MATHS IN FOCUS

MATHEMATICS ADVANCED

4TH EDITION

Margaret Grove

Series editor: Robert Yen

Maths in Focus 11 Mathematics Advanced
4th Edition
Margaret Grove
ISBN 9780170498197

Product manager: Robert Yen
Content developer: Helen Kaminski
Content project manager: Ananya Sarkar
Project designer: Nikita Bansal
Text designer: Cengage Creative Studio
Cover designer: Leigh Ashforth (Watershed Art & Design) and Cengage Creative Studio
Editor: Anne Eastaugh, Nikki M Group Pty Ltd
Proofreader: Jane Fitzpatrick, Nikki M Group Pty Ltd
Permissions/Photo researcher: Liz McShane/Lumina Datamatics
Cover: Nikita Bansal
Typeset by: KGL

National Library of Australia Cataloguing-in-Publication Data
A catalogue record for this book is available from the National Library of Australia

Cengage Learning Australia
Level 7, 80 Dorcas Street
South Melbourne, Victoria Australia 3205

For learning solutions, visit **cengage.com.au**

Printed in China by 1010 Printing International Limited.
1 2 3 4 5 6 7 29 28 27 26 25

PREFACE

This 4th edition of *Maths in Focus 11 Mathematics Advanced* has been written for the new Mathematics Advanced Syllabus (2024). Based on feedback from teachers and experience from the 2019 syllabus and subsequent HSC exams, this new edition now contains additional application and problem-solving questions, including more detailed and complex HSC-style problems.

This book retains the clear and abundant explanations and syllabus coverage of previous editions, which allows students to achieve self-paced learning and develop a strong understanding of the theory behind each syllabus topic. Exercise questions are now colour-coded according to level of difficulty and cross-referenced to the relevant worked examples.

This book covers Year 11 Mathematics Advanced. The theory follows a logical order, although some topics may be learned in any order. Syllabus coverage is shown by the table of contents and the syllabus reference grid on the next pages.

Maths in Focus has always provided straightforward explanations, examples and graded exercises to support all students and teachers. Each exercise set has a wide range of questions, and students are not required to complete every question. For example, the most able students could skip the Foundation and revision questions.

The *Nelson MindTap* online learning platform contains an eBook, worksheets, videos, topic tests and worked solutions. We have designed this book to be user-friendly and uncomplicated so you can pick it up and use it straight away. We wish all students and teachers using *Maths in Focus* every success in embracing this revised senior mathematics course.

About the author

Margaret Grove OAM started writing *Maths in Focus* in 1992, a breakthrough series that explained HSC mathematics in a clearer and more concise way. She has spent over 30 years teaching mathematics, mostly specialising in HSC Mathematics at TAFE. She completed a Master's degree in Pure Mathematics at the University of Sydney. Margaret has written several HSC texts and study guides, including the bestselling *Maths in Focus* series for Mathematics Advanced and Extension 1.

In 2023, Margaret was awarded a Medal of the Order of Australia for her services to the community: as a long-term president of the Australian Breastfeeding Association, an executive leader of several community choirs and as a mathematics teacher and author.

Margaret thanks her family, especially her husband Geoff, for their support in writing this book.

Contributing authors

Gaspare Carrozza and **Haroon Ha** wrote many of the *Nelson MindTap* worksheets.

Scott Smith, Ilhea Yen and **Cherylanne Saywell** created the video tutorials.

Rachel Eastcott, Tania Eastcott and **Elizabeth Nabhan** wrote the topic tests.

Roger Walter wrote the testbank questions.

Andrew Siu wrote the worked solutions to the exercise sets, with contributions from **Shane Scott, Anna Ma, George Dimitriadis, George Higgins, Bela Adash** and **Brandon Pettis**.

CONTENTS

CONTENTS

SYLLABUS REFERENCE GRID

Topic and subtopic		*Maths in Focus 11: Mathematics Advanced* chapter
FUNCTIONS		
MAV-11-01	**Working with functions** Algebraic techniques	1 Algebraic techniques
MAV-11-02	**Working with functions** Introduction to functions and relations Linear functions Quadratic and cubic functions Reciprocal functions Constructing and using functions Direct and inverse variation Circles and semicircles Properties of functions, relations and graphs Piecewise-defined functions Absolute value functions	2 Equations and inequalities 3 Functions and graphs 5 Further functions and graphs
MAV-11-03	**Graph transformations**	5 Further functions and graphs 8 Transformations of functions
TRIGONOMETRIC FUNCTIONS		
MAV-11-04	**Trigonometry and measure of angles** Trigonometry with acute angles Trigonometry with angles of any magnitude Radians	4 Trigonometry 9 Trigonometric identities and functions
MAV-11-05	**Trigonometric identities and equations**	9 Trigonometric identities and functions
CALCULUS		
MAV-11-06	**Introduction to differentiation** Estimating change The derivative Calculations with the derivative Graphical applications of the derivative The derivative as a rate of change	3 Functions and graphs 6 Introduction to differentiation
EXPONENTIAL AND LOGARITHMIC FUNCTIONS		
MAV-11-07, -08	**Exponential and logarithmic functions** Exponential functions Logarithmic functions	7 Exponential and logarithmic functions
STATISTICAL ANALYSIS		
MAV-11-09	**Probability and data** Sets and set notation Probability Conditional probability	10 Probability and data
MAV-11-10	**Probability and data** Data	10 Probability and data

ABOUT THIS BOOK

At the beginning of each chapter

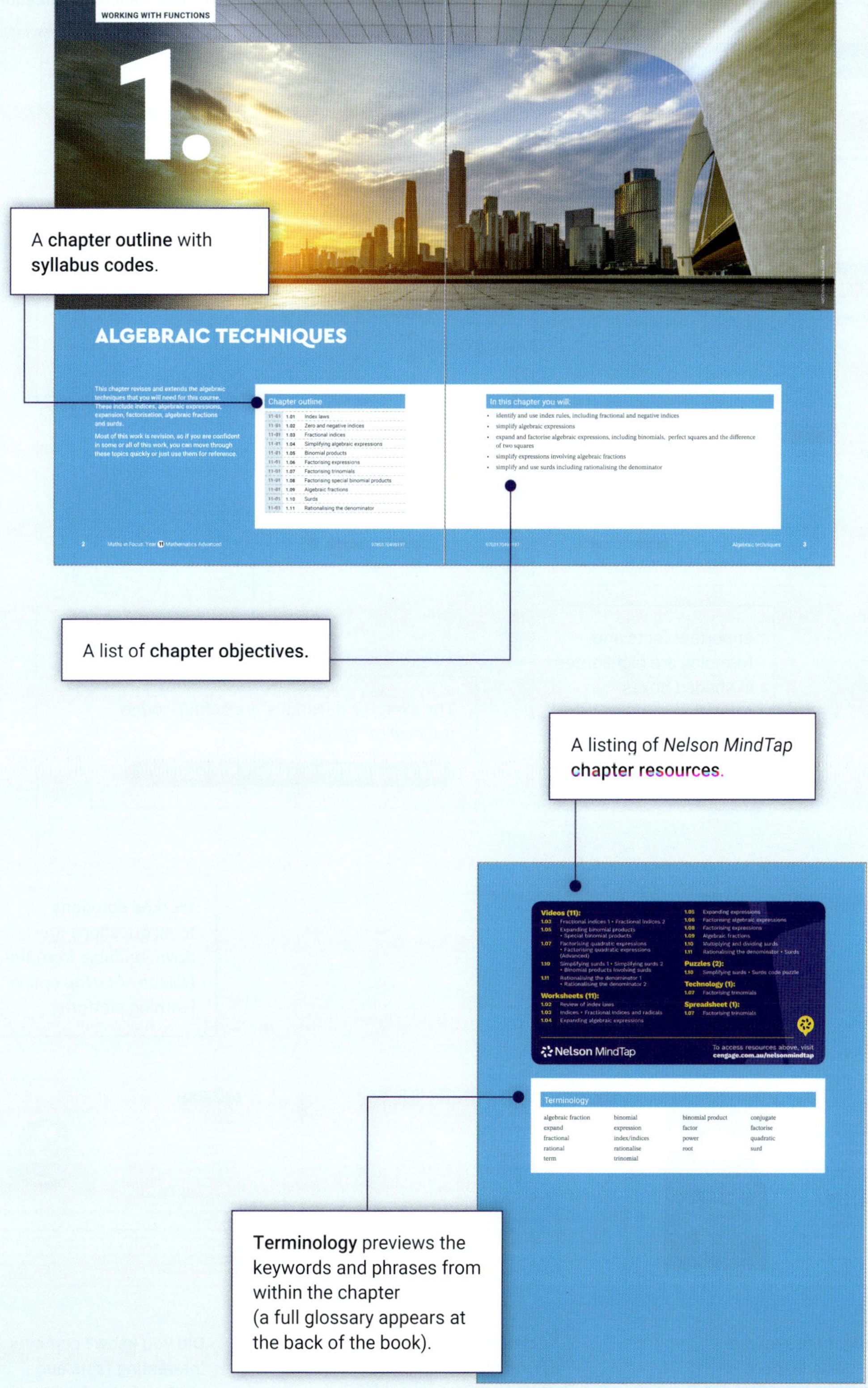

In each chapter

Glossary terms are printed in **red**.

Graded exercises are linked to worked examples and include HSC exam-style problems and realistic applications.

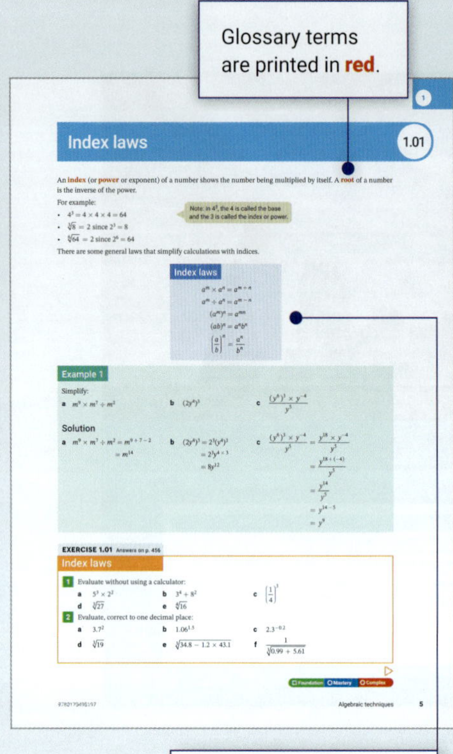

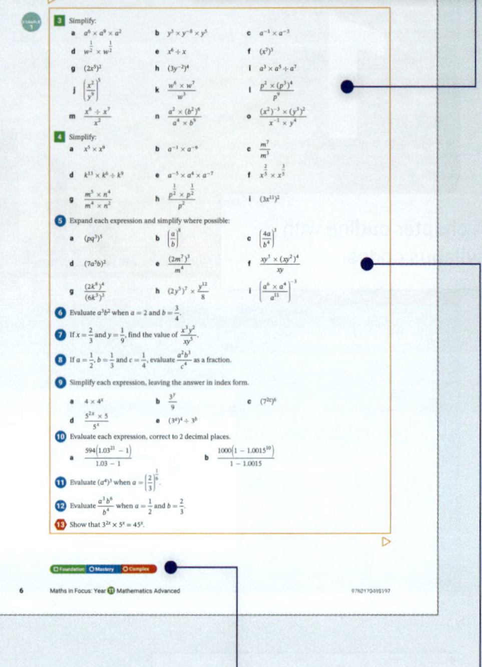

Important facts and formulas are highlighted in shaded boxes.

The exercise questions are colour-coded for level of difficulty:

Worked solutions to all questions are downloadable from the *Nelson MindTap* online learning platform

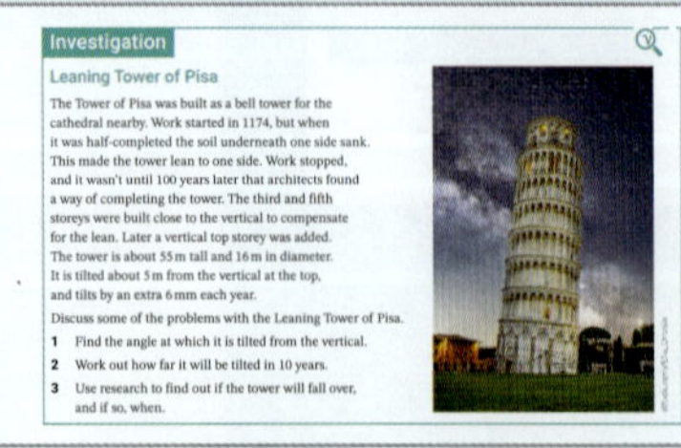

Investigations explore the syllabus in more detail, providing ideas for modelling activities and assessment tasks.

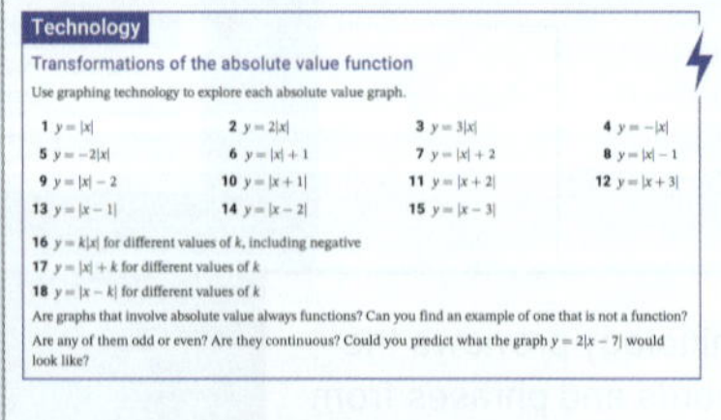

Technology integrates ICT, using spreadsheets, graphing software and the internet.

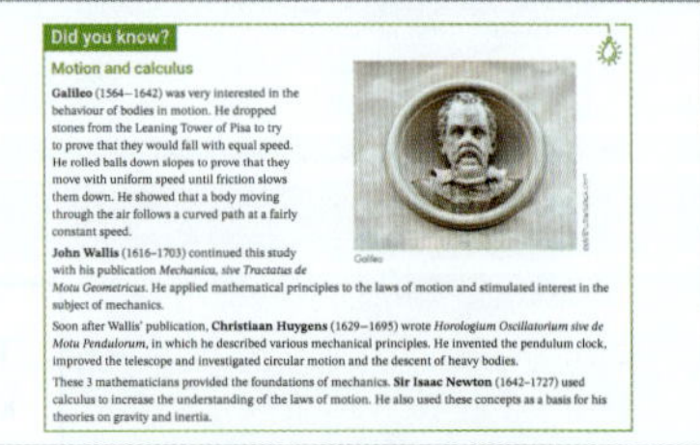

Did you know? contains interesting facts and applications of the mathematics learned in the chapter.

At the end of each chapter

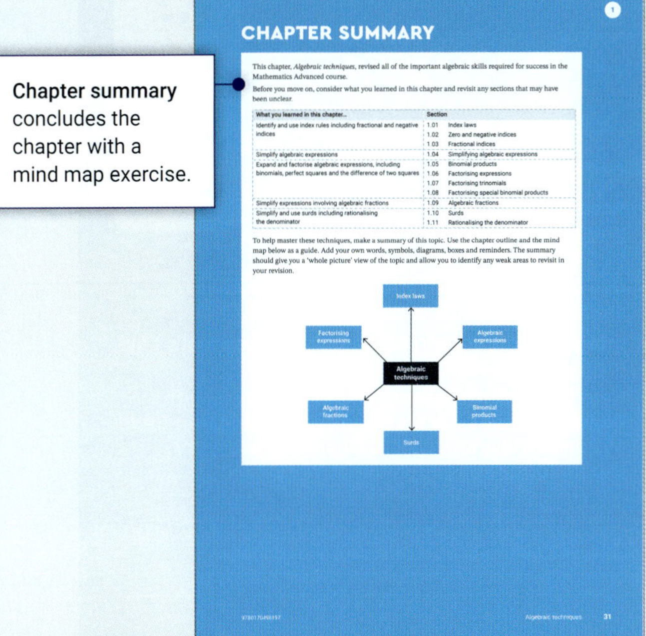

Chapter summary concludes the chapter with a mind map exercise.

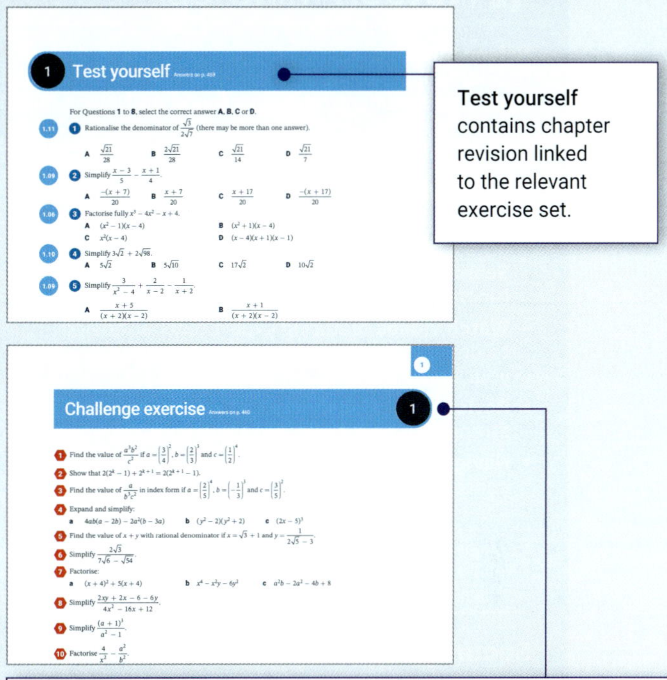

Test yourself contains chapter revision linked to the relevant exercise set.

Challenge exercise contains chapter extension questions. Attempt these only if you have mastered the *Test yourself* exercises, because these questions are beyond the level of the syllabus and HSC exams.

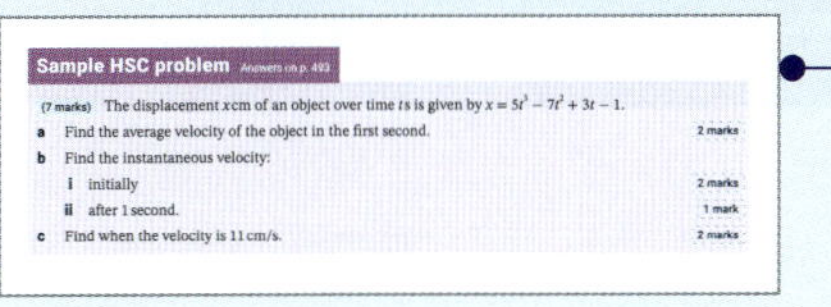

Sample HSC problem demonstrates a typical HSC-style question using knowledge learned from the chapter.

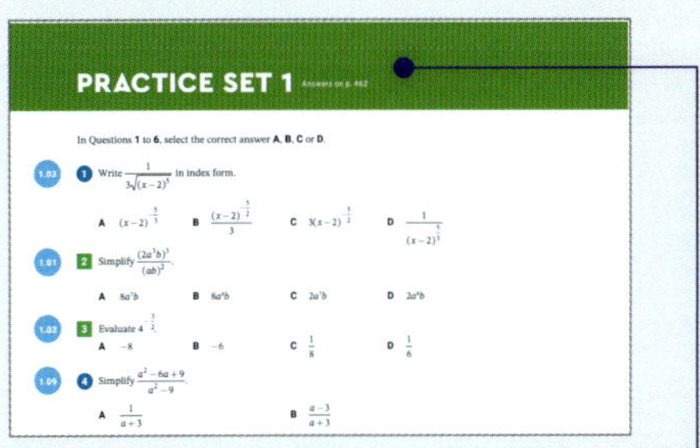

Practice sets after every 2–3 chapters revise the skills and knowledge of those chapters.

At the end of the book

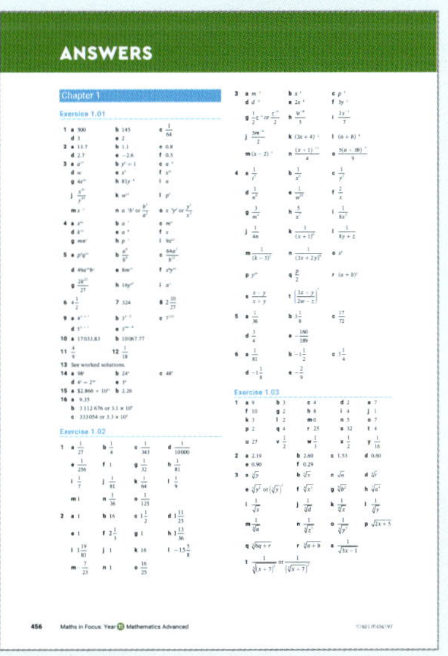

Answers (worked solutions, including answers that are proofs, are provided on *Nelson MindTap* for teachers to allocate to students).

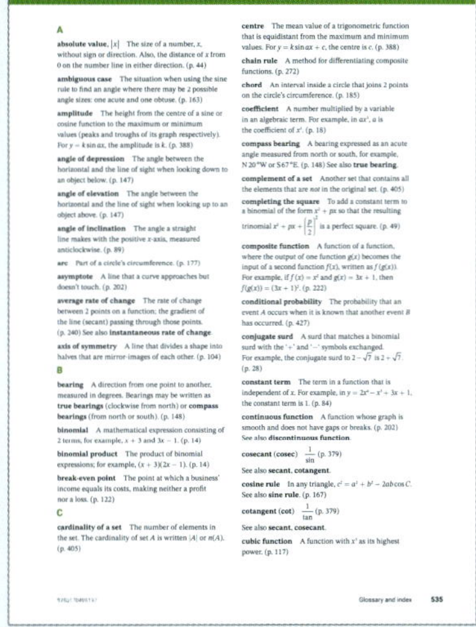

Glossary and index includes a comprehensive dictionary of course terminology.

ABOUT THIS BOOK

Nelson MindTap

Nelson MindTap is an online learning space that provides students and teachers with access to additional learning and teaching resources. Margin links in the student book signpost multimedia student resources found on Nelson MindTap*.

Videos

Skillsheets

Worksheets

For students:

- **Engage** with the **online eBook** by adding notes, highlights and bookmarks, and using the **Search** and **Read Aloud** (in Australian voice) functions.
- **Watch videos** featuring expert teacher advice to unpack worked examples and deepen your understanding.
- **Revise** using **skillsheets** and **worksheets** to strengthen your skills and build your confidence.
- **Navigate** your own learning path, accessing the content and support as you need it.

For teachers:

- **Access topic tests, teaching programs** and **worked solutions** to each exercise set.
- **Use course customisation** to tailor content to groups of students or the whole class.
- **Integrate** content directly within your school's LMS for ease of access.
- Help build your students' exam readiness with **Cognero Assess** – a test bank containing hundreds of questions and answers to create, assign or export formative and summative tests.

* Complimentary access to these resources is only available to schools that use this book as part of a class set, book hire or booklist. Not available for single purchases. Contact your Cengage Learning Consultant for information about access and conditions.

Security & privacy:

Nelson MindTap joined the Safer Technologies 4 Schools (ST4S) Product Badge Program in 2024. The annual ST4S assessment is part of our commitment to supporting the online security and safety of students and schools. Learn more at st4s.edu.au

STUDY SKILLS

There is no right or wrong way to learn. Different styles of learning suit different people. There is also no magical number of hours a week that you should study, because this will be different for every student. But just listening in class and taking notes is not enough, especially when you are learning material that is totally new.

If a skill is not practised within the first 24 hours, up to 50% can be forgotten.

If it is not practised within 72 hours, up to 90% can be forgotten!

So it is really important that, whatever your study routine, you study new work soon after it is presented to you.

With a constant flow of new work to learn and retain, this is a challenge.

But the good news is that you don't have to study for hours on end!

In class

Leave distractions and issues outside the classroom	Listen to what you're being taught	Take notes	Highlight main points	Ask questions

At home

To best remember, revise

Today

Tomorrow

In a week

In a month.

Set a realistic timetable	Revise when you are fresh and have energy	Study in small chunks instead of marathons

Create an environment and routine that helps you focus	Revise previous topics as well as new work	Balance study with other activities and commitments

Learning skills

Try different learning styles:

Seeing

Hearing

Doing.

Summarise on cue cards

Draw mind maps or pictures

Use colours to highlight main points

Read notes out loud

Discuss work with a friend

Create songs or rhymes to remember facts and formulas

Revise by teaching someone else

Once you have mastered a topic, do practice exams under exam conditions

During exams

During reading time, skim the whole paper to know how many questions there are

During the exam, read each question carefully

Underline or highlight key words

Divide time so you can answer each question

Allow time at the end to check answers or complete harder questions

Write legibly

Don't spend too much time on one question: move on if you are genuinely stuck

Show all working out and include diagrams and formulas

Cross out mistakes with a single line so that they can still be read

Finally…

Study involves knowing what you don't know, and putting in a lot of time focusing on these areas. This is a positive way to learn. Rather than just saying, 'I can't do this', say 'I can't do this *yet*', and use your teachers, friends, textbooks and other ways of finding out.

Some students hardly find time to study while others give up their outside lives to devote their time to study.

The ideal situation is to balance study with other aspects of your life, including going out with friends, working, and keeping up with sport and other activities that you enjoy.

Good luck with your studies!

STUDY SKILLS

New Century Maths / Maths in Focus 7–12 series

MATHEMATICAL VERBS

A glossary of 'doing words' commonly found in mathematics problems

analyse: study and state in detail the relationship of parts of a situation

apply: use knowledge or a procedure in a given situation

calculate: find a numerical value

classify: identify and sort into a category

comment: express an observation or opinion about a statement or result

compare: show how 2 or more things are similar or different

construct: draw an accurate diagram or logically arrange items or ideas

convert: change from one form to another

define: give the meaning of or identify in exact terms

demonstrate: show to be correct

describe: state the features of a situation, object, pattern, event, etc.

differentiate: find the derivative of a function

estimate: make an educated guess for a number, measurement or solution; to find roughly or approximately

evaluate: find the value of or state the application, strengths and limitations of a solution

examine: investigate the details and assumptions of a situation

expand: remove brackets in an algebraic expression by multiplying; for example, expanding $3(2y + 1)$ gives $6y + 3$

explain: state the meaning in logical detail; describe why or how

explore: examine or state the details and assumptions of a situation

factorise: take out the highest common factor (HCF) in an algebraic expression and insert brackets; for example, factorising $5x - 20$ gives $5(x - 4)$. The opposite of **expand**

give reasons: show the rules or thinking used when solving a problem

graph: display on a number line, number plane or statistical graph

hence find/prove: find an answer or prove a result using previous answers or information supplied

identify: state the type, name or distinguishing feature of an item or situation

interpret: find meaning in a mathematical result

justify: give reasons or evidence to support your argument or conclusion

prove: use logical steps to establish the truth of a result, statement or assumption

rationalise the denominator: simplify a fraction involving a surd by making its denominator rational (that is, not a surd)

recall: remember (and state)

show that: in questions where the answer is given, use calculation, procedure or reasoning to prove that an answer or result is true

show working: show the steps you used to find the answer

simplify: reduce a value or equation to give a result in its most basic, shortest, neatest form; for example, simplifying a ratio or algebraic expression

sketch: draw a diagram that shows the general shape and includes relevant features

solve: find the answer or explanation for a problem, particularly the value(s) of variables in equations or inequalities

substitute: replace a variable by a number and evaluate

verify: check a solution or result, usually by substituting back into the equation or referring back to the problem

write/state: give the answer, formula or result without showing any working or explanation (This usually means that the answer can be found mentally, or in one step)

SYMBOLS AND ABBREVIATIONS

$=$	is equal to	$\varnothing, \{\ \}$	empty set
$\neq$	is not equal to	U	universal set
$\approx$	is approximately equal to	$\bar{A}, A'$	not A, the complement of A
$<$	is less than	$\cup$	union
$>$	is greater than	$\cap$	intersection
$\leq$	is less than or equal to	S	sample space
$\geq$	is greater than or equal to	$P(A)$	the probability of A
$\pm$	plus or minus	$P(A \mid B)$	the probability of A given B
π	pi $= 3.141...$	$\lim\limits_{h=0}$	the limit as $h \to 0$
e	Euler's number, $e = 2.718...$	$\frac{dy}{dx}, y', f'(x)$	the derivative of $y, f(x)$
$\circ$	degree	$f(a)$	the value of $f(x)$ at $x = a$
α	alpha	$f'(a)$	the derivative of $f(x)$ at $x = a$
θ	theta	$f(g(x))$	the composite function 'f of g of x'
∞	infinity	$[a, b]$	the interval $a \leq x \leq b$
Δ	discriminant	(a, b)	the interval $a < x < b$
$\sqrt{\ }$	square root, radical sign	$\log_{10}$	common logarithm
$\|x\|$	absolute value of x	$\ln, \log_e$	natural logarithm (base e)
LHS	left-hand side	$\log_a b$	thc logarithm of b to the base a
RHS	right-hand side	$\sin^2 x$	$(\sin x)^2$
$\|\|$	is parallel to	$\angle A$	angle A
$\perp$	is perpendicular to	$\triangle ABC$	triangle ABC
$\therefore$	therefore	S 37°W	a compass bearing
$\|A\|, n(A)$	cardinality of set A	217°T	a true bearing

ALGEBRAIC TECHNIQUES

This chapter revises and extends the algebraic techniques that you will need for this course. These include indices, algebraic expressions, expansion, factorisation, algebraic fractions and surds.

Most of this work is revision, so if you are confident in some or all of this work, you can move through these topics quickly or just use them for reference.

Chapter outline

ABCDstock/Adobe Stock Photos

In this chapter you will:

- identify and use index rules, including fractional and negative indices
- simplify algebraic expressions
- expand and factorise algebraic expressions, including binomials, perfect squares and the difference of two squares
- simplify expressions involving algebraic fractions
- simplify and use surds including rationalising the denominator

Videos (11):

1.03 Fractional indices 1 • Fractional indices 2

1.05 Expanding binomial products • Special binomial products

1.07 Factorising quadratic expressions • Factorising quadratic expressions (Advanced)

1.10 Simplifying surds 1 • Simplifying surds 2 • Binomial products involving surds

1.11 Rationalising the denominator 1 • Rationalising the denominator 2

Worksheets (11):

1.02 Review of index laws

1.03 Indices • Fractional indices and radicals

1.04 Expanding algebraic expressions

1.05 Expanding expressions

1.06 Factorising algebraic expressions

1.08 Factorising expressions

1.09 Algebraic fractions

1.10 Multiplying and dividing surds

1.11 Rationalising the denominator • Surds

Puzzles (2):

1.10 Simplifying surds • Surds code puzzle

Technology (1):

1.07 Factorising trinomials

Spreadsheet (1):

1.07 Factorising trinomials

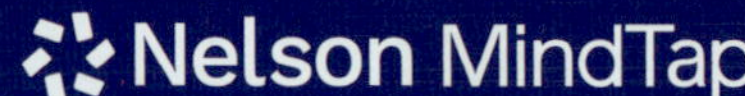

To access resources above, visit **cengage.com.au/nelsonmindtap**

Terminology

algebraic fraction	binomial	binomial product	conjugate
expand	expression	factor	factorise
fractional	index/indices	power	quadratic
rational	rationalise	root	surd
term	trinomial		

Index laws

1.01

An **index** (or **power** or exponent) of a number shows the number being multiplied by itself. A **root** of a number is the inverse of the power.

For example:

- $4^3 = 4 \times 4 \times 4 = 64$
- $\sqrt[3]{8} = 2$ since $2^3 = 8$
- $\sqrt[6]{64} = 2$ since $2^6 = 64$

Note: In 4^3, the 4 is called the base and the 3 is called the index or power.

There are some general laws that simplify calculations with indices.

Index laws

$$a^m \times a^n = a^{m+n}$$
$$a^m \div a^n = a^{m-n}$$
$$(a^m)^n = a^{mn}$$
$$(ab)^n = a^n b^n$$
$$\left(\frac{a}{b}\right)^n = \frac{a^n}{b^n}$$

Example 1

Simplify:

a $m^9 \times m^7 \div m^2$ **b** $(2y^4)^3$ **c** $\dfrac{(y^6)^3 \times y^{-4}}{y^5}$

Solution

a $m^9 \times m^7 \div m^2 = m^{9+7-2}$
$= m^{14}$

b $(2y^4)^3 = 2^3(y^4)^3$
$= 2^3y^{4 \times 3}$
$= 8y^{12}$

c $\dfrac{(y^6)^3 \times y^{-4}}{y^5} = \dfrac{y^{18} \times y^{-4}}{y^5}$
$= \dfrac{y^{18+(-4)}}{y^5}$
$= \dfrac{y^{14}}{y^5}$
$= y^{14-5}$
$= y^9$

EXERCISE 1.01 Answers on p. 456

Index laws

1 Evaluate without using a calculator:

a $5^3 \times 2^2$ **b** $3^4 + 8^2$ **c** $\left(\frac{1}{4}\right)^3$

d $\sqrt[3]{27}$ **e** $\sqrt[4]{16}$

2 Evaluate, correct to one decimal place:

a 3.7^2 **b** $1.06^{1.5}$ **c** $2.3^{-0.2}$

d $\sqrt[3]{19}$ **e** $\sqrt[3]{34.8 - 1.2 \times 43.1}$ **f** $\dfrac{1}{\sqrt[3]{0.99} + 5.61}$

Foundation Mastery Complex

EXAMPLE 1

3 Simplify:

a $a^6 \times a^9 \times a^2$ b $y^3 \times y^{-8} \times y^5$ c $a^{-1} \times a^{-3}$

d $w^{\frac{1}{2}} \times w^{\frac{1}{2}}$ e $x^6 \div x$ f $(x^7)^3$

g $(2x^5)^2$ h $(3y^{-2})^4$ i $a^3 \times a^5 \div a^7$

j $\left(\frac{x^2}{y^9}\right)^5$ k $\frac{w^6 \times w^7}{w^3}$ l $\frac{p^2 \times (p^3)^4}{p^9}$

m $\frac{x^6 \div x^7}{x^2}$ n $\frac{a^2 \times (b^2)^6}{a^4 \times b^9}$ o $\frac{(x^2)^{-3} \times (y^3)^2}{x^{-1} \times y^4}$

4 Simplify:

a $x^5 \times x^9$ b $a^{-1} \times a^{-6}$ c $\frac{m^7}{m^3}$

d $k^{13} \times k^6 \div k^9$ e $a^{-5} \times a^4 \times a^{-7}$ f $x^{\frac{2}{5}} \times x^{\frac{3}{5}}$

g $\frac{m^5 \times n^4}{m^4 \times n^2}$ h $\frac{p^{\frac{1}{2}} \times p^{\frac{1}{2}}}{p^2}$ i $(3x^{11})^2$

5 Expand each expression and simplify where possible:

a $(pq^3)^5$ b $\left(\frac{a}{b}\right)^8$ c $\left(\frac{4a}{b^4}\right)^3$

d $(7a^5b)^2$ e $\frac{(2m^7)^3}{m^4}$ f $\frac{xy^3 \times (xy^2)^4}{xy}$

g $\frac{(2k^8)^4}{(6k^3)^3}$ h $(2y^5)^7 \times \frac{y^{12}}{8}$ i $\left(\frac{a^6 \times a^4}{a^{11}}\right)^{-3}$

6 Evaluate a^3b^2 when $a = 2$ and $b = \frac{3}{4}$.

7 If $x = \frac{2}{3}$ and $y = \frac{1}{9}$, find the value of $\frac{x^3y^2}{xy^5}$.

8 If $a = \frac{1}{2}$, $b = \frac{1}{3}$ and $c = \frac{1}{4}$, evaluate $\frac{a^2b^3}{c^4}$ as a fraction.

9 Simplify each expression, leaving the answer in index form.

a 4×4^x b $\frac{3^y}{9}$ c $(7^{2z})^6$

d $\frac{5^{2x} \times 5}{5^x}$ e $(3^a)^4 \div 3^b$

10 Evaluate each expression, correct to 2 decimal places.

a $\frac{594(1.03^{21} - 1)}{1.03 - 1}$ b $\frac{1000(1 - 1.0015^{10})}{1 - 1.0015}$

11 Evaluate $(a^4)^3$ when $a = \left(\frac{2}{3}\right)^{\frac{1}{6}}$.

12 Evaluate $\frac{a^3b^6}{b^4}$ when $a = \frac{1}{2}$ and $b = \frac{2}{3}$.

13 Show that $3^{2x} \times 5^x = 45^x$.

☐ Foundation ◯ Mastery ⬡ Complex

14 Write each expression with a single index, as in Question **13**.

a $2^y \times 7^{2y}$ **b** $3^a \times 2^{3a}$ **c** $2^{4b} \times 3^b$

d $\dfrac{2^{3x} \times 3^x}{6^x}$ **e** $\dfrac{2^{4a} \times 3^{2a}}{6^a \times 2^{3a}}$

15 In the USA, damage by Hurricane Sandy cost 8.78×10^{10} and damage by Hurricane Katrina cost 1.988×10^{11}.

a What was the total cost from the 2 hurricanes?

b How many times more than Hurricane Sandy were the costs from Hurricane Katrina?

16 The mass of Earth is around 5.972×10^{24} kg, the mass of Mars is around 6.39×10^{23} kg and the mass of the Sun is around 1.989×10^{30} kg. How many times heavier:

a than Mars is Earth?

b than Mars is the Sun?

c than Earth is the Sun?

Zero and negative indices 1.02

Worksheet Review of index laws

Zero and negative indices

$$x^0 = 1$$

$$x^{-n} = \frac{1}{x^n}$$

$$\left(\frac{a}{b}\right)^{-n} = \left(\frac{b}{a}\right)^n$$

Example 2

a Simplify $\left(\dfrac{ab^5c}{abc^4}\right)^0$.

b Evaluate 2^{-3}.

c Write in index form:

i $\dfrac{1}{x^2}$ **ii** $\dfrac{3}{x^5}$ **iii** $\dfrac{1}{5x}$ **iv** $\dfrac{1}{x+1}$

d Write a^{-3} with a positive index.

e Evaluate $\left(\dfrac{2}{3}\right)^{-3}$.

Solution

a $\left(\dfrac{ab^5c}{abc^4}\right)^0 = 1$

b $2^{-3} = \dfrac{1}{2^3} = \dfrac{1}{8}$

c **i** $\dfrac{1}{x^2} = x^{-2}$

ii $\dfrac{3}{x^5} = 3 \times \dfrac{1}{x^5} = 3x^{-5}$

iii $\dfrac{1}{5x} = \dfrac{1}{5} \times \dfrac{1}{x} = \dfrac{1}{5}x^{-1}$

iv $\dfrac{1}{x+1} = \dfrac{1}{(x+1)^1} = (x+1)^{-1}$

Foundation Mastery Complex

d $a^{-3} = \dfrac{1}{a^3}$

e $\left(\dfrac{2}{3}\right)^{-3} = \left(\dfrac{3}{2}\right)^{3}$

$= \dfrac{27}{8}$

$= 3\dfrac{3}{8}$

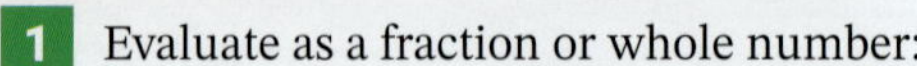

EXERCISE 1.02 Answers on p. 456

Zero and negative indices

EXAMPLE 2

1 Evaluate as a fraction or whole number:

a 3^{-3} **b** 4^{-1} **c** 7^{-3} **d** 10^{-4}

e 2^{-8} **f** 6^{0} **g** 2^{-5} **h** 3^{-4}

i 7^{-1} **j** 9^{-2} **k** 2^{-6} **l** 3^{-2}

m 4^{0} **n** 6^{-2} **o** 5^{-3}

2 Evaluate:

a 2^0 **b** $\left(\dfrac{1}{2}\right)^{-4}$ **c** $\left(\dfrac{2}{3}\right)^{-1}$ **d** $\left(\dfrac{5}{6}\right)^{-2}$

e $\left(\dfrac{x+2y}{3x-y}\right)^{0}$ **f** $\left(\dfrac{3}{7}\right)^{-1}$ **g** $\left(\dfrac{8}{9}\right)^{0}$ **h** $\left(\dfrac{6}{7}\right)^{-2}$

i $\left(\dfrac{9}{10}\right)^{-2}$ **j** $\left(\dfrac{6}{11}\right)^{0}$ **k** $\left(-\dfrac{1}{4}\right)^{-2}$ **l** $\left(-\dfrac{2}{5}\right)^{-3}$

m $\left(-3\frac{2}{7}\right)^{-1}$ **n** $\left(-\dfrac{3}{8}\right)^{0}$ **o** $\left(-1\frac{1}{4}\right)^{-2}$

3 Write in index form:

a $\dfrac{1}{m^3}$ **b** $\dfrac{1}{x}$ **c** $\dfrac{1}{p^7}$ **d** $\dfrac{1}{d^9}$

e $\dfrac{2}{x^4}$ **f** $\dfrac{3}{y^2}$ **g** $\dfrac{1}{2z^6}$ **h** $\dfrac{3}{5t^8}$

i $\dfrac{2}{7x}$ **j** $\dfrac{5}{2m^6}$ **k** $\dfrac{1}{(3x+4)^2}$ **l** $\dfrac{1}{(a+b)^8}$

m $\dfrac{1}{x-2}$ **n** $\dfrac{1}{4(x+1)^{11}}$ **o** $\dfrac{5}{9(a+3b)^7}$

☐ Foundation ○ Mastery ○ Complex

4 Write each term with a positive index.

- **a** t^{-5}
- **b** x^{-6}
- **c** y^{-3}
- **d** n^{-8}
- **e** w^{-10}
- **f** $2x^{-1}$
- **g** $3m^{-4}$
- **h** $5x^{-7}$
- **i** $(2x)^{-3}$
- **j** $(4n)^{-1}$
- **k** $(x+1)^{-6}$
- **l** $(8y+z)^{-1}$
- **m** $(k-3)^{-2}$
- **n** $(3x+2y)^{-9}$
- **o** $\left(\frac{1}{x}\right)^{-5}$
- **p** $\left(\frac{1}{y}\right)^{-10}$
- **q** $\left(\frac{2}{p}\right)^{-1}$
- **r** $\left(\frac{1}{a+b}\right)^{-2}$
- **s** $\left(\frac{x+y}{x-y}\right)^{-1}$
- **t** $\left(\frac{2w-z}{3x+y}\right)^{-7}$

5 Evaluate as a fraction:

- **a** $3^{-2} \times 4^{-1}$
- **b** $2^{-3} \div 5^{-2}$
- **c** $2^{-3} + 3^{-2}$
- **d** $\frac{3^{-1} - 2^{-2}}{3^{-2}}$
- **e** $\frac{7^{-1} + 3^{-2}}{5^{-1} - 2^{-1}}$

6 If $a = 3$ and $b = -2$, evaluate as a fraction:

- **a** a^{-4}
- **b** $\left(\frac{b}{a}\right)^{-1}$
- **c** $a + b^{-2}$
- **d** $\frac{b^{-3}}{a^{-2}}$
- **e** $\frac{a^{-1} + b^{-1}}{ab^{-2}}$

Fractional indices 1.03

Investigation

Fractional indices

Consider the following example.

$$\left(x^{\frac{1}{2}}\right)^2 = x^1 \text{ (by index laws)}$$
$$= x$$

and

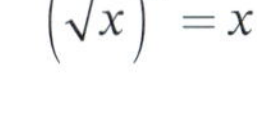
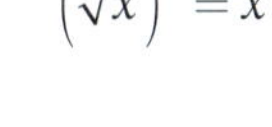

$$\left(\sqrt{x}\right)^2 = x$$

$$\text{So} \left(x^{\frac{1}{2}}\right)^2 = \left(\sqrt{x}\right)^2$$
$$= x$$

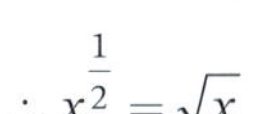

$$\therefore x^{\frac{1}{2}} = \sqrt{x}$$

Now, simplify these expressions.

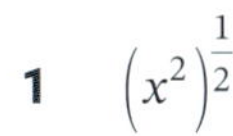

- **1** $\left(x^2\right)^{\frac{1}{2}}$
- **2** $\sqrt{x^2}$
- **3** $\left(x^{\frac{1}{3}}\right)^3$
- **4** $\left(x^3\right)^{\frac{1}{3}}$
- **5** $\left(\sqrt[3]{x}\right)^3$
- **6** $\sqrt[3]{x^3}$
- **7** $\left(x^{\frac{1}{4}}\right)^4$
- **8** $\left(x^4\right)^{\frac{1}{4}}$
- **9** $\left(\sqrt[4]{x}\right)^4$
- **10** $\sqrt[4]{x^4}$

Use your results to complete:

$$x^{\frac{1}{n}} =$$

Videos
Fractional indices 1
Fractional indices 2

Worksheets
Indices
Fractional indices and radicals

Foundation Mastery Complex

Fractional indices

$$a^{\frac{1}{n}} = \sqrt[n]{a}$$

$$a^{-\frac{1}{n}} = \frac{1}{\sqrt[n]{a}}$$

$$a^{\frac{m}{n}} = \sqrt[n]{a^m} \text{ or } \left(\sqrt[n]{a}\right)^m$$

Example 3

a Evaluate:

i $49^{\frac{1}{2}}$ **ii** $27^{\frac{1}{3}}$

b Write $\sqrt{3x-2}$ in index form.

c Write $(a+b)^{\frac{1}{7}}$ in root form.

Solution

a i $49^{\frac{1}{2}} = \sqrt{49} = 7$ **ii** $27^{\frac{1}{3}} = \sqrt[3]{27} = 3$

b $\sqrt{3x-2} = (3x-2)^{\frac{1}{2}}$

c $(a+b)^{\frac{1}{7}} = \sqrt[7]{a+b}$

Example 4

a Evaluate:

i $8^{\frac{4}{3}}$ **ii** $125^{-\frac{1}{3}}$

b Write in index form:

i $\sqrt{x^5}$ **ii** $\dfrac{1}{\sqrt[3]{\left(4x^2-1\right)^2}}$

c Write $r^{-\frac{3}{5}}$ in root form.

Solution

a i
$$\begin{aligned} 8^{\frac{4}{3}} &= \left(\sqrt[3]{8}\right)^4 \left(\text{or } \sqrt[3]{8^4}\right) \\ &= 2^4 \\ &= 16 \end{aligned}$$

ii
$$\begin{aligned} 125^{-\frac{1}{3}} &= \frac{1}{125^{\frac{1}{3}}} \\ &= \frac{1}{\sqrt[3]{125}} \\ &= \frac{1}{5} \end{aligned}$$

b i $\sqrt{x^5} = x^{\frac{5}{2}}$

ii
$$\begin{aligned} \frac{1}{\sqrt[3]{(4x^2-1)^2}} &= \frac{1}{(4x^2-1)^{\frac{2}{3}}} \\ &= (4x^2-1)^{-\frac{2}{3}} \end{aligned}$$

c $r^{-\frac{3}{5}} = \dfrac{1}{r^{\frac{3}{5}}} = \dfrac{1}{\sqrt[5]{r^3}}$

Did you know?

Fractional indices

Nicole Oresme (1323–82) was the first mathematician to use fractional indices.

John Wallis (1616–1703) was the first person to explain the significance of zero, negative and fractional indices. He also introduced the symbol ∞ for infinity.

Research these mathematicians and find out more about their work and backgrounds. You could use keywords such as indices and infinity, as well as their names to find this information.

EXERCISE 1.03 Answers on p. 456

Fractional indices

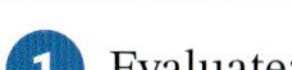

EXAMPLES 3, 4

1 Evaluate:

a $81^{\frac{1}{2}}$ **b** $27^{\frac{1}{3}}$ **c** $16^{\frac{1}{2}}$ **d** $8^{\frac{1}{3}}$ **e** $49^{\frac{1}{2}}$

f $1000^{\frac{1}{3}}$ **g** $16^{\frac{1}{4}}$ **h** $64^{\frac{1}{2}}$ **i** $64^{\frac{1}{3}}$ **j** $1^{\frac{1}{7}}$

k $81^{\frac{1}{4}}$ **l** $32^{\frac{1}{5}}$ **m** $0^{\frac{1}{8}}$ **n** $125^{\frac{1}{3}}$ **o** $343^{\frac{1}{3}}$

p $128^{\frac{1}{7}}$ **q** $256^{\frac{1}{4}}$ **r** $125^{\frac{2}{3}}$ **s** $4^{\frac{5}{2}}$ **t** $8^{\frac{2}{3}}$

u $9^{\frac{3}{2}}$ **v** $8^{-\frac{1}{3}}$ **w** $9^{-\frac{1}{2}}$ **x** $16^{-\frac{1}{4}}$ **y** $64^{-\frac{2}{3}}$

2 Evaluate, correct to 2 decimal places:

a $23^{\frac{1}{4}}$ **b** $\sqrt[4]{45.8}$ **c** $\sqrt[7]{1.24 + 4.3^2}$

d $\dfrac{1}{\sqrt[5]{12.9}}$ **e** $\sqrt[8]{\dfrac{3.6 - 1.4}{1.5 + 3.7}}$ **f** $\dfrac{\sqrt[4]{5.9 \times 3.7}}{8.79 - 1.4}$

3 Write in root form:

a $y^{\frac{1}{3}}$ **b** $x^{\frac{1}{6}}$ **c** $a^{\frac{1}{2}}$ **d** $t^{\frac{1}{9}}$

e $y^{\frac{2}{3}}$ **f** $x^{\frac{3}{4}}$ **g** $b^{\frac{2}{5}}$ **h** $a^{\frac{4}{7}}$

i $x^{-\frac{1}{2}}$ **j** $d^{-\frac{1}{3}}$ **k** $x^{-\frac{1}{8}}$ **l** $y^{\frac{1}{3}}$

m $a^{-\frac{1}{4}}$ **n** $z^{-\frac{3}{4}}$ **o** $y^{-\frac{3}{5}}$ **p** $(2x + 5)^{\frac{1}{2}}$

q $(6q + r)^{\frac{1}{3}}$ **r** $(a + b)^{\frac{1}{9}}$ **s** $(3x - 1)^{-\frac{1}{2}}$ **t** $(x + 7)^{-\frac{2}{5}}$

4 Write in index form:

a $\sqrt{t}$ **b** $\sqrt[5]{y}$ **c** $\sqrt{x^3}$

d $\sqrt[3]{9 - x}$ **e** $\sqrt{4s + 1}$ **f** $\sqrt{(3x + 1)^5}$

g $\dfrac{1}{\sqrt{2t + 3}}$ **h** $\dfrac{1}{\sqrt{(5x - y)^3}}$ **i** $\dfrac{1}{\sqrt[3]{(x - 2)^2}}$

j $\dfrac{1}{2\sqrt{y + 7}}$ **k** $\dfrac{5}{\sqrt[3]{x+4}}$ **l** $\dfrac{1}{3\sqrt{y^2 - 1}}$

m $\dfrac{3}{5\sqrt[4]{(x^2 + 2)^3}}$

Foundation | Mastery | Complex

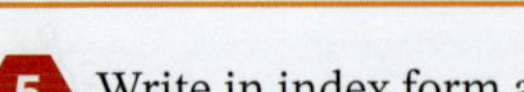

5 Write in index form and simplify:

a $x\sqrt{x}$ **b** $\frac{\sqrt{x}}{x}$ **c** $\frac{x}{\sqrt[3]{x}}$

d $\frac{x^2}{\sqrt[3]{x}}$ **e** $x\sqrt[4]{x}$

6 Write in root form:

a $(a - 2b)^{-\frac{1}{3}}$ **b** $(y - 3)^{-\frac{2}{3}}$ **c** $4(6a + 1)^{-\frac{4}{7}}$

d $\frac{(x + y)^{-\frac{5}{4}}}{3}$ **e** $\frac{6(3x + 8)^{-\frac{2}{9}}}{7}$

Did you know?

The beginnings of algebra

One of the earliest mathematicians to use algebra was **Diophantus of Alexandria** in Greece. It is not known when he lived, but it is thought this may have been around 250 CE.

In Persia around 700–800 CE, a mathematician named **Muhammad ibn Musa al-Khwarizmi** wrote books on algebra and Hindu numerals. One of his books was titled *Al-Jabr wa'l Muqabala*, and the word *algebra* comes from the first word in this title.

1.04 Simplifying algebraic expressions

Example 5

Simplify:

a $-5x \times 3y \times 2x$

b $\frac{5a^3b}{15ab^2}$

Solution

a $-5x \times 3y \times 2x = -30xyx$

$= -30x^2y$

b $\frac{5a^3b}{15ab^2} = \frac{1}{3}a^{3-1}b^{1-2}$

$= \frac{1}{3}a^2 b^{-1}$

$= \frac{a^2}{3b}$

Foundation | Mastery | Complex

When we remove grouping symbols, we say that we are **expanding** an expression.

Worksheet
Expanding algebraic expressions

Expanding expressions

To expand an expression, use the distributive law:

$$a(b+c)=ab+ac$$

Example 6

Expand and simplify:

a $5a^2(4+3ab-c)$ **b** $5-2(y+3)$ **c** $2(b-5)-(b+1)$

Solution

a $5a^2(4+3ab-c)=5a^2\times 4+5a^2\times 3ab-5a^2\times c$
$=20a^2+15a^3b-5a^2c$

b $5-2(y+3)=5-2\times y-2\times 3$
$=5-2y-6$
$=-2y-1$

c $2(b-5)-(b+1)=2\times b+2\times -5-1\times b-1\times 1$
$=2b-10-b-1$
$=b-11$

EXERCISE 1.04 Answers on p. 457

Simplifying algebraic expressions

1 Simplify:

a $6x-5y-y$ **b** $8a+b-4b-7a$ **c** $xy+2y+3xy$
d $2ab^2-5ab^2-3ab^2$ **e** $m^2-5m-m+12$ **f** $p^2-7p+5p-6$
g $ab+2b-3ab+8b$ **h** $ab+bc-ab-ac+bc$ **i** $a^5-7x^3+a^5-2x^3+1$
j $x^3-3xy^2+4x^2y-x^2y+xy^2+2y^3$

2 Simplify: EXAMPLE 5

a $2ab^3\times 3a$ **b** $5a^2b\times -2ab$ **c** $7pq^2\times 3p^2q^2$
d $5ab\times a^2b^2$ **e** $4h^3\times -2h^7$ **f** $k^3p\times p^2$
g $(-3t^3)^4$ **h** $7m^6\times -2m^5$ **i** $-2x^2\times 3x^3y\times -4xy^2$

3 Simplify:

a $\frac{8a^2}{a}$ **b** $\frac{8a^2}{2a}$ **c** $\frac{xy}{2x}$
d $12p^3\div 4p^2$ **e** $\frac{3a^2b^2}{6ab}$ **f** $\frac{20x}{15xy}$
g $-15ab\div -5b$ **h** $\frac{2ab}{6a^2b^3}$ **i** $\frac{-8p}{4pqs}$
j $\frac{2xy^2z^3}{4x^3y^2z}$ **k** $\frac{42p^5q^4}{7pq^3}$ **l** $5a^9b^4c^{-2}\div 20a^5b^{-3}c^{-1}$
m $\frac{2(a^{-5})^2b^4}{4a^{-9}(b^2)^{-1}}$ **n** $-5x^4y^7z\div 15xy^8z^{-2}$ **o** $-9(a^4b^{-1})^3\div -18a^{-1}b^3$

□ Foundation ○ Mastery ○ Complex

4 Expand and simplify each expression.

a	$2(x-4)$	**b**	$3(2h+3)$	**c**	$-5(a-2)$
d	$x(2y+3)$	**e**	$x(x-2)$	**f**	$2a(3a-8b)$
g	$ab(2a+b)$	**h**	$5n(n-4)$	**i**	$3x^2y(xy+2y^2)$
j	$3+4(k+1)$	**k**	$2(t-7)-3$	**l**	$y(4y+3)+8y$
m	$9-5(b+3)$	**n**	$3-(2x-5)$	**o**	$5(3-2m)+7(m-2)$
p	$2(h+4)+3(2h-9)$	**q**	$3(2d-3)-(5d-3)$	**r**	$a(2a+1)-(a^2+3a-4)$
s	$x(3x-4)-5(x+1)$	**t**	$2ab(3-a)-b(4a-1)$	**u**	$5x-(x-2)-3$
v	$8-4(2y+1)+y$	**w**	$(a+b)-(a-b)$	**x**	$2(3t-4)-(t+1)+3$

1.05 Binomial products

Video Expanding binomial products

Worksheet Expanding expressions

A **binomial** expression consists of 2 **terms**; for example, $x+3$.

A set of 2 binomial expressions multiplied together is called a **binomial product**; for example, $(x+3)(x-2)$.

To expand, each term in the first bracket is multiplied by each term in the second bracket.

Binomial product

$$(x+a)(x+b)=x^2+bx+ax+ab$$

Example 7

Expand and simplify:

a $(p+3)(q-4)$ **b** $(a+5)^2$ **c** $(x+4)(2x-3y-1)$

Solution

a $(p+3)(q-4)=pq-4p+3q-12$

b $(a+5)^2=(a+5)(a+5)$
$=a^2+5a+5a+25$
$=a^2+10a+25$

c $(x+4)(2x-3y-1)=2x^2-3xy-x+8x-12y-4$
$=2x^2-3xy+7x-12y-4$

Special binomial products

Video Special binomial products

Difference of two squares

$$(a+b)(a-b)=a^2-b^2$$

Perfect squares

$$(a+b)^2=a^2+2ab+b^2$$
$$(a-b)^2=a^2-2ab+b^2$$

☐ Foundation ◯ Mastery ⬡ Complex

Example 8

Expand and simplify:

a $(2x-3)^2$ **b** $(3y-4)(3y+4)$

Solution

a $(2x-3)^2 = (2x)^2 - 2(2x)3 + 3^2$
$= 4x^2 - 12x + 9$

b $(3y-4)(3y+4) = (3y)^2 - 4^2$
$= 9y^2 - 16$

EXERCISE 1.05 Answers on p. 457

Binomial products

1 Expand and simplify:

a	$(a+5)(a+2)$	**b**	$(x+3)(x-1)$	**c**	$(2y-3)(y+5)$
d	$(m-4)(m-2)$	**e**	$(x+4)(x+3)$	**f**	$(y+2)(y-5)$
g	$(2x-3)(x+2)$	**h**	$(h-7)(h-3)$	**i**	$(x+5)(x-5)$
j	$(5a-4)(3a-1)$	**k**	$(2y+3)(4y-3)$	**l**	$(x-4)(y+7)$
m	$(x^2+3)(x-2)$	**n**	$(n+2)(n-2)$	**o**	$(a+2b)(a-2b)$
p	$(3x-4y)(3x+4y)$	**q**	$(x+9)(x-2y+2)$	**r**	$(b-3)(2a+2b-1)$
s	$(x+2)(x^2-2x+4)$	**t**	$(a-3)(a^2+3a+9)$	**u**	$(a+9)^2$
v	$(k-4)^2$	**w**	$(3a+4b)^2$	**x**	$(x-5y)^2$

2 Prove that:

a $(a+b)(a-b) = a^2 - b^2$ **b** $(a+b)^2 = a^2 + 2ab + b^2$

c $(a-b)^2 = a^2 - 2ab + b^2$

3 Expand and simplify:

a $\left(x-\frac{2}{3}\right)\left(x+\frac{1}{3}\right)$ **b** $\left(\frac{y}{2}-1\right)\left(y+\frac{1}{3}\right)$ **c** $\left(m+\frac{1}{m}\right)\left(m^2-3\right)$

d $\left(x+\frac{1}{x}\right)\left(x-\frac{1}{x}\right)$ **e** $\left(x^2-\frac{1}{x}\right)\left(x-\frac{1}{x^2}\right)$

EXAMPLE 8

4 Expand and simplify:

a	$(t+4)^2$	**b**	$(z-6)^2$	**c**	$(x-1)^2$
d	$(y+8)^2$	**e**	$(q+3)^2$	**f**	$(k-7)^2$
g	$(n+1)^2$	**h**	$(2b+5)^2$	**i**	$(3-x)^2$
j	$(3y-1)^2$	**k**	$(x+y)^2$	**l**	$(3a-b)^2$
m	$(4d+5e)^2$	**n**	$(t+4)(t-4)$	**o**	$(x-3)(x+3)$
p	$(p+1)(p-1)$	**q**	$(r+6)(r-6)$	**r**	$(x-10)(x+10)$
s	$(2a+3)(2a-3)$	**t**	$(x-5y)(x+5y)$	**u**	$(4a+1)(4a-1)$
v	$(7-3x)(7+3x)$	**w**	$(x^2+2)(x^2-2)$	**x**	$(x^2+5)^2$

Foundation Mastery Complex

5 Expand and simplify.

a $(3ab - 4c)(3ab + 4c)$

b $\left(x + \frac{2}{x}\right)^2$

c $\left(a - \frac{1}{a}\right)\left(a + \frac{1}{a}\right)$

d $[x + (y - 2)][x - (y - 2)]$

e $[(a + b) + c]^2$

f $[(x + 1) - y]^2$

g $(a + 3)^2 - (a - 3)^2$

h $16 - (z - 4)(z + 4)$

i $2x + (3x + 1)^2 - 4$

j $(x + y)^2 - x(2 - y)$

k $(4n - 3)(4n + 3) - 2n^2 + 5$

l $(x - 4)^3$

m $\left(x - \frac{1}{x}\right)^2 - \left(\frac{1}{x}\right)^2 + 2$

n $(x^2 + y^2)^2 - 4x^2y^2$

o $(2a + 5)^3$

6 Show that $(a + b)^3 = a^3 + 3a^2b + 3ab^2 + b^3$.

7 Show that $(a - b)^3 = a^3 - 3a^2b + 3ab^2 - b^3$.

8 Expand and simplify:

a $3(2y + 1)^2 - (y + 4)^2$

b $(h - 2)^2 - 3(h - 7)^2$

c $(2x - 5)^2 - (x - 3)(x + 3)$

d $(p + 5)^2 + (3p - 1)^2$

e $(x + 2y)^2 + (y - 2x)^2 - (2y - x)(2y + x)$

1.06 Factorising expressions

Worksheet
Factorising algebraic expressions

Factors divide exactly into an equal or larger number or term, without leaving a remainder.

Factorising

To **factorise** an expression, we use the distributive law in the opposite way from when we expand brackets.

$$ax + bx = x(a + b)$$

$$-ax - bx = -x(a + b)$$

Example 9

Factorise:

a $x^3 - 2x^2$

b $5(x + 3) + 2y(x + 3)$

c $-8a^3b^2 - 2ab^3$

Solution

a x and x^2 are both common factors. Take out the highest common factor, which is x^2.

$x^3 - 2x^2 = x^2(x - 2)$

b The highest common factor is $x + 3$.

$5(x + 3) + 2y(x + 3) = (x + 3)(5 + 2y)$

c The highest common factor is $-2ab^2$.

$-8a^3b^2 - 2ab^3 = -2ab^2(4a^2 + b)$

☐ Foundation ○ Mastery ○ Complex

Factorising by grouping in pairs

If an expression has 4 terms, it can sometimes be factorised in pairs.

$$\begin{aligned} ax + bx + ay + by &= x(a + b) + y(a + b) \\ &= (a + b)(x + y) \end{aligned}$$

Example 10

Factorise:

a $x^2 - 2x + 3x - 6$ **b** $2x - 4 + 6y - 3xy$

Solution

a $\begin{aligned} x^2 - 2x + 3x - 6 &= x(x - 2) + 3(x - 2) \\ &= (x - 2)(x + 3) \end{aligned}$

b $\begin{aligned} 2x - 4 + 6y - 3xy &= 2(x - 2) + 3y(2 - x) \\ &= 2(x - 2) - 3y(x - 2) \\ &= (x - 2)(2 - 3y) \end{aligned}$

EXERCISE 1.06 Answers on p. 457

Factorising expressions

1 Factorise: EXAMPLE 9

a	$m^2 - 3m$	**b**	$2y^2 + 4y$	**c**	$-15a - 3a^2$
d	$ab^2 + ab$	**e**	$4x^2y - 2xy$	**f**	$3mn^3 + 9mn$
g	$-8x^2z - 2xz^2$	**h**	$6ab + 3a - 2a^2$	**i**	$5x^2 - 2x + xy$
j	$3q^5 - 2q^2$	**k**	$5b^3 + 15b^2$	**l**	$-6a^2b^3 - 3a^3b^2$
m	$x(m + 5) + 7(m + 5)$	**n**	$2(y - 1) - y(y - 1)$	**o**	$4(7 + y) - 3x(7 + y)$
p	$6x(a - 2) + 5(a - 2)$	**q**	$-x(2t + 1) - y(2t + 1)$	**r**	$35m^3n^4 - 25m^2n$
s	$24a^2b^5 + 16ab^2$	**t**	$2\pi r^2 + 2\pi rh$	**u**	$(x - 3)^2 + 5(x - 3)$
v	$y^2(x + 4) + 2(x + 4)$	**w**	$-a(a + 1) - (a + 1)^2$		

2 Factorise: EXAMPLE 10

a	$2x + 8 + bx + 4b$	**b**	$ay - 3a + by - 3b$	**c**	$x^2 + 5x + 2x + 10$
d	$m^2 - 2m + 3m - 6$	**e**	$ad - ac + bd - bc$	**f**	$x^3 + x^2 + 3x + 3$
g	$5ab - 3b + 10a - 6$	**h**	$2xy - x^2 + 2y^2 - xy$	**i**	$ay + a + y + 1$
j	$x^2 + 5x - x - 5$	**k**	$y + 3 + ay + 3a$	**l**	$m - 2 + 4y - 2my$
m	$2x^2 + 10xy - 3xy - 15y^2$	**n**	$a^2b + ab^3 - 4a - 4b^2$	**o**	$5x - x^2 - 3x + 15$
p	$x^4 + 7x^3 - 4x - 28$	**q**	$7x - 21 - xy + 3y$	**r**	$4d + 12 - de - 3e$
s	$xy + 7x - 4y - 28$	**t**	$x^4 - 4x^3 - 5x + 20$	**u**	$4x^3 - 6x^2 + 8x - 12$
v	$3a^2 + 9a + 6ab + 18b$	**w**	$5y - 15 + 10xy - 30x$	**x**	$\pi r^2 + 2\pi r - 3r - 6$

3 Factorise and simplify each expression.

a $3^{x+2} + 3^x$ **b** $5^{n+1} + 5^n$ **c** $2^{n-1} - 2^{n-2}$ **d** $\dfrac{5^{n+3} + 5^{n+1}}{5^{n+1} - 5^n}$

4 Factorise and simplify each expression.

a $3^{2x+1} + 3^{x+1} + 7 \times 3^x + 7$

b $5^{2b-1} + 4 \times 5^{b-1} + 3 \times 5^b + 12$

c $7^{2b+1} - 2 \times 7^b + 5 \times 7^{b+1} - 10$

d $3^{2a+1} - 2 \times 3^a - 7 \times 3^{a+1} + 14$

e $2^{2n+1} + 2^{n+1} - 3 \times 2^n - 3$

□ Foundation ○ Mastery ⬡ Complex

1.07 Factorising trinomials

Video
Factorising quadratic expressions

A **trinomial** is an expression with 3 terms; for example, $x^2 - 4x + 3$. Factorising a trinomial usually gives a **binomial product**.

We know that: $(x + a)(x + b) = x^2 + bx + ax + ab$

$$= x^2 + (a + b)x + ab$$

Factorising trinomials

$$x^2 + (a + b)x + ab = (x + a)(x + b)$$

Find values for a and b so that the sum $a + b$ is the coefficient of the middle term and the product ab is the last term.

Example 11

Factorise each trinomial.

a $m^2 - 5m + 6$ **b** $y^2 + y - 2$

Solution

a $a + b = -5$ and $ab = 6$

For $ab = 6$, a and b have the same sign: they are both positive or both negative.

For $a + b = -5$, at least one must be negative.

So both are negative.

For $ab = 6$: we could have $-6 \times (-1)$ or $-3 \times (-2)$.

$-3 + (-2) = -5$, so $a = -3$ and $b = -2$.

So $m^2 - 5m + 6 = (m - 3)(m - 2)$.

Check this is correct by expanding $(m - 3)(m - 2)$.

b $a + b = 1$ and $ab = -2$

For $ab = -2$, a and b must have opposite signs: one is positive and one is negative.

For $ab = -2$: we could have -2×1 or -1×2.

$-1 + 2 = 1$, so $a = -1$ and $b = 2$.

So $y^2 + y - 2 = (y - 1)(y + 2)$.

Check this is correct by expanding $(y - 1)(y + 2)$.

Video
Factorising quadratic expressions (Advanced)

When the **coefficient** of the first term is not 1, for example, $5x^2 - 13x + 6$, we need to use a different method to factorise the trinomial.

This method still involves finding 2 numbers that give a required sum and product, but it also involves grouping in pairs.

The coefficient of the first term is the number in front of the x^2.

Example 12

Factorise each trinomial

a $5x^2 - 13x + 6$ **b** $4y^2 + 4y - 3$

Solution

a First, multiply the coefficient of the first term by the last term: $5 \times 6 = 30$

$a + b = -13$ and $ab = 30$.

For $ab = 30$, a and b have the same sign: they are both positive or both negative.

For $a + b = -13$, at least one must be negative.

So both are negative.

For $ab = 30$, we could have -10 and -3.

$-10 + (-3) = -13$, so $a = -10$ and $b = -3$.

Now write the trinomial with the middle term split into 2 terms $-10x$ and $-3x$, and then factorise by grouping in pairs.

$$
\begin{aligned}
5x^2 - 13x + 6 &= 5x^2 - 10x - 3x + 6 \\
&= 5x(x-2) - 3(x-2) \\
&= (x-2)(5x-3)
\end{aligned}
$$

If you factorise correctly, you should always find a common factor remaining, such as (x – 2) here.

b First, multiply the coefficient of the first term by the last term: $4(-3) = -12$

$a + b = 4$ and $ab = -12$.

For $ab = -12$, a and b have opposite signs: one is positive, one is negative.

We could have 6 and -2.

$6 + (-2) = 4$, so $a = 6$ and $b = -2$.

Now write the trinomial with the middle term split into 2 terms $6y$ and $-2y$, and then factorise by grouping in pairs.

$$
\begin{aligned}
4y^2 + 4y - 3 &= 4y^2 + 6y - 2y - 3 \\
&= 2y(2y+3) - 1(2y+3) \\
&= (2y+3)(2y-1)
\end{aligned}
$$

There are other ways of factorising these trinomials. Your teacher may show you some of these.

Technology
Factorising trinomials

Spreadsheet
Factorising trinomials

EXERCISE 1.07 Answers on p. 458

Factorising trinomials

1 Factorise: EXAMPLE 11

a $x^2 + 4x + 3$ **b** $y^2 + 7y + 12$ **c** $m^2 + 2m + 1$

d $t^2 + 8t + 16$ **e** $z^2 + z - 6$ **f** $x^2 - 5x - 6$

g $v^2 - 8v + 15$ **h** $t^2 - 6t + 9$ **i** $x^2 + 9x - 10$

j $y^2 - 10y + 21$ **k** $m^2 - 9m + 18$ **l** $y^2 + 9y - 36$

m $x^2 - 5x - 24$ **n** $a^2 - 4a + 4$ **o** $x^2 + 14x - 32$

p $y^2 - 5y - 36$ **q** $n^2 - 10n + 24$ **r** $x^2 - 10x + 25$

2 Factorise each trinomial by first taking out the common factor.

a $3p^2 - 3p - 36$ **b** $2x^2 - 10x + 8$ **c** $5y^2 + 15y - 50$

d $11a^2 - 55a + 66$ **e** $x^3 + 6x^2 + 5x$ **f** $x^3 - 3x^2 - 10x$

3 Factorise each trinomial.

a $2^{2x} + 10 \times 2^x + 21$ **b** $7^{2k} - 8 \times 7^k + 15$ **c** $3^{2m} + 2 \times 3^m - 8$

d $5^{2h} + 3 \times 5^h - 4$ **e** $3^{2n} - 11 \times 3^n + 28$

Foundation Mastery Complex

EXAMPLE 12

4 Factorise:

a	$2a^2 + 11a + 5$	**b**	$5y^2 + 7y + 2$	**c**	$3x^2 + 10x + 7$
d	$3x^2 + 8x + 4$	**e**	$2b^2 - 5b + 3$	**f**	$7x^2 - 9x + 2$
g	$3y^2 + 5y - 2$	**h**	$2x^2 + 11x + 12$	**i**	$5p^2 + 13p - 6$
j	$6x^2 - x - 12$	**k**	$6y^2 + 47y - 8$	**l**	$4n^2 - 11n + 6$
m	$8t^2 + 18t - 5$	**n**	$12q^2 + 23q + 10$	**o**	$4r^2 + 11r - 3$
p	$4x^2 - 4x - 15$	**q**	$6y^2 - 13y + 2$	**r**	$16y^2 + 24y + 9$
s	$25k^2 - 20k + 4$	**t**	$36a^2 - 12a + 1$	**u**	$49m^2 + 84m + 36$

5 Factorise each trinomial by first taking out the common factor.

a	$6a^2 + 39a + 60$	**b**	$15y^2 + 35y - 30$	**c**	$20m^2 - 34m + 6$
d	$6x^2 + 8x - 8$	**e**	$18x^2 + 33x - 30$	**f**	$2x^3 + x^2 - 36x$
g	$3y^3 - 5y^2 + 2y$	**h**	$6n^4 + 19n^3 + 10n^2$	**i**	$63b^3 - 147b^2 + 70b$

6 Factorise each trinomial.

a	$3 \times 2^{2n} + 13 \times 2^n + 12$	**b**	$5 \times 3^{2x} + 19 \times 3^x - 4$
c	$2 \times 7^{2k} - 13 \times 7^k + 15$	**d**	$6 \times 5^{2m} - 8 \times 5^m - 8$
e	$15 \times 2^{2x} + 26 \times 2^x - 21$		

1.08 Factorising special binomial products

Worksheet Factorising expressions

You have looked at expanding $(a + b)^2 = a^2 + 2ab + b^2$ and $(a - b)^2 = a^2 - 2ab + b^2$. These are called **perfect squares**.

When factorising, use these results the other way around.

Perfect squares

$$a^2 + 2ab + b^2 = (a + b)^2$$
$$a^2 - 2ab + b^2 = (a - b)^2$$

Example 13

Factorise:

a $x^2 - 8x + 16$ **b** $4a^2 + 20a + 25$

Solution

a $x^2 - 8x + 16 = x^2 - 2(4)x + 4^2$
$= (x - 4)^2$

b $4a^2 + 20a + 25 = (2a)^2 + 2(2a)(5) + 5^2$
$= (2a + 5)^2$

☐ Foundation ○ Mastery ⬡ Complex

Difference of two squares

$$a^2 - b^2 = (a + b)(a - b)$$

Example 14

Factorise:

a $d^2 - 36$ **b** $1 - 9b^2$ **c** $(a + 3)^2 - (b - 1)^2$

Solution

a $d^2 - 36 = d^2 - 6^2$
$= (d + 6)(d - 6)$

b $1 - 9b^2 = 1^2 - (3b)^2$
$= (1 + 3b)(1 - 3b)$

c $(a + 3)^2 - (b - 1)^2 = [(a + 3) + (b - 1)][(a + 3) - (b - 1)]$
$= (a + 3 + b - 1)(a + 3 - b + 1)$
$= (a + b + 2)(a - b + 4)$

EXERCISE 1.08 Answers on p. 458

Factorising special binomial products

1 Factorise: EXAMPLE 13

a $y^2 - 2y + 1$ **b** $x^2 + 6x + 9$ **c** $m^2 + 10m + 25$
d $t^2 - 4t + 4$ **e** $x^2 - 12x + 36$ **f** $4x^2 + 12x + 9$
g $16b^2 - 8b + 1$ **h** $9a^2 + 12a + 4$ **i** $25x^2 - 40x + 16$
j $49y^2 + 14y + 1$ **k** $9y^2 - 30y + 25$ **l** $16k^2 - 24k + 9$
m $3x^2 + 6x + 3$ **n** $5a^2 - 30a + 45$ **o** $7y^2 - 70y + 175$

2 Factorise: EXAMPLE 14

a $a^2 - 4$ **b** $x^2 - 9$ **c** $y^2 - 1$
d $x^2 - 25$ **e** $4x^2 - 49$ **f** $16y^2 - 9$
g $1 - 4z^2$ **h** $25t^2 - 1$ **i** $9t^2 - 4$
j $9 - 16x^2$ **k** $x^2 - 4y^2$ **l** $36x^2 - y^2$
m $4a^2 - 9b^2$ **n** $x^2 - 100y^2$ **o** $4a^2 - 81b^2$
p $(x + 2)^2 - y^2$ **q** $(a - 1)^2 - (b - 2)^2$ **r** $z^2 - (1 + w)^2$
s $x^2 - \frac{1}{4}$ **t** $\frac{y^2}{9} - 1$ **u** $(x + 2)^2 - (2y + 1)^2$
v $x^4 - 1$ **w** $9x^6 - 4y^2$ **x** $x^4 - 16y^4$

3 Factorise:

a $4a^3 - 36a$ **b** $2x^2 - 18$ **c** $3p^2 - 3p - 36$
d $5y^2 - 5$ **e** $5a^2 - 10a + 5$ **f** $3z^3 + 27z^2 + 60z$
g $9ab - 4a^3b^3$ **h** $x^3 - x$ **i** $6x^2 + 8x - 8$
j $y^2(y + 5) - 16(y + 5)$ **k** $x^4 + 8x^3 - x^2 - 8x$ **l** $y^6 - 4$
m $x^3 - 3x^2 - 10x$ **n** $x^3 - 3x^2 - 9x + 27$ **o** $4x^2y^3 - y$
p $24 - 6b^2$ **q** $18x^2 + 33x - 30$ **r** $3x^2 - 6x + 3$
s $x^3 + 2x^2 - 25x - 50$ **t** $z^3 + 6z^2 + 9z$ **u** $3y^2 + 30y + 75$
v $ab^2 - 9a$ **w** $4k^3 + 40k^2 + 100k$ **x** $3x^3 + 9x^2 - 3x - 9$
y $4a^3b + 8a^2b^2 - 4ab^2 - 2a^2b$

☐ Foundation ◯ Mastery ⬡ Complex

1.09 Algebraic fractions

Example 15

Simplify:

a $\dfrac{4x+2}{2}$

b $\dfrac{2x^2-3x-2}{x^2-4}$

Solution

a $\dfrac{4x+2}{2} = \dfrac{2(2x+1)}{2}$

$= 2x+1$

b Factorise both top and bottom.

$$\frac{2x^2-3x-2}{x^2-4} = \frac{(2x+1)(x-2)}{(x-2)(x+2)}$$

$$= \frac{2x+1}{x+2}$$

Worksheet Algebraic fractions

Example 16

Simplify:

a $\dfrac{x-1}{5} - \dfrac{x+3}{4}$

b $\dfrac{2a^2b+10ab}{b^2-9} \div \dfrac{a^2-25}{4b+12}$

c $\dfrac{2}{x-5} + \dfrac{1}{x+2}$

d $\dfrac{2}{x+1} - \dfrac{1}{x^2-1}$

Solution

a $\dfrac{x-1}{5} - \dfrac{x+3}{4} = \dfrac{4(x-1)-5(x+3)}{20}$

$$= \frac{4x-4-5x-15}{20}$$

$$= \frac{-x-19}{20}$$

b $\dfrac{2a^2b+10ab}{b^2-9} \div \dfrac{a^2-25}{4b+12} = \dfrac{2a^2b+10ab}{b^2-9} \times \dfrac{4b+12}{a^2-25}$

$$= \frac{2ab(a+5)}{(b+3)(b-3)} \times \frac{4(b+3)}{(a+5)(a-5)}$$

$$= \frac{8ab}{(a-5)(b-3)}$$

c $\dfrac{2}{x-5} + \dfrac{1}{x+2} = \dfrac{2(x+2)+1(x-5)}{(x-5)(x+2)}$

$$= \frac{2x+4+x-5}{(x-5)(x+2)}$$

$$= \frac{3x-1}{(x-5)(x+2)}$$

d $\frac{2}{x+1} - \frac{1}{x^2-1} = \frac{2}{x+1} - \frac{1}{(x+1)(x-1)}$

$= \frac{2(x-1)}{(x+1)(x-1)} - \frac{1}{(x+1)(x-1)}$

$= \frac{2x-2}{(x+1)(x-1)} - \frac{1}{(x+1)(x-1)}$

$= \frac{2x-2-1}{(x+1)(x-1)}$

$= \frac{2x-3}{(x+1)(x-1)}$

EXERCISE 1.09 Answers on p. 458

Algebraic fractions

EXAMPLE 15

1 Simplify:

a $\frac{5a+10}{5}$ **b** $\frac{6t-3}{3}$ **c** $\frac{8y+2}{6}$

d $\frac{8}{4d-2}$ **e** $\frac{x^2}{5x^2-2x}$ **f** $\frac{y-4}{y^2-8y+16}$

g $\frac{2ab-4a^2}{a^2-3a}$ **h** $\frac{s^2+s-2}{s^2+5s+6}$ **i** $\frac{b^4-1}{b^2-1}$

j $\frac{2p^2+7p-15}{6p-9}$ **k** $\frac{a^2-1}{a^2+2a-3}$ **l** $\frac{3(x-2)+y(x-2)}{x^2-4}$

m $\frac{x^3+3x^2-9x-27}{x^2+6x+9}$ **n** $\frac{2p^2-3p-2}{2p^2+p}$ **o** $\frac{ay-ax+by-bx}{2ay-by-2ax+bx}$

2 Simplify:

a $\frac{x}{2} + \frac{3x}{4}$ **b** $\frac{y+1}{5} + \frac{2y}{3}$ **c** $\frac{a+2}{3} - \frac{a}{4}$

d $\frac{p-3}{6} + \frac{p+2}{2}$ **e** $\frac{x-5}{2} - \frac{x-1}{3}$

3 Simplify:

a $\frac{3x+6}{5} \times \frac{10}{x+2}$ **b** $\frac{a^2-4}{3} \times \frac{5b}{a+2}$

c $\frac{t^2+3t-10}{xy^2} \div \frac{5t-10}{2xy}$ **d** $\frac{2a-6}{2x+4} \times \frac{5x+10}{4}$

e $\frac{5x+10-xy-2y}{15} \div \frac{7x+14}{3}$ **f** $\frac{3}{b+2} \times \frac{b^2+2b}{6a-3}$

g $\frac{3ab^2}{5xy} \div \frac{12ab-6a}{x^2y+2xy^2}$ **h** $\frac{ax-ay+bx-by}{x^2-y^2} \times \frac{x^2y+xy^2}{ab^2+a^2b}$

i $\frac{x^2-6x+9}{x^2-25} \div \frac{x^2-5x+6}{x^2+4x-5}$ **j** $\frac{p^2-4}{q^2+2q+1} \times \frac{5q+5}{3p+6}$

☐ Foundation ○ Mastery ○ Complex

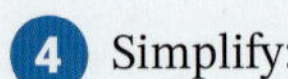

4 Simplify:

a $\frac{2}{x}+\frac{3}{x}$ b $\frac{1}{x-1}-\frac{2}{x}$ c $1+\frac{3}{a+b}$

d $x-\frac{x^2}{x+2}$ e $p-q+\frac{1}{p+q}$ f $\frac{1}{x+1}+\frac{1}{x-3}$

g $\frac{2}{x^2-4}-\frac{3}{x+2}$ h $\frac{1}{a^2+2a+1}+\frac{1}{a+1}$

5 Simplify:

a $\frac{a^2-5a}{y^2-4y+4}\div\frac{3a-15}{y^2-4}\times\frac{y^2-y-2}{5ay}$ b $\frac{3}{x-3}+\frac{2x+8}{x^2-9}\times\frac{x^2+3x}{4x-16}$

c $\frac{5b}{2b+6}\div\frac{b^2}{b^2+b-6}-\frac{b}{b+1}$ d $\frac{x^2-8x+15}{5x^2+10x}\div\frac{x^2-9}{10x^2}\times\frac{x^2+5x+6}{2x-10}$

6 Simplify:

a $\frac{5}{x^2-4}-\frac{3}{x-2}-\frac{2}{x+2}$ b $\frac{2}{p^2+pq}+\frac{3}{pq-q^2}$ c $\frac{a}{a+b}-\frac{b}{a-b}+\frac{1}{a^2-b^2}$

1.10 Surds

Videos Simplifying surds 1

Simplifying surds 2

Binomial products involving surds

Worksheet Multiplying and dividing surds

Puzzles Simplifying surds

Surds code puzzle

An **irrational number** is a number that cannot be written as a fraction $\frac{a}{b}$, where a and b are integers. **Surds** such as $\sqrt{2}$, $\sqrt{3}$ and $\sqrt{5}$ are special types of irrational numbers.

If a question involving surds asks for an exact answer, then leave it as a surd.

Properties of surds

$$\sqrt{a}\times\sqrt{b}=\sqrt{ab}$$

$$\frac{\sqrt{a}}{\sqrt{b}}=\sqrt{\frac{a}{b}}$$

$$\left(\sqrt{x}\right)^2=\sqrt{x^2}=x \text{ for } x\geq 0$$

Example 17

a Express $\sqrt{45}$ in simplest surd form.

b Simplify $3\sqrt{40}$.

c Write $5\sqrt{2}$ as a single surd.

Solution

a
$$\begin{aligned}\sqrt{45}&=\sqrt{9\times 5}\\&=\sqrt{9}\times\sqrt{5}\\&=3\times\sqrt{5}\\&=3\sqrt{5}\end{aligned}$$

b
$$\begin{aligned}3\sqrt{40}&=3\times\sqrt{4}\times\sqrt{10}\\&=3\times 2\times\sqrt{10}\\&=6\sqrt{10}\end{aligned}$$

c
$$\begin{aligned}5\sqrt{2}&=\sqrt{25}\times\sqrt{2}\\&=\sqrt{50}\end{aligned}$$

Foundation Mastery Complex

Example 18

Simplify $\sqrt{3} - \sqrt{12}$.

Solution

First, simplify the surds.

$$\begin{aligned}\sqrt{3} - \sqrt{12} &= \sqrt{3} - \sqrt{4} \times \sqrt{3} \\ &= \sqrt{3} - 2\sqrt{3} \\ &= -\sqrt{3}\end{aligned}$$

Multiplication and division, as in algebra, are easier to do than adding and subtracting.

Example 19

Simplify:

a $4\sqrt{2} \times 5\sqrt{18}$ **b** $\dfrac{2\sqrt{14}}{4\sqrt{2}}$ **c** $\left(\sqrt{\dfrac{10}{3}}\right)^2$

Solution

a $$\begin{aligned}4\sqrt{2} \times 5\sqrt{18} &= 20\sqrt{36} \\ &= 20 \times 6 \\ &= 120\end{aligned}$$

b $$\begin{aligned}\frac{2\sqrt{14}}{4\sqrt{2}} &= \frac{2 \times \sqrt{7}}{4} \\ &= \frac{\sqrt{7}}{2}\end{aligned}$$

c $$\begin{aligned}\left(\sqrt{\frac{10}{3}}\right)^2 &= \frac{10}{3} \\ &= 3\frac{1}{3}\end{aligned}$$

Example 20

Expand and simplify:

a $3\sqrt{7}(2\sqrt{3} - 3\sqrt{2})$ **b** $(\sqrt{2} + 3\sqrt{5})(\sqrt{3} - \sqrt{2})$ **c** $(\sqrt{5} + 2\sqrt{3})(\sqrt{5} - 2\sqrt{3})$

Solution

a $$\begin{aligned}3\sqrt{7}(2\sqrt{3} - 3\sqrt{2}) &= 3\sqrt{7} \times 2\sqrt{3} - 3\sqrt{7} \times 3\sqrt{2} \\ &= 6\sqrt{21} - 9\sqrt{14}\end{aligned}$$

b $$\begin{aligned}(\sqrt{2} + 3\sqrt{5})(\sqrt{3} - \sqrt{2}) &= \sqrt{2} \times \sqrt{3} - \sqrt{2} \times \sqrt{2} + 3\sqrt{5} \times \sqrt{3} - 3\sqrt{5} \times \sqrt{2} \\ &= \sqrt{6} - 2 + 3\sqrt{15} - 3\sqrt{10}\end{aligned}$$

c Using the difference of two squares: $$\begin{aligned}(\sqrt{5} + 2\sqrt{3})(\sqrt{5} - 2\sqrt{3}) &= \left(\sqrt{5}\right)^2 - \left(2\sqrt{3}\right)^2 \\ &= 5 - 4 \times 3 \\ &= -7\end{aligned}$$

EXERCISE 1.10 Answers on p. 458

Surds

1 Simplify each surd.

a $\sqrt{12}$ **b** $\sqrt{63}$ **c** $\sqrt{24}$ **d** $\sqrt{50}$ **e** $\sqrt{72}$

f $\sqrt{200}$ **g** $\sqrt{48}$ **h** $\sqrt{75}$ **i** $\sqrt{32}$ **j** $\sqrt{54}$

k $\sqrt{112}$ **l** $\sqrt{300}$ **m** $\sqrt{128}$ **n** $\sqrt{243}$ **o** $\sqrt{245}$

p $\sqrt{108}$ **q** $\sqrt{99}$ **r** $\sqrt{125}$

2 Simplify:

a $2\sqrt{27}$ **b** $5\sqrt{80}$ **c** $4\sqrt{98}$ **d** $2\sqrt{28}$ **e** $8\sqrt{20}$

f $4\sqrt{56}$ **g** $8\sqrt{405}$ **h** $15\sqrt{8}$ **i** $7\sqrt{40}$ **j** $8\sqrt{45}$

3 Write as a single surd:

a $3\sqrt{2}$ **b** $2\sqrt{5}$ **c** $4\sqrt{11}$ **d** $8\sqrt{2}$ **e** $5\sqrt{3}$

f $4\sqrt{10}$ **g** $3\sqrt{13}$ **h** $7\sqrt{2}$ **i** $11\sqrt{3}$ **j** $12\sqrt{7}$

EXAMPLE 18

4 Simplify:

a $\sqrt{5} + 4\sqrt{5} + 3\sqrt{5}$ **b** $\sqrt{2} - 2\sqrt{2} - 3\sqrt{2}$ **c** $\sqrt{5} + \sqrt{45}$

d $\sqrt{8} - \sqrt{2}$ **e** $\sqrt{3} + \sqrt{48}$ **f** $\sqrt{12} - \sqrt{27}$

g $\sqrt{50} - \sqrt{32}$ **h** $\sqrt{28} + \sqrt{63}$ **i** $2\sqrt{8} - \sqrt{18}$

j $3\sqrt{54} + 2\sqrt{24}$ **k** $\sqrt{90} - 5\sqrt{40} - 2\sqrt{10}$ **l** $4\sqrt{48} + 3\sqrt{147} + 5\sqrt{12}$

m $3\sqrt{2} + \sqrt{8} - \sqrt{12}$ **n** $\sqrt{63} - \sqrt{28} - \sqrt{50}$ **o** $\sqrt{12} - \sqrt{45} - \sqrt{48} - \sqrt{5}$

5 Simplify:

a $\sqrt{7} \times \sqrt{3}$ **b** $\sqrt{3} \times \sqrt{5}$ **c** $\sqrt{2} \times 3\sqrt{3}$

d $5\sqrt{7} \times 2\sqrt{2}$ **e** $-3\sqrt{3} \times 2\sqrt{2}$ **f** $5\sqrt{3} \times 2\sqrt{3}$

g $-4\sqrt{5} \times 3\sqrt{11}$ **h** $2\sqrt{7} \times \sqrt{7}$ **i** $2\sqrt{3} \times 5\sqrt{12}$

j $\sqrt{6} \times \sqrt{2}$ **k** $\left(\sqrt{2}\right)^2$ **l** $\left(2\sqrt{7}\right)^2$

m $\sqrt{3} \times \sqrt{5} \times \sqrt{2}$ **n** $2\sqrt{3} \times \sqrt{7} \times -\sqrt{5}$ **o** $\sqrt{2} \times \sqrt{6} \times 3\sqrt{3}$

6 Simplify:

a $\dfrac{4\sqrt{12}}{2\sqrt{2}}$ **b** $\dfrac{12\sqrt{18}}{3\sqrt{6}}$ **c** $\dfrac{5\sqrt{8}}{10\sqrt{2}}$ **d** $\dfrac{16\sqrt{2}}{2\sqrt{12}}$

e $\dfrac{10\sqrt{30}}{5\sqrt{10}}$ **f** $\dfrac{2\sqrt{2}}{6\sqrt{20}}$ **g** $\dfrac{4\sqrt{2}}{8\sqrt{10}}$ **h** $\dfrac{\sqrt{3}}{3\sqrt{15}}$

i $\dfrac{\sqrt{2}}{\sqrt{8}}$ **j** $\dfrac{3\sqrt{15}}{6\sqrt{10}}$ **k** $\left(\sqrt{\dfrac{2}{3}}\right)^2$ **l** $\left(\sqrt{\dfrac{5}{7}}\right)^2$

EXAMPLE 20

7 Expand and simplify:

a $\sqrt{2}\left(\sqrt{5} + \sqrt{3}\right)$ **b** $\sqrt{3}\left(2\sqrt{2} - \sqrt{5}\right)$ **c** $4\sqrt{3}\left(\sqrt{3} + 2\sqrt{5}\right)$

d $\sqrt{7}\left(5\sqrt{2} - 2\sqrt{3}\right)$ **e** $-\sqrt{3}\left(\sqrt{2} - 4\sqrt{6}\right)$ **f** $\sqrt{3}\left(5\sqrt{11} + 3\sqrt{7}\right)$

g $-3\sqrt{2}\left(\sqrt{2} + 4\sqrt{3}\right)$ **h** $\sqrt{5}\left(\sqrt{5} - 5\sqrt{3}\right)$ **i** $\sqrt{3}\left(\sqrt{12} + \sqrt{10}\right)$

j $2\sqrt{3}\left(\sqrt{18} + \sqrt{3}\right)$ **k** $-4\sqrt{2}\left(\sqrt{2} - 3\sqrt{6}\right)$ **l** $-7\sqrt{5}\left(-3\sqrt{20} + 2\sqrt{3}\right)$

m $10\sqrt{3}\left(\sqrt{2} - 2\sqrt{12}\right)$ **n** $-\sqrt{2}\left(\sqrt{5} + 2\right)$ **o** $2\sqrt{3}\left(2 - \sqrt{12}\right)$

☐ Foundation ◯ Mastery ⬡ Complex

8 Expand and simplify:

a $\left(\sqrt{2}+3\right)\left(\sqrt{5}+3\sqrt{3}\right)$

b $\left(\sqrt{5}-\sqrt{2}\right)\left(\sqrt{2}-\sqrt{7}\right)$

c $\left(\sqrt{2}+5\sqrt{3}\right)\left(2\sqrt{5}-3\sqrt{2}\right)$

d $\left(3\sqrt{10}-2\sqrt{5}\right)\left(4\sqrt{2}+6\sqrt{6}\right)$

e $\left(2\sqrt{5}-7\sqrt{2}\right)\left(\sqrt{5}-3\sqrt{2}\right)$

f $\left(\sqrt{5}+6\sqrt{2}\right)\left(3\sqrt{5}-\sqrt{3}\right)$

g $\left(\sqrt{7}+\sqrt{3}\right)\left(\sqrt{7}-\sqrt{3}\right)$

h $\left(\sqrt{2}-\sqrt{3}\right)\left(\sqrt{2}+\sqrt{3}\right)$

i $\left(\sqrt{6}+3\sqrt{2}\right)\left(\sqrt{6}-3\sqrt{2}\right)$

j $\left(2\sqrt{2}-\sqrt{3}\right)^2$

k $\left(3\sqrt{2}+\sqrt{7}\right)^2$

l $\left(2\sqrt{3}+3\sqrt{5}\right)^2$

m $\left(\sqrt{7}-2\sqrt{5}\right)^2$

n $\left(2\sqrt{8}-3\sqrt{5}\right)^2$

o $\left(3\sqrt{5}+2\sqrt{2}\right)^2$

9 If $a=3\sqrt{2}$, simplify:

a a^2

b $2a^3$

c $(2a)^3$

d $(a+1)^2$

e $(a+3)(a-3)$

10 Expand and simplify:

a $\left(\sqrt{a+3}-2\right)\left(\sqrt{a+3}+2\right)$

b $\left(\sqrt{p-1}-\sqrt{p}\right)^2$

11 Simplify $\left(2\sqrt{x}+\sqrt{y}\right)\left(\sqrt{x}-3\sqrt{y}\right)$.

12 If $\left(2\sqrt{3}-\sqrt{5}\right)^2=a-\sqrt{b}$, evaluate a and b.

Santhosh Varghese/Shutterstock.com

1.11 Rationalising the denominator

Rationalising the denominator of a fractional surd means writing it with a rational number (not a surd) in the denominator. For example, after rationalising the denominator $\frac{3}{\sqrt{5}}$ becomes $\frac{3\sqrt{5}}{5}$.

Video
Rationalising the denominator 1

To rationalise the denominator, multiply top and bottom by the same surd as in the denominator:

Worksheets
Rationalising the denominator

Surds

Rationalising the denominator

$$\frac{a}{\sqrt{b}} \times \frac{\sqrt{b}}{\sqrt{b}} = \frac{a\sqrt{b}}{b}$$

Example 21

Rationalise the denominator of $\frac{2}{5\sqrt{3}}$.

Solution

$$\frac{2}{5\sqrt{3}} \times \frac{\sqrt{3}}{\sqrt{3}} = \frac{2\sqrt{3}}{5\sqrt{9}}$$

$$= \frac{2\sqrt{3}}{5 \times 3}$$

$$= \frac{2\sqrt{3}}{15}$$

When there is a binomial denominator, we use the difference of two squares to rationalise it.

Rationalising a binomial denominator

To rationalise the denominator of $\frac{a}{\sqrt{b} + \sqrt{c}}$, multiply by $\frac{\sqrt{b} - \sqrt{c}}{\sqrt{b} - \sqrt{c}}$.

To rationalise the denominator of $\frac{a}{\sqrt{b} - \sqrt{c}}$, multiply by $\frac{\sqrt{b} + \sqrt{c}}{\sqrt{b} + \sqrt{c}}$.

When there is a binomial denominator, multiply the top and bottom by the **conjugate surd** of the denominator.

The conjugate of $a + \sqrt{b}$ is $a - \sqrt{b}$. It is simply the same binomial surd with the '+' replaced by a '–' (or the other way around).

The conjugate of $\sqrt{x} - \sqrt{y}$ is $\sqrt{x} + \sqrt{y}$.

The conjugate of $\sqrt{2} - 3$ is $\sqrt{2} + 3$.

Multiplying the denominator by the conjugate surd will rationalise it because of the difference of two squares:

$$\left(a + \sqrt{b}\right)\left(a - \sqrt{b}\right) = a^2 - \left(\sqrt{b}\right)^2 = a^2 - b$$

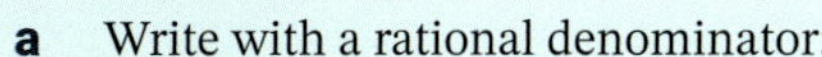

Example 22

a Write with a rational denominator:

i $\dfrac{\sqrt{5}}{\sqrt{2}-3}$ **ii** $\dfrac{2\sqrt{3}+\sqrt{5}}{\sqrt{3}+4\sqrt{2}}$

b Evaluate a and b if $\dfrac{3\sqrt{3}}{\sqrt{3}-\sqrt{2}} = a+\sqrt{b}$.

c Evaluate $\dfrac{2}{\sqrt{3}+2}+\dfrac{\sqrt{5}}{\sqrt{3}-2}$ as a fraction with a rational denominator.

Video Rationalising the denominator 2

Solution

a i
$$\begin{aligned}\frac{\sqrt{5}}{\sqrt{2}-3}\times\frac{\sqrt{2}+3}{\sqrt{2}+3} &= \frac{\sqrt{5}\left(\sqrt{2}+3\right)}{\left(\sqrt{2}\right)^2-3^2}\\ &= \frac{\sqrt{10}+3\sqrt{5}}{2-9}\\ &= -\frac{\sqrt{10}+3\sqrt{5}}{7}\end{aligned}$$

ii
$$\begin{aligned}\frac{2\sqrt{3}+\sqrt{5}}{\sqrt{3}+4\sqrt{2}}\times\frac{\sqrt{3}-4\sqrt{2}}{\sqrt{3}-4\sqrt{2}} &= \frac{\left(2\sqrt{3}+\sqrt{5}\right)\left(\sqrt{3}-4\sqrt{2}\right)}{\left(\sqrt{3}\right)^2-\left(4\sqrt{2}\right)^2}\\ &= \frac{2\times 3-8\sqrt{6}+\sqrt{15}-4\sqrt{10}}{3-16\times 2}\\ &= \frac{6-8\sqrt{6}+\sqrt{15}-4\sqrt{10}}{-29}\\ &= \frac{-6+8\sqrt{6}-\sqrt{15}+4\sqrt{10}}{29}\end{aligned}$$

b
$$\begin{aligned}\frac{3\sqrt{3}}{\sqrt{3}-\sqrt{2}}\times\frac{\sqrt{3}+\sqrt{2}}{\sqrt{3}+\sqrt{2}} &= \frac{3\sqrt{3}\left(\sqrt{3}+\sqrt{2}\right)}{\left(\sqrt{3}-\sqrt{2}\right)\left(\sqrt{3}+\sqrt{2}\right)}\\ &= \frac{3\sqrt{9}+3\sqrt{6}}{\left(\sqrt{3}\right)^2-\left(\sqrt{2}\right)^2}\\ &= \frac{3\times 3+3\sqrt{6}}{3-2}\\ &= \frac{9+3\sqrt{6}}{1}\\ &= 9+3\sqrt{6}\\ &= 9+\sqrt{9}\times\sqrt{6}\\ &= 9+\sqrt{54}\end{aligned}$$

So $a = 9$ and $b = 54$.

c
$$\begin{aligned}\frac{2}{\sqrt{3}+2}+\frac{\sqrt{5}}{\sqrt{3}-2} &= \frac{2\left(\sqrt{3}-2\right)+\sqrt{5}\left(\sqrt{3}+2\right)}{\left(\sqrt{3}+2\right)\left(\sqrt{3}-2\right)}\\ &= \frac{2\sqrt{3}-4+\sqrt{15}+2\sqrt{5}}{\left(\sqrt{3}\right)^2-2^2}\\ &= \frac{2\sqrt{3}-4+\sqrt{15}+2\sqrt{5}}{3-4}\\ &= \frac{2\sqrt{3}-4+\sqrt{15}+2\sqrt{5}}{-1}\\ &= -2\sqrt{3}+4-\sqrt{15}-2\sqrt{5}\end{aligned}$$

EXERCISE 1.11 Answers on p. 459

Rationalising the denominator

EXAMPLE 21

1 Express with a rational denominator:

a $\frac{1}{\sqrt{7}}$ **b** $\frac{\sqrt{3}}{2\sqrt{2}}$ **c** $\frac{2\sqrt{3}}{\sqrt{5}}$ **d** $\frac{6\sqrt{7}}{5\sqrt{2}}$

e $\frac{1+\sqrt{2}}{\sqrt{3}}$ **f** $\frac{\sqrt{6}-5}{\sqrt{2}}$ **g** $\frac{\sqrt{5}+2\sqrt{2}}{\sqrt{5}}$ **h** $\frac{3\sqrt{2}-4}{2\sqrt{7}}$

i $\frac{8+3\sqrt{2}}{4\sqrt{5}}$ **j** $\frac{4\sqrt{3}-2\sqrt{2}}{7\sqrt{5}}$

EXAMPLE 22

2 Express with a rational denominator:

a $\frac{4}{\sqrt{3}+\sqrt{2}}$ **b** $\frac{\sqrt{3}}{\sqrt{2}-7}$ **c** $\frac{2\sqrt{3}}{\sqrt{5}+2\sqrt{6}}$

d $\frac{\sqrt{3}-4}{\sqrt{3}+4}$ **e** $\frac{\sqrt{2}+5}{\sqrt{3}-\sqrt{2}}$ **f** $\frac{3\sqrt{3}+\sqrt{2}}{2\sqrt{5}+3\sqrt{2}}$

3 Express as a single fraction with a rational denominator:

a $\frac{1}{\sqrt{2}+1}+\frac{1}{\sqrt{2}-1}$ **b** $\frac{\sqrt{2}}{\sqrt{2}-\sqrt{3}}-\frac{3}{\sqrt{2}+\sqrt{3}}$

c $t+\frac{1}{t}$, where $t=\sqrt{3}-2$ **d** $z^2-\frac{1}{z^2}$, where $z=1+\sqrt{2}$

e $\frac{\sqrt{2}+3}{\sqrt{2}}+\frac{1}{\sqrt{3}}$ **f** $\frac{\sqrt{3}}{\sqrt{2}+3}+\frac{\sqrt{2}}{\sqrt{3}}$

g $\frac{\sqrt{5}}{\sqrt{6}+2}-\frac{2}{5\sqrt{3}}$ **h** $\frac{\sqrt{2}+7}{4+\sqrt{3}}-\frac{\sqrt{2}}{4-\sqrt{3}}$

i $\frac{\sqrt{5}-\sqrt{2}}{\sqrt{3}-\sqrt{2}}-\frac{2+\sqrt{3}}{\sqrt{3}+1}$

4 Find a and b if:

a $\frac{3}{2\sqrt{5}}=\frac{\sqrt{a}}{b}$ **b** $\frac{\sqrt{3}}{4\sqrt{2}}=\frac{a\sqrt{6}}{b}$ **c** $\frac{2}{\sqrt{5}+1}=a+b\sqrt{5}$

d $\frac{2\sqrt{7}}{\sqrt{7}-4}=a+b\sqrt{7}$ **e** $\frac{\sqrt{2}+3}{\sqrt{2}-1}=a+\sqrt{b}$

5 Show that $\frac{\sqrt{2}-1}{\sqrt{2}+1}+\frac{4}{\sqrt{2}}$ is rational.

6 If $x=\sqrt{3}+2$, simplify:

a $x+\frac{1}{x}$ **b** $x^2+\frac{1}{x^2}$ **c** $\left(x+\frac{1}{x}\right)^2$

Sample HSC problem Answers on p. 459

(2 marks) Find the value of a and b if $\left(7\sqrt{2}-3\right)^2=a+b\sqrt{2}$.

Foundation Mastery Complex

CHAPTER SUMMARY

This chapter, *Algebraic techniques*, revised all of the important algebraic skills required for success in the Mathematics Advanced course.

Before you move on, consider what you learned in this chapter and revisit any sections that may have been unclear.

What you learned in this chapter...	Section	
Identify and use index rules including fractional and negative indices	1.01	Index laws
	1.02	Zero and negative indices
	1.03	Fractional indices
Simplify algebraic expressions	1.04	Simplifying algebraic expressions
Expand and factorise algebraic expressions, including binomials, perfect squares and the difference of two squares	1.05	Binomial products
	1.06	Factorising expressions
	1.07	Factorising trinomials
	1.08	Factorising special binomial products
Simplify expressions involving algebraic fractions	1.09	Algebraic fractions
Simplify and use surds including rationalising the denominator	1.10	Surds
	1.11	Rationalising the denominator

To help master these techniques, make a summary of this topic. Use the chapter outline and the mind map below as a guide. Add your own words, symbols, diagrams, boxes and reminders. The summary should give you a 'whole picture' view of the topic and allow you to identify any weak areas to revisit in your revision.

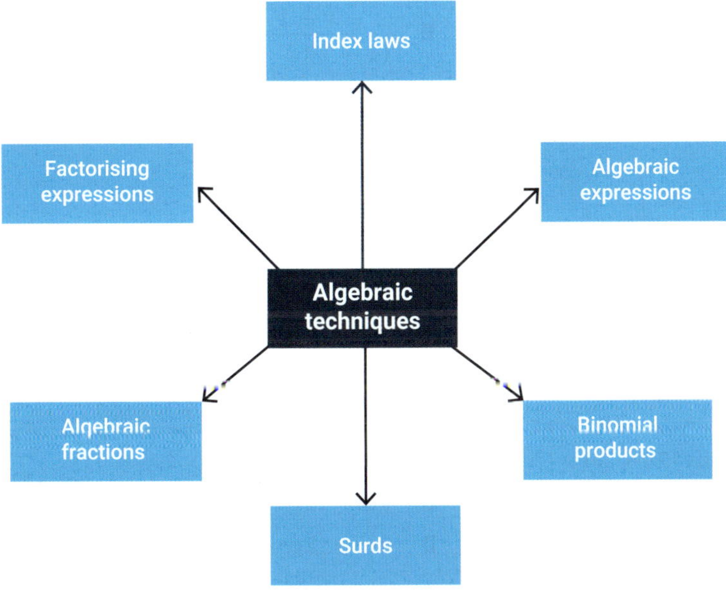

1 Test yourself Answers on p. 459

For Questions **1** to **8**, select the correct answer **A**, **B**, **C** or **D**.

1.11 **1** Rationalise the denominator of $\frac{\sqrt{3}}{2\sqrt{7}}$ (there may be more than one answer).

A $\frac{\sqrt{21}}{28}$ **B** $\frac{2\sqrt{21}}{28}$ **C** $\frac{\sqrt{21}}{14}$ **D** $\frac{\sqrt{21}}{7}$

1.09 **2** Simplify $\frac{x-3}{5} - \frac{x+1}{4}$.

A $\frac{-(x+7)}{20}$ **B** $\frac{x+7}{20}$ **C** $\frac{x+17}{20}$ **D** $\frac{-(x+17)}{20}$

1.06 **3** Factorise fully $x^3 - 4x^2 - x + 4$.

A $(x^2-1)(x-4)$ **B** $(x^2+1)(x-4)$

C $x^2(x-4)$ **D** $(x-4)(x+1)(x-1)$

1.10 **4** Simplify $3\sqrt{2} + 2\sqrt{98}$.

A $5\sqrt{2}$ **B** $5\sqrt{10}$ **C** $17\sqrt{2}$ **D** $10\sqrt{2}$

1.09 **5** Simplify $\frac{3}{x^2-4} + \frac{2}{x-2} - \frac{1}{x+2}$.

A $\frac{x+5}{(x+2)(x-2)}$ **B** $\frac{x+1}{(x+2)(x-2)}$

C $\frac{x+9}{(x+2)(x-2)}$ **D** $\frac{x-3}{(x+2)(x-2)}$

1.04 **6** Simplify $5ab - 2a^2 - 7ab - 3a^2$.

A $2ab + a^2$ **B** $-2ab - 5a^2$ **C** $-13a^3b$ **D** $-2ab + 5a^2$

1.10 **7** Simplify $\sqrt{\frac{80}{27}}$.

A $\frac{4\sqrt{5}}{3\sqrt{3}}$ **B** $\frac{4\sqrt{5}}{9\sqrt{3}}$ **C** $\frac{8\sqrt{5}}{9\sqrt{3}}$ **D** $\frac{8\sqrt{5}}{3\sqrt{3}}$

1.05 **8** Expand and simplify $(3x - 2y)^2$.

A $3x^2 - 12xy - 2y^2$ **B** $9x^2 - 12xy - 4y^2$

C $3x^2 - 6xy + 2y^2$ **D** $9x^2 - 12xy + 4y^2$

1.02 **9** Evaluate as a fraction:

a 7^{-2} **b** 5^{-1} **c** $9^{-\frac{1}{2}}$

1.01 **10** Simplify:

a $x^5 \times x^7 \div x^3$ **b** $(5y^3)^2$ **c** $\frac{(a^5)^4 b^7}{a^9 b}$ **d** $\left(\frac{2x^6}{3}\right)^3$ **e** $\left(\frac{ab^4}{a^5b^6}\right)^0$

1.02, 1.03 **11** Evaluate:

a $36^{\frac{1}{2}}$ **b** 4^{-3} as a fraction **c** $8^{\frac{2}{3}}$

d $49^{-\frac{1}{2}}$ as a fraction **e** $16^{\frac{1}{4}}$ **f** $(-3)^0$

☐ Foundation ○ Mastery ○ Complex

12 Simplify: 1.01

a $a^{14} \div a^9$ b $(x^5y^3)^6$ c $p^6 \times p^5 \div p^2$

d $(2b^9)^4$ e $\dfrac{(2x^7)^3 y^2}{x^{10} y}$

13 Write in index form: 1.03

a $\sqrt{n}$ b $\dfrac{1}{x^5}$ c $\dfrac{1}{x+y}$ d $\sqrt[4]{x+1}$

e $\sqrt[7]{a+b}$ f $\dfrac{2}{x}$ g $\dfrac{1}{2x^3}$ h $\sqrt[3]{x^4}$

i $\sqrt[7]{(5x+3)^9}$ j $\dfrac{1}{\sqrt[4]{m^3}}$

14 Write without fractional or negative indices: 1.02, 1.03

a a^{-5} b $n^{\frac{1}{4}}$ c $(x+1)^{\frac{1}{2}}$ d $(x-y)^{-1}$

e $(4t-7)^{-4}$ f $(a+b)^{\frac{1}{5}}$ g $x^{-\frac{1}{3}}$ h $b^{\frac{3}{4}}$

i $(2x+3)^{\frac{4}{3}}$ j $x^{-\frac{3}{2}}$

15 Rationalise the denominator of: 1.11

a $\dfrac{3}{\sqrt{7}}$ b $\dfrac{\sqrt{2}}{5\sqrt{3}}$ c $\dfrac{2}{\sqrt{5}-1}$

d $\dfrac{2\sqrt{2}}{3\sqrt{2}+\sqrt{3}}$ e $\dfrac{\sqrt{5}+\sqrt{2}}{4\sqrt{5}-3\sqrt{3}}$

16 Simplify: 1.09, 1.11

a $\dfrac{3x}{5} - \dfrac{x-2}{2}$ b $\dfrac{a+2}{7} + \dfrac{2a-3}{3}$ c $\dfrac{1}{x^2-1} - \dfrac{2}{x+1}$

d $\dfrac{4}{k^2+2k-3} + \dfrac{1}{k+3}$ e $\dfrac{\sqrt{3}}{\sqrt{2}+\sqrt{5}} - \dfrac{5}{\sqrt{3}-\sqrt{2}}$

17 Evaluate a^2b^4 when $a = \dfrac{9}{25}$ and $b = 1\dfrac{2}{3}$. 1.01

18 If $a = \left(\dfrac{1}{3}\right)^4$ and $b = \dfrac{3}{4}$, evaluate ab^3 as a fraction. 1.01

19 Write in index form: 1.02, 1.03

a $\sqrt{x}$ b $\dfrac{1}{y}$ c $\sqrt[6]{x+3}$

d $\dfrac{1}{(2x-3)^{11}}$ e $\sqrt[3]{y^7}$

20 Write using a positive index: 1.02

a x^{-3} b $(2a+5)^{-1}$ c $\left(\dfrac{a}{b}\right)^{-5}$

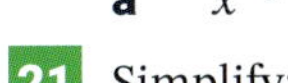

21 Simplify: 1.04, 1.10

a $5y - 7y$ b $\dfrac{3a+12}{3}$ c $-2k^3 \times 3k^2$

d $\dfrac{x}{3} + \dfrac{y}{5}$ e $4a - 3b - a - 5b$ f $\sqrt{8} + \sqrt{32}$

g $3\sqrt{5} - \sqrt{20} + \sqrt{45}$

22 Factorise: 1.07, 1.08

a $x^2 - 36$ b $a^2 + 2a - 3$ c $4ab^2 - 8ab$

d $5y - 15 + xy - 3x$ e $4n - 2p + 6$

□ Foundation ○ Mastery ⬡ Complex

1.05, 1.10 **23** Expand and simplify:

a $b + 3(b - 2)$ b $(2x - 1)(x + 3)$ c $5(m + 3) - (m - 2)$

d $(4x - 3)^2$ e $(p - 5)(p + 5)$ f $7 - 2(a + 4) - 5a$

g $\sqrt{3}(2\sqrt{2} - 5)$ h $(3 + \sqrt{7})(\sqrt{3} - 2)$ i $(\sqrt{7} + 2)^2$

1.09 **24** Simplify:

a $\frac{4a - 12}{5b^3} \times \frac{10b}{a^2 - 9}$ b $\frac{5m + 10}{m^2 - m - 2} \div \frac{m^2 - 4}{3m + 3}$

1.11 **25** Simplify $\frac{3}{\sqrt{5} + 2} - \frac{\sqrt{2}}{2\sqrt{2} - 1}$, writing your answer with a rational denominator.

1.11 **26** a Expand and simplify $(2\sqrt{5} + \sqrt{3})(2\sqrt{5} - \sqrt{3})$.

b Rationalise the denominator of $\frac{3\sqrt{3}}{2\sqrt{5} + \sqrt{3}}$.

1.09 **27** Simplify $\frac{3}{x - 2} + \frac{1}{x + 3} - \frac{2}{x^2 + x - 6}$.

1.10, 1.04 **28** Simplify:

a $3\sqrt{8}$ b $-2\sqrt{2} \times 4\sqrt{3}$ c $\sqrt{108} - \sqrt{48}$

d $\frac{8\sqrt{6}}{2\sqrt{18}}$ e $5a \times -3b \times -2a$ f $\frac{2m^3n}{6m^2n^5}$

g $3x - 2y - x - y$

1.10 **29** Simplify:

a $\frac{3\sqrt{12}}{6\sqrt{15}}$ b $\frac{4\sqrt{32}}{2\sqrt{2}}$

1.11 **30** If $x = \sqrt{3} + 1$, simplify $x + \frac{1}{x}$ and give your answer with a rational denominator.

1.10 **31** Expand and simplify:

a $2\sqrt{2}(\sqrt{3} + \sqrt{2})$ b $(5\sqrt{7} - 3\sqrt{5})(2\sqrt{2} - \sqrt{3})$ c $(3 + \sqrt{2})(3 - \sqrt{2})$

d $(4\sqrt{3} - \sqrt{5})(4\sqrt{3} + \sqrt{5})$ e $(3\sqrt{7} - \sqrt{2})^2$

1.10 **32** Evaluate n if:

a $\sqrt{108} - \sqrt{12} = \sqrt{n}$ b $\sqrt{112} + \sqrt{7} = \sqrt{n}$ c $2\sqrt{8} + \sqrt{200} = \sqrt{n}$

d $4\sqrt{147} + 3\sqrt{75} = \sqrt{n}$ e $2\sqrt{245} + \frac{\sqrt{180}}{2} = \sqrt{n}$

1.07, 1.08 **33** Factorise fully:

a $3x^2 - 27$ b $6x^2 - 12x - 18$ c $5y^2 - 30y + 45$

1.10 **34** Simplify:

a $(3\sqrt{11})^2$ b $(2\sqrt{3})^3$

1.05 **35** Expand and simplify:

a $(a + b)(a - b)$ b $(a + b)^2$

☐ Foundation ○ Mastery ⬡ Complex

Challenge exercise 1

Answers on p. 460

1 Find the value of $\frac{a^3b^2}{c^2}$ if $a = \left(\frac{3}{4}\right)^2$, $b = \left(\frac{2}{3}\right)^3$ and $c = \left(\frac{1}{2}\right)^4$.

2 Show that $2(2^k - 1) + 2^{k+1} = 2(2^{k+1} - 1)$.

3 Find the value of $\frac{a}{b^3c^2}$ in index form if $a = \left(\frac{2}{5}\right)^4$, $b = \left(-\frac{1}{3}\right)^3$ and $c = \left(\frac{3}{5}\right)^2$.

4 Expand and simplify:

a $4ab(a - 2b) - 2a^2(b - 3a)$ **b** $(y^2 - 2)(y^2 + 2)$ **c** $(2x - 5)^3$

5 Find the value of $x + y$ with rational denominator if $x = \sqrt{3} + 1$ and $y = \frac{1}{2\sqrt{5} - 3}$.

6 Simplify $\frac{2\sqrt{3}}{7\sqrt{6} - \sqrt{54}}$.

7 Factorise:

a $(x + 4)^2 + 5(x + 4)$ **b** $x^4 - x^2y - 6y^2$ **c** $a^2b - 2a^2 - 4b + 8$

8 Simplify $\frac{2xy + 2x - 6 - 6y}{4x^2 - 16x + 12}$.

9 Simplify $\frac{(a + 1)^3}{a^2 - 1}$.

10 Factorise $\frac{4}{x^2} - \frac{a^2}{b^2}$.

11 **a** Expand $(2x - 1)^3$.

b Hence, or otherwise, simplify $\frac{6x^2 + 5x - 4}{8x^3 - 12x^2 + 6x - 1}$.

12 Simplify each expression.

a $2^{x+1} + 2^x$ **b** $\frac{3^n - 3^{n-2}}{3^{n+1} + 3^n}$ **c** $\frac{3}{x - 1} - \frac{1}{x} + \frac{x}{x + 1}$

d $3^{n+1} + 5 \times 3^n - 2 \times 3^{n-1}$ **e** $\frac{\frac{1}{x} - \frac{1}{x + 2}}{\frac{1}{x - 2} + \frac{1}{x}}$

13 Expand and simplify, and write in index form.

a $\left(\sqrt{x} + x\right)^2$ **b** $\left(\sqrt[3]{a} + \sqrt[3]{b}\right)\left(\sqrt[3]{a} - \sqrt[3]{b}\right)$

c $\left(p + \frac{1}{\sqrt{p}}\right)^2$ **d** $\left(\sqrt{x} + \frac{1}{\sqrt{x}}\right)^2$

Foundation Mastery Complex

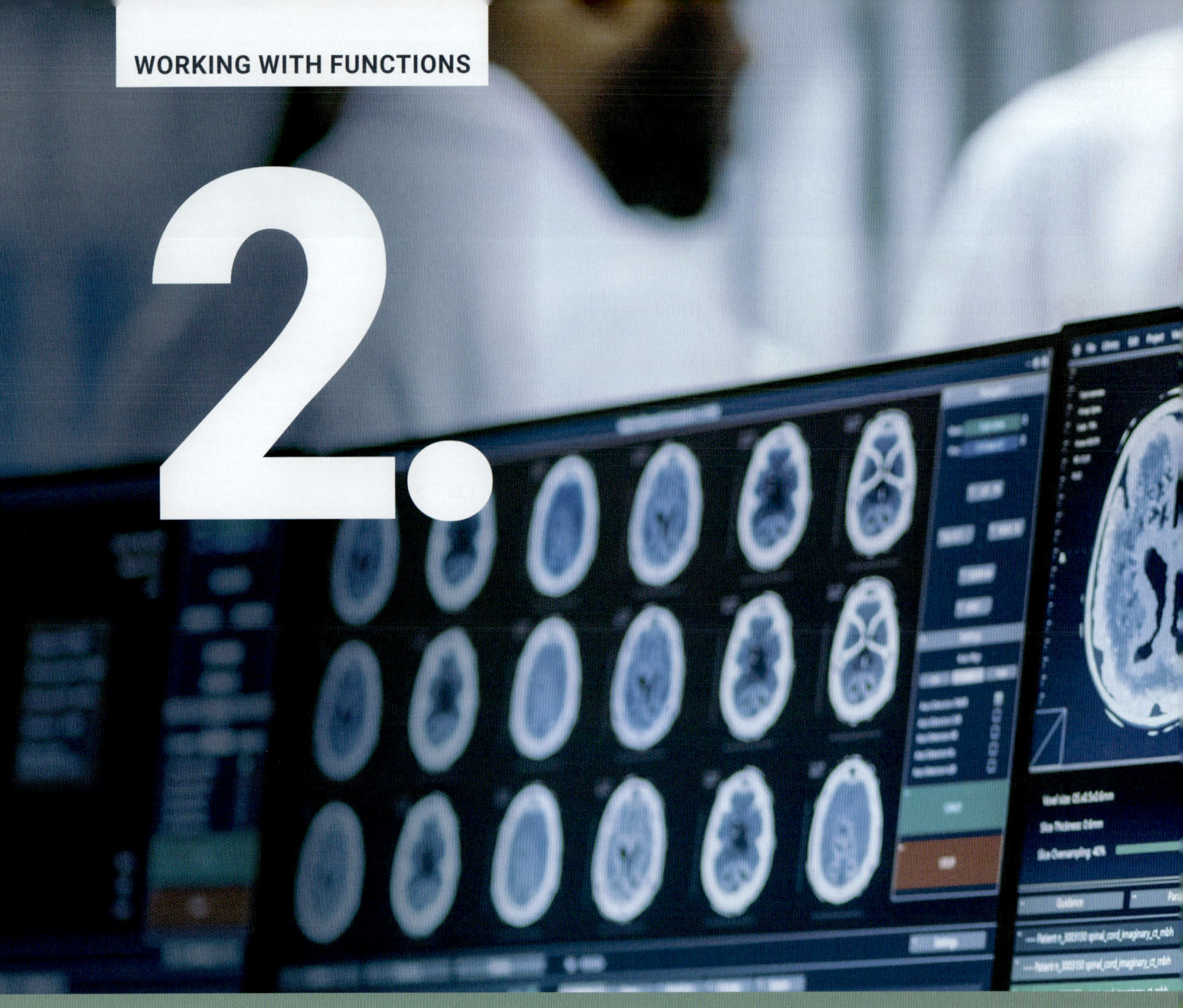

2. EQUATIONS AND INEQUALITIES

Equations are found in most branches of mathematics. They are also important in many other fields, such as science, economics, statistics and engineering. In this chapter you will revise basic equations and inequalities, including absolute value equations, quadratic equations and simultaneous equations.

Most of this topic is revision, with the new content including quadratic inequalities and absolute values. You can work through the revision content at a faster pace or just use it for reference, so that you can focus on learning the new algebraic skills.

Chapter outline

* REVISION

Gorodenkoff/Adobe Stock Photos

In this chapter you will:

- solve linear equations and inequalities
- substitute values into formulas and solve equations
- understand and use absolute values, and solve equations involving absolute value
- solve quadratic equations by factorisation, completing the square and the quadratic formula
- solve quadratic inequalities
- solve linear and non-linear simultaneous equations

Videos (11):

2.02 Graphing inequalities on a number line

2.04 Absolute value equations

2.05 Quadratic equations by factorising

2.07 The quadratic formula
• Solving quadratic equations

2.09 Simultaneous equations 1
• Simultaneous equations 2
• The elimination method
• The substitution method

2.10 Non-linear simultaneous equations 1
• Non-linear simultaneous equations 2

Worksheets (11):

2.01 Equations

2.02 Inequalities on a number line

2.06 Completing the square

2.07 Quadratic formula • Quadratic equations
• Problems involving quadratic equations
• Solving algebraic equations

2.09 Solving simultaneous equations
• Simultaneous equations order activity
• Simultaneous equations problems

2.10 Simultaneous equations

Puzzles (2):

2.05 Factorising quadratic equations

2.09 Simultaneous equations puzzle

Technology (1):

2.07 The quadratic formula

Spreadsheet (1):

2.07 The quadratic formula

Nelson MindTap

To access resources above, visit
cengage.com.au/nelsonmindtap

Terminology

absolute value	completing the square	elimination method	equation
inequality	quadratic equation	quadratic formula	quadratic inequality
simultaneous equations	substitution method		

Equations and formulas

2.01

Example 1

Solve each equation.

a $4y - 3 = 8y + 21$ **b** $2(3x + 7) = 6 - (x - 1)$

Solution

a

$$4y - 3 = 8y + 21$$
$$4y - 4y - 3 = 8y - 4y + 21$$
$$-3 = 4y + 21$$
$$-3 - 21 = 4y + 21 - 21$$
$$-24 = 4y$$
$$\frac{-24}{4} = \frac{4y}{4}$$
$$-6 = y$$
$$y = -6$$

b

$$2(3x + 7) = 6 - (x - 1)$$
$$6x + 14 = 6 - x + 1$$
$$= 7 - x$$
$$6x + x + 14 = 7 - x + x$$
$$7x + 14 = 7$$
$$7x + 14 - 14 = 7 - 14$$
$$7x = -7$$
$$\frac{7x}{7} = \frac{-7}{7}$$
$$x = -1$$

When an **equation** involves fractions, multiply both sides of the equation by the common denominator of the fractions.

Example 2

Solve:

a $\frac{m}{3} - 4 = \frac{1}{2}$ **b** $\frac{x+1}{3} + \frac{x}{4} = 5$

Solution

a

$$\frac{m}{3} - 4 = \frac{1}{2}$$
$$6\left(\frac{m}{3}\right) - 6(4) = 6\left(\frac{1}{2}\right)$$
$$2m - 24 = 3$$
$$2m - 24 + 24 = 3 + 24$$
$$2m = 27$$
$$\frac{2m}{2} = \frac{27}{2}$$
$$m = \frac{27}{2}$$
$$= 13\tfrac{1}{2}$$

b

$$\frac{x+1}{3} + \frac{x}{4} = 5$$
$$12\left(\frac{x+1}{3}\right) + 12\left(\frac{x}{4}\right) = 12(5)$$
$$4(x + 1) + 3x = 60$$
$$4x + 4 + 3x = 60$$
$$7x + 4 = 60$$
$$7x + 4 - 4 = 60 - 4$$
$$7x = 56$$
$$\frac{7x}{7} = \frac{56}{7}$$
$$x = 8$$

Did you know?

History of algebra

Algebra was known in ancient civilisations. Many equations were known in Babylon, although general solutions were difficult because symbols were not used in those times.

Diophantus, around 250 CE, first used algebraic notation and symbols (e.g. the minus sign). He wrote a treatise on algebra in his *Arithmetica*, comprising 13 books. Only 6 of these books survived. About 400 CE, Hypatia of Alexandria wrote a commentary on them.

Hypatia was the first female mathematician on record, and was a philosopher and teacher. She was the daughter of Theon, who was also a mathematician and who ensured that she had the best education.

In 1799, **Carl Friedrich Gauss** proved the Fundamental Theorem of Algebra: that every algebraic equation involving a power of x has at least one solution, which may be a real number or a non-real number.

Example 3

a The formula for the surface area of a rectangular prism is given by $S = 2(lb + bh + lh)$. Find the value of b when $S = 180$, $l = 9$ and $h = 6$.

b Make y the subject of the equation $3x - 2y + 1 = 0$.

Solution

a

$$\begin{aligned} S &= 2(lb + bh + lh) \\ 180 &= 2(9b + 6b + 9 \times 6) \\ &= 2(15b + 54) \\ &= 30b + 108 \\ 72 &= 30b \\ \frac{72}{30} &= \frac{30b}{30} \\ 2.4 &= b \\ b &= 2.4 \end{aligned}$$

b

$$\begin{aligned} 3x - 2y + 1 &= 0 \\ 3x + 1 &= 2y \\ \frac{3x+1}{2} &= y \\ y &= \frac{3x+1}{2} \end{aligned}$$

EXERCISE 2.01 Answers on p. 460

Equations and formulas

Solve each equation.

EXAMPLE 1

1 **a** $4a + 7 = -21$ **b** $7y - 1 = 20$ **c** $3(x + 2) = 15$

d $-2(3a + 1) = 8$ **e** $7t + 4 = 3t - 12$ **f** $x - 3 = 6x - 9$

g $2(a - 2) = 4 - 3a$ **h** $5b + 2 = -3(b - 1)$ **i** $3(t + 7) = 2(2t - 9)$

j $2 + 5(p - 1) = 5p - (p - 2)$ **k** $3.7x + 1.2 = 5.4x - 6.3$ **l** $\frac{b}{5} = \frac{2}{3}$

m $\frac{5x}{4} = \frac{11}{7}$ **n** $\frac{x}{3} - 4 = 8$ **o** $\frac{5+x}{7} = \frac{2}{7}$

EXAMPLE 2

2 **a** $\frac{y}{2} = -\frac{3}{5}$ **b** $\frac{x}{9} - \frac{2}{3} = 7$ **c** $\frac{w-3}{2} = 5$

d $\frac{2t}{5} - \frac{t}{3} = 2$ **e** $\frac{x}{4} + \frac{1}{2} = 4$ **f** $\frac{x}{5} - \frac{x}{2} = \frac{3}{10}$

g $\frac{x+4}{3} + \frac{x}{2} = 1$ **h** $\frac{p-3}{2} + \frac{2p}{3} = 2$ **i** $\frac{t+3}{7} + \frac{t-1}{3} = 4$

j $\frac{x+5}{9} - \frac{x+2}{5} = 1$ **k** $\frac{q-1}{3} - \frac{q-2}{4} = 2$ **l** $\frac{x+3}{5} + 2 = \frac{x+7}{2}$

□ Foundation ○ Mastery ⬡ Complex

EXAMPLE 3

3 Given that $v = u + at$ is the formula for the velocity of a particle at time t, find the value of t when $u = 17.3$, $v = 100.6$ and $a = 9.8$.

4 The formula for finding the area of a triangle is $A = \frac{1}{2}bh$. Find b when $A = 36$ and $h = 9$.

5 The area of a trapezium is given by $A = \frac{1}{2}h(a + b)$. Find the value of a when $A = 120$, $h = 5$ and $b = 7$.

6 Find the value of y when $x = 3$, given the straight line equation $5x - 2y - 7 = 0$.

7 The area of a circle is given by $A = \pi r^2$. Find r, correct to 3 significant figures, if $A = 140$.

8 The area of a rhombus is given by the formula $A = \frac{1}{2}xy$, where x and y are its diagonals.
Find the value of x, correct to 2 decimal places, when $y = 7.8$ and $A = 25.1$.

9 The gradient of a straight line is given by $m = \frac{y_2 - y_1}{x_2 - x_1}$. Find y_1 when $m = -\frac{5}{6}$, $y_2 = 7$, $x_2 = -3$ and $x_1 = 1$.

10 The surface area of a cylinder is given by the formula $S = 2\pi r(r + h)$.
Evaluate h, correct to one decimal place, if $S = 232$ and $r = 4.5$.

11 The formula for body mass index is $\text{BMI} = \frac{w}{h^2}$. Evaluate:

- **a** the BMI when $w = 65$ and $h = 1.6$
- **b** w when $\text{BMI} = 21.5$ and $h = 1.8$
- **c** h when $\text{BMI} = 19.7$ and $w = 73.8$.

12 Change the subject of:

- **a** $y = 2x - 1$ to x
- **b** $x + 5y - 4 = 0$ to y
- **c** $S = lbh$ to h
- **d** $V = \frac{4}{3}\pi r^3$ to
  - **i** r^3
  - **ii** r
- **e** $m = \frac{y_2 - y_1}{x_2 - x_1}$ to
  - **i** y_2
  - **ii** x_2
  - **iii** x_1
- **f** $S = \frac{n}{2}[2a + (n - 1)d]$ to
  - **i** a
  - **ii** d

13 The x value of the midpoint is given by $x = \frac{x_1 + x_2}{2}$. Find x_1 when $x = -2$ and $x_2 = 5$.

14 Given $y = \sqrt{2x + 5}$, evaluate x when $y = 4$.

15 The volume of a sphere is $V = \frac{4}{3}\pi r^3$. Evaluate r, to one decimal place, when $V = 150$.

16 Solve each equation.

- **a** $\frac{x}{3} - \frac{2}{5} = \frac{7}{10}$
- **b** $\frac{n + 1}{4} + \frac{2}{3} - \frac{n}{2} = 0$
- **c** $\frac{y - 3}{5} - \frac{y + 1}{6} + \frac{y}{4} = 0$ (to one decimal place)

Did you know?

Diophantus' riddle

The age of Greek mathematician, Diophantus at his death can be calculated from his epitaph:

Diophantus passed one-sixth of his life in childhood, one-twelfth in youth, and one-seventh more as a bachelor; five years after his marriage a son was born who died four years before his father at half his father's final age.

How old was Diophantus when he died?

Foundation Mastery Complex

2.02 Inequalities

Video
Graphing inequalities on a number line

Worksheet
Inequalities on a number line

We can show the relative size of numbers on a number line.

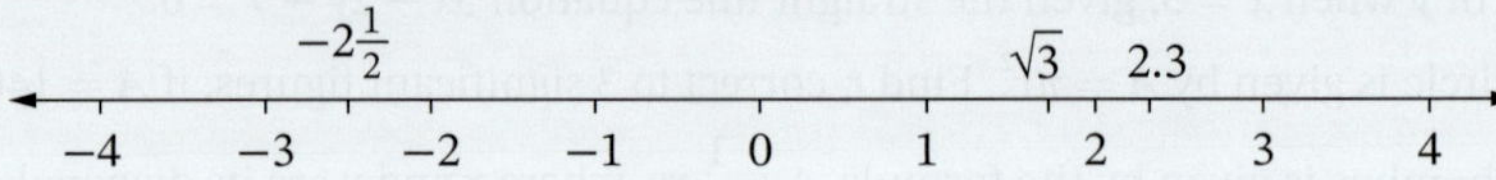

We can also show inequalities such as $x < 3$ on a number line.

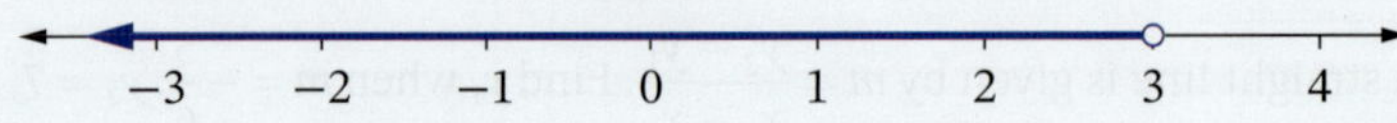

The open circle at 3 means that 3 is excluded (not included) in the **inequality** $x < 3$.

If there was a closed circle at 3, then that would mean that 3 is included and the inequality would be $x \leq 3$.

Inequalities on a number line

Less than	$<$	←—○	Less than or equal to	$\leq$	←—●
Greater than	$>$	○—→	Greater than or equal to	$\geq$	●—→

Inequalities between values

Inequalities between 2 values are written in a special way, using the $<$ and $\leq$ signs only, with the variable x in between the 2 values.

$1 < x < 3$ means x is between 1 and 3 exclusive (that is, excluding 1 and 3).

$1 \leq x \leq 3$ means x is between 1 and 3 inclusive (that is, including 1 and 3).

$1 < x \leq 3$ means x is between 1 and 3, not including 1 but including 3.

$1 \leq x < 3$ means x is between 1 and 3, not including 3 but including 1. On the number line, it looks like this:

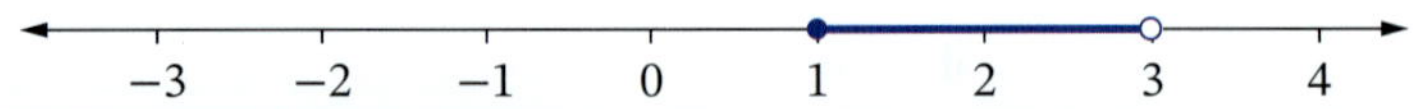

Solving inequalities

When solving inequalities, solve like equations, but when multiplying or dividing by a negative number, reverse the inequality sign.

Example 4

Solve each inequality and show its solution on a number line.

a $5x + 7 \geq 17$ **b** $3t - 2 > 5t + 4$ **c** $1 < 2z + 7 \leq 11$

Solution

a

$$5x + 7 \geq 17$$
$$5x + 7 - 7 \geq 17 - 7$$
$$5x \geq 10$$
$$\frac{5x}{5} \geq \frac{10}{5}$$
$$x \geq 2$$

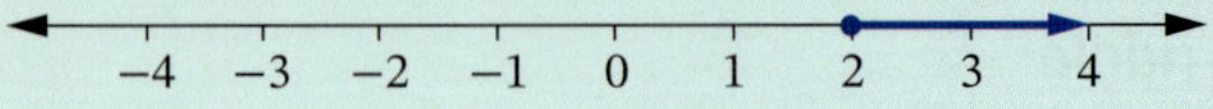

(Number line from −4 to 4 with closed circle at 2 and arrow to the right.)

b

$$3t - 2 > 5t + 4$$
$$3t - 5t - 2 > 5t - 5t + 4$$
$$-2t - 2 > 4$$
$$-2t - 2 + 2 > 4 + 2$$
$$-2t > 6$$
$$\frac{-2t}{-2} < \frac{6}{-2}$$
$$t < -3$$

Remember to change the inequality sign when dividing by −2.

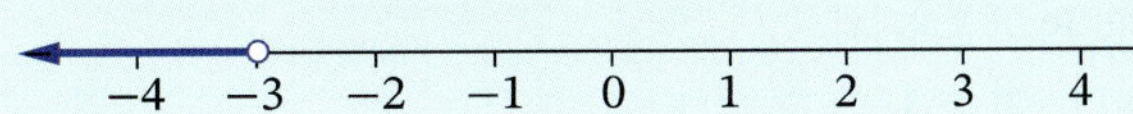

c

$$1 < 2z + 7 \leq 11$$
$$1 - 7 < 2z + 7 - 7 \leq 11 - 7$$
$$-6 < 2z \leq 4$$
$$-3 < z \leq 2$$

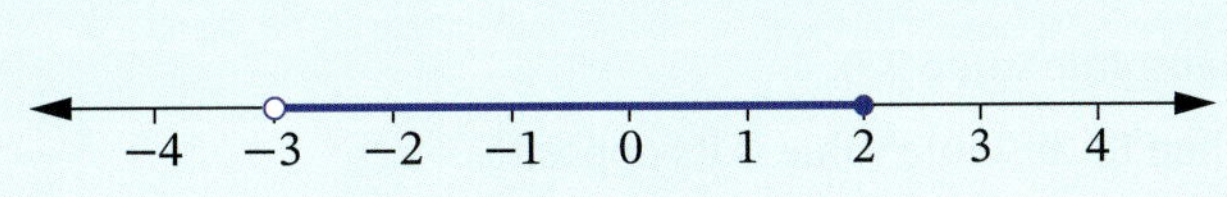

EXERCISE 2.02 Answers on p. 460

Inequalities

EXAMPLE 4

1 Solve each inequality and plot the solution on a number line.

a $3x + 16 > 25$ **b** $2y - 10 \leq -2$

2 Solve:

a $5t > 35$ **b** $3x - 7 \geq 2$ **c** $2(p + 5) > 8$

d $4 - (x - 1) \leq 7$ **e** $3y + 5 > 2y - 4$ **f** $2a - 6 \leq 5a - 3$

g $3 + 4y \geq -2(1 - y)$ **h** $2x + 9 < 1 - 4(x + 1)$ **i** $\frac{a}{2} \leq -3$

j $8 > \frac{2y}{3}$ **k** $\frac{b}{2} + 5 < -4$ **l** $\frac{x}{3} - 4 > 6$

3 Solve:

a $\frac{1}{4} + \frac{x}{5} \leq 1$ **b** $\frac{m}{4} - 3 > \frac{2}{3}$ **c** $\frac{2b}{5} - \frac{1}{2} \geq 6$

d $\frac{r-3}{2} \leq -6$ **e** $\frac{z+1}{9} + 2 > 3$ **f** $\frac{w}{6} + \frac{2w+5}{3} < 4$

g $\frac{x+1}{2} - \frac{x-2}{3} \geq 7$ **h** $\frac{t+2}{7} - \frac{t+3}{2} \leq 2$ **i** $\frac{q-2}{3} < 2 + \frac{3q}{4}$

j $\frac{2x}{3} - \frac{x-1}{2} > \frac{2}{9}$ **k** $\frac{2b-5}{8} + 3 \leq \frac{b+6}{12}$

4 Solve and plot each solution on a number line.

a $3 < x + 2 < 9$ **b** $-4 \leq 2p < 10$ **c** $2 < 3x - 1 < 11$

d $-6 \leq 5y + 9 \leq 34$ **e** $-2 < 3(2y - 1) < 7$

5 By writing each problem as an inequality involving the variable x, solve the problem.

a The product of 10 and a number is greater than 500.

b A number plus 17 is less than or equal to 30.

c A number divided by 3 is less than 20.

d Subtracting 5 from the product of a number and 2 is greater than or equal to 13.

e Multiplying a number by 3 and adding 7 gives an answer less than the product of the number and 5.

f A number divided by 2 is greater than the sum of the number and 10.

g The total of 7 and the product of 9 and a number is less than or equal to the sum of 12 and the product of 3 and the number.

Foundation Mastery Complex

2.03 Absolute value

The **absolute value** of a number is the size of the number without the sign or direction. It is also called the **magnitude** of the number and its value is always positive (or zero).

We write the absolute value of x as $|x|$.

For example, $|4| = 4$ and $|-3| = 3$.

The absolute value of x is also the distance of x from 0 on the number line.

- If x is positive, then its absolute value is itself.
- If $x = 0$, then its absolute value is 0.
- If x is negative, then its absolute value is its opposite, $-x$.

$|4| = 4$

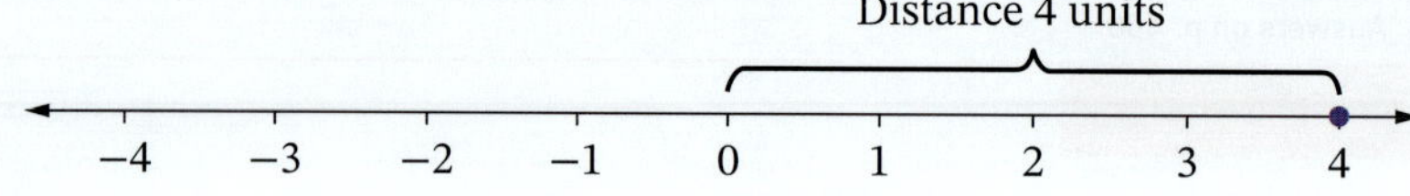

$|-3| = 3$

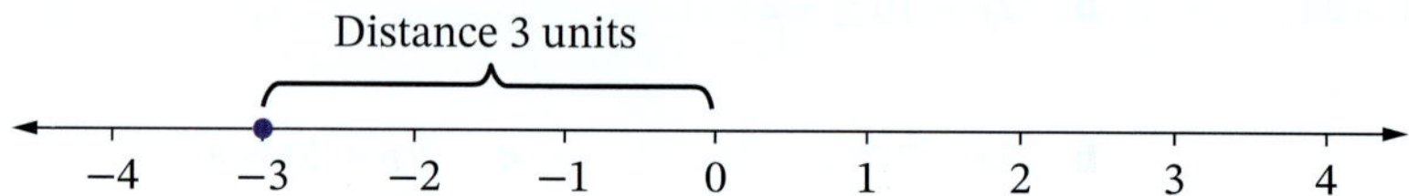

$|4| = 4$ since $4 \geq 0$.

$$\begin{aligned}|-3| &= -(-3) \text{ since } -3 < 0\\ &= 3\end{aligned}$$

Properties of absolute value

Property	Example
$\|ab\| = \|a\| \times \|b\|$	$\|2 \times -3\| = \|2\| \times \|-3\| = 6$
$\|a\|^2 = a^2$	$\|-3\|^2 = (-3)^2 = 9$
$\sqrt{a^2} = \|a\|$	$\sqrt{(-5)^2} = \|-5\| = 5$
$\|-a\| = \|a\|$	$\|-7\| = \|7\| = 7$
$\|a - b\| = \|b - a\|$	$\|2 - 3\| = \|3 - 2\| = 1$
$\|a + b\| \leq \|a\| + \|b\|$	$\|2 + 3\| = \|2\| + \|3\|$ but $\|-3 + 4\| < \|-3\| + \|4\|$

Example 5

a Evaluate $|2| - |-1| + |-3|^2$.

b Show that $|a + b| \leq |a| + |b|$ when $a = -2$ and $b = 3$.

c Write an expression for $|x - 4|$ when:

i $x \geq 4$ **ii** $x < 4$

Solution

a
$$\begin{aligned}|2| - |-1| + |-3|^2 &= 2 - 1 + 3^2\\ &= 10\end{aligned}$$

b LHS $= |a + b|$

$= |-2 + 3|$

$= |1|$

$= 1$

RHS $= |a| + |b|$

$= |-2| + |3|$

$= 2 + 3$

$= 5$

Note: LHS means left-hand side and RHS means right-hand side.

Since $1 < 5$, $|a + b| \leq |a| + |b|$

c $|x - 4| = x - 4$ when $x - 4 \geq 0$ that is, when $x \geq 4$

$|x - 4| = -(x - 4)$ when $x - 4 < 0$ that is, when $x < 4$

$= -x + 4$

EXERCISE 2.03 Answers on p. 461

Absolute value

1 Evaluate:

a $|7|$ **b** $|-5|$ **c** $|-6|$ **d** $|0|$ **e** $|2|$

f $|-11|$ **g** $|-2||3|$ **h** $3|-8|$ **i** $|-5|^2$ **j** $|-5|^3$

2 Evaluate:

a $|3| + |-2|$ **b** $|-3| - |4|$ **c** $|-5 + 3|$

d $|2 \times -7|$ **e** $|-3| + |-1|$ **f** $5 - |-2| \times |6|^2$

g $|-2 + 5 \times -1|$ **h** $3|-4|$ **i** $2|-3| - 3|-4|$

j $|5 - 7| + 4|-2|$

3 Evaluate $|a - b|$ if:

a $a = 5$ and $b = 2$ **b** $a = -1$ and $b = 2$ **c** $a = -2$ and $b = -3$

d $a = 4$ and $b = 7$ **e** $a = -1$ and $b = -2$

4 Write an expression for:

a $|a|$ when $a > 0$ **b** $|a|$ when $a < 0$ **c** $|a|$ when $a = 0$

d $|3a|$ when $a > 0$ **e** $|3a|$ when $a < 0$ **f** $|3a|$ when $a = 0$

g $|a + 1|$ when $a > -1$ **h** $|a + 1|$ when $a < -1$ **i** $|x - 2|$ when $x > 2$

5 Show that $|a + b| \leq |a| + |b|$ when:

a $a = 2$ and $b = 4$ **b** $a = -1$ and $b = -2$ **c** $a = -2$ and $b = 3$

d $a = -4$ and $b = 5$ **e** $a = -7$ and $b = -3$

6 Show that $\sqrt{x^2} = |x|$ when:

a $x = 5$ **b** $x = -2$ **c** $x = -3$ **d** $x = 4$ **e** $x = -9$

7 Find values of x for which $|x| = 3$.

8 Simplify $\frac{|n|}{n}$, where $n \neq 0$.

9 Simplify $\frac{x-2}{|x-2|}$ and state which value x cannot be.

Foundation | Mastery | Complex

Investigation

Absolute value

Are these statements true? If so, are there some values for which the expression is undefined (values of x or y that the expression cannot have)?

1 $\frac{x}{|x|}=1$ **2** $|2x|=2x$ **3** $|2x|=2|x|$

4 $|x|+|y|=|x+y|$ **5** $|x|^2=x^2$ **6** $|x|^3=x^3$

7 $|x+1|=|x|+1$ **8** $\frac{|3x-2|}{3x-2}=1$ **9** $\frac{|x|}{x^2}=1$

10 $|x|\geq 0$

Discuss in groups absolute value and its definition in relation to these statements.

2.04 Equations involving absolute values

Video
Absolute value equations

On a number line, $|x|$ means the distance of x from 0 in either direction.

Example 6

Solve $|x|=2$.

Solution

$|x|=2$ means the distance of x from 0 is 2 (in either direction).

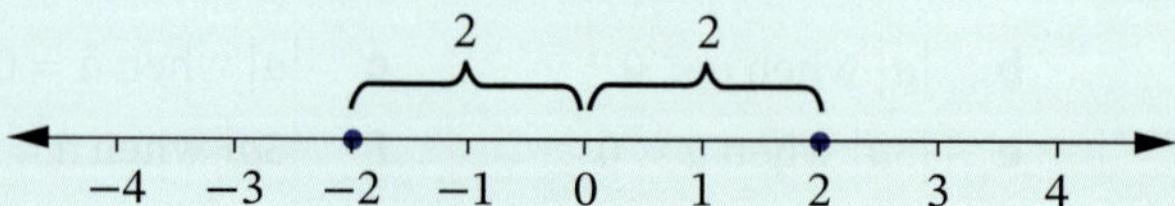

$x=\pm 2$

$|x|=a$

The equation $|x|=a$, where $a\geq 0$, has solutions $x=a$ and $x=-a$.

Investigation

Absolute value and the number line

What does $|a-b|$ mean as a distance along the number line?

Discuss your response with the class. Select different values of a and b to help with this discussion.

Also look at your answers to Question **3** from Exercise 2.03.

Video
Absolute value equations

Example 7

Solve:

a $|x+4|=7$ **b** $|2x-3|=9$

Solution

a This means that the distance from $x+4$ to 0 is 7 in either direction.

So $x+4=\pm 7$.

$x+4=7$ or $x+4=-7$

$x=3$ $x=-11$

So $x=3$ or -11.

Checking your answer:

$$\begin{aligned} \text{LHS} &= |3+4| \\ &= |7| \\ &= 7 \\ &= \text{RHS} \end{aligned} \qquad \begin{aligned} \text{LHS} &= |-11+4| \\ &= |-7| \\ &= 7 \\ &= \text{RHS} \end{aligned}$$

b $|2x-3|=9$

$2x-3=9$ or $2x-3=-9$

$2x=12$ $2x=-6$

$x=6$ $x=-3$

So $x=6$ or -3.

Checking your answer:

$$\begin{aligned} \text{LHS} &= |2\times 6-3| \\ &= |9| \\ &= 9 \\ &= \text{RHS} \end{aligned} \qquad \begin{aligned} \text{LHS} &= |2\times(-3)-3| \\ &= |-9| \\ &= 9 \\ &= \text{RHS} \end{aligned}$$

EXERCISE 2.04 Answers on p. 461

Equations involving absolute values

EXAMPLE 6

1 Solve:

a $|x|=5$ **b** $|y|=8$ **c** $|x|=0$

EXAMPLE 7

2 Solve:

a $|x+2|=7$ **b** $|n-1|=3$ **c** $9=|2x+3|$

d $|7x-1|=34$ **e** $\left|\frac{x}{3}\right|=4$

3 Solve:

a $|8x-5|=11$ **b** $|5-3n|=1$ **c** $16=|5t+4|$

d $21=|9-2y|$ **e** $|3x+2|-7=0$

4 Solve:

a $|3h+1|=5$ **b** $\left|\frac{k+2}{5}\right|=3$

c $\left|\frac{2t-5}{3}\right|=\frac{3}{4}$ **d** $\frac{|y+2|}{4}-\frac{1}{2}=0$

☐ Foundation ○ Mastery ○ Complex

Investigation

Faulty proof

Here is a proof that $1 = 2$. Can you see the fault in the proof?

$$x^2 - x^2 = x^2 - x^2$$
$$x(x - x) = (x + x)(x - x)$$
$$x = x + x$$
$$x = 2x$$
$$\therefore 1 = 2$$

2.05 Quadratic equations by factorisation

Video Quadratic equations by factorising

Puzzle Factorising quadratic equations

A **quadratic equation** is an equation involving a square. For example, $x^2 - 4 = 0$.

When solving quadratic equations by factorising, we use a property of 0 called the **null factor law** or the **zero product law** ('null' means 0).

The null factor law

For any 2 numbers a and b, if $ab = 0$ then $a = 0$ or $b = 0$.

Example 8

Solve:

a $x^2 + x - 6 = 0$ **b** $y^2 - 7y = 0$ **c** $3a^2 - 14a = -8$

Solution

a

$$x^2 + x - 6 = 0$$
$$(x + 3)(x - 2) = 0$$
$$\therefore x + 3 = 0 \text{ or } x - 2 = 0$$
$$x = -3 \text{ or } x = 2$$

So the solution is $x = -3$ or 2.

b

$$y^2 - 7y = 0$$
$$y(y - 7) = 0$$
$$\therefore y = 0 \text{ or } y - 7 = 0$$
$$y = 7$$

So the solution is $y = 0$ or 7.

c First we make the equation equal to zero so we can factorise and use the rule for zero.

$$3a^2 - 14a = -8$$
$$3a^2 - 14a + 8 = -8 + 8$$
$$3a^2 - 14a + 8 = 0$$
$$(3a - 2)(a - 4) = 0$$
$$\therefore 3a - 2 = 0 \text{ or } a - 4 = 0$$
$$3a = 2 \text{ or } a = 4$$
$$\frac{3a}{3} = \frac{2}{3}$$
$$a = \frac{2}{3}$$

So the solution is $a = \frac{2}{3}$ or 4.

EXERCISE 2.05 Answers on p. 461

Quadratic equations by factorisation

EXAMPLE 8

1 Solve each quadratic equation.

a	$y^2 + y = 0$	**b**	$b^2 - b - 2 = 0$	**c**	$p^2 + 2p - 15 = 0$
d	$t^2 - 5t = 0$	**e**	$x^2 + 9x + 14 = 0$	**f**	$q^2 - 9 = 0$
g	$x^2 - 1 = 0$	**h**	$a^2 + 3a = 0$	**i**	$2x^2 + 8x = 0$
j	$4x^2 - 1 = 0$	**k**	$3x^2 + 7x + 4 = 0$	**l**	$2y^2 + y - 3 = 0$
m	$8b^2 - 10b + 3 = 0$	**n**	$x^2 - 3x = 10$	**o**	$3x^2 = 2x$
p	$2x^2 = 7x - 5$	**q**	$5x - x^2 = 0$	**r**	$y^2 = y + 2$
s	$8n = n^2 + 15$	**t**	$12 = 7x - x^2$	**u**	$m^2 = 6 - 5m$

2 The sum of a number squared together with the product of the number and 7 gives a total of 8. Find the values of the number.

3 Double a number is 15 less than its square. Find its value(s).

4 Solve each equation.

a $x - \frac{4}{x} = 0$ **b** $x + 2 = \frac{3}{x}$ **c** $\frac{6}{x} + x = 5$ **d** $x + \frac{16}{x} = 8$ **e** $3x + 14 - \frac{5}{x} = 0$

5 For the formula $S = \frac{n}{2}[2a + (n - 1)d]$, find the values of n if:

a $S = 60$, $a = 25$ and $d = -5$ **b** $S = 44$, $a = 23$ and $d = -2$.

6 A rectangle has one side 3 cm longer than the other. Its area is 108 cm^2. Find the dimensions of its sides.

Quadratic equations by completing the square 2.06

Not all trinomials will factorise, so other methods need to be used to solve quadratic equations.

Worksheet
Completing the square

Example 9

Solve:

a $(x + 3)^2 = 11$ **b** $(y - 2)^2 = 7$

Solution

a
$$\begin{aligned}(x+3)^2 &= 11\\ x+3 &= \pm\sqrt{11}\\ x+3-3 &= \pm\sqrt{11}-3\\ x &= \pm\sqrt{11}-3\\ x &= -3\pm\sqrt{11}\end{aligned}$$

b
$$\begin{aligned}(y-2)^2 &= 7\\ y-2 &= \pm\sqrt{7}\\ y-2+2 &= \pm\sqrt{7}+2\\ y &= \pm\sqrt{7}+2\\ y &= 2\pm\sqrt{7}\end{aligned}$$

If a quadratic equation cannot be factorised, then its solutions are surds (irrational). To solve a quadratic equation such as $x^2 - 6x + 3 = 0$, which will not factorise, we can use the method of **completing the square**.

We use the perfect square:

$$a^2 + 2ab + b^2 = (a + b)^2$$

Notice that the middle term $2ab$ is double ab and the last term b^2 is the square of b.

☐ Foundation ○ Mastery ⬡ Complex

Completing the square

To complete the square on $a^2 + pa$, divide p by 2 and square it.

$$a^2 + pa + \left(\frac{p}{2}\right)^2 = \left(a + \frac{p}{2}\right)^2$$

In this formula, $p = 2b$.

Example 10

Complete the square on $x^2 - 10x + 3$.

Solution

$p = -10$, the coefficient of x.

Halve the 10 to get 5 and square 5 to get 25.

$$\begin{aligned} x^2 - 10x + 3 &= x^2 - 10x + 25 + 3 - 25 \\ &= (x-5)^2 + 3 - 25 \\ &= (x-5)^2 - 22 \end{aligned}$$

Example 11

Solve by completing the square:

a $x^2 - 6x + 3 = 0$

b $y^2 + 2y - 7 = 0$ (correct to 3 significant figures)

Solution

The 3rd line shows the 'completing the square' step in both solutions.

a

$$\begin{aligned} x^2 - 6x + 3 &= 0 \\ x^2 - 6x &= -3 \\ x^2 - 6x + 9 &= -3 + 9 \qquad \left(\frac{6}{2}\right)^2 = 3^2 = 9 \\ (x-3)^2 &= 6 \\ \therefore x - 3 &= \pm\sqrt{6} \\ x &= 3 \pm \sqrt{6} \end{aligned}$$

Notice that the solutions are surds.

b

$$\begin{aligned} y^2 + 2y - 7 &= 0 \\ y^2 + 2y &= 7 \\ y^2 + 2y + 1 &= 7 + 1 \qquad \left(\frac{2}{2}\right)^2 = 1^2 = 1 \\ (y+1)^2 &= 8 \\ \therefore y + 1 &= \pm\sqrt{8} \\ y &= -1 \pm \sqrt{8} \\ y &\approx 1.83 \text{ or } -3.83 \end{aligned}$$

EXERCISE 2.06 Answers on p. 461

Quadratic equations by completing the square

1 Solve and give exact solutions:

a $(x+1)^2 = 7$ **b** $(y+5)^2 = 5$ **c** $(a-3)^2 = 6$

d $(x-2)^2 = 13$ **e** $(2y+3)^2 = 2$

2 Solve and give solutions correct to one decimal place:

a $(h+2)^2 = 15$ **b** $(a-1)^2 = 8$ **c** $(x-4)^2 = 17$

d $(y+7)^2 = 21$ **e** $(3x-1)^2 = 12$

EXAMPLE 10

3 Complete the square on each quadratic expression.

a $y^2 + 2y - 7$ **b** $a^2 - 16a + 3$ **c** $n^2 + 12n - 1$

d $a^2 - 10a + 5$ **e** $g^2 + 20g - 19$ **f** $x^2 - 2x - 11$

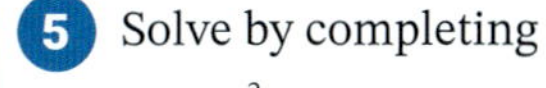

4 Solve by completing the square, giving exact solutions in simplest surd form:

a $x^2 + 4x - 1 = 0$ **b** $a^2 - 6a + 2 = 0$ **c** $y^2 - 8y - 7 = 0$

d $x^2 + 2x - 12 = 0$ **e** $p^2 + 14p + 5 = 0$ **f** $x^2 - 10x - 3 = 0$

EXAMPLE 11

5 Solve by completing the square and writing answers correct to 3 significant figures:

a $x^2 - 2x - 5 = 0$ **b** $x^2 + 12x + 34 = 0$ **c** $q^2 + 18q - 1 = 0$

d $x^2 - 4x - 2 = 0$ **e** $b^2 + 16b + 50 = 0$ **f** $x^2 - 24x + 112 = 0$

6 Solve by completing the square, giving exact solutions in simplest surd form:

a $2x^2 - 12x - 2 = 0$ **b** $3y^2 + 12y + 1 = 0$ **c** $2a^2 - 16a - 3 = 0$

d $5y^2 + 40y = 4$ **e** $3x^2 = 36x + 2$ **f** $x^2 + 3x - 2 = 0$

The quadratic formula 2.07

Completing the square is difficult with harder quadratic equations, such as $2x^2 - x - 5 = 0$. It is often easier to use the **quadratic formula**, which is proved by completing the square.

Videos
The quadratic formula
Solving quadratic equations

Worksheets
Quadratic formula
Quadratic equations
Problems involving quadratic equations
Solving algebraic equations

Technology
The quadratic formula

Spreadsheet
The quadratic formula

The quadratic formula

For the equation $ax^2 + bx + c = 0$, where a, b and c are real numbers:

$$x = \frac{-b \pm \sqrt{b^2 - 4ac}}{2a}$$

Proof

Solve $ax^2 + bx + c = 0$ by completing the square.

$$ax^2 + bx + c = 0$$

$$x^2 + \frac{bx}{a} + \frac{c}{a} = 0$$

$$x^2 + \frac{bx}{a} = -\frac{c}{a}$$

Completing the square:

$$x^2 + \frac{bx}{a} + \frac{b^2}{4a^2} = -\frac{c}{a} + \frac{b^2}{4a^2}$$

$$\left(\frac{b}{2a}\right)^2 = \frac{b^2}{4a^2}$$

$$\left(x + \frac{b}{2a}\right)^2 = -\frac{c}{a} + \frac{b^2}{4a^2}$$

$$= \frac{-4ac + b^2}{4a^2}$$

$$x + \frac{b}{2a} = \pm\sqrt{\frac{-4ac + b^2}{4a^2}}$$

$$= \pm\frac{\sqrt{b^2 - 4ac}}{2a}$$

$$x = \frac{-b}{2a} \pm \frac{\sqrt{b^2 - 4ac}}{2a}$$

$$= \frac{-b \pm \sqrt{b^2 - 4ac}}{2a}$$

Example 12

a Solve $x^2 - x - 2 = 0$ by using the quadratic formula.

b Solve $2y^2 - 9y + 3 = 0$ by using the quadratic formula and give your answer correct to 2 decimal places.

Solution

a $a = 1, b = -1, c = -2$

$$\begin{aligned} x &= \frac{-b \pm \sqrt{b^2 - 4ac}}{2a} \\ &= \frac{-(-1) \pm \sqrt{(-1)^2 - 4(1)(-2)}}{2(1)} \\ &= \frac{1 \pm \sqrt{1+8}}{2} \\ &= \frac{1 \pm \sqrt{9}}{2} \\ &= \frac{1 \pm 3}{2} \\ &= 2 \text{ or } -1 \end{aligned}$$

b $a = 2, b = -9, c = 3$

$$\begin{aligned} x &= \frac{-b \pm \sqrt{b^2 - 4ac}}{2a} \\ &= \frac{-(-9) \pm \sqrt{(-9)^2 - 4(2)(3)}}{2(2)} \\ &= \frac{9 \pm \sqrt{81 - 24}}{4} \\ &= \frac{9 \pm \sqrt{57}}{4} \\ &\approx 4.14 \text{ or } 0.36 \end{aligned}$$

EXERCISE 2.07 Answers on p. 461

The quadratic formula

EXAMPLE 12

1 Solve each equation using the quadratic formula.

a $x^2 + x - 4 = 0$ **b** $3x^2 - 5x + 1 = 0$ **c** $q^2 - 4q - 3 = 0$

d $4h^2 + 12h + 1 = 0$ **e** $3s^2 - 8s + 2 = 0$ **f** $x^2 + 11x - 3 = 0$

g $6d^2 + 5d - 2 = 0$ **h** $x^2 - 2x = 7$ **i** $t^2 = t + 1$

2 Solve each quadratic equation, correct to 3 significant figures where necessary.

a $y^2 + 6y + 2 = 0$ **b** $2x^2 - 5x + 3 = 0$ **c** $b^2 - b - 9 = 0$

d $2x^2 - x - 1 = 0$ **e** $-8x^2 + x + 3 = 0$ **f** $n^2 + 8n - 2 = 0$

g $m^2 + 7m + 10 = 0$ **h** $x^2 - 7x = 0$ **i** $x^2 + 5x = 6$

3 A number squared minus the product of the number and 2 gives a total of 4. Find the exact value of the number.

4 The product of a number and 9 is 5 more than its square. Find its value, correct to 2 decimal places.

5 Solve each equation correct to one decimal place.

a $x + \frac{1}{x} = 5$ **b** $y - \frac{8}{y} + 6 = 0$ **c** $n - 1 - \frac{7}{n} = 0$

d $3x - \frac{1}{x} + 4 = 0$ **e** $2b + \frac{1}{b} = 4$

□ Foundation ○ Mastery ⬡ Complex

Quadratic inequalities

2.08

Solving **quadratic inequalities** is similar to solving quadratic equations, but you need to check the inequality on a number line.

There is also a graphical method of solving quadratic inequalities by sketching a parabola, which you can find in Chapter 3 (3.09) on page 103.

Example 13

Solve:

a $x^2 + x - 6 > 0$

b $9 - x^2 \geq 0$

Solution

a First solve $x^2 + x - 6 = 0$

$(x - 2)(x + 3) = 0$

$\therefore x = 2$ or -3

This can also be solved using the quadratic formula.

Plot 2 and -3 on the number line as open circles because the inequality involves $>$.

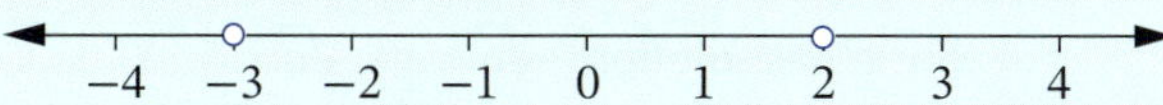

With quadratic inequalities, the solution is either the interval between these 2 values, $-3 < x < 2$, or the interval beyond these 2 values, $x < -3$ and $x > 2$. We need to test which interval it is.

Choose a number between -3 and 2, say $x = 0$.

Substitute $x = 0$ into the inequality.

$x^2 + x - 6 > 0$

$0 + 0 - 6 > 0$

$-6 > 0$ (false)

So, the solution is not between -3 and 2.

$\therefore$ the solution lies either side of -3 and 2.

An alternative to testing points is to sketch a parabola of $y = x^2 + x - 6$ with x-intercepts at -3 and 2 and look at where the graph is above the y-axis. You can find this method in Chapter 3 (3.09) on page 103.

Check by choosing a number, say $x = 3$.

Substitute $x = 3$ into the inequality.

$3^2 + 3 - 6 > 0$

$6 > 0$ (true)

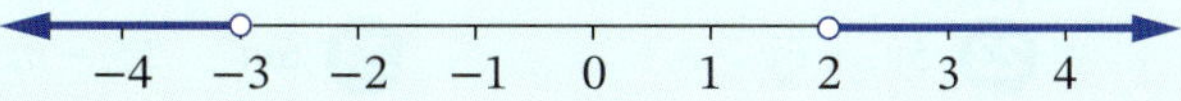

The solution is $x < -3$, $x > 2$.

b First solve $9 - x^2 = 0$

$9 = x^2$

$x = \pm\sqrt{9}$

$x = \pm 3$

Plot 3 and -3 on the number line as closed circles because the inequality involves $\geq$.

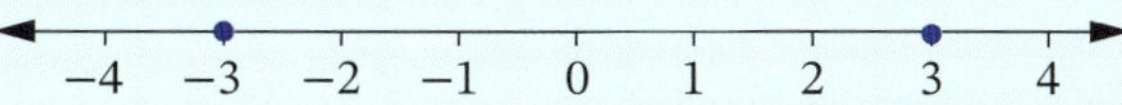

Choose a number between -3 and 3, say $x = 0$.

Substitute $x = 0$ into the inequality.

$9 - x^2 \geq 0$

$9 - 0^2 \geq 0$

$9 \geq 0$ (true)

So, the solution is between -3 and 3, that is $-3 \leq x \leq 3$ on the number line:

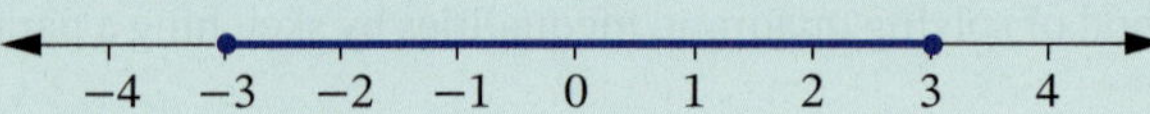

This inequality can also be solved algebraically:

$9 - x^2 \geq 0$

$9 \geq x^2$

$x^2 \leq 9$

$-3 \leq x \leq 3$

$x^2 < a$ and $x^2 > a$ inequalities

For $a > 0$:

$x^2 = a$ has solution $x = \pm\sqrt{a}$

$x^2 < a$ has solution $-\sqrt{a} < x < \sqrt{a}$

$x^2 > a$ has solution $x < -\sqrt{a}$, $x > \sqrt{a}$

EXERCISE 2.08 Answers on p. 461

Quadratic inequalities

Solve each quadratic inequality.

EXAMPLE 13

1 $x^2 + 3x < 0$
2 $y^2 - 4y < 0$
3 $n^2 - n \geq 0$
4 $x^2 - 4 \geq 0$
5 $1 - n^2 < 0$
6 $n^2 + 2n - 15 \leq 0$
7 $c^2 - c - 2 > 0$
8 $x^2 + 6x + 8 \leq 0$
9 $x^2 - 9x + 20 < 0$
10 $2b^2 + 5b + 2 \geq 0$
11 $1 - 2a - 3a^2 < 0$
12 $2y^2 - y - 6 > 0$
13 $3x^2 - 5x + 2 \geq 0$
14 $6 - 13b - 5b^2 < 0$
15 $6x^2 + 11x + 3 \leq 0$
16 $y^2 + y \leq 12$
17 $x^2 > 16$
18 $a^2 \leq 1$
19 $x^2 < x + 6$
20 $x^2 \geq 2x + 3$
21 $x^2 < 2x$
22 $2a^2 \leq 5a - 3$
23 $5y^2 + 6y \geq 8$
24 $6m^2 > 15 - m$

Linear simultaneous equations 2.09

Investigation

School musical

A group of 39 people went to see a school musical. The total cost of the tickets was $939, with children paying $17 each and adults paying $29 each. How many in the group were adults and how many were children? (Hint: let x be the number of adults and y the number of children.)

Videos
Simultaneous equations 1
Simultaneous equations 2
The elimination method
The substitution method

You can solve 2 equations together to find one solution that satisfies both equations. Such equations are called **simultaneous equations** and there are 2 ways of solving them. The **elimination method** adds or subtracts the equations. The **substitution method** substitutes one equation into the other.

Worksheets
Solving simultaneous equations
Simultaneous equations order activity
Simultaneous equations problems

Puzzle
Simultaneous equations puzzle

Example 14

Solve $3a + 2b = 5$ and $2a - b = -6$ simultaneously using:

a the elimination method

b the substitution method

Solution

a

$3a + 2b = 5$ [1]

$2a - b = -6$ [2]

[2] × 2 to make $-2b$ to eliminate with [1]: $4a - 2b = -12$ [3]

[1] + [3]: $7a = -7$

$a = -1$

Substitute $a = -1$ in [1] to find b:

$$3(-1) + 2b = 5$$
$$-3 + 2b = 5$$
$$2b = 8$$
$$b = 4$$

Check that the solution is correct by substituting back into *both* equations.

∴ the solution is $a = -1$, $b = 4$.

b

$3a + 2b = 5$ [1]

$2a - b = -6$ [2]

Make b the subject of equation [2]: $2a + 6 = b$ [3]

Substitute [3] into [1]:

$$3a + 2(2a + 6) = 5$$
$$3a + 4a + 12 = 5$$
$$7a + 12 = 5$$
$$7a = -7$$
$$a = -1$$

Substitute $a = -1$ into [3]:

$$b = 2(-1) + 6$$
$$b = 4$$

∴ the solution is $a = -1$, $b = 4$.

EXERCISE 2.09 Answers on p. 462

Linear simultaneous equations

EXAMPLE 14

1 Solve each pair of simultaneous equations using either the elimination or substitution method.

a $a - b = -2$ and $a + b = 4$

b $5x + 2y = 12$ and $3x - 2y = 4$

c $4p - 3q = 11$ and $5p + 3q = 7$

d $y = 3x - 1$ and $y = 2x + 5$

e $2x + 3y = -14$ and $x + 3y = -4$

f $7t + v = 22$ and $4t + v = 13$

g $4x + 5y + 2 = 0$ and $4x + y + 10 = 0$

h $2x - 4y = 28$ and $2x - 3y = -11$

i $5x - y = 19$ and $2x + 5y = -14$

j $5m + 4n = 22$ and $m - 5n = -13$

k $3a - 2b = -6$ and $a - 3b = -2$

l $3k - 2h = -14$ and $2k - 5h = -13$

2 The sum of 2 numbers is 13 and their difference is 5. Find the numbers.

3 A group of parents and their children all go to the cinema. Adult tickets cost \$20 and children's tickets cost \$12. If there are 17 people altogether and the total cost for tickets is \$260, how many are parents and how many are children?

4 Vi pays \$255 for some shirts and dresses. Each shirt costs \$15 and each dress costs \$35. Vi buys 9 items altogether. How many shirts and how many dresses does she buy?

5 Paola is twice the age of Natalia. In 3 years' time, Paola will be 5 years older than Natalia. What are their current ages?

6 The sum of 2 numbers is 21 and one number is twice the other. What are the 2 numbers?

7 Solve simultaneously:

$a - b + c = 7$

$a + 2b - c = -4$

$3a - b - c = 3$

□ Foundation ○ Mastery ⬡ Complex

Non-linear simultaneous equations 2.10

In simultaneous equations involving **non-linear equations,** there may be more than one set of solutions. When solving these, you need to use the substitution method only as the elimination method will not work.

Videos Non-linear simultaneous equations 1

Non-linear simultaneous equations 2

Worksheet Simultaneous equations

Example 15

Solve each pair of equations simultaneously using the substitution method:

a $xy = 6$ and $x + y = 5$

b $x^2 + y^2 = 16$ and $3x - 4y - 20 = 0$

Solution

a $xy = 6$ [1]

$x + y = 5$ [2]

From [2], make y the subject. $y = 5 - x$ [3]

Substitute [3] in [1].

$$x(5 - x) = 6$$
$$5x - x^2 = 6$$
$$0 = x^2 - 5x + 6$$
$$0 = (x - 2)(x - 3)$$
$$\therefore x = 2 \text{ or } x = 3$$

Substitute $x = 2$ in [3] to find y: $y = 5 - 2 = 3$

Substitute $x = 3$ in [3]: $y = 5 - 3 = 2$

There are 2 solutions.

Solutions are $x = 2, y = 3$ and $x = 3, y = 2$.

b $x^2 + y^2 = 16$ [1]

$3x - 4y - 20 = 0$ [2]

From [2] make y the subject:

$$3x - 20 = 4y$$
$$\frac{3x - 20}{4} = y \quad [3]$$

Substitute [3] into [1]:

$$x^2 + \left(\frac{3x - 20}{4}\right)^2 = 16$$
$$x^2 + \left(\frac{9x^2 - 120x + 400}{16}\right) = 16$$
$$16x^2 + 9x^2 - 120x + 400 = 256$$
$$25x^2 - 120x + 144 = 0$$

Factorising to solve:

$$(5x - 12)^2 = 0$$
$$\therefore 5x - 12 = 0$$
$$x = 2.4$$

Substitute $x = 2.4$ into [3] to find y:

$$y = \frac{3(2.4) - 20}{4}$$
$$= -3.2$$

Only one solution.

So the solution is $x = 2.4, y = -3.2$.

EXERCISE 2.10 Answers on p. 462

Non-linear simultaneous equations

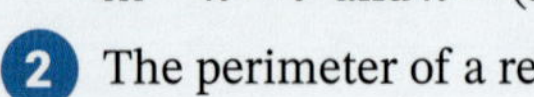

EXAMPLE 15

1. Solve each pair of simultaneous equations.

a $y = x^2$ and $y = x$

b $y = x^2$ and $2x + y = 0$

c $x^2 + y^2 = 9$ and $x + y = 3$

d $x - y = 7$ and $xy = -12$

e $y = x^2 + 4x$ and $2x - y - 1 = 0$

f $y = x^2$ and $6x - y - 9 = 0$

g $xy = 2$ and $y = 2x$

h $y = x^3$ and $y = x^2$

i $y = x - 1$ and $y = x^2 - 3$

j $y = x^2 + 1$ and $y = 1 - x^2$

k $y = x^2 - 3x + 7$ and $y = 2x + 3$

l $xy = 1$ and $4x - y + 3 = 0$

m $h = t^2$ and $h = (t + 1)^2$

n $y = x^2 - 7x + 6$ and $24x + 4y - 23 = 0$

2. The perimeter of a rectangle is 16 cm and its area is 12 cm^2. Find the length of each side.

Sample HSC problem

Answers on p. 462

(3 marks) Solve these simultaneous equations:

$$y = x^2$$
$$2x - y + 3 = 0$$

☐ Foundation ○ Mastery ⬡ Complex

CHAPTER SUMMARY

This chapter, *Equations and inequalities*, revised and extended skills in solving equations and inequalities algebraically.

Before you move on, consider what you learned in this chapter and revisit any sections and examples that may have been unclear.

What you learned in this chapter…	Section
Solve linear equations and inequalities	2.01 Equations and formulas 2.02 Inequalities
Substitute values into formulas and solve equations	2.01 Equations and formulas 2.02 Inequalities
Understand and use absolute values and solve equations involving absolute value	2.03 Absolute value 2.04 Equations involving absolute values
Solve quadratic equations by factorisation, completing the square and the quadratic formula	2.05 Quadratic equations by factorisation 2.06 Quadratic equations by completing the square 2.07 The quadratic formula
Solve quadratic inequalities	2.08 Quadratic inequalities
Solve linear and non-linear simultaneous equations	2.09 Linear simultaneous equations 2.10 Non-linear simultaneous equations

To help master these techniques, make a summary of this topic. Use the chapter outline and the mind map below as a guide. Add your own words, symbols, diagrams, boxes and reminders. The summary should give you a 'whole picture' view of the topic and allow you to identify any weak areas to revisit in your revision.

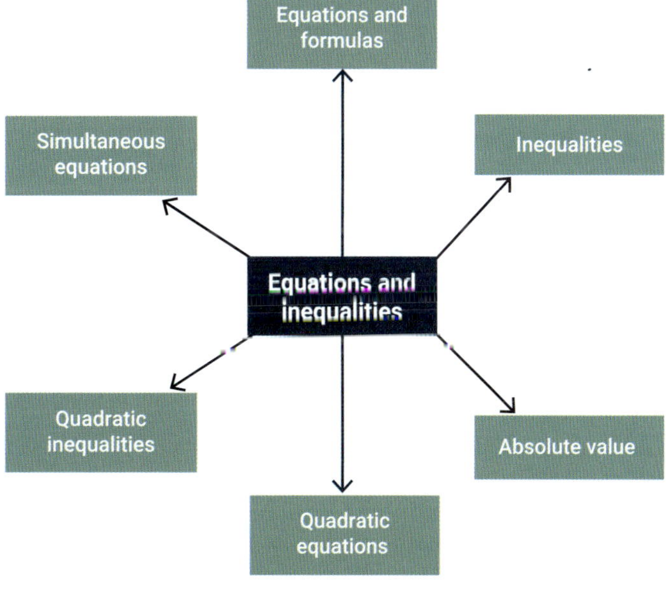

2 Test yourself

Answers on p. 462

For Questions **1** to **4**, select the correct answer **A**, **B**, **C** or **D**.

2.06, 2.07 **1** Find the exact solution of $x^2 - 5x - 1 = 0$.

A $\frac{-5 \pm \sqrt{29}}{2}$ **B** $\frac{5 \pm \sqrt{21}}{2}$ **C** $\frac{5 \pm \sqrt{29}}{2}$ **D** $\frac{-5 \pm \sqrt{21}}{2}$

2.01 **2** If $S = 4\pi r^2$, find the value of r when $S = 200$ (there may be more than one answer).

A $5\sqrt{\frac{2}{\pi}}$ **B** $\sqrt{\frac{200}{\pi}}$ **C** $10\sqrt{\frac{2}{\pi}}$ **D** $\sqrt{\frac{50}{\pi}}$

2.09 **3** Solve the simultaneous equations $x - y = 7$ and $x + 2y = 1$.

A $x = 5, y = 2$ **B** $x = 5, y = -2$ **C** $x = -5, y = -2$ **D** $x = -5, y = 2$

2.08 **4** Which is the graph of the solution of $x^2 + 4x - 5 \geq 0$?

A −7 −6 −5 −4 −3 −2 −1 0 1 2

B −7 −6 −5 −4 −3 −2 −1 0 1 2

C −7 −6 −5 −4 −3 −2 −1 0 1 2

D −7 −6 −5 −4 −3 −2 −1 0 1 2

2.03 **5** Evaluate:

a $|-15|$ **b** $|5| - |7|$ **c** $4|-3|$ **d** $|4|^2 - 2|3| + |-1|$

2.01, 2.02 **6** Solve:

a $\frac{a}{4} - \frac{a+2}{3} = 9$ **b** $4(3x + 1) = 11x - 3$ **c** $3p + 1 \leq p + 9$

2.09, 2.10 **7** Solve each pair of simultaneous equations.

a $x - y + 7 = 0$ and $3x - 4y + 26 = 0$ **b** $xy = 4$ and $2x - y - 7 = 0$

2.04 **8** Solve $|3b - 1| = 5$

2.01 **9** The area of a trapezium is given by $A = \frac{1}{2}h(a + b)$. Find:

a A when $h = 6$, $a = 5$ and $b = 7$

b b when $A = 40$, $h = 5$ and $a = 4$.

2.05 **10** Solve $2x^2 - 3x + 1 = 0$.

2.02 **11** Solve $-2 < 3y + 1 \leq 10$ and plot the solution on a number line.

2.06, 2.07 **12** Solve, correct to 3 significant figures:

a $x^2 + 7x + 2 = 0$ **b** $y^2 - 2y - 9 = 0$ **c** $3n^2 + 2n - 4 = 0$

2.01 **13** The surface area of a sphere is given by $A = 4\pi r^2$. Evaluate to one decimal place:

a A when $r = 7.8$ **b** r when $A = 102.9$

2.02 **14** Solve $\frac{x-3}{7} - \frac{3}{4} > 9$.

□ Foundation ○ Mastery ○ Complex

15 Solve $x^2 - 11x + 18 > 0$. 2.08

16 Solve the simultaneous equations $x^2 + y^2 = 16$ and $3x + 4y - 20 = 0$. 2.10

17 Solve: 2.02, 2.04

a $2(3y - 5) > y + 5$ **b** $n^2 + 3n \le 0$ **c** $|5x - 4| = 11$

d $x^2 + 2x - 8 \le 0$ **e** $y^2 - 4 > 0$ **f** $1 - x^2 \le 0$ 2.08

g $x^2 < 2x + 3$ **h** $m^2 + m \ge 6$

18 For each equation, decide if it has: 2.04, 2.07

A 2 solutions **B** one solution **C** no solutions.

a $x^2 - 6x + 9 = 0$ **b** $|2x - 3| = 7$ **c** $x^2 - x - 5 = 0$

d $2x^2 - x + 4 = 0$ **e** $3x + 2 = 7$

Challenge exercise 2

Answers on p. 462

1 Find the value of y if $a^{3y-5} = \frac{1}{a^2}$.

2 Solve $x^2 > a^2$, where $a > 0$.

3 The product of a number and 5 less 11 is greater than the product of the number and 3 plus 7. Find the values of the number.

4 Solve $\frac{2}{x-1} - \frac{1}{x+1} = 1$, correct to 3 significant figures.

5 **a** Factorise $x^5 - 9x^3 - 8x^2 + 72$.

b Hence or otherwise solve $x^5 - 9x^3 - 8x^2 + 72 = 0$.

6 Solve the simultaneous equations $y = x^3 + x^2$ and $y = x + 1$.

7 Find the value of b if $x^2 - 8x + b$ is a perfect square. Hence solve $x^2 - 8x - 1 = 0$ by completing the square.

8 Considering the definition of absolute value, solve $\frac{|x-3|}{3-x} = x$, where $x \neq 3$.

9 Solve $(x - 4)(x - 1) \le 28$.

10 Solve $x^{\frac{3}{2}} = \frac{1}{8}$.

11 Find the solutions of $x^2 - 2ax - b = 0$ by completing the square.

12 Solve $3x^2 = 8(2x - 1)$ and write the solution in its simplest surd form.

13 Solve $|2x - 1| = 5 - x$ and check solutions.

14 Solve $\frac{2}{x-1} - \frac{1}{x+1} + \frac{3}{x} = 0$, correct to one decimal place.

15 Large drinks cost \$5 each and small ones \$3. Pham buys 11 drinks altogether for her friends and spends \$43. How many of each drink does she buy?

16 Solve each equation.

a $n + \frac{4}{n} = 4$ **b** $n - 3 - \frac{7}{n} = 0$ **c** $3n + 9 + \frac{1}{n} = 0$

Foundation | Mastery | Complex

PRACTICE SET 1

Answers on p. 462

In Questions **1** to **6**, select the correct answer **A**, **B**, **C** or **D**.

1.03 **1** Write $\dfrac{1}{3\sqrt{(x-2)^5}}$ in index form.

A $(x-2)^{-\frac{5}{3}}$ **B** $\dfrac{(x-2)^{-\frac{5}{2}}}{3}$ **C** $3(x-2)^{-\frac{5}{2}}$ **D** $\dfrac{1}{(x-2)^{\frac{5}{3}}}$

1.01 **2** Simplify $\dfrac{(2a^3b)^3}{(ab)^2}$.

A $8a^7b$ **B** $8a^8b$ **C** $2a^7b$ **D** $2a^8b$

1.03 **3** Evaluate $4^{-\frac{3}{2}}$.

A -8 **B** -6 **C** $\dfrac{1}{8}$ **D** $\dfrac{1}{6}$

1.09 **4** Simplify $\dfrac{a^2-6a+9}{a^2-9}$.

A $\dfrac{1}{a+3}$ **B** $\dfrac{a-3}{a+3}$

C $\dfrac{a+3}{a-3}$ **D** $\dfrac{-6a+9}{a-9}$

1.08 **5** Factorise $a^2 - \dfrac{b^2}{4}$.

A $\left(a-\dfrac{b}{2}\right)^2$ **B** $\left(a+\dfrac{b}{4}\right)\left(a-\dfrac{b}{4}\right)$

C $\left(a+\dfrac{b}{2}\right)^2$ **D** $\left(a+\dfrac{b}{2}\right)\left(a-\dfrac{b}{2}\right)$

2.06, 2.07 **6** The solution to $x^2 + 2x - 6 = 0$ is:

A $x = -1 \pm 2\sqrt{7}$ **B** $x = \dfrac{2 \pm \sqrt{28}}{2}$

C $x = \dfrac{-2 \pm \sqrt{-20}}{2}$ **D** $x = -1 \pm \sqrt{7}$

2.01, 2.04, 2.05 **7** Solve each equation.

a $3x - 7 = 23$ **b** $5(b - 3) = 15$ **c** $\dfrac{x}{3} + 4 = 5$

d $4y - 7 = 3y + 9$ **e** $8z + 1 = 11z - 17$ **f** $x^2 = 32$

g $x^2 - 3x = 0$ **h** $|x + 2| = 5$ **i** $|5a - 2| = 8$

2.01 **8** Solve $\dfrac{p-3}{2} - \dfrac{p+1}{5} = 1$.

1.10 **9** Simplify $2\sqrt{12}$.

1.06 **10** Factorise fully $10x + 2xy - 10y - 2y^2$.

1.02, 1.03 **11** Write in index form:

a $\dfrac{1}{x}$ **b** $\sqrt[3]{x^4}$

Foundation Mastery Complex

12 Simplify the expression $8y - 2(y + 5)$. 1.04

13 Rationalise the denominator of $\dfrac{5}{5-\sqrt{2}}$. 1.11

14 Solve $2x^2 - 3x - 1 = 0$, correct to 3 significant figures. 2.07

15 Simplify $\dfrac{x+1}{5} \div \dfrac{x^2-2x-3}{10}$. 1.09

16 Simplify $2\sqrt{3} - \sqrt{27}$. 1.10

17 Expand and simplify $(x - 3)(x^2 + 5x - 1)$. 1.04

18 Expand and simplify $\sqrt{2}(3\sqrt{5} - 2\sqrt{2})$. 1.10

19 Solve $4a - 5 < 7a + 4$. 2.02

20 Solve this pair of simultaneous equations. 2.09

$3a - b = 7$ and $2a + b = 8$

21 Solve $5 - 2x < 3$ and show the solution on a number line. 2.02

22 Solve the equation $x^2 - 4x + 1 = 0$, giving exact solutions in simplest surd form. 2.06, 2.07

23 Write 7^{-2} as a fraction. 1.02

24 Solve the simultaneous equations $y = 3x - 1$ and $y = x^2 - 5$. 2.10

25 Find integers x and y such that $\dfrac{\sqrt{3}}{2\sqrt{3}+3} = x + y\sqrt{3}$. 1.11

26 Evaluate $|-2|^2 - |-1| + |4|$. 2.03

27 Factorise $8x^2 - 32$. 1.08

28 Rationalise the denominator of $\dfrac{2\sqrt{3}}{3\sqrt{5}-\sqrt{2}}$. 1.11

29 Simplify $2|-4| - |3| + |-2|$. 2.03

30 Rationalise the denominator of $\dfrac{\sqrt{5}+1}{2\sqrt{2}+3}$. 1.11

31 Simplify $\dfrac{(a^{-4})^3 \times b^6}{a^9 \times (b^{-1})^4}$. 1.01

32 Evaluate $4^{-\frac{3}{2}}$ as a rational number. 1.03

33 Simplify $2(x - 5) - 3(x - 1)$. 1.04

34 Write $\dfrac{1}{x+3}$ in index form. 1.02

35 Write $(3x+2)^{-\frac{1}{2}}$ without an index. 1.03

□ Foundation ○ Mastery ○ Complex

PRACTICE SET 1

1.04, 1.09 **36** Simplify:

a $8x - 7y - y + 4x$

b $\sqrt{124}$

1.10, 1.11 c $\dfrac{x^2 - 9}{2x^2 + 5x - 3}$

d $\dfrac{1}{\sqrt{2}+1} + \dfrac{2}{\sqrt{2}-1}$

e $\dfrac{3}{x+1} + \dfrac{2}{x^2-1} - \dfrac{4}{x-1}$

f $\dfrac{(x^{-2})^5 y^4 z^{-3}}{x^4 (y^3)^{-1} (z^{-4})^{-2}}$

g $\dfrac{a+b}{5a-20ab^2} \div \dfrac{a^2+2ab+b^2}{3-6b}$

h $8\sqrt{5} - 3\sqrt{20} + 2\sqrt{45}$

2.01 **37** The volume of a sphere is given by the formula $V = \dfrac{4}{3}\pi r^3$.

Find the exact radius r if the volume V is $10\dfrac{2}{3}$ cm³.

1.05 **38** Find the value of k if $(2x + 5)^2 = 4x^2 + kx + 25$.

1.03 **39** Simplify $\sqrt{81x^2 y^3}$.

1.08 **40** Factorise:

a $5(a - 2)^2 + 40(a - 2)$

b $(2a - b + c)^2 - (a + 5b - c)^2$

2.02 **41** Solve $-2 \le \dfrac{8x-1}{5} < 9$.

1.09 **42** Simplify $\dfrac{x+1}{5} - \dfrac{x+2}{3}$.

2.05 **43** Solve $x^2 - 5x = 0$.

2.07 **44** Solve $x^2 - 5x - 1 = 0$ and write the solutions, correct to 2 decimal places.

1.10 **45** Simplify $\sqrt{8} + \sqrt{98}$.

1.09 **46** Write $\dfrac{3}{x^2+5x} - \dfrac{4}{x} + \dfrac{2}{x+5}$ as a single fraction.

1.07, 1.08 **47** Factorise:

a $x^2 - 2x - 8$

b $a^2 - 9$

c $y^2 + 6y + 9$

d $t^2 + 8t + 16$

e $3x^2 - 11x + 6$

2.08 **48** Solve each inequality:

a $a^2 - 1 < 0$

b $y^2 + 3y \ge 0$

c $y^2 - y - 2 \le 0$

d $x^2 > 9$

e $2d - d^2 \ge 0$

2.04 **49** Solve:

a $y^2 - 2y - 13 = 0$ (correct to 2 decimal places)

2.07, 2.08 b $|2b + 3| = 7$

c $m^2 \le 9$

☐ Foundation ◯ Mastery ⬡ Complex

FUNCTIONS AND GRAPHS

Functions and their graphs are used in many areas such as mathematics, science and economics. In this chapter you will explore what functions are and how to sketch some types of graphs, namely lines, parabolas and cubics, as well as solve problems and inequalities involving them. There are many graphing applications you can use to explore functions and their graphs.

Functions and graphs make up a large portion of this HSC course, so this is an important foundation topic. However, you may know these graphs from Year 10. If you are confident that you know them well and want to focus on the new concepts, you can move through these quickly.

Chapter outline

In this chapter you will:

- understand the definition of a function and test a function using the vertical line test
- use function notation
- identify and use piecewise functions
- find the domain and range of functions, including using interval notation
- identify and prove even and odd functions
- identify and use linear, quadratic and cubic functions, their graphs and properties
- find the gradient and angle of inclination of a line
- find the equation of a line, including parallel and perpendicular lines, using the gradient-intercept, general and point-gradient forms on the linear equation
- solve simultaneous linear equations graphically
- solve quadratic inequalities
- find the equation of a parabola, including its axis of symmetry, vertex and intercepts
- find coefficients in identical quadratic expressions
- use the discriminant to classify and number the roots of a quadratic equation
- solve simultaneous equations involving linear and quadratic equations graphically
- apply linear and quadratic functions, including cost and revenue

Videos (18):

3.02 Piecewise functions

3.03 Interval notation • Domain and range 1 • Domain and range 2 • Odd and even functions

3.04 The general equation $ax + by + c = 0$

3.05 Gradient and y-intercept of a line

3.06, 3.14 Linear modelling

3.07 Equations of parallel lines • Equations of perpendicular lines

3.09 Quadratic inequalities

3.10 Graphing the parabola $y = a(x - b)^2 + c$ • The turning point form of a parabola

3.12 The discriminant

3.13 The intercept form of a cubic curve

3.14 Break-even points • Linear modelling • Applying linear functions • Applying quadratic functions

Worksheets (22):

3.01 Functions and relations

3.02 Function notation 1

3.03 Function notation 2

3.04 Graphing linear functions 1 • Graphing linear functions 2 • x- and y-intercepts • Intersection of lines

3.05 $y = mx + c$

3.06 Finding the equation of a line • Equations of a line

3.07 Linear functions

3.08 Graphing quadratic functions • Graphing quadratics

3.09 Quadratic inequalities

3.10 Quadratic functions • Sketching quadratic functions • Features of a parabola • Matching parabolas with their equations

3.11 A page of parabolas

3.12 The discriminant

3.13 Cubic functions • Graphing cubics • Graphing cubics 2

Graph paper (1):

3.04 A page of number planes

Puzzles (2):

3.06 Linear functions code puzzle

3.07 Parallel and perpendicular lines

Nelson MindTap

To access resources above, visit **cengage.com.au/nelsonmindtap**

Terminology

axis of symmetry	break-even point	cubic function	dependent variable
discriminant	domain	even function	function
function notation	gradient	independent variable	intercept
interval notation	linear function	odd function	parabola
piecewise function	point of inflection	quadratic function	range
root	vertex	vertical line test	zero

Functions and relations 3.01

A **relation** is an association or mapping between the elements of one set of numbers and the elements of a second set of numbers. It can also be described as a set of **ordered pairs** (x, y), where the **variables** x and y are related according to some pattern or rule.

Worksheet
Functions and relations

This diagram shows a mapping between the x values in set A and the y values in set B.

The x value 1 in set A is mapped to the y values 3 and 4 in set B, giving the ordered pairs $(1, 3)$ and $(1, 4)$.

The x value 2 is mapped to the y value 1, giving the ordered pair $(2, 1)$.

The x is called the **independent variable**, and the y is called the **dependent variable** because the value of y depends on the value of x.

A relation can also be described by an algebraic formula, a table of values, or a graph on the number plane.

Functions

A **function** is a special type of relation of ordered pairs (x, y) where each element of the first set X is associated with exactly one element from the second set Y.

Every value of x corresponds to exactly one value of y.

The set of all x values of a function is called the **domain**.

The set of all y values of a function is called the **range**.

This diagram and graph show a mapping between the x and y values of a function.

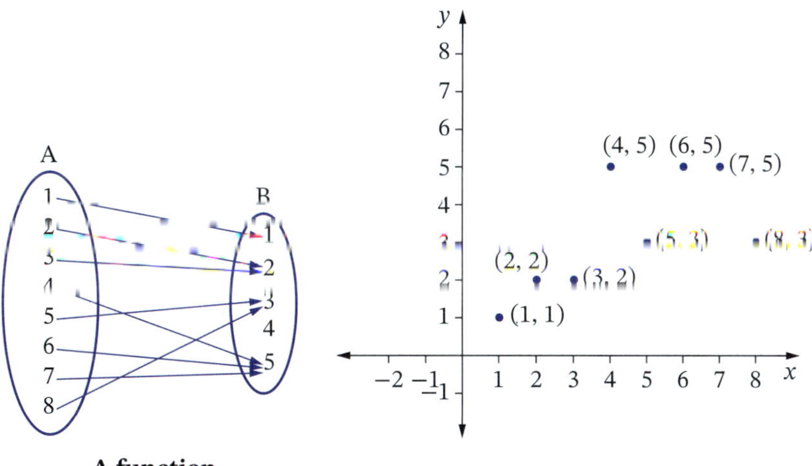

A function

Each x value is related to one y value: the ordered pairs are $(1, 1)$, $(2, 2)$, $(3, 2)$, $(4, 5)$, $(5, 3)$, $(6, 5)$, $(7, 5)$, $(8, 3)$.

The domain is $\{1, 2, 3, 4, 5, 6, 7, 8\}$.

The range is $\{1, 2, 3, 5\}$.

For a function, each x value must have only one y value, but different x values can have the same y value. In the above diagram, $(2, 2)$ and $(3, 2)$ have the same y value, 2.

This diagram and graph show a relation that is *not* a function.

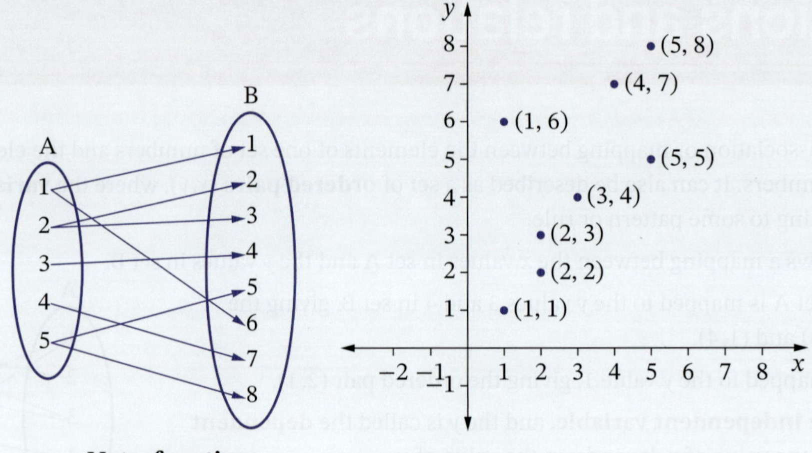

Not a function

For example, $x = 1$ is related to $y = 1$ and $y = 6$, giving $(1, 1)$ and $(1, 6)$.

$x = 1$ is mapped to 2 values of y, not one. So are $x = 2$ and $x = 5$.

For a function, you cannot have 2 different points in the same column on a number plane, that is, 2 different points can't have the same *x* value.

The vertical line test

If a relation is graphed on a number plane, you can use the **vertical line test** to test if the graph represents a function.

If any vertical line crosses a graph at most once, the graph represents a function. This shows that, for every value of *x*, there is only one value of *y*.

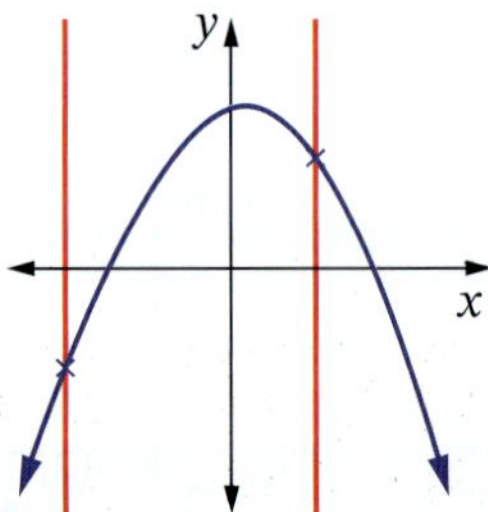

If any vertical line crosses a graph at more than one point, the graph does not represent a function. This shows that, for some value of *x*, there is more than one value of *y*.

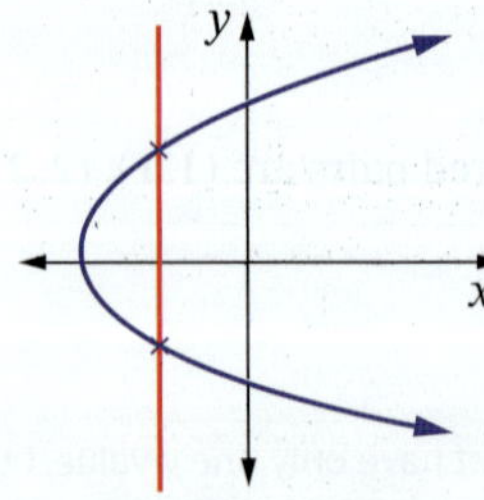

Example 1

Does each graph or set of ordered pairs represent a function?

a

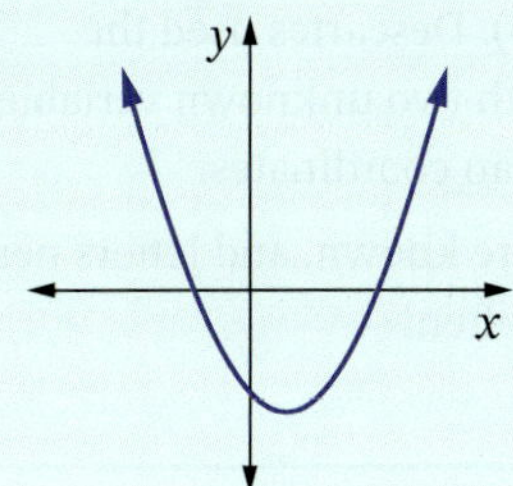

b

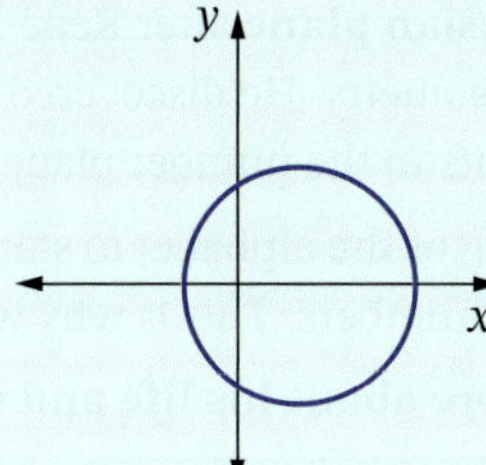

c (−2, 3), (−1, 4), (0, 5), (1, 3), (2, 4)

d

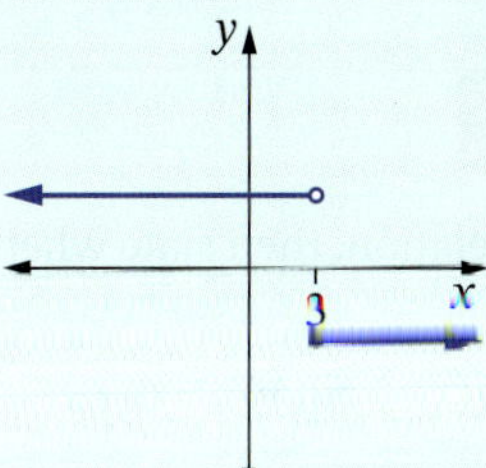

Solution

a

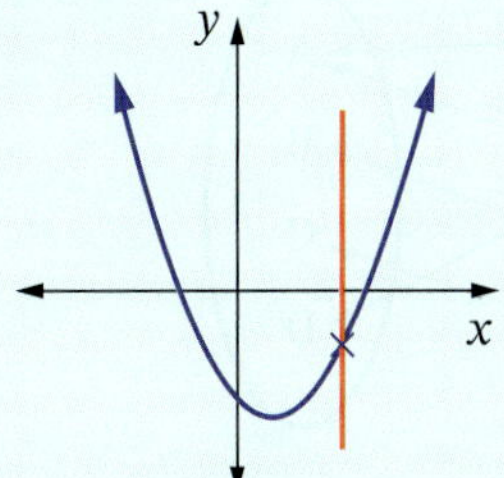

A vertical line only cuts the graph once. So the graph represents a function.

b

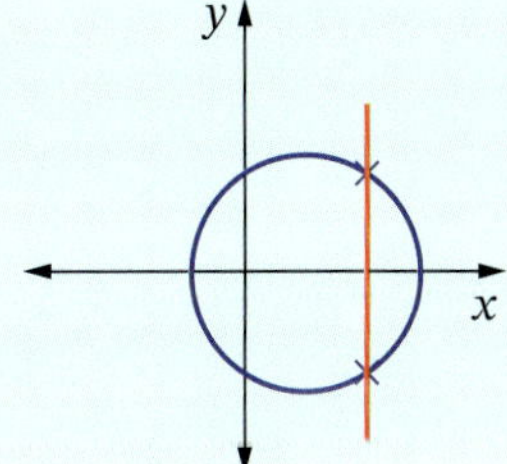

A vertical line can cut the curve in more than one place. So the circle does not represent a function.

c For each x value there is only one y value, so this set of ordered pairs is a function.

d

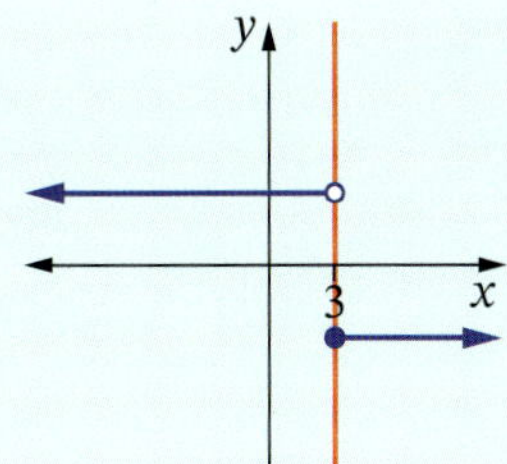

The open circle at $x = 3$ on the top line means that $x = 3$ is not included, while the closed circle on the bottom line means that $x = 3$ is included on this line.

So, a vertical line only touches the graph once at $x = 3$.

The graph represents a function.

Did you know?

René Descartes

The number plane is called the **Cartesian plane** after René Descartes (1596–1650). Descartes used the number plane to develop analytical geometry. He discovered that any equation with two unknown variables can be represented by a line. The points in the number plane can be called Cartesian coordinates.

Descartes used letters at the beginning of the alphabet to stand for numbers that are known, and letters near the end of the alphabet for unknown numbers. This is why we still use x and y so often!

Research Descartes to find out more about his life and work.

EXERCISE 3.01 Answers on p. 463

Functions and relations

1 List the ordered pairs for each relation, then state whether the relation is a function.

a

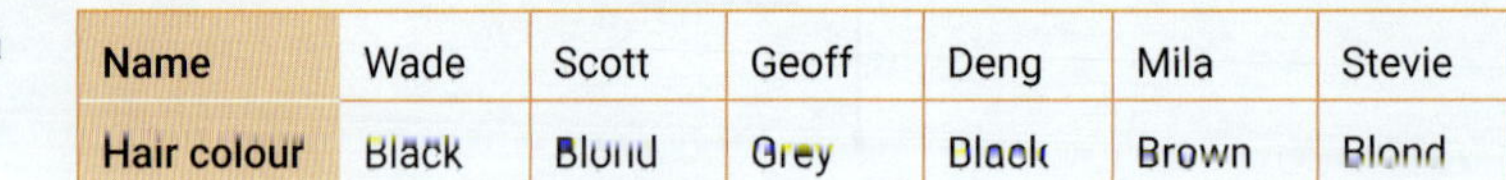

Name	Wade	Scott	Geoff	Deng	Mila	Stevie
Hair colour	Black	Blond	Grey	Black	Brown	Blond

b

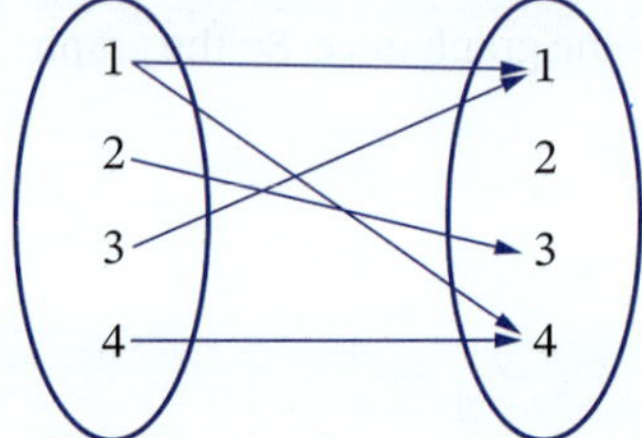

c

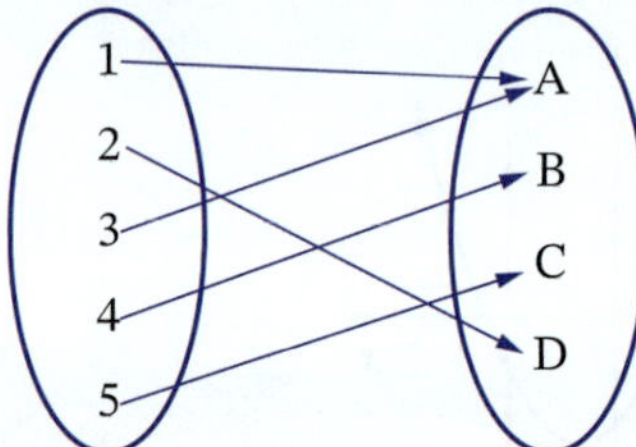

d

x	3	5	8	9	5	8
y	5	± 2	-7	3	6	0

e

x	y
1	9
2	15
3	27
4	33
5	45

Foundation Mastery Complex

EXAMPLE 1

2 Does each graph or set of ordered pairs represent a function?

a

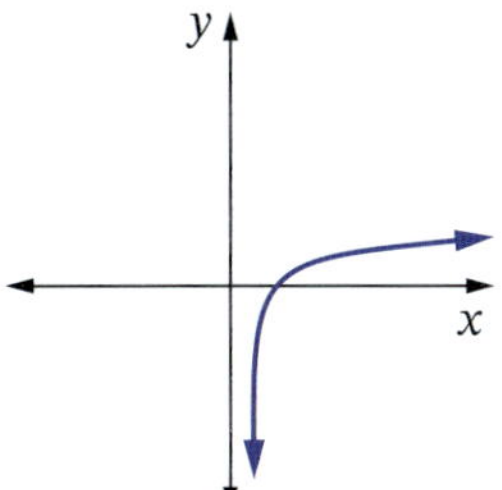

b

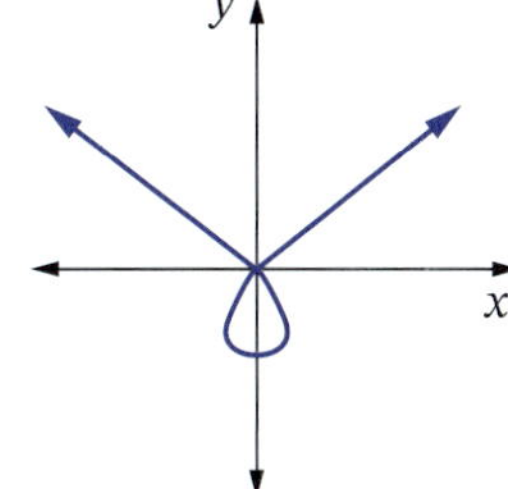

c

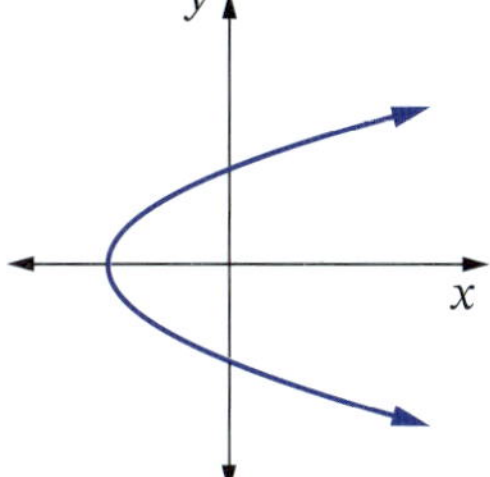

d

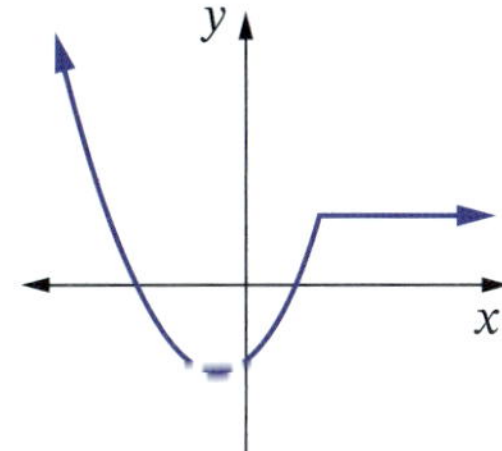

e

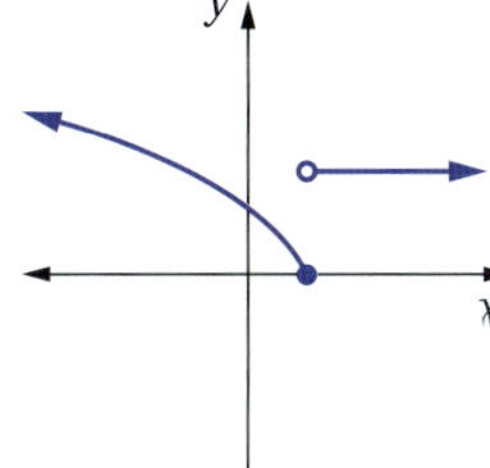

f

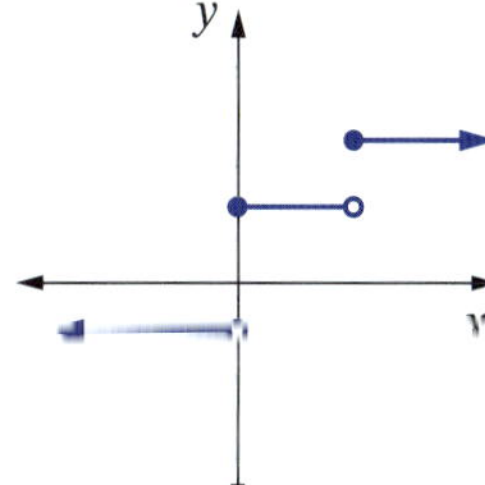

g

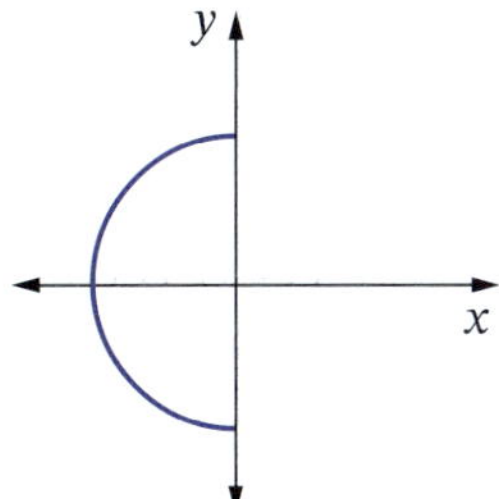

h

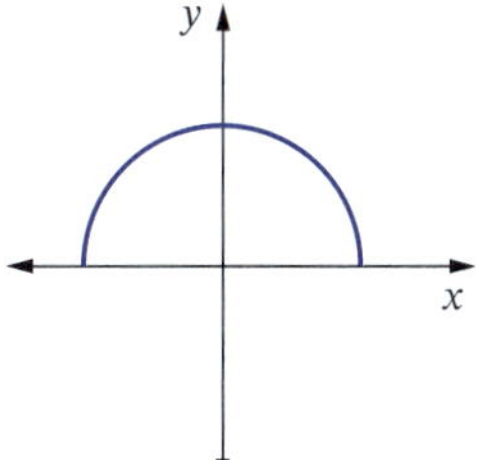

i (1, 3), (2, −1), (3, 3), (4, 0)

j (1, 3), (2, −1), (2, 7), (4, 0)

k

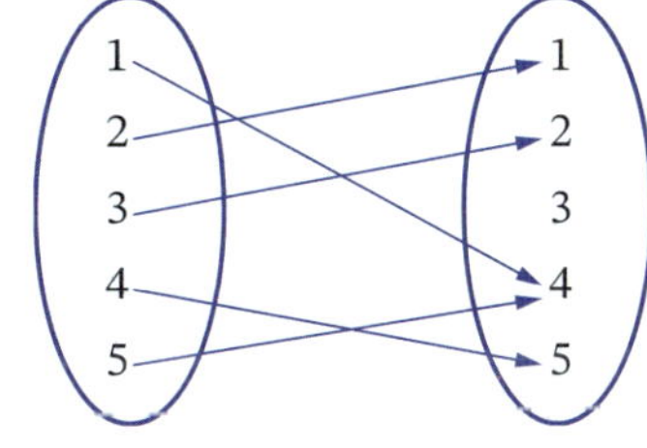

l

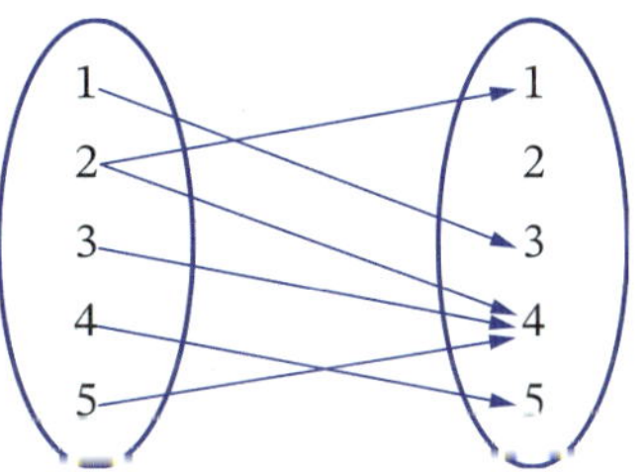

m (2, 5), (3, −1), (4, 0), (−1, 3), (−2, 7)

n

Person	Ben	Paula	Pierre	Hamish	Jacob	Leanne	Pierre	Lien
Sport	Tennis	Football	Tennis	Football	Football	Badminton	Football	Badminton

o

Input	Output
A	3
B	4
C	7
D	3
E	5
F	7
G	4

☐ Foundation ◯ Mastery ⬡ Complex

3 A relation consists of the ordered pairs $(-3, 4)$, $(-1, 5)$, $(0, -2)$, $(1, 4)$ and $(6, 8)$.

a List the values of the domain.

b List the values of the range.

c Is the relation a function?

4 A relation is represented by this table of values.

x	2	-1	0	3	-1	-2
y	7	5	-2	-6	0	1

a Is this relation a function?

b List the values of the independent variable.

c List the values of the dependent variable.

3.02 Function notation and piecewise functions

Worksheet Function notation 1

Since the value of y depends on the value of x, we say that y is a function of x. We write this using **function notation** as $y = f(x)$, where x is the independent variable and y is the dependent variable.

For different values of x we can find values of f at x. For example, the value of f at $x = 5$ can be written in function notation as $f(5)$.

Example 2

a If $f(x) = x^2 + 3x + 1$, find $f(-2)$.

b If $f(x) = x^3 - x^2$, find the value of $f(-1)$.

c Find the values of x for which $f(x) = 0$, given that $f(x) = x^2 + 3x - 10$.

Solution

a
$$f(x) = x^2 + 3x + 1$$
$$f(-2) = (-2)^2 + 3(-2) + 1$$
$$= 4 - 6 + 1$$
$$= -1$$

b
$$f(x) = x^3 - x^2$$
$$f(-1) = (-1)^3 - (-1)^2$$
$$= -1 - 1$$
$$= -2$$

c
$$f(x) = 0$$
$$x^2 + 3x - 10 = 0$$
$$(x + 5)(x - 2) = 0$$
$$x = -5, x = 2$$

You can also substitute algebraic expressions instead of numbers into functions.

Example 3

Find $f(h+1)$ given $f(x) = 5x + 4$.

Solution

Substitute $h + 1$ for x:

$$f(h+1) = 5(h+1) + 4$$
$$= 5h + 5 + 4$$
$$= 5h + 9$$

Piecewise functions

A **piecewise function** is a function made up of 2 or more different functions defined for different intervals in its domain.

Example 4

a $f(x) = \begin{cases} 3x+4 & \text{when } x \geq 2 \\ -2x & \text{when } x < 2 \end{cases}$

Find $f(3), f(2), f(0)$ and $f(-4)$.

b $g(x) = \begin{cases} x^2 & \text{when } x > 2 \\ 2x-1 & \text{when } -1 \leq x \leq 2 \\ 5 & \text{when } x < -1 \end{cases}$

Find $g(1) + g(-2) - g(3)$.

Solution

a $f(3) = 3(3) + 4$ since $3 \geq 2$

$= 13$

$f(2) = 3(2) + 4$ since $2 \geq 2$

$= 10$

$f(0) = -2(0)$ since $0 < 2$

$= 0$

$f(-4) = -2(-4)$ since $-4 < 2$

$= 8$

b $g(1) = 2(1) - 1$ since $-1 \leq 1 \leq 2$

$= 1$

$g(-2) = 5$ since $-2 < -1$

$g(3) = 3^2$ since $3 > 2$

$= 9$

So $g(1) + g(-2) - g(3) = 1 + 5 - 9$

$= -3$

Did you know?

Leonhard Euler

Leonhard Euler (1707–83), from Switzerland, studied functions and invented the function notation $f(x)$. He studied theology, astronomy, anatomy, physics and oriental languages, as well as mathematics, and published more than 800 articles and over 25 books on mathematics. He found time between writing to marry and have 13 children, and even when he went blind he kept on having his work published.

EXERCISE 3.02 Answers on p. 463

Answers on p. 463

Function notation and piecewise functions

1. Given $f(x) = x + 3$, find $f(1)$ and $f(-3)$.
2. If $h(x) = x^2 - 2$, find $h(0)$, $h(2)$ and $h(-4)$.
3. If $f(x) = -x^2$, find $f(5)$, $f(-1)$, $f(3)$ and $f(-2)$.
4. Find the value of $f(0) + f(-2)$ if $f(x) = x^4 - x^2 + 1$.
5. Find $f(-3)$ if $f(x) = 2x^3 - 5x + 4$.
6. If $f(x) = 2x - 5$, find x when $f(x) = 13$.
7. Given $f(x) = x^2 + 3$, find any values of x for which $f(x) = 28$.
8. If $f(x) = 3^x$, find x when $f(x) = \frac{1}{27}$.
9. Find values of z for which $f(z) = 5$, given $f(z) = |2z + 3|$.

EXAMPLE 3

10. If $f(x) = 2x - 9$, find $f(p)$ and $f(x + h)$.
11. Find $g(x - 1)$ when $g(x) = x^2 + 2x + 3$.
12. If $f(x) = x^2 - 1$, find $f(k)$ as a product of factors.
13. Given $f(t) = t^2 - 2t + 1$, find:
    - **a** t when $f(t) = 0$
    - **b** any values of t for which $f(t) = 9$.
14. Given $f(t) = t^4 + t^2 - 5$, find the value of $f(b) - f(-b)$.

15. $f(x) = \begin{cases} x^3 & \text{for } x > 1 \\ x & \text{for } x \le 1 \end{cases}$

    Find $f(5)$, $f(1)$ and $f(-1)$.
16. Find the value of $f(3) - f(2) + 2f(-3)$ when $f(x) = \begin{cases} x & \text{for } x > 2 \\ x^2 & \text{for } -2 \le x \le 2 \\ 4 & \text{for } x < -2 \end{cases}$
17. Find the value of $f(-1) - f(3)$ if $f(x) = \begin{cases} x^3 - 1 & \text{for } x \ge 2 \\ 2x^2 + 3x - 1 & \text{for } x < 2 \end{cases}$
18. If $f(x) = x^2 - 5x + 4$, find $f(x + h) - f(x)$ in its simplest form.
19. Simplify $\frac{f(x+h) - f(x)}{h}$, where $f(x) = 2x^2 + x$.
20. If $f(x) = 5x - 4$, find $f(x) - f(c)$ in its simplest form.
21. Find the value of $f(k^2)$ if $f(x) = \begin{cases} 3x + 5 & \text{for } x \ge 0 \\ x^2 & \text{for } x < 0 \end{cases}$
22. If $f(x) = \begin{cases} x^3 & \text{when } x \ge 3 \\ 5 & \text{when } 0 < x < 3 \\ x^2 - x + 2 & \text{when } x \le 0 \end{cases}$

    evaluate:
    - **a** $f(0)$
    - **b** $f(2) - f(1)$
    - **c** $f(-n^2)$

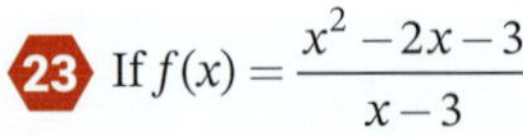

23. If $f(x) = \frac{x^2 - 2x - 3}{x - 3}$:
    - **a** evaluate $f(2)$
    - **b** explain why the function does not exist for $x = 3$
    - **c** by taking several x values close to 3, find the value of y that the function is moving towards as x moves towards 3.

□ Foundation ○ Mastery ○ Complex

Domain, range, even and odd functions 3.03

When you are not sure how to sketch the graph of a function, you can draw up a table of values first. If you know what its basic shape should be, such as a line or parabola, you can use their properties to sketch the graph.

Worksheet Function notation 2

Intercepts

The x-intercept of a graph is the value of x where the graph crosses the x-axis.

The y-intercept of a graph is the value of y where the graph crosses the y-axis.

Intercepts of the graph of a function

For x-intercept(s), substitute $y = 0$.

For y-intercept, substitute $x = 0$.

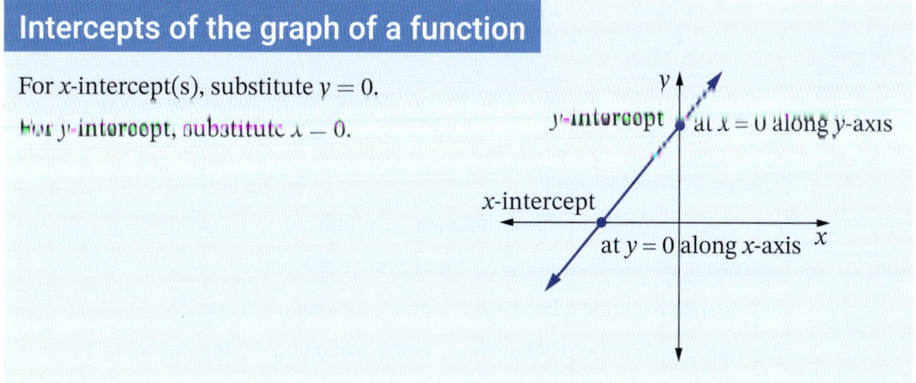

For the graph of $y = f(x)$, solving $f(x) = 0$ gives the x-intercepts and evaluating $f(0)$ gives the y-intercept.

The solutions of $f(x) = 0$ are also called the **zeroes** of the function because they are the value(s) of x for which $f(x) = 0$.

Example 5

Find the x- and y-intercepts of the function $f(x) = x^2 + 7x - 8$.

Solution

For x-intercepts, $f(x) = 0$.

$0 = x^2 + 7x - 8$

$= (x + 8)(x - 1)$

$x = -8, x = 1$

So x-intercepts are -8 and 1.

-8 and 1 are also called the **zeroes** of $f(x)$.

For y-intercept, $x = 0$.

$f(0) = 0^2 + 7(0) - 8$

$= -8$

So the y-intercept is -8.

Domain and range

The domain of a function $y = f(x)$ is the set of all x values for which $f(x)$ is defined.

The range of a function $y = f(x)$ is the set of all y values for which $f(x)$ is defined.

We can describe the domain and range in different ways, for example, 'all real x', '$y \geq 0$', '$0 < x < 2$'.

We can also use **interval notation**. This uses square brackets [ ] to include the end values and round brackets ( ) to exclude the end values.

As long as you describe the domain and range accurately, it doesn't matter which notation you choose.

Interval notation

- $[a, b]$ means the interval is between a and b, including a and b.
- (a, b) means the interval is between a and b, excluding a and b.
- $[a, b)$ means the interval is between a and b, including a but excluding b.
- $(a, b]$ means the interval is between a and b, excluding a but including b.
- $(-\infty, \infty)$ means that the interval includes the set of all real numbers $\mathbb{R}$.

Video
Interval notation

Example 6

Write each inequality in interval notation.

a $x > 1$ **b** $x \geq 0$ **c** all real numbers y

d $1 < y < 5$ **e** $-3 \leq x \leq 2$ **f** $-2 < y \leq 3$

Solution

a $(1, \infty)$ We don't include ∞ because infinity isn't a number.

b $[0, \infty)$ '[' means that we include 0.

c $(-\infty, \infty)$ This means all (real) numbers (between $-\infty$ and ∞).

d $(1, 5)$ This means all numbers between 1 and 5, but not 1 and 5.

e $[-3, 2]$ This means all numbers between -3 and 2, including -3 and 2.

f $(-2, 3]$ This means all numbers between -2 and 3, excluding -2 but including 3.

Videos
Domain and range 1

Domain and range 2

Example 7

Find the domain and range of each function.

a $f(x) = x^2$ **b** $y = \sqrt{x-1}$

Solution

a You can find the domain and range from the equation or the graph.

For $f(x) = x^2$, you can substitute any value for x. The y values will be 0 or positive.

So the domain is all real values of x and the range is $y \geq 0$.

We can write this using interval notation:

Domain: $(-\infty, \infty)$

Range: $[0, \infty)$

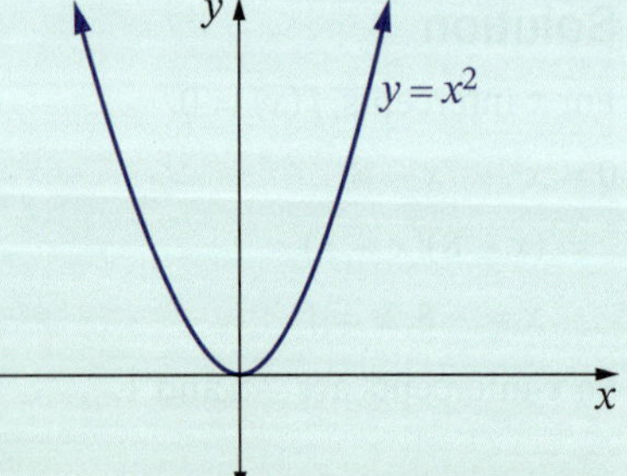

b The function $y = \sqrt{x-1}$ is only defined if $x - 1 \geq 0$ because we can only evaluate the square root of a positive number or 0.

For example, $x = 0$ gives $y = \sqrt{-1}$, which is undefined for real numbers.

So $x - 1 \geq 0$

$x \geq 1$

Domain: $[1, \infty)$

The value of $\sqrt{x-1}$ is always positive or zero. So $y \geq 0$.

Range: $[0, \infty)$

Increasing and decreasing graphs

When you draw a graph, it helps to know whether the function is increasing or decreasing on an interval.

If a graph is increasing, y increases as x increases, and the graph is moving upwards.

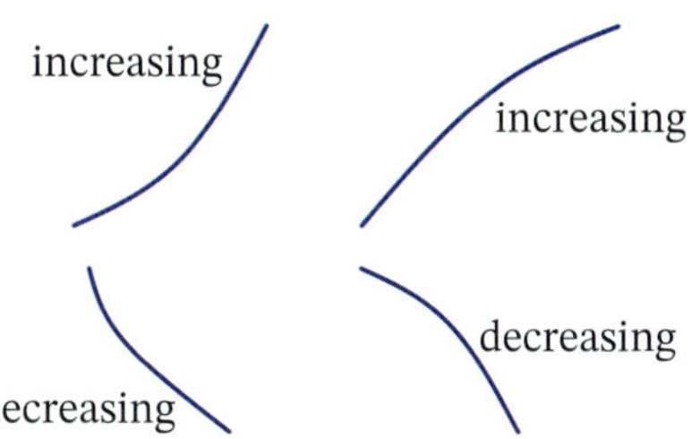

If a graph is decreasing, then y decreases as x increases, and the curve moves downwards.

Example 8

State the domain over which each curve is increasing.

a

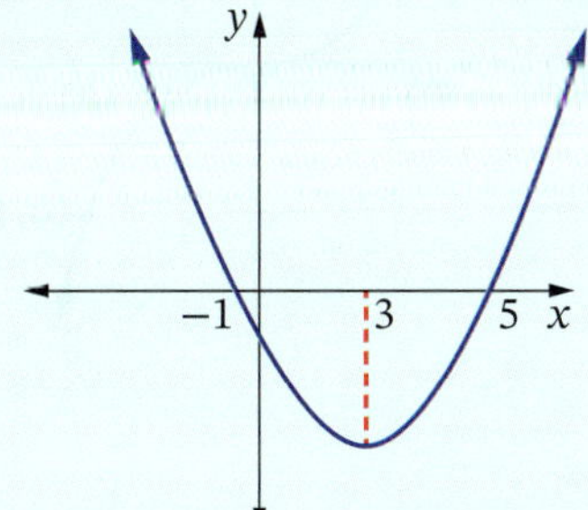

b

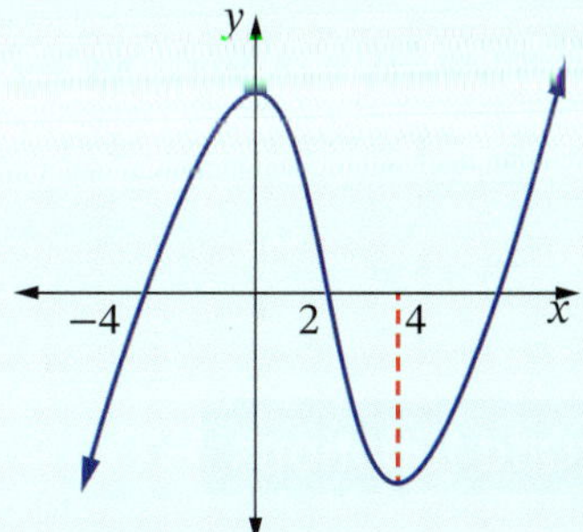

Solution

a The curve is decreasing to the left of $x = 3$ and increasing to the right of $x = 3$, that is, when $x > 3$.

So, the domain over which the graph is increasing is $(3, \infty)$.

b The curve is increasing on the left of the y-axis ($x = 0$), decreasing from $x = 0$ to $x = 4$, then increasing again from $x = 4$.

So the curve is increasing for $x < 0, x > 4$.

So the domain over which the graph is increasing is $(-\infty, 0) \cup (4, \infty)$.

The symbol $\cup$ is for 'union' and stands for the union or joining of 2 separate parts. You will meet this symbol again in Chapter 10, *Probability and data*.

Even and odd functions

Even functions have graphs that are symmetrical about the y-axis. The graph doesn't change when it is reflected in the y-axis. The left and right halves are mirror images of each other. This means the value of y is the same for $x = a$ and $x = -a$.

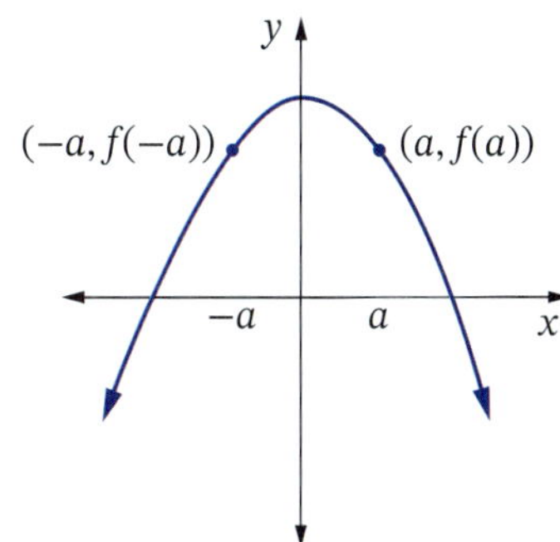

Even functions

A function is even if $f(-x) = f(x)$ for all values of x in the domain.

Video
Odd and even functions

Example 9

Show that $f(x) = x^2 + 3$ is an even function.

Solution

$f(-x) = (-x)^2 + 3$

$= x^2 + 3$

$= f(x)$

So $f(x) = x^2 + 3$ is an even function.

Odd functions have graphs that have point symmetry about the origin. A graph rotated 180° about the origin gives the original graph. This means the values of y are the same but opposite in sign for $x = a$ and $x = -a$.

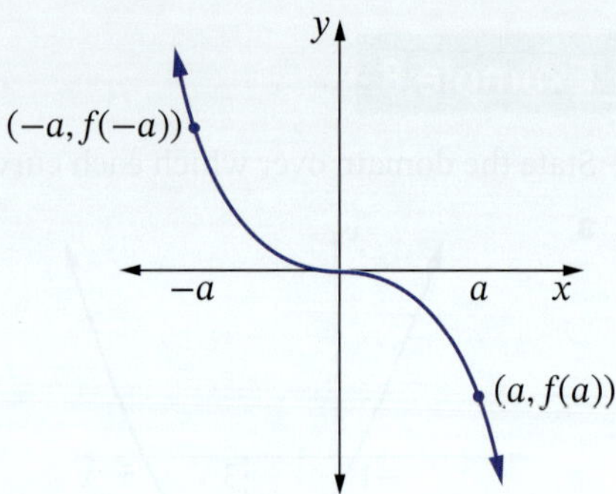

Odd functions

A function is odd if $f(-x) = -f(x)$ for all values of x in the domain.

Video
Odd and even functions

Example 10

Show that $f(x) = x^3 - x$ is an odd function.

Solution

$f(-x) = (-x)^3 - (-x)$

$= -x^3 + x$

$= -(x^3 - x)$

$= -f(x)$

So $f(x) = x^3 - x$ is an odd function.

Investigation

Even and odd functions

Explore the family of graphs of $f(x) = kx^n$, the **power functions**.

For what values of n is the function even?

For what values of n is the function odd?

Does the value of k change this?

Are the families of functions below even or odd? Does the value of k change this?

1 $f(x) = x^n + k$

2 $f(x) = (x + k)^n$

EXERCISE 3.03 Answers on p. 463

Domain, range, even and odd functions

EXAMPLE 5

1 Find the x- and y-intercepts of each function.

a $y = 3x - 2$ **b** $2x - 5y + 20 = 0$ **c** $x + 3y - 12 = 0$

d $f(x) = x^2 + 3x$ **e** $f(x) = x^2 - 4$ **f** $p(x) = x^2 + 5x + 6$

g $y = x^2 - 8x + 15$ **h** $p(x) = x^3 + 5$ **i** $y = \dfrac{x+3}{x}$

j $g(x) = 9 - x^2$

2 $f(x) = 3x - 6$.

a Find the zero of $f(x)$.

b Find the x- and y-intercepts.

3 State the domain and range for each function.

a $f(x) = x^2 + 1$ **b** $y = x^3$ **c** $y = \sqrt{x}$

d $f(x) = \sqrt{x+5}$ **e** $y = -\sqrt{2x-6}$

4 For the functions below, state:

i the domain over which the graph is increasing

ii the domain over which the graph is decreasing

iii whether the graph is odd, even or neither.

a

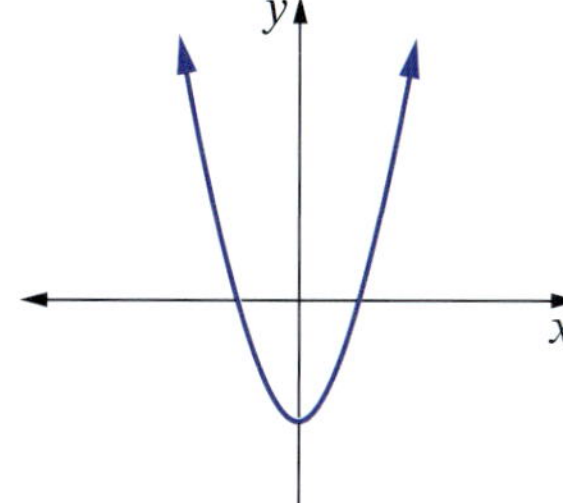

b

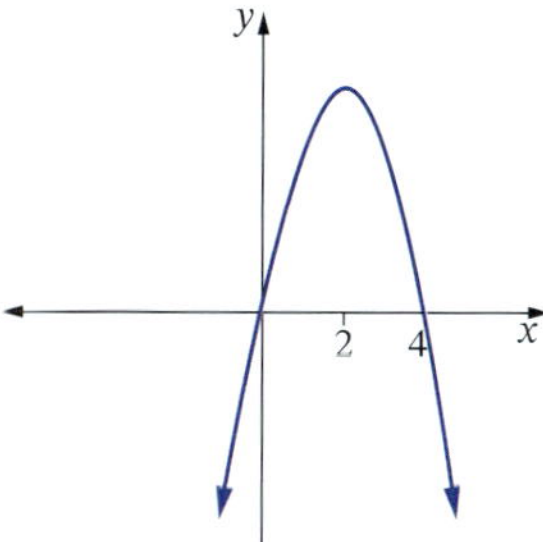

c

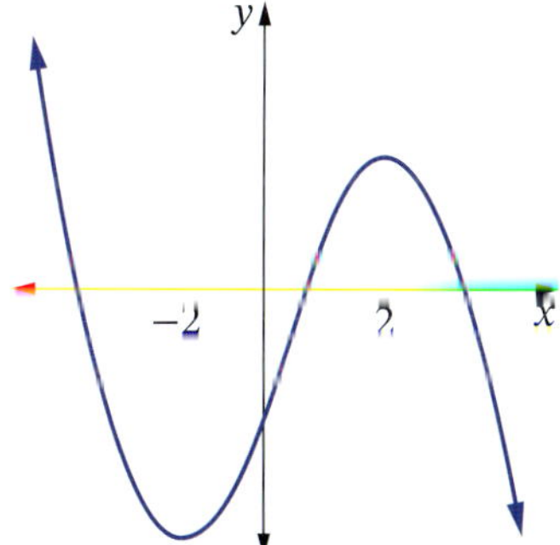

d

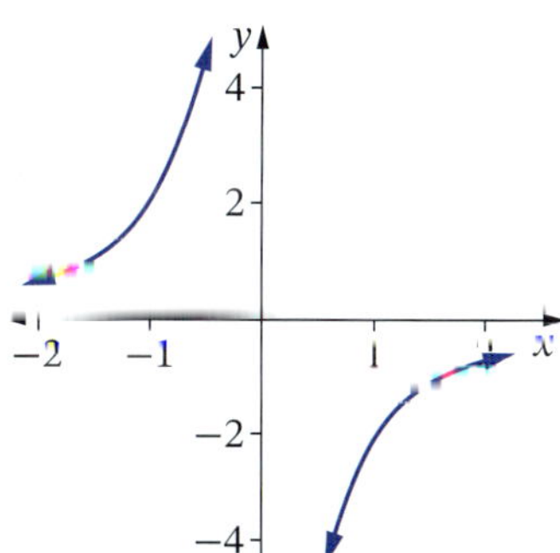

e

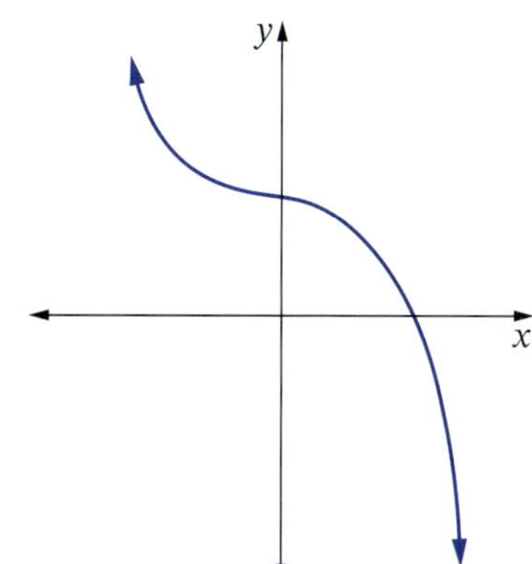

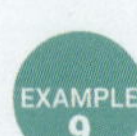

EXAMPLE 9

5 Show that $f(-x) = f(x)$ when $f(x) = x^2 - 2$. What type of function is it?

6 Let $f(x)$ be the function $f(x) = x^3 + 1$.

- **a** Find $f(x^2)$.
- **b** Find $[f(x)]^2$.
- **c** Find $f(-x)$.
- **d** Is $f(x) = x^3 + 1$ an even or odd function, or neither?
- **e** Solve $f(x) = 0$.
- **f** Find the intercepts of the function.

7 Show that $g(x) = x^8 + 3x^4 - 2x^2$ is an even function.

EXAMPLE 10

8 Show that $f(x)$ is odd, given $f(x) = x$.

9 Show that $f(x) = x^2 - 1$ is an even function.

10 Show that $f(x) = 4x - x^3$ is an odd function.

11 **a** Prove that $f(x) = x^4 + x^2$ is an even function.

- **b** Find $f(x) - f(-x)$.

12 Are these functions even, odd or neither?

- **a** $y = \dfrac{x^3}{x^4 - x^2}$
- **b** $f(x) = \dfrac{1}{x^3 - 1}$
- **c** $f(x) = \dfrac{3}{x^2 - 4}$
- **d** $y = \dfrac{x - 3}{x + 3}$
- **e** $f(x) = \dfrac{x^3}{x^5 - x^2}$

13 If n is a positive integer, for what values of n is the power function $f(x) = kx^n$:

- **a** even?
- **b** odd?

14 Can the function $f(x) = x^n + x$ ever be:

- **a** even?
- **b** odd?

15 $f(x) = (x - 2)^2$

- **a** Find $f(3)$.
- **b** Find $f(-5)$.
- **c** Find the zeroes of $f(x)$.
- **d** Find the x- and y-intercepts.
- **e** State the domain and range of $f(x)$.
- **f** Find $f(-x)$.
- **g** Is $f(x)$ even, odd or neither?

16 The incomplete graph of $y = f(x)$ shown below is an even function. Sketch the part of the graph in the domain $x < 0$.

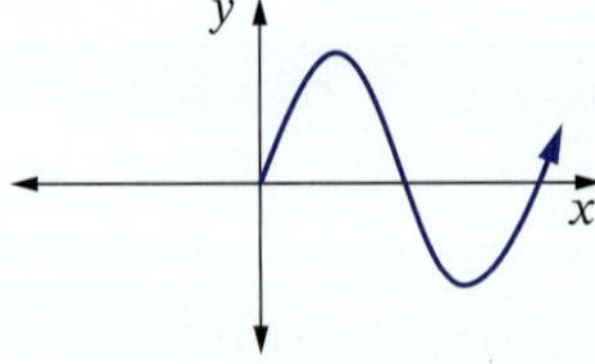

☐ Foundation ◯ Mastery ⬡ Complex

17 The incomplete graph of $y = g(x)$ shown below is an odd function. Sketch the part of the graph in the domain $x < 0$.

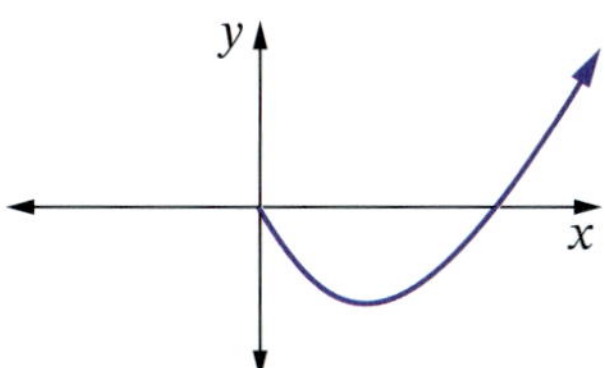

18 Copy and complete this graph in the domain $x < 0$ if:

a $y = f(x)$ is an even function.

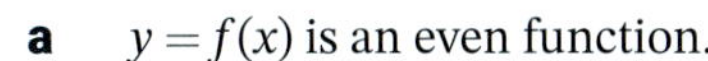

b $y = f(x)$ is an odd function.

19 Find the domain and range of each function.

a $y = 3^x$

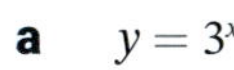

b $f(x) = \dfrac{1}{x-2}$

c $y = \dfrac{1}{\sqrt{x+3}}$

d $f(x) = \dfrac{1}{\sqrt{x^2-9}}$

20 **a** Find the domain and range of $y = \sqrt{4-x^2}$.

b Find the domain and range of $y = -\sqrt{4-x^2}$.

c Change the subject of $x^2 + y^2 = 4$ to y.

d Find the domain and range of $x^2 + y^2 = 4$.

e Is $x^2 + y^2 = 4$ a function?

21 Find the domain and range of:

$$f(x) = \begin{cases} x+1, & [2, \infty) \\ x^2, & (-\infty, 2) \end{cases}$$

22 Given $f(x) = x + 3$ and $g(x) = \dfrac{1}{x+3}$, find the domain and range of:

a $f(x)\,g(x)$

b $\dfrac{f(x)}{g(x)}$

23 Given $f(x) = x^2$ and $g(x) = 2x + 1$,

a Find the domain and range of:

i $f(x) + g(x)$

ii $f(x) - g(x)$

iii $f(x)\,g(x)$

iv $\dfrac{f(x)}{g(x)}$ (domain only).

b Verify your answers using graphing technology.

24 Prove whether each function is even, odd or neither if $f(x)$ is:

i even

ii odd.

a $y = [f(x)]^2$

b $y = f(x^2)$

c $y = \dfrac{f(-x)}{f(x)}$

d $y = f(-x)\,f(x)$

e $y = x\,f(x)$

Foundation | Mastery | Complex

3.04 Linear functions

Video
The general equation
$ax + by + c = 0$

Graph paper
A page of number planes

Worksheets
Graphing linear functions 1

Graphing linear functions 2

x- and y-intercepts

Linear functions

A **linear function** is a function whose graph is a straight line.

Its equation can be written in **gradient–intercept form** $y = mx + c$, where m is the **gradient** and c is the y-intercept.

It can also be written in **general form** $ax + by + c = 0$, where a, b and c are **constants**, usually integers.

Graphing linear functions

Example 11

a Find the x- and y-intercepts of the graph of $y = 2x - 4$ and draw its graph on the number plane.

b Find the x- and y-intercepts of the line with equation $x + 2y + 6 = 0$ and draw its graph.

Solution

a For x-intercept, $y = 0$.

$0 = 2x - 4$

$4 = 2x$

$2 = x$

So the x-intercept is 2.

For y-intercept, $x = 0$.

$y = 2(0) - 4$

$= -4$

So the y-intercept is -4.

Use the intercepts to graph the line.

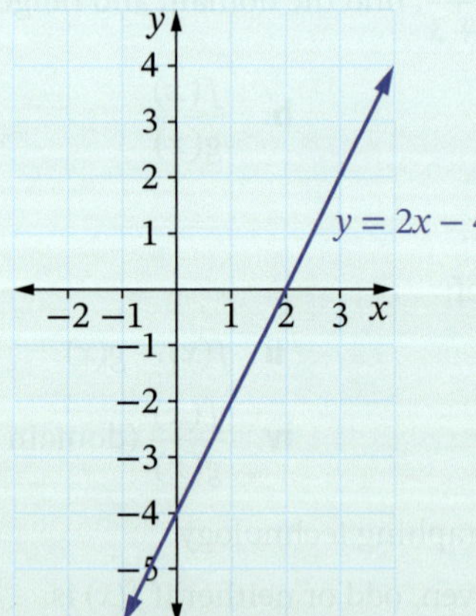

b For x-intercept, $y = 0$.

$x + 2(0) + 6 = 0$

$x + 6 = 0$

$x = -6$

So the x-intercept is -6.

For y-intercept, $x = 0$.

$0 + 2y + 6 = 0$

$2y = -6$

$y = -3$

So the y-intercept is -3.

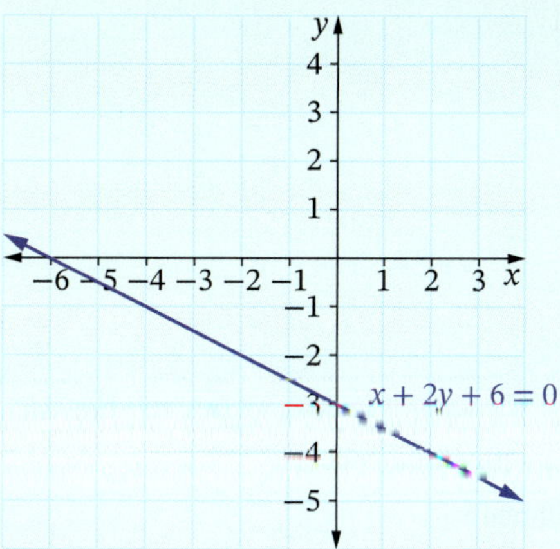

Domain and range of linear functions

The domain of a linear function is $(-\infty, \infty)$ or all real numbers.

The range of a linear function is $(-\infty, \infty)$ or all real numbers.

Horizontal and vertical lines

Example 12

a Sketch the graph of $y = 2$ on a number plane. What is its domain and range?

b Sketch the graph of $x = -1$ on a number plane and state its domain and range.

Solution

a x can have any value and y is always 2.

Some of the points on the line will be (0, 2), (1, 2) and (2, 2).

This gives a horizontal line with y-intercept 2.

The domain is all real x and the range is $y = 2$.

Domain: $(-\infty, \infty)$, range: [2, 2]

'$y = 2$' in interval notation is [2, 2].

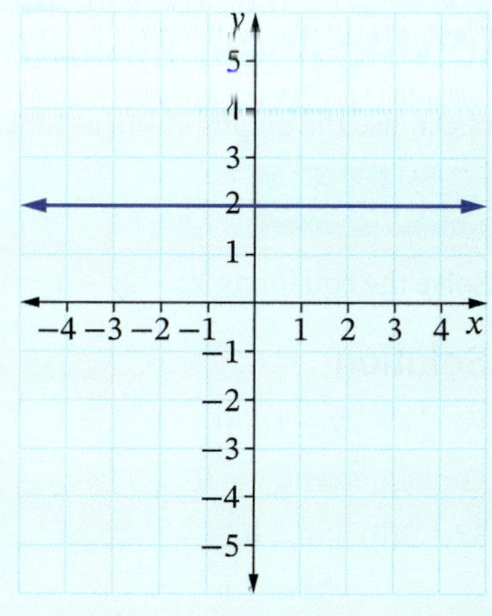

b y can have any value and x is always -1.

Some of the points on the line will be $(-1, 0)$, $(-1, 1)$ and $(-1, 2)$.

This gives a vertical line with x-intercept -1.

Domain: $[-1, -1]$, range: $(-\infty, \infty)$.

'$x = 1$' in interval notation is $[1,1]$.

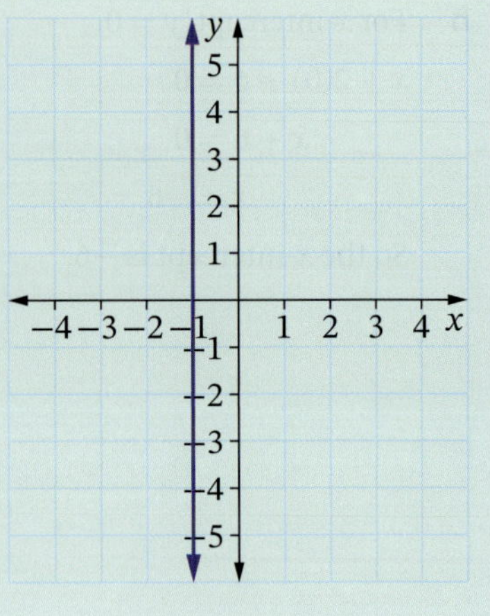

Horizontal lines

$y = c$ is a horizontal line with y-intercept c.

Domain: $(-\infty, \infty)$ or all real x

Range: $[c, c]$ or $y = c$

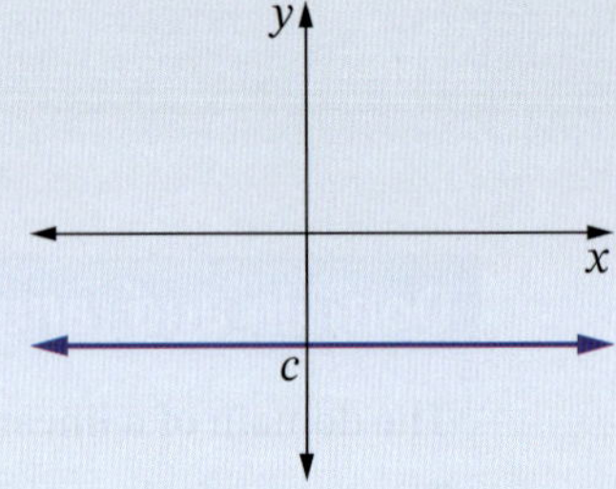

Vertical lines

$x = c$ is a vertical line with x-intercept c.

$x = c$ is not a function and its gradient is undefined.

Domain: $[c, c]$ or $x = c$

Range: $(-\infty, \infty)$ or all real y

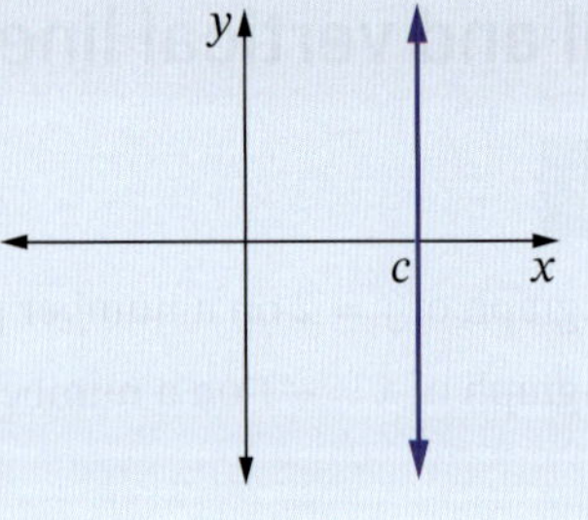

Solving linear simultaneous equations graphically

Worksheet Intersection of lines

We can use the graphs of linear functions to solve simultaneous equations graphically.

Example 13

Solve the equations $3x - 2y + 1 = 0$ and $2x + y + 3 = 0$ simultaneously by sketching their graphs.

Solution

For $3x - 2y + 1 = 0$

For x-intercept, $y = 0$.

$3x - 2(0) + 1 = 0$

$3x = -1$

$x = -\frac{1}{3}$

For y-intercept, $x = 0$.

$3(0) - 2y + 1 = 0$

$-2y = -1$

$y = \frac{1}{2}$

For $2x + y + 3 = 0$

For x-intercept, $y = 0$.

$2x + 0 + 3 = 0$

$2x = -3$

$x = -1\frac{1}{2}$

For y-intercept, $x = 0$.

$2(0) + y + 3 = 0$

$y = -3$

Sketch both lines on the same number plane.

From the graph, the 2 lines intersect at $(-1, -1)$.

So, the solution of simultaneous equations $3x - 2y + 1 = 0$ and $2x + y + 3 = 0$ is $x = -1$, $y = -1$.

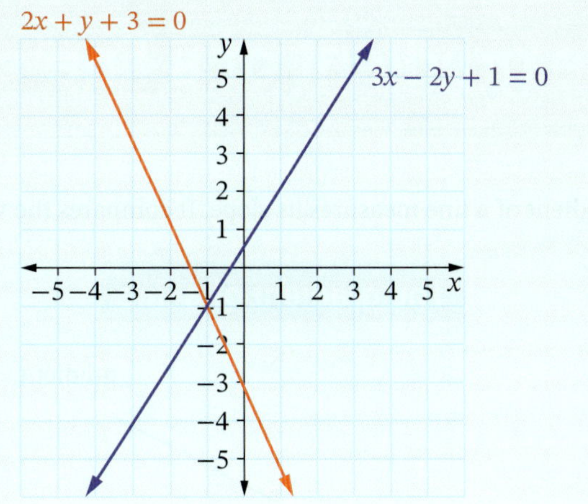

EXERCISE 3.04 Answers on p. 464

Linear functions

1 Write an equation for:

- **a** the number of months (N) in x years
- **b** the amount of juice (A) in n lots of 2-litre bottles
- **c** the cost (c) of x litres of petrol at \$2.50 per litre
- **d** the number (y) of people in x debating teams if there are 4 people in each team
- **e** the weight (w) of x lots of 400 g cans of peaches.

2 Find the equation and draw the graph of the cost (c) of x refrigerators if each refrigerator costs \$850.

3 Find the x- and y-intercepts of the graph of each function. EXAMPLE 11

- **a** $y = x - 2$
- **b** $y = 3x + 9$
- **c** $y = 4 - 2x$
- **d** $f(x) = 2x + 3$
- **e** $f(x) = 5x - 4$
- **f** $f(x) = 10x + 5$
- **g** $x + y - 2 = 0$
- **h** $2x - y + 4 = 0$
- **i** $x - y + 3 = 0$
- **j** $3x - 6y - 2 = 0$

4 Draw the graph of each linear function.

- **a** $y = x + 4$
- **b** $f(x) = 2x - 1$
- **c** $f(x) = 3x + 2$
- **d** $x + y = 3$
- **e** $x - y - 1 = 0$

5 Find the domain and range of each equation. 

- **a** $3x - 2y + 7 = 0$
- **b** $y = 2$
- **c** $x = -4$
- **d** $x - 2 = 0$
- **e** $3 - y = 0$

6 Sketch each equation's graph and state its domain and range.

- **a** $x = 4$
- **b** $x - 3 = 0$
- **c** $y = 5$
- **d** $y + 1 = 0$

7 By sketching the graphs of $x - y - 4 = 0$ and $2x + 3y - 3 = 0$ on the same set of axes, solve the simultaneous equations graphically. 

8 **a** Solve the simultaneous equations $y = 2x + 1$ and $y = x - 2$ graphically.

b Hence solve the inequality $2x + 1 < x - 2$.

9 Solve each pair of simultaneous equations graphically.

- **a** $4x + 3y = 11$ and $3x + y = 2$
- **b** $3a - 4b = -16$ and $2a + 3b = 12$
- **c** $5p + 2q + 18 = 0$ and $2p - 3q + 11 = 0$
- **d** $7x + 3y = 4$ and $3x + 5y = -2$
- **e** $9x - 2y = -1$ and $7x - 4y = 9$

Foundation Mastery Complex

3.05 The gradient of a line

Video
Gradient and y-intercept of a line

The gradient of a line measures its slope. It compares the vertical rise with the horizontal run.

The gradient of a line

$$\text{gradient} = \frac{\text{rise}}{\text{run}}$$

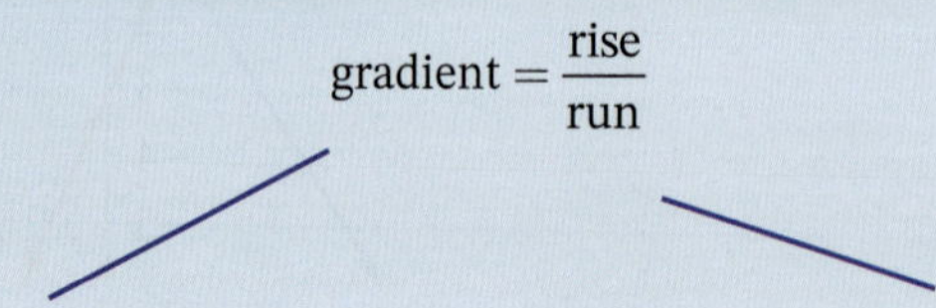

Positive gradient leans to the right. Negative gradient leans to the left.

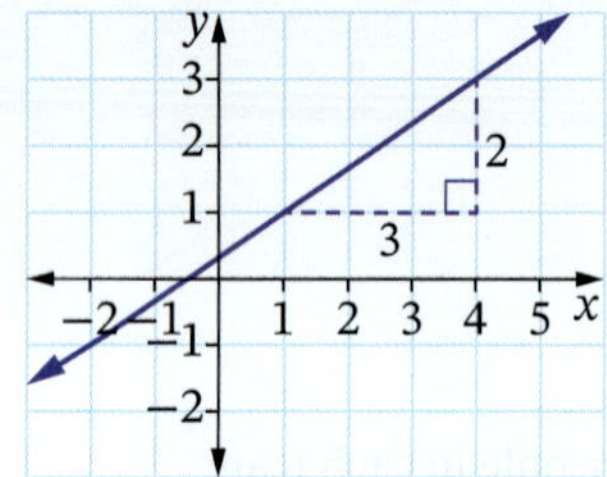

$$\text{gradient} = \frac{\text{rise}}{\text{run}} = \frac{2}{3}$$

On the number plane, gradient is a measure of the rate of change of y with respect to x.

Gradient formula

The gradient of the line joining points (x_1, y_1) and (x_2, y_2) is:

$$m = \frac{y_2 - y_1}{x_2 - x_1}$$

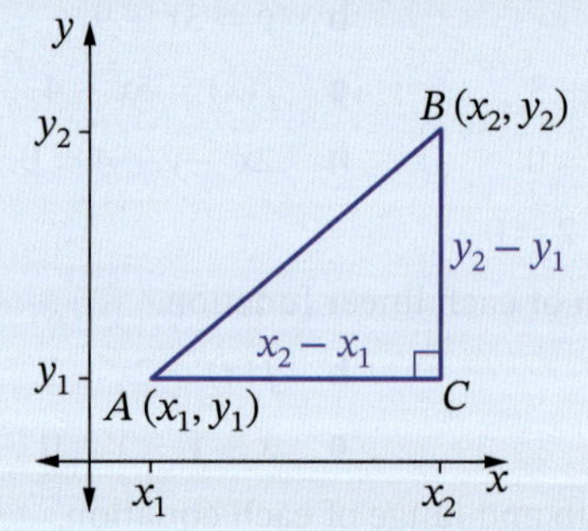

Example 14

Find the gradient of the line joining points (2, 3) and (−3, 4).

Solution

$$\text{gradient, } m = \frac{y_2 - y_1}{x_2 - x_1}$$

$$= \frac{4-3}{-3-2}$$

$$= \frac{1}{-5}$$

$$= -\frac{1}{5}$$

The angle of inclination of a line

The **angle of inclination**, θ, is the angle a straight line makes with the positive x-axis, measured anticlockwise.

$$m = \frac{\text{rise}}{\text{run}}$$
$$= \frac{\text{opposite}}{\text{adjacent}}$$
$$= \tan\theta$$

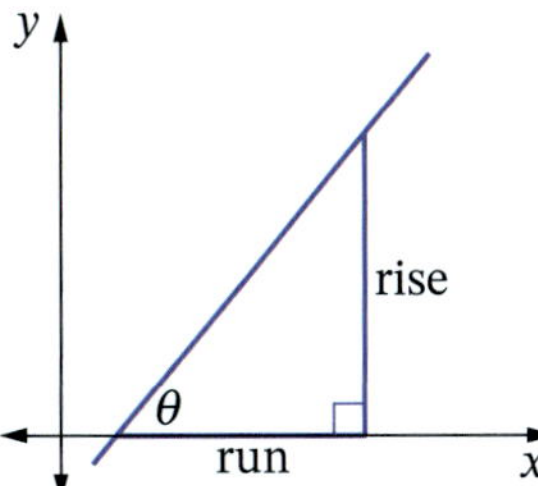

Gradient and angle of inclination of a line

$$m = \tan\theta$$

where m is the gradient and θ is the **angle of inclination**.

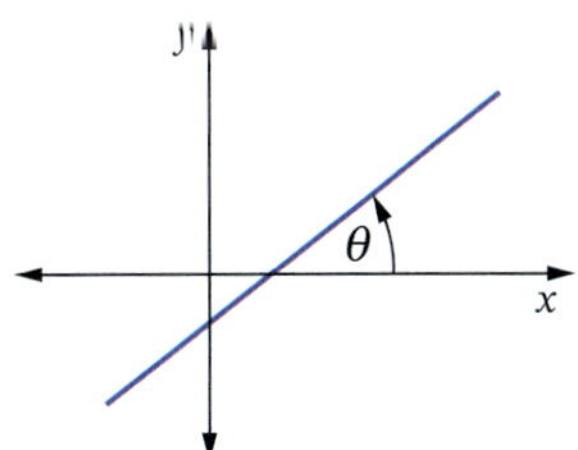

For an acute angle, $\tan\theta > 0$.

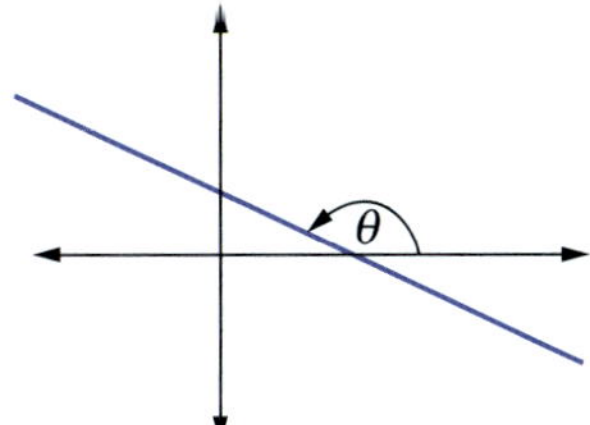

For an obtuse angle, $\tan\theta < 0$.

Investigation

Angles and gradients

1 What type of angles give a positive gradient?

2 What type of angles give a negative gradient? Why?

3 What is the gradient of a horizontal line? What angle does it make with the x-axis?

4 What angle does a vertical line make with the x-axis? Can you find its gradient?

Example 15

a Find the gradient of the line that makes an angle of inclination of 135°.

b Find, correct to the nearest minute, the angle of inclination of a straight line whose gradient is:

i 0.5 **ii** -3

Solution

a $m = \tan\theta$

$= \tan 135°$

$= -1$

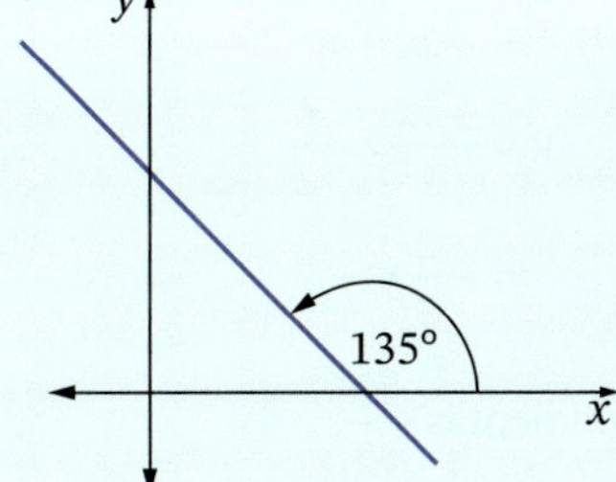

b **i** $m = \tan\theta$

$\therefore \tan\theta = 0.5$

$\theta = \tan^{-1}(0.5)$

$= 26°33'54.18''$

$\approx 26°34'$

Operation	Casio scientific	Sharp scientific
Enter data.	SHIFT tan 0.5 =	2nd F tan 0.5 =
Change to degrees and minutes.	° ′ ″	2nd F D°M′S

ii $m = \tan\theta$

$\therefore \tan\theta = -3$

$\theta = \tan^{-1}(-3)$

$= -71°33'54.18''$

$\approx -71°34'$

This negative angle describes the angle of the line under the x-axis, but we want its angle above the x-axis.

y
x
180° − 71°34
71°34′

A negative gradient means that the angle of inclination is obtuse.

To find this angle, subtract the acute angle from 180°.

$\theta = 180° - 71°34'$

$= 108°26'$

Worksheet
$y = mx + c$

The gradient–intercept form of a linear equation

The linear function with equation $y = mx + c$ has gradient m and y-intercept c.

Example 16

a Find the gradient and y-intercept of the linear function $y = 7x - 5$.

b Find the gradient of the straight line with equation $2x + 3y - 6 = 0$.

Solution

a gradient = 7, y-intercept = −5

b First, change the equation into the form $y = mx + c$.

$$2x + 3y - 6 = 0$$
$$2x + 3y = 6$$
$$3y = 6 - 2x$$
$$= -2x + 6$$
$$y = \frac{-2x}{3} + \frac{6}{3}$$
$$= -\frac{2}{3}x + 2$$

The gradient of the line $ax + by + c = 0$ is $-\frac{a}{b}$.

So the gradient is $-\frac{2}{3}$.

EXERCISE 3.05 Answers on p. 465

The gradient of a line

EXAMPLE 14

1 Find the gradient of the line joining the points:

a (3, 2) and (1, −2) **b** (0, 2) and (3, 6) **c** (−2, 3) and (4, −5)

d (2, −5) and (−3, 7) **e** (2, 3) and (−1, 1) **f** (−5, 1) and (3, 0)

g (−2, −3) and (−4, 6) **h** (−1, 3) and (−7, 7) **i** (1, −4) and (5, 5)

2 Find the gradient of the straight line, correct to one decimal place, whose angle of inclination is:

a 25° **b** 82° **c** 68°

d 100° **e** 130° **f** 164°

3 For each linear function, find:

i the gradient **ii** the y-intercept.

a $y = 3x + 5$ **b** $f(x) = 2x + 1$ **c** $y = 6x - 7$

d $y = -x$ **e** $y = -4x + 3$ **f** $y = x - 2$

g $f(x) = 6 - 2x$ **h** $y = 1 - x$ **i** $y = 9x$

4 Find the gradient of the linear function:

a with x-intercept 3 and y-intercept −1

b passing through (2, 4) and has x-intercept 5

c passing through (1, 1) and (−2, 7)

d with x-intercept −3 and passing through (2, 3)

e passing through the origin and (−3, −1).

5 Find the angle of inclination, to the nearest minute, of a line with gradient:

a 2 **b** 1.7 **c** 6

d −5 **e** −0.85 **f** −1.2

6 For each linear function, find:

i the gradient **ii** the y-intercept.

a $2x + y - 3 = 0$ **b** $5x + y + 6 = 0$ **c** $6x - y - 1 = 0$

d $x - y + 4 = 0$ **e** $4x + 2y - 1 = 0$ **f** $6x - 2y + 3 = 0$

7 Find the gradient of each linear function.

a $y = -2x - 1$ **b** $y = 2$ **c** $x + y + 1 = 0$

d $3x + y = 8$ **e** $2x - y + 5 = 0$ **f** $x + 4y - 12 = 0$

g $y = \frac{x}{5} - 1$ **h** $y = \frac{2x}{7} + 5$ **i** $y = -\frac{3x}{5} - 2$

j $2y = -\frac{x}{7} + \frac{1}{3}$ **k** $3x - \frac{y}{5} = 8$ **l** $\frac{x}{2} + \frac{y}{3} = 1$

8 If the gradient of the line joining $(8, y_1)$ and (−1, 3) is 2, find the value of y_1.

9 The gradient of the line through (2, −1) and $(x, 0)$ is −5. Find the value of x.

10 The gradient of a line is −1 and the line passes through the points (4, 2) and $(x, -3)$. Find the value of x.

11 The number of phones, n, produced in a factory is given by the linear function $n = 350t + 1500$, where t is the time in days.

a Find the gradient of this function.

b Explain what the gradient represents with regard to the number of phones produced.

12 The points $A(-1, 2)$, $B(1, 5)$, $C(6, 5)$ and $D(4, 2)$ form a parallelogram. Find the gradients of all 4 sides of the parallelogram. What do you notice?

Foundation Mastery Complex

3.06 The point-gradient formula

Video
Linear modelling

Worksheets
Finding the equation of a line

Equations of a line

Puzzle
Linear functions code puzzle

Example 17

Find the equation of the line with gradient 3 and y-intercept -1.

Solution

Use the gradient–intercept equation $y = mx + c$, where m represents the gradient and c represents the y-intercept.

$m = 3$ and $c = -1$.

The equation is $y = 3x - 1$.

There is a formula you can use if you know the gradient and the coordinates of a point on the line.

The point-gradient form of a linear equation

The linear function with equation $y - y_1 = m(x - x_1)$ has gradient m and the point (x_1, y_1) lies on the line.

Proof

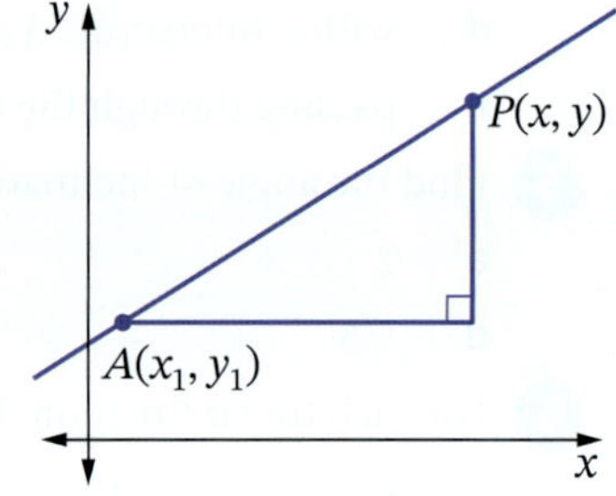

Let $P(x, y)$ be a general point on the line with gradient m that passes through $A(x_1, y_1)$.

Then line AP has gradient

$$m = \frac{y - y_1}{x - x_1}$$

$$m(x - x_1) = y - y_1$$

Example 18

Find the equation of the line:

a with gradient -4 and x-intercept 1

b passing through (2, 3) and $(-1, 4)$.

Solution

a The x-intercept of 1 means the line passes through the point (1, 0).

Substituting $m = -4$, $x_1 = 1$ and $y_1 = 0$ into the formula:

$$y - y_1 = m(x - x_1)$$

$$y - 0 = -4(x - 1)$$

$$y = -4x + 4$$

b First, find the gradient.

$$m = \frac{y_2 - y_1}{x_2 - x_1}$$
$$= \frac{4-3}{-1-2}$$
$$= -\frac{1}{3}$$

Substitute the gradient and one of the points, say (2, 3), into the formula.

$$y - y_1 = m(x - x_1)$$
$$y - 3 = -\frac{1}{3}(x - 2)$$
$$3 \times (y - 3) = 3 \times -\frac{1}{3}(x - 2)$$
$$3y - 9 = -(x - 2)$$
$$= -x + 2$$
$$x + 3y - 9 = 2$$

Writing the equation in general form:

$$x + 3y - 11 = 0$$

EXERCISE 3.06 Answers on p. 466

The point–gradient formula

EXAMPLES 17, 18

1 Find the equation of the straight line:

- **a** with gradient 4 and y-intercept -1
- **b** with gradient -3 and passing through (0, 4)
- **c** passing through the origin with gradient 5
- **d** with gradient 4 and x-intercept -5
- **e** with x-intercept 1 and y-intercept 3
- **f** with x-intercept 3, y-intercept -4.

2 Find the equation of the straight line passing through the points:

a (2, 5) and (−1, 1) **b** (0, 1) and (−4, −2) **c** (−2, 1) and (3, 5)
d (3, 4) and (−1, 7) **e** (−4, −1) and (−2, 0)

3 What is **a** the gradient and **b** the equation of the line with x-intercept 2 that passes through (3, −4)?

4 Find the equation of the line:

- **a** parallel to the x-axis and passing through (2, 3)
- **b** parallel to the y-axis and passing through (−1, 2).

5 A straight line passing through the origin has a gradient of −2. Find:

a the y-intercept **b** its equation.

6 What could be the equation of this line?
Select the correct answer **A**, **B**, **C** or **D**.

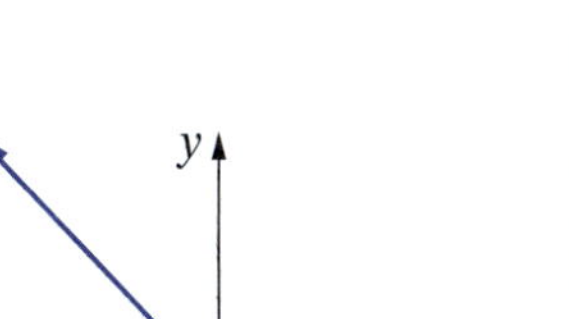

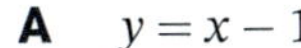

A $y = x - 1$ **B** $y = -x + 1$
C $y = -x - 1$ **D** $y = x + 1$

7 Find the x- and y-intercepts of the line with equation:

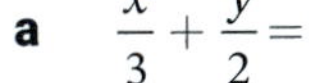

a $\frac{x}{3} + \frac{y}{2} = 1$ **b** $\frac{x}{4} - \frac{y}{5} = 1$ **c** $\frac{x}{a} + \frac{y}{b} = 1$

8 Find the equation of the linear function with x-intercept 2 and y-intercept 7.

☐ Foundation ◯ Mastery ◯ Complex

9 Find the equation of the straight line passing through $(-1, 3)$ and the point of intersection of $3x + 2y - 5 = 0$ and $x - 2y + 1 = 0$.

10 A line passes through the points (x_1, y_1) and (x_2, y_2). Show that its equation can be written as $\frac{y - y_1}{x - x_1} = \frac{y_2 - y_1}{x_2 - x_1}$.

3.07 Parallel and perpendicular lines

Videos
Equations of parallel lines
Equations of perpendicular lines

Worksheet
Linear functions

Puzzle
Parallel and perpendicular lines

Gradients of parallel lines

If two lines are parallel, then they have the same gradient, i.e., $m_1 = m_2$.

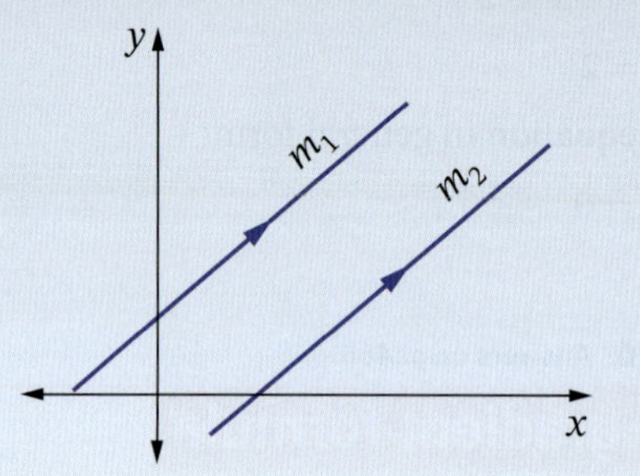

Example 19

a Prove that the straight lines with equations $5x - 2y - 1 = 0$ and $5x - 2y + 7 = 0$ are parallel.

b Find the equation of a straight line parallel to the line $2x - y - 3 = 0$ and passing through $(1, -5)$.

Solution

a First, change the equation into the form $y = mx + c$.

Line 1:

$$5x - 2y - 1 = 0$$
$$5x - 1 = 2y$$
$$\frac{5}{2}x - \frac{1}{2} = y$$
$$\therefore m_1 = \frac{5}{2}$$

Line 2:

$$5x - 2y + 7 = 0$$
$$5x + 7 = 2y$$
$$\frac{5}{2}x + \frac{7}{2} = y$$
$$\therefore m_2 = \frac{5}{2}$$

$$m_1 = m_2 = \frac{5}{2}$$

$\therefore$ the lines are parallel.

b $2x - y - 3 = 0$

$$2x - 3 = y$$
$$\therefore m_1 = 2$$

For parallel lines $m_1 = m_2$.

$$\therefore m_2 = 2$$

Substitute this and $(1, -5)$ into $y - y_1 = m(x - x_1)$:

$$y - (-5) = 2(x - 1)$$
$$y + 5 = 2x - 2$$
$$y = 2x - 7$$

☐ Foundation ◯ Mastery ◯ Complex

Investigation

Perpendicular lines

Sketch each pair of straight lines on the same number plane or using graphing technology.

1 $3x - 4y + 12 = 0$ and $4x + 3y - 8 = 0$

2 $2x + y + 4 = 0$ and $x - 2y + 2 = 0$

What do you notice about each pair of lines?

Find the gradient of each line. What do you notice?

Gradients of perpendicular lines

If two lines with gradients m_1 and m_2 are perpendicular,

then $m_1 m_2 = -1$, i.e., $m_2 = -\dfrac{1}{m_1}$.

Example 20

a Show that the lines with equations $3x + y - 11 = 0$ and $x - 3y + 1 = 0$ are perpendicular.

b Find the equation of the straight line through (2, 3) that is perpendicular to the line passing through (−1, 7) and (3, 3).

Solution

a Line 1:

$3x + y - 11 = 0$

$y = -3x + 11$

$\therefore m_1 = -3$

Line 2:

$x - 3y + 1 = 0$

$x + 1 = 3y$

$\dfrac{1}{3}x + \dfrac{1}{3} = y$

$\therefore m_2 = \dfrac{1}{3}$

$m_1 m_2 = -3 \times \dfrac{1}{3} = -1$

$\therefore$ the lines are perpendicular.

b Line through (−1, 7) and (3, 3):

$$m_1 = \frac{y_2 - y_1}{x_2 - x_1}$$

$$= \frac{3 - 7}{3 - (-1)}$$

$$= \frac{-4}{4}$$

$$= -1$$

For perpendicular lines, $m_1 m_2 = -1$:

$-1 \times m_2 = -1$

$m_2 = 1$

Substitute $m = 1$ and the point (2, 3) into $y - y_1 = m(x - x_1)$:

$y - 3 = 1(x - 2)$

$y - 3 = x - 2$

$y = x + 1$

EXERCISE 3.07 Answers on p. 466

Parallel and perpendicular lines

1 Find the gradient of the straight line:

- **a** parallel to the line $3x + y - 4 = 0$
- **b** perpendicular to the line $3x + y - 4 = 0$
- **c** parallel to the line joining $(3, 5)$ and $(-1, 2)$
- **d** perpendicular to the line with x-intercept 3 and y-intercept 2
- **e** perpendicular to the line that has an angle of inclination of $135°$
- **f** perpendicular to the line $6x - 5y - 4 = 0$
- **g** parallel to the line $x - 3y - 7 = 0$
- **h** perpendicular to the line passing through $(4, -2)$ and $(3, 3)$.

2 Find the equation of the straight line:

- **a** passing through $(2, 3)$ and parallel to the line $y = x + 6$
- **b** through $(-1, 5)$ and parallel to the line $x - 3y - 7 = 0$
- **c** with x-intercept 5 and parallel to the line $y = 4 - x$
- **d** through $(3, -4)$ and perpendicular to the line $y = 2x$
- **e** through $(-2, 1)$ and perpendicular to the line $2x + y + 3 = 0$
- **f** through $(3, 7)$ and parallel to the line $5x - y - 2 = 0$
- **g** through $(0, -2)$ and perpendicular to the line $x - 2y = 9$
- **h** perpendicular to the line $3x + 2y - 1 = 0$ and passing through the point $(-2, 4)$.

3 Show that the lines with equations $y = 3x - 2$ and $6x - 2y - 9 = 0$ are parallel.

4 Show that lines $x + 5y = 0$ and $y = 5x + 3$ are perpendicular.

5 Show that lines $6x - 5y + 1 = 0$ and $6x - 5y - 3 = 0$ are parallel.

6 Show that lines $7x + 3y + 2 = 0$ and $3x - 7y = 0$ are perpendicular.

7 If the lines $3x - 2y + 5 = 0$ and $y = kx - 1$ are perpendicular, find the value of k.

8 Show that the line joining $(3, -1)$ and $(2, -5)$ is parallel to the line $8x - 2y - 3 = 0$.

9 Show that the points $A(-3, -2)$, $B(-1, 4)$, $C(7, -1)$ and $D(5, -7)$ are the vertices of a parallelogram.

10 The points $A(-2, 0)$, $B(1, 4)$, $C(6, 4)$ and $D(3, 0)$ form a rhombus. Show that the diagonals are perpendicular.

11 Find the equation of the straight line passing through $(6, -3)$ that is perpendicular to the line joining $(2, -1)$ and $(-5, -7)$.

12 The vertices of a quadrilateral are $A(-2, 1)$, $B(1, 4)$, $C(5, 6)$ and $D(-5, -4)$. What type of quadrilateral is $ABCD$?

☐ Foundation ◯ Mastery ⬡ Complex

Quadratic functions 3.08

Worksheets
Graphing quadratic functions

Graphing quadratics

Quadratic functions

A **quadratic function** has an equation in the form $y = ax^2 + bx + c$, where a, b and c are constants, $a \neq 0$, and the highest power of x is 2.

The graph of a quadratic function is a **parabola**, a curve that has a **vertex** (**turning point**) and an axis of symmetry that goes through the vertex.

Graphing quadratic functions

Example 21

Graph the quadratic function $y = x^2 - x$.

Solution

Draw up a table of values for $y = x^2 - x$.

x	−3	−2	−1	0	1	2	3
y	12	6	2	0	0	2	6

Plot (−3, 12), (−2, 6), (−1, 2), (0, 0), (1, 0), (2, 2) and (3, 6) and draw a parabola through them.

Label the graph with its equation.

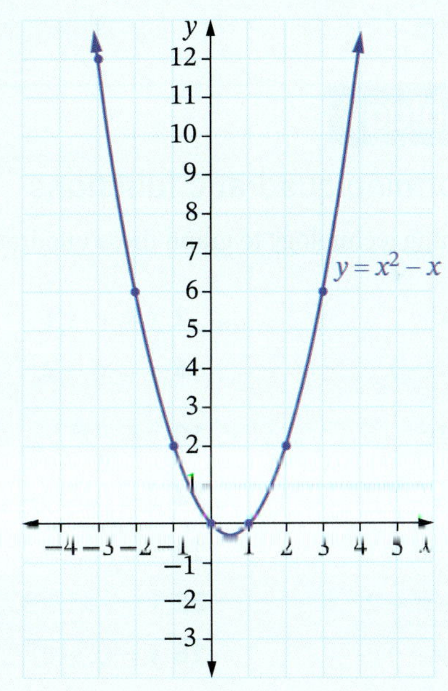

Example 22

Sketch the graphs of $y = x^2$, $y = x^2 + 1$ and $y = (x + 1)^2$ on the same number plane.

Describe the relationships between these graphs.

Solution

$y = x^2$

x	−3	−2	−1	0	1	2	3
y	9	4	1	0	1	4	9

$y = x^2 + 1$

x	−3	−2	−1	0	1	2	3
y	10	5	2	1	2	5	10

$y = (x + 1)^2$

x	−3	−2	−1	0	1	2	3
y	4	1	0	1	4	9	16

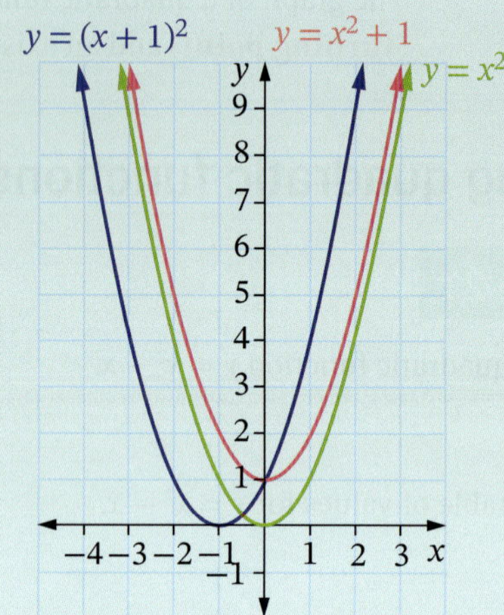

The graph of $y = x^2 + 1$ is 1 unit up from $y = x^2$.

The graph of $y = (x + 1)^2$ is 1 unit to the left of $y = x^2$.

Technology

Transforming quadratic functions

Use graphing technology to graph these quadratic functions. Look for any patterns.

1 $y = x^2$ **2** $y = x^2 + 1$ **3** $y = x^2 + 2$ **4** $y = x^2 + 3$

5 $y = x^2 - 1$ **6** $y = x^2 - 2$ **7** $y = x^2 - 3$ **8** $y = 2x^2$

9 $y = 3x^2$ **10** $y = \frac{1}{2}x^2$ **11** $y = (x + 2)^2$ **12** $y = (x + 3)^2$

13 $y = (x - 1)^2$ **14** $y = (x - 2)^2$ **15** $y = (x - 3)^2$ **16** $y = -x^2$

17 $y = -x^2 + 1$ **18** $y = -x^2 + 2$ **19** $y = -x^2 + 3$ **20** $y = -x^2 - 1$

21 $y = -x^2 - 2$ **22** $y = -x^2 - 3$ **23** $y = -2x^2$ **24** $y = -3x^2$

25 $y = (x + 1)^2$ **26** $y = (x + 2)^2$ **27** $y = (x - 1)^2$ **28** $y = (x - 2)^2$

Could you predict where the graphs $y = x^2 + 9$, $y = 5x^2$ or $y = (x + 6)^2$ would lie?
Is a parabola always a function? Can you find an example of a parabola that is not a function?

9780170498197

Did you know?

The parabola

The parabola shape has special properties that are very useful. For example, if a light is placed inside a parabolic mirror at a special place called the focus, then all light rays coming from this point and reflecting off the parabola shape will radiate out parallel to each other, giving a strong light. This is how car headlights work. The dishes of radio telescopes also use this property of the parabola, because radio signals coming in to the dish will reflect back to the focus.

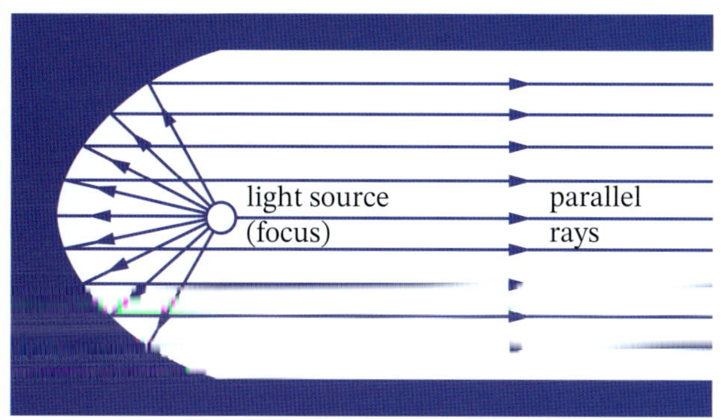

Zack Frank/Shutterstock.com

Concavity and turning points

For the parabola $y = ax^2 + bx + c$

- if $a > 0$ the parabola is **concave upwards** and has a **minimum turning point**
- if $a < 0$ the parabola is **concave downwards** and has a **maximum turning point**.

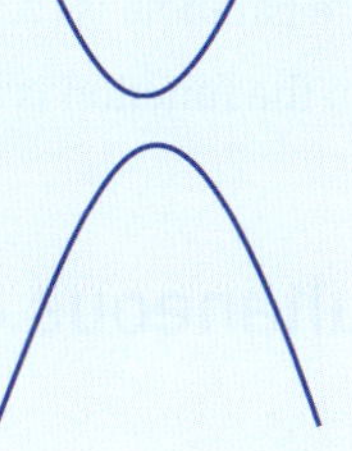

Example 23

a Find the x- and y-intercepts of the quadratic function $f(x) = -x^2 + 4x + 5$.

b Sketch a graph of the function.

c Find the maximum value of the function.

d State the domain and range.

e Write the coordinates of the vertex of the parabola.

f For what values of x is the function increasing?

Solution

a For x-intercepts, $f(x) = 0$.

$$0 = -x^2 + 4x + 5$$
$$x^2 - 4x - 5 = 0$$
$$(x - 5)(x + 1) = 0$$
$$x = 5, x = -1$$

For y-intercept, $x = 0$.

$$f(0) = -(0)^2 + 4(0) + 5$$
$$= 5$$

b Since $a < 0$, the quadratic function is concave downwards.

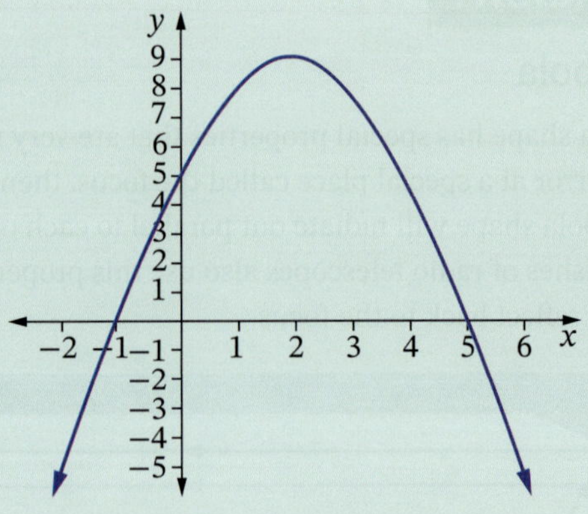

c The turning point is halfway between $x = -1$ and $x = 5$.

$x = \frac{-1+5}{2}$

$= 2$

$f(2) = -(2)^2 + 4(2) + 5$

$= 9$

The maximum value of $f(x)$ is 9.

d For the domain, the function can take on all real numbers for x.

Domain: $(-\infty, \infty)$

For the range, $y \leq 9$.

Range: $(-\infty, 9]$

e The vertex of the parabola is $(2, 9)$.

f From the graph, the function is increasing when $x < 2$.

Solving simultaneous equations graphically

Example 24

Solve the simultaneous equations $y = x^2 - 2x$ and $x + y - 6 = 0$ graphically.

Solution

Sketch the parabola $y = x^2 - 2x$.

For x-intercepts, $y = 0$.

$x^2 - 2x = 0$

$x(x - 2) = 0$

$x = 0, 2$

For y-intercepts, $x = 0$.

$y = 0^2 - 2(0)$

$= 0$

Sketch the line $x + y - 6 = 0$.

For x-intercepts, $y = 0$.

$x + 0 - 6 = 0$

$x - 6 = 0$

$x = 6$

For y-intercepts, $x = 0$.

$0 + y - 6 = 0$

$y = 6$

Sketch both graphs on the same number plane and find their points of intersection.

The points of intersection are $(-2, 8)$ and $(3, 3)$.

So the solutions of the simultaneous equations are $x = -2, y = 8$ and $x = 3, y = 3$.

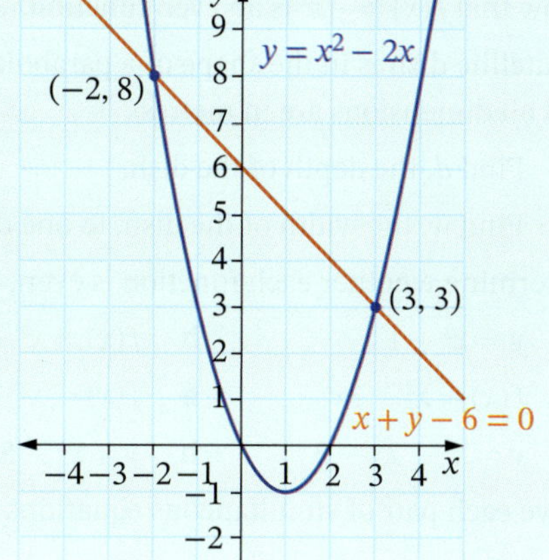

EXERCISE 3.08 Answers on p. 466

Quadratic functions

1 Graph each quadratic function and find its x- and y-intercepts.

a $y = x^2 + 2x$ **b** $y = -x^2 + 3x$ **c** $f(x) = x^2 - 1$

d $y = x^2 - x - 2$ **e** $y = x^2 - 9x + 8$

2 Sketch the graphs of $y = x^2$, $y = x^2 + 5$ and $y = x^2 - 2$ on paper or using technology, and describe the relationship between the graphs.

3 Sketch the graphs of $y = x^2$, $y = 4x^2$ and $y = \frac{x^2}{4}$ on paper or using technology, and describe the relationship between the graphs.

4 Sketch the graphs of $y = x^2$, $y = (x + 3)^2$ and $y = (x - 4)^2$ on paper or using technology, and describe the relationship between the graphs.

5 Graph each parabola and find its maximum or minimum value.

a $y = x^2 + 2$ **b** $y = -x^2 + 1$ **c** $f(x) = x^2 - 4$

d $y = x^2 + 2x$ **e** $y = -x^2 - x$ **f** $f(x) = (x - 3)^2$

g $f(x) = (x + 1)^2$ **h** $y = x^2 + 3x - 4$ **i** $f(x) = -x^2 + 3x - 2$

6 For each parabola, find:

i the x- and y-intercepts **ii** the domain and range.

a $y = x^2 - 7x + 12$ **b** $f(x) = x^2 + 4x$ **c** $y = x^2 - 2x - 8$

d $y = x^2 - 6x + 9$ **e** $f(x) = 4 - x^2$

7 Find the domain and range of the following quadratic functions.

a $y = x^2 - 5$ **b** $f(x) = x^2 - 6x$ **c** $f(x) = x^2 - x - 2$

d $y = -x^2$ **e** $f(x) = (x - 7)^2$

8 Sketch each parabola, showing its intercepts and the coordinates of its vertex.

a $y = x^2 - 6x + 8$ **b** $f(x) = -x^2 + 3x$ **c** $y = x^2 - 6x + 9$

d $y = -x^2 - x + 12$ **e** $f(x) = 2x^2 - x - 3$

9 Find the domain over which each function is increasing.

a $y = x^2$ **b** $y = -x^2$ **c** $f(x) = x^2 - 9$

d $y = -x^2 + 4x$ **e** $f(x) = (x + 5)^2$

Foundation | Mastery | Complex

10 Show that $f(x) = -x^2$ is an even function.

11 A satellite dish is in the shape of a parabola with equation $y = -3x^2 + 6$, and all dimensions are in metres.

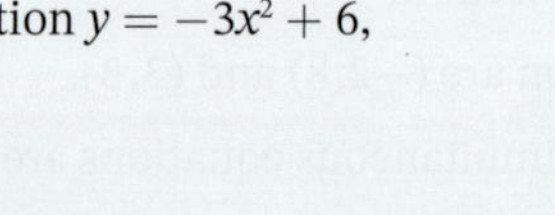

a Find d, the depth of the dish.

b Find w, the width of the dish, to one decimal place.

12 Determine whether each function is even, odd or neither.

a $y = x^2 + 1$ **b** $f(x) = x^2 - 3$ **c** $y = -2x^2$

d $f(x) = x^2 - 3x$ **e** $f(x) = x^2 + x$ **f** $y = x^2 - 4$

g $y = x^2 - 2x - 3$ **h** $y = x^2 - 5x + 4$ **i** $p(x) = (x + 1)^2$

13 Solve each pair of simultaneous equations graphically.

a $y = x - 1$ and $y = x^2 - 3$

b $y = x^2 + 1$ and $y = 1 - x^2$

c $y = x^2 - 3x + 7$ and $y = 2x + 3$

14 **a** Draw the graphs of $f(x) = x^2$ and $f(x) = (x - 2)^2$ on the same number plane.

b From the graph, find the number of points of intersection of the functions.

c From the graph or by using algebra, find any points of intersection.

15 By graphing the parabola $y = -x^2 + x$ and the line $y = -2$, solve the equation $-x^2 + x = -2$.

16 Solve $x^2 - 5x + 6 = 2$ by graphing the functions $y = x^2 - 5x + 6$ and $y = 2$.

17 Graph $f(x) = (x - 2)(x + 1)$ and $g(x) = 4 - x^2$ on the same number plane to solve $(x - 2)(x + 1) = 4 - x^2$.

18 The graph of $y = x^2$ meets the line $y = c$, where $c > 0$, at points A and B as shown. The length of the interval AB is k units.

The graph of $y = \frac{x^2}{n^2}$, where n is a positive integer, meets the line $y = c$ at points C and D (not shown on the graph).
Find the length of CD in terms of n and k.

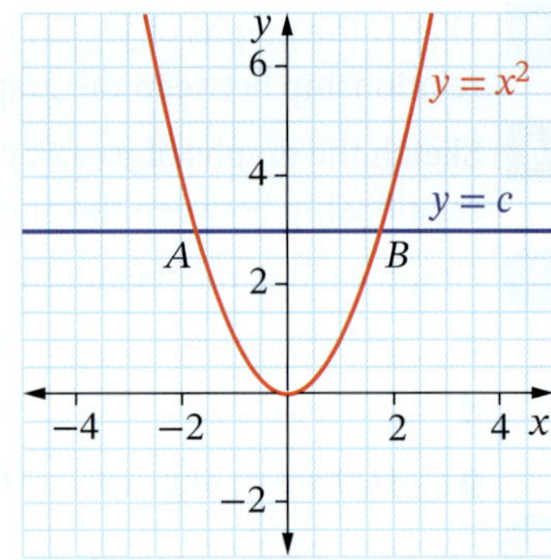

Quadratic inequalities

3.09

In Chapter 2, *Equations and inequalities*, you solved quadratic inequalities using the number line. You can also solve quadratic inequalities using the graph of a parabola.

Worksheet Quadratic inequalities

Solving quadratic inequalities graphically

For the graph of the quadratic function $y = ax^2 + bx + c$:

- $ax^2 + bx + c = 0$ is on the x-axis
- $ax^2 + bx + c > 0$ is above the x-axis
- $ax^2 + bx + c < 0$ is below the x-axis.

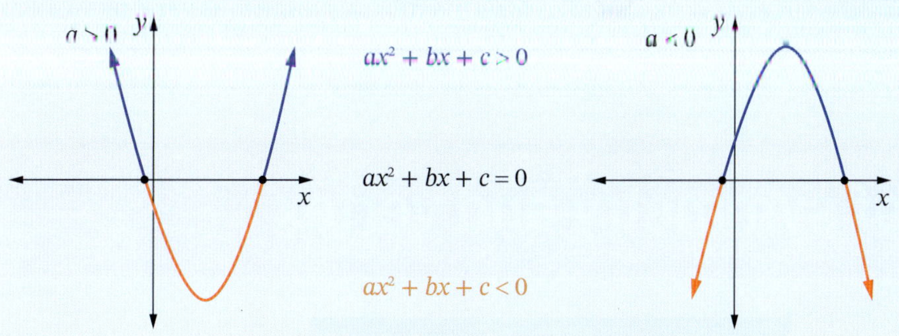

Example 25

Video Quadratic inequalities

Solve:

a $x^2 - 3x + 2 \geq 0$

b $4x - x^2 > 0$

Solution

a Sketch the graph of $y = x^2 - 3x + 2$, showing x-intercepts.

$a > 0$ so it is concave upwards.

For x-intercepts, $y = 0$.

$0 = x^2 - 3x + 2$

$= (x - 2)(x - 1)$

$x = 2, x = 1$

$x^2 - 3x + 2 \geq 0$ is on and above the x-axis.

$\therefore x \leq 1, x \geq 2$

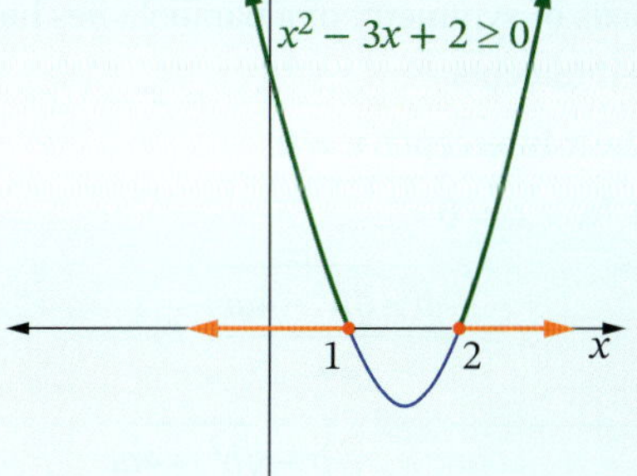

b For $y = 4x - x^2$, $a < 0$ so its graph is concave downwards.

For x-intercepts, $y = 0$.

$0 = 4x - x^2$

$= x(4 - x)$

$x = 0, x = 4$

$4x - x^2 > 0$ is above the x-axis.

$\therefore 0 < x < 4$

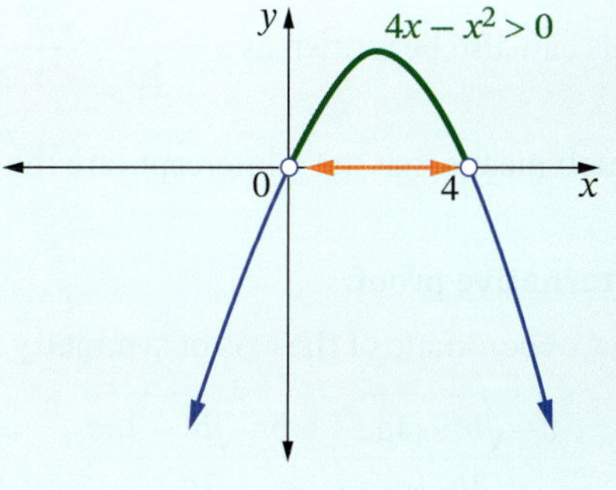

EXERCISE 3.09 Answers on p. 469

Quadratic inequalities

EXAMPLE 25

Solve each quadratic inequality.

1 $x^2 - 9 > 0$
2 $x^2 + x \le 0$
3 $x^2 - 2x \ge 0$
4 $4 - x^2 < 0$
5 $y^2 - 6y \le 0$
6 $2t - t^2 > 0$
7 $x^2 + 2x - 8 > 0$
8 $p^2 + 4p + 3 \ge 0$
9 $m^2 - 6m + 8 > 0$
10 $6 - x - x^2 \le 0$
11 $2h^2 - 7h + 6 < 0$
12 $x^2 - x - 20 \le 0$
13 $35 + 9k - 2k^2 \ge 0$
14 $q^2 - 9q + 18 > 0$
15 $(x + 2)^2 \ge 0$
16 $12 - n - n^2 \le 0$
17 $x^2 - 2x < 15$
18 $-t^2 \ge 4t - 12$
19 $3y^2 > 14y + 5$
20 $(x - 3)(x + 1) \ge 5$

3.10 Axis of symmetry and vertex

Worksheets
Quadratic functions

Sketching quadratic functions

Features of a parabola

Matching parabolas with their equations

Axis of symmetry of a parabola

The **axis of symmetry** of a parabola with the equation $y = ax^2 + bx + c$ is the vertical line with equation

$$x = -\frac{b}{2a}$$

Proof

The axis of symmetry of a parabola lies halfway between the x-intercepts.

For the x-intercepts, $y = 0$.

$ax^2 + bx + c = 0$

$$x = \frac{-b \pm \sqrt{b^2 - 4ac}}{2a}$$

$\therefore$ x-intercepts at $x = \frac{-b + \sqrt{b^2 - 4ac}}{2a}$ and $x = \frac{-b - \sqrt{b^2 - 4ac}}{2a}$.

This can also be written as $x = \frac{-b}{2a} + \frac{\sqrt{b^2 - 4ac}}{2a}$ and $x = \frac{-b}{2a} - \frac{\sqrt{b^2 - 4ac}}{2a}$,

which means that the x-intercepts are the same distance right and left from $x = \frac{-b}{2a}$ respectively.

$x = -\frac{b}{2a}$, y, x, $\frac{-b - \sqrt{b^2 - 4ac}}{2a}$, $\frac{-b + \sqrt{b^2 - 4ac}}{2a}$

Alternative proof:

The x-coordinate of the axis of symmetry is the average of the x-intercepts.

$$x = \frac{\frac{-b - \sqrt{b^2 - 4ac}}{2a} + \frac{-b + \sqrt{b^2 - 4ac}}{2a}}{2} = \frac{\frac{-2b}{2a}}{2} = \frac{-2b}{4a} = -\frac{b}{2a}$$

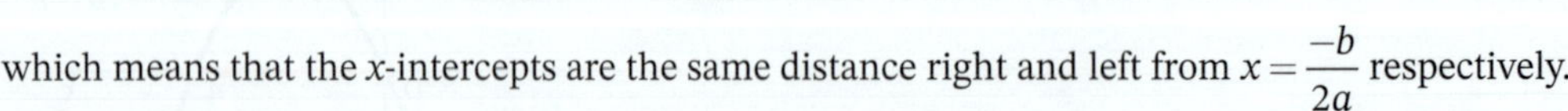

Vertex of a parabola

The quadratic function $f(x) = ax^2 + bx + c$ has a minimum value if $a > 0$ and a maximum value if $a < 0$.

The minimum or maximum value of the quadratic function is $f\left(-\frac{b}{2a}\right)$.

The turning point or vertex of a parabola is $\left(-\frac{b}{2a}, f\left(-\frac{b}{2a}\right)\right)$.

Example 26

a Find the equation of the axis of symmetry and the minimum value of the quadratic function $y = x^2 - 5x + 1$.

b Find the equation of the axis of symmetry, the maximum value and the vertex of the quadratic function $y = -3x^2 + x - 5$.

Solution

a $x = -\frac{b}{2a}$

$= -\frac{(-5)}{2(1)}$

$= \frac{5}{2}$

$= 2\frac{1}{2}$

$\therefore$ axis of symmetry is the line $x = 2\frac{1}{2}$.

Minimum value $y = \left(\frac{5}{2}\right)^2 - 5\left(\frac{5}{2}\right) + 1$

$= \frac{25}{4} - \frac{25}{2} + 1$

$= -5\frac{1}{4}$

b $x = -\frac{b}{2a}$

$= -\frac{1}{2(-3)}$

$= \frac{1}{6}$

$\therefore$ axis of symmetry is the line $x = \frac{1}{6}$.

Maximum value $y = -3\left(\frac{1}{6}\right)^2 + \left(\frac{1}{6}\right) - 5$

$= -\frac{1}{12} + \frac{1}{6} - 5$

$= -4\frac{11}{12}$

The vertex is $\left(\frac{1}{6}, -4\frac{11}{12}\right)$.

The vertex form of a quadratic function $y = a(x - b)^2 + c$

- The graph of $y = a(x - b)^2$ is a parabola with a vertex at $(b, 0)$.
- The parabola $y = a(x - b)^2$ is the graph of $y = ax^2$ translated b units to the right (or left if b is negative).
- The graph of $y = a(x - b)^2 + c$ is the parabola $y = a(x - b)^2$ translated c units up (or down if c is negative), so its vertex is at (b, c).

Videos
Graphing the parabola $y = a(x - b)^2 + c$

The turning point form of a parabola

Example 27

Express $y = x^2 + 2x + 3$ in the form $y = (x - b)^2 + c$ and determine the coordinates of the vertex. Graph the parabola showing its y-intercept.

Solution

Method 1: By completing the square

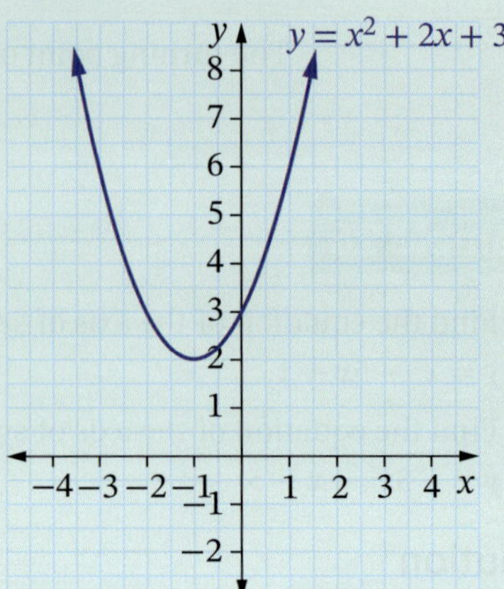

$x^2 + 2x + 3 = x^2 + 2x + 1 + 2$

$= (x + 1)^2 + 2$

$= (x - (-1))^2 + 2$

$\therefore$ the vertex of the graph of $y = x^2 + 2x + 3$ is $(-1, 2)$.

For y-intercept, $x = 0$.

$y = 0^2 + 2(0) + 3$

$= 3$

$a = 1 > 0$, so parabola is concave up.

There are no x intercepts because the vertex is above the x-axis.

Method 2: By expanding the RHS and comparing coefficients

$y = (x - b)^2 + c = x^2 - 2bx + b^2 + c$

$\therefore x^2 + 2x + 3 = x^2 - 2bx + b^2 + c$

$\therefore -2b = 2$ [1]

$b^2 + c = 3$ [2]

From [1]:

$b = -1$

Substitute into [2]:

$(-1)^2 + c = 3$

$1 + c = 3$

$c = 2$

$\therefore x^2 + 2x + 3 = (x - b)^2 + c$

$= (x - (-1))^2 + 2$

$\therefore$ the vertex of the graph of $y = x^2 + 2x + 3$ is $(-1, 2)$.

Continue as in Method 1.

EXERCISE 3.10 Answers on p. 469

Axis of symmetry and vertex

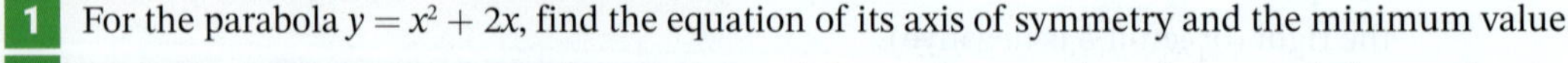

EXAMPLE 26

1 For the parabola $y = x^2 + 2x$, find the equation of its axis of symmetry and the minimum value.

2 Find the equation of the axis of symmetry and the minimum value of the parabola $y = x^2 - 4$.

3 Find the equation of the axis of symmetry and the vertex of the parabola $y = 4x^2 - 3x + 1$.

4 Find the equation of the axis of symmetry and the maximum value of the parabola $y = -x^2 + 2x - 7$.

5 Find the equation of the axis of symmetry and the vertex of the parabola $y = -2x^2 - 4x + 5$.

6 Find the equation of the axis of symmetry and the minimum value of the parabola $y = x^2 + 3x + 2$.

Foundation Mastery Complex

7 Find the equation of the axis of symmetry and the coordinates of the vertex for each parabola:

a $y = x^2 + 6x - 3$ **b** $y = -x^2 - 8x + 1$ **c** $y = 3x^2 + 18x + 4$

d $y = -2x^2 + 5x$ **e** $y = 4x^2 + 10x - 7$

8 For each parabola, find:

i the equation of the axis of symmetry

ii the minimum or maximum value

iii the vertex.

a $y = x^2 + 2x - 2$ **b** $y = -2x^2 + 4x - 1$

9 Find the vertex of the parabola for each quadratic function. EXAMPLE 27

a $y = (x - 3)^2$ **b** $y = (x + 2)^2$ **c** $y = 2(x + 1)^2 + 3$

d $y = -(x - 3)^2$ **e** $y = -4(x + 1)^2 + 8$ **f** $y = -(x - 5)^2 - 1$

g $y = 3(x + 4)^2$ **h** $y = -2(x - 1)^2 - 5$ **i** $y = -\frac{1}{2}(x + 6)^2 + 2$

10 Match each quadratic function to its graph.

a $y = (x + 4)^2$ **b** $y = -(x - 1)^2$ **c** $y = (x + 2)^2$

d $y = (x - 3)^2 + 4$ **e** $y = -(x + 5)^2$ **f** $y = -2(x + 1)^2$

g $y = -0.5(x - 4)^2$ **h** $y = 4(x - 1)^2 - 3$

A

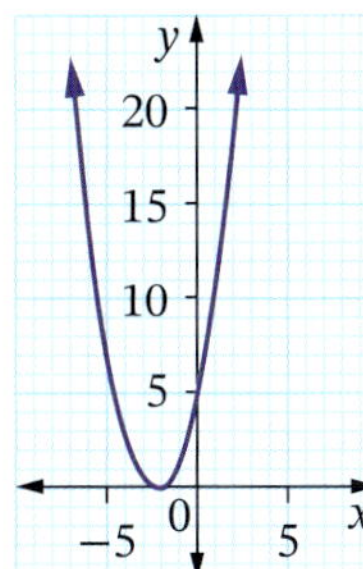

B

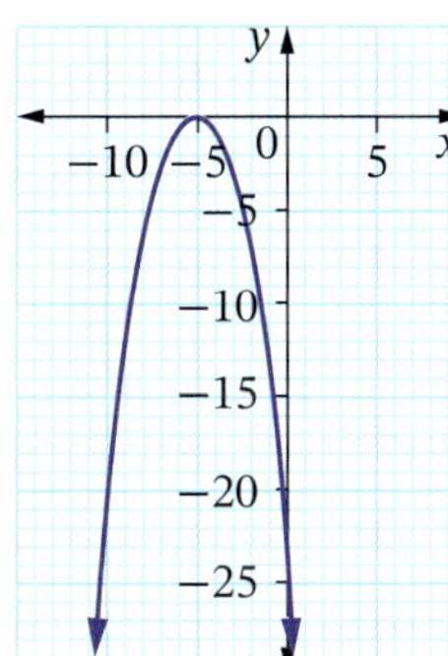

C

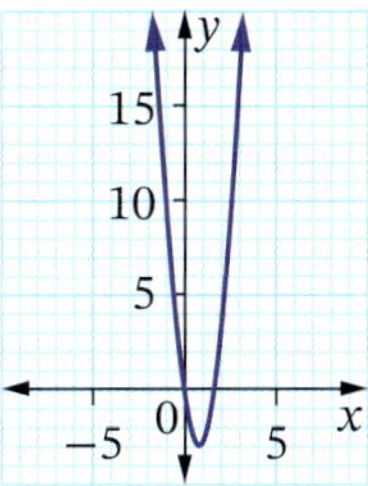

D

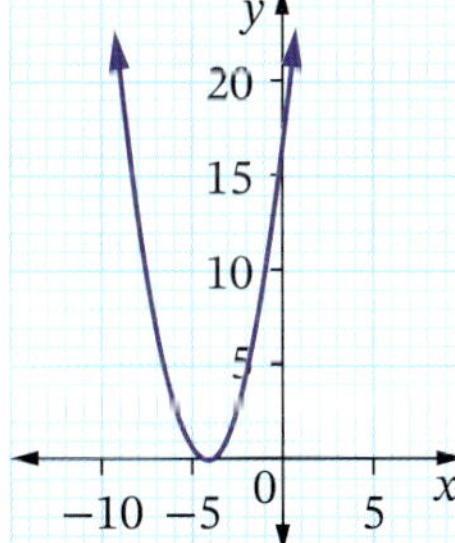

E

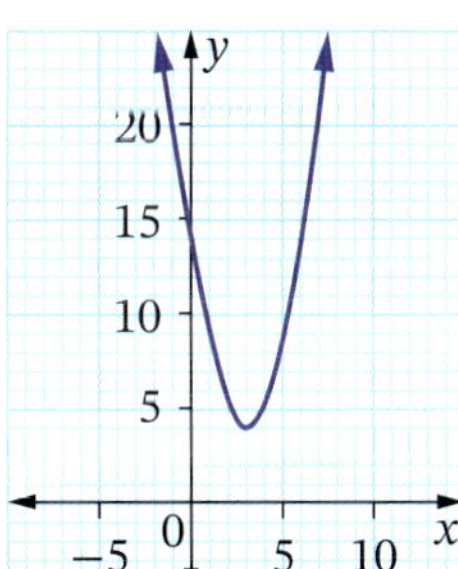

F

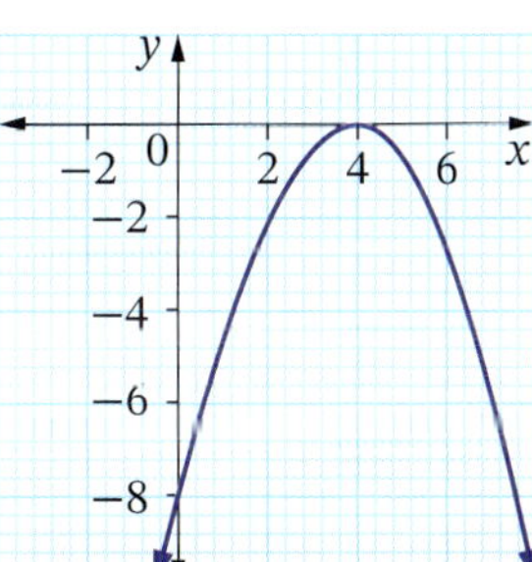

G

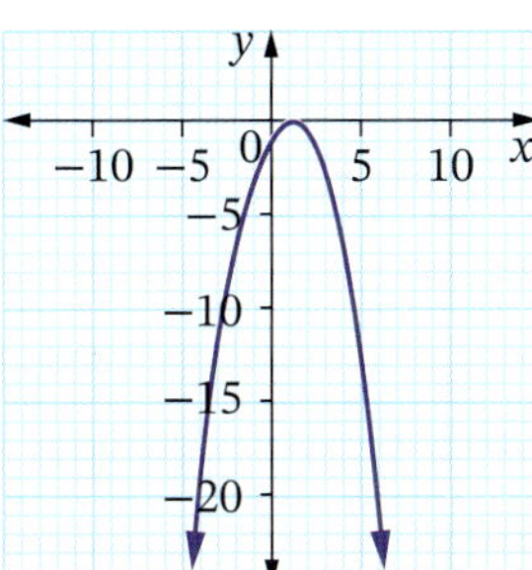

H

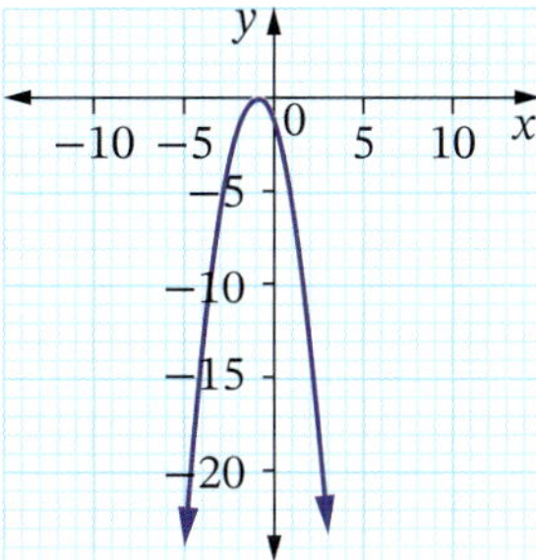

Foundation Mastery Complex

11 Express each quadratic function in the form $y = a(x - b)^2 + c$, find its vertex, and state whether it is a maximum or minimum.

a $y = x^2 + 2x + 1$ **b** $y = x^2 - 8x - 7$ **c** $f(x) = x^2 + 4x - 3$

d $y = -x^2 - 2x + 5$ **e** $y = -2x^2 + 8x + 3$ **f** $f(x) = -3x^2 + 3x + 7$

12 **a** Write the quadratic function $y = x^2 - 4x + 1$ in the form $y = (x - b)^2 + c$.

b Sketch the graph of the function showing its axis of symmetry, vertex and y-intercept.

c Find the domain and range of the function.

13 For each quadratic function:

i find x-intercepts using the quadratic formula

ii state whether the function has a maximum or minimum value and find this value

iii sketch the graph of the function on a number plane

iv find the zeroes of $f(x)$ graphically.

a $f(x) = x^2 + 4x + 4$ **b** $f(x) = x^2 - 2x - 3$ **c** $y = x^2 - 6x + 1$

d $f(x) = -x^2 - 2x + 6$ **e** $f(x) = -x^2 - x + 3$

14 **a** Find the minimum value of the parabola with equation $y = x^2 - 2x + 5$.

b How many solutions does the quadratic equation $x^2 - 2x + 5 = 0$ have?

c Sketch the parabola.

15 **a** Find the maximum value of the quadratic function $f(x) = -2x^2 + x - 4$.

b How many solutions are there to the quadratic equation $-2x^2 + x - 4 = 0$?

c Sketch the graph of the quadratic function.

16 Which parabola could represent $y = x^2 + bx - 2$, given that $b > 0$? Select the correct answer **A**, **B**, **C** or **D**.

A

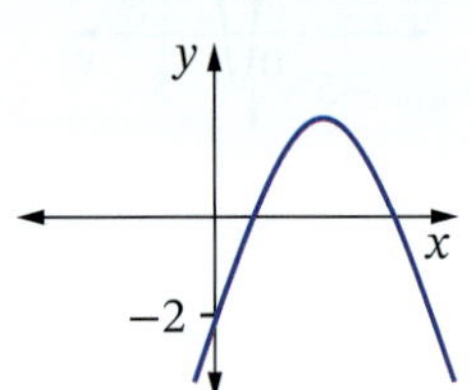

B

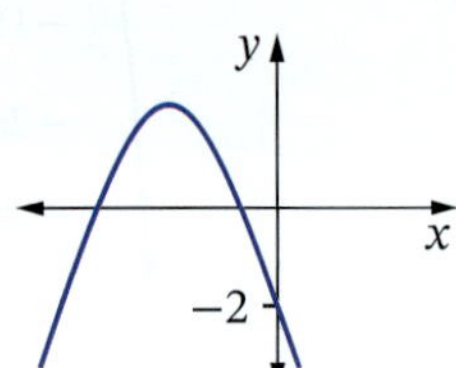

C

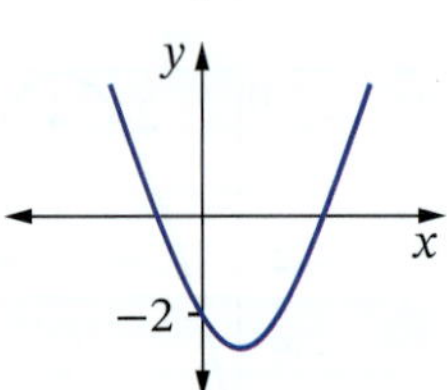

D

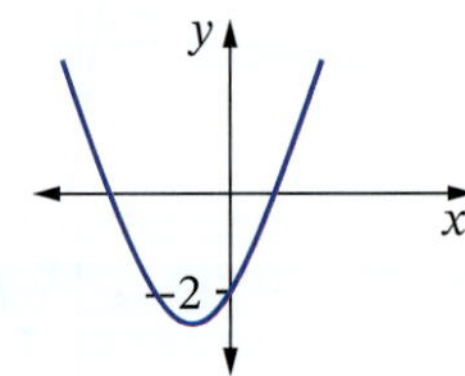

17 Draw a rough sketch of:

a $y = -ax^2 + 3x + 5$ if $a > 0$

b $f(x) = 3x^2 - 6x + c$ given $c < 0$

c $y = -x^2 + bx + 2$ if $b > 0$.

☐ Foundation ○ Mastery ○ Complex

Quadratic identities

3.11

If 2 quadratic functions are equal for *all* values of x, then they are **identical** and their corresponding coefficients must be equal.

Worksheet
A page of parabolas

Identity of quadratic expressions

If $ax^2 + bx + c$ and $dx^2 + ex + f$ are equal for all values of x, then $a = d$, $b = e$ and $c = f$.

Example 28

Write $2x^2 - 3x + 5$ in the form $A(x - 1)^2 + B(x - 1) + C$.

Solution

$$A(x - 1)^2 + B(x - 1) + C = A(x^2 - 2x + 1) + Bx - B + C$$
$$= Ax^2 - 2Ax + A + Bx - B + C$$
$$= Ax^2 + (-2A + B)x + A - B + C$$

Let $2x^2 - 3x + 5 = Ax^2 + (-2A + B)x + A - B + C$

Equating corresponding coefficients:

$A = 2$ [1]

$-2A + B = -3$ [2]

$A - B + C = 5$ [3]

Substitute [1] into [2]:

$-2(2) + B = -3$

$-4 + B = -3$

$B = 1$

Substitute $A = 2$ and $B = 1$ into [3]:

$2 - 1 + C = 5$

$1 + C = 5$

$C = 4$

$\therefore 2x^2 - 3x + 5 = 2(x - 1)^2 + (x - 1) + 4$

Example 29

Find values for a, b and c if $x^2 - x \equiv a(x + 3)^2 + bx + c - 1$.

Solution

$$a(x + 3)^2 + bx + c - 1 = a(x^2 + 6x + 9) + bx + c - 1$$
$$= ax^2 + 6ax + 9a + bx + c - 1$$
$$= ax^2 + (6a + b)x + 9a + c - 1$$

Let $x^2 - x = ax^2 + (6a + b)x + 9a + c - 1$

Equating coefficients:

$a = 1$ [1]

$6a + b = -1$ [2]

$9a + c - 1 = 0$ [3]

Substitute [1] into [2]:

$6(1) + b = -1$

$6 + b = -1$

$b = -7$

Substitute [1] into [3]:

$9(1) + c - 1 = 0$

$8 + c = 0$

$c = -8$

$\therefore a = 1, b = -7, c = -8$

Finding the equation of a parabola

Example 30

a Find the equation of the parabola that passes through the points $(-1, -3)$, $(0, 3)$ and $(2, 21)$.

b A parabolic satellite dish is built so it is 30 cm deep and 80 cm wide, as shown.

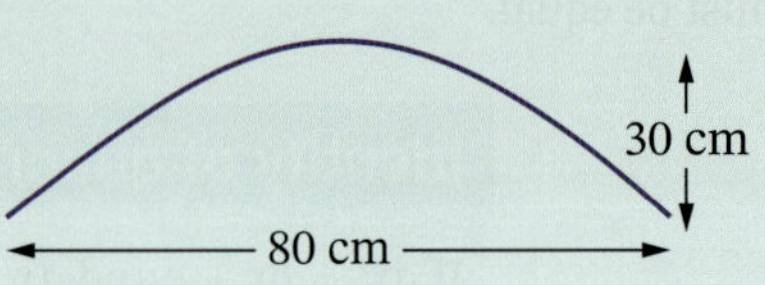

i Find an equation for the parabola.

ii Find the depth of the dish 10 cm out from the vertex.

Solution

a The parabola has equation in the form $y = ax^2 + bx + c$.

Substitute the points into the equation.

$(-1, -3)$:

$-3 = a(-1)^2 + b(-1) + c$

$= a - b + c$

$\therefore a - b + c = -3$ [1]

$(0, 3)$:

$3 = a(0)^2 + b(0) + c$

$= c$

$\therefore c = 3$ [2]

$(2, 21)$:

$21 = a(2)^2 + b(2) + c$

$= 4a + 2b + c$

$\therefore 4a + 2b + c = 21$ [3]

Solve simultaneous equations to find a, b and c.

Substitute [2] into [1]:

$a - b + 3 = -3$

$a - b = -6$ [4]

Substitute [2] into [3]:

$4a + 2b + 3 = 21$

$4a + 2b = 18$ [5]

[4] × 2:

$2a - 2b = -12$ [6]

[5] + [6]:

$6a = 6$

$a = 1$

Substitute $a = 1$ into [5]:

$4(1) + 2b = 18$

$4 + 2b = 18$

$2b = 14$

$b = 7$

$\therefore a = 1, b = 7, c = 3$

Thus, the parabola has equation $y = x^2 + 7x + 3$.

b i We can put the dish onto a number plane as shown. Since the parabola is symmetrical, the width of 80 cm means 40 cm either side of the y-axis.

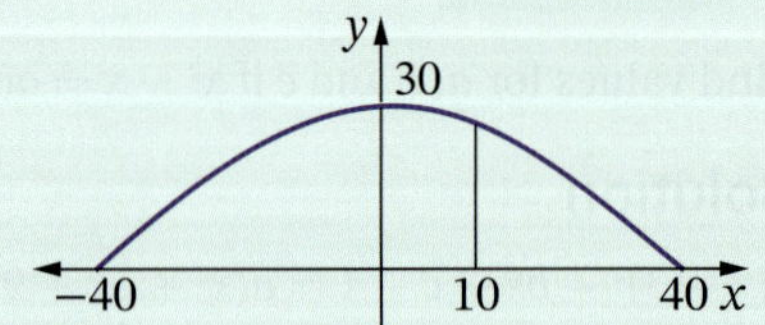

The parabola passes through points $(0, 30)$, $(40, 0)$ and $(-40, 0)$.

Substitute these points into $y = ax^2 + bx + c$.

$(0, 30)$:

$30 = a(0)^2 + b(0) + c$

$30 = c$

So $y = ax^2 + bx + 30$.

Substitute $(40, 0)$ into $y = ax^2 + bx + 30$:

$0 = a(40)^2 + b(40) + 30$

$0 = 1600a + 40b + 30$ [1]

Substitute $(-40, 0)$ into $y = ax^2 + bx + 30$:

$0 = a(-40)^2 + b(-40) + 30$

$0 = 1600a - 40b + 30$ [2]

[1] + [2]:

$0 = 3200a + 60$

$a = \frac{-60}{3200}$

$= -\frac{3}{160}$

Substitute a into [1]:

$0 = 1600\left(-\frac{3}{160}\right) + 40b + 30$

$= -30 + 40b + 30$

$= 40b$

$0 = b$

So $y = -\frac{3}{160}x^2 + 30$.

ii Substitute $x = 10$:

$y = -\frac{3}{160}(10)^2 + 30$

$= 28.125$

So the depth of the dish at 10 cm is 28.125 cm.

EXERCISE 3.11 Answers on p. 471

Quadratic identities

1. Find values of a, b and c for which:
   - **a** $x^2 + 4x - 3 = a(x + 1)^2 + b(x + 1) + c$
   - **b** $2x^2 - 3x + 1 = a(x + 2)^2 + b(x + 2) + c$
   - **c** $x^2 - x - 2 = a(x - 1)^2 + b(x - 1) + c$
   - **d** $x^2 + x + 6 = a(x - 3)^2 + b(x - 3) + c$
   - **e** $3x^2 - 5x - 2 = a(x + 1)^2 + b(x - 1) + c$
   - **f** $4x^2 + x - 7 = a(x - 2)^2 + b(x - 2) + c$
2. Find values of m, p and q for which $2x^2 - x - 1 = m(x + 1)^2 + p(x + 1) + q$.
3. Express $x^2 - 4x + 5$ in the form $Ax(x - 2) + B(x + 1) + C + 4$.
4. Show that $x^2 + 2x + 9$ can be written in the form $a(x - 2)(x + 3) + b(x - 2) + c$, where $a = 1$, $b = 1$ and $c = 17$.
5. Find values of A, B and C if $x^2 + x - 2 = A(x - 2)^2 + Bx + C$.
6. Find values of a, b and c for which $3x^2 + 5x - 1 = ax(x + 3) + bx^2 + c(x + 1)$.
7. Evaluate K, L and M if $x^2 = K(x - 3)^2 + L(x + 1) - 2M$.
8. Express $4x^2 + 2$ in the form $a(x + 5) + b(2x - 3)^2 + c - 2$.
9. Find the values of a, b and c if $20x - 17 = a(x - 4)^2 - b(5x + 1) + c$.

Foundation Mastery Complex

10 Find the equation of the parabola that passes through the points:

a $(0, -5)$, $(2, -3)$ and $(-3, 7)$ **b** $(1, -2)$, $(3, 0)$ and $(-2, 10)$

c $(-2, 21)$, $(1, 6)$ and $(-1, 12)$ **d** $(2, 3)$, $(1, -4)$ and $(-1, -12)$

e $(0, 1)$, $(-2, 1)$ and $(2, -7)$

11 Grania throws a ball off a 10 m high cliff. After 1 s it is 22.5 m above ground and it reaches the ground after 4 s.

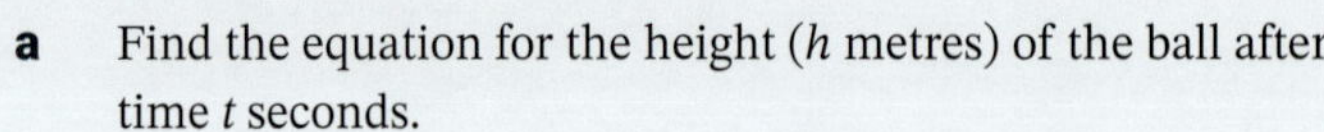

a Find the equation for the height (h metres) of the ball after time t seconds.

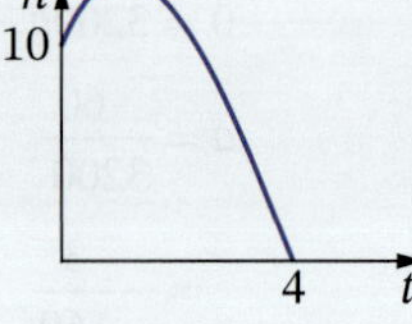

b Find the height of the ball after 2 seconds.

c Find when the ball is in line with the cliff.

12 A parabolic headlight is 15 cm wide and 8 cm deep as shown.

a Find an equation for the parabola.

b Find the depth of the headlight at a point 3 cm out from its axis of symmetry.

c At what width from the axis of symmetry does the headlight have a depth of 5 cm?

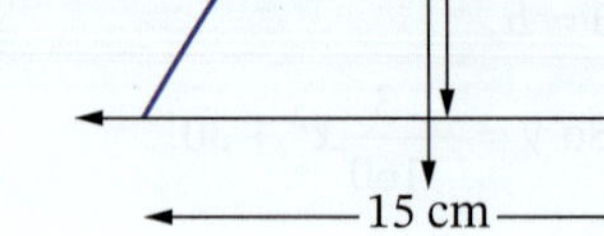

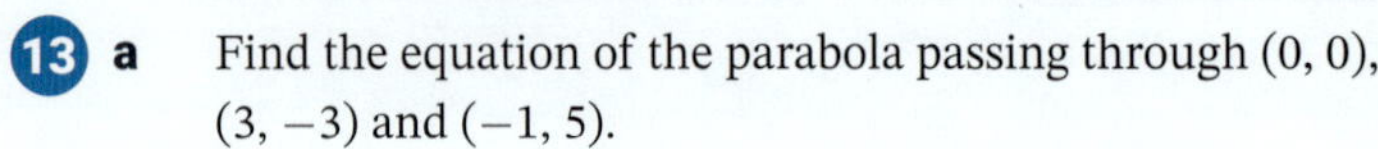

13 **a** Find the equation of the parabola passing through $(0, 0)$, $(3, -3)$ and $(-1, 5)$.

b Find the value of y when:

i $x = 5$ **ii** $x = -4$.

c Find the value/s of x when $y = -4$.

d Find the exact values of x when $y = 2$.

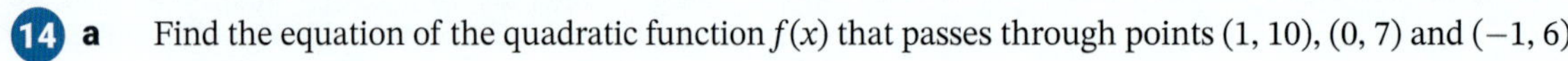

14 **a** Find the equation of the quadratic function $f(x)$ that passes through points $(1, 10)$, $(0, 7)$ and $(-1, 6)$.

b Evaluate $f(-5)$.

c Show that $f(x) > 0$ for all x.

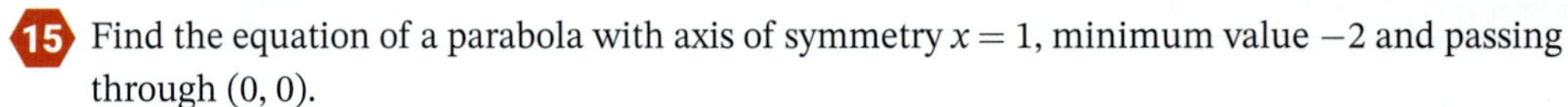

15 Find the equation of a parabola with axis of symmetry $x = 1$, minimum value -2 and passing through $(0, 0)$.

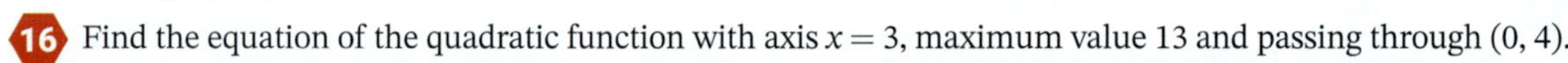

16 Find the equation of the quadratic function with axis $x = 3$, maximum value 13 and passing through $(0, 4)$.

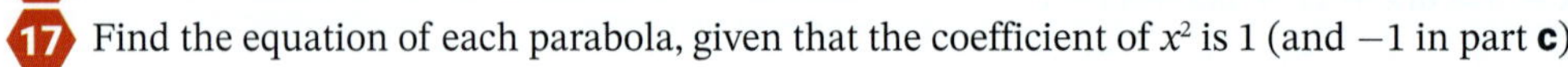

17 Find the equation of each parabola, given that the coefficient of x^2 is 1 (and -1 in part **c**).

a

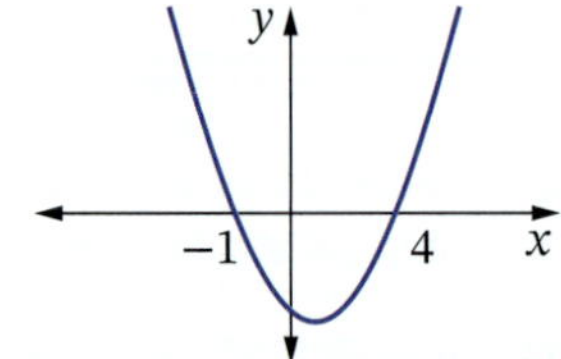

b

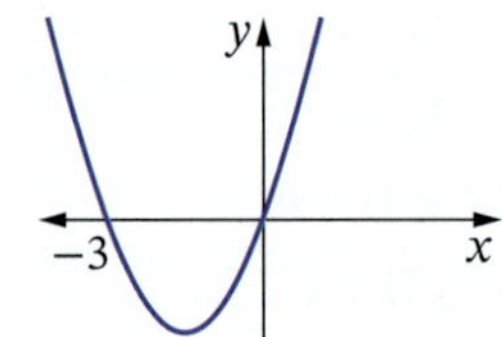

c

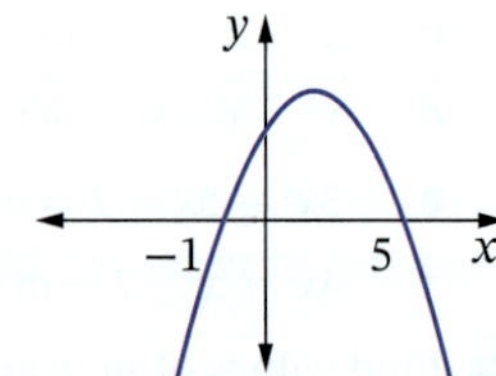

18 Find the equation of each parabola.

a

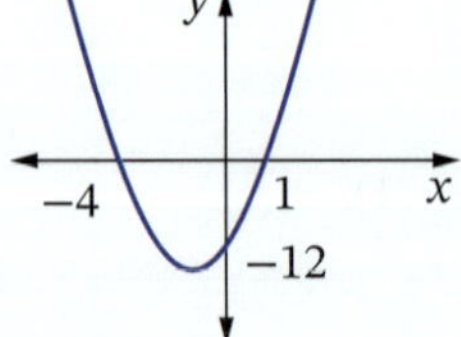

b

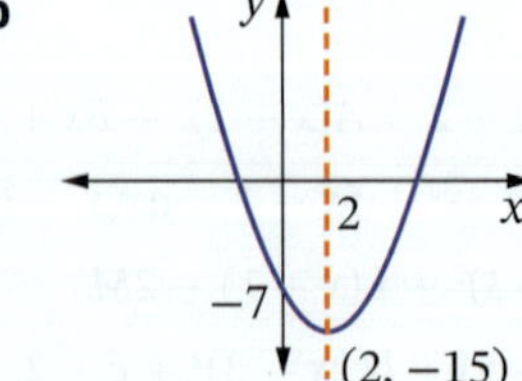

c

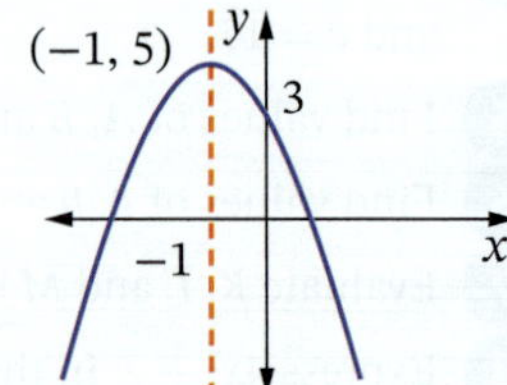

Foundation Mastery Complex

The discriminant

3.12

The solutions of an equation are also called the **roots** of the equation.

In the quadratic formula $x = \frac{-b \pm \sqrt{b^2 - 4ac}}{2a}$, the expression $b^2 - 4ac$ is called the **discriminant**.

It gives us information about the roots of the quadratic equation $ax^2 + bx + c = 0$.

Video
The discriminant

Worksheet
The discriminant

Example 31

'real roots' means solutions that are real numbers.

Use the quadratic formula to find how many real roots each quadratic equation has.

a $x^2 + 5x - 3 = 0$ **b** $x^2 - x + 4 = 0$ **c** $x^2 - 2x + 1 = 0$

Solution

a $x = \frac{-b \pm \sqrt{b^2 - 4ac}}{2a}$

$= \frac{-5 \pm \sqrt{5^2 - 4 \times 1 \times (-3)}}{2 \times 1}$

$= \frac{-5 \pm \sqrt{25 + 12}}{2}$

$= \frac{-5 \pm \sqrt{37}}{2}$

There are 2 real roots:

$x = \frac{-5 + \sqrt{37}}{2}, \frac{-5 - \sqrt{37}}{2}$

b $x = \frac{-b \pm \sqrt{b^2 - 4ac}}{2a}$

$= \frac{-(-1) \pm \sqrt{(-1)^2 - 4 \times 1 \times 4}}{2 \times 1}$

$= \frac{1 \pm \sqrt{-15}}{2}$

There are no real roots since $\sqrt{-15}$ has no real value.

c $x = \frac{-b \pm \sqrt{b^2 - 4ac}}{2a}$

$= \frac{-(-2) \pm \sqrt{(-2)^2 - 4 \times 1 \times 1}}{2 \times 1}$

$= \frac{2 \pm \sqrt{0}}{2}$

$= 1$

There are 2 real roots:

$x = 1, 1$

However, these are equal roots.

The discriminant

The value of the discriminant $\Delta = b^2 - 4ac$ tells us information about the roots of the quadratic equation $ax^2 + bx + c = 0$.

When $\Delta \geq 0$, there are 2 real roots.

- If $\Delta > 0$, the roots are distinct (different, unequal).
- If Δ is a perfect square, the roots are rational.
- If Δ is not a perfect square, the roots are irrational.

When $\Delta = 0$, there are 2 equal rational roots (or 1 rational root).

When $\Delta < 0$, there are no real roots.

Example 32

a Show that the equation $2x^2 + x + 4 = 0$ has no real roots.

b Describe the roots of the equation:

i $2x^2 - 7x - 1 = 0$ **ii** $x^2 + 6x + 9 = 0$

c Find the values of k for which the quadratic equation $5x^2 - 2x + k = 0$ has real roots.

Solution

a
$$\begin{aligned}\Delta &= b^2 - 4ac\\ &= 1^2 - 4(2)(4)\\ &= -31\\ &< 0\end{aligned}$$

$\Delta < 0$, so the equation has no real roots.

b **i**
$$\begin{aligned}\Delta &= b^2 - 4ac\\ &= (-7)^2 - 4(2)(-1)\\ &= 57\\ &> 0\end{aligned}$$

$\Delta > 0$, so there are 2 real irrational roots.

The roots are irrational because 57 is not a perfect square.

ii
$$\begin{aligned}\Delta &= b^2 - 4ac\\ &= (6)^2 - 4(1)(9)\\ &= 0\end{aligned}$$

$\Delta = 0$, so there are 2 real equal rational roots.

c For real roots, $\Delta \geq 0$.

$$\begin{aligned}b^2 - 4ac &\geq 0\\ (-2)^2 - 4(5)(k) &\geq 0\\ 4 - 20k &\geq 0\\ 4 &\geq 20k\\ k &\leq \frac{1}{5}\end{aligned}$$

The discriminant and the parabola

The roots of the quadratic equation $ax^2 + bx + c = 0$ give the x-intercepts of the parabola $y = ax^2 + bx + c$.

If $\Delta > 0$, then the quadratic equation has 2 real distinct roots and the parabola has 2 x-intercepts.

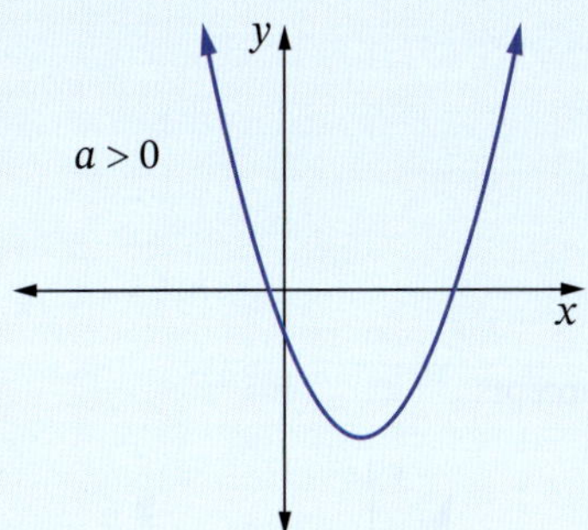

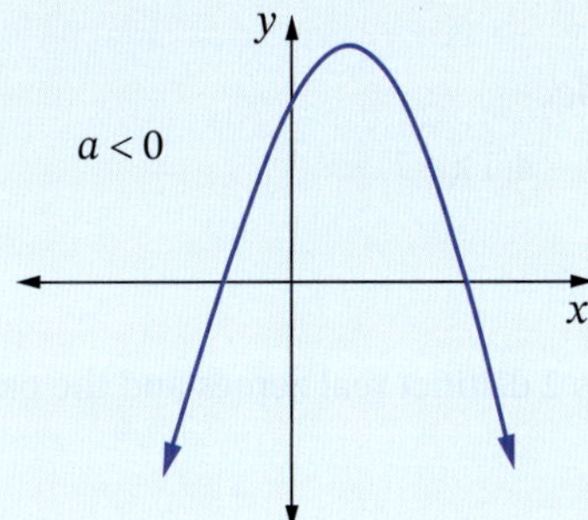

If $\Delta = 0$, then the quadratic equation has 1 real root or 2 equal roots and the parabola has one x-intercept ('touches' the x-axis at only one point)

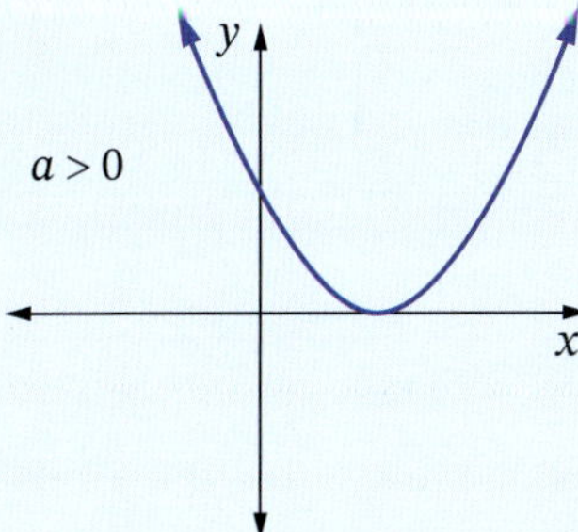

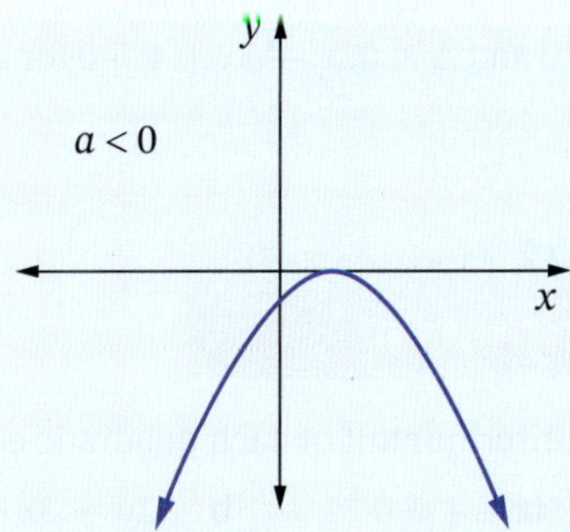

If $\Delta < 0$, then the quadratic equation has no real roots and the parabola has no x-intercepts.

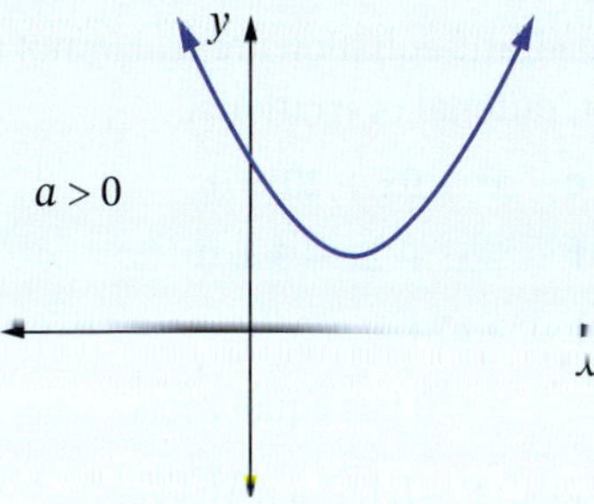

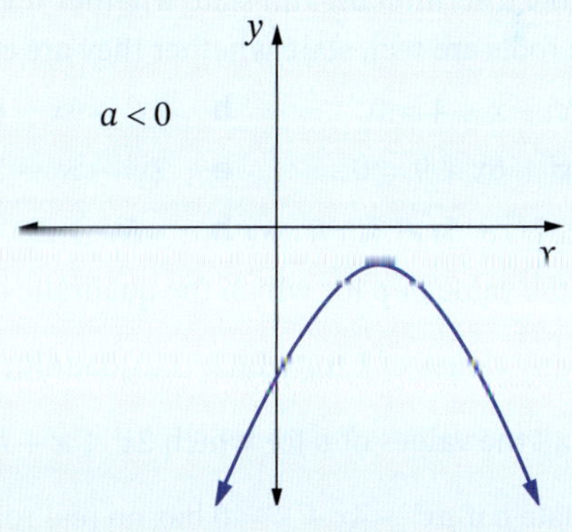

If $\Delta < 0$ and $a > 0$, then the parabola is completely above the x-axis and $ax^2 + bx + c > 0$ for all x.

If $\Delta < 0$ and $a < 0$, then the parabola is completely below the x-axis and $ax^2 + bx + c < 0$ for all x.

Example 33

a Show that the parabola $f(x) = x^2 - x - 2$ has 2 x-intercepts.

b Show that $x^2 - 2x + 4 > 0$ for all x.

Solution

a $\Delta = b^2 - 4ac$

$= (-1)^2 - 4(1)(-2)$

$= 9$

> 0

So $f(x)$ has 2 distinct real zeroes and the parabola has 2 x-intercepts.

b $a = 1 > 0$

$\Delta = b^2 - 4ac$

$= (-2)^2 - 4(1)(4)$

$= -12$

< 0

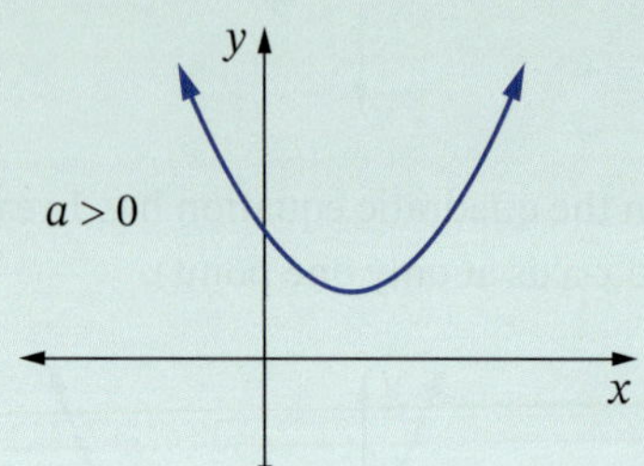

Since $a > 0$ and $\Delta < 0$, $x^2 - 2x + 4 > 0$ for all x.

EXERCISE 3.12 Answers on p. 471

The discriminant

EXAMPLE 32

1 Find the discriminant of each quadratic equation.

a	$x^2 - 4x - 1 = 0$	**b**	$2x^2 + 3x + 7 = 0$	**c**	$-4x^2 + 2x - 1 = 0$
d	$6x^2 - x - 2 = 0$	**e**	$-x^2 - 3x = 0$	**f**	$x^2 + 4 = 0$
g	$x^2 - 2x + 1 = 0$	**h**	$-3x^2 - 2x + 5 = 0$	**i**	$-2x^2 + x + 2 = 0$

2 Find the discriminant and state whether the roots of the quadratic equation are real or not real. If the roots are real, state whether they are equal or distinct, rational or irrational.

a	$x^2 - x - 4 = 0$	**b**	$2x^2 + 3x + 6 = 0$	**c**	$x^2 - 9x + 20 = 0$
d	$x^2 + 6x + 9 = 0$	**e**	$2x^2 - 5x - 1 = 0$	**f**	$-x^2 + 2x - 5 = 0$
g	$-2x^2 - 5x + 3 = 0$	**h**	$-5x^2 + 2x - 6 = 0$	**i**	$-x^2 + x = 0$

3 Find the value of p for which the quadratic equation $x^2 + 2x + p = 0$ has equal roots.

4 Find any values of k for which the quadratic equation $x^2 + kx + 1 = 0$ has equal roots.

5 Find all the values of b for which $2x^2 + x + b + 1 = 0$ has real roots.

6 Evaluate p if $px^2 + 4x + 2 = 0$ has no real roots.

7 Find all values of k for which $(k + 2)x^2 + x - 3 = 0$ has 2 real distinct roots.

8 Prove that $3x^2 - x + 7 > 0$ for all real x.

EXAMPLE 33

9 Show that the line $y = 2x + 6$ cuts the parabola $y = x^2 + 3$ in 2 points.

10 Show that the line $3x + y - 4 = 0$ cuts the parabola $y = x^2 + 5x + 3$ in 2 points.

11 Show that the line $y = -x - 4$ does not touch the parabola $y = x^2$.

12 Show that the line $y = 5x - 2$ is a tangent to the parabola $y = x^2 + 3x - 1$.

A tangent is a line that 'touches' a curve at one point only.

13 Find the sign of the discriminant of $y = x^2 - 2x + 7$ and hence describe where its graph lies in relation to the x-axis.

14 Find the values of k for which $x^2 + (k + 1)x + 4 = 0$ has real roots.

Foundation Mastery Complex

15 Find values of k for which the expression $kx^2 + 3kx + 9 > 0$ for all real x.

16 Find the values of m for which the quadratic equation $x^2 - 2mx + 9 = 0$ has real and distinct roots.

17 **a** How many points of intersection are possible between a line and a parabola?

b Find the number of points of intersection of the line $3x - y + 2 = 0$ and the parabola $y = 2x^2 - 5x - 3$.

c From part **b**, if there are points of intersection, find their exact x-coordinates.

Cubic functions 3.13

A **cubic function** has an equation where the highest power of x is 3, such as $f(x) = kx^3$, $f(x) = k(x - b)^3 + c$ and $f(x) = k(x - a)(x - b)(x - c)$, where a, b, c and k are constants.

Worksheets
Cubic functions

Graphing cubics

Graphing cubics 2

Example 34

a Sketch the graph of the cubic function $f(x) = x^3 + 2$.

b State its domain and range.

c Solve the equation $x^3 + 2 = 0$ graphically.

Solution

a Draw up a table of values and use it to sketch the graph.

x	−3	−2	−1	0	1	2	3
y	−25	−6	1	2	3	10	29

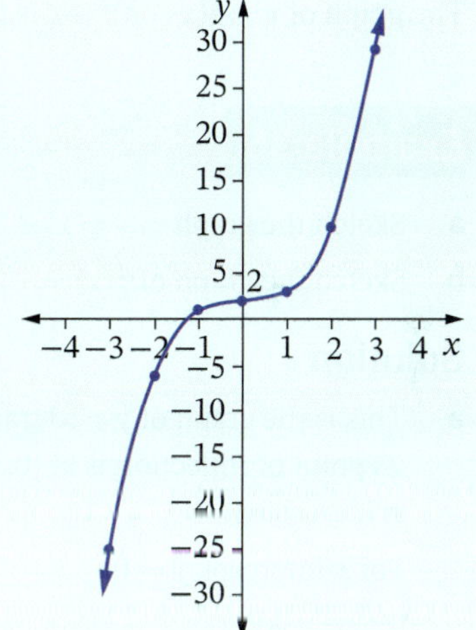

b The function can have any real x or y value.

Domain: $(-\infty, \infty)$

Range: $(-\infty, \infty)$

c From the graph, the x-intercept is approximately -1.3.

So, the root of $x^3 + 2 = 0$ is approximately -1.3.

Technology

Transforming cubic functions

Use graphing technology to sketch the graphs of some cubic functions, such as:

1 $y = x^3$ **2** $y = x^3 + 1$ **3** $y = x^3 + 3$ **4** $y = x^3 - 1$

5 $y = x^3 - 2$ **6** $y = 2x^3$ **7** $y = 3x^3$ **8** $y = -x^3$

9 $y = -2x^3$ **10** $y = -3x^3$ **11** $y = 2x^3 + 1$ **12** $y = (x + 1)^3$

13 $y = (x + 2)^3$ **14** $y = (x - 1)^3$ **15** $y = -2(x - 2)^3$ **16** $y = 3(x + 2)^3 + 1$

17 $y = (x - 1)(x - 2)(x - 3)$ **18** $y = x(x + 1)(x + 4)$ **19** $y = -2(x + 1)(x - 2)(x + 5)$

Can you see any patterns? Can you describe the shape of the cubic function? Can you predict where the graphs of different cubic functions would lie?

Is the cubic graph always a function? Can you find an example of a cubic that is not a function?

Foundation Mastery Complex

Point of inflection

The flat turning point of the cubic function $y = kx^3$ is called a **point of inflection**, which is where the concavity of the curve changes.

The graph of $y = kx^3$

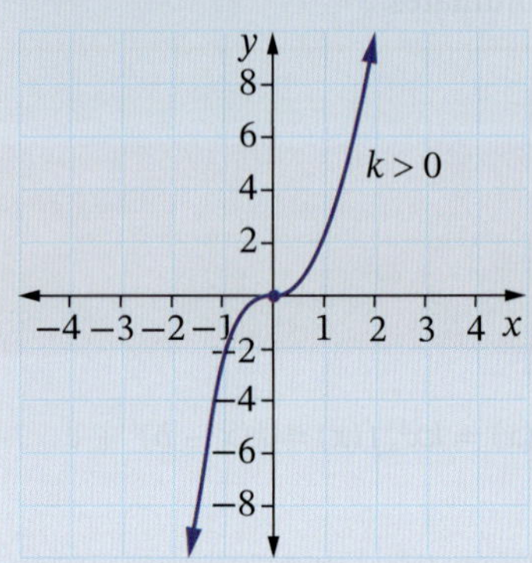

This cubic curve is increasing and has a point of inflection at (0, 0) where the curve changes from concave downwards to concave upwards.

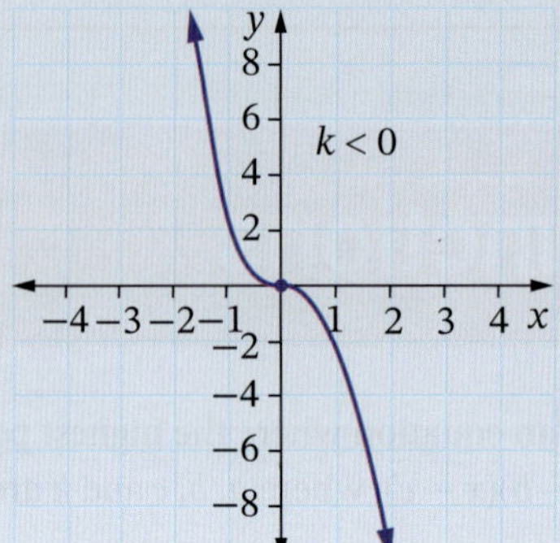

This cubic curve is decreasing and has a point of inflection at (0, 0) where the curve changes from concave upwards to concave downwards.

The graph of $y = k(x - b)^3 + c$

The graph of $y = k(x - b)^3 + c$ is the graph of $y = kx^3$ translated so that its point of inflection is at (b, c).

Example 35

a Sketch the graph of $y = x^3 - 8$, showing intercepts.

b Sketch the graph of $f(x) = -2(x - 3)^3 + 2$.

Solution

a This is the graph of $y = x^3$ translated downwards 8 units so that its point of inflection is at $(0, -8)$. Since $k > 0$, the function is increasing.

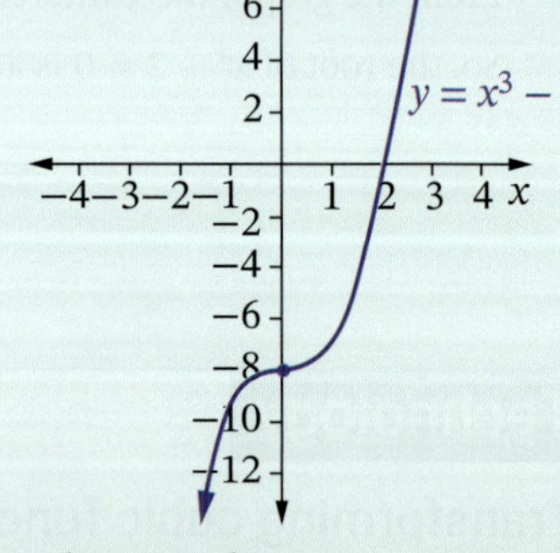

For x-intercept, $y = 0$.

$0 = x^3 - 8$

$8 = x^3$

$x = 2$

For y-intercept, $x = 0$.

$y = 0^3 - 8$

$= -8$

The point of inflection is at $(0, -8)$, where the curve changes from concave downwards to concave upwards.

b Since $k < 0$, $f(x)$ is decreasing.

This is the graph of $y = -2x^3$ translated upwards and to the right, so that its point of inflection is at (3, 2).

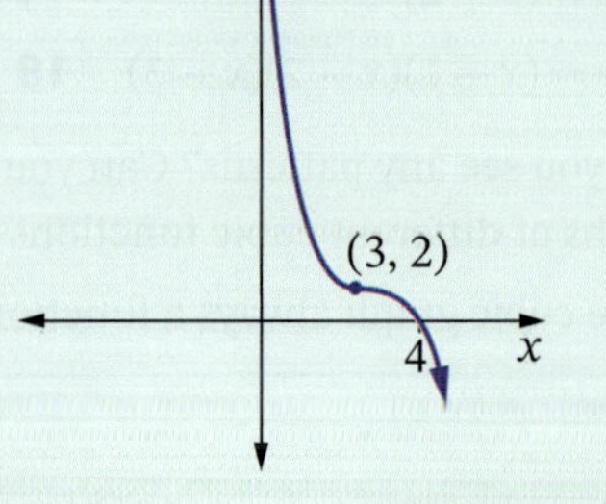

For x-intercept, $f(x) = 0$.

$0 = -2(x - 3)^3 + 2$

$-2 = -2(x - 3)^3$

$1 = (x - 3)^3$

$1 = x - 3$

$x = 4$

For y-intercept, $x = 0$.

$y = -2(0 - 3)^3 + 2$

$= -2(-27) + 2$

$= 56$

Example 36

Show that $y = 2x^3$ is an odd function.

Solution

Let $f(x) = 2x^3$.

$f(-x) = 2(-x)^3$

$= -2x^3$

$= -f(x)$

So $y = 2x^3$ is an odd function.

A cubic function has one y-intercept and up to 3 x-intercepts. We can sketch the graph of a more general cubic function using intercepts. This will not give a very accurate graph but it will show the shape and important features.

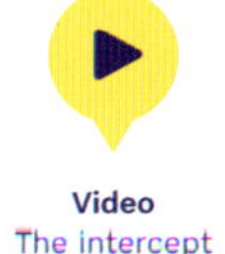

The graph of $y = k(x - a)(x - b)(x - c)$

The graph of $y = k(x - a)(x - b)(x - c)$ has x-intercepts at a, b and c.

Example 37

a **i** Sketch the graph of the cubic function $f(x) = x(x + 3)(x - 2)$.

ii Describe the shape of the graph and state its domain and range.

b Sketch the graph of the cubic function $f(x) = (x - 3)(x + 1)^2$ and describe its shape.

Solution

a **i** For x-intercepts, $f(x) = 0$.

$0 = x(x + 3)(x - 2)$

$x = 0, -3, 2$

Plot x-intercepts on graph.

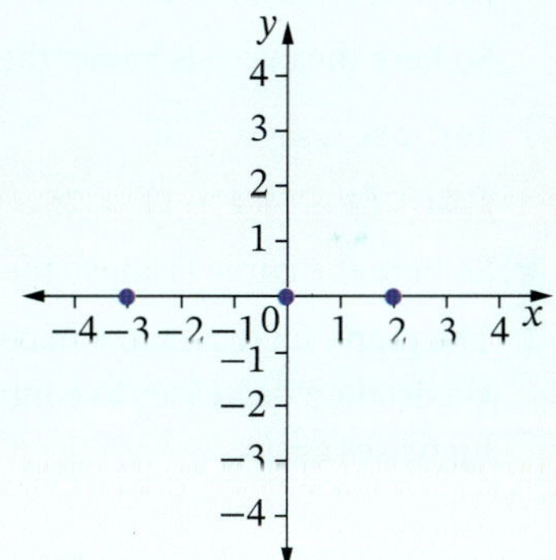

For y-intercept, $x = 0$.

$f(0) = 0(0 + 3)(0 - 2)$

$= 0$

So y-intercept is 0.

We look at which parts of the graph are above and which are below the x-axis between the x-intercepts.

Test $x < -3$, say $x = -4$:

$f(-4) = -4(-4 + 3)(-4 - 2) = -24 < 0$

So here the curve is below the x-axis.

Test $-3 < x < 0$, say $x = -1$:

$f(-1) = -1(-1 + 3)(-1 - 2) = 6 > 0$

So here the curve is above the x-axis.

Test $0 < x < 2$, say $x = 1$:

$f(1) = 1(1 + 3)(1 - 2) = -4 < 0$

So here the curve is below the x-axis.

Test $x > 2$, say $x = 3$:

$f(3) = 3(3 + 3)(3 - 2) = 18 > 0$

So here the curve is above the x-axis.

We can sketch the cubic curve as shown.

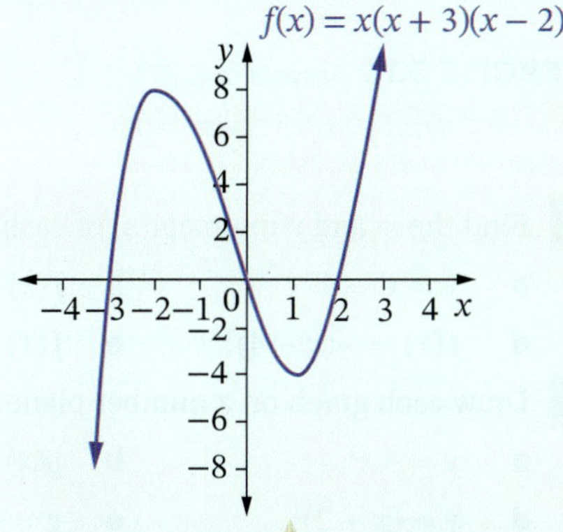

Notice that $k = 1 > 0$, so the curve is mostly increasing. If $k < 0$, then the curve is mostly decreasing.

ii The graph increases to a maximum turning point, then decreases to a minimum turning point. Then it increases again.

Domain: $(-\infty, \infty)$

Range: $(-\infty, \infty)$

b For x-intercepts, $f(x) = 0$.

$0 = (x - 3)(x + 1)^2$

$x = 3, x = -1$

So x-intercepts are -1 and 3.

For y-intercept, $x = 0$.

$f(0) = (0 - 3)(0 + 1)^2$
$= (-3)(1)$
$= -3$

So y-intercept is -3.

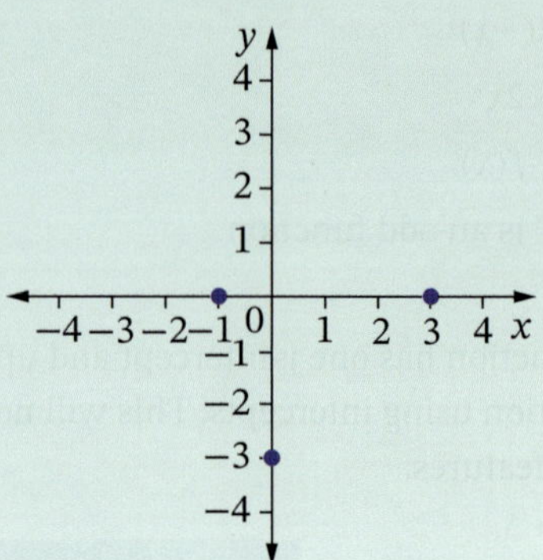

We look at which parts of the graph are above and below the x-axis.

Test $x < -1$, say $x = -2$:

$f(-2) = (-2 - 3)(-2 + 1)^2 = -5 < 0$

So here the curve is below the x-axis.

Test $-1 < x < 3$, say $x = 0$:

$f(0) = (0 - 3)(0 + 1)^2 = -3 < 0$

So here the curve is below the x-axis.

Test $x > 3$, say $x = 4$:

$f(4) = (4 - 3)(4 + 1)^2 = 25 > 0$

So here the curve is above the x-axis.

The graph increases to a maximum turning point at the x-axis, then decreases to a minimum turning point, then increases again.

Again, $k = 1 > 0$, so the curve is mostly increasing.

We can sketch the cubic curve as shown.

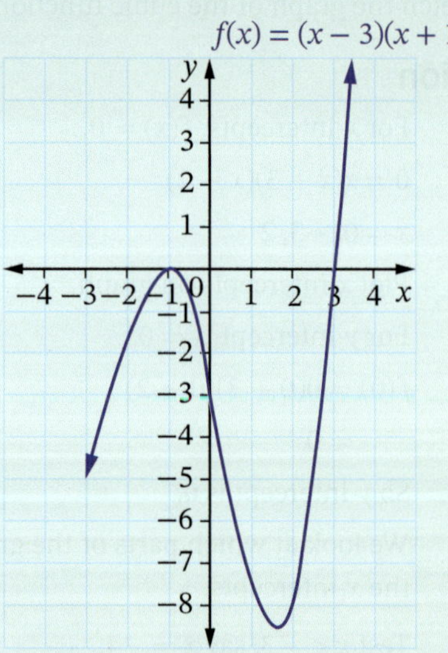

EXERCISE 3.13 Answers on p. 471

Cubic functions

1 Find the x- and y-intercept(s) of each cubic function.

a $y = x^3 - 1$ **b** $f(x) = -x^3 + 8$ **c** $y = (x + 5)^3$

d $f(x) = -(x - 4)^3$ **e** $f(x) = 3(x + 7)^3 - 3$ **f** $y = (x - 2)(x - 1)(x + 5)$

EXAMPLE 34

2 Draw each graph on a number plane.

a $y = -x^3$ **b** $p(x) = 2x^3$ **c** $g(x) = x^3 + 1$

d $y = (x + 2)^3$ **e** $y = -(x - 3)^3 + 1$ **f** $f(x) = -x(x + 2)(x - 4)$

g $y = (x + 2)(x - 3)(x + 6)$ **h** $y = x^2(x - 2)$ **i** $f(x) = (x - 1)(x + 3)^2$

Foundation Mastery Complex

3 Find the point of inflection of each cubic function by sketching its graph. EXAMPLE 35

a $y = 8x^3 + 1$ **b** $y = -x^3 + 27$ **c** $f(x) = (x + 2)^3$

d $y = 2(x - 1)^3 - 16$ **e** $f(x) = -(x + 1)^3 + 1$

4 Find the x-intercept of each cubic function correct to one decimal place.

a $y = 2x^3 - 5$ **b** $f(x) = (x - 1)^3 + 2$ **c** $f(x) = -3x^3 + 1$

d $y = 2(x + 3)^3 - 3$ **e** $y = -3(2x - 1)^3 + 2$

5 Show that $f(x) = -x^3$ is an odd function. EXAMPLE 36

6 Determine which of the following are odd functions.

a $y = 3x^3$ **b** $y = (x + 1)^3$ **c** $f(x) = -2x^3 - 1$

d $y = -5x^3$ **e** $y = (x - 2)^3 + 3$

7 Describe the shape of each cubic function. EXAMPLE 37

a $y = x^3 - 64$ **b** $f(x) = -(x - 3)^3$ **c** $y = x(x + 2)(x + 4)$

d $f(x) = -2(x + 3)(x + 1)(x - 4)$ **e** $y = x(x + 5)^2$

8 Solve each of the following graphically. Approximate your answer correct to one decimal place.

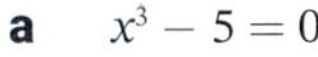

a $x^3 - 5 = 0$ **b** $x^3 + 2 = 0$ **c** $2x^3 - 9 = 0$

d $3x^3 + 4 = 0$ **e** $(x - 1)^3 + 6 = 0$ **f** $x(x + 2)(x - 1) = 0$

9 Which cubic function could describe this graph?
Select the correct answer **A**, **B**, **C** or **D**.

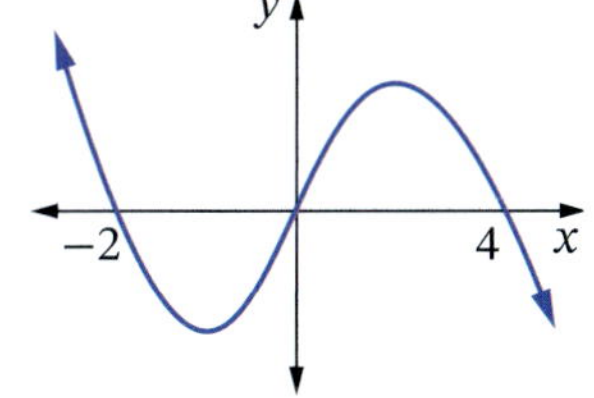

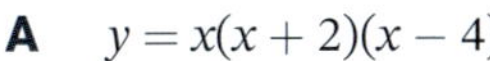

A $y = x(x + 2)(x - 4)$

B $y = -x(x + 2)(x - 4)$

C $y = x(x - 2)(x + 4)$

D $y = -x(x - 2)(x + 4)$

10 Which of the following is the graph of $y = (x - 2)(x - 5)(x + 3)$? Select **A**, **B**, **C** or **D**.

A

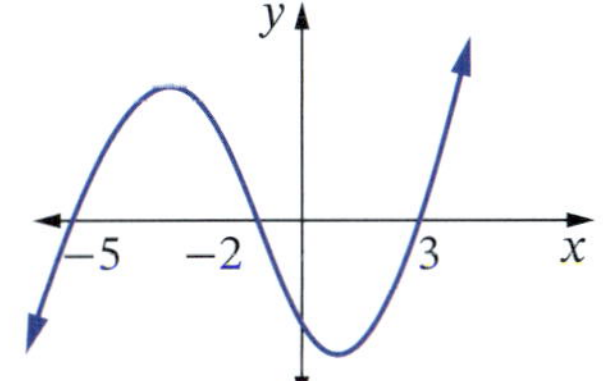

B

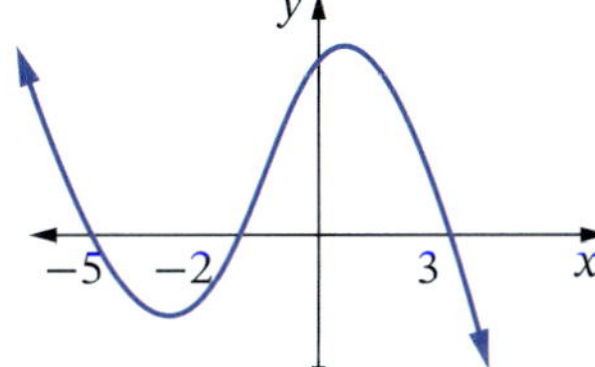

C

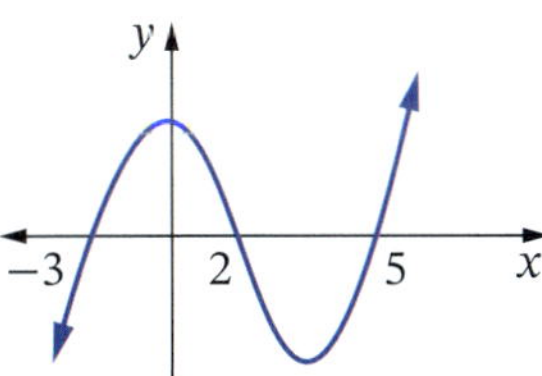

D

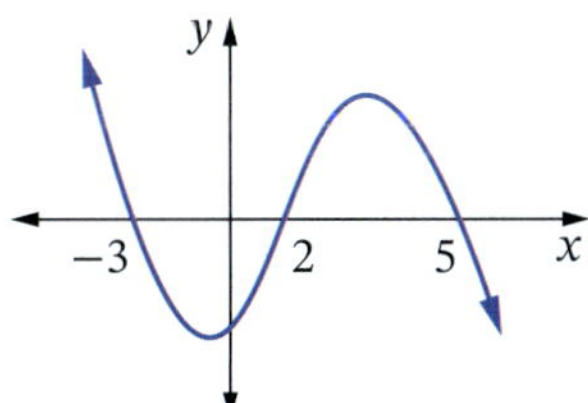

11 Graph $y = x^3$, $y = x^3 + 3$ and $y = x^3 - 6$ on paper or using technology, then describe the relationship between the graphs.

12 Graph $y = x^3$, $y = 3x^3$ and $y = \frac{x^3}{2}$ on paper or using technology, then describe the relationship between the graphs.

13 Graph $y = x^3$, $y = (x + 1)^3$ and $y = (x - 2)^3$ on paper or using technology, then describe the relationship between the graphs.

Foundation | Mastery | Complex

3.14 Applications of functions

Break-even points

Videos
Break-even points
Linear modelling
Applying linear functions
Applying quadratic functions

Example 38

Greenway company sells solar panels at \$850 each. Its costs are \$6500 per day and each panel costs \$200.

a Find the formula for the daily revenue $\$R$ (income) Greenway earns from selling x panels each day.

b Find the formula for the daily costs $\$C$ from selling x panels each day.

c Draw the graphs of both functions on the same number plane.

d Find the **break-even point** (where revenue and costs are equal).

e From the graphs, for what values of x will Greenway make

i a profit? **ii** a loss?

Solution

a The company earns \$850 for each solar panel so its revenue from selling x panels is given by

$R = 850x$.

b Each panel costs \$200 to make so selling x panels costs $200x$.

Also, fixed costs (called overheads) each day are \$6500.

So $C = 200x + 6500$.

c Both are linear functions.

First, graph $R = 850x$:

For x-intercept, $R = 0$.

$0 = 850x$

$0 = x$

For R-intercept, $x = 0$.

$R = 850(0)$

$= 0$

Find another point on the line for an accurate graph.

When $x = 20$:

$R = 850(20)$

$= 17\,000$

$(20, 17\,000)$

Next, graph $C = 200x + 6500$:

For x-intercept, $C = 0$.

$0 = 200x + 6500$

$-6500 = 200x$

$-32.5 = x$

The company costs are never negative (can you see why?).

So we can find another point on the graph.

When $x = 20$:

$C = 200(20) + 6500$

$= 10\,500$

$(20, 10\,500)$

For C-intercept, $x = 0$.

$C = 200(0) + 6500$

$= 6500$

Sketch both graphs using a scale that fits the information well.

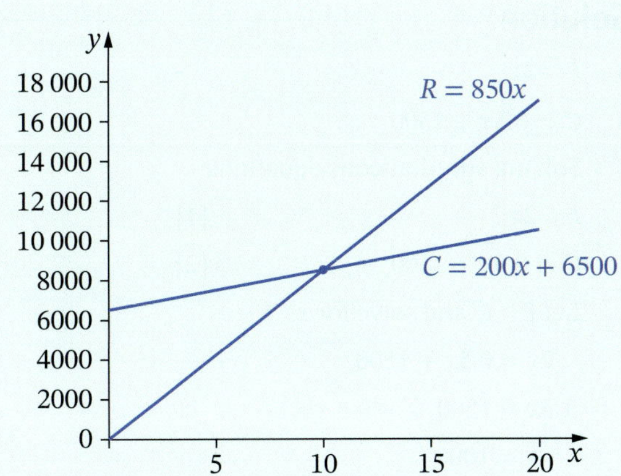

d For break-even, revenue = costs, that is, $R = C$.

Solve $R = 850x$ and $C = 200x + 6500$ simultaneously, either algebraically or graphically.

From the graph, the break-even point is where $x = 10$.

So, the company needs to sell 10 panels each day to break even.

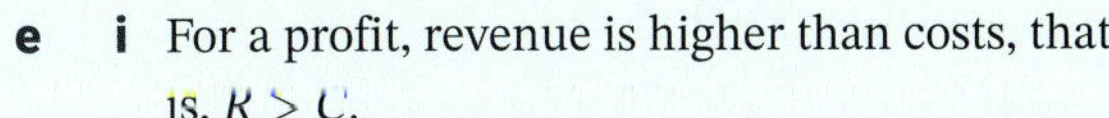

e **i** For a profit, revenue is higher than costs, that is, $R > C$.

Solve the inequality $850x > 200x + 6500$ or read the answer from the graph.

On the number plane, this means that the revenue line is higher than the cost line.

On the graph, this happens when $x > 10$.

The company makes a profit when it sells more than 10 panels each day.

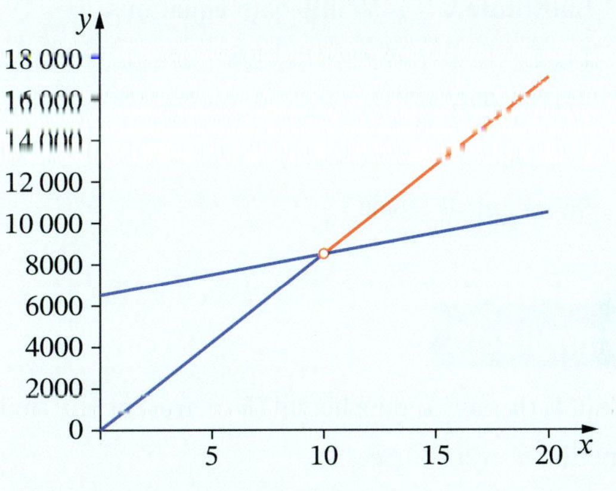

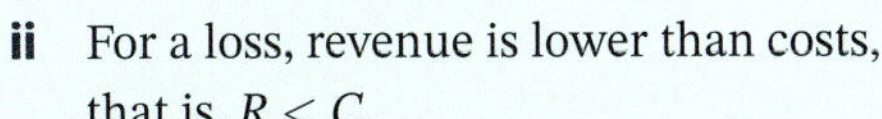

ii For a loss, revenue is lower than costs, that is, $R < C$.

This happens when $x < 10$.

So, the company makes a loss when it sells fewer than 10 panels a day.

Of course, once you know the break-even point, it is easy to work out when the company will make a profit or a loss.

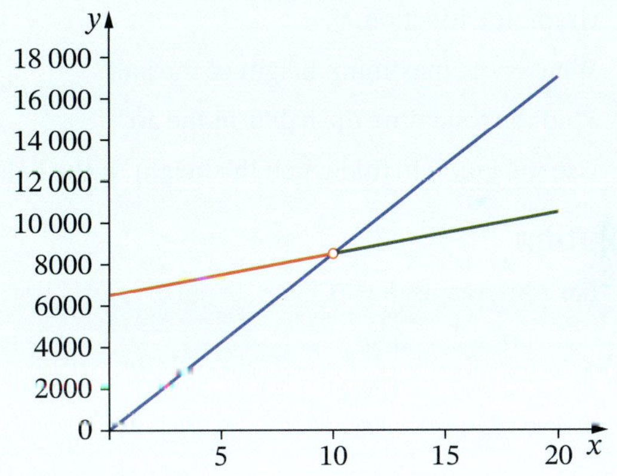

Break-even points

Profit: revenue > costs

Loss: revenue < costs

Break-even point: revenue = costs

Example 39

A company that manufactures cables sells them for $2 each. It costs 50 cents to produce each cable and the company has fixed costs of $1500 per week.

a Find the equation for the income, $*I*, on *x* cables per week.

b Find the equation for the costs, $*C*, of manufacturing *x* cables per week.

c Find the break-even point (where income = costs).

d Find the profit on 1450 cables.

Solution

a $I = 2x$

b $C = 0.5x + 1500$

c Solving simultaneous equations:

$I = 2x$ [1]

$C = 0.5x + 1500$ [2]

Let $I = C$ and solve for x.

$2x = 0.5x + 1500$

$1.5x = 1500$

$x = 1000$

1000 cables is where income = costs.

Substitute $x = 1000$ into [1] (or [2]):

$I = 2(1000)$

$= 2000$

So the break-even point is (1000, 2000).

1000 cables gives an income and cost of \$2000.

d profit = income − costs = $I - C$

Substitute $x = 1450$ into both equations.

$I = 2x$

$= 2(1450)$

$= 2900$

So income is \$2900.

$C = 0.5x + 1500$

$= 0.5(1450) + 1500$

$= 2225$

So costs are \$2225.

profit = \$2900 − \$2225

= \$675

Example 40

A ball is thrown and its height (h metres) at any time (t seconds) is given by the quadratic function $h = 2t - 0.5t^2$.

a Graph the function.

b What is the maximum height of the ball?

c Find the total time the ball is in the air.

d Use the graph to find when the height of the ball is 1.5 m.

Solution

a For t-intercepts, $h = 0$.

$0 = 2t - 0.5t^2$

$0 = t(2 - 0.5t)$

$t = 0$

$2 - 0.5t = 0$

$2 = 0.5t$

$t = 4$

For h-intercept, $t = 0$.

$h = 2(0) - 0.5(0)^2$

$= 0$

Axis of symmetry is $x = \frac{0+4}{2} = 2$.

OR

$x = -\frac{b}{2a} = -\frac{2}{2(-0.5)} = -\frac{2}{-1} = 2$.

When $t = 2$, $h = 2(2) - 0.5(2)^2 = 2$.

Vertex is at (2, 2).

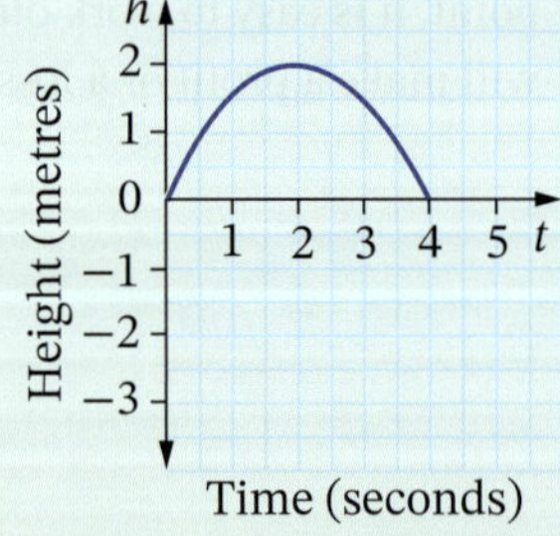

This is not the path of the ball. It is the graph of the height of the ball at different *times*, t seconds. We can ignore that part of the graph after $t = 4$ below the t-axis because height cannot be negative.

b Maximum height is 2 m. reading from vertex

c Ball is in the air from 0 to 4 seconds, so for 4 seconds.

d The height of the ball is 1.5 m at 1 s and 3 s. reading from graph

EXERCISE 3.14 Answers on p. 472

Applications of functions

1 In a game, each person starts with 20 points, then earns 15 points for every level completed.

a Write an equation for the number of points earned (P) for x levels completed.

b Find the number of points earned for completing:

i 24 levels **ii** 55 levels **iii** 247 levels

c Find the number of levels completed if the number of points earned is:

i 2195 **ii** 7700 **iii** 12 665

2 A TV manufacturing business has fixed costs of $1500 rental, $3000 wages and other costs of $2500 each week. It costs $250 to produce each TV.

a Write an equation for the cost (c) of producing n TVs each week.

b From the equation, find the cost of producing:

i 100 TVs **ii** 270 TVs **iii** 1200 TVs

c From the equation find the number of TVs produced if the cost is:

i $52 000 **ii** $78 250 **iii** $367 000

d If each TV sells for $950, find the number of TVs needed to sell to break even.

3 There are 450 litres of water in a pond, and 8 litres of water evaporate out of the pond every hour.

a Write an equation for the amount of water in the pond (A) after h hours.

b Find the amount of water in the pond after

i 3 hours **ii** a day.

c After how many hours will the pond be empty?

4 Geordie has a $20 credit to spend on his online music store account. He uses the credit to buy songs at $1.69 each.

a Write an equation for the amount of credit (C) left if Geordie buys x songs.

b How many songs can Geordie buy before his credit runs out?

5 Emily-Rose owes $20 000 and she pays back $320 a month.

a Write an equation for the amount of money she owes (A) after x months.

b How much does Emily-Rose owe after

i 5 months **ii** 1 year **iii** 5 years?

c How long will it take for Emily-Rose to pay all the money back?

6 Acme Party Supplies earns $5 for every helium balloon it sells.

a If overhead costs are $100 each day, find an equation for the profit (P) of selling x balloons.

b How much profit does Acme make in a day if it sells 300 balloons today?

c How many balloons does it sell if it makes a profit of $1055?

d What is the break-even point for this business?

7 Find the number of calculators that a company needs to sell to break even each week if it costs $3 to make each calculator and they are sold for $15 each. Fixed overheads are $852 a week.

8 Cupcakes Online sells cupcakes at $5 each. The cost of making each cupcake is $1 and the company has fixed overheads of $264 a day.

a Find the equations for daily income and costs.

b Find how many cupcakes the company needs to sell daily to break even.

c What is the profit on 250 cupcakes?

d What is the loss on 50 cupcakes?

9 The braking distance of a car travelling at 100 km/h is 40 metres. The formula for braking distance (d) in metres is $d = kx^2$ where k, is a constant and x is the speed in km/h.

- **a** Find the value of k.
- **b** Find the braking distance at 80 km/h.
- **c** A dog runs out onto the road 15 m in front of a car travelling at 50 km/h. Will the car be able to stop in time without hitting the dog?
- **d** If the dog is 40 m in front of a car travelling at 110 km/h, will the car stop in time?

10 A rectangle has sides x and $3 - x$.

- **a** Write an equation for its area.
- **b** Draw the graph of the area in terms of x.
- **c** Find the value of x that gives the maximum area.
- **d** Find the maximum area of the rectangle.

11 A company that manufactures nails sells them for \$2 per packet. It costs 50 cents to produce each packet, and the company has fixed costs of \$1500 per week.

- **a** Find the equation for the income \$$I$ on x packets per week.
- **b** Find the equation for weekly cost \$$C$ of manufacturing x packets.
- **c** Find the break-even point.
- **d** Find the profit on 1450 packets of nails.

12 Mila makes toys and sells them for \$15 each at a market each day. The materials for each toy cost her \$3 and the fixed costs to sell at the market are \$120 per day.

- **a** Write equations for Mila's revenue r and costs c for selling n toys.
- **b** Graph both functions on a number plane.
- **c** How many toys would Mila need to sell to make a profit?
- **d** What is the break-even point?
- **e** Mila sells 21 toys on Wednesday and 25 toys on Thursday. What is her combined profit for these 2 days?

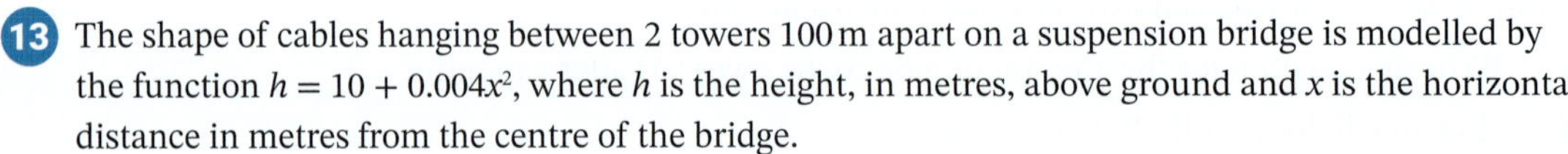

EXAMPLE 40

13 The shape of cables hanging between 2 towers 100 m apart on a suspension bridge is modelled by the function $h = 10 + 0.004x^2$, where h is the height, in metres, above ground and x is the horizontal distance in metres from the centre of the bridge.

- **a** Graph this function for the domain $[-50, 50]$.
- **b** How tall are the towers at either end of the cables?
- **c** What is the height of the cables at their lowest point?
- **d** How far from the centre are the cables when the height is 15 m? Give your answer to the nearest metre.

14 A ball is thrown upwards and its height (y metres) after x seconds is given by the equation $y = 8 - 0.2(x - 2)^2$.

- **a** What is the y-intercept and what does it represent?
- **b** What is the maximum height of the ball and when does it occur?
- **c** When is the ball at a height of 6 metres? Approximate your answer to one decimal place.
- **d** Find the height of the ball after
  - **i** 4.5 seconds
  - **ii** 2 seconds.
- **e** How far does the ball travel in the
  - **i** 4th second?
  - **ii** 7th second?
- **f** After how many seconds did the ball hit the ground? Approximate your answer to one decimal place.

Foundation | Mastery | Complex

15 The path of a golf ball being hit off the tee is given by the quadratic equation $h = -0.004(x - 100)^2 + 40$, where h metres is the height of the ball and x metres is the horizontal distance the ball has travelled.

- **a** What is the h-intercept and what does it represent?
- **b** What is the maximum height of the ball?
- **c** How far from the tee does the ball hit the ground?
- **d** At what distance, rounded to the nearest metre, does the golf ball reach a height of
  - **i** 25 m
  - **ii** 8 m?

Sample HSC problem

Answers on p. 473

(8 marks) A bridge has a parabolic span as shown, with equation $d = -\dfrac{w^2}{800} + 200$

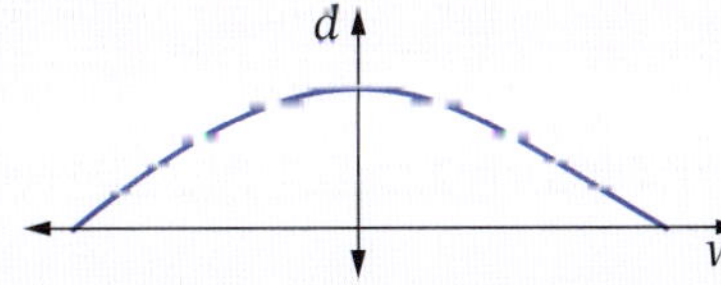

where d is the depth of the arch in metres and w is the width of the span in metres.

- **a** Show that the quadratic function is even. 2 marks
- **b** Find the depth of the arch from the top of the span. 2 marks
- **c** Find the total width of the span. 2 marks
- **d** Find the depth of the arch at a point 10 m from its widest span. 1 mark
- **e** Find the width across the span at a depth of 100 m. 1 mark

Foundation Mastery Complex

CHAPTER SUMMARY

This chapter, *Functions and graphs*, introduced the concept of a function, function notation, domain and range, interval notation, and even and odd functions. You revised and extended your knowledge of linear, quadratic and cubic functions; their equations, graphs, properties and applications. You also looked at graphing techniques, solving equations graphically including simultaneous equations, and solving inequalities. When revising the quadratic function, make sure that you master the theory behind the formulas for the vertex of a parabola, the discriminant and quadratic identities.

Before you move on, consider what you learned in this chapter and revisit any sections that may have been unclear.

What you learned in this chapter…	Section	
Understand the definition of a function and test a function using the vertical line test	3.01	Functions and relations
Use function notation	3.02	Function notation and piecewise functions
Identify and use piecewise functions	3.02	Function notation and piecewise functions
Find the domain and range of functions, including using interval notation	3.03	Domain, range, even and odd functions
Identify and prove even and odd functions	3.03	Domain, range, even and odd functions
Identify and use linear, quadratic and cubic functions, their graphs and properties	3.04 3.08 3.13	Linear functions Quadratic functions Cubic functions
Find the gradient and angle of inclination of a line	3.05 3.06	The gradient of a line The point-gradient formula
Find the equation of a line, including parallel and perpendicular lines, using the gradient-intercept, general and point-gradient forms on the linear equation	3.07	Parallel and perpendicular lines
Solve simultaneous linear equations graphically	3.04	Linear functions
Solve quadratic inequalities	3.09	Quadratic inequalities
Find coefficients in identical quadratic expressions	3.11	Quadratic identities
Use the discriminant to classify and number the roots of a quadratic equation	3.12	The discriminant
Find the equation of a parabola, including its axis of symmetry, vertex and intercepts	3.10 3.12	Axis of symmetry and vertex The discriminant
Solve simultaneous equations involving linear and quadratic equations graphically	3.11	Quadratic identities
Apply linear and quadratic functions, including break-even problems involving cost and revenue	3.14	Applications of functions

To help master these techniques, make a summary of this topic. Use the chapter outline and the mind map below as a guide. Add your own words, symbols, diagrams, boxes and reminders. The summary should give you a 'whole picture' view of the topic and allow you to identify any weak areas to revisit in your revision.

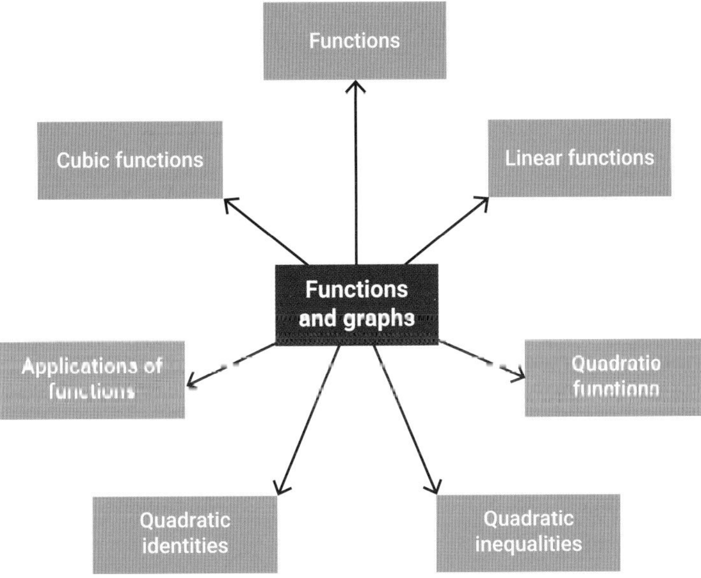

3 Test yourself

Answers on p. 473

For Questions **1** to **5**, select the correct answer **A**, **B**, **C** or **D**.

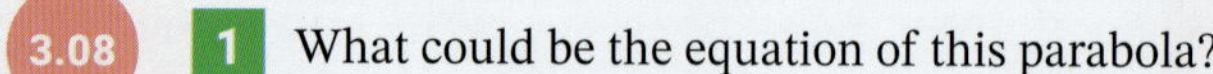

3.08 **1** What could be the equation of this parabola?

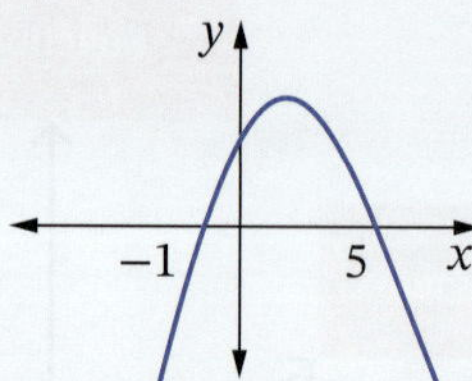

A $y = (x - 1)(x + 5)$

B $y = -(x - 1)(x + 5)$

C $y = -(x + 1)(x - 5)$

D $y = (x - 1)(x + 5)$

3.10 **2** The axis of symmetry and vertex of the quadratic function $f(x) = 1 + 2x - x^2$ are, respectively:

A $x = 1, (1, 2)$ **B** $x = -1, (-1, 4)$ **C** $x = 2, (2, 5)$ **D** $x = -2, (-2, 5)$

3.04 **3** The linear function $2x - 3y - 6 = 0$ has x- and y-intercepts, respectively:

A -3 and 2 **B** 3 and -2 **C** -3 and -2 **D** 3 and 2

3.03 **4** The domain and range of the straight line with equation $x = -2$ are:

A domain: $(-\infty, \infty)$, range: $[-2, -2]$ **B** domain: $[-2, -2]$, range: $(-\infty, \infty)$

C domain: $(-\infty, \infty)$, range: $(-\infty, \infty)$ **D** domain: $[-2, -2]$, range: $[-2, -2]$

3.13 **5** Which cubic function has this graph?

A $y = x(x + 1)(x - 2)$

B $y = -x(x - 1)(x + 2)$

C $y = x(x - 1)(x + 2)$

D $y = -x(x + 1)(x - 2)$

3.02 **6** If $f(x) = x^2 - 3x - 4$, find:

a $f(-2)$ **b** $f(a)$ **c** the zeroes of $f(x)$.

3.03 **7** Sketch each graph and find its domain and range.

a $y = x^2 - 3x - 4$ **b** $f(x) = x^3$ **c** $2x - 5y + 10 = 0$

d $x = 3$ **e** $y = (x + 1)^3$ **f** $y = -2$

g $f(x) = -x^2 + x$ **h** $f(x) = x^2 + 4x + 4$

3.02 **8** If $f(x) = 3x - 4$, find:

a $f(2)$ **b** x when $f(x) = 7$ **c** x when $f(x) = 0$.

3.09 **9** Solve each inequality.

a $x^2 - 3x \le 0$ **b** $n^2 - 9 > 0$ **c** $4 - y^2 \ge 0$

3.05, 3.07 **10** Find the gradient of the straight line:

a passing through $(3, -1)$ and $(-2, 5)$ **b** with equation $2x - y + 1 = 0$

c perpendicular to the line $5x + 3y - 8 = 0$ **d** making an angle of inclination of 45°.

3.10 **11** For the parabola $y = x^2 - 4x + 1$, find:

a the equation of the axis of symmetry **b** the minimum value.

3.13 **12** Sketch the graph of $f(x) = (x - 2)(x + 3)^2$, showing the intercepts.

3.13 **13** A function has equation $f(x) = x^3 - x^2 - 4x + 4$.

a Solve $f(x) = 0$.

b Find its x- and y-intercepts.

c Sketch the graph of the function.

d From the graph, state how many solutions there are for:

i $f(x) = 1$ **ii** $f(x) = -5$

Foundation Mastery Complex

14 Find the x- and y-intercepts of: 3.03

a $2x - 5y + 20 = 0$ **b** $y = x^2 - 5x - 14$

c $y = (x + 2)^3$ **d** $2x - 5y - 10 = 0$

15 Find the point of intersection of the lines $y = 2x + 3$ and $x - 5y + 6 = 0$. 3.04

16 For the quadratic function $y = -2x^2 - x + 6$, find: 3.11

a the equation of the axis of symmetry

b the maximum value.

17 Find the domain and range of $y = -2x^2 - x + 6$. 3.03

18 For each quadratic equation, select the correct property of its roots **A**, **B**, **C** or **D**. 3.12

a $2x^2 - x + 3 = 0$ **b** $x^2 - 10x - 25 = 0$ **c** $x^2 - 10x + 25 = 0$

d $3x^2 + 7x - 2 = 0$ **e** $6x^2 - x - 2 = 0$

A real, distinct and rational **B** real, distinct and irrational

C equal **D** no real roots

19 Find the equation of the line 3.07

a passing through (2, 3) and with gradient 7

b parallel to the line $5x + y - 3 = 0$ and passing through (1, 1)

c through the origin, and perpendicular to the line $2x - 3y + 6 = 0$

d through (3, 1) and (−2, 4)

e with x-intercept 3 and y-intercept −1.

20 The function $f(x) = ax^2 + bx + c$ has zeroes 4 and 5, and $f(-1) = 60$. Evaluate a, b and c. 3.11

21 Determine whether each function is even, odd or neither. 3.03

a $y = x^2 - 1$ **b** $y = x + 1$ **c** $y = x^3$

d $y = (x + 1)^2$ **e** $y = -5x^3$

22 Show that $f(x) = x^3 - x$ is odd. 3.03

23 Prove that the line between (−1, 4) and (3, 3) is perpendicular to the line $4x - y - 6 = 0$. 3.07

24 Show that $-4 + 3x - x^2 < 0$ for all x. 3.12

25 A large aquarium holds 500 L of water. It starts out full but has a leak and loses 50 L every hour. The volume V of water in the aquarium is modelled by $V = 500 - 50t$, where t is time in hours. Another aquarium is initially empty but after 3 hours it is filled by a hose that adds water at a constant rate of 100 L per hour. Its formula is given by $V = 100(t - 3)$. 3.14

a Draw both graphs on the same axes.

b From the graph, estimate when both aquariums will have the same volume of water.

c From the graph, estimate when the total volume of water in both aquariums will be 600 L.

d Use algebraic methods to find

i when both aquariums will have the same volume of water

ii what the volume of water will be when both aquariums have the same volume

iii when the total volume will be 600 L.

e The second aquarium holds 750 L of water. When will it be full?

3.07 **26** Prove that the lines with equations $y = 5x - 7$ and $10x - 2y + 1 = 0$ are parallel.

3.08 **27** Find the zeroes of $g(x) = -x^2 + 9x - 20$.

3.13 **28** Sketch the graph of $P(x) = 2x(x - 3)(x + 5)$, showing intercepts.

3.13 **29** Solve $P(x) = 0$ when $P(x) = x^3 - 4x^2 + 4x$.

3.05 **30** Find x if the gradient of the line through $(3, -4)$ and $(x, 2)$ is -5.

3.02 **31** If $f(x) = \begin{cases} 2x & \text{if } x \geq 1 \\ x^2 - 3 & \text{if } x < 1 \end{cases}$, find $f(5) - f(0) + f(1)$.

3.02 **32** Given $f(x) = \begin{cases} 3 & \text{if } x > 3 \\ x^2 & \text{if } 1 \leq x \leq 3 \\ 2 - x & \text{if } x < 1 \end{cases}$

find:

a $f(2)$ **b** $f(-3)$ **c** $f(3)$ **d** $f(5)$ **e** $f(0)$

3.12 **33** Find the equation of the parabola:

a that passes through the points $(-2, 18)$, $(3, -2)$ and $(1, 0)$

b with x-intercepts 3 and -2 and y-intercept 12.

3.12 **34** Find the value of k, l and m if $(x - 2)^2 + 3x - 4 = kx^2 + lx + m$.

3.01 **35** For each graph and set of ordered pairs, state whether it represents a function.

a

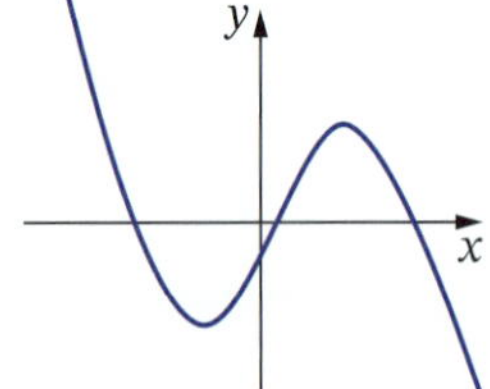

b

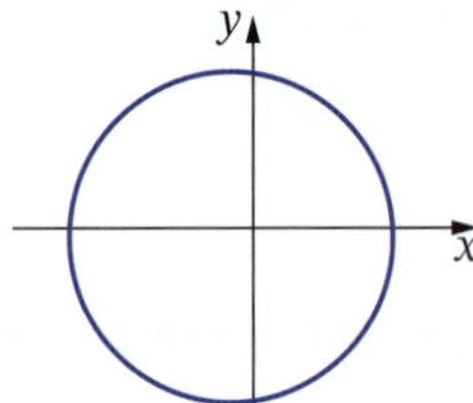

c

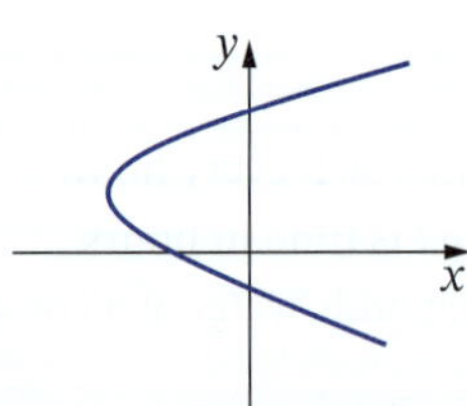

d

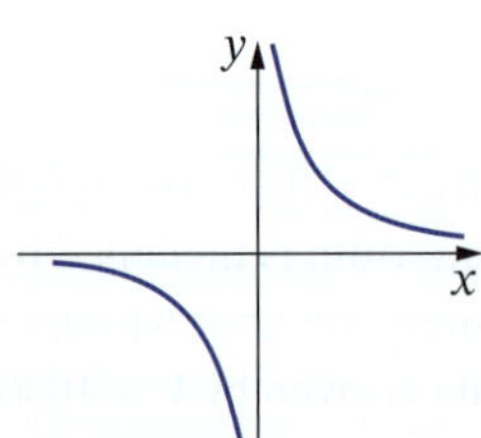

e $(1, 2), (2, 5), (-1, 4), (1, 3), (3, 4)$

3.13 **36** Find the equation of a cubic function $f(x) = kx^3 + c$ if it passes through the point $(1, 2)$ and has y-intercept 5.

3.14 **37** A company has costs given by $y = 7x + 15$ and income $y = 12x$. Find the break-even point.

3.07 **38** **a** Find the equation of the straight line that is perpendicular to the line $y = \frac{1}{2}x - 3$ and passes through $(1, -1)$.

b Find the x-intercept of this line.

Foundation Mastery Complex

39 Find values of m such that $mx^2 + 3x - 4 < 0$ for all x. 3.11

40 Find any points of intersection of the graphs of:

a $y = 3x - 4$ and $y = 1 - 2x$

b $y = x^2 - x$ and $y = 2x - 2$

c $y = x^2$ and $y = 2x^2 - 9$.

41 Solve each inequality.

a $x^2 \geq x + 2$ **b** $x^2 + 2x < 3$

42 The cost of producing n wooden toys made in a factory is given by $y = 5n + 300$ and the income on the toys is given by $y = 20n$. 3.14

a Graph both linear functions on the same axes, using a scale of 0, 5, 10, 15, 20, ... on the n-axis and a scale of 0, 100, 200, 300, 400, ... on the y-axis.

b Find the point of intersection of the 2 functions graphically.

c Find the point of intersection of the 2 functions algebraically.

d How many wooden toys must be produced in order for the factory to break even?

43 A park contains 5 tonnes of contaminated soil.

a If 200 kg of soil are removed each day, how many days will it take to clear all the contaminated soil?

b Find a formula for the mass m kg of the contaminated soil in the park after t days.

c Starting on the 9th day of the contaminated soil removal, 500 kg of clean soil is brought into a storage shed every day to replace it. Find a formula for the mass of the clean soil after t days.

d Draw the graph of both formulas and use it to find when the mass of contaminated soil is the same as the mass of clean soil.

e Use algebra to find on which day the mass of contaminated and clean soils are equal.

f Find the day when the total mass of soil is 7 tonnes.

g The park needs 10 tonnes of clean soil. How many days will it take to bring this amount of soil in?

44 A parabolic arch is designed with equation $h = 50 - 2x^2$ where all dimensions are in metres. 3.14

a Find the height of the arch.

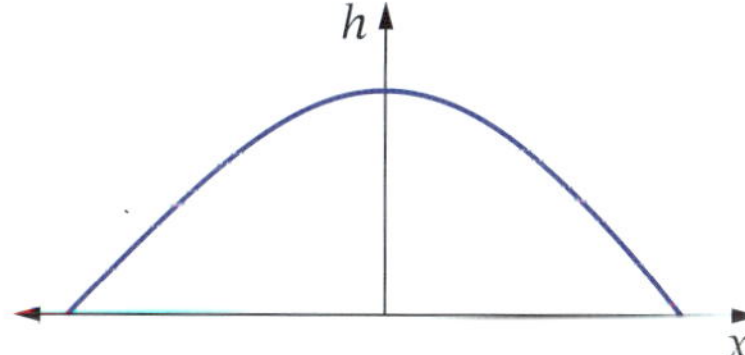

b Find the width of the arch.

3 Challenge exercise

Answers on p. 474

1. Find the values of b if $f(x) = 3x^2 - 7x + 1$ and $f(b) = 7$.
2. Sketch the graph of $y = (x + 2)^2 - 1$ in the domain $[-3, 0]$.
3. If points $(-3k, 1)$, $(k - 1, k - 3)$ and $(k - 4, k - 5)$ are collinear (lie on the same line), find the value of k.
4. Find the equation of the line that passes through the point of intersection of the lines $2x + 5y + 19 = 0$ and $4x - 3y - 1 = 0$ and is perpendicular to the line $3x - 2y + 1 = 0$.
5. If $ax - y - 2 = 0$ and $bx - 5y + 11 = 0$ intersect at the point $(3, 4)$, find the values of a and b.
6. By writing each as a quadratic equation, solve:
   - **a** $(3x - 2)^2 - 2(3x - 2) - 3 = 0$
   - **b** $5^{2x} - 26(5^x) + 25 = 0$
   - **c** $2^{2x} - 10(2^x) + 16 = 0$
   - **d** $2^{2x+1} - 5(2^x) + 2 = 0$
   - **e** $\left(x+\dfrac{1}{x}\right)^2 - 5\left(x+\dfrac{1}{x}\right) + 6 = 0$
7. Find the equation of the straight line through $(1, 3)$ that passes through the intersection of the lines $2x - y + 5 = 0$ and $x + 2y - 5 = 0$.
8. $f(x) = \begin{cases} 2x+3 & \text{when } x > 2 \\ 1 & \text{when } -2 \le x \le 2 \\ x^2 & \text{when } x < -2 \end{cases}$

   Find $f(3)$, $f(-4)$, $f(0)$ and sketch the graph of the function.
9. If $h(t) = \begin{cases} 1-t^2 & \text{if } t > 1 \\ t^2 - 1 & \text{if } t \le 1 \end{cases}$, find the value of $h(2) + h(-1) - h(0)$ and sketch the curve.
10. If $f(x) = 2x^3 - 2x^2 - 12x$, find x when $f(x) = 0$.
11. Show that the quadratic equation $2x^2 - kx + k - 2 = 0$ has real rational roots.
12. Find the values of p for which $x^2 - x + 3p - 2 > 0$ for all x.
13. Find the equation of the straight line through $(3, -4)$ that is perpendicular to the line with x-intercept -2 and y-intercept 5.
14. Find any points of intersection between $y = x^2$ and $y = x^3$.
15. Find the equation of a cubic function $y = ax^3 + bx^2 + cx + d$ if it passes through $(0, 1)$, $(1, 3)$, $(-1, 3)$ and $(2, 15)$.
16. Show that the quadratic equation $x^2 - 2px + p^2 = 0$ has equal roots.
17. Solve $x^2 + 1 + \dfrac{25}{x^2 + 1} = 10$.
18. Find exact values of k for which $x^2 + 2kx + k + 5 = 0$ has real roots.

Foundation Mastery Complex

19 The shape of the Sydney Harbour Bridge is based on 2 concave-down parabolas as shown.

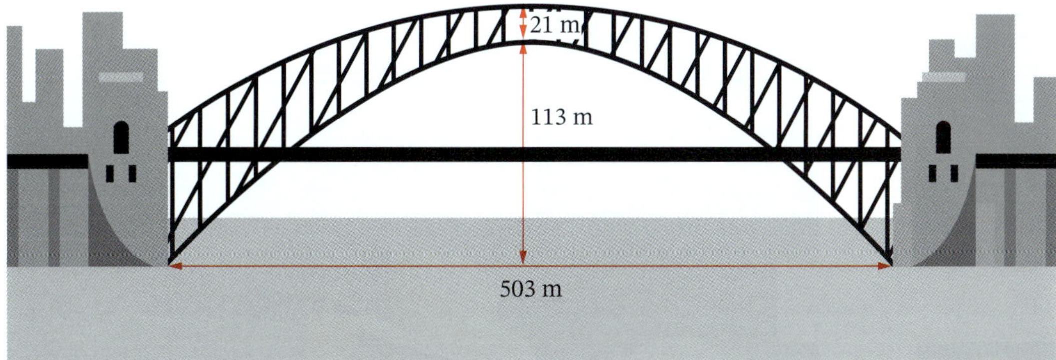

a By placing the x-axis across the base of the pylons and the y-axis at the centre of the bridge, show that the equation of the lower arch is $h = -0.00179x^2 + 113$ and the equation of the upper arch is $h = -0.00212x^2 + 134$.

b A crane on a boat is set up 100 m from the centre of the bridge to hang a new sign from the upper arch. How high up must it reach?

c How far out from a pylon is a boat if the height from the boat up to the lower arch is 48 m?

d Sam is doing the bridge climb and walks along the upper arch until he is 125 m above the water. How far out from the pylon is he?

20 Vanna is paid $35 an hour to drive a truck between Sydney and Melbourne. The cost of petrol and other expenses per hour is $6000 + v^2$ cents, where v is the average speed in km/h.

a Find the equation for the total cost per hour of driving the truck.

b Find the equation for c, the total cost, C, if the trip from Sydney to Melbourne is 970 km at v km/h.

c If Vanna's average speed is 70 km/h, find the total cost for the trip.

d Vanna drives a long way back to Sydney through country Victoria and her average speed is 56 km/h. If the return trip is 1248 km, what is the total cost of the return trip from Sydney to Melbourne and back?

□ Foundation ○ Mastery ○ Complex

TRIGONOMETRY

Trigonometry is used in many fields, such as building, surveying and navigating. It is the geometry and measurement of triangles.

This chapter covers the trigonometry of right-angled and non-right-angled triangles, and applies it to problems and real-life situations, including the use of angles of elevation and depression, and bearings. This chapter also introduces trigonometry with angles greater than 90° and radians, an alternative to degrees for measuring the size of an angle. We will apply radians to circle measurement by finding the length of an arc and the area of a sector and segment.

Chapter outline

* REVISION

In this chapter you will:

- identify the trigonometric ratios
- solve right-angled triangle problems
- apply trigonometry to angles of elevation and depression, and bearings
- evaluate trigonometric ratios for angles of any magnitude, including the exact ratios and multiples of 90°
- understand and apply the sine and cosine rules
- find the area of a triangle given the length of two sides and the size of their included angle
- understand radians and convert between degrees and radians
- find the length of an arc and area of a sector and segment of a circle

Videos (17):

4.01 Finding an unknown side • Finding an unknown angle

4.02 Angles of elevation and depression • Angle of depression 1 • Angle of depression 2 • Compass bearings • Bearings • True bearings

4.03 Exact trigonometric ratios • Negative angles

4.05 The cosine rule for angles

4.06 The sine area formula

4.07 The sine and cosine rules

4.08 Degrees and radians • Exact trigonometric values • The exact trigonometric ratios

4.09 Arc length

Worksheets (16):

4.01 Trigonometric ratios 2 • Trigonometry calculations

4.02 Right-angled trigonometry • A page of bearings • Elevations and bearings

4.03 Angles of any magnitude • Exact ratios

4.04 Sine rule problems

4.05 Cosine rule problems • Finding an unknown side • Finding an unknown angle

4.06 Areas of triangles

4.07 The sine and cosine rules

4.08 Radians • Converting degrees and radians • Radians of any magnitude

Puzzles (3):

4.02 Bearings match-up

4.08 Exact values • Exact values 2

Nelson MindTap

To access resources above, visit **cengage.com.au/nelsonmindtap**

Terminology

acute angle	ambiguous case	angle of depression	angle of elevation
arc	bearing	compass bearing	cosine (cos)
cosine rule	exact ratio	included angle	major segment
minor segment	obtuse angle	quadrant	radian
reflex angle	sector	segment	sine (sin)
sine rule	tangent (tan)	true bearing	unit circle

Right-angled trigonometry 4.01

Worksheets
Trigonometric ratios 2
Trigonometry calculations

The sides of a right-angled triangle

- The **hypotenuse** is the longest side, and is always opposite the right angle.
- The **opposite** side is opposite the angle marked in the triangle.
- The **adjacent** side is next to the angle marked.

The opposite and adjacent sides vary according to where the angle is marked. For example:

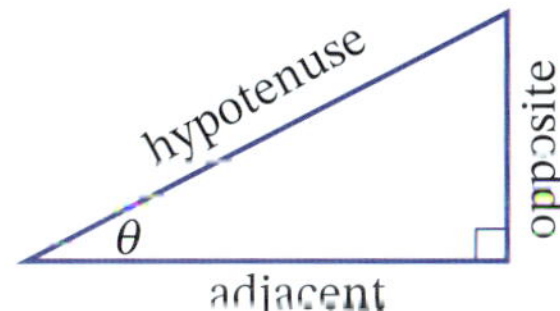

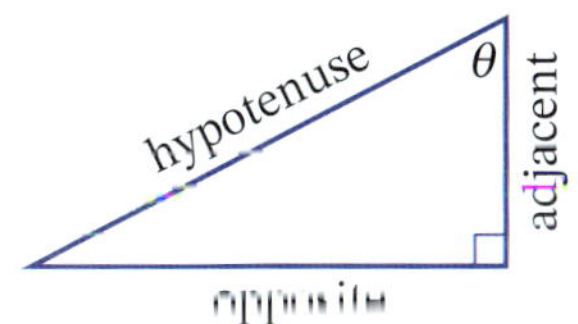

The trigonometric ratios

Sine $\sin\theta = \dfrac{\text{opposite}}{\text{hypotenuse}}$

Cosine $\cos\theta = \dfrac{\text{adjacent}}{\text{hypotenuse}}$

Tangent $\tan\theta = \dfrac{\text{opposite}}{\text{adjacent}}$

Did you know?

The origins of trigonometry

Trigonometry, or **triangle measurement**, progressed from the study of geometry in ancient Greece. Trigonometry was seen as applied mathematics. It gave a tool for the measurement of planets and their motion. It was also used extensively in navigation, surveying and mapping, and it is still used in these fields today.

Trigonometry was crucial in setting up an accurate calendar, since this involved measuring the distances between Earth, the Sun and the Moon.

saiko3p/Shutterstock.com

Astronomical instruments at Jantar Mantar observatory, Jaipur, India

Degrees, minutes, seconds

Angles are measured in degrees, minutes and seconds.

> **Degrees, minutes, seconds**
>
> 60 minutes = 1 degree ($60' = 1°$)
>
> 60 seconds = 1 minute ($60'' = 1'$)

With angles, you round up to the next degree if there are 30 minutes or more.

Similarly, round angles up to the nearest minute if there are 30 seconds or more.

Example 1

a Round to the nearest degree:

i $54°17'45''$ **ii** $29°32'52''$

b Round to the nearest minute:

i $23°12'22''$ **ii** $84°19'30''$

Solution

a **i** $17'$ is less than $30'$ so rounding gives $54°$.

ii $32'$ is more than $30'$ so rounding gives $30°$.

b **i** $22''$ is less than $30''$ so rounding gives $23°12'$.

ii $30''$ is exactly halfway so round up to $84°20'$.

Decimal degrees and degrees-minutes-seconds

Scientific calculators have a [° ′ ″] or [D°M′S] key for converting between decimal degrees and degrees, minutes, seconds.

Example 2

Change 45.236° into degrees and minutes.

Solution

Operation	Casio scientific	Sharp scientific
Enter data.	45.236 [=]	45.236 [=]
Change to degrees and minutes.	[° ′ ″]	[2nd F] [D°M′S]

So $45.236° = 45°14'9.6'' \approx 45°14'$.

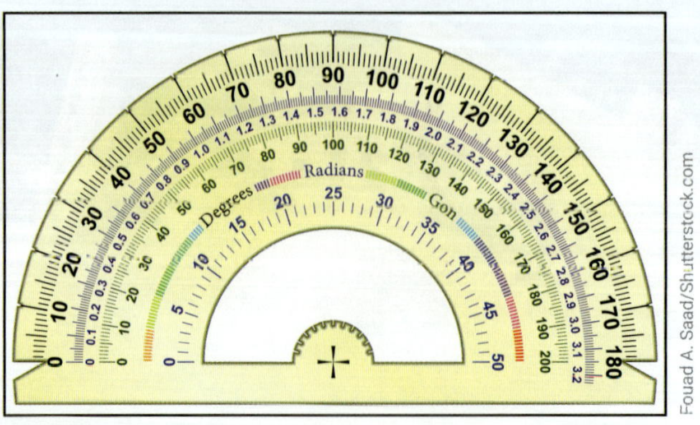

Fouad A. Saad/Shutterstock.com

Example 3

a Find $\cos 58°19'$, correct to 3 decimal places.

b If $\tan\theta = 0.348$, find θ in degrees and minutes.

Solution

a

Operation	Casio scientific	Sharp scientific
Enter data.	cos 58 ° ' " 19 ° ' " =	cos 58 D°M'S 19 D°M'S =

So $\cos 58°19' = 0.52522\ldots \approx 0.525$.

b To find the angle given the ratio, use the inverse key ($\tan^{-1}$).

Operation	Casio scientific	Sharp scientific
Enter data.	SHIFT tan 0.348 =	2nd F tan 0.348 =
Change to degrees and minutes.	° ' "	2nd F D°M'S

So $\theta = 19°11'16.43'' \approx 19°11'$

Example 4

a Find the value of x, correct to one decimal place.

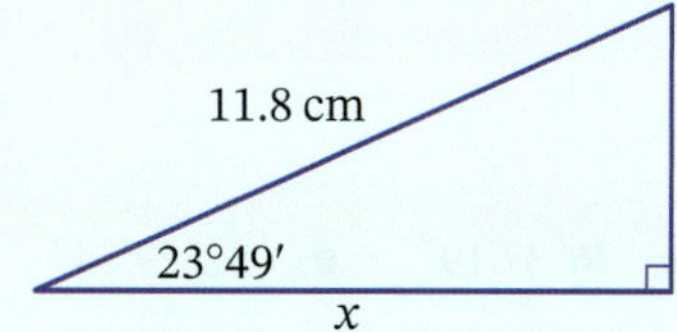

b Find the value of y, correct to 3 significant figures.

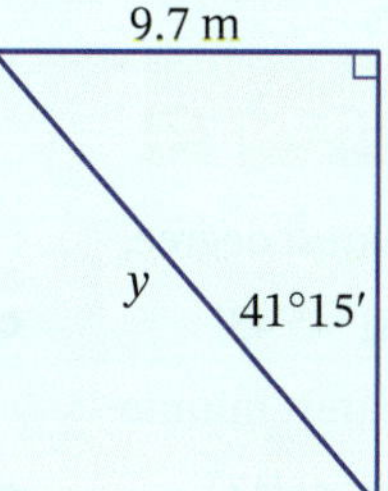

Video
Finding an unknown side

Solution

a

$$\cos\theta = \frac{\text{adjacent}}{\text{hypotenuse}}$$

$$\cos 23°49' = \frac{x}{11.8}$$

$$11.8\cos 23°49' = x$$

$$x \approx 10.8\text{ cm}$$

b

$$\sin\theta = \frac{\text{opposite}}{\text{hypotenuse}}$$

$$\sin 41°15' = \frac{9.7}{y}$$

$$y\sin 41°15' = 9.7$$

$$y = \frac{9.7}{\sin 41°15'}$$

$$\approx 14.7\text{ m}$$

Video
Finding an unknown angle

Example 5

Find the size of each unknown angle, in degrees and minutes.

a

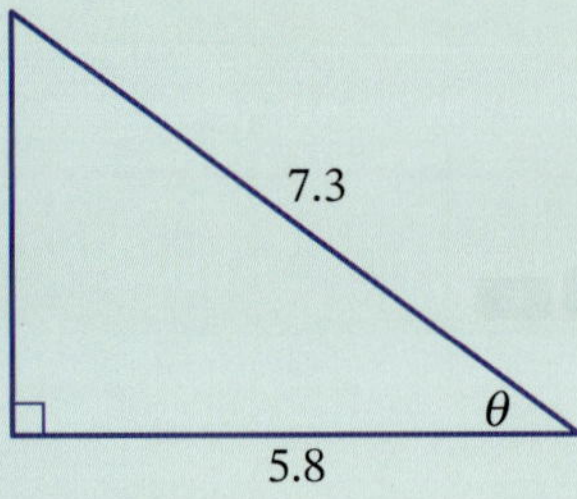

b

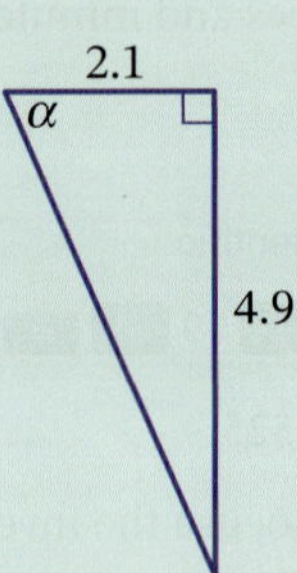

Solution

a

$$\cos\theta = \frac{\text{adjacent}}{\text{hypotenuse}}$$

$$= \frac{5.8}{7.3}$$

$$\therefore \theta = \cos^{-1}\left(\frac{5.8}{7.3}\right)$$

$$\approx 37°23'$$

b

$$\tan\theta = \frac{\text{opposite}}{\text{adjacent}}$$

$$= \frac{4.9}{2.1}$$

$$\therefore \alpha = \tan^{-1}\left(\frac{4.9}{2.1}\right)$$

$$= 66°48'$$

EXERCISE 4.01 Answers on p. 475

Right-angled trigonometry

EXAMPLE 1

1 Round each angle to the nearest degree.

a $47°13'12''$ **b** $81°45'43''$ **c** $19°25'34''$ **d** $76°37'19''$ **e** $52°29'54''$

2 Round each angle to the nearest minute.

a $47°13'12''$ **b** $81°45'43''$ **c** $19°25'34''$ **d** $76°37'19''$ **e** $52°29'54''$

EXAMPLE 2

3 Change into degrees and minutes.

a $59.53°$ **b** $72.231°$ **c** $85.887°$ **d** $46.9°$ **e** $73.213°$

EXAMPLE 3

4 Find, correct to 3 decimal places:

a $\sin 39°25'$ **b** $\cos 45°51'$ **c** $\tan 18°43'$ **d** $\sin 68°06'$ **e** $\tan 54°20'$

5 Find θ in degrees and minutes if:

a $\sin\theta = 0.298$ **b** $\tan\theta = 0.683$ **c** $\cos\theta = 0.827$

d $\tan\theta = 1.056$ **e** $\cos\theta = 0.188$

EXAMPLE 4

6 Find each unknown side, correct to one decimal place.

a

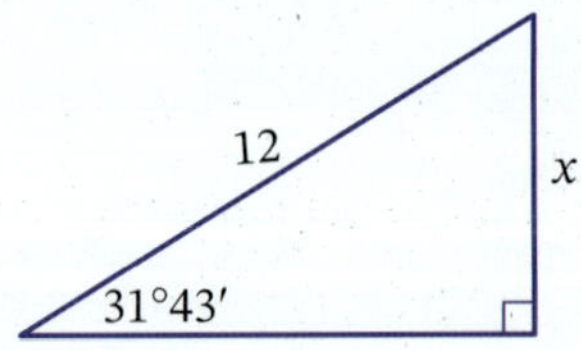

b

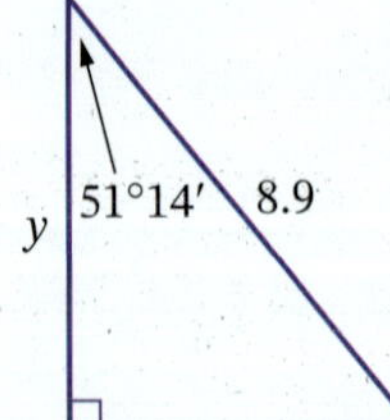

c

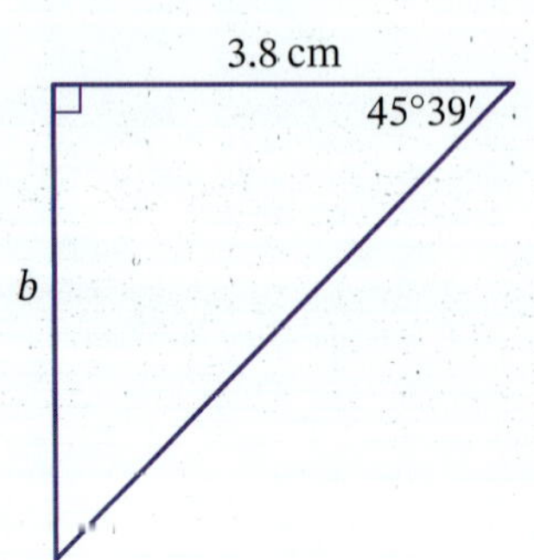

Foundation Mastery Complex

d

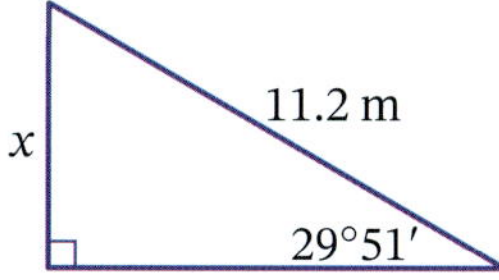

e

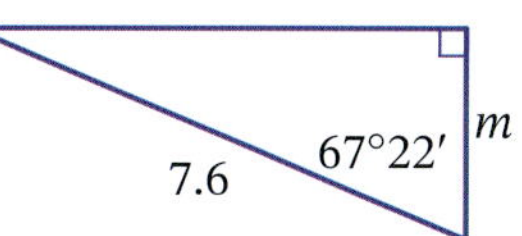

f

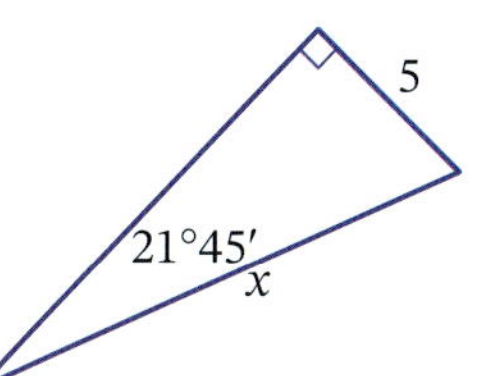

g

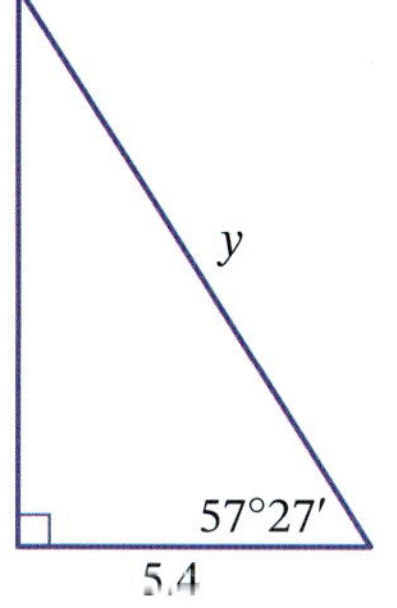

h

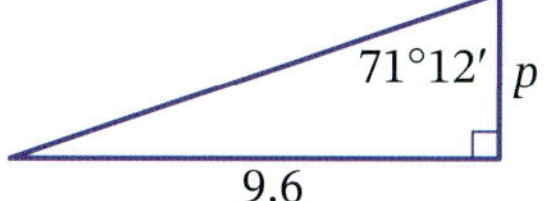

i

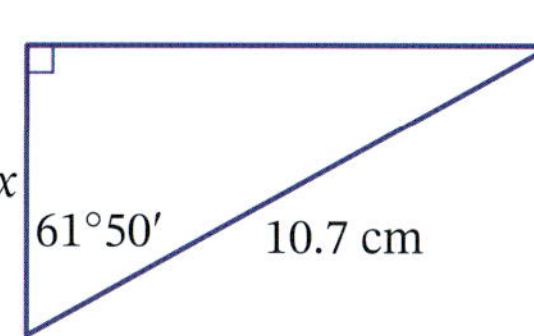

j

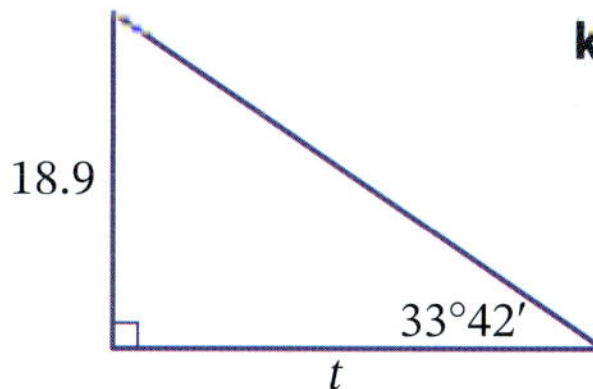

k

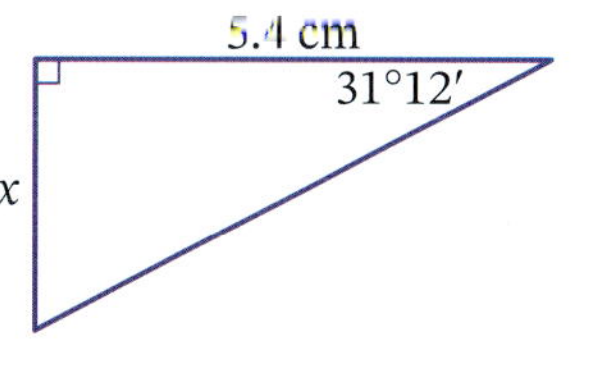

l

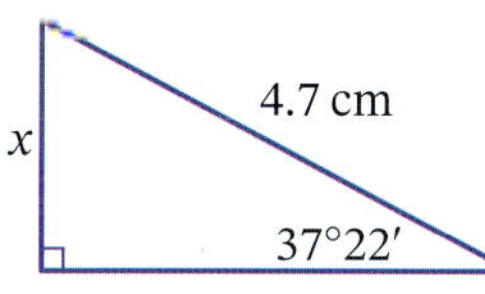

7 A roof is pitched at 60°. A room built inside the roof space is to have a 2.7 m high ceiling. How far in from the side of the roof will the wall for the room go?

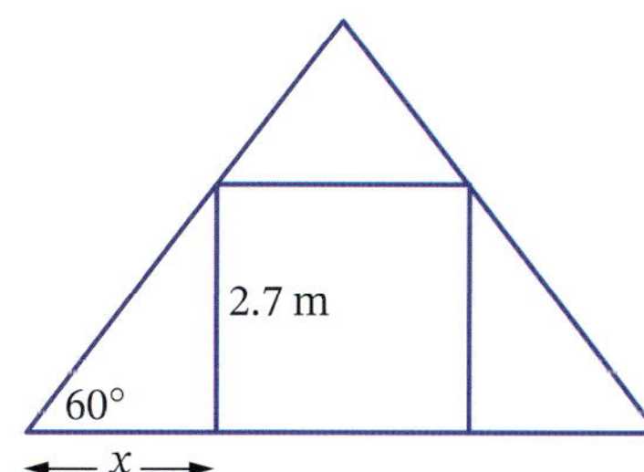

8 Hamish is standing on the sideline of a soccer field, and the goal is at an angle of 67° from his position as shown. The goal is 12.8 m from the corner of the field. How far does he need to kick a ball for it to reach the goal?

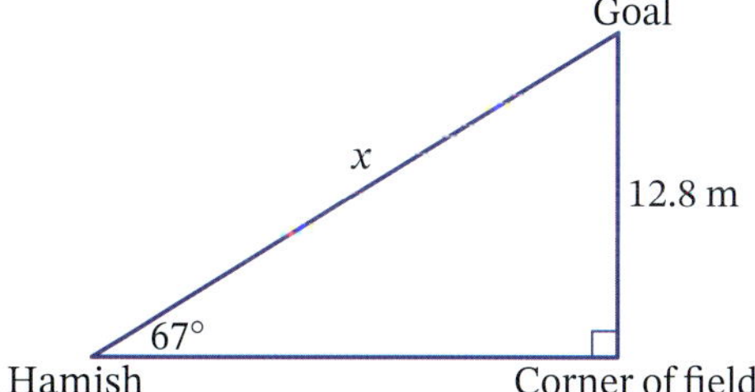

9 Find the size of each unknown angle, in degrees and minutes.

EXAMPLE 5

a

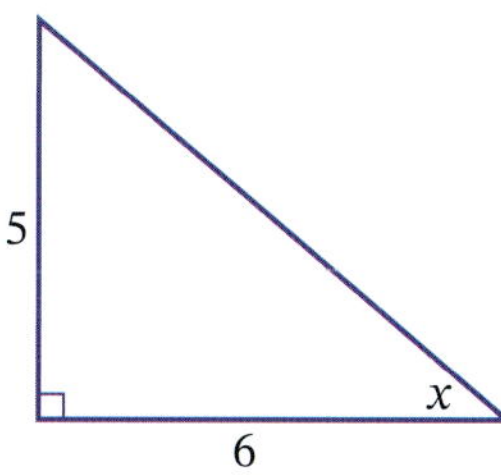

b

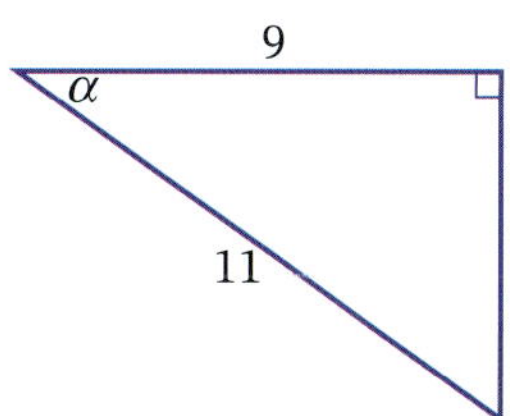

c

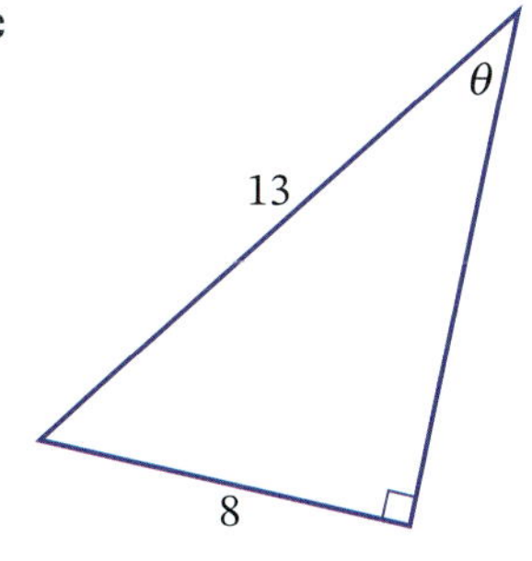

☐ Foundation ○ Mastery ⬡ Complex

d

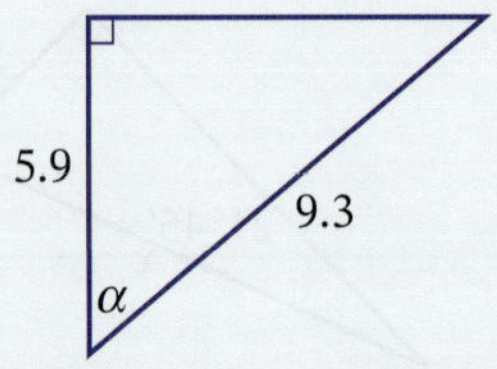

5.9, 9.3, α

e

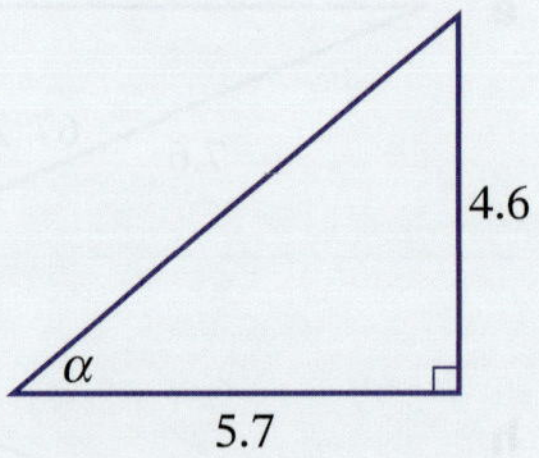

4.6, 5.7, α

f

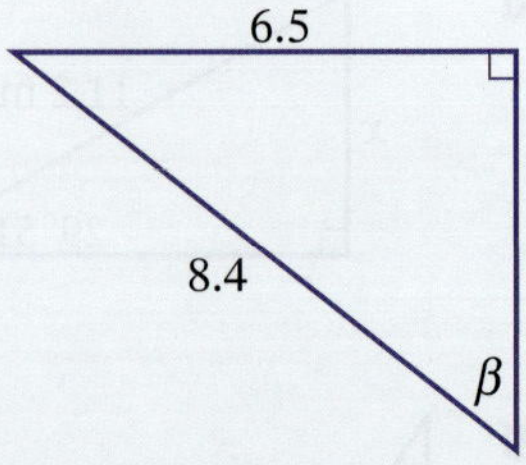

6.5, 8.4, β

g

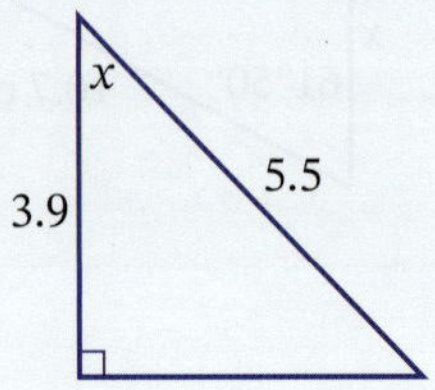

x, 3.9, 5.5

h

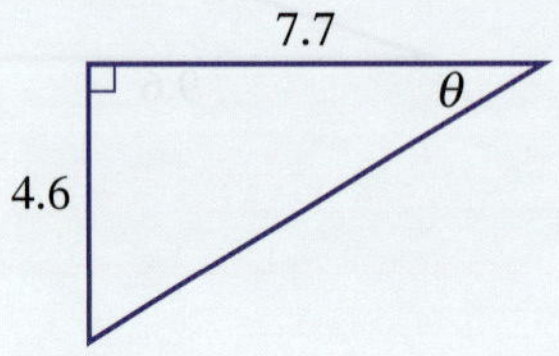

7.7, 4.6, θ

i

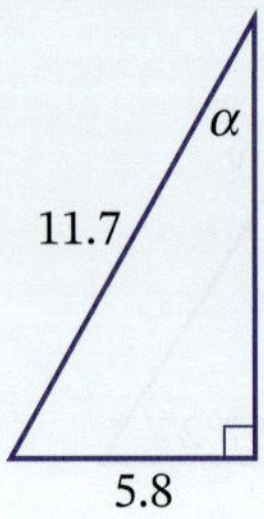

α, 11.7, 5.8

j

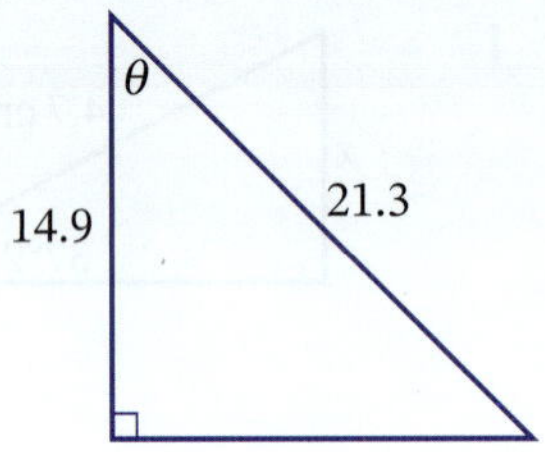

θ, 14.9, 21.3

k

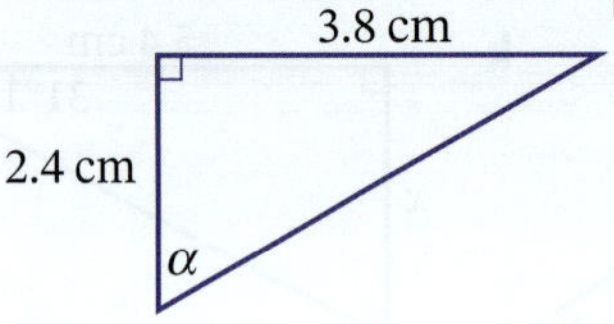

3.8 cm, 2.4 cm, α

l

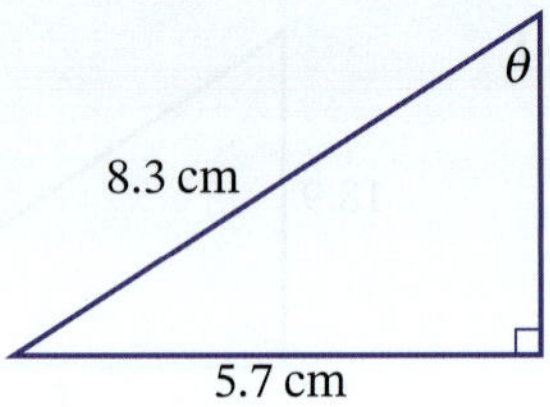

θ, 8.3 cm, 5.7 cm

10 A kite is flying at an angle of θ above the ground as shown. If the kite is 12.3 m above the ground and has 20 m of string, find angle θ.

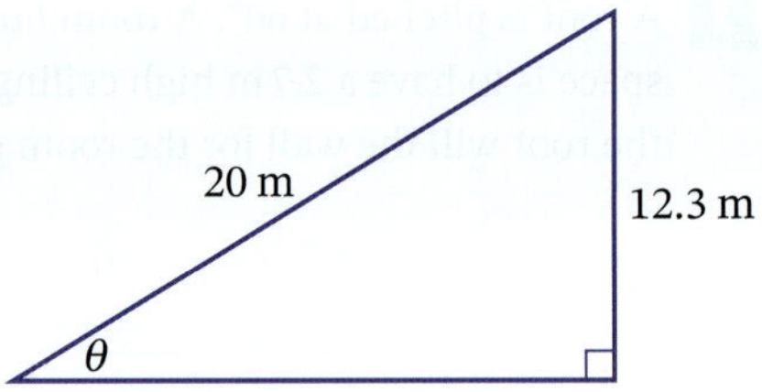

20 m, 12.3 m, θ

11 A field is 13.7 m wide and Andre is on one side. There is a gate on the opposite side and 5.6 m along from where Andre is. At what angle will he walk to get to the gate?

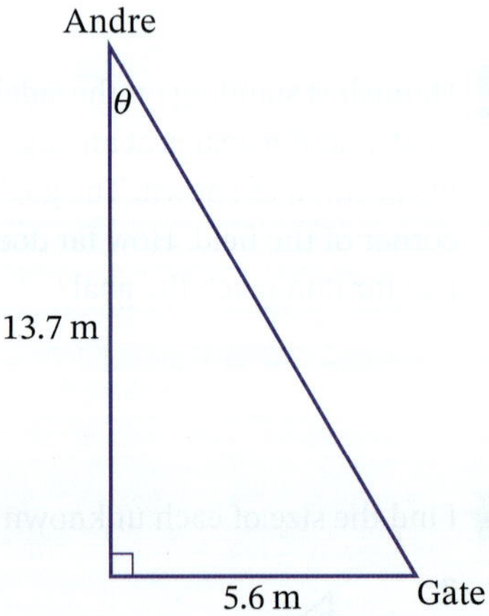

Andre, θ, 13.7 m, 5.6 m, Gate

12 A 60 m long bridge has an opening in the middle and both sides open up to let boats pass underneath. The two parts of the bridge floor rise up to a height of 18 m. Through what angle do they move?

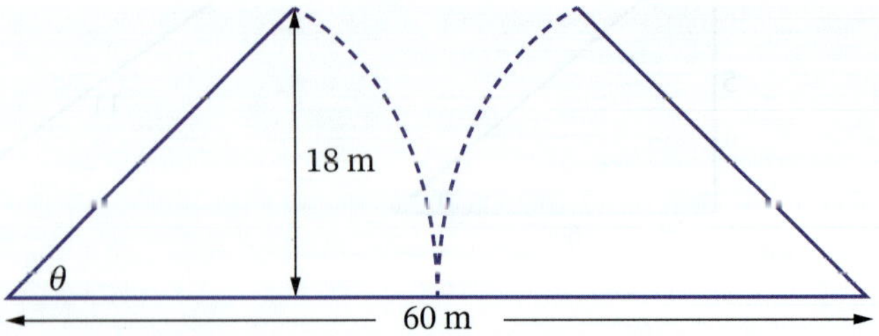

18 m, θ, 60 m

Foundation Mastery Complex

13 Square $ABCD$ with side 6 cm has line CD produced to E as shown so that $\angle EAD = 64°12'$. Evaluate the length, correct to one decimal place, of:

a CE **b** AE

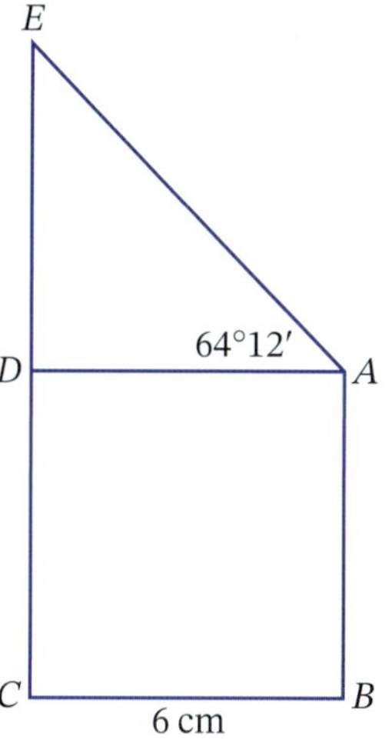

14 A triangular fence is made for a garden inside a park. Three holes A, B and C for fence posts are made at the corners so that A and B are 10.2 m apart, AB and CB are perpendicular, and angle CAB is $59°51'$. How far apart are A and C?

15 A 52 m tall tower has wire stays on either side to minimise wind movement. One stay is 61.3 m long and the other is 74.5 m long, as shown. Find the angles that the tower makes with each stay.

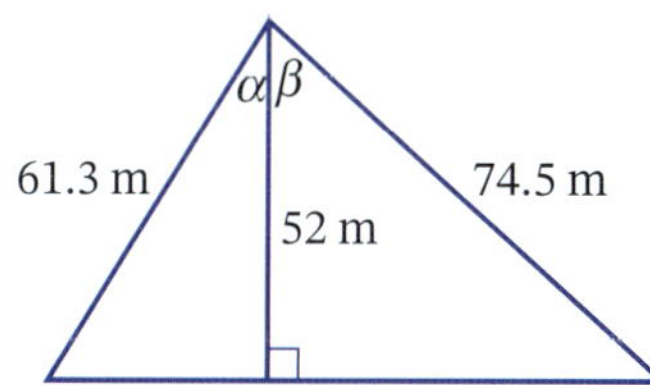

16 Triangle ABC has $\angle BAC = 46°$ and $\angle ABC = 54°$. An altitude (perpendicular line) is drawn from C to meet AB at point D. If the altitude is 5.3 cm long, find, correct to one decimal place, the length of:

a AC **b** BC **c** AB

17 The angle up from the ground to the top of a pole is 41° from a position 15 m to one side.

a Find the height h of the pole, to the nearest metre.

b If Sarah stands 6 m away on the other side, find the angle of elevation θ from Sarah to the top of the pole.

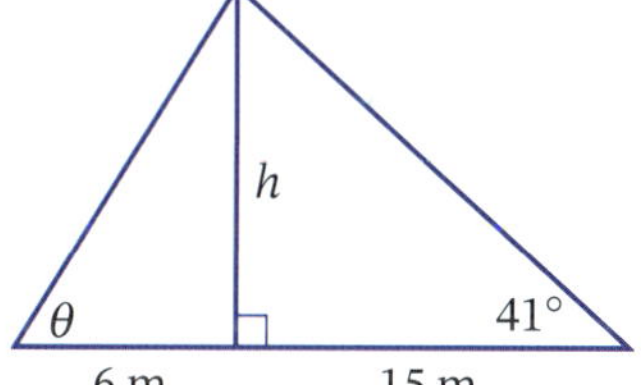

18 A rhombus has one diagonal 12 cm long and the other diagonal makes an angle of $28°23'$ with the side of the rhombus.

a Find the length of the side of the rhombus.

b Find the length of the other diagonal.

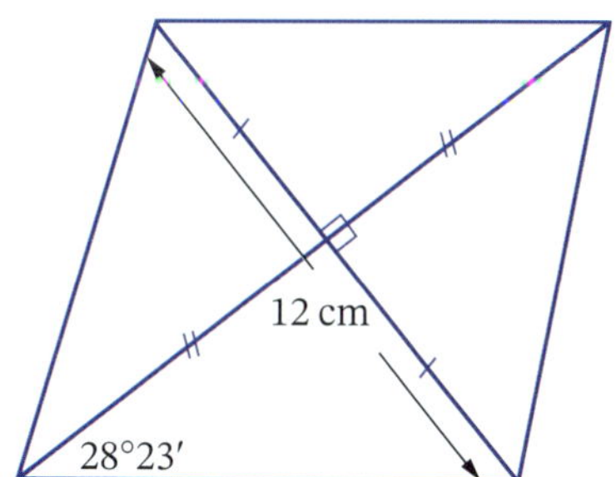

19 **a** Frankie is standing at the side of a road at point A, 15.9 m away from an intersection. She is at an angle of 39° from point B on the other side of the road. What is the width w of the road?

b Frankie walks 7.4 m to point C. At what angle is she from point B?

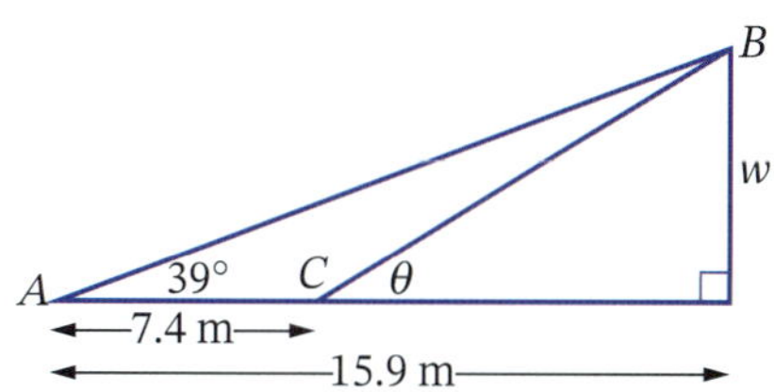

□ Foundation ○ Mastery ○ Complex

20 Kite $ABCD$ has diagonal $BD = 15.8$ cm as shown. If $\angle ABD = 57°29'$ and $\angle DBC = 72°51'$, find the length of the other diagonal AC.

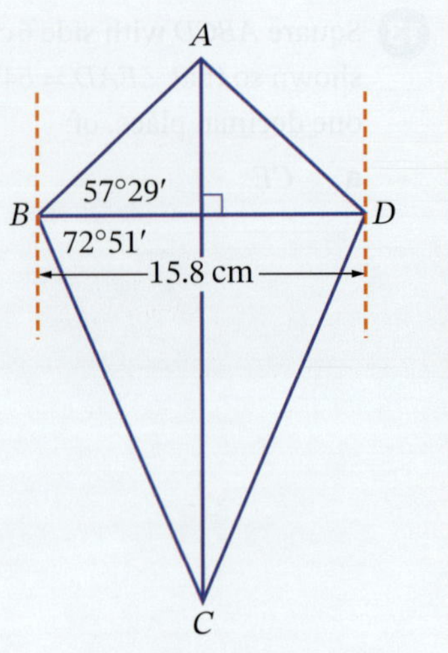

Investigation

Leaning Tower of Pisa

The Tower of Pisa was built as a bell tower for the cathedral nearby. Work started in 1174, but when it was half-completed the soil underneath one side sank. This made the tower lean to one side. Work stopped, and it wasn't until 100 years later that architects found a way of completing the tower. The third and fifth storeys were built close to the vertical to compensate for the lean. Later a vertical top storey was added. The tower is about 55 m tall and 16 m in diameter. It is tilted about 5 m from the vertical at the top, and tilts by an extra 6 mm each year.

iStock.com/Eloi_Omella

Discuss some of the problems with the Leaning Tower of Pisa.

1. Find the angle at which it is tilted from the vertical.
2. Work out how far it will be tilted in 10 years.
3. Use research to find out if the tower will fall over, and if so, when.

☐ Foundation ◯ Mastery ◯ Complex

Bearings and angles of elevation 4.02

Angle of elevation

The **angle of elevation** can be used to measure the height of tall objects that cannot be measured directly; for example, a tree, cliff, tower or building. Stand outside a tall building and look up to the top of the building. Think about what angle your eyes pass through to look up to the top of the building.

Video
Angles of elevation and depression

Worksheet
Right-angled trigonometry

Angle of elevation

The angle of elevation, θ, is the angle measured when looking from the ground up to the top of the object. We assume that the ground is horizontal.

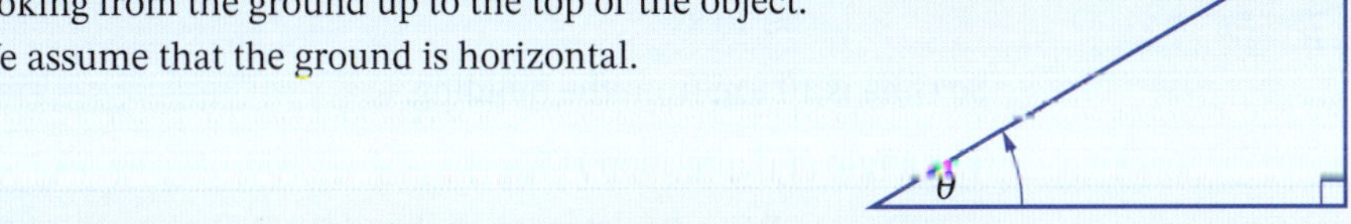

Example 6

The angle of elevation of a tree from a point 50 m out from its base is $38°14'$.

Find the height of the tree to the nearest metre.

Solution

We assume that the tree is vertical.

$$\tan 38°14' = \frac{h}{50}$$

$$50 \tan 38°14' = h$$

$$39 \approx h$$

So the tree is 39 m tall.

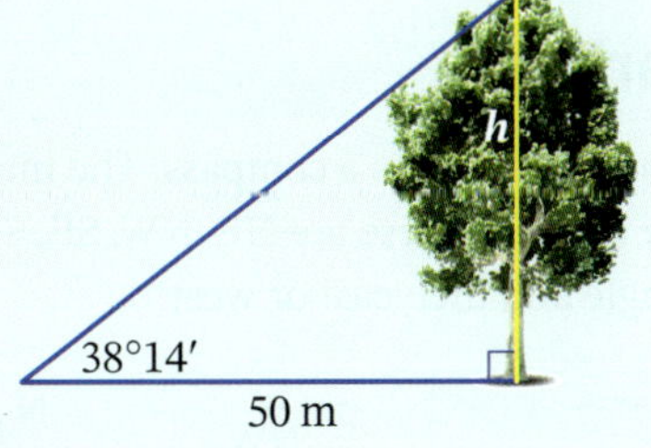

Angle of depression

The **angle of depression** is the angle formed when looking down from a high place to an object below. Find a tall building, hill or other high place, and look down to something below. Through what angle do your eyes pass as you look down?

Videos
Angle of depression 1

Angle of depression 2

Angle of depression

The angle of depression, θ, is the angle measured when looking down from the horizontal to an object below.

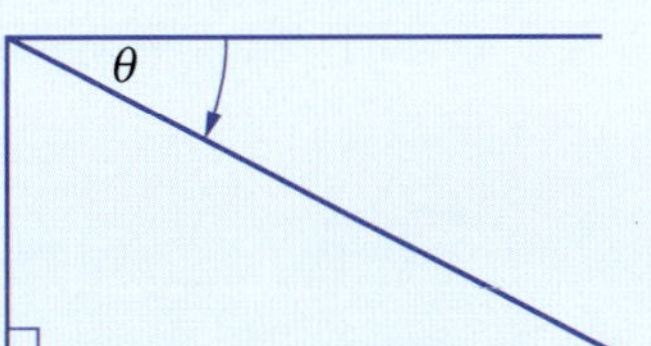

Example 7

a The angle of depression from the top of a 20 m building to Gina below is 61°39′. How far is Gina from the building, to one decimal place?

b A bird sitting on top of an 8 m tall tree looks down at a possum 3.5 m out from the base of the tree. Find the angle of depression to the nearest minute.

Solution

a

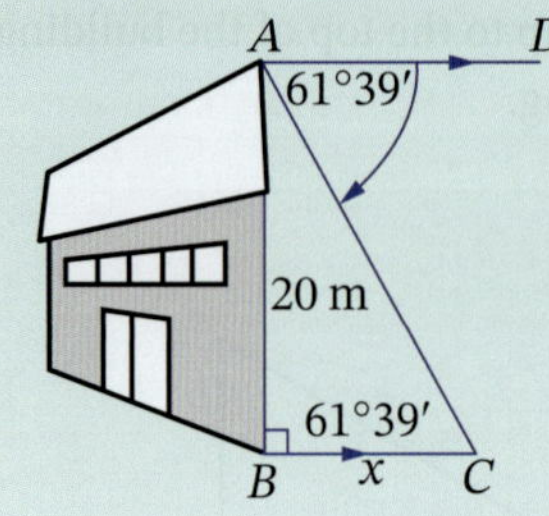

$\angle DAC = \angle ACB = 61°39'$ (alternate angles, $AD \parallel BC$)

$$\tan 61°39' = \frac{20}{x}$$

$$x \tan 61°39' = 20$$

$$x = \frac{20}{\tan 61°39'}$$

$$\approx 10.8$$

So Gina is 10.8 m from the building.

b

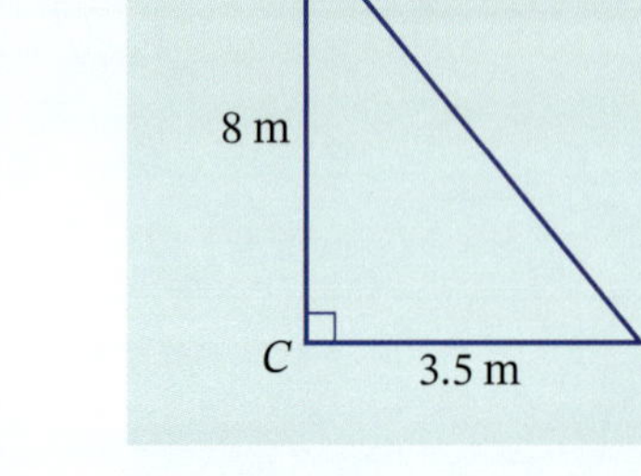

The angle of depression is θ.

$\angle ABD = \angle BDC = \theta$ (alternate angles, $AB \parallel DC$).

$$\tan\theta = \frac{8}{3.5}$$

$$\therefore \theta \approx 66°22'$$

Compass bearings

Video Compass bearings

A **bearing** is a direction according to a compass. The main points on a compass are north (N), south (S), east (E) and west (W). Halfway between these are NE, NW, SE, SW. We write **compass bearings** with north or south first, followed by an angle and then east or west.

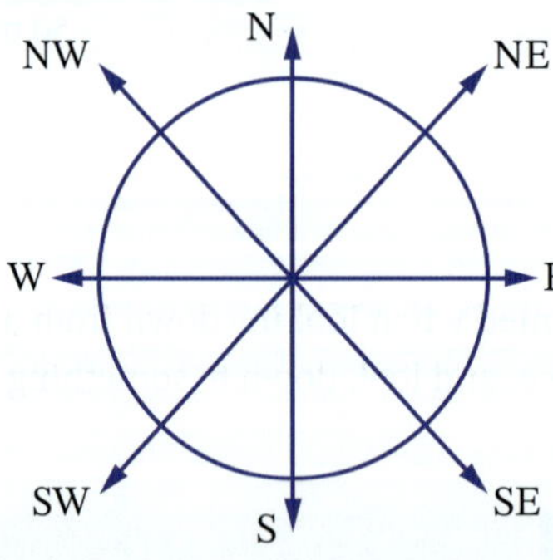

Example 8

a Draw a compass bearing of N 70° W.

b Eli walks from his house on a bearing of S 25° E. If he walks 5.7 km, how far south is he from his house?

Solution

a Start at north and turn 70° towards west.

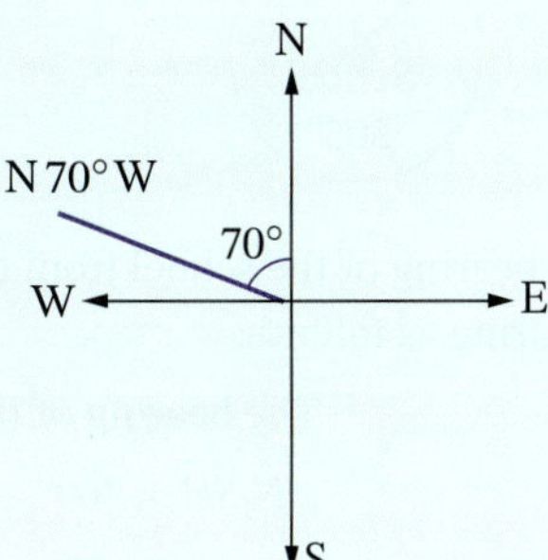

b Start at south and turn 25° towards east.

The hypotenuse is 5.7 and we want to measure the adjacent side (x).

$$\cos\theta = \frac{\text{adjacent}}{\text{hypotenuse}}$$

$$\cos 25° = \frac{x}{5.7}$$

$$5.7\cos 25° = x$$

$$x \approx 5.2$$

So Eli is 5.2 km south of his house.

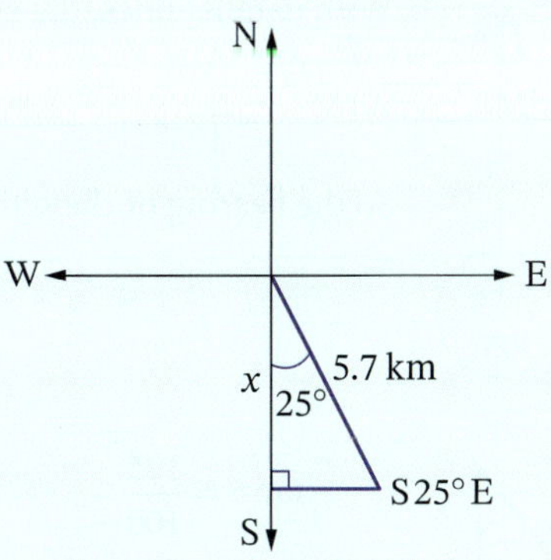

True bearings

True bearings measure angles clockwise from north.

We say B is on a bearing of θ from A.

A true bearing uses 3 digits from 000° to 360°.

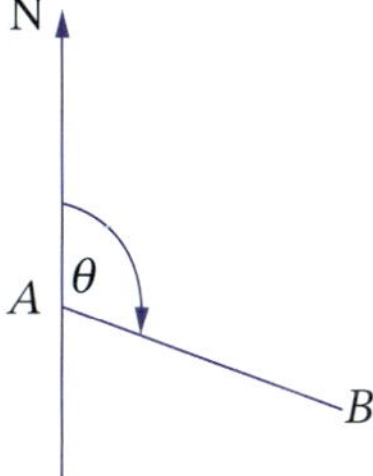

Videos
Bearings
True bearings

Worksheets
A page of bearings
Bearings match-up
Elevations and bearings

Example 9

a X is on a bearing of 030° from Y. Sketch this diagram.

b A house is on a bearing of 305° from a school. What is the bearing of the school from the house?

c A plane leaves Sydney and flies 100 km due east, then 125 km due north. Find the bearing of the plane from Sydney, to the nearest degree.

d A ship sails on a bearing of 140° from Sydney for 250 km. How far east of Sydney is the ship now, to the nearest km?

Solution

a North

X

30°

Y

Note: Bearings use 3 digits so a bearing of 030° is a 30° angle.

b The diagram shows the bearing of the house from the school.

North

House

School

305°

To find the bearing of the school from the house, draw in north from the house and use geometry to find the bearing as follows:

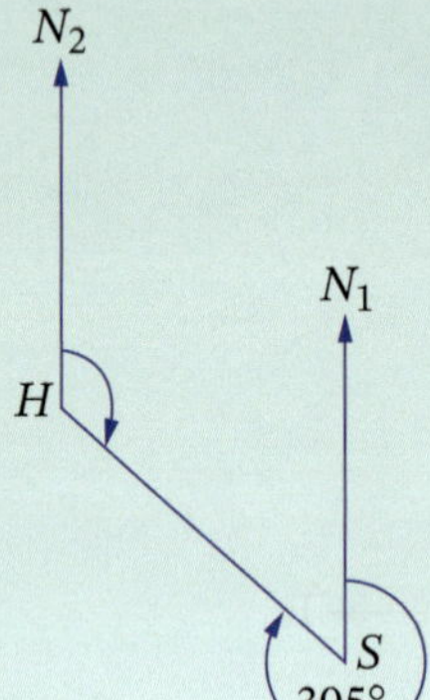

The bearing of the school from the house is $\angle N_2HS$.

$$\angle N_1SH = 360° - 305° \text{ (angles in a revolution)}$$
$$= 55°$$

With parallel lines, the sum of cointerior angles is 180°.

$$\angle N_2HS = 180° - 55°$$
$$= 125°$$

So the bearing of the school from the house is 125°.

c

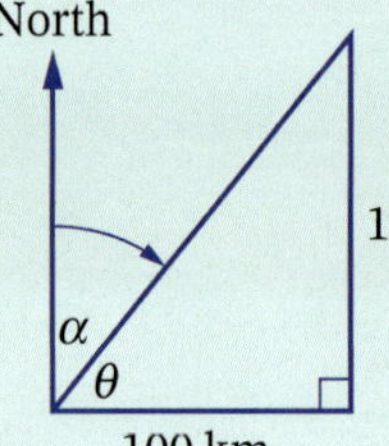

$$\tan\theta = \frac{125}{100}$$
$$\theta \approx 51°$$
$$\alpha = 90° - 51° = 39°$$

So the bearing of the plane from Sydney is 039°.

d

$$\theta = 140° - 90° = 50°$$
$$\cos 50° = \frac{x}{250}$$
$$250 \cos 50° = x$$
$$x \approx 161$$

So the ship is 161 km east of Sydney.

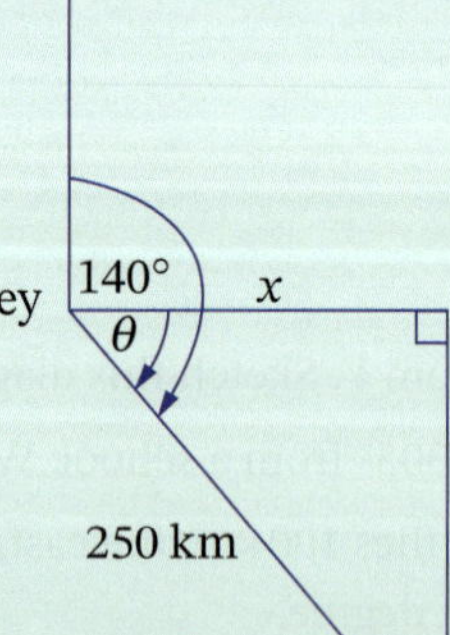

Note: A navigator on a ship uses a sextant to measure angles. A clinometer measures angles of elevation and depression.

EXERCISE 4.02 Answers on p. 475

Bearings and angles of elevation

1 The angle of elevation from a point 11.5 m away from the base of a tree up to the top of the tree is 42°12′. Find the height of the tree to one decimal place.

2 Fineas stands 25.8 m away from the base of a tower and measures the angle of elevation as 39°20′. Find the height of the tower to the nearest metre.

3 A wire is suspended from the top of a 100 m tall bridge tower down to the bridge at an angle of elevation of 52°. How long is the wire, to one decimal place?

4 A cat crouches at the top of a 4.2 m high cliff and looks down at a mouse 1.3 m out from the foot (base) of the cliff. What is the angle of depression, to the nearest minute?

5 The angle of depression from the top of an 8 m tree down to a rabbit is 43°52′. If an eagle is perched in the top of the tree, how far does it need to fly to reach the rabbit, to the nearest metre?

6 The angle of depression from the top of a cliff down to a boat 100 m out from the foot of the cliff is 59°42′. How high is the cliff, to the nearest metre?

EXAMPLES 0, 9

7 Draw a diagram to show the bearing in each question.

a N 50° E **b** S 60° W **c** S 80° E **d** N 40° W

e A boat is on a bearing of 100° from a beach house.

f Jamie is on a bearing of 320° from a campsite.

g A seagull is on a bearing of 200° from a jetty.

h Gena is on a bearing of 050° from the bus stop.

i A plane is on a bearing of 285° from Broken Hill.

j A farmhouse is on a bearing of 012° from a dam.

k Mohammed is on a bearing of 160° from his house.

8 Find the bearing of X from Y in each question using:

i compass bearings **ii** true bearings.

a

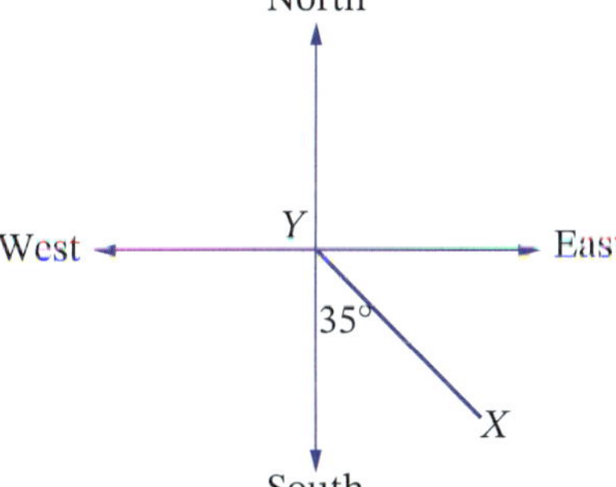

b

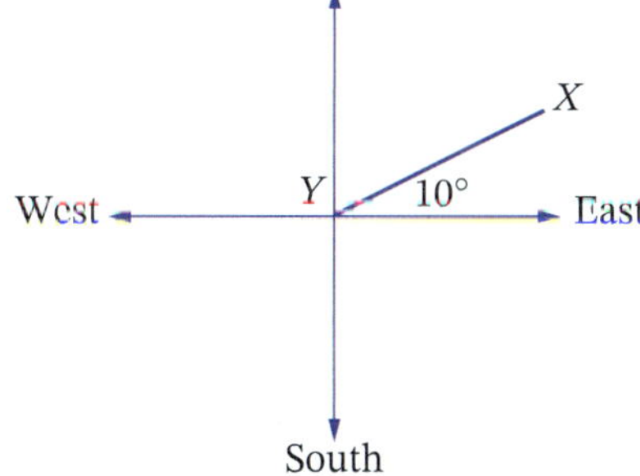

c

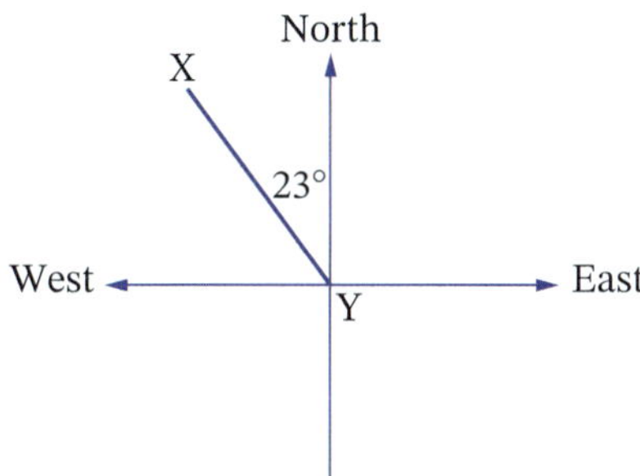

d

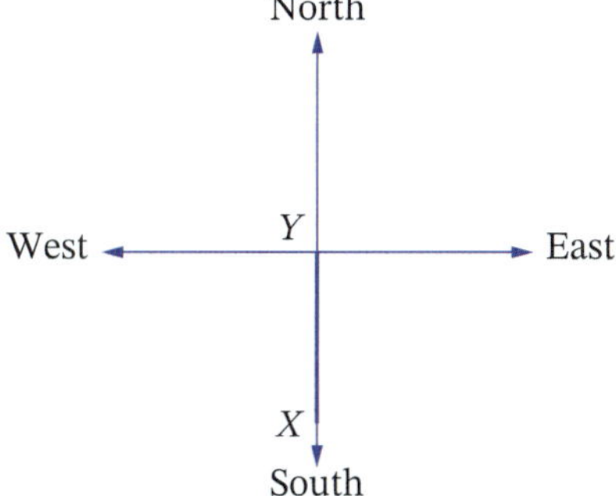

9 Jack is on a bearing of 260° from Jill. What is Jill's bearing from Jack?

10 A tower is on a bearing of 030° from a house. What is the bearing of the house from the tower?

11 Tamworth is on a bearing of 340° from Newcastle. What is the bearing of Newcastle from Tamworth?

12 A plane leaves Melbourne and flies on a bearing of 065° for 2500 km.

a How far north of Melbourne is the plane?

b How far east of Melbourne is it?

c What is the bearing of Melbourne from the plane?

13 The angle of elevation of a tower is 39°44′ when measured at a point 100 m from its base. Find the height of the tower, to one decimal place.

14 Kim leaves her house and walks for 2 km on a bearing of 155°. How far south is Kim from her house now, to one decimal place?

15 Sanjay rides a motorbike through his property, starting at his house. If he rides south for 1.3 km, then rides west for 2.4 km, what is his bearing from the house, to the nearest degree?

16 A plane flies north from Sydney for 560 km, then turns and flies east for 390 km. What is its bearing from Sydney, to the nearest degree?

17 Find the height of a pole, correct to one decimal place, if a 10 m rope tied to it at the top and stretched out straight to reach the ground makes an angle of elevation of 67°13′.

18 A group of students are bushwalking. They walk north from their camp for 7.5 km, then walk west until their bearing from camp is 320°. How far are they from camp, to one decimal place?

19 A 20 m tall tower casts a shadow 15.8 m long at a certain time of day. What is the angle of elevation from the edge of the shadow up to the top of the tower at this time?

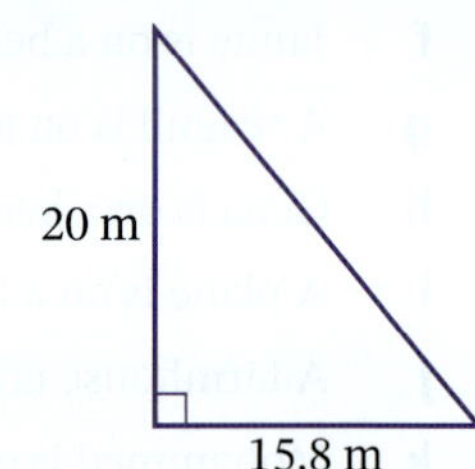

20 A flat verandah roof 1.8 m long is 2.6 m above the ground. At a certain time of day, the sun makes an angle of elevation of 72°25′. How much shade is provided on the ground by the verandah roof at that time, to one decimal place?

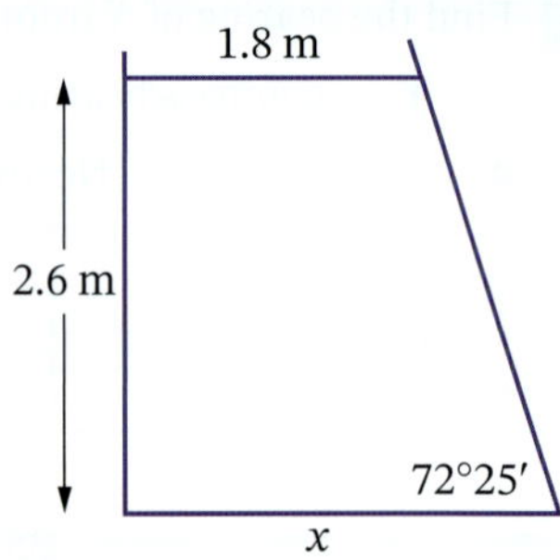

21 Find the angle of elevation of a cliff 15.9 m high from a point 100 m out from its base.

22 A plane leaves Sydney and flies for 2000 km on a bearing of 195°. How far due south of Sydney is it?

23 The angle of depression from the top of a tree 15 m tall down to a pond is 25°41′. If a bird is perched in the top of the tree, how far does it need to fly to reach the pond, to the nearest metre?

24 Robin starts at her house, walks south for 2.7 km then walks east for 1.6 km. What is her bearing from the house, to the nearest degree?

25 The angle of depression from the top of a tower down to a car 250 m out from the foot of the tower is 38°19′. How high is the tower, to the nearest metre?

26 A hot air balloon flies south for 3.6 km then turns and flies east until it is on a bearing of 127° from where it started. How far east does it fly?

27 A 24 m wire is attached to the top of a pole and runs down to the ground where the angle of elevation is 22°32′. Find the height of the pole.

28 A train depot has train tracks running north for 7.8 km where they meet another set of tracks going east for 5.8 km into a station. What is the bearing of the depot from the station, to the nearest degree?

Foundation | Mastery | Complex

29 Jessica leaves home and walks for 4.7 km on a bearing of 075°. She then turns and walks for 2.9 km on a bearing of 115° and she is then due east of her home.

a What is the furthest north that Jessica walks?

b How far is she from home?

30 Builder Jo stands 4.5 m out from the foot of a building and looks up to the top of the building where the angle of elevation is 71°. Builder Ben stands at the top of the building looking down at his wheelbarrow that is 10.8 m out from the foot of the building on the opposite side from where Jo is standing.

a Find the height of the building.

b Find the angle of depression from Ben down to his wheelbarrow.

Angles of any magnitude 4.03

The sin, cos and tan of angles greater than 90° give some interesting results.

Investigation

Obtuse angles

1 Use your calculator to find the sin, cos and tan of some angles greater than 90°. What do you notice?

2 Can you see a pattern for angles between 90° and 180° for:

i sin? **ii** cos? **iii** tan?

We can use a circle to show angles, starting with 0° at the x-axis and turning anticlockwise to show other angles. We divide the number plane into 4 **quadrants** as shown:

1st quadrant:	0° to 90°
2nd quadrant:	90° to 180°
3rd quadrant:	180° to 270°
4th quadrant:	270° to 360°

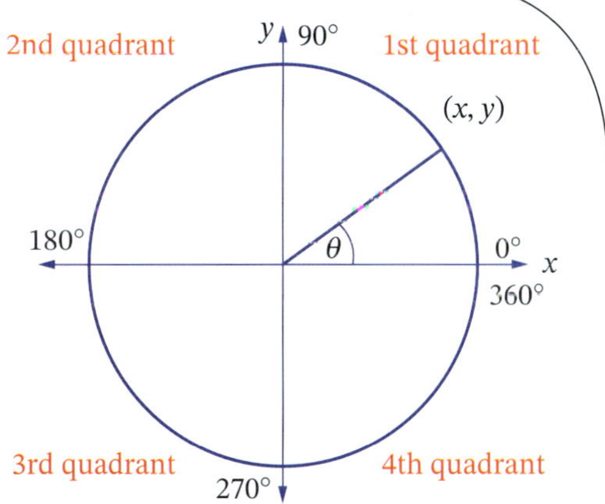

To make it easier to explore these results, we use a **unit circle** with radius 1.

We can find the trigonometric ratios for angle θ.

The trigonometric ratios on a unit circle

$$\sin\theta = \frac{y}{1} = y$$

$$\cos\theta = \frac{x}{1} = x$$

$$\tan\theta = \frac{y}{x}$$

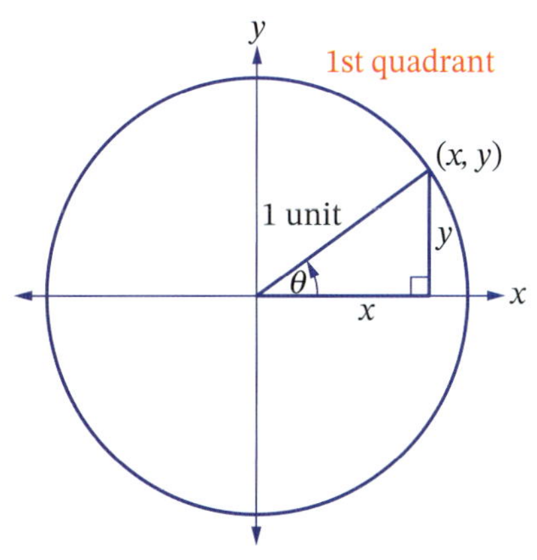

Foundation Mastery Complex

2nd quadrant: Obtuse angles (between 90° and 180°)

The signs of the x and y values are different in each of the 4 quadrants. We can examine each quadrant separately. In the 2nd quadrant, notice that x values are negative and y values are positive.

So the point in the 2nd quadrant will be $(-x, y)$.

Since $\sin\theta = y$, sin will be **positive** in the 2nd quadrant.

Since $\cos\theta = -x$, cos will **negative** in the 2nd quadrant.

Since $\tan\theta = \frac{y}{-x}$, tan will be **negative** in the 2nd quadrant (positive divided by negative).

To have an angle of θ in the triangle, the obtuse angle in the 2nd quadrant is $180° - \theta$.

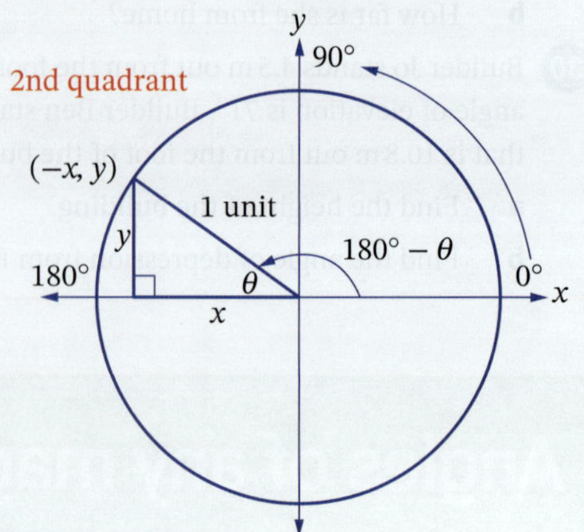

Trigonometric ratios of obtuse angles

$$\sin(180° - \theta) = \sin\theta$$
$$\cos(180° - \theta) = -\cos\theta$$
$$\tan(180° - \theta) = -\tan\theta$$

Example 10

a If $\cos 80° = 0.174$, evaluate $\cos 100°$.

b If $\sin 55° = 0.819$, find the value of $\sin 125°$.

Solution

a
$$\cos(180° - \theta) = -\cos\theta$$
So $\cos(180° - 80°) = -\cos 80°$
$$\cos 100° = -\cos 80°$$
$$= -0.174$$

b
$$\sin(180° - \theta) = \sin\theta$$
So $\sin(180° - 55°) = \sin 55°$
$$\sin 125° = \sin 55°$$
$$= 0.819$$

You can check that this is true by finding both ratios on the calculator.

3rd quadrant: angles between 180° and 270°

$\sin\theta = -y$ (negative)

$\cos\theta = -x$ (negative)

$\tan\theta = \frac{-y}{-x} = \frac{y}{x}$ (positive)

The angle that gives θ in the triangle is $180° + \theta$.

3rd quadrant

$$\sin(180° + \theta) = -\sin\theta$$
$$\cos(180° + \theta) = -\cos\theta$$
$$\tan(180° + \theta) = \tan\theta$$

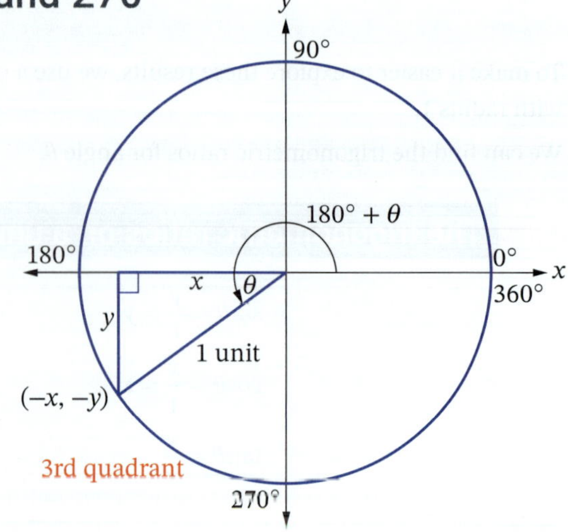

4th quadrant: angles between 270° and 360°

$\sin\theta = -y$ (negative)

$\cos\theta = x$ (positive)

$\tan\theta = \frac{-y}{x}$ (negative)

The angle that gives θ in the triangle is $360° - \theta$.

4th quadrant

$\sin(360° - \theta) = -\sin\theta$

$\cos(360° - \theta) = \cos\theta$

$\tan(360° - \theta) = -\tan\theta$

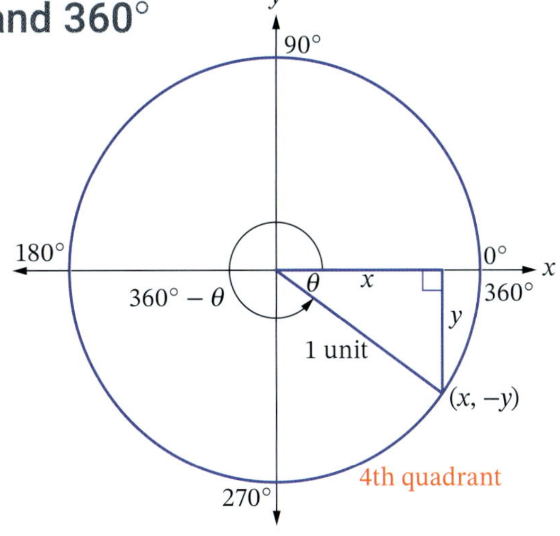

Putting all of these results together gives a rule for all 4 quadrants that we usually call the **ASTC rule**.

ASTC rule

A: All ratios are positive in the 1st quadrant.

S: sin is positive in the 2nd quadrant (cos and tan are negative).

T: tan is positive in the 3rd quadrant (sin and cos are negative).

C: cos is positive in the 4th quadrant (sin and tan are negative).

ASTC can be remembered using the phrase '**A**ll **S**tations **T**o **C**entral'.

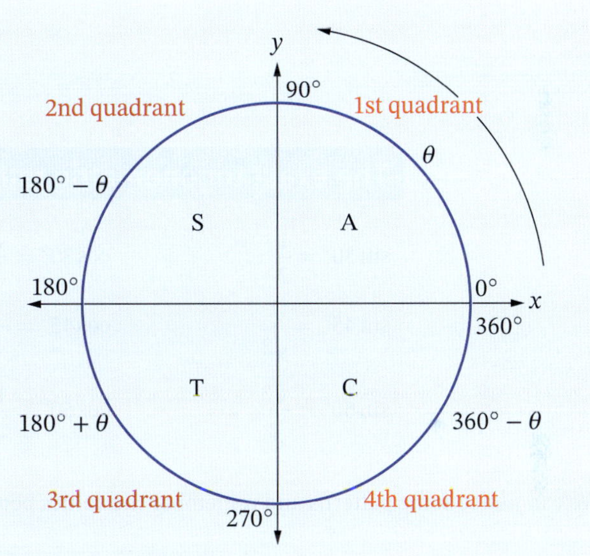

The trigonometric ratios of 0°, 90°, 180°, 270°, 360°

The point (x, y) on the unit circle gives us the values of the trigonometric ratios of the angle at the origin.

$\sin\theta = y$, $\cos\theta = x$, $\tan\theta = \frac{y}{x}$

We can use this to find the trigonometric ratios of the boundary angles on the edges of the quadrants: 0°, 90°, 180°, 270° and 360°.

θ	0°	90°	180°	270°	360° (same as 0°)
Point (x, y)	(1, 0)	(0, 1)	(−1, 0)	(0, −1)	(1, 0)
$\sin\theta = y$	0	1	0	−1	0
$\cos\theta = x$	1	0	−1	0	1
$\tan\theta = \frac{y}{x}$	$\frac{0}{1} = 0$	$\frac{1}{0}$ is undefined, no value	$\frac{0}{-1} = 0$	$\frac{-1}{0}$ is undefined, no value	$\frac{0}{1} = 0$

Video
Exact trigonometric ratios

Worksheet
Exact ratios

The exact trigonometric ratios of 30°, 45°, 60°

The trigonometric ratios of 30°, 45° and 60° are **exact ratios**. This can be shown and remembered using 2 special right-angled triangles.

The 45° isosceles triangle on the left has equal sides of length 1 unit. The 30°–60° triangle on the right is half of an equilateral triangle (below it) with sides of 2 units, so its base is 1 unit.

These 2 special triangles appear on the HSC exam reference sheet.

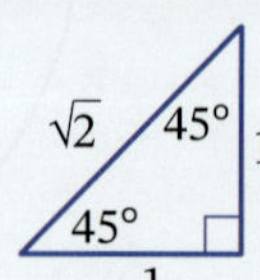

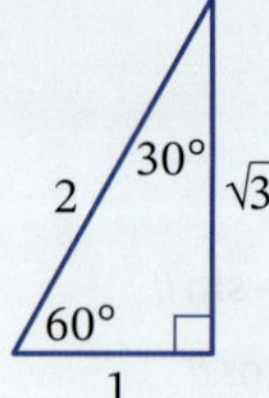

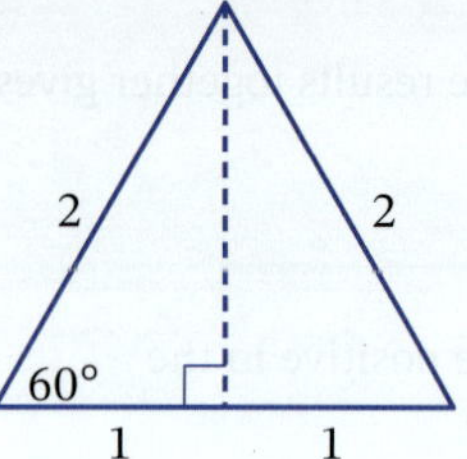

The exact trigonometric ratios

$\sin 30° = \frac{1}{2}$ $\qquad \cos 30° = \frac{\sqrt{3}}{2}$ $\qquad \tan 30° = \frac{1}{\sqrt{3}}$

$\sin 45° = \frac{1}{\sqrt{2}}$ $\qquad \cos 45° = \frac{1}{\sqrt{2}}$ $\qquad \tan 45° = 1$

$\sin 60° = \frac{\sqrt{3}}{2}$ $\qquad \cos 60° = \frac{1}{2}$ $\qquad \tan 60° = \sqrt{3}$

We can notice some patterns and remember the values better when we write the exact ratios in a table.

	30°	45°	60°
sin	$\frac{1}{2}$	$\frac{1}{\sqrt{2}}$	$\frac{\sqrt{3}}{2}$
cos	$\frac{\sqrt{3}}{2}$	$\frac{1}{\sqrt{2}}$	$\frac{1}{2}$
tan	$\frac{1}{\sqrt{3}}$	1	$\sqrt{3}$

- $\sin 30° = \cos 60° = \frac{1}{2}$
- $\sin 60° = \cos 30° = \frac{\sqrt{3}}{2}$
- $\sin 45° = \cos 45° = \frac{1}{\sqrt{2}}$
- The values of cos are the values of sin in reverse order.

Example 11

a Find all quadrants where:

i $\sin\theta > 0$ **ii** $\cos\theta < 0$ **iii** $\tan\theta < 0$ and $\cos\theta > 0$

b Find the exact value of:

i $\tan 330°$ **ii** $\sin 225°$

c Simplify $\cos(180° + x)$.

d If $\sin x = -\frac{3}{5}$ and $\cos x > 0$, find the value of $\tan x$.

Solution

a **i** Using the ASTC rule, $\sin\theta > 0$ in the 1st and 2nd quadrants.

ii $\cos\theta > 0$ in the 1st and 4th quadrants, so $\cos\theta < 0$ in the 2nd and 3rd quadrants.

iii $\tan\theta > 0$ in the 1st and 3rd quadrants so $\tan\theta < 0$ in the 2nd and 4th quadrants. Also $\cos\theta > 0$ in the 1st and 4th quadrants.

So, $\tan\theta < 0$ and $\cos\theta > 0$ in the 4th quadrant.

b **i** 330° lies in the 4th quadrant.

The angle inside the triangle in the 4th quadrant is $360° - 330° = 30°$ and tan is negative in the 4th quadrant.

$$\tan 330° = -\tan 30° = -\frac{1}{\sqrt{3}}$$

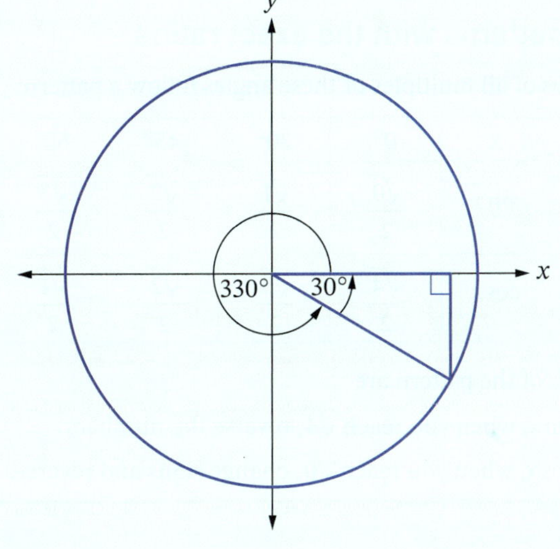

ii 225° is in the 3rd quadrant.
The angle in the triangle in the 3rd quadrant is $225° - 180° = 45°$ and sin is negative in the 3rd quadrant.

$$\sin 225° = -\sin 45° = -\frac{1}{\sqrt{2}}$$

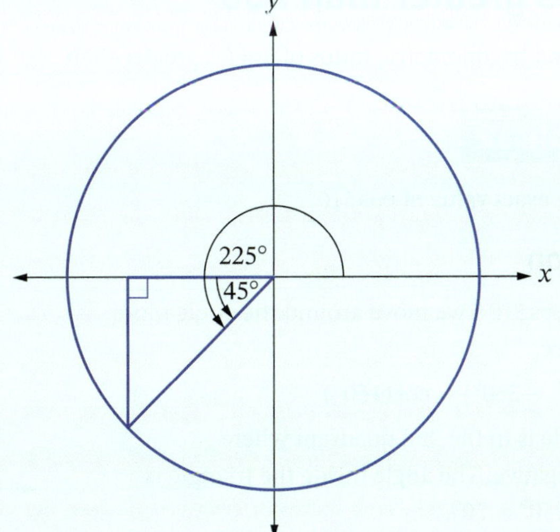

c $180° + x$ is in the 3rd quadrant, where $\cos x$ is negative.

So, $\cos(180° + x) = -\cos x$.

d $\sin x < 0$ and $\cos x > 0$ so x is in the 4th quadrant.

$\sin x = -\frac{3}{5}$

So, the opposite side is 3 and the hypotenuse is 5.

By Pythagoras' theorem, the adjacent side is 4 (3, 4, 5 triangle).

$\tan x < 0$ in the 4th quadrant.

So, $\tan x = -\frac{3}{4}$.

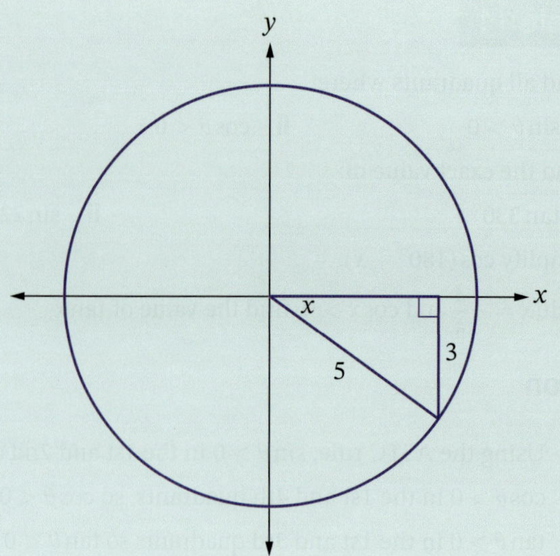

Did you know?

More patterns with the exact ratios

The ratios of all multiples of these angles follow a pattern:

x	0°	30°	45°	60°	90°	120°	135°	150°
$\sin x$	$\frac{\sqrt{0}}{2}$	$\frac{\sqrt{1}}{2}$	$\frac{\sqrt{2}}{2}$	$\frac{\sqrt{3}}{2}$	$\frac{\sqrt{4}}{2}$	$\frac{\sqrt{3}}{2}$	$\frac{\sqrt{2}}{2}$	$\frac{\sqrt{1}}{2}$
$\cos x$	$\frac{\sqrt{4}}{2}$	$\frac{\sqrt{3}}{2}$	$\frac{\sqrt{2}}{2}$	$\frac{\sqrt{1}}{2}$	$\frac{\sqrt{0}}{2}$	$\frac{-\sqrt{1}}{2}$	$\frac{-\sqrt{2}}{2}$	$\frac{-\sqrt{3}}{2}$

The rules of the pattern are

- for $\sin x$, when you reach $\sqrt{4}$, reverse the numbers
- for $\cos x$, when you reach $\sqrt{0}$, change signs and reverse.

Angles greater than 360°

We can find trigonometric ratios of angles greater than 360° by turning around the circle more than once.

Example 12

Find the exact value of cos 510°.

Solution

To find cos 510°, we move around the circle more than once.

$\cos(510° - 360°) = \cos(150°)$

The angle is in the 2nd quadrant where cos is negative. The angle inside the triangle is $180° - 150° = 30°$.

$$\begin{aligned} \text{So, } \cos 510° &= \cos 150° \\ &= -\cos 30° \\ &= -\frac{\sqrt{3}}{2} \end{aligned}$$

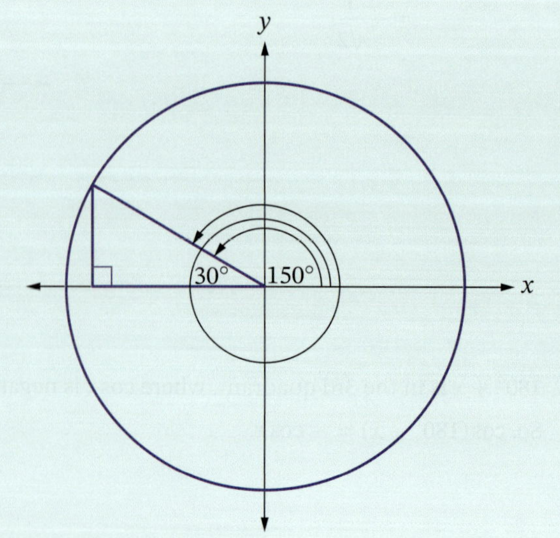

Negative angles

The ASTC rule also works for negative angles. These are measured in the opposite direction (clockwise) from positive angles as shown. Notice the relationship between the negative angle and the matching positive angle as the negative angles move through the quadrants in reverse order.

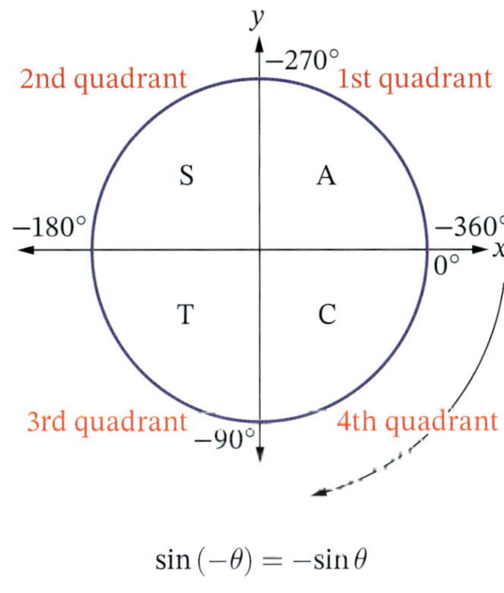

$$\sin(-\theta) = -\sin\theta$$
$$\cos(-\theta) = \cos\theta$$
$$\tan(-\theta) = -\tan\theta$$

	4th quadrant	3rd quadrant	2nd quadrant	1st quadrant	Relationship
$\sin(-\theta)$	Negative	Negative	Positive	Positive	$\sin(-\theta) = -\sin\theta$
$\sin\theta$	Positive (1st quadrant)	Positive (2nd quadrant)	Negative (3rd quadrant)	Negative (4th quadrant)	
$\cos(-\theta)$	Positive	Negative	Negative	Positive	$\cos(-\theta) = \cos\theta$
$\cos\theta$	Positive (1st quadrant)	Negative (2nd quadrant)	Negative (3rd quadrant)	Positive (4th quadrant)	
$\tan(-\theta)$	Negative	Positive	Negative	Positive	$\tan(-\theta) = -\tan\theta$
$\tan\theta$	Positive (1st quadrant)	Positive (2nd quadrant)	Negative (3rd quadrant)	Negative (4th quadrant)	

Example 13

Find the exact value of $\tan(-120°)$.

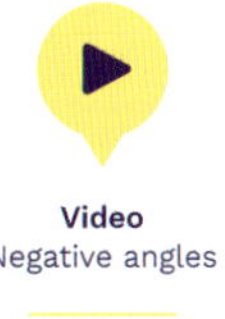

Solution

Moving clockwise around the circle, the angle is in the 3rd quadrant, with $180° - 120° = 60°$ in the triangle.

tan is positive in the 3rd quadrant.

$$\begin{aligned}\tan(-120°) &= \tan 60° \\ &= \sqrt{3}\end{aligned}$$

or using the rule $\tan(-\theta) = -\tan\theta$:

$$\begin{aligned}\tan(-120°) &= -\tan 120° \\ &= -[-\tan(180° - 120°)] \\ &= \tan 60° \\ &= \sqrt{3}\end{aligned}$$

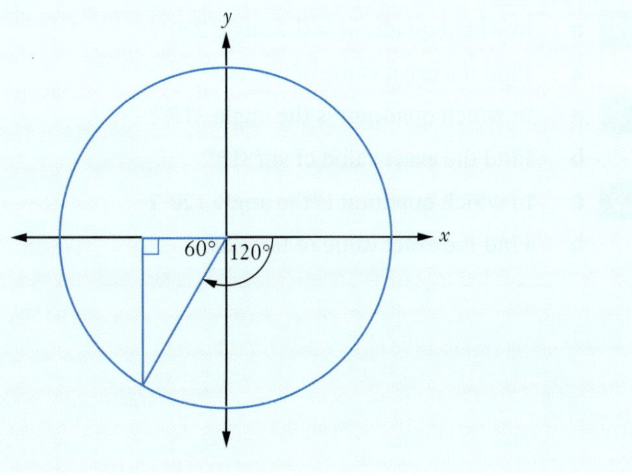

EXERCISE 4.03 Answers on p. 476

Angles of any magnitude

1 $\triangle ABC$ is a right-angled isosceles triangle with $\angle B = 90°$ and $AB = BC = 1$.

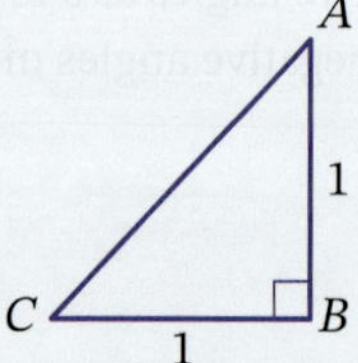

a Find the exact length of AC.

b Find $\angle BAC$.

c From the triangle, write down the exact ratios of $\sin 45°$, $\cos 45°$ and $\tan 45°$.

2 $\triangle ABC$ is half of an equilateral triangle with $\angle C = 90°$, $AB = 2$ and $BC = 1$.

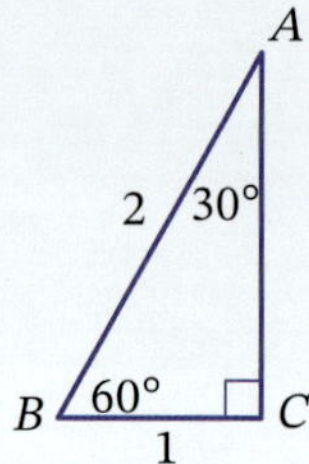

a Find the exact length of AC.

b Use the triangle to find the exact ratios of $\sin 30°$, $\cos 30°$ and $\tan 30°$.

c Use the triangle to find the exact ratios of $\sin 60°$, $\cos 60°$ and $\tan 60°$.

3 Use the coordinates (x, y) on the unit circle to evaluate each trigonometric ratio.

a	$\tan 360°$	**b**	$\cos 90°$	**c**	$\sin 180°$	**d**	$\sin 0°$
e	$\tan 180°$	**f**	$\cos 0°$	**g**	$\cos 270°$	**h**	$\sin 360°$
i	$\tan 270°$	**j**	$\cos 180°$	**k**	$\tan 90°$	**l**	$\cos 540°$
m	$\sin 720°$	**n**	$\cos 630°$	**o**	$\sin(-90°)$	**p**	$\cos(-360°)$

EXAMPLE 10

4 **a** If $\sin 20° = 0.342$, then evaluate $\sin 160°$, $\sin 200°$ and $\sin 340°$.

b If $\cos 55° = 0.574$, then evaluate $\cos 305°$, $\cos 235°$ and $\cos 125°$.

c If $\tan 78° = x$, then evaluate $\tan 258°$, $\tan 102°$ and $\tan(-78°)$.

5 Find θ if it is in the first quadrant (acute).

a	$\tan 205° = \tan\theta$	**b**	$\cos 140° = -\cos\theta$
c	$\sin 300° = -\sin\theta$	**d**	$\tan 110° = -\tan\theta$
e	$\cos 185° = -\cos\theta$	**f**	$\sin 130° = \sin\theta$
g	$\tan 250° = -\tan\theta$	**h**	$\cos 295° = \cos\theta$
i	$\sin 420° = \sin\theta$	**j**	$\tan(-80°) = -\tan\theta$
k	$\cos 600° = -\cos\theta$	**l**	$\sin(-320°) = \sin\theta$

EXAMPLE 11

6 Find all quadrants where:

a $\cos\theta > 0$ **b** $\tan\theta > 0$ **c** $\sin\theta > 0$

d $\tan\theta < 0$ **e** $\sin\theta < 0$ **f** $\cos\theta < 0$

g $\sin\theta < 0$ and $\tan\theta > 0$ **h** $\cos\theta < 0$ and $\tan\theta < 0$

i $\cos\theta > 0$ and $\tan\theta < 0$ **j** $\sin\theta < 0$ and $\tan\theta < 0$

7 **a** In which quadrant is the angle 240°?

b Find the exact value of $\cos 240°$.

8 **a** In which quadrant is the angle 315°?

b Find the exact value of $\sin 315°$.

9 **a** In which quadrant is the angle 120°?

b Find the exact value of $\tan 120°$.

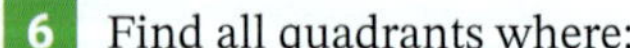

10 a In which quadrant is the angle $-225°$?

b Find the exact value of $\sin(-225°)$.

11 a In which quadrant is the angle $-330°$?

b Find the exact value of $\cos(-330°)$.

12 Find the exact value of:

a $\tan 225°$ b $\cos 315°$ c $\tan 300°$ d $\sin 150°$

e $\cos 120°$ f $\sin 210°$ g $\cos 330°$ h $\tan 150°$

i $\sin 300°$ j $\cos 135°$

13 Find the exact value of:

a $\cos 570°$ b $\tan 420°$ c $\sin 480°$ d $\cos 660°$

e $\sin 690°$ f $\tan 600°$ g $\sin 495°$ h $\cos 405°$

i $\tan 675°$ j $\sin 390°$

14 Find the exact value of:

a $\cos(-225°)$ b $\cos(-210°)$ c $\tan(-300°)$ d $\cos(-150°)$

e $\sin(-60°)$ f $\tan(-240°)$ g $\cos(-300°)$ h $\tan(-30°)$

i $\cos(-45°)$ j $\sin(-135°)$

15 If $\tan\theta = \frac{3}{4}$ and $\cos\theta < 0$, find $\sin\theta$ and $\cos\theta$ as fractions.

16 Given $\sin\theta = \frac{4}{7}$ and $\tan\theta < 0$, find the exact value of $\cos\theta$ and $\tan\theta$.

17 If $\sin x < 0$ and $\tan x = -\frac{5}{8}$, find the exact value of $\cos x$.

18 Given $\cos x = \frac{2}{5}$ and $\tan x < 0$, find the exact value of $\sin x$ and $\tan x$.

19 If $\cos x < 0$ and $\sin x > 0$, find $\cos x$ and $\sin x$ in surd form if $\tan x = \frac{5}{7}$.

20 If $\sin\theta = -\frac{4}{9}$ and $270° < \theta < 360°$, find the exact value of $\tan\theta$ and $\cos\theta$.

21 If $\cos x = -\frac{3}{8}$ and $180° < x < 270°$, find the exact value of $\tan x$ and $\sin x$.

22 Simplify:

a $\sin(180° - \theta)$ b $\cos(360° - x)$ c $\tan(180° + \beta)$

d $\sin(180° + \alpha)$ e $\tan(360° - \theta)$ f $\sin(-\theta)$

g $\cos(-\alpha)$ h $\tan(-x)$ i $\cos(180° - \alpha)$

Foundation Mastery Complex

4.04 The sine rule

Worksheet
Sine rule problems

Naming the sides and angles of a triangle

Side a is opposite angle A, side b is opposite angle B and side c is opposite angle C.

The shortest side is opposite the smallest angle.

The longest side is opposite the largest angle.

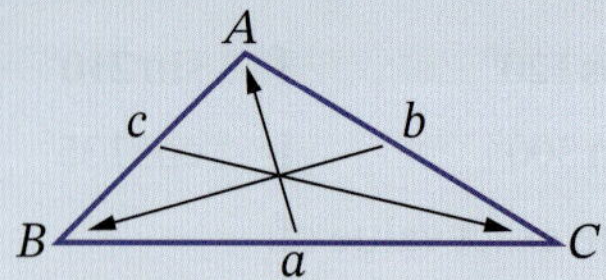

The sine rule

The **sine rule** relates each side of a triangle to its opposite side.

The rule is used to find unknown sides and angles in non-right-angled triangles.

$$\frac{a}{\sin A} = \frac{b}{\sin B} = \frac{c}{\sin C}$$

OR $$\frac{\sin A}{a} = \frac{\sin B}{b} = \frac{\sin C}{c}$$

Proof

In $\triangle ABC$ draw perpendicular AD and call it h.

From $\triangle ABD$, $\sin B = \frac{h}{c}$

$\therefore h = c \sin B$ [1]

From $\triangle ACD$, $\sin C = \frac{h}{b}$

$\therefore h = b \sin C$ [2]

From [1] and [2], $c \sin B = b \sin C$

$$\frac{\sin B}{b} = \frac{\sin C}{c}$$

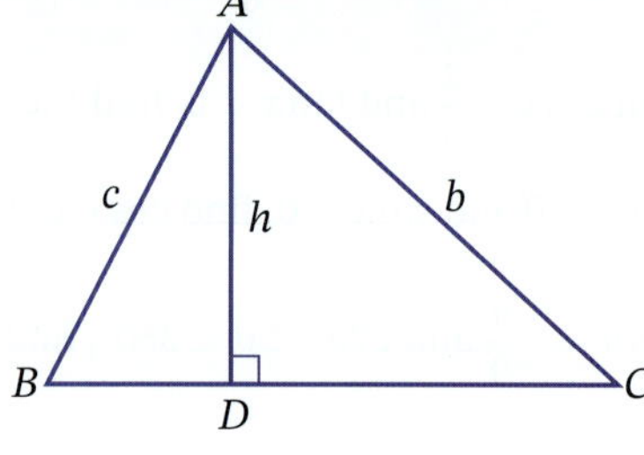

Similarly, by drawing a perpendicular from C, it can be proved that $\frac{\sin A}{a} = \frac{\sin B}{b}$

Example 14

a Find the value of x, correct to one decimal place.

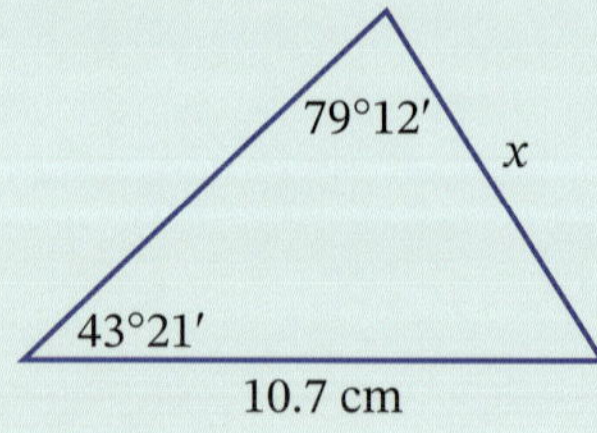

b Find the value of y, to the nearest whole number.

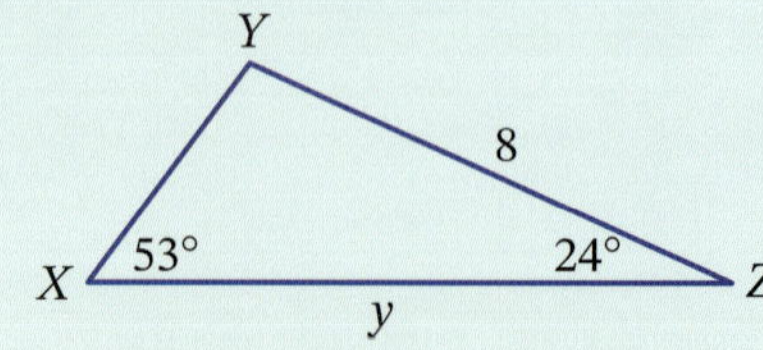

c Find the value of θ, in degrees and minutes, given θ is acute.

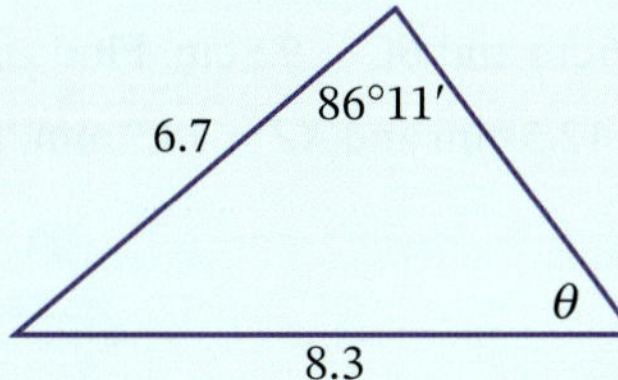

Solution

a Name the sides a and b, and opposite angles A and B.

$$\frac{a}{\sin A} = \frac{b}{\sin B}$$

Look for 2 sides paired with their opposite angles.

$$\frac{x}{\sin 43°21'} = \frac{10.7}{\sin 79°12'}$$

$$x = \frac{10.7 \sin 43°21'}{\sin 79°12'}$$

$$\approx 7.5 \text{ cm}$$

b First, we need to find angle Y since it is opposite side y.

$\angle Y = 180° - (53° + 24°) = 103°$

$$\frac{a}{\sin A} = \frac{b}{\sin B}$$

$$\frac{y}{\sin 103°} = \frac{8}{\sin 53°}$$

$$y = \frac{8 \sin 103°}{\sin 53°}$$

$$\approx 10$$

c

$$\frac{\sin A}{a} = \frac{\sin B}{b}$$

$$\frac{\sin\theta}{6.7} = \frac{\sin 86°11'}{8.3}$$

Look for 2 angles paired with their opposite sides.

$$\sin\theta = \frac{6.7 \sin 86°11'}{8.3}$$

$$= 0.8054...$$

$$\theta = \sin^{-1}(0.8054...)$$ SHIFT sin Ans

$$\approx 53°39'$$

The ambiguous case when finding an angle

When using the sine rule to find an unknown angle, there are 2 possible solutions: one acute and one obtuse. This is because sine has the same positive value in both the first and second quadrants. (The calculator will only show the acute solution but from that you can work out the obtuse solution). This is called the **ambiguous case** of the sine rule.

Example 15

a Triangle ABC has $\angle B = 53°$, $AC = 7.6$ cm and $BC = 9.5$ cm. Find $\angle A$ to the nearest degree.

b In triangle XYZ, $\angle Y = 118°35'$, $YZ = 12.5$ mm and $XZ = 14.3$ mm. Find $\angle X$ in degrees and minutes.

Solution

a Draw a diagram.

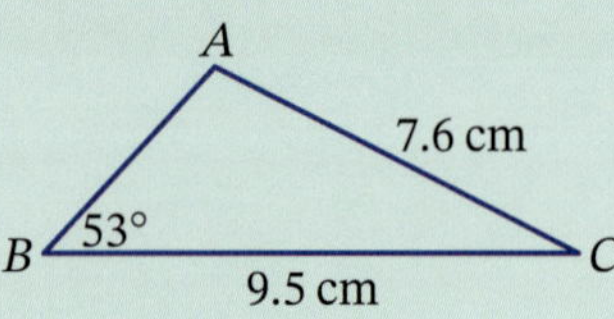

$$\frac{\sin A}{a} = \frac{\sin B}{b}$$

$$\frac{\sin A}{9.5} = \frac{\sin 53°}{7.6}$$

$$\sin A = \frac{9.5 \sin 53°}{7.6}$$

$$= 0.998$$

$$A = \sin^{-1}(0.998)$$

$$\approx 87°$$

But $\angle A$ could be obtuse.

So $\angle A = 180° - 87°$

$= 93°$

Checking angle sum of a triangle:

$53° + 87° = 140° < 180°$, so $87°$ is a possible answer.

$53° + 93° = 146° < 180°$, so $93°$ is a possible answer.

So $\angle A = 87°$ or $93°$.

b

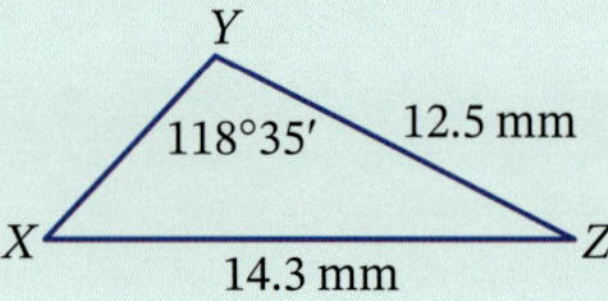

$$\frac{\sin A}{a} = \frac{\sin B}{b}$$

$$\frac{\sin X}{12.5} = \frac{\sin 118°35'}{14.3}$$

$$\sin X = \frac{12.5 \sin 118°35'}{14.3}$$

$$= 0.768$$

$$X = \sin^{-1}(0.768)$$

$$\approx 50°8'$$

But $\angle X$ could be obtuse.

So $\angle X = 180° - 50°8'$

$= 129°52'$.

Checking angle sum of a triangle:

$118°35' + 50°8' = 168°43' < 180°$, so a possible answer.

$118°35' + 129°52' = 248°27' > 180°$, so an impossible answer.

So $\angle X = 50°8'$.

Example 16

Find the value of θ, in degrees and minutes, given θ is obtuse.

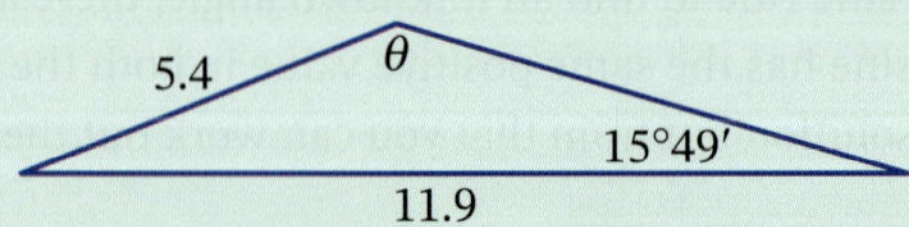

Solution

$$\frac{\sin A}{a} = \frac{\sin B}{b}$$

$$\frac{\sin\theta}{11.9} = \frac{\sin 15°49'}{5.4}$$

$$\sin\theta = \frac{11.9 \sin 15°49'}{5.4}$$

$$= 0.6006\ldots$$

$$\theta = \sin^{-1}(0.6006\ldots)$$

$$\approx 36°55'$$

But θ is obtuse.

$\therefore \theta = 180° - 36°55'$

$= 143°05'$

EXERCISE 4.04 Answers on p. 477

The sine rule

1 Evaluate each unknown, correct to one decimal place.

a
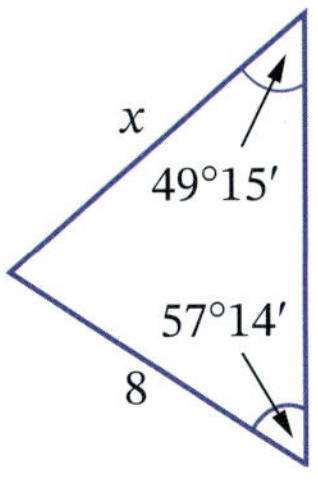

b
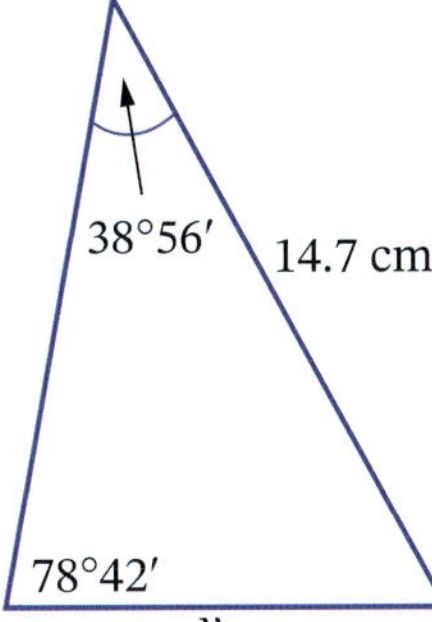

c
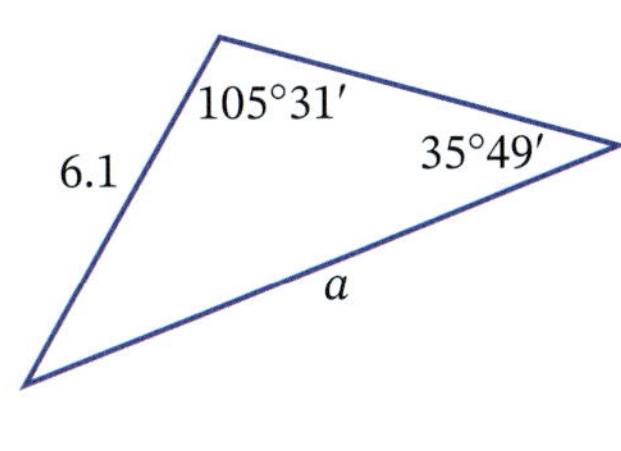

d
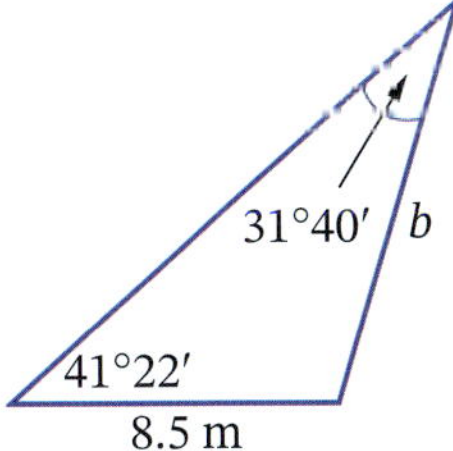

e
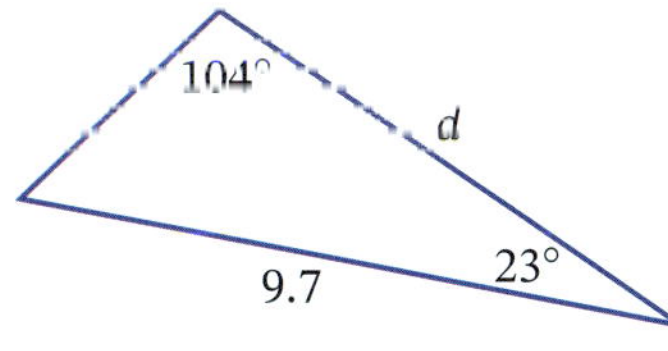

2 Find all possible values for each unknown angle, in degrees and minutes.

a
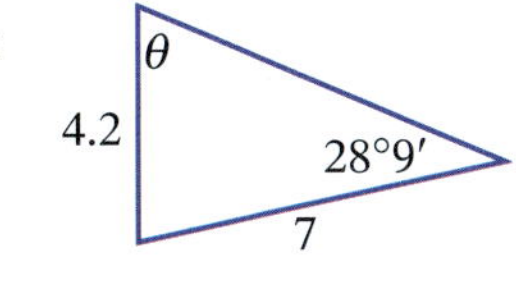

b
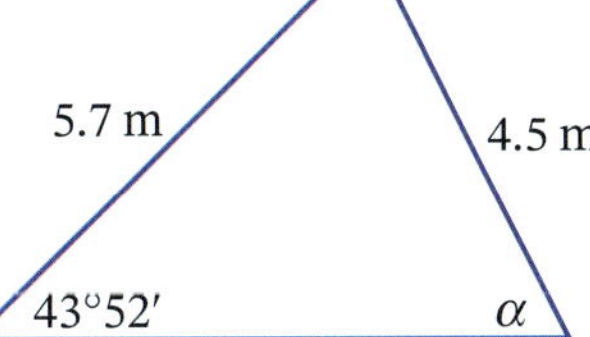

c
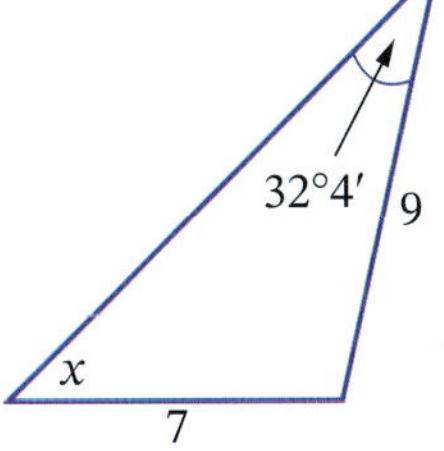

d
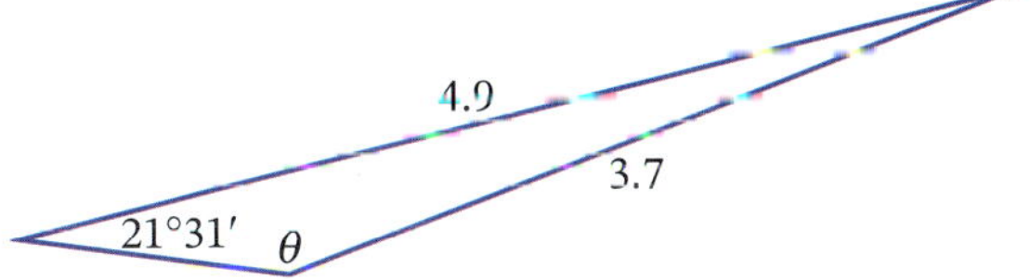

e
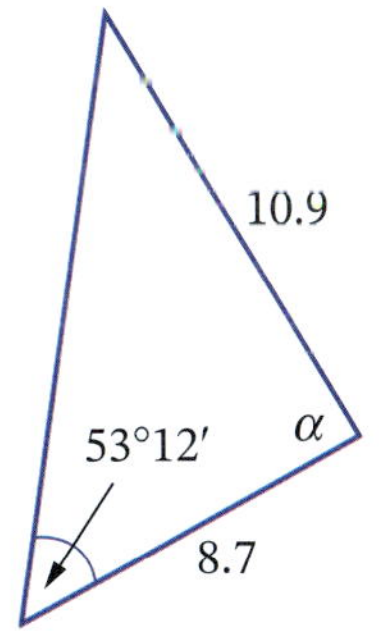

3 Triangle XYZ has $XY = 5.4$ cm, $\angle ZXY = 48°$ and $\angle XZY = 63°$. Find the length of XZ.

4 Triangle ABC has an obtuse angle at A. Evaluate this angle to the nearest minute if $AB = 3.2$ cm, $BC = 4.6$ cm and $\angle ACB = 33°47'$.

5 Triangle ABC has $BC = 12.7$ m, $\angle ABC = 47°$ and $\angle ACB = 53°$ as shown. Find the length of:

a AB

b AC

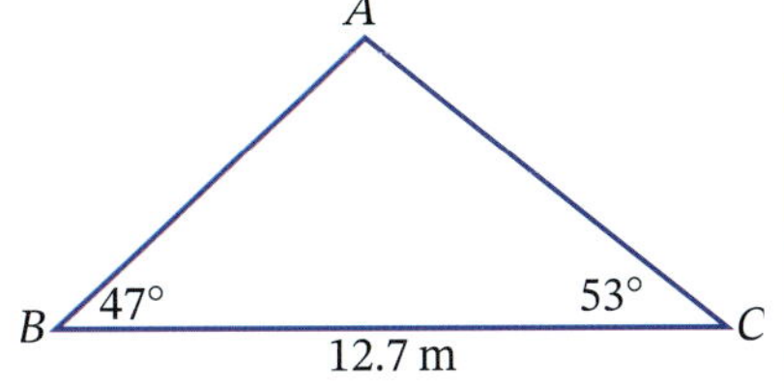

Foundation Mastery Complex

6 Triangle EFG has $\angle FEG = 48°$, $\angle EGF = 32°$ and $FG = 18.9$ mm. Find the length of:

a the shortest side **b** the longest side.

7 Manuel starts from home and walks for 2.7 km on a bearing of 054°.
He then turns and walks on a bearing of 165° until he is due east of his house.

a How far is Manuel from his house?

b How far does Manuel walk on the second part of his walk?

8 A plane flies from Sydney for 495 km on a bearing of 315° then turns and flies on a bearing of 200° until it is 550 km from Sydney. On what bearing is Sydney from the plane to the nearest degree?

9 The city of Silvertown is on a bearing of 293° from Bayview. A plane leaves Bayview to fly to Silvertown at a constant speed of 210 km/h. There is a constant wind blowing at 38 km/h from the south. On what bearing should the plane fly to reach Silvertown? Answer to the nearest degree.

10 Triangle ABC is isosceles with $AB = AC$.
BC is produced to D as shown.

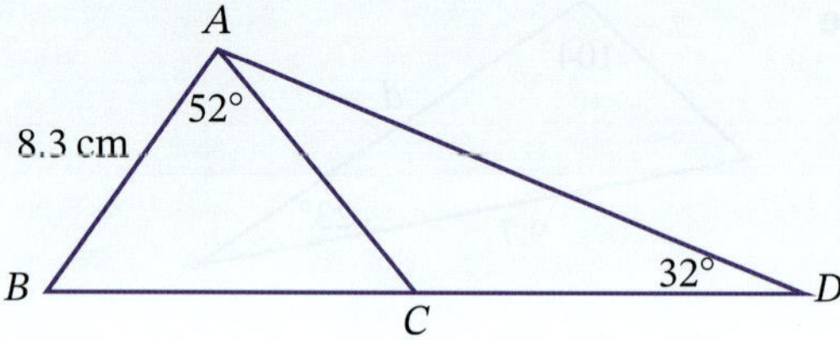

If $AB = 8.3$ cm, $\angle BAC = 52°$ and $\angle ADC = 32°$, find the length of:

a AD **b** BD

11 Triangle ABC is equilateral with side 63 mm.
A line is drawn from A to BC, where it meets BC at D and $\angle DAB = 26°15'$.
Find the length of:

a AD **b** DC

12 In triangle ABC, find $\angle B$ to the nearest degree given:

a $\angle C = 67°$, $AB = 7.2$, $AC = 7.5$

b $\angle A = 92°$, $BC = 10.7$, $AC = 8.4$

c $\angle A = 29°$, $BC = 4.9$, $AC = 8.3$

13 In the diagram, AB and DC are parallel, and $BD = 10$ cm.

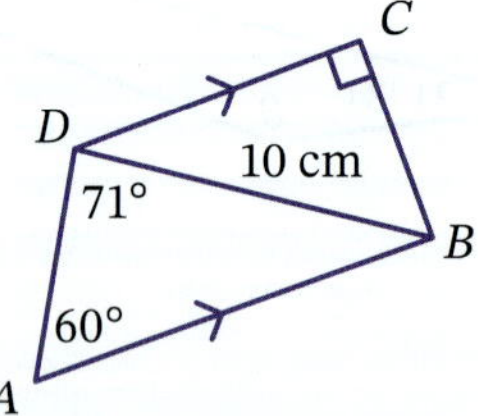

a Show that $CB = 10 \sin 49°$.

b Show that $AD = \dfrac{10 \sin 49°}{\sin 60°}$.

c Show $AD = \dfrac{2\sqrt{3}CB}{3}$ cm.

Foundation Mastery Complex

The cosine rule 4.05

The cosine rule

The **cosine rule** relates the 3 sides of a triangle to one of its angles. The rule is also used to find unknown sides and angles in non-right-angled triangles.

$$c^2 = a^2 + b^2 - 2ab\cos C$$

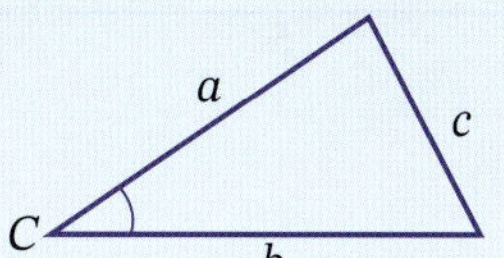

Worksheets
Cosine rule problems

Finding an unknown side

Finding an unknown angle

Proof

In triangle ABC, draw perpendicular AD with length p and let $CD = x$.

Since $BC = a$, $BD = a - x$.

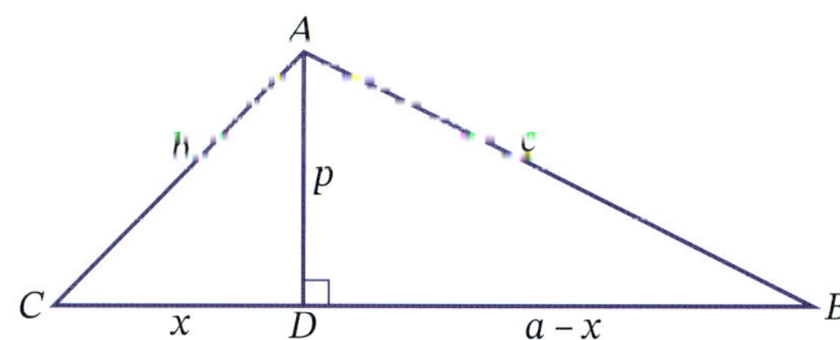

From triangle ACD:

$b^2 = x^2 + p^2$ [1]

$\cos C = \dfrac{x}{b}$

$\therefore b\cos C = x$ [2]

From triangle DAB:

$c^2 = p^2 + (a - x)^2$

$= p^2 + a^2 - 2ax + x^2$

$= p^2 + x^2 + a^2 - 2ax$ [3]

Substitute [1] into [3]:

$c^2 = b^2 + a^2 - 2ax$ [4]

Substituting [2] into [4]:

$c^2 = b^2 + a^2 - 2ab\cos C$

Example 17

Find the value of x, correct to the nearest whole number.

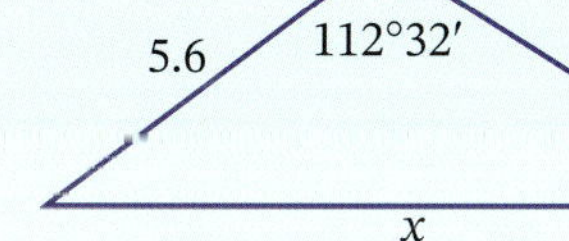

Solution

$c^2 = a^2 + b^2 - 2ab\cos C$

$x^2 = 5.6^2 + 6.4^2 - 2(5.6)(6.4)\cos 112°32'$

$= 99.7892\ldots$

$x = \sqrt{99.7892\ldots}$

$= 9.9894\ldots$

≈ 10

Look for 3 sides and the angle opposite the unknown side.

When using the cosine rule to find an unknown angle, it may be more convenient to change the subject of this formula to $\cos C$.

$c^2 = a^2 + b^2 - 2ab\cos C$

$2ab\cos C = a^2 + b^2 - c^2$

$\cos C = \dfrac{a^2 + b^2 - c^2}{2ab}$

The cosine rule for angles

$$\cos C = \frac{a^2 + b^2 - c^2}{2ab}$$

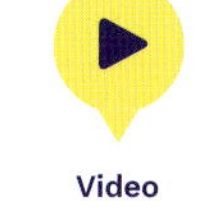

Video
The cosine rule for angles

Example 18

a Find θ, in degrees and minutes.

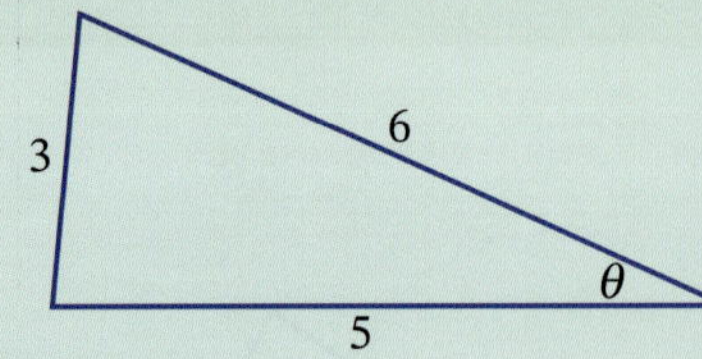

b Evaluate $\angle BCA$ in degrees and minutes.

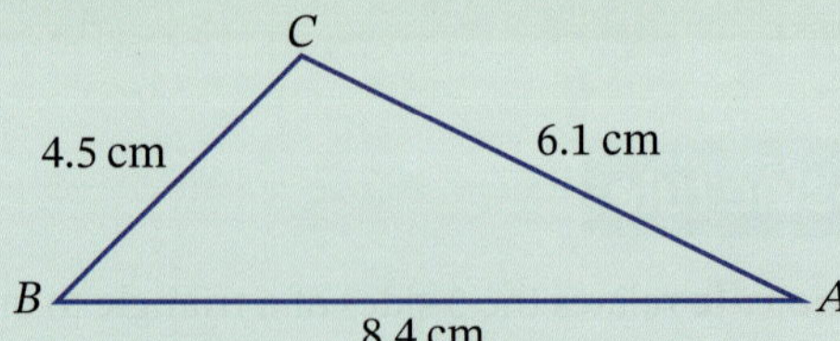

Solution

Naming sides and opposite angles, side c is opposite the unknown angle C.

a

$$\cos C = \frac{a^2 + b^2 - c^2}{2ab}$$

$$\cos\theta = \frac{5^2 + 6^2 - 3^2}{2(5)(6)}$$

$$= \frac{52}{60}$$

$$\theta = \cos^{-1}\left(\frac{52}{60}\right)$$

$$\approx 29°56'$$

b

$$\cos C = \frac{a^2 + b^2 - c^2}{2ab}$$

$$\cos\angle BCA = \frac{4.5^2 + 6.1^2 - 8.4^2}{2(4.5)(6.1)}$$

$$= -0.2386...$$

$$\angle BCA = \cos^{-1}(-0.2386...)$$

$$\approx 103°48'$$

Look for 3 sides and the side opposite the unknown angle.

Did you know?

The cosine rule for right-angled triangles

Pythagoras' theorem is a special case of the cosine rule when the triangle is right-angled.

$$c^2 = a^2 + b^2 - 2ab\cos C$$

When $C = 90°$:

$$c^2 = a^2 + b^2 - 2ab\cos 90°$$

$$= a^2 + b^2 - 2ab \times 0$$

$$= a^2 + b^2$$

EXERCISE 4.05 Answers on p. 477

The cosine rule

1 Find each unknown side, correct to one decimal place.

a

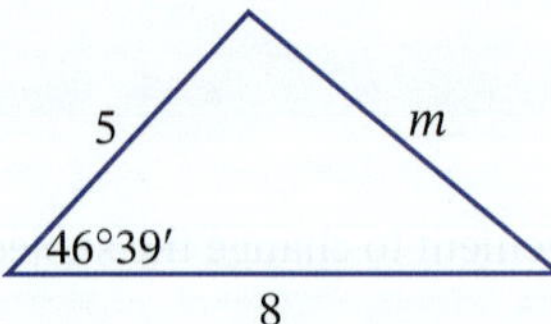

b

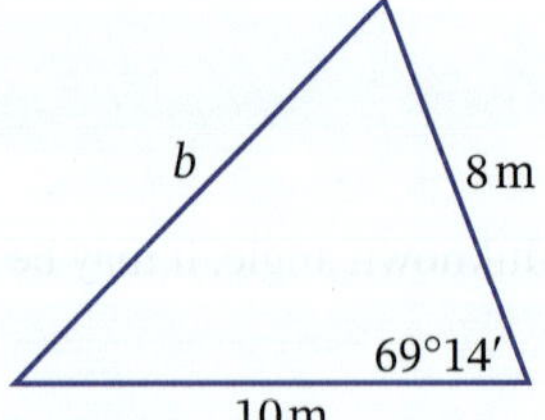

c

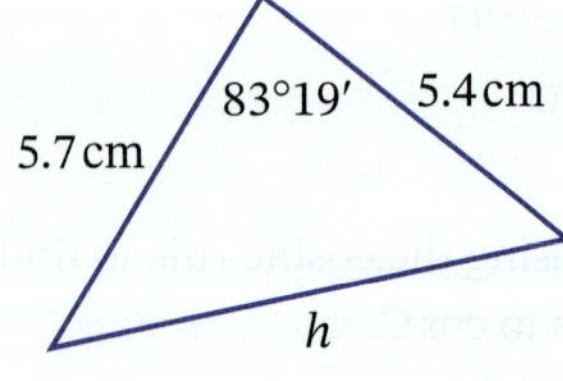

d

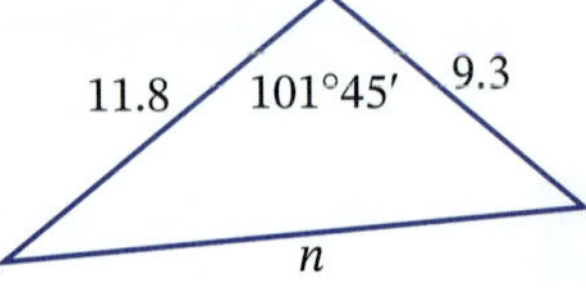

e

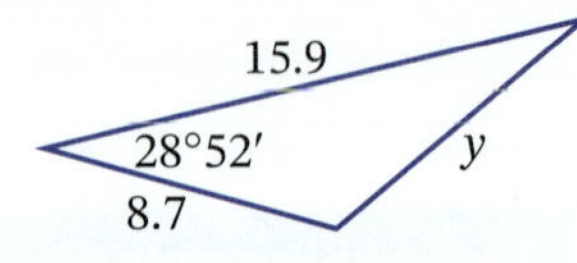

Foundation Mastery Complex

EXAMPLE 18

2 Find each unknown angle, correct to the nearest minute.

a

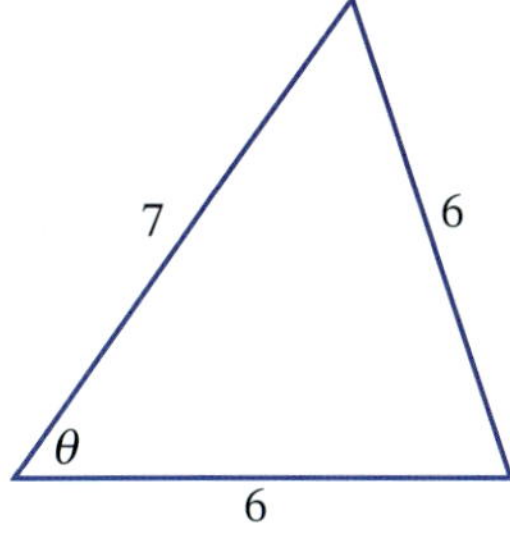

b

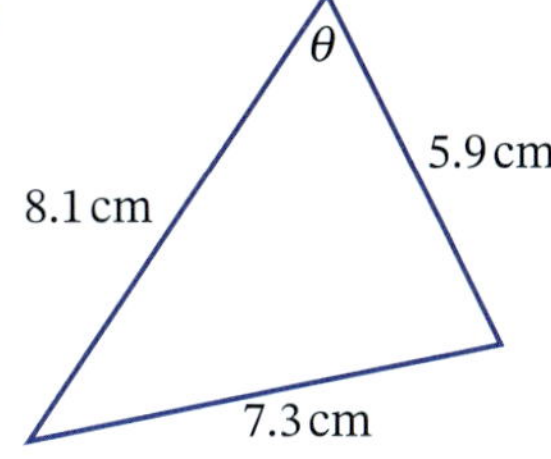

c

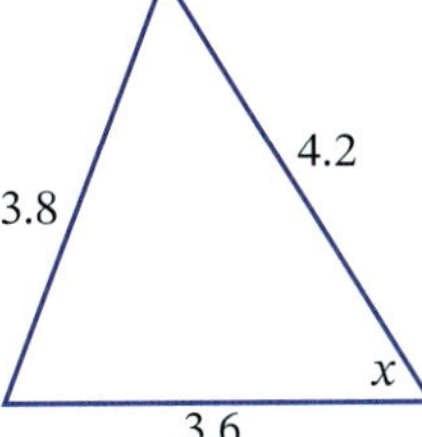

d

5.3 mm β 6.1 mm

10.4 mm

e

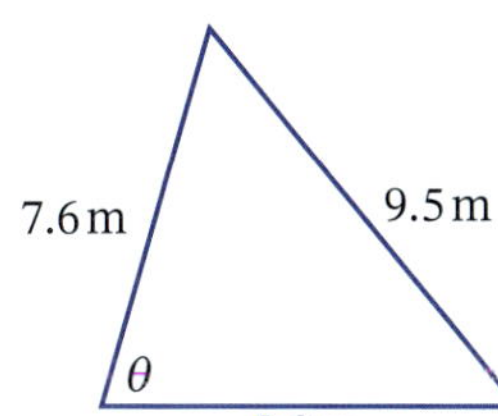

3 Kite $ABCD$ has $AB = 12.9$ mm, $CD = 23.8$ mm and $\angle ABC = 125°$ as shown.
Find the length of diagonal AC.

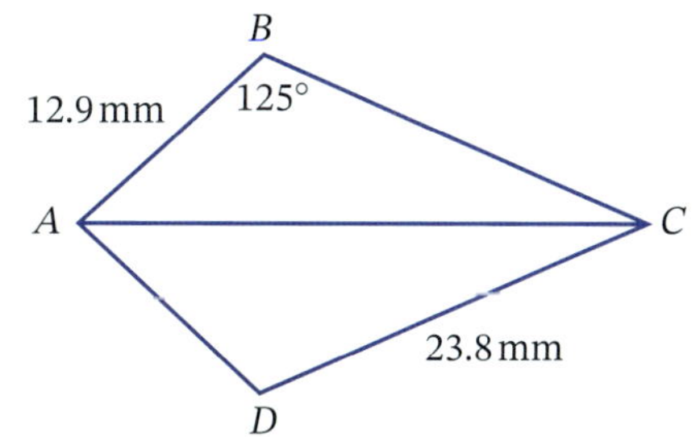

4 Parallelogram $ABCD$ has sides 11 cm and 5 cm, and one interior angle 79°25′. Find the length of the diagonals.

5 Quadrilateral $ABCD$ has sides $AB = 12$ cm, $BC = 10.4$ cm, $CD = 8.4$ cm and $AD = 9.7$ cm with $\angle ABC = 63°57'$. Find:

a the length of diagonal AC **b** $\angle DAC$ **c** $\angle ADC$.

6 Triangle XYZ is isosceles with $XY = XZ = 7.3$ cm and $YZ = 5.9$ cm. Find the value of all angles, to the nearest minute.

7 Quadrilateral $MNOP$ has $MP = 12$ mm, $NO = 12.7$ mm, $MN = 8.9$ mm, $OP = 15.6$ mm and $\angle NMP = 119°15'$. Find:

a the length of diagonal NP **b** $\angle NOP$.

8 Given the figure, find the length of:

a AC

b AD

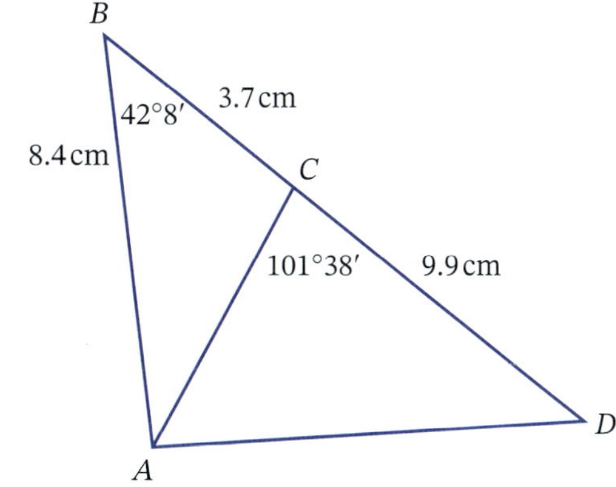

Foundation Mastery Complex

9 In a regular pentagon $ABCDE$ with sides 8 cm, find the length of diagonal AD.

10 A regular hexagon $ABCDEF$ has sides 5.5 cm. Find:

a the length of AD **b** $\angle ADF$.

11 Two people on motorbikes start at the same position. One travels for 81.3 km on a bearing of 225°, while the other travels for 59.7 km on a bearing of 305°. How far apart are they now?

12 A regular pentagon has equal sides of 3 cm. It forms an isosceles triangle AOE where O is the centre of the pentagon, as shown.

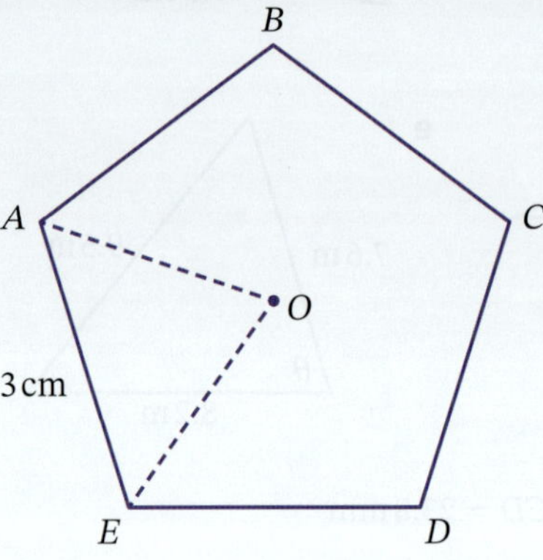

a Find the size of angle AOE.

b Find the length of AO, correct to one decimal place.

4.06 Area of a triangle

Video
The sine area formula

Worksheet
Areas of triangles

Trigonometry allows us to find the area of a triangle if we know 2 sides and their **included angle**.

Sine formula for the area of a triangle

$$A = \frac{1}{2}ab\sin C$$

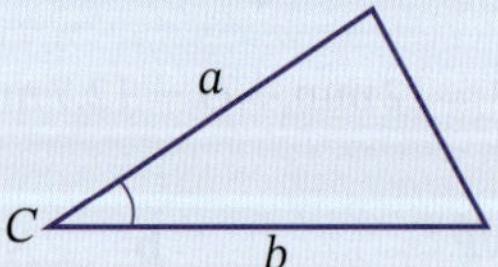

Proof

From $\triangle BCD$,

$$\sin C = \frac{h}{a}$$

$$\therefore h = a\sin C$$

$$A = \frac{1}{2}bh$$

$$= \frac{1}{2}ba\sin C$$

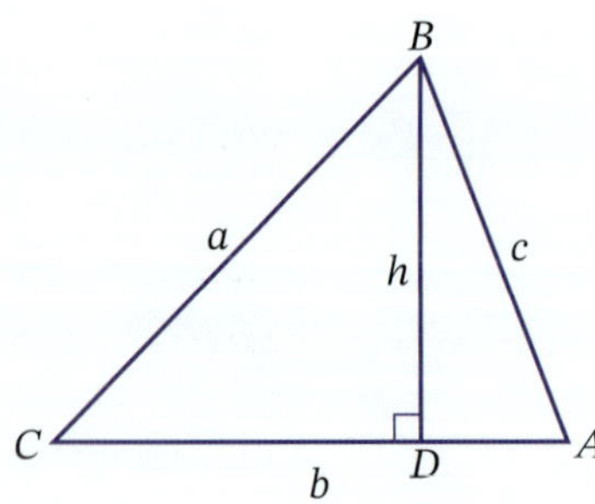

Foundation Mastery Complex

Example 19

Find the area of $\triangle ABC$, correct to 2 decimal places.

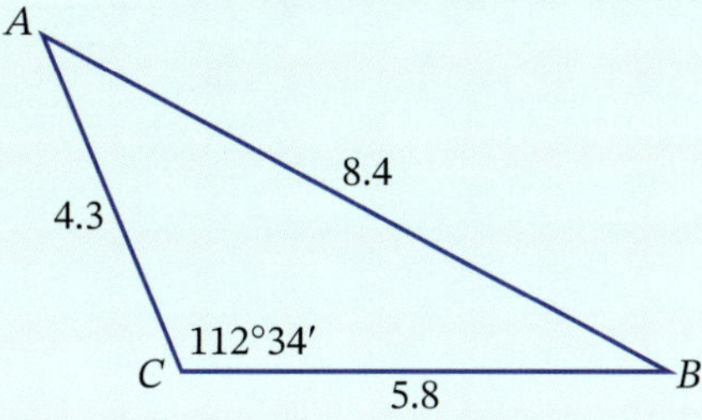

Solution

$$A = \frac{1}{2}ab\sin C$$

$$= \frac{1}{2}(4.3)(5.8)\sin 112°34'$$

$$\approx 11.52 \text{ units}^2$$

EXERCISE 4.06 Answers on p. 477

Area of a triangle

EXAMPLE 19

1 Find the area of each triangle, correct to one decimal place.

a

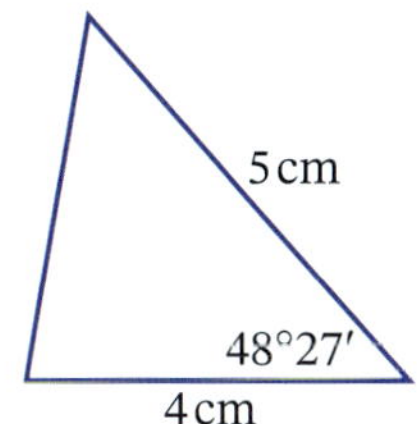

b

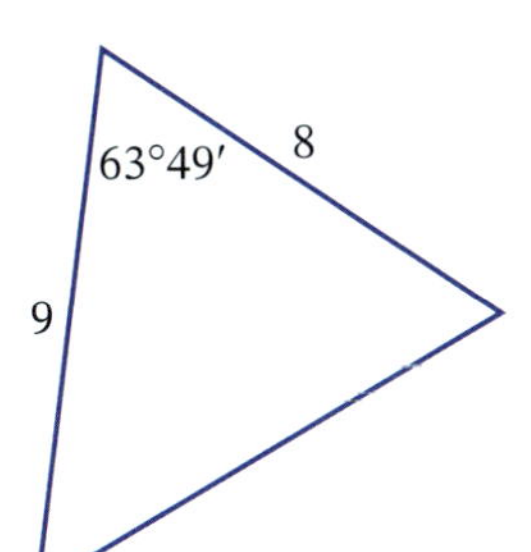

c

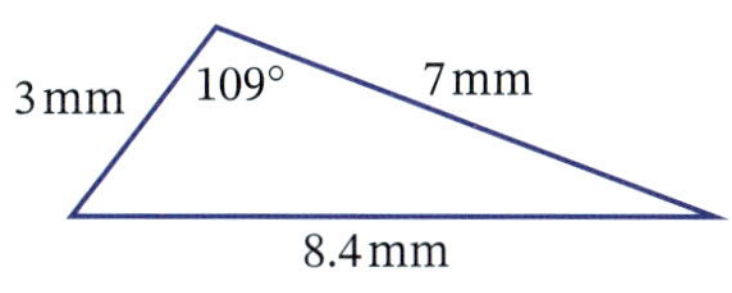

d

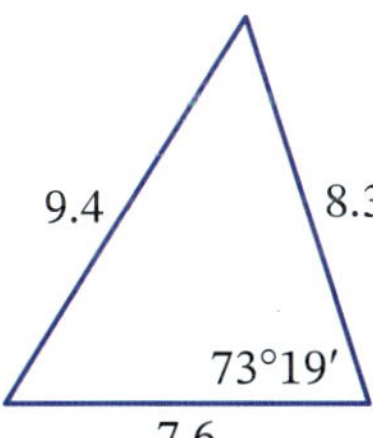

e

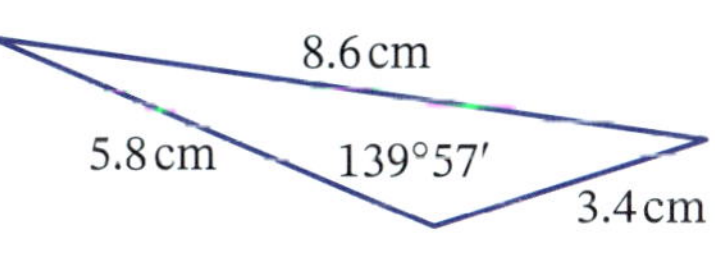

2 Find the area of $\triangle OAB$, correct to one decimal place (O is the centre of the circle).

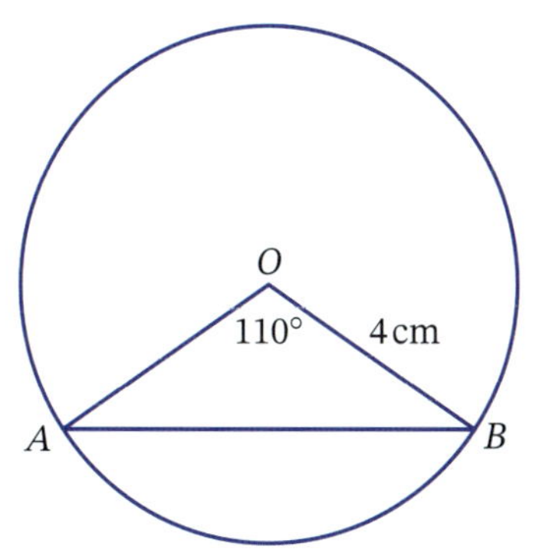

□ Foundation ○ Mastery ⬡ Complex

3 Find the area of a parallelogram with sides 3.5 cm and 4.8 cm and with one of its internal angles $67°13'$, correct to one decimal place.

4 Find the area of kite $ABCD$, correct to 3 significant figures.

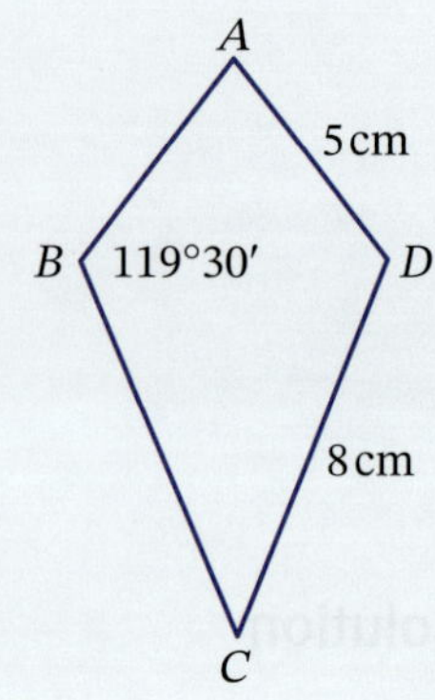

5 This pentagon is made from a rectangle and isosceles triangle with $AE = AB$, as shown. Find:

a the length of AE

b the area of the figure.

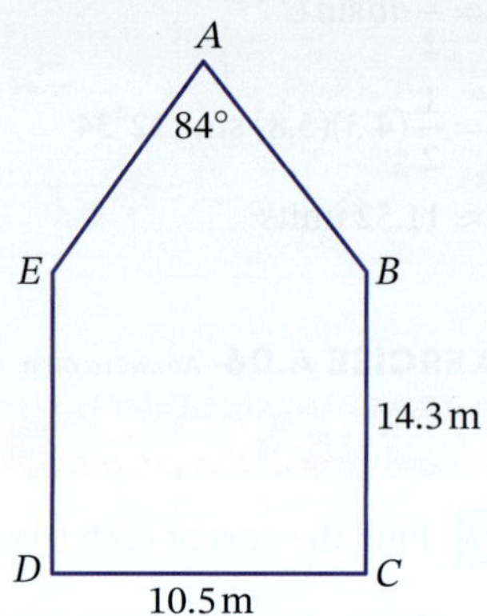

6 For this figure, find:

a the length of AC

b the area of triangle ACD

c the area of triangle ABC.

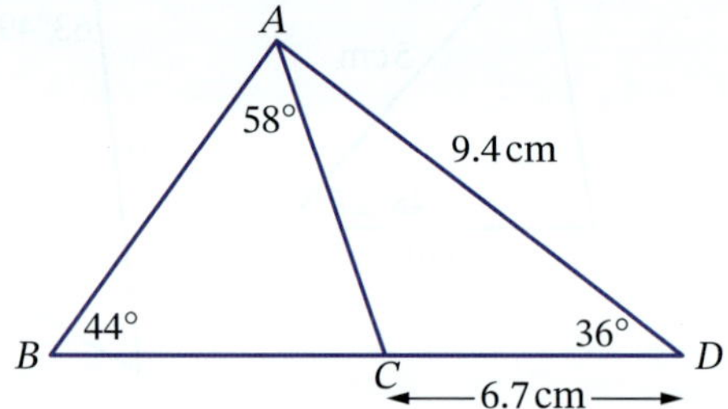

7 Find the exact area of an equilateral triangle with sides 5 cm.

8 Triangle MNO is cut from a semicircle with diameter 15 mm as shown. Angle $NMO = 51°$.

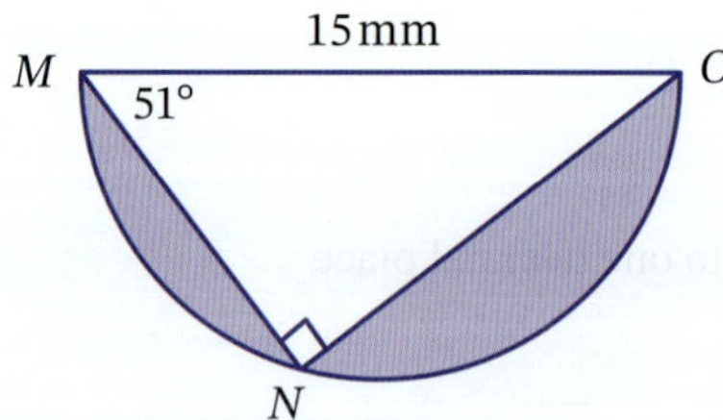

a Calculate the length of MN.

b Find the area of the shaded region, correct to one decimal place.

Foundation Mastery Complex

9 A regular hexagon has sides of length 2.5 mm. It forms an isosceles triangle EOD with O at the center of the hexagon.

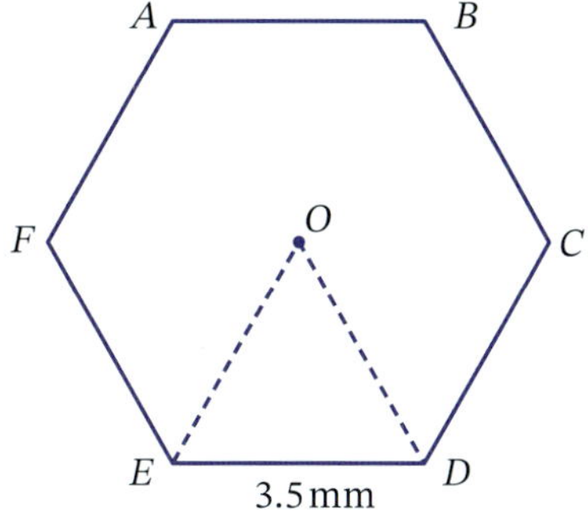

a Find the size of angle EOD.

b Find the length of EO, correct to one decimal place.

c Find the area of triangle EOD, correct to one decimal place.

10 An isosceles triangle with equal sides of 10 cm has an area of 42 cm^2.

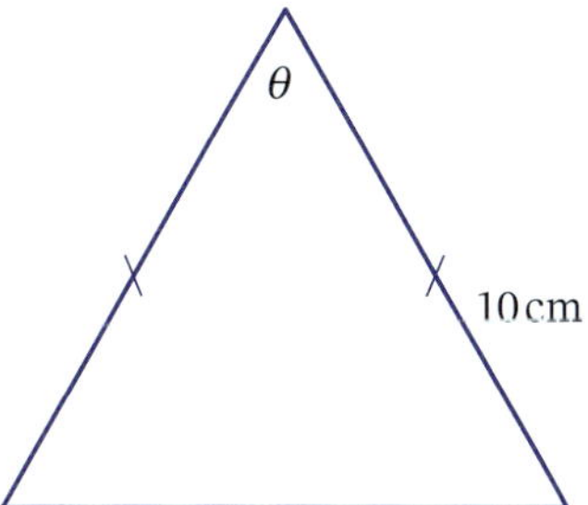

If θ is obtuse, evaluate its size correct to the nearest minute.

11 A triangular garden bed with side lengths 1.3 m and 1.7 m has an area of 1 m^2. Find, correct to the nearest degree, 2 possible values for the size of the angle between the 2 sides.

Further applications of trigonometry 4.07

The sine and cosine rules

Use the **sine rule** to find:

- a side, given one side and 2 angles
- an angle, given 2 sides and one angle.

Use the **cosine rule** to find:

- a side, given 2 sides and the included angle
- an angle, given 3 sides.

Video The sine and cosine rules

Worksheet The sine and cosine rules

☐ Foundation ◯ Mastery ⬡ Complex

Example 20

a The angle of elevation of a tower from point A is 72°. From point B, 50 m further away from the tower than A, the angle of elevation is 47°.

i Find the exact length of AT, the distance from A to the top of the tower.

ii Hence, or otherwise, find the height h of the tower, correct to one decimal place.

b A ship sails from Sydney for 200 km on a bearing of 040° then sails on a bearing of 157° for 345 km.

i How far from Sydney is the ship, to the nearest km?

ii What is the bearing of the ship from Sydney, to the nearest degree?

Solution

a

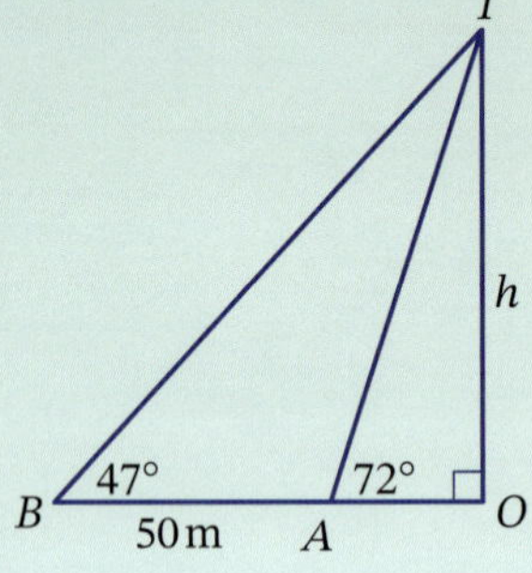

i $\angle BAT = 180° - 72° = 108°$ (straight angle)

$\angle BTA = 180° - (47° + 108°)$ (angle sum of △BTA)

$= 25°$

$$\frac{a}{\sin A} = \frac{b}{\sin B}$$

$$\frac{AT}{\sin 47°} = \frac{50}{\sin 25°}$$

$$\therefore AT = \frac{50 \sin 47°}{\sin 25°}$$

ii $\sin 72° = \dfrac{h}{AT}$

$\therefore h = AT \sin 72°$

$= \dfrac{50 \sin 47°}{\sin 25°} \times \sin 72°$

≈ 82.3 m

b

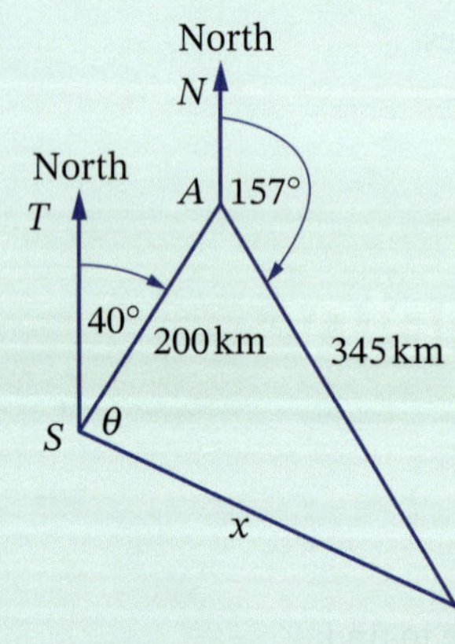

i $\angle SAN = 180° - 40° = 140°$ (cointerior angles)

$\therefore \angle SAB = 360° - (140° + 157°)$ (angle of revolution)

$= 63°$

$c^2 = a^2 + b^2 - 2ab \cos C$

$x^2 = 200^2 + 345^2 - 2(200)(345) \cos 63°$

$= 96\ 374.3110...$

$x = \sqrt{96\,374.3110...}$

$= 310.4421...$

≈ 310

So the ship is 310 km from Sydney.

ii $\dfrac{\sin A}{a} = \dfrac{\sin B}{b}$

$\dfrac{\sin \theta}{345} = \dfrac{\sin 63°}{310.4421...}$

$\therefore \sin \theta = \dfrac{345 \sin 63°}{310.4421...}$

$= 0.9901...$

$\theta \approx 82°$

The bearing from Sydney $= 40° + 82° = 122°$.

EXERCISE 4.07 Answers on p. 477

Further applications of trigonometry

1 A car is broken down to the north of 2 towns. The car is 39 km from town A and 52 km from town B. If A is due west of B and the 2 towns are 68 km apart, what is the bearing, to the nearest degree, of the car from:

a town A **b** town B?

2 The angle of elevation to the top of a tower is 54°37′ from a point 12.8 m out from its base. The tower is leaning at an angle of 85°58′ as shown. Find the height of the tower.

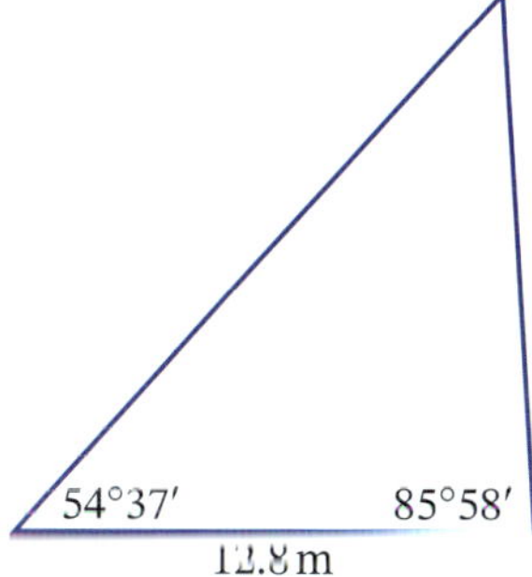

3 Rugby league goal posts are 5.5 m apart. If a footballer is standing 8 m from one post and 11 m from the other, find the angle within which the ball must be kicked to score a goal, to the nearest degree.

4 A boat is sinking 1.3 km out to sea from a marina. Its bearing is 041° from the marina and 324° from a rescue boat. The rescue boat is due east of the marina.

a How far, correct to 2 decimal places, is the rescue boat from the sinking boat?

b How long will it take the rescue boat, to the nearest minute, to reach the other boat if it travels at 80 km/h?

5 The angle of elevation of the top of a flagpole is 20° from where Thuy stands a certain distance away from its base. After walking 80 m towards the flagpole, Thuy finds the angle of elevation is 75°. Find the height of the flagpole, correct to the nearest metre.

6 A triangular field ABC has sides $AB = 85$ m and $AC = 50$ m. If B is on a bearing of 065° from A and C is on a bearing of 166° from A, find the length of BC, correct to the nearest metre.

7 Find the value of h, correct to one decimal place.

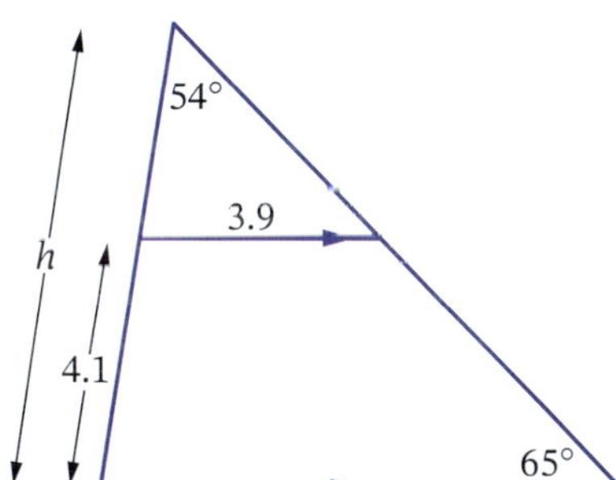

8 A motorbike and a car leave a service station at the same time. The motorbike travels on a bearing of 080° and the car travels for 15.7 km on a bearing of 108° until the bearing of the motorbike from the car is 310°. How far, correct to one decimal place, has the motorbike travelled?

9 A submarine is being followed by two ships, A and B, 3.8 km apart, with A due east of B. If A is on a bearing of 165° from the submarine and B is on a bearing of 205° from the submarine, find the distance from the submarine to both ships.

10 A plane flies from Dubbo on a bearing of 139° for 852 km, then turns and flies on a bearing of 285° until it is due west of Dubbo. How far from Dubbo is the plane, correct to the nearest km?

Foundation Mastery Complex

11. Rhombus $ABCD$ with side 8 cm has diagonal BD 11.3 cm long. Find $\angle DAB$.
12. Zeke leaves school and runs for 8.7 km on a bearing of $338°$, then turns and runs on a bearing of $061°$ until he is due north of school. How far north of school is he?
13. A car drives due east for 83.7 km then turns and travels for 105.6 km on a bearing of $029°$. How far is the car from its starting point?
14. A plane leaves Sydney and flies for 1280 km on a bearing of $050°$. It then turns and flies for 3215 km on a bearing of $149°$. How far is the plane from Sydney, to the nearest km?
15. Trapezium $ABCD$ has $AD \| BC$, with $AB = 4.6$ cm, $BC = 11.3$ cm, $CD = 6.4$ cm, $\angle DAC = 23°30'$ and $\angle ABC = 78°$. Find:
    - **a** the length of AC
    - **b** $\angle ADC$ to the nearest minute.
16. A plane leaves Adelaide and flies for 875 km on a bearing of $056°$. It then turns and flies on a bearing of θ for 630 km until it is due east of Adelaide. Evaluate θ, correct to the nearest degree.
17. Quadrilateral $ABCD$ has $AB = AD = 7.2$ cm, $BC = 8.9$ cm and $CD = 10.4$ cm, with $\angle DAB = 107°$. Find:
    - **a** the length of diagonal BD
    - **b** $\angle BCD$.
18. A wall leans inwards and makes an angle of $88°$ with the floor.
    - **a** A 4 m long ladder leans against the wall with its base 2.3 m out from the wall. Find the angle that the top of the ladder makes with the wall.
    - **b** A longer ladder is placed the same distance out from the wall and its top makes an angle of $31°$ with the wall.
        - **i** How long is this ladder?
        - **ii** How much further does it reach up the wall than the first ladder?
19. A group of Year 11 students walk 1.8 km from the school to the local pool on a bearing of $074°$ while a group of Year 12 students walk 2.3 km to the tennis courts on a bearing of $296°$.
    - **a** Find the distance between the pool and the tennis courts, correct to one decimal place.
    - **b** Find the bearing of the pool from the tennis courts, correct to the nearest degree.
20. A regular octagon has all sides and angles equal. This regular octagon has a perimeter of 120 cm.

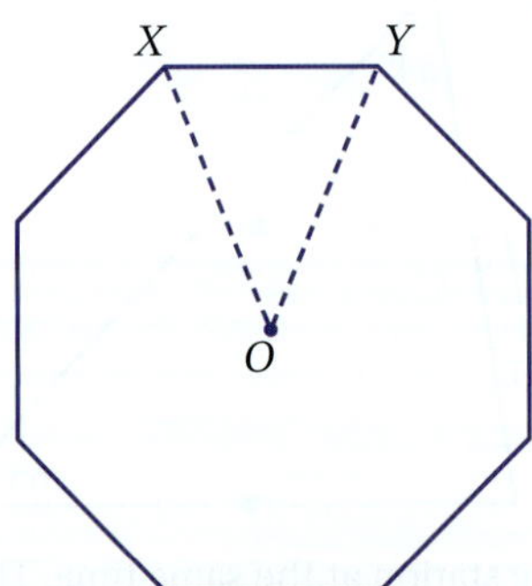

Calculate the area of the octagon, correct to one decimal place.

Foundation Mastery Complex

Radians

4.08

We use degrees to measure angles in geometry and trigonometry, but there are other units for measuring angles, such as radians and grads.

A **radian** is a unit for measuring angles based on the length of an **arc** in a circle. It is the arc length as a multiple of the radius of the circle, that is, the ratio of the arc length to the radius.

1 radian is the angle subtended by an arc with length r in a circle (of radius r).

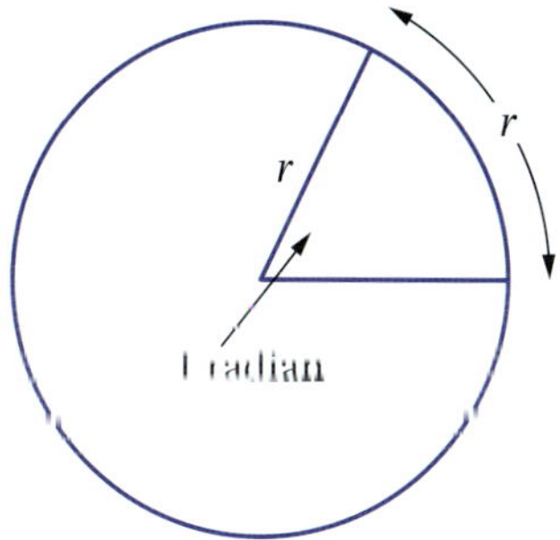

Can you estimate how many degrees are in a radian?

Since we know how many degrees and radians are in a revolution (full circle), we can use this to make conversions between degrees and radians.

The circumference of a circle is $C = 2\pi r$.

$\therefore$ the number of radians in a whole circle is $\dfrac{2\pi r}{r} = 2\pi$.

But there are 360° in a whole circle (angle of revolution).

So $2\pi = 360°$

$\pi = 180°$.

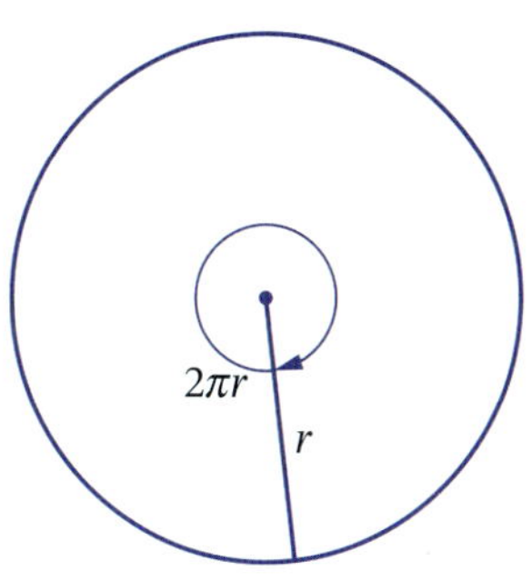

Radians and degrees

π radians = 180°

To convert x degrees to radians:

$180° = \pi$ radians

$1° = \dfrac{\pi}{180}$

$x° = x \times \dfrac{\pi}{180}$

To convert x radians to degrees:

π radians $= 180°$

1 radian $= \dfrac{180}{\pi}$

x radians $= x \times \dfrac{180}{\pi}$

Converting between radians and degrees

To change from radians to degrees: multiply by $\dfrac{180}{\pi}$.

To change from degrees to radians: multiply by $\dfrac{\pi}{180}$.

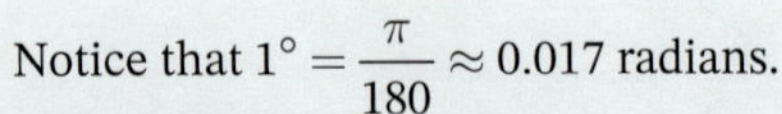

Notice that $1° = \frac{\pi}{180} \approx 0.017$ radians.

Also 1 radian $= \frac{180}{\pi} \approx 57°18'$.

Video
Degrees and radians

Example 21

a Convert $\frac{3\pi}{2}$ into degrees.

b Change 60° to radians, leaving your answer in terms of π.

c Convert 50° into radians, correct to 2 decimal places.

d Change 1.145 radians into degrees, correct to the nearest minute.

e Convert 38°41′ into radians, correct to 3 decimal places.

f Evaluate cos 1.145, correct to 2 decimal places.

Solution

a Since $\pi = 180°$,

$$\frac{3\pi}{2} = \frac{3(180°)}{2} = 270°$$

b $180° = \pi$ radians

So $1° = \frac{\pi}{180}$ radians

$$60° = \frac{\pi}{180} \times 60$$
$$= \frac{60\pi}{180}$$
$$= \frac{\pi}{3}$$

You could skip the above steps and go straight to this step.

c $180° = \pi$ radians

So $1° = \frac{\pi}{180}$ radians

$$50° = \frac{\pi}{180} \times 50$$
$$= \frac{50\pi}{180}$$
$$\approx 0.87$$

d π radians $= 180°$

$\therefore$ 1 radian $= \frac{180°}{\pi}$

$$1.145 \text{ radians} = \frac{180°}{\pi} \times 1.145$$
$$\approx 65.6°$$
$$= 65°36'$$

You could skip the above steps and go straight to this step.

e $180° = \pi$ radians

$$1° = \frac{\pi}{180°}$$

$$38°41' = \frac{\pi}{180°} \times 38°41'$$
$$= 0.675$$

f

Operation	Casio scientific	Sharp scientific
Make sure the calculator is in radians.	SHIFT MODE Rad	Press DRG until rad is on the screen.
Enter data.	cos 1.145 =	cos 1.145 =

$\cos 1.145 = 0.4130...$

≈ 0.41

If $180° = \pi$, then it's easy to remember the following frequently-used angles:

Special angles

$30° = \frac{\pi}{6}$ $\quad 45° = \frac{\pi}{4}$ $\quad 60° = \frac{\pi}{3}$ $\quad 90° = \frac{\pi}{2}$

Videos
Exact trigonometric values

The exact trigonometric ratios

Puzzles
Exact values

Exact values 2

The exact ratios

$\sin 30^\circ = \frac{1}{2}$ $\quad \sin 45^\circ = \frac{1}{\sqrt{2}}$ $\quad \sin 60^\circ = \frac{\sqrt{3}}{2}$

$\cos 30^\circ = \frac{\sqrt{3}}{2}$ $\quad \cos 45^\circ = \frac{1}{\sqrt{2}}$ $\quad \cos 60^\circ = \frac{1}{2}$

$\tan 30^\circ = \frac{1}{\sqrt{3}}$ $\quad \tan 45^\circ = 1$ $\quad \tan 60^\circ = \sqrt{3}$

The exact ratios in radians

$\sin\frac{\pi}{6} = \frac{1}{2}$ $\quad \sin\frac{\pi}{4} = \frac{1}{\sqrt{2}}$ $\quad \sin\frac{\pi}{3} = \frac{\sqrt{3}}{2}$

$\cos\frac{\pi}{6} = \frac{\sqrt{3}}{2}$ $\quad \cos\frac{\pi}{4} = \frac{1}{\sqrt{2}}$ $\quad \cos\frac{\pi}{3} = \frac{1}{2}$

$\tan\frac{\pi}{6} = \frac{1}{\sqrt{3}}$ $\quad \tan\frac{\pi}{4} = 1$ $\quad \tan\frac{\pi}{3} = \sqrt{3}$

Example 22

a **i** Convert $\frac{\pi}{3}$ to degrees.

ii Find the exact value of $\tan\frac{\pi}{3}$.

b Find the exact value of $\cos\frac{\pi}{4}$.

Solution

a **i** $\frac{\pi}{3} = \frac{180^\circ}{3}$

$= 60^\circ$

ii $\tan\frac{\pi}{3} = \tan 60^\circ$

$= \sqrt{3}$

b $\cos\frac{\pi}{4} = \cos 45^\circ$

$= \frac{1}{\sqrt{2}}$

Frequently-used angles in radians

As $\frac{\pi}{6} = 30^\circ$, $\frac{\pi}{4} = 45^\circ$, $\frac{\pi}{3} = 60^\circ$ and $\frac{\pi}{2} = 90^\circ$, then multiples of these angles in the quadrants 2, 3 and 4 are shown below.

$\frac{\pi}{6}$	$\frac{2\pi}{6} = \frac{\pi}{3}$	$\frac{3\pi}{6} = \frac{\pi}{2}$	$\frac{4\pi}{6} = \frac{2\pi}{3}$	$\frac{5\pi}{6}$	$\frac{6\pi}{6} = \pi$
30°	60°	90°	120°	150°	180°

$\frac{7\pi}{6}$	$\frac{8\pi}{6} = \frac{4\pi}{3}$	$\frac{9\pi}{6} = \frac{3\pi}{2}$	$\frac{10\pi}{6} = \frac{5\pi}{3}$	$\frac{11\pi}{6}$	$\frac{12\pi}{6} = 2\pi$
210°	240°	270°	300°	330°	360°

$\frac{\pi}{4}$	$\frac{2\pi}{4} = \frac{\pi}{2}$	$\frac{3\pi}{4}$	$\frac{4\pi}{4} = \pi$	$\frac{5\pi}{4}$	$\frac{6\pi}{4} = \frac{3\pi}{2}$	$\frac{7\pi}{4}$	$\frac{8\pi}{4} = 2\pi$
45°	90°	135°	180°	225°	270°	315°	360°

The rules and formulas learned in this chapter can also be expressed in radians.

Worksheet
Radians of any magnitude

ASTC rule

Positive angles

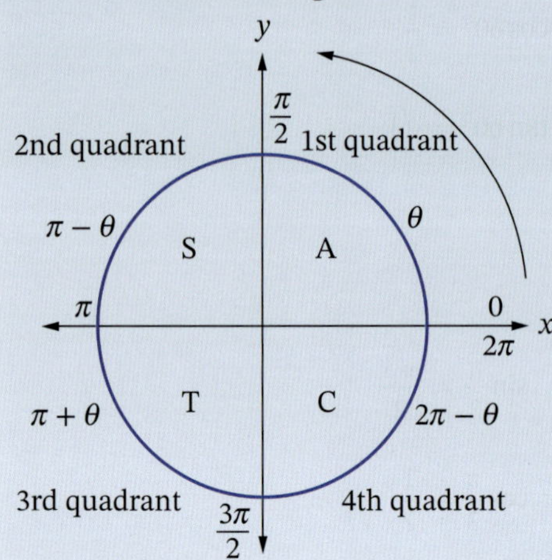

Negative angles

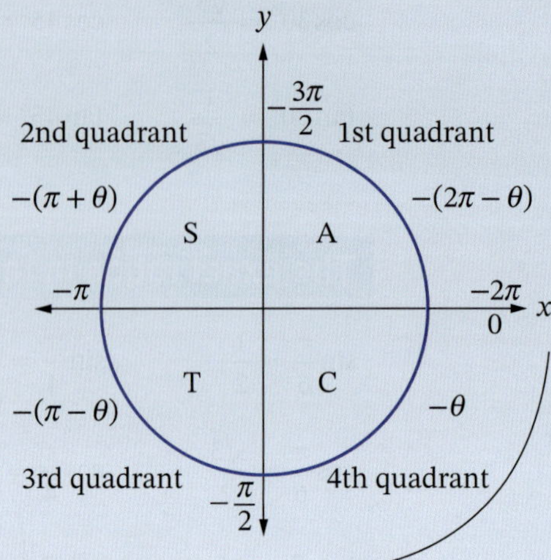

2nd quadrant:

$\sin(\pi - \theta) = \sin\theta$

$\cos(\pi - \theta) = -\cos\theta$

$\tan(\pi - \theta) = -\tan\theta$

4th quadrant:

$\sin(2\pi - \theta) = -\sin\theta$

$\cos(2\pi - \theta) = \cos\theta$

$\tan(2\pi - \theta) = -\tan\theta$

3rd quadrant:

$\sin(\pi + \theta) = -\sin\theta$

$\cos(\pi + \theta) = -\cos\theta$

$\tan(\pi + \theta) = \tan\theta$

Negative angles:

$\sin(-\theta) = -\sin\theta$

$\cos(-\theta) = \cos\theta$

$\tan(-\theta) = -\tan\theta$

Example 23

Find the exact value of:

a $\sin\left(\frac{5\pi}{4}\right)$

b $\cos\left(\frac{11\pi}{6}\right)$

Solution

a
$$\frac{5\pi}{4} = \frac{4\pi}{4} + \frac{\pi}{4}$$
$$= \pi + \frac{\pi}{4}$$

Angle is in the 3rd quadrant, so $\sin\theta < 0$.

$$\sin\left(\frac{5\pi}{4}\right) = \sin\left(\pi + \frac{\pi}{4}\right)$$
$$= -\sin\frac{\pi}{4}$$
$$= -\frac{1}{\sqrt{2}}$$

This question can also be completed by converting $\frac{5\pi}{4}$ to degrees first.

b
$$\frac{11\pi}{6} = \frac{12\pi}{6} - \frac{\pi}{6}$$
$$= 2\pi - \frac{\pi}{6}$$

Angle is in the 4th quadrant, so $\cos\theta > 0$.

$$\cos\left(\frac{11\pi}{6}\right) = \cos\left(2\pi - \frac{\pi}{6}\right)$$
$$= \cos\frac{\pi}{6}$$
$$= \frac{\sqrt{3}}{2}$$

EXERCISE 4.08 Answers on p. 477

Radians

EXAMPLE 21

1 Convert to degrees:

a $\frac{\pi}{5}$ **b** $\frac{2\pi}{3}$ **c** $\frac{5\pi}{4}$ **d** $\frac{7\pi}{6}$ **e** 3π

f $\frac{7\pi}{9}$ **g** $\frac{4\pi}{3}$ **h** $\frac{7\pi}{3}$ **i** $\frac{\pi}{9}$ **j** $\frac{5\pi}{18}$

2 Convert to radians in terms of π:

a 135° **b** 30° **c** 150° **d** 240° **e** 300°

f 63° **g** 15° **h** 450° **i** 225° **j** 120°

3 Change to radians, correct to 2 decimal places:

a 56° **b** 68° **c** 127° **d** 289° **e** 312°

4 Change to radians, correct to 2 decimal places:

a 18°34′ **b** 35°12′ **c** 101°56′ **d** 88°29′ **e** 50°39′

5 Convert each radian measure into degrees and minutes, to the nearest minute.

a 1.09 **b** 0.768 **c** 1.16 **d** 0.99 **e** 0.32

f 3.2 **g** 2.7 **h** 4.31 **i** 5.6 **j** 0.11

6 Find correct to 2 decimal places:

a $\sin 0.342$ **b** $\cos 1.5$ **c** $\tan 0.056$ **d** $\cos 0.589$ **e** $\tan 2.29$

f $\sin 2.8$ **g** $\tan 5.3$ **h** $\cos 4.77$ **i** $\cos 3.9$ **j** $\sin 2.98$

7 Find the exact value of:

a $\sin\frac{\pi}{4}$ **b** $\cos\frac{\pi}{3}$ **c** $\tan\frac{\pi}{6}$ **d** $\sin\frac{\pi}{3}$ **e** $\tan\frac{\pi}{4}$

f $\sin\frac{\pi}{6}$ **g** $\cos\frac{\pi}{4}$ **h** $\cos\frac{\pi}{6}$ **i** $\tan\frac{\pi}{3}$ **j** $\tan\pi$

8 **a** Show that $\frac{3\pi}{4} = \pi - \frac{\pi}{4}$.

b In which quadrant is $\frac{3\pi}{4}$?

c Find the exact value of $\cos\frac{3\pi}{4}$.

9 **a** Show that $\frac{5\pi}{6} = \pi - \frac{\pi}{6}$.

b In which quadrant is $\frac{5\pi}{6}$?

c Find the exact value of $\sin\frac{5\pi}{6}$.

Foundation Mastery Complex

10 **a** Show that $\frac{7\pi}{4} = 2\pi - \frac{\pi}{4}$.

b In which quadrant is $\frac{7\pi}{4}$?

c Find the exact value of $\tan \frac{7\pi}{4}$.

11 **a** Show that $\frac{4\pi}{3} = \pi + \frac{\pi}{3}$.

b In which quadrant is the angle $\frac{4\pi}{3}$?

c Find the exact value of $\cos \frac{4\pi}{3}$.

12 **a** Show that $\frac{5\pi}{3} = 2\pi - \frac{\pi}{3}$.

b In which quadrant is $\frac{5\pi}{3}$?

c Find the exact value of $\sin \frac{5\pi}{3}$.

13 **a** **i** Show that $\frac{13\pi}{6} = 2\pi + \frac{\pi}{6}$.

ii In which quadrant is $\frac{13\pi}{6}$?

iii Find the exact value of $\cos \frac{13\pi}{6}$.

b Find the exact value of:

i $\sin \frac{9\pi}{4}$ **ii** $\tan \frac{8\pi}{3}$ **iii** $\cos \frac{11\pi}{4}$

iv $\tan \frac{19\pi}{6}$ **v** $\sin \frac{10\pi}{3}$ **vi** $\cos \frac{8\pi}{3}$

14 Copy and complete each table with exact values.

a

	$\frac{\pi}{3}$	$\frac{2\pi}{3}$	$\frac{4\pi}{3}$	$\frac{5\pi}{3}$	$\frac{7\pi}{3}$	$\frac{8\pi}{3}$	$\frac{10\pi}{3}$	$\frac{11\pi}{3}$
sin								
cos								
tan								

b

	$\frac{\pi}{4}$	$\frac{3\pi}{4}$	$\frac{5\pi}{4}$	$\frac{7\pi}{4}$	$\frac{9\pi}{4}$	$\frac{11\pi}{4}$	$\frac{13\pi}{4}$	$\frac{15\pi}{4}$
sin								
cos								
tan								

Foundation Mastery Complex

c

	$\frac{\pi}{6}$	$\frac{5\pi}{6}$	$\frac{7\pi}{6}$	$\frac{11\pi}{6}$	$\frac{13\pi}{6}$	$\frac{17\pi}{6}$	$\frac{19\pi}{6}$	$\frac{23\pi}{6}$
sin								
cos								
tan								

15 Copy and complete the table where possible.

	0	$\frac{\pi}{2}$	π	$\frac{3\pi}{2}$	2π	$\frac{5\pi}{2}$	3π	$\frac{7\pi}{2}$	4π
sin									
cos									
tan									

The length of an arc 4.09

Since radian measure is defined from the length of an arc of a circle, we can use radians to find the arc length of a circle. You can find formulas for these using degrees, but they are not as simple. All the work on circles in this chapter uses radians.

Video
Arc length

The length of an arc

$$l = r\theta$$

where r is the radius of the circle and θ is the central angle subtended by the arc, measured in radians.

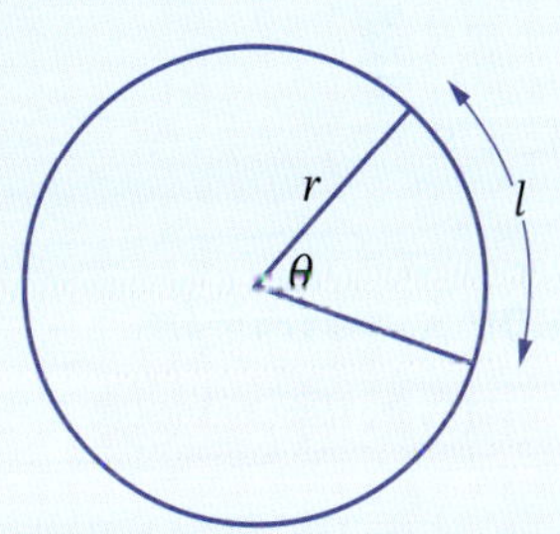

This formula can also be proven by comparing the arc length to the circumference, and the central angle to the whole revolution.

$$\frac{\text{arc length } l}{\text{circumference}} = \frac{\text{angle } \theta}{\text{whole revolution}}$$

$$\frac{l}{2\pi r} = \frac{\theta}{2\pi}$$

$$\therefore l = \frac{\theta 2\pi r}{2\pi}$$

$$= r\theta$$

☐ Foundation ○ Mastery ⬡ Complex

Example 24

a Find the length of the arc formed if it subtends an angle of $\frac{\pi}{4}$ at the centre of a circle of radius 5 m.

b Find the length of the arc formed given the central angle is 30° and the radius is 9 cm.

c The area of a circle is 450 cm². Find, in degrees and minutes, the angle subtended at the centre of the circle by a 2.7 cm arc.

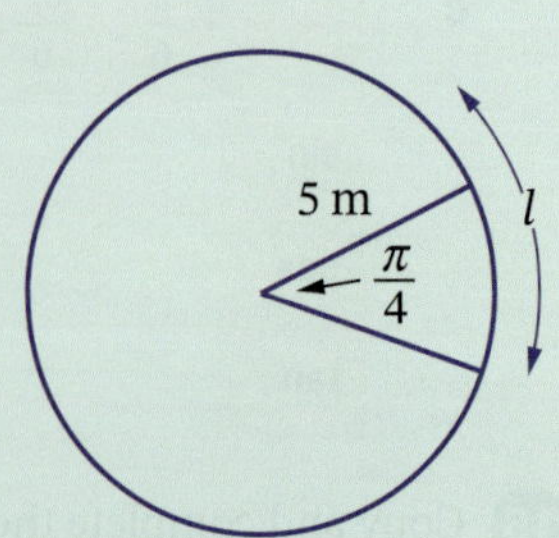

Solution

a
$$l = r\theta$$
$$= 5\left(\frac{\pi}{4}\right)$$
$$= \frac{5\pi}{4}\text{ m}$$

b First change 30° into radians.
$$\theta = \frac{\pi}{6}$$
$$l = r\theta$$
$$= 9\left(\frac{\pi}{6}\right)$$
$$= \frac{3\pi}{2}\text{ cm}$$

This formula works only if θ is in radians.

c
$$A = \pi r^2$$
$$450 = \pi r^2$$
$$\frac{450}{\pi} = r^2$$
$$\sqrt{\frac{450}{\pi}} = r$$
$$11.9682... = r$$
Now, $l = r\theta$.
$$2.7 = 11.9682...\theta$$
$$\frac{2.7}{11.9682...} = \theta$$
$$0.2255... = \theta$$

$$\pi \text{ radians} = 180°$$
$$1 \text{ radian} = \frac{180°}{\pi}$$
$$0.2255... \text{ radians} = \frac{180°}{\pi} \times 0.2255...$$
$$= 12.9257°$$
$$\approx 12°56'$$
So, $\theta = 12°56'$.

EXERCISE 4.09 Answers on p. 478

The length of an arc

EXAMPLE 24

1 Find the exact arc length of a circle with:

a radius 4 cm and subtended angle π

b radius 3 m and subtended angle $\frac{\pi}{3}$

c radius 10 cm and central angle $\frac{5\pi}{6}$

d radius 3 cm and central angle 30°

e radius 7 mm and subtended angle 45°.

2 Find the arc length, correct to 2 decimal places, given:

a radius 1.5 m and central angle 0.43

b radius 3.21 cm and angle 1.22

c radius 7.2 mm and angle 55°

d radius 5.9 cm and subtended angle 23°12′

e radius 2.1 m and angle 82°35′.

Foundation Mastery Complex

3 The angle at the centre of a circle of radius 3.4 m is 29°51′. Find the length of the arc subtending this angle, correct to one decimal place.

4 The arc length subtending an angle of $\frac{\pi}{5}$ at the centre is $\frac{3\pi}{2}$ m. Find the radius of the circle.

5 The radius of a circle is 3 cm and an arc is $\frac{2\pi}{7}$ cm. Find the angle subtended at the centre of the circle by the arc.

6 The circumference of a circle is 300 mm. Find the length of the arc that is formed by an angle of $\frac{\pi}{6}$ at the centre of the circle.

7 A circle with area 60 cm^2 has an arc 8 cm long. Find the angle that is subtended at the centre of the circle by the arc.

8 A circle with circumference 124 mm has an arc cut off that subtends an angle of 40° at the centre. Find the length of the arc.

9 A circle has a chord of 25 mm with an angle of $\frac{\pi}{6}$ subtended at the centre. Find, to one decimal place:

a the radius

b the length of the arc cut off by the chord.

A **chord** is an interval that joins 2 points on a circle.

10 A sector of a circle with radius 5 cm and an angle of $\frac{\pi}{3}$ subtended at the centre is curved around to form an open cone. Find the exact volume of the cone.

11 In this circle with radius 2.8 cm, $\angle AOB = 40°$.

a Find the length of the arc AB, correct to 2 decimal places.

b Find the length of the major arc AB (the rest of the circumference of the circle).

c Show that the area of the sector OAB is approximately 2.74 cm^2.

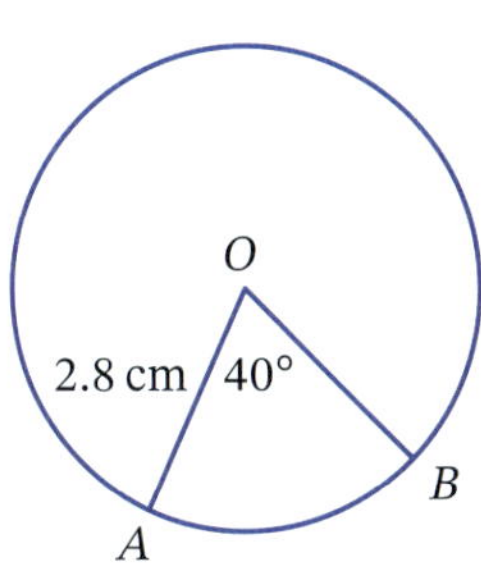

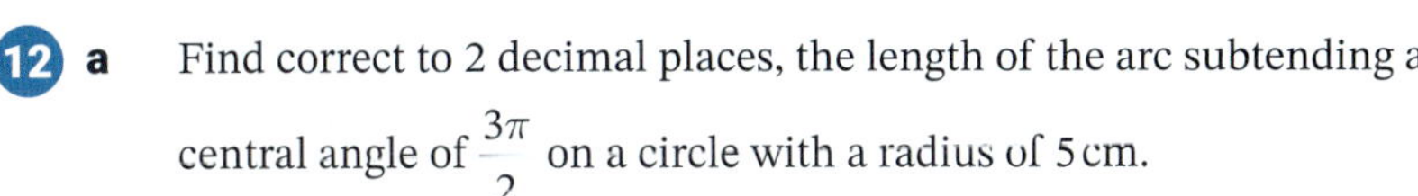

12 **a** Find correct to 2 decimal places, the length of the arc subtending a central angle of $\frac{3\pi}{2}$ on a circle with a radius of 5 cm.

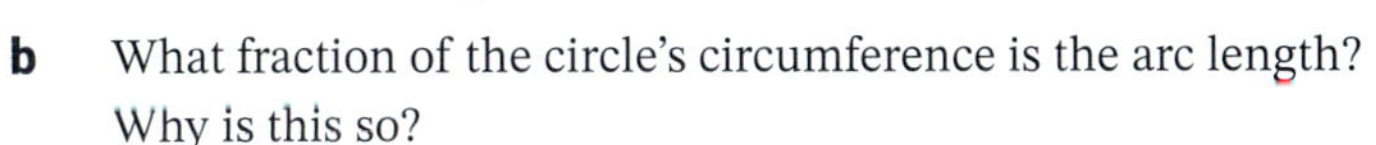

b What fraction of the circle's circumference is the arc length? Why is this so?

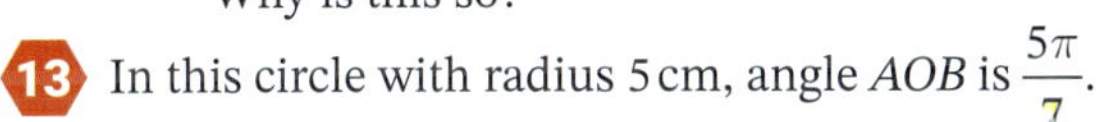

13 In this circle with radius 5 cm, angle AOB is $\frac{5\pi}{7}$.

a Find the exact length of the arc cut off by the chord AB.

b Find the length of the chord AB, correct to the nearest centimetre.

c Find the shaded area, correct to 2 decimal places.

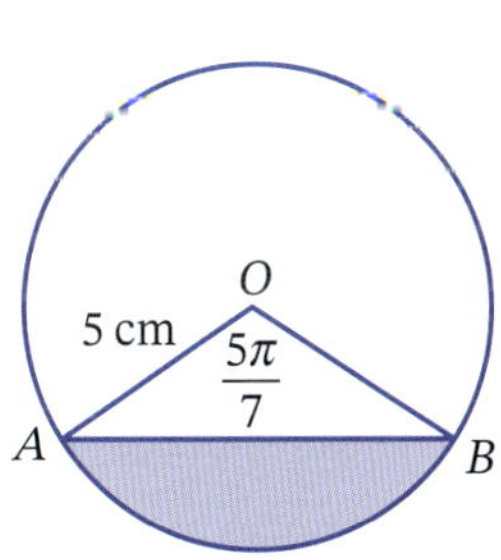

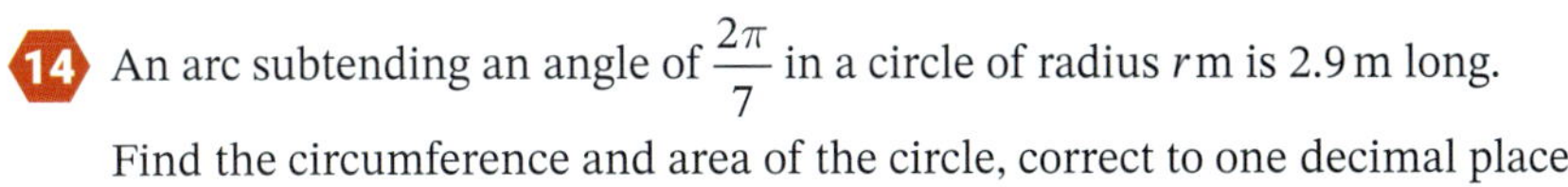

14 An arc subtending an angle of $\frac{2\pi}{7}$ in a circle of radius r m is 2.9 m long. Find the circumference and area of the circle, correct to one decimal place.

☐ Foundation ○ Mastery ⬡ Complex

4.10 Area of a sector

Area of a sector

$$A = \frac{1}{2}r^2\theta$$

where r is the radius of the circle and θ is the central angle, measured in radians.

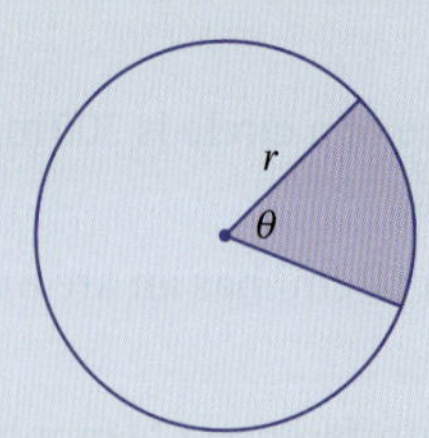

Proof

We can find the area of a **sector** by equating the ratio of the area of a sector to the area of the whole circle and the ratio of its central angle to a whole revolution.

$$\frac{\text{area of sector, } A}{\text{area of circle}} = \frac{\text{angle } \theta}{\text{whole revolution}}$$

$$\frac{A}{\pi r^2} = \frac{\theta}{2\pi}$$

$$\therefore A = \frac{\theta \pi r^2}{2\pi}$$

$$= \frac{1}{2}r^2\theta$$

OR

area of a sector $= \frac{x}{360} \times \pi r^2$, where x is measured in degrees

To convert θ radians to degrees, multiply by $\frac{180}{\pi}$:

$$\therefore \text{area of a sector} = \frac{\theta}{360} \times \frac{180}{\pi} \times \pi r^2$$

$$= \frac{1}{2}r^2\theta$$

Dahin/Shutterstock.com

Example 25

a Find the area of the sector with an angle of $\frac{\pi}{4}$ at the centre of a circle of radius 5 m.

b The area of the sector of a circle with radius 4 cm is $\frac{6\pi}{5}$ cm^2.

Find the angle, in degrees, at the centre of the circle.

Solution

a

$$A = \frac{1}{2}r^2\theta$$

$$= \frac{1}{2}(5)^2\left(\frac{\pi}{4}\right)$$

$$= \frac{25\pi}{8} \text{ m}^2$$

b

$$A = \frac{1}{2}r^2\theta$$

$$\frac{6\pi}{5} = \frac{1}{2}(4)^2\theta$$

$$= 8\theta$$

$$\frac{6\pi}{40} = \theta$$

$$\theta = \frac{3\pi}{20}$$

$$= \frac{3(180^\circ)}{20}$$

$$= 27^\circ$$

EXERCISE 4.10 Answers on p. 478

Area of a sector

EXAMPLE 25

1 Find the exact area of the sector of a circle whose radius is:

a 3 m and the central angle is $\frac{\pi}{3}$ **b** 10 cm and the central angle is $\frac{5\pi}{6}$

c 3 cm and the central angle is 30° **d** 7 mm and the central angle is 225°.

2 Find the area of the sector, correct to 2 decimal places, given the radius is:

a 1.5 m and the central angle is 0.43 **b** 3.21 cm and the central angle is 1.22

c 7.2 mm and the central angle is 55° **d** 5.9 cm and the central angle is 23°12′

e 2.1 m and the central angle is 82°35′.

3 Find the area, correct to 3 significant figures, of the sector of a circle with radius 4.3 m and an angle of 1.8 at the centre.

4 The area of a sector of a circle is 20 cm². If the radius of the circle is 3 cm, find the angle at the centre of the circle, correct to one decimal place.

5 **a** Find, correct to 2 decimal places, the area of a sector with radius 7.5 m and central angle π.

b What fraction of a circle with radius 7.5 m is the area of the above sector? Why is this so?

6 A circle with radius 7 cm has a sector cut off by an angle of 150° subtended at the centre of the circle. Find the exact value of:

a the arc length **b** the area of the sector.

7 A circle has a circumference of 185 mm. Find the area of the sector cut off by an angle of $\frac{8\pi}{5}$ at the centre. What fraction of the circle's area is the area of this sector?

8 If the area of a circle is 200 cm² and a sector is cut off by an angle of $\frac{3\pi}{4}$ at the centre, find the area of the sector.

9 Find the area of the sector of a circle with radius 5.7 cm if the length of the arc formed by this sector is 4.2 cm.

10 The area of a sector is $\frac{3\pi}{10}$ cm² and the arc length cut off by the sector is $\frac{\pi}{5}$ cm. Find the angle subtended at the centre of the circle and the radius of the circle.

11 If an angle of $\frac{\pi}{7}$ is at the centre of a circle with radius 3 cm, find the exact value of:

a the arc length **b** the area of the sector.

12 An angle of $\frac{7\pi}{6}$ is at the centre of a circle with radius 5 cm. Find:

a the length of the arc **b** the area of the sector **c** the length of the chord.

13 A chord 8 mm long subtends an angle of 45° at the centre of a circle. Find, correct to one decimal place,

a the radius of the circle

b the area of the sector cut off by the angle and the percentage it is of the circle's area.

14 **a** Find the area of the sector of a circle with radius 4 cm if the angle at the centre is $\frac{\pi}{4}$.

b Find the length of BC, correct to one decimal place.

c Find the exact area of triangle ABC.

d Hence find the exact area of the shaded minor segment of the circle.

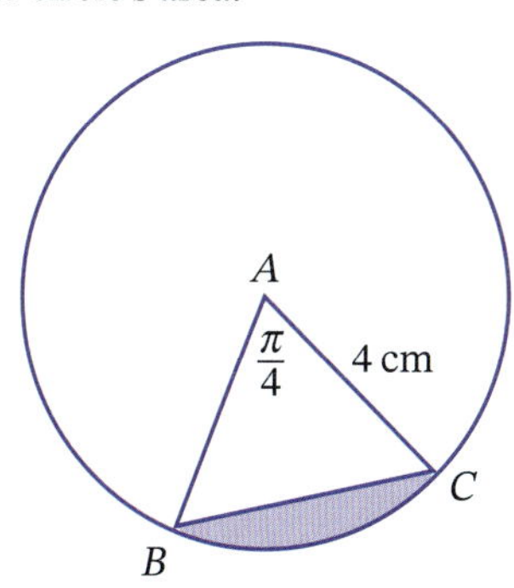

Foundation Mastery Complex

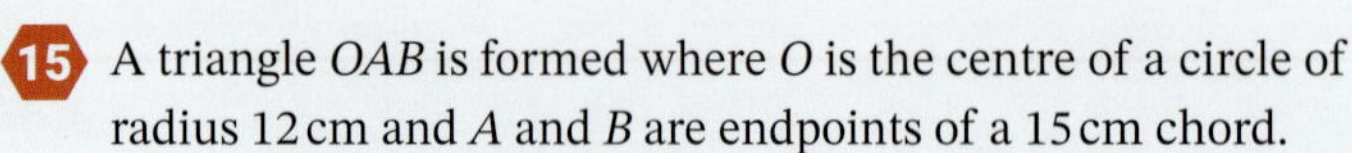

15 A triangle OAB is formed where O is the centre of a circle of radius 12 cm and A and B are endpoints of a 15 cm chord.

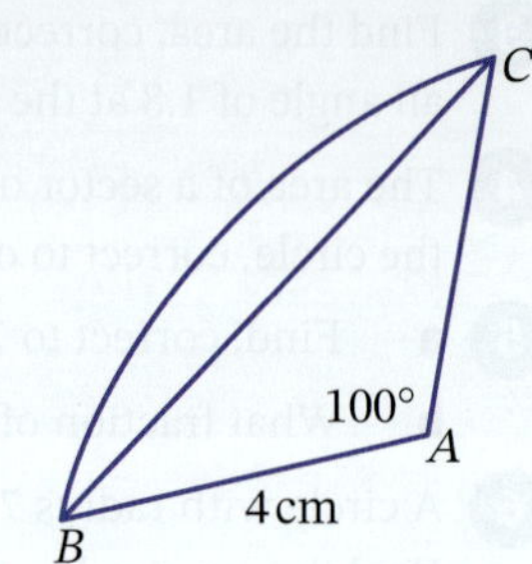

a Find the angle subtended at the centre of the circle, in degrees and minutes.

b Find the area of $\triangle OAB$, correct to one decimal place.

c Find the area of the minor segment cut off by the chord, correct to 2 decimal places.

d Find the area of the major segment cut off by the chord, correct to 2 decimal places.

> A chord divides a circle into 2 segments. The major **segment** is the larger segment.

16 Arc BC subtends an angle of $100°$ at the centre A of a circle with radius 4 cm. Find the perimeter of sector ABC.

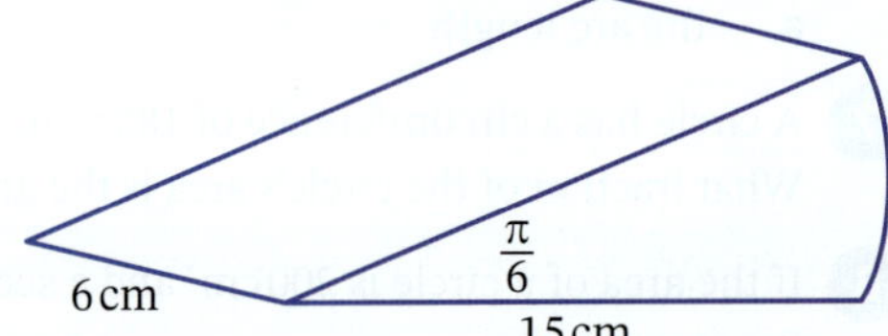

17 A wedge is cut so that its cross-sectional area is a sector of a circle with radius 15 cm and subtending an angle of $\dfrac{\pi}{6}$ at the centre. Find the exact volume of the wedge.

18 A sector in a circle has radius r and central angle θ radians.

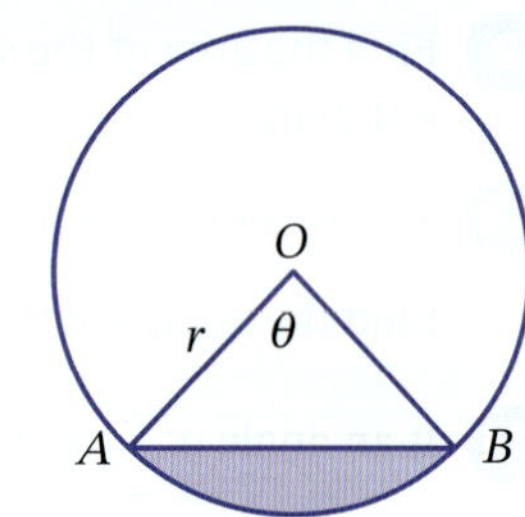

a Write the formula for the area of:

i the circle

ii $\triangle OAB$

iii sector OAB

iv the shaded minor segment.

b Find the exact area of the segment cut off by an angle of $\dfrac{\pi}{3}$ in a circle with radius 2 cm.

4.11 Area of a segment

A chord splits a circle into 2 segments as shown.

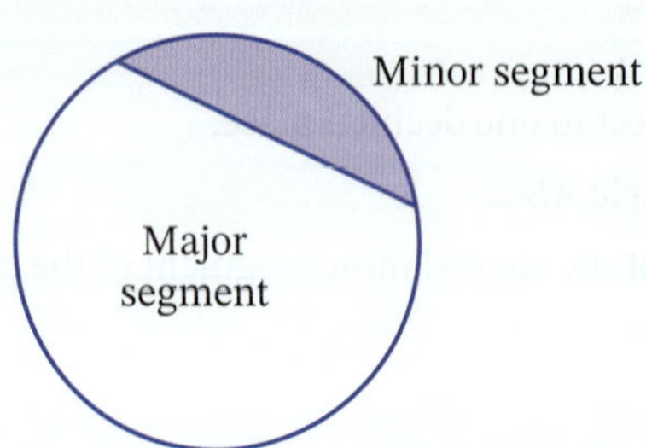

Consider a segment cut off by a chord in a sector of a circle, with radius r and central angle θ.

We can find the area of the segment.

The segment is minor if the central angle is less than 180° or π (acute or obtuse).

The segment is major if the central angle is more than 180° or π (reflex).

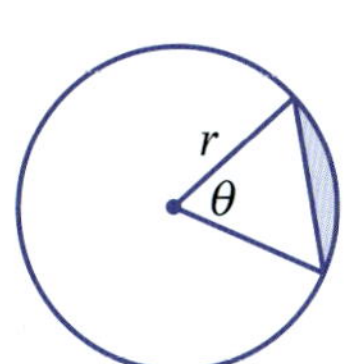

Minor segment

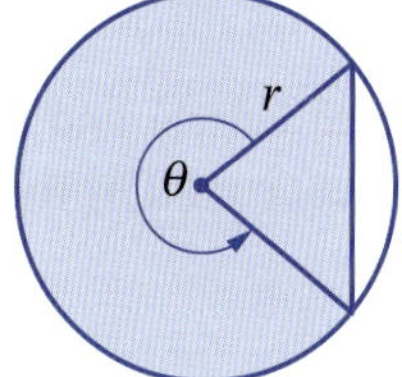

Major segment

Area of a segment

$$A=\frac{1}{2}r^2(\theta-\sin\theta)$$

where r is the radius of the circle and θ is the central angle, measured in radians.

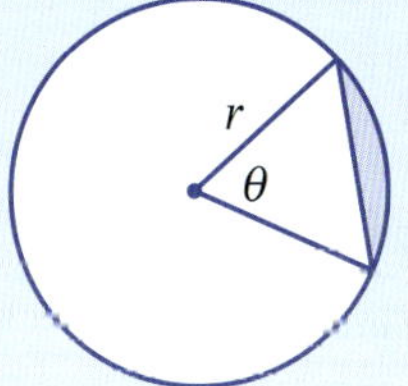

Proof

Area of segment = area of sector − area of triangle

Area of triangle $=\frac{1}{2}\times r\times r\times\sin\theta$ using $\frac{1}{2}ab\sin C$

$=\frac{1}{2}r^2\sin\theta$

$\therefore$ Area of segment $=\frac{1}{2}r^2\theta-\frac{1}{2}r^2\sin\theta$

$=\frac{1}{2}r^2(\theta-\sin\theta)$ factorising

If $\theta>\pi$ (reflex) and a major segment results, the diagram is different but the proof is similar.

The area of a major segment is calculated by *adding* the area of the major sector and the area of the triangle with central angle $2\pi-\theta$.

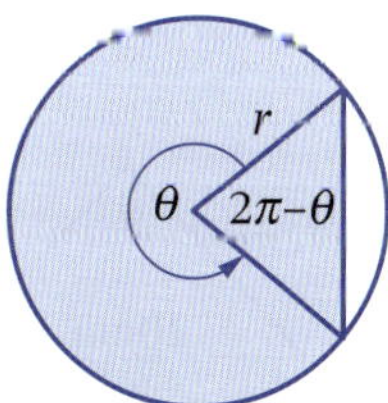

Area of segment = area of sector + area of triangle

Area of triangle $=\frac{1}{2}\times r\times r\times\sin(2\pi-\theta)$ using $\frac{1}{2}ab\sin C$

$=\frac{1}{2}r^2\sin(2\pi-\theta)$

$=-\frac{1}{2}r^2\sin\theta$

Using $\sin(2\pi-\theta)=-\sin\theta$:

Area of segment $=\frac{1}{2}r^2\theta+\left(-\frac{1}{2}r^2\sin\theta\right)$

$=\frac{1}{2}r^2\theta-\frac{1}{2}r^2\sin\theta$

$=\frac{1}{2}r^2(\theta-\sin\theta)$ giving the same formula as above.

Example 26

a Find the exact area of the minor segment formed if it subtends an angle of $\frac{\pi}{4}$ at the centre of a circle of radius 5 m.

b An 80 cm piece of wire is bent into a circle, and a 10 cm piece of wire is joined to it to form a chord. Find the area of the minor segment cut off by the chord, correct to 2 decimal places.

Solution

a
$$\begin{aligned}A &= \frac{1}{2}r^2(\theta - \sin\theta)\\ &= \frac{1}{2}(5)^2\left(\frac{\pi}{4} - \sin\frac{\pi}{4}\right)\\ &= \frac{25}{2}\left(\frac{\pi}{4} - \frac{1}{\sqrt{2}}\right)\\ &= \frac{25\pi}{8} - \frac{25}{2\sqrt{2}}\\ &= \frac{25\pi}{8} - \frac{25\sqrt{2}}{4}\\ &= \frac{25\pi - 50\sqrt{2}}{8}\ \text{m}^2\end{aligned}$$

b We can work out the radius, r, of the circle from its circumference.

$$\begin{aligned}C &= 2\pi r = 80\\ \therefore r &= \frac{80}{2\pi}\\ &= \frac{40}{\pi}\\ &\approx 12.7324\end{aligned}$$

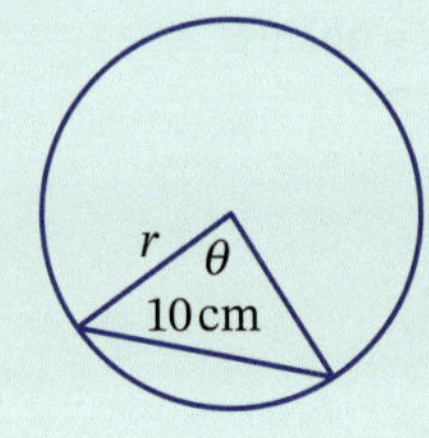

Don't round off the radius.

Use the full value in your calculator's display for the rest of the solution.

We can work out the central angle, θ, using the cosine rule.

$$\cos C = \frac{a^2 + b^2 - c^2}{2ab}$$

$$\cos\theta = \frac{12.73...^2 + 12.73...^2 - 10^2}{2(12.73...)(12.73...)}$$

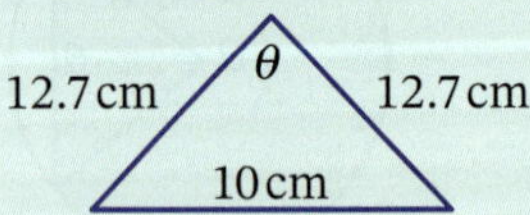

$$\begin{aligned}&= \frac{2(12.73...)^2 - 100}{2(12.73...)^2}\\ &= 0.6918\\ \theta &= 0.8068\end{aligned}$$

Substitute r and θ into $A = \frac{1}{2}r^2(\theta - \sin\theta)$:

$$\begin{aligned}A &= \frac{1}{2}(12.73...^2)(0.8068... - \sin 0.8068...)\\ &= 6.8673\\ &= 6.87\ \text{cm}^2\end{aligned}$$

EXERCISE 4.11 Answers on p. 479

Area of a segment

1 Find the exact area of the segment in a circle with radius and central angle, respectively:

a 4 cm and π **b** 3 m and $\frac{\pi}{3}$ **c** 10 cm and $\frac{5\pi}{6}$

d 3 cm and 30° **e** 7 mm and 45°.

2 Find, correct to 2 decimal places, the area of the segment in a circle with radius and central angle, respectively:

a 1.5 m and 0.43 **b** 3.21 cm and 1.22 **c** 7.2 mm and 0.960

d 5.9 cm and 23°12′ **e** 2.1 m and 82°35′.

3 Find the area of the segment formed by an angle of 2.443 subtended at the centre of a circle with radius 2.82 cm, correct to 2 significant figures.

4 Find for a central angle of $\frac{\pi}{7}$ of a circle with radius 3 cm:

a the exact arc length

b the exact area of the sector

c the area of the segment, correct to 2 decimal places.

5 The area of the segment cut off by an angle of $\frac{2\pi}{9}$ is 500 cm^2. Find the radius of the circle, correct to one decimal place.

6 Find for an angle of $\frac{\pi}{6}$ at the centre of a circle with radius 5 cm:

a the length of the chord, correct to one decimal place

b the length of the arc

c the area of the minor segment, correct to 2 decimal places.

7 A chord 8 mm long is formed by an angle of 315° subtended at the centre of a circle. Find:

a the radius of the circle

b the area of the major segment cut off by the angle, correct to one decimal place.

8 An angle of $\frac{\pi}{8}$ is subtended by an arc of length $\frac{5\pi}{4}$ cm. Find:

a the exact area of the sector

b the area of the minor segment formed, correct to one decimal place.

9 $\triangle OAB$, is formed where O is the centre of a circle of radius 12 cm, and A and B are endpoints of a 15 cm chord.

O
θ
12 cm
15 cm
A
B

a Find the angle (θ) subtended at the centre of the circle, in degrees and minutes.

b Find the area of $\triangle OAB$, correct to one decimal place.

c Find the area of the minor segment cut off by the chord, correct to 2 decimal places.

d Find the area of the major segment cut off by the chord, correct to 2 decimal places.

e Show that the sum of the areas of the segments gives the area of the circle.

10 A major segment has a central angle of $\frac{5\pi}{4}$ in a circle of radius 4 cm. Find, correct to 2 decimal places, the area of this segment:

a using the formula

b by finding the area of the minor segment and subtracting it from the area of the circle.

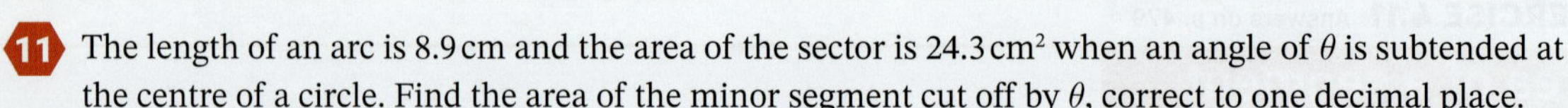

11 The length of an arc is 8.9 cm and the area of the sector is 24.3 cm^2 when an angle of θ is subtended at the centre of a circle. Find the area of the minor segment cut off by θ, correct to one decimal place.

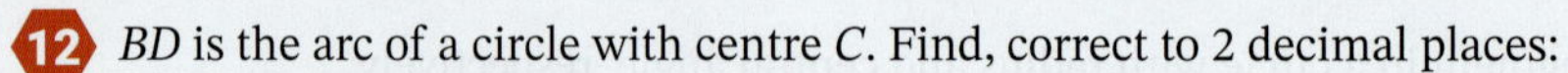

12 BD is the arc of a circle with centre C. Find, correct to 2 decimal places:

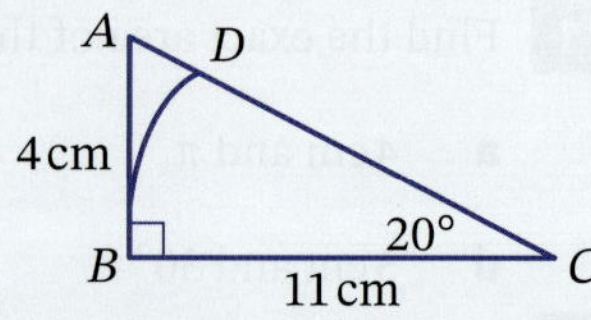

- **a** the length of arc BD
- **b** the area of region ABD
- **c** the perimeter of sector BDC.

13 The dartboard has a radius of 23 cm. The shaded area has a height of 7 mm. If a player must hit the shaded area to win, find:

- **a** the area of the shaded part, to the nearest square centimetre
- **b** the percentage of the shaded part in relation to the whole area of the dartboard, to one decimal place
- **c** the perimeter of the shaded area.

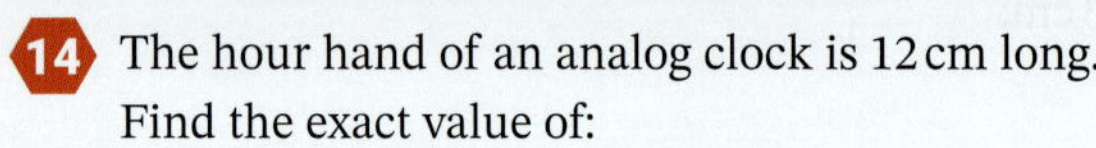

14 The hour hand of an analog clock is 12 cm long. Find the exact value of:

- **a** the length of the arc through which the hand would turn in 5 hours
- **b** the area through which the hand would pass in 2 hours.

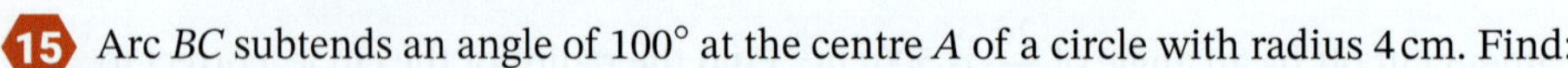

15 Arc BC subtends an angle of 100° at the centre A of a circle with radius 4 cm. Find:

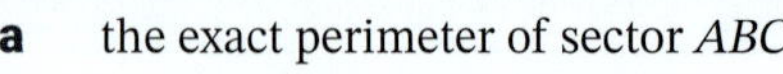

- **a** the exact perimeter of sector ABC
- **b** the approximate ratio of the area of the segment to the area of the sector.

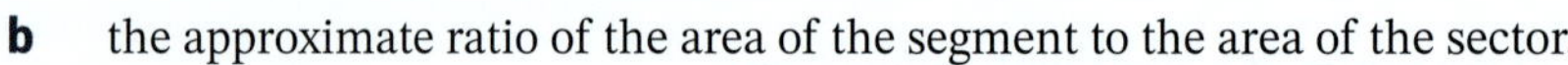

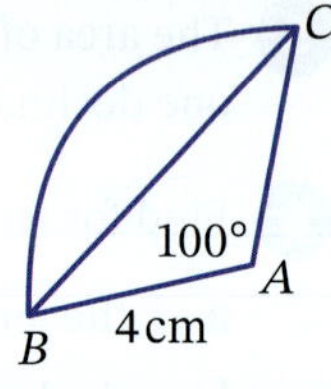

Sample HSC problem

Answers on p. 479

(6 marks) Two cars leave from the same intersection I.

Taylor drives for 85 km along a road on a bearing of 043° to P while Nate drives for 50 km on a bearing of 110° to Q.

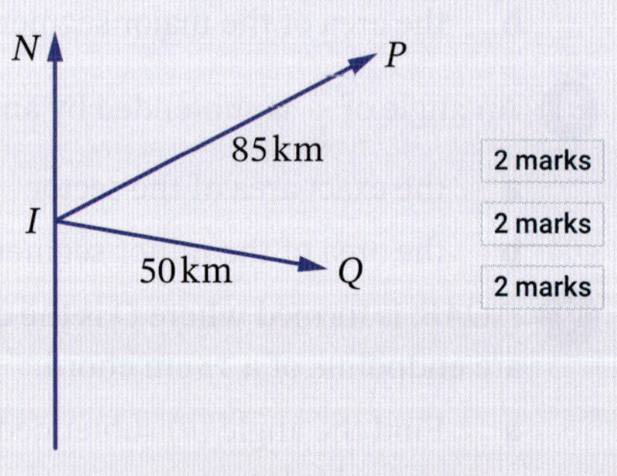

- **a** Show that $\angle PIQ$ is 67°. **2 marks**
- **b** Find PQ, the distance the cars are apart. **2 marks**
- **c** Find the true bearing of Nate from Taylor, to the nearest degree. **2 marks**

Foundation Mastery Complex

CHAPTER SUMMARY

This chapter, *Trigonometry*, revised and extended concepts and skills in trigonometry.

You were introduced to the trigonometry of angles greater than 90° and radian measure, so make sure that you master converting between degrees and radians, and know the relationships between angles in all the 4 quadrants. The formulas for the sine rule, cosine rule and area of a triangle can be applied to all types of triangles, while the formulas for arc length, areas of sectors and segments only work for angles expressed in radians.

Before you move on, consider what you learned in this chapter and revisit any sections that may have been unclear.

What you learned in this chapter...	Section	
Solve right-angled triangle problems	4.01	Right-angled trigonometry
Apply trigonometry to angles of elevation and depression and bearings	4.02	Bearings and angles of elevation
Evaluate trigonometric ratios for angles of any magnitude including the exact ratios and multiples of 90°	4.03	Angles of any magnitude
Understand and apply the sine and cosine rules	4.04	The sine rule
	4.05	The cosine rule
	4.07	Further applications of trigonometry
Find the area of a triangle given the length of two sides and the size of their included angle	4.06	Area of a triangle
Understand radians and convert between degrees and radians	4.08	Radians
Find the length of an arc and area of a sector and segment of a circle	4.09	The length of an arc
	4.10	Area of a sector
	4.11	Area of a segment

To help master these techniques, make a summary of this topic. Use the chapter outline and the mind map below as a guide. Add your own words, symbols, diagrams, boxes and reminders. The summary should give you a 'whole picture' view of the topic and allow you to identify any weak areas to revisit in your revision.

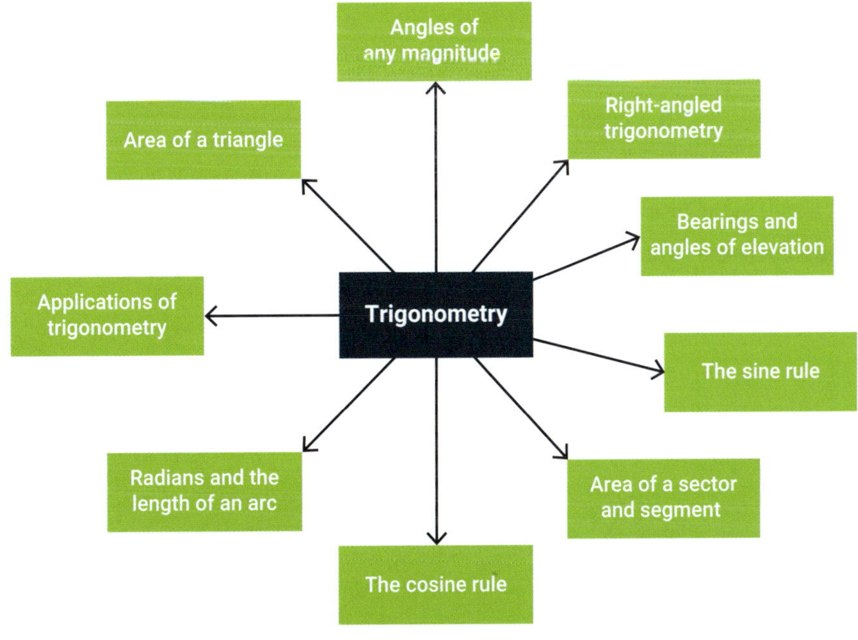

4 Test yourself

Answers on p. 479

For Questions **1** to **4**, select the correct answer **A**, **B**, **C** or **D**.

4.09 **1** Find the exact length of the radius of a circle if the arc length cut off by an angle of $\frac{5\pi}{4}$ is $\frac{25\pi}{8}$ cm.

A 5π cm **B** 5 cm **C** 2.5 cm **D** $\frac{5\pi}{2}$ cm

4.05 **2** The cosine rule is (there is more than one answer):

A $c^2 = a^2 + b^2 - 2ab\cos C$

B $c^2 = a^2 + c^2 - 2ac\cos C$

C $a^2 = b^2 + c^2 - 2bc\cos A$

D $a^2 = b^2 + c^2 - 2ab\cos A$

4.02 **3** What bearing is shown on the diagram (there may be more than one answer)?

A 035°

B W 35° S

C S 35° W

D 215°

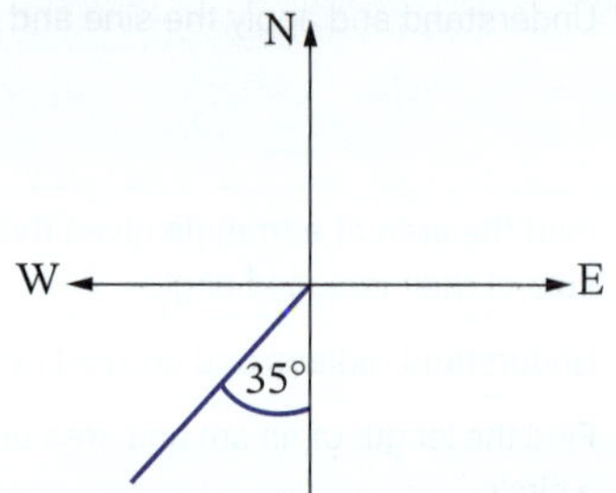

4.08 **4** The exact value of $\cos\frac{2\pi}{3}$ is:

A $\frac{1}{2}$ **B** $-\frac{\sqrt{3}}{2}$ **C** $\frac{\sqrt{3}}{2}$ **D** $-\frac{1}{2}$

4.03, 4.08 **5** Simplify each expression.

a $\cos(180° + \theta)$ **b** $\tan(-\theta)$ **c** $\sin(\pi - \theta)$

4.01 **6** Find θ to the nearest minute if:

a $\sin\theta = 0.72$ **b** $\cos\theta = 0.286$ **c** $\tan\theta = \frac{5}{7}$

4.02 **7** A ship sails on a bearing of 215° from port until it is 100 km due south of port. How far does it sail, to the nearest km?

4.03 **8** Find the length of AB as a surd.

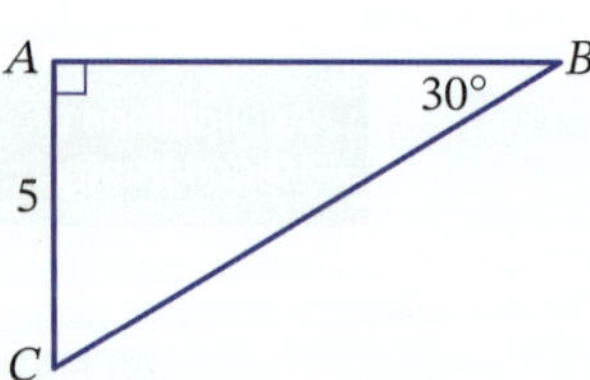

4.04, 4.05 **9** Evaluate x, correct to 2 significant figures.

a

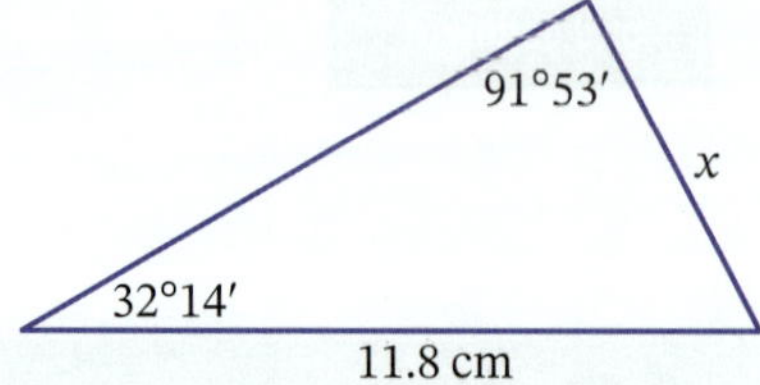

b

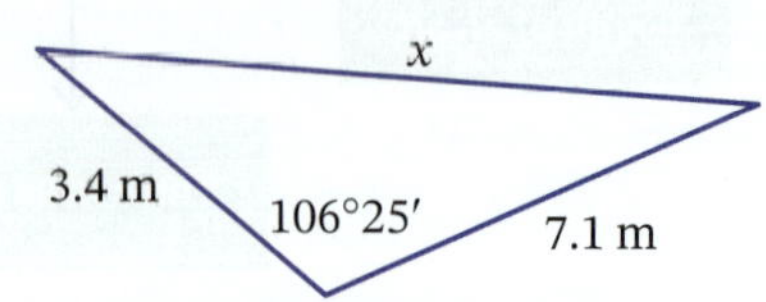

Foundation Mastery Complex

10 Convert each radian measure to degrees and minutes. 4.08

a 0.75 **b** 1.3 **c** 3.95 **d** 4.2 **e** 5.66

11 Evaluate θ, correct to the nearest minute. 4.04, 4.05

a

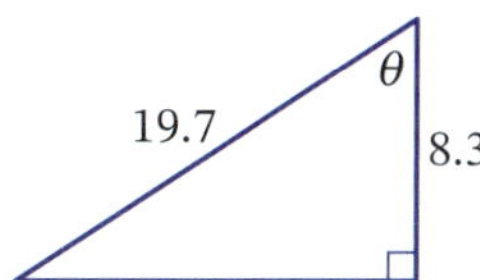

b

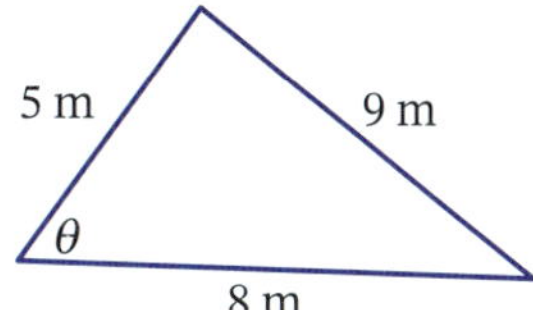

c

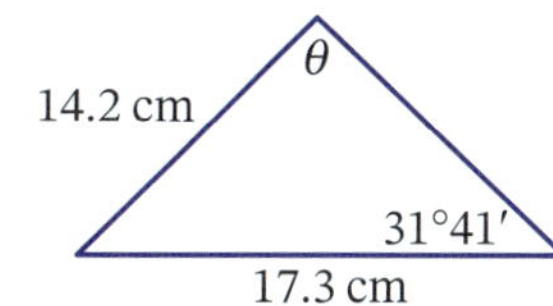

12 Find the area of this triangle. 4.06

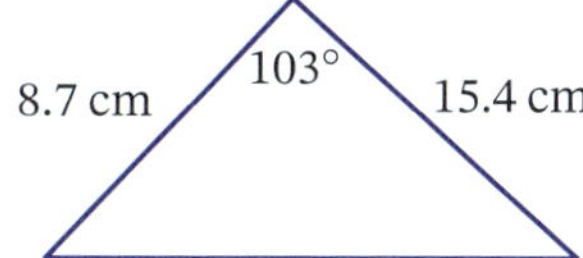

13 Paris walks south from home for 3.2 km, then turns and walks west for 1.8 km. Find the true bearing, to the nearest degree, of: 4.02

a Paris from her home

b her home from where Paris is now.

14 The angle of elevation from point B to the top of a pole AC is 39° and the angle of elevation from D, on the other side of the pole, is 42°. B and D are 20 m apart.

a Find an expression for the length of AD.

b Find the height of the pole, correct to one decimal place.

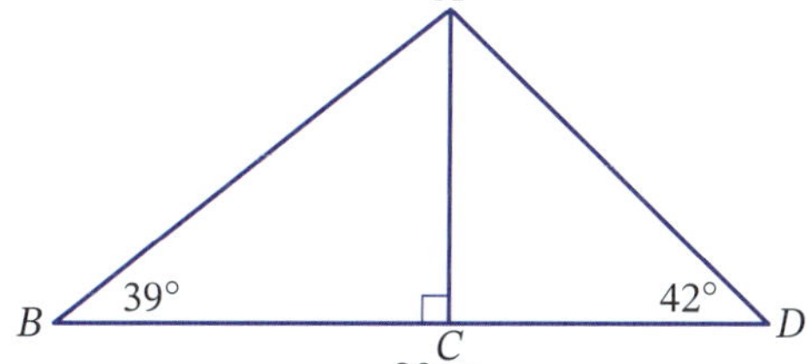

15 A plane flies from Orange for 1800 km on a bearing of 300°. It then turns and flies for 2500 km on a bearing of 205°. How far is the plane from Orange, to the nearest km?

16 Convert to radians, leaving in terms of π: 4.08

a 60° **b** 45° **c** 150° **d** 180° **e** 20°

17 A sector with radius 5 cm has an angle of $\frac{\pi}{6}$ at the centre. Find the exact value of:

a the arc length

b the area of the sector. 4.11

c the area of the minor segment inside the sector.

18 Find the exact value of: 4.08

a $\tan\frac{\pi}{3}$ **b** $\cos\frac{\pi}{6}$ **c** $\sin\frac{\pi}{4}$ **d** $\tan\frac{\pi}{6}$ **e** $\cos\frac{\pi}{4}$

f $\sin\frac{\pi}{6}$ **g** $\tan\frac{\pi}{4}$ **h** $\cos\frac{\pi}{3}$ **i** $\sin\frac{\pi}{3}$

19 A circle has a circumference of 8π cm. If an angle of $\frac{\pi}{7}$ is at the centre of a sector, find:

a the exact area of the sector

b the area of the minor segment, correct to 2 decimal places.

☐ Foundation ○ Mastery ⬡ Complex

4.04 **20** In triangle MNP, $NP = 14.9$ cm, $MP = 12.7$ cm and $\angle N = 43°49'$. Find $\angle M$ in degrees and minutes.

4.03 **21** If $\tan 52° = p$, then evaluate each expression.

a $\tan 412°$ **b** $\tan(-52°)$ **c** $\tan 128°$ **d** $\tan 232°$

4.03, 4.08 **22** Find the exact value of:

a $\cos 315°$ **b** $\sin(-60°)$ **c** $\tan 180°$

d $\sin \frac{5\pi}{4}$ **e** $\cos \frac{5\pi}{6}$ **f** $\tan \frac{4\pi}{3}$

g $\cos 0$ **h** $\sin(-315°)$ **i** $\tan\left(-\frac{\pi}{6}\right)$

4.06 **23** Find the exact area of $\triangle ABC$.

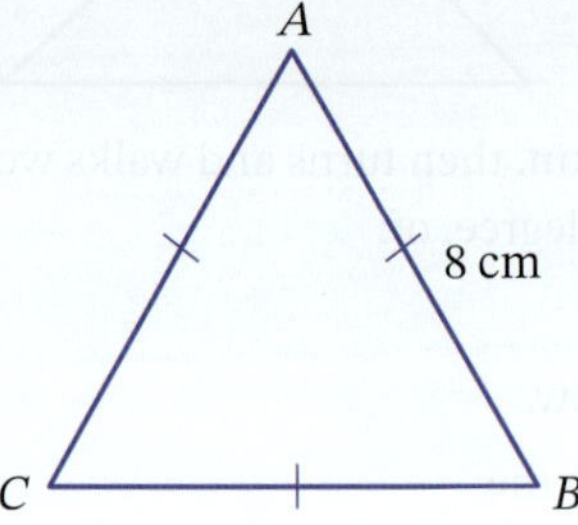

4.03 **24** Find θ if it is in the first quadrant (acute).

a $\cos 150° = -\cos\theta$ **b** $\sin 225° = -\sin\theta$

c $\tan 318° = -\tan\theta$ **d** $\sin(-300°) = \sin\theta$

Iain Masterton/Alamy Stock Photo

☐ Foundation ○ Mastery ○ Complex

Challenge exercise 4

Answers on p. 479

1 Two cars leave an intersection at the same time, one travelling at 70 km/h along one straight road and the other car travelling at 80 km/h along another straight road. After 2 hours they are 218 km apart. At what angle, to the nearest minute, do the roads meet at the intersection?

2 Evaluate x, correct to 3 significant figures.

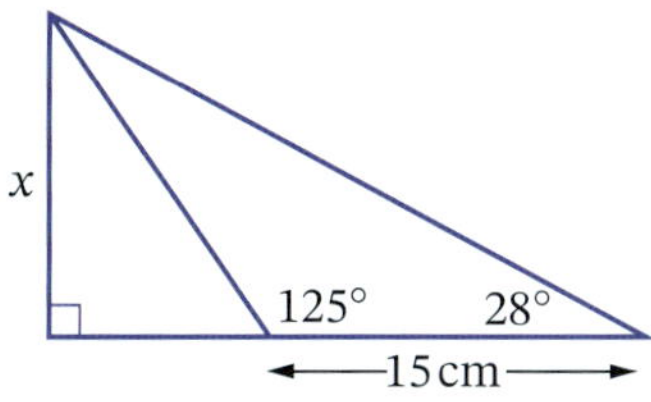

3 **a** Find an exact expression for the length of AC.

b Hence, or otherwise, find the value of h correct to one decimal place.

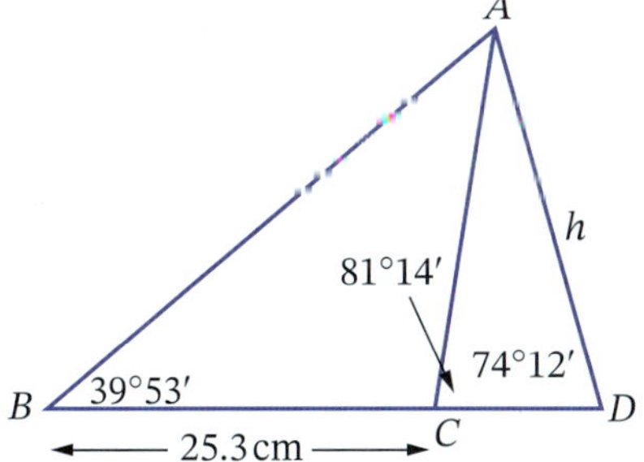

4 From the top of a vertical pole the angle of depression to Ian standing at the foot of the pole is 43°. Liam is on the other side of the pole, and the angle of depression from the top of the pole to Liam is 52°. The boys are standing 58 m apart. Find the height of the pole, to the nearest metre.

5 Find the area of a regular hexagon with sides 4 cm, correct to the nearest cm².

6 Calculate correct to one decimal place, the area of a regular pentagon with sides 12 mm.

7 Find the exact area of the minor segment cut off by a quarter of a circle with radius 3 cm.

8 Find the exact value of each expression.

a $\sin 600°$ **b** $\tan(-405°)$ **c** $\cos 810°$

9 The area of a sector in a circle of radius 4 cm is $\frac{8\pi}{3}$ cm². Find the area of the minor segment in exact form.

10 This composite shape consists of an isosceles triangle with $AB = AC = 8.7$ cm and $\angle BAC = 25°$ and a segment of a circle of radius 5 cm and central angle 84°.

Find the perimeter and area of this shape, correct to 2 decimal places.

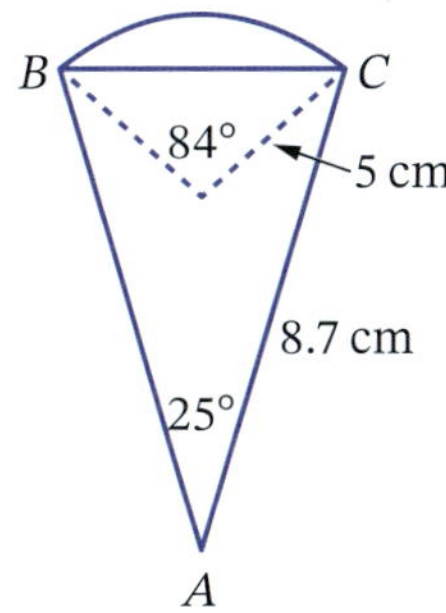

Foundation Mastery Complex

5. FURTHER FUNCTIONS AND GRAPHS

In this chapter, we look at more complex functions and relations, including the hyperbola, absolute value function, circles and semicircles. We will also study relationships between functions, including composite functions. This chapter builds upon concepts learned in Year 10 and Chapter 3, *Functions and graphs*.

Chapter outline

Deemerwha studio/Adobe Stock Photos

In this chapter you will:

- identify values of a function where it is continuous and discontinuous
- identify characteristics of a hyperbola, piecewise and absolute value functions, including domain and range
- solve absolute value equations graphically
- understand direct and inverse variations and solve problems involving them
- sketch graphs of circles and semicircles and find their equations
- work with composite functions

Videos (7):

5.01 Graphing hyperbolas

5.03 Inverse proportion • Inverse variation

5.04 Graphing piecewise functions • Absolute value graphs

5.05 Identifying graphs • Equation of a circle

Worksheets (7):

5.01 The hyperbola • Graphing hyperbolas • Graphing $y = \frac{10}{x}$

5.04 Absolute value functions

5.05 Equations of circles • Advanced graphs

5.06 Circles and composite functions

Puzzles (2):

5.05 Circle equations • Matching graphs (Advanced)

Nelson MindTap

To access resources above, visit **cengage.com.au/nelsonmindtap**

Terminology

absolute value function	asymptote	circle	composite function
constant of variation	continuous	direct variation	discontinuous
even function	hyperbola	inverse variation	piecewise function
proportion	radius	semicircle	symmetry
variation			

The hyperbola

5.01

The graph of the function $y = \frac{k}{x}$, where k is a constant, is a **hyperbola**.

Video
Graphing hyperbolas

Worksheets
The hyperbola

Graphing hyperbolas

Graphing $y = \frac{10}{x}$

Example 1

Sketch the graph of $y = \frac{1}{x}$. What is the function's domain and range?

Solution

x	-3	-2	-1	$-\frac{1}{2}$	$-\frac{1}{4}$	0	$\frac{1}{4}$	$\frac{1}{2}$	1	2	3
y	$-\frac{1}{3}$	$-\frac{1}{2}$	-1	-2	-4	–	4	2	1	$\frac{1}{2}$	$\frac{1}{3}$

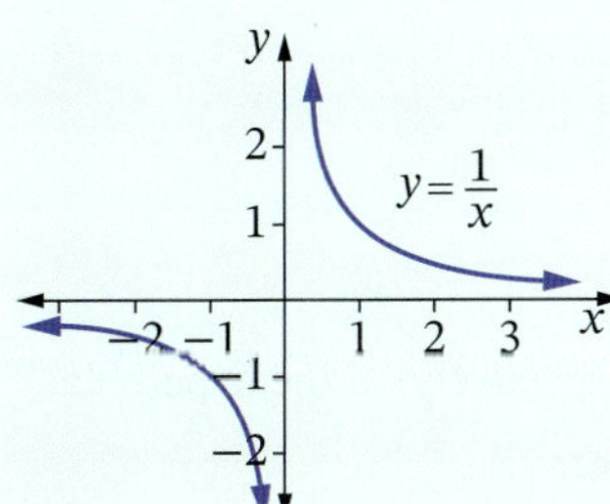

When $x = 0$ the value of y is undefined.

Domain: All real x, $x \neq 0$.
Or in interval notation: $(-\infty, 0) \cup (0, \infty)$

Range: All real y, $y \neq 0$.
In interval notation: $(-\infty, 0) \cup (0, \infty)$

Technology

Transforming the hyperbola

Use graphing technology to sketch each hyperbola. What do you notice?

a $y = \frac{3}{x}$ **b** $y = \frac{10}{x}$ **c** $y = \frac{7}{x}$

d $y = -\frac{2}{x}$ **e** $y = -\frac{4}{x}$ **f** $y = \frac{1}{2x}$

g $y = \frac{1}{5x}$ **h** $y = -\frac{1}{7x}$ **i** $y = \frac{k}{x}$ for different values of k

j $y = \frac{1}{kx}$ for different values of k

What happens to the graph as x becomes closer to 0? What happens as x becomes very large in both positive and negative directions? The value of y is never 0. Why?

Is a hyperbola always a function? Is it an even or odd function?

Use graphing technology to sketch each hyperbola. What do you notice?

a $y = \frac{1}{x} + 4$ **b** $y = \frac{1}{x} + 9$ **c** $y = \frac{1}{x} - 1$

d $y = \frac{1}{x} - 4$ **e** $y = \frac{1}{x} + k$ for different values of k

Use graphing technology to sketch each hyperbola. What do you notice?

a $y = \frac{1}{x+2}$ **b** $y = \frac{1}{x-1}$ **c** $y = \frac{1}{x-6}$

d $y = \frac{1}{x+5}$ **e** $y = \frac{1}{x-k}$ for different values of k

Continuity and asymptotes

Most functions have graphs that are smooth unbroken curves (or lines). They are called **continuous functions** and their graphs can be drawn without lifting the pen off the paper. However, some functions have discontinuities, meaning that their graphs have gaps or breaks. These are called **discontinuous functions**.

The hyperbola is discontinuous at $x = 0$ because there is a gap in the graph at that point and it has two separate parts. The graph of $y = \frac{1}{x}$ cannot be drawn without lifting the pen off the paper at $x = 0$. The hyperbola also does not touch the x- or y-axes, but it gets closer and closer to them. We call the x- and y-axes **asymptotes**: lines that the curve approaches but never touches. For the function $y = \frac{1}{x}$ the x-axis is a **horizontal asymptote** and the y-axis is a **vertical asymptote**.

To find the shape of the graph close to the asymptotes or as $x \to \pm\infty$, we can check points nearby.

Example 2

Find the domain and range of $f(x) = \frac{3}{x-3}$ and sketch the graph of the function.

Solution

To find the domain, we notice that $x - 3 \neq 0$. So $x \neq 3$.

Domain: all real x, $x \neq 3$

Also, y cannot be zero: $y \neq 0$.

Range: all real y, $y \neq 0$

The lines $x = 3$ and $y = 0$ (the x-axis) are the asymptotes (vertical and horizontal respectively) of the hyperbola.

To find the limiting behaviour of the graph, look at what is happening as $x \to \pm\infty$.

As x increases and approaches ∞, $\frac{3}{x-3}$ becomes closer to 0 and is positive.

Substitute large values of x into the function, for example, $x = 1000$.

As $x \to \infty$, $y \to 0^+$ (as x approaches infinity, y approaches 0 from above, the positive side).

Similarly, as x decreases and approaches $-\infty$, y becomes closer to 0 and is negative.
Substitute $x = -1000$, for example.

As $x \to -\infty$, $y \to 0^-$ (as x approaches negative infinity, y approaches 0 from below, the negative side).

To see the behaviour of the function near the asymptote $x = 3$, we can test values either side.

LHS: When $x = 2.999$,

$$\frac{3}{2.999-3} = -3000$$
$$< 0$$

As $x \to 3^-$, $y \to -\infty$

RHS: When $x = 3.001$,

$$\frac{3}{3.001-3} = 3000$$
$$> 0$$

As $x \to 3^+$, $y \to \infty$

For y-intercept, $x = 0$.

$$y = \frac{3}{0-3}$$
$$= -1$$

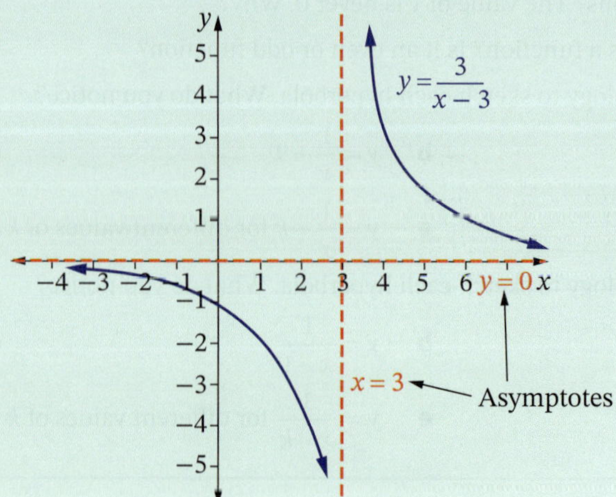

Example 3

Sketch the graph of $y = -\dfrac{1}{2x+4}$.

Solution

To find the discontinuity, notice that:

$$\begin{aligned} 2x + 4 &\neq 0 \\ 2x &\neq -4 \\ x &\neq -2 \end{aligned}$$

Vertical asymptote at $x = -2$.

Domain: All real x, $x \neq -2$. Range: All real y, $y \neq 0$.

Horizontal asymptote at $y = 0$.

Limiting behaviour:

As $x \to \infty$, $y \to 0^-$.

As $x \to -\infty$, $y \to 0^+$.

Substitute, say, $x = 5000$ and $x = -5000$.

To see the shape of the graph near the asymptote $x = -2$, we can test values either side.

LHS: When $x = -2.0001$,

$$y = -\frac{1}{2(-2.0001)+4} = 10\,000 > 0$$

As $x \to -2^-$, $y \to \infty$

RHS: When $x = -1.9999$,

$$y = -\frac{1}{2(-1.9999)+4} = -10\,000 < 0$$

As $x \to -2^+$, $y \to -\infty$

For y-intercept, $x = 0$.

$$\begin{aligned} y &= -\frac{1}{2(0)+4} \\ &= -\frac{1}{4} \end{aligned}$$

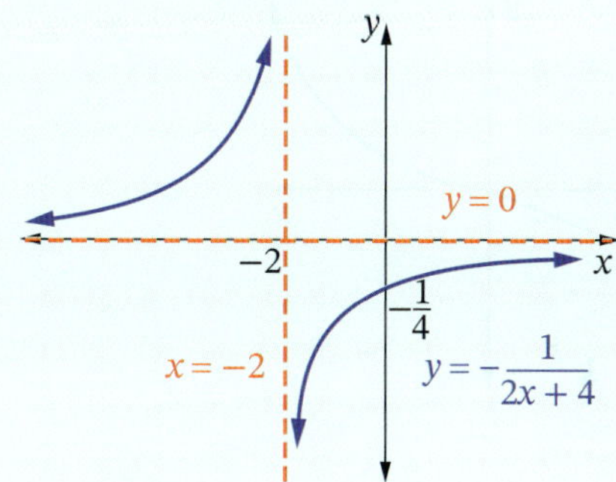

The hyperbola $y = \frac{k}{x}$

- The graph of $y = \dfrac{k}{x}$ is a hyperbola with a centre of symmetry at $(0, 0)$, with rotational symmetry of 180° about $(0, 0)$.
- It has a discontinuity at $x = 0$ and has two separate branches in different quadrants.
- The horizontal and vertical asymptotes are the x- and y-axes respectively.
- domain: All real x, $x \neq 0$, or $(-\infty, 0) \cup (0, \infty)$
- range: All real y, $y \neq 0$, or $(-\infty, 0) \cup (0, \infty)$
- If k is positive, the graph is in the 1st and 3rd quadrants.
- If k is negative, the graph is in the 2nd and 4th quadrants.
- The higher the value of k, the further the hyperbola is from the x- and y-axes.

EXERCISE 5.01 Answers on p. 480

The hyperbola

1 For each function:

- **i** state the domain and range
- **ii** find the y-intercept if it exists
- **iii** sketch the graph.

a $y=\frac{2}{x}$ **b** $y=-\frac{1}{x}$ **c** $f(x)=\frac{1}{x+1}$

d $f(x)=\frac{3}{x-2}$ **e** $y=\frac{1}{3x+6}$ **f** $f(x)=-\frac{2}{x-3}$

g $f(x)=\frac{4}{x-1}$ **h** $y=-\frac{2}{x+1}$ **i** $f(x)=\frac{2}{6x-3}$

2 Show that $f(x)=\frac{2}{x}$ is an odd function.

3 **a** Is the hyperbola $y=-\frac{2}{x+1}$

i a function? **ii** even, odd or neither? **iii** continuous?

b What are the equations of the asymptotes?

c State its domain and range.

4 State whether each graph is continuous or discontinuous.

a

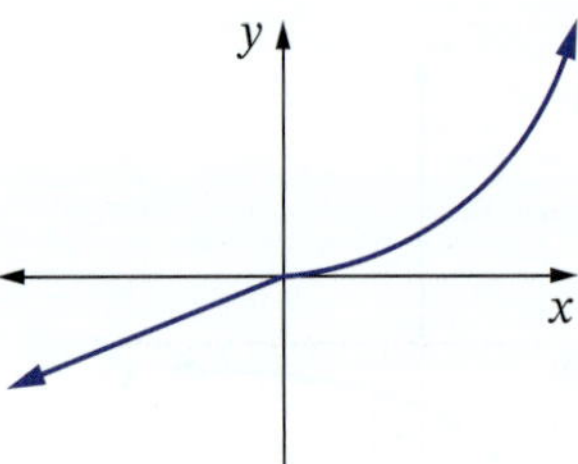

b

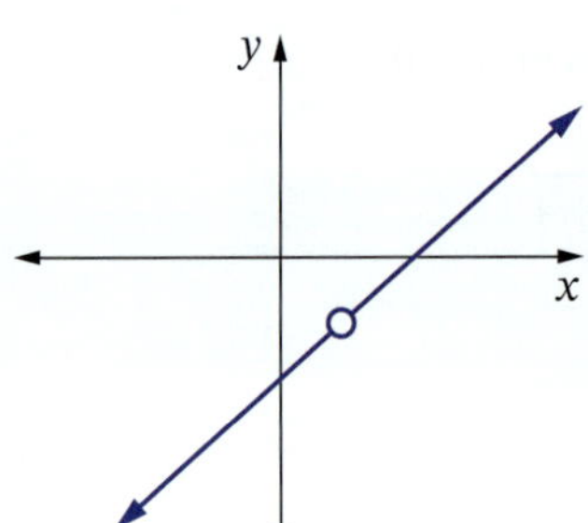

c

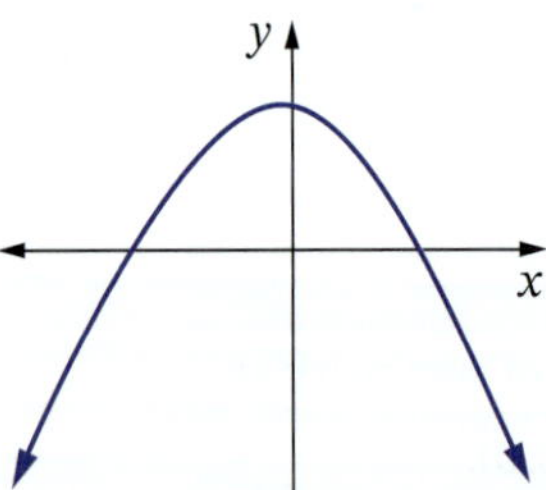

d

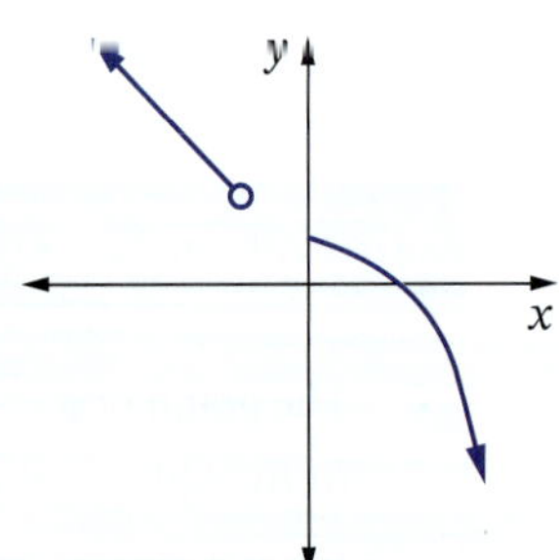

5 Is the function $f(x)=\begin{cases} x^2 & \text{for } x \le 0 \\ x^3+1 & \text{for } x > 0 \end{cases}$ continuous? If not, where is the discontinuity?

6 Determine whether each function is discontinuous and if so, find the x values for which they are discontinuous.

a $y=x^2-3$ **b** $y=\frac{1}{x+1}$ **c** $f(x)=\sqrt{x-1}$

d $y=\frac{1}{x^2+4}$ **e** $y=\frac{1}{x^2-4}$ **f** $y=\frac{x-1}{x+1}$

g $f(x)=\frac{2x^2-1}{3}$ **h** $f(x)=\begin{cases} 5x-1 & \text{for } x<-2 \\ x^2-3 & \text{for } x \ge -2 \end{cases}$ **i** $f(x)=\begin{cases} 2x^2 & \text{when } x \le 1 \\ \frac{2}{x} & \text{when } x \ge 1 \end{cases}$

☐ Foundation ◯ Mastery ◯ Complex

7 The average cost per day of producing x phones is given by $\frac{1000}{x} + 150 + x$.

a If average cost $= \frac{\text{total cost}}{x}$, find the total cost of producing x phones each day.

b Find the number of phones that would need to be produced if the total cost is \$8400.

8 Amie travels to her grandparents' home 800 km away at an average speed of v km/h. Her car costs $v^2 + 2000$ cents per hour to run.

a Show that the cost of the trip to Amie's grandparents' home is given by $C = 800\left(v + \frac{2000}{v}\right)$.

b If Amie's average speed is 65 km/h, find the cost of the trip.

c If the trip costs \$798.22, what is Amie's average speed?

Direct variation

5.02

When one variable is in **direct variation** (or **direct proportion**) with another variable, one is a constant multiple of the other. This means that as one increases, so does the other.

Direct variation

If y is in **direct variation** to x, then $y = kx$, where k is a constant called the **constant of variation** or **constant of proportionality**.

- A direct linear relationship exists between x and y.
- If x increases (or decreases), y increases (or decreases).
- If x is doubled (or halved), y is doubled (or halved).
- Another way of describing this relationship is 'y is directly proportional to x' or 'y varies directly with x (or as x)'.
- The graph of direct variation is a straight line going through $(0, 0)$ with gradient k, in the first quadrant only (positive values of x only).

Example 4

Huang earns \$20 an hour. Find an equation for Huang's income (I) for working x hours and draw its graph.

Solution

Income for 1 hour is \$20.

Income for 2 hours is \$20 × 2 or \$40.

Income for 3 hours is \$20 × 3 or \$60.

Income for x hours is \$20 × x or \$20x.

We can write the equation as $I = 20x$.

We can graph the equation using a table of values.

x	1	2	3
I	20	40	60

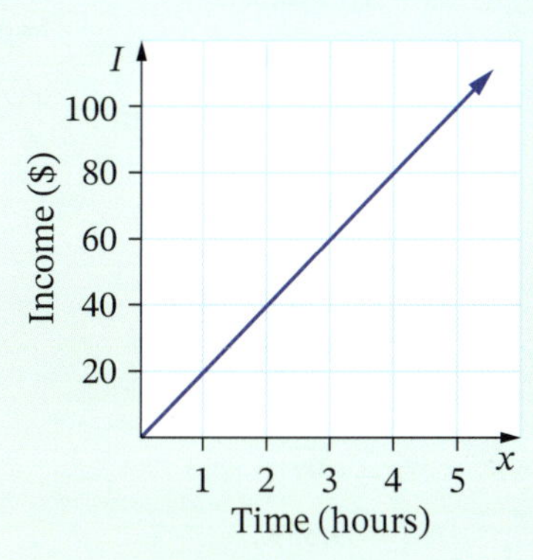

□ Foundation ○ Mastery ○ Complex

Direct variation is not always linear. For example, one variable could vary directly with the square or cube of another variable.

Direct variation with a power of x

If y is in **direct variation** to a power of x, then $y = kx^n$.

Variation with the square	$y = kx^2$
Variation with the cube	$y = kx^3$
Variation with the square root	$y = k\sqrt{x}$

Example 5

a The area A of a pentagon varies directly with the square of its length x. When the length is 7 cm, its area is 84 cm^2.

i Find a formula for A.

ii Find, correct to one decimal place, the area when its length is 13 cm.

iii Find, correct to one decimal place, the length when the area is 156 cm^2.

b As a seedpod grows, its volume V varies directly as the cube of its diameter d. When its diameter is 5 mm, its volume is 327 mm^3.

i Find the equation for volume in terms of d.

ii Find the volume when the diameter is 10 mm.

iii Find, correct to one decimal place, the diameter when the volume is 500 mm^3.

Solution

a i For variation with the square: $A = kx^2$

Substitute $x = 7, A = 84$ to find the value of k.

$$84 = k(7^2)$$
$$= 49k$$
$$k = \frac{84}{49}$$
$$= \frac{12}{7}$$
$$= 1.7142...$$

So $A = 1.7142...x^2$.

Do not round the value of k or your final answer will be inaccurate. Alternatively, leave k in fraction form and write as $A = \frac{12x^2}{7}$

ii When $x = 13$:

$$A = 1.7142...(13^2)$$
$$= 289.7142...$$

So the area is 289.7 cm^2.

iii When $A = 156$:

$$156 = 1.7142...x^2$$
$$x^2 = \frac{156}{1.7142...}$$
$$= 91$$
$$x = \sqrt{91}$$
$$= 9.539...$$

So the length is 9.5 cm.

b i For direct variation with the cube: $V = kd^3$

When $d = 5, V = 327$.

$$327 = k(5^3)$$
$$= 125k$$
$$k = \frac{327}{125}$$
$$= 2.616$$

So $V = 2.616d^3$.

ii When $d = 10$:

$$V = 2.616(10^3)$$
$$= 2616$$

So the volume is 2616 mm^3.

iii When $V = 500$:

$$500 = 2.616d^3$$
$$d^3 = \frac{500}{2.616}$$
$$d^3 = 191.1314...$$
$$\sqrt[3]{191.14...} = d$$
$$5.7602... = d$$

So the diameter is 5.8 mm.

EXERCISE 5.02 Answers on p. 481

Direct variation

1 Write an equation for:

- **a** the number of months (N) in x years
- **b** the amount of juice (A) in n lots of 2-litre bottles
- **c** the cost (C) of x litres of petrol at \$1.50 per litre
- **d** the number (y) of people in x debating teams if there are 4 people in each team
- **e** the weight (w) of x lots of 400 g cans of peaches.

2 Find the equation and draw the graph of the cost (c) of x refrigerators if each refrigerator costs \$850.

3 A supermarket has boxes containing cans of dog food. The number of cans of dog food is directly proportional to the number of boxes.

- **a** If there are 144 cans in 4 boxes, find an equation for the number of cans (N) in x boxes.
- **b** How many cans are in 28 boxes?
- **c** How many boxes would be needed for 612 cans of dog food?

4 The number of frequent flyer points that Matteo earns on his credit card varies directly with the number of dollars he spends on his card.

- **a** If Matteo earns 150 points when he spends \$450, find an equation for the number of points (P) he earns when spending d dollars.
- **b** Find the number of points Matteo earns when he spends \$840.
- **c** If Matteo earns 57 points, how much did he spend?

5 The braking distance of a car in metres varies as the square of its speed in km/h. When travelling at 60 km/h, its braking distance is 14.4 metres.

- **a** Find the formula for d, the braking distance in metres, in terms of x, the speed in km/h.
- **b** Find the braking distance at 100 km/h.
- **c** A person runs out onto the road 40 m in front of a car travelling at 80 km/h. Will the car be able to stop in time without hitting the person?
- **d** If the person was 40 m in front of a car travelling at 110 km/h, could the car stop in time?
- **e** If a car had a braking distance of 33 m, how fast was it going? Answer to the nearest km/h.

6 The area (A) of an ellipse is directly proportional to the square of its length (x).
When its length is 5 cm, its area is 125 cm^2.

- **a** Find the equation for the area.
- **b** Find the area when the length is 4.2 cm.
- **c** Find the length correct to one decimal place when the area is 250 cm^2.

7 The capacity of a water bottle, C, in mL, varies directly with the cube of its diameter, d, in cm. A bottle with diameter 7 cm has a capacity of 1232 mL.

- **a** Find the equation for C in terms of d.
- **b** Find the capacity of a water bottle with a radius of 5 cm. Round your answer to the nearest mL.
- **c** Find the diameter of a water bottle with a capacity of 250 mL. Round your answer to the nearest cm.

8
- **a** The perimeter P of a polygon varies directly with its side length x. Find an equation for the perimeter if it is 90 cm when its side is 5 cm long.
- **b** The area A of the polygon varies directly to the square of its side length x. If its area is 105.8 cm^2 when its side is 23 cm long, find its equation.
- **c** Find any non-zero x values of the side length for which the perimeter and area have the same value.

Foundation Mastery Complex

9 The volume, V, in cm^3, of a square pyramid varies with the cube of its base length, x, in cm.

a Find the equation for V if $V = 120$ when $x = 3.5$.

b Find the volume when $x = 6$. Round your answer to the nearest cm^3.

c Find x when $V = 250$.

10 Find the equation for each variation statement.

a d is in direct variation with the cube of t and when $t = 2$, $d = 72$.

b P varies directly as the 6th power of l and when $l = 1$, $P = 5$.

5.03 Inverse variation

Videos
Inverse proportion
Inverse variation

When one variable is in **inverse variation** (or inverse proportion) with another variable, one is a constant multiple of the reciprocal of the other. This means that as one variable increases, the other decreases and when one decreases, the other increases.

For example:

- the more slices you cut a pizza into, the smaller the size of each slice
- the more workers there are on a project, the less time it takes to complete
- the fewer people sharing a house, the higher the rent each person pays.

Inverse variation

If y is in **inverse variation** to x, then $y = \frac{k}{x}$, where k is a constant called the **constant of variation** or **constant of proportionality**.

- If x increases (or decreases), y decreases (or increases).
- If x is doubled (or halved), y is halved (or doubled).
- Another way of describing this relationship is 'y is inversely proportional to x' or 'y varies inversely with x'.
- The graph of inverse variation is a hyperbola in the first quadrant only (positive values of x only).

Example 6

a Building a shed in 12 hours requires 3 builders. If the number of builders, N, is in inverse variation to the amount of time, t hours:

i find the equation for N in terms of t

ii find the number of builders it would take to build the shed in 9 hours

iii find how long it would take 2 builders to build the shed.

b The faster a car travels, the less time it takes to travel a certain distance. It takes the car 2 hours to travel to Nelson Bay at a speed of 80 km/h. If the time taken, t hours, varies inversely with the speed, s km/h, then,

i find the equation for t

ii find the time it would take if travelling at 100 km/h

iii find the speed at which the trip would take $2\frac{1}{2}$ hours

iv graph the equation.

Foundation | Mastery | Complex

Solution

a **i** For inverse variation, the equation is in the form $N = \frac{k}{t}$.

Substitute $t = 12$, $N = 3$ to find the value of k:

$$3 = \frac{k}{12}$$
$$36 = k$$
$$\therefore N = \frac{36}{t}$$

ii Substitute $t = 9$:

$$N = \frac{36}{9}$$
$$= 4$$

So it takes 4 builders to build the shed in 9 hours.

iii Substitute $N = 2$:

$$2 = \frac{36}{t}$$
$$2t = 36$$
$$t = 18$$

It takes 2 builders 18 hours to build the shed.

b **i** For inverse variation, the equation is in the form $t = \frac{k}{s}$.

Substitute $s = 80$, $t = 2$ to find k:

$$2 = \frac{k}{80}$$
$$k = 160$$
$$\therefore t = \frac{160}{s}$$

ii Substitute $s = 100$:

$$t = \frac{160}{100}$$
$$= 1.6 \text{ hours}$$
$$= 1 \text{ h } 36 \text{ min}$$

So the car takes 1 h 36 min to travel the distance if travelling at 100 km/h.

iii Substitute $t = 2.5$:

$$2.5 = \frac{160}{s}$$
$$2.5s = 160$$
$$s = \frac{160}{2.5}$$
$$= 64$$

So the car travels at 64 km/h if the trip takes $2\frac{1}{2}$ hours.

iv To graph this function, complete a table of values for $t = \frac{160}{s}$.

s	20	40	60	80	100	120
t	8	4	2.667	2	1.6	1.333

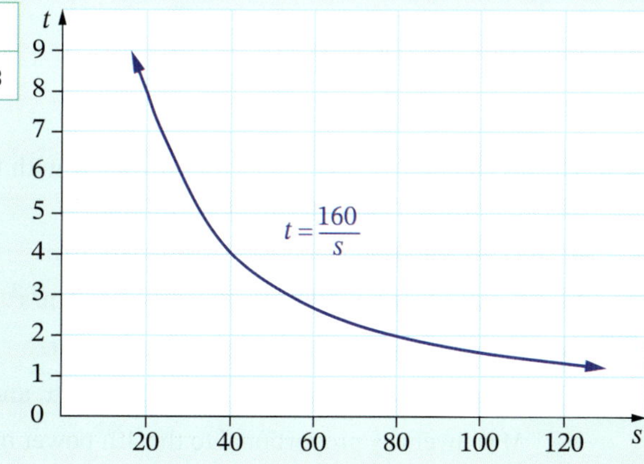

Inverse variation with a power of x

If y is in **inverse variation** with a power of x, then $y = \frac{k}{x^n}$.

Inverse variation with the square	$y = \frac{k}{x^2}$
Inverse variation with the cube	$y = \frac{k}{x^3}$
Inverse variation with the square root	$y = \frac{k}{\sqrt{x}}$

EXERCISE 5.03 Answers on p. 481

Inverse variation

EXAMPLE 6

1 The diameter of a balloon varies inversely with the thickness of the rubber. The diameter of the balloon is 80 mm when the rubber is 2 mm thick.

- **a** Find an equation for the diameter D in terms of the thickness x.
- **b** Find the diameter when the thickness is 0.8 mm.
- **c** Find the thickness, correct to one decimal place, when the diameter is 115.3 mm.
- **d** Sketch the graph showing this information.

2 The more boxes a factory produces, the less it costs to produce each box. When 128 boxes are produced, it costs \$2 per box. The cost per box is inversely proportional to the number of boxes produced.

- **a** Write an equation for the cost c to produce each box when manufacturing n boxes.
- **b** Find the cost of each box when 100 boxes are produced.
- **c** Find how many boxes must be produced for the cost for each box to be 50 cents.
- **d** Sketch the graph of this information.

3 The volume V of a gas varies inversely with the pressure p on it.

- **a** If the volume of the gas is 300 cm^3 under pressure of 35 kg/cm^2, find a formula for V in terms of p.
- **b** Find the volume of the gas under 40 kg/cm^2 of pressure.
- **c** Find the pressure that needs to be applied to the gas for its volume to be 500 cm^3.

4 The intensity of light I lumens is inversely proportional to the square of the distance d m from the light source. At a distance of 0.6 m, the intensity is 8 lumens. Find:

- **a** the equation for I in terms of d
- **b** the light intensity when the distance is 1.5 m from the light source
- **c** the distance from the light source when the intensity is 15 lumens. Round your answer to 2 decimal places.

5 $y = f(x)$ where y varies inversely as the square root of x. The point $(4, 6)$ lies on the graph of $y = f(x)$.

- **a** Find the equation in terms of y and x.
- **b** Find the value of y when $x = 8$.
- **c** Find the value of x when $y = 4$.
- **d** Sketch the graph of the function.

6 The acceleration of an object varies inversely with the square of its displacement. If the acceleration is 300 m/s^2 when its displacement is 5 m, find:

- **a** its acceleration when displacement is 10 m
- **b** the displacement when acceleration is 4 m/s^2. Round your answer to one decimal place.

7 Find the equation for each variation statement.

- **a** y varies inversely with the square root of x, and when $x = 4$, $y = 7$.
- **b** M is inversely proportional to the 4th power of x, and when $x = 2$, $M = 3$.

Foundation Mastery Complex

8 The number of hoses (n) needed to fill a pool with water varies inversely with the cube root of time (t) in minutes. If 10 hoses fill the pool in 27 minutes, find the number of hoses needed to fill the pool in 2 hours and 5 minutes.

9 y varies inversely with a power of x. When $x = 2$, $y = 1215$ and when $x = 3$, $y = 160$. Find the formula for y.

Absolute value and piecewise functions 5.04

Piecewise functions

You were introduced to piecewise functions in Chapter 3, *Functions and graphs*. We can graph them by looking at the different functions over their specified domains.

Video Graphing piecewise functions

Worksheet Absolute value functions

Example 7

Graph each piecewise function.

a $f(x)=\begin{cases} 2x+1 & \text{when } x<-1 \\ 3 & \text{when } x\geq -1 \end{cases}$

b $f(x)=\begin{cases} x^2 & \text{for } x\leq 0 \\ x^3+1 & \text{for } x>0 \end{cases}$

Solution

a $f(x)$ is made up of 2 linear functions. Graph each one separately.

1st function:

$f(x) = 2x + 1$ in the domain $x < -1$

$f(-2) = 2(-2) + 1 = -3$ $\quad (-2, -3)$

$f(-1) = 2(-1) + 1 = -1$ $\quad (-1, -1)$

Draw the line with an open circle at $(-1, -1)$ to exclude $x = -1$.

2nd function:

$f(x) = 3$ in the domain $x \geq -1$

$f(-1) = 3$

Draw a horizontal line with a closed circle at $(-1, 3)$ to include $x = -1$.

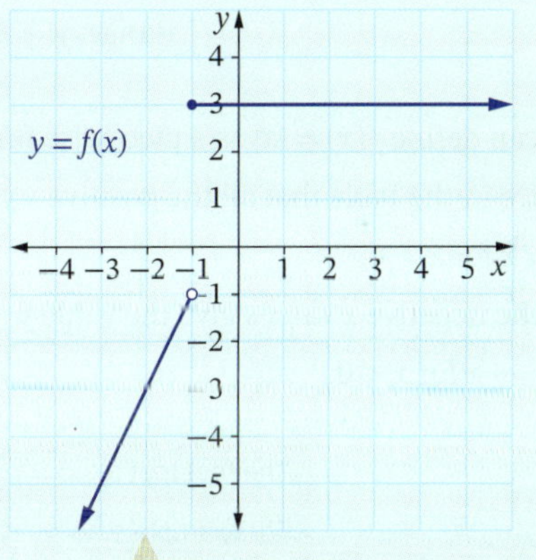

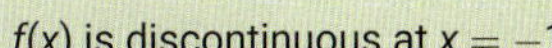
$f(x)$ is discontinuous at $x = -1$

b $f(x)$ is made up of a quadratic function (parabola) and a cubic function.

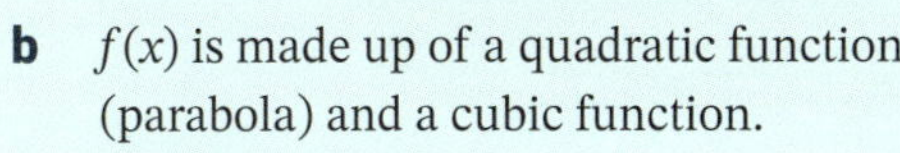

1st function:

$f(x) = x^2$ for $x \leq 0$

$f(0) = 0^2 = 0$ $\quad (0, 0)$

$f(-1) = (-1)^2 = 1$ $\quad (-1, 1)$

Draw the parabola with a closed circle at $(0, 0)$.

2nd function:

$f(x) = x^3 + 1$ for $x > 0$

$f(0) = 0^3 + 1 = 1$ $\quad (0, 1)$

$f(1) = 1^3 + 1 = 2$ $\quad (1, 2)$

Draw the curve with an open circle at $(0, 1)$.

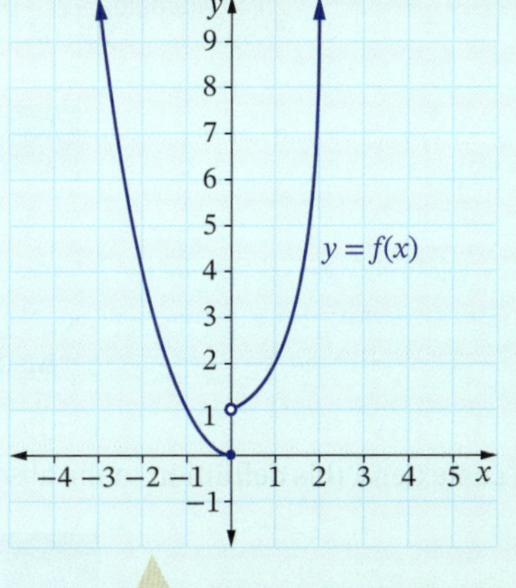

$f(x)$ is discontinuous at $x = 0$

Foundation Mastery Complex

Absolute value functions

You were introduced to absolute value in Chapter 2, *Equations and inequalities*. It is the **magnitude** of a number or the distance of the number from 0 on a number line, so its value is always positive (or zero).

We can graph the absolute value function $y = |x|$ by completing a table of values.

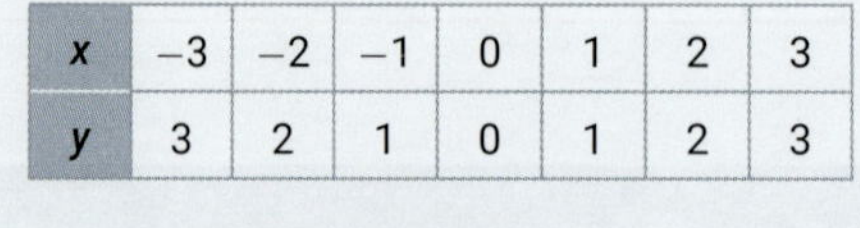

x	-3	-2	-1	0	1	2	3
y	3	2	1	0	1	2	3

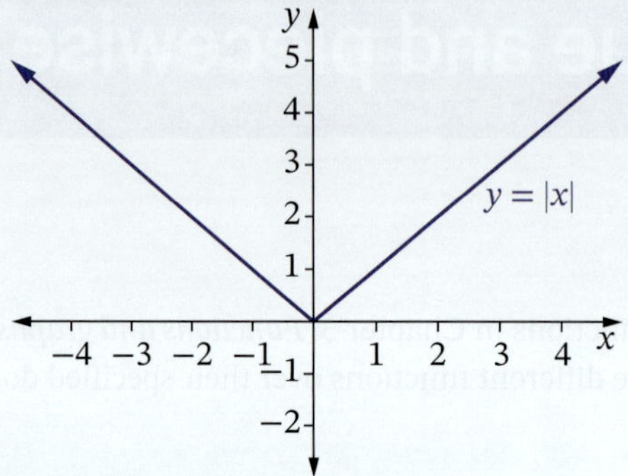

$y = |x|$ is a continuous function

The absolute value function $y = |x|$

- The graph of $y = |x|$ is V-shaped, consisting of 2 lines with gradients -1 and 1 that meet at $(0, 0)$.
- It is symmetrical about the y-axis so it is an even function.
- Domain: All real x, or $(-\infty, \infty)$
- Range: $y \geq 0$, or $[0, \infty)$

We can define $f(x) = |x|$ as a piecewise function.

Notice in the table that all the positive x values and 0 give the value $y = x$.

$|x| = x$ for $x \geq 0$.

All the negative x values give the value $y = -x$.

$|x| = -x$ for $x < 0$.

The fact that $f(-x) = f(x)$ for all x actually proves that $y = |x|$ is an even function.

For example, $|-2| = -(-2) = 2$.

A further definition of absolute value is $|x| = \sqrt{x^2}$.

For example, $|-2| = \sqrt{(-2)^2} = \sqrt{4} = 2$.

$y = |x|$ as a piecewise function

$$|x| = \begin{cases} x & \text{if } x \geq 0 \\ -x & \text{if } x < 0 \end{cases}$$

Also, $|x| = \sqrt{x^2}$.

We can extend this definition to the absolute value of any function, $f(x)$:

$y = |f(x)|$ as a piecewise function

$$|f(x)| = \begin{cases} f(x) & \text{if } f(x) \geq 0 \\ -f(x) & \text{if } f(x) < 0 \end{cases}$$

Video
Absolute value graphs

Example 8

a Sketch the graph of $f(x) = |x| - 1$ and state its domain and range.

b Sketch the graph of $y = |x + 2|$.

Solution

a Using the piecewise function definition of $|x|$:

$$y = \begin{cases} x - 1 & \text{for } x \geq 0 \\ -x - 1 & \text{for } x < 0 \end{cases}$$

Draw $y = x - 1$ for $x \geq 0$ and $y = -x - 1$ for $x < 0$.

For x-intercepts, $y = 0$.

$$\begin{aligned} y &= x - 1 \\ 0 &= x - 1 \\ 1 &= x \end{aligned} \qquad \begin{aligned} y &= -x - 1 \\ 0 &= -x - 1 \\ x &= -1 \end{aligned}$$

For y-intercept, $x = 0$.

$$\begin{aligned} y &= x - 1 \text{ for } x \geq 0 \\ &= 0 - 1 \\ &= -1 \end{aligned}$$

From the graph, notice that x can be any real number while $y \geq -1$.

Domain: $(-\infty, \infty)$ (or all real x)

Range: $[-1, \infty)$ (or $y \geq -1$)

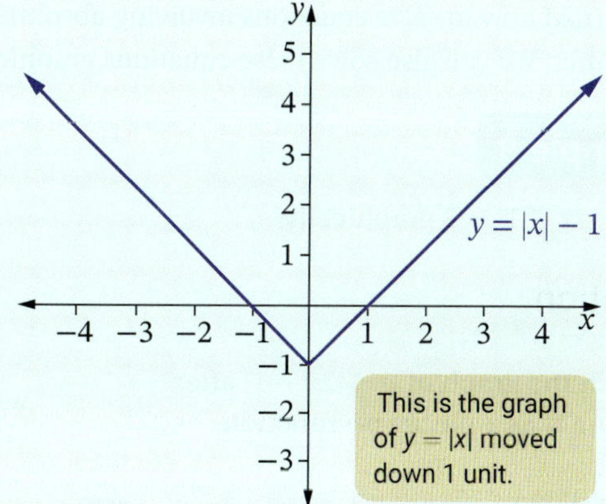

b Using the piecewise function definition of $|f(x)|$:

$$y = \begin{cases} x + 2 & \text{for } x + 2 \geq 0 \\ -(x + 2) & \text{for } x + 2 < 0 \end{cases}$$

Simplifying this gives:

$$y = \begin{cases} x + 2 & \text{for } x \geq -2 \\ -x - 2 & \text{for } x < -2 \end{cases}$$

For x-intercepts, $y = 0$.

$$\begin{aligned} y &= x + 2 \\ 0 &= x + 2 \\ -2 &= x \end{aligned} \qquad \begin{aligned} y &= -x - 2 \\ 0 &= -x - 2 \\ x &= -2 \end{aligned}$$

For y-intercept, $x = 0$.

$$\begin{aligned} y &= x + 2 \text{ for } x \geq 0 \\ &= 0 + 2 \\ &= 2 \end{aligned}$$

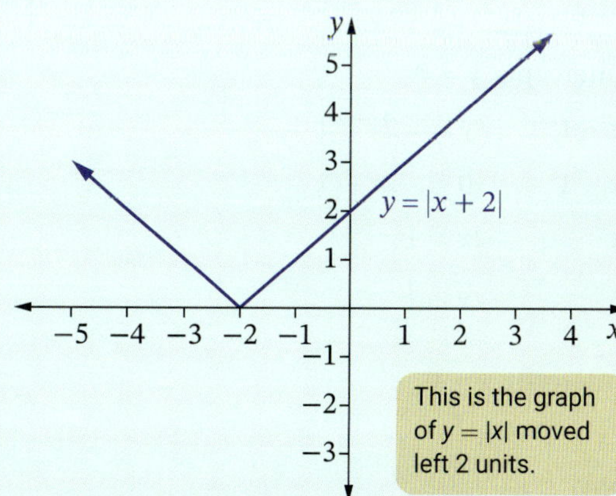

Technology

Transformations of the absolute value function

Use graphing technology to explore each absolute value graph.

1 $y = |x|$ **2** $y = 2|x|$ **3** $y = 3|x|$ **4** $y = -|x|$

5 $y = -2|x|$ **6** $y = |x| + 1$ **7** $y = |x| + 2$ **8** $y = |x| - 1$

9 $y = |x| - 2$ **10** $y = |x + 1|$ **11** $y = |x + 2|$ **12** $y = |x + 3|$

13 $y = |x - 1|$ **14** $y = |x - 2|$ **15** $y = |x - 3|$

16 $y = k|x|$ for different values of k, including negative

17 $y = |x| + k$ for different values of k

18 $y = |x - k|$ for different values of k

Are graphs that involve absolute value always functions? Can you find an example of one that is not a function?

Are any of them odd or even? Are they continuous? Could you predict what the graph $y = 2|x - 7|$ would look like?

Equations involving absolute values

You learned how to solve equations involving absolute values using algebra in Chapter 2 *Equations and inequalities*. We can also solve these equations graphically.

Example 9

Solve $|2x - 1| = 3$ graphically.

Solution

Sketch the graph of $y = |2x - 1|$ after writing it as a piecewise function.

$$y = \begin{cases} 2x-1 & \text{for } 2x-1 \geq 0 \\ -(2x-1) & \text{for } 2x-1 < 0 \end{cases}$$

Simplifying:

$$y = \begin{cases} 2x-1 & \text{for } x \geq \frac{1}{2} \\ -2x+1 & \text{for } x < \frac{1}{2} \end{cases}$$

For x-intercepts, $y = 0$.

$$\begin{aligned} y &= 2x - 1 & y &= -2x + 1 \\ 0 &= 2x - 1 & 0 &= -2x + 1 \\ 1 &= 2x & 2x &= 1 \\ \frac{1}{2} &= x & x &= \frac{1}{2} \end{aligned}$$

For y-intercept, $x = 0$.

$$\begin{aligned} y &= -2x + 1 \text{ for } x\frac{1}{2} \\ &= -2(0) + 1 \\ &= 1 \end{aligned}$$

The graph of $y = 3$ is a horizontal line through 3 on the y-axis.

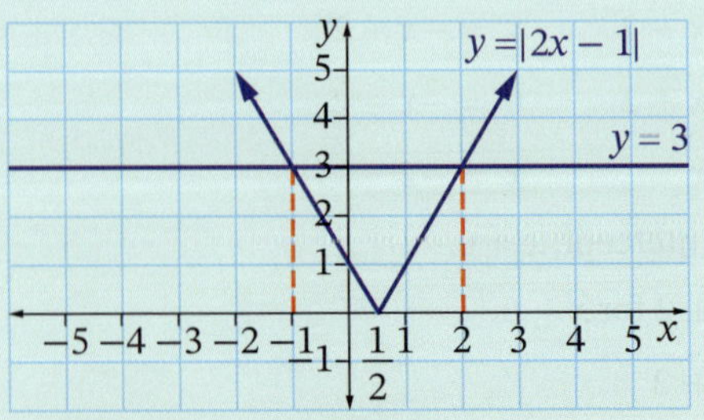

The solutions of $|2x - 1| = 3$ are the values of x at the points of intersection of the graphs of $y = |2x - 1|$ and $y = 3$.

$x = -1, 2$.

We can check that our solutions are correct by substituting them back into the equation.

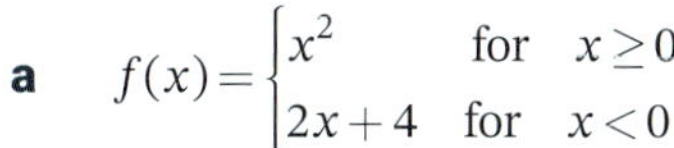

EXERCISE 5.04 Answers on p. 482

Absolute value and piecewise functions

EXAMPLE 7

1 Graph each piecewise function, state whether it is continuous or discontinuous and find its domain and range.

a $f(x)=\begin{cases} x^2 & \text{for} \quad x\geq 0 \\ 2x+4 & \text{for} \quad x<0 \end{cases}$　　**b** $f(x)=\begin{cases} 2x & \text{for} \quad x>0 \\ x^3 & \text{for} \quad x\leq 0 \end{cases}$

c $f(x)=\begin{cases} |x| & \text{for} \quad x\geq 1 \\ x^2 & \text{for} \quad x<1 \end{cases}$　　**d** $y=\begin{cases} x-3 & \text{for} \quad x\geq -1 \\ 2 & \text{for} \quad x<-1 \end{cases}$

e $f(x)=\begin{cases} -1 & \text{for} \quad x>2 \\ x+1 & \text{for} \quad x\leq 2 \end{cases}$

EXAMPLE 8

2 Sketch the graph of each function.

a $y = |x|$　　**b** $f(x) = |x| + 1$　　**c** $f(x) = |x| - 3$

d $y = 2|x|$　　**e** $f(x) = -|x|$　　**f** $y = |x + 1|$

g $f(x) = |x - 1|$　　**h** $y = |2x - 3|$　　**i** $f(x) = |3x| + 1$

3 Prove that $f(x) = 2|x|$ is an even function.

4 Sketch $f(x)=\begin{cases} x & \text{for} \quad x>3 \\ x-1 & \text{for} \quad -1\leq x\leq 3 \\ 2 & \text{for} \quad <-1 \end{cases}$

5 By sketching $y = |x|$, $y = |x| + 2$ and $y = |x + 2|$, describe their relationship to each other.

6 Use a table of values or graphing technology to graph each function.

a $y = |x + 1| - x$　　**b** $y = x^2 + |x|$　　**c** $y=\dfrac{|x|}{x}$

d $f(x)=\dfrac{|x|}{x^2}$　　**e** $f(x)=|3-||x|-2||$

EXAMPLE 9

7 Solve each equation graphically.

a $|x| = 3$　　**b** $|x + 2| = 1$　　**c** $|x - 3| = 0$

d $|2x - 3| = 1$　　**e** $|2x + 3| = 11$　　**f** $|5b - 2| = 8$

g $|3x + 1| = 2$　　**h** $5 = |2x + 1|$　　**i** $0 = |6t - 3|$

8 By sketching $y = |x|$ and $y = 2x + 6$ on the same number plane, solve $|x| = 2x + 6$.

9 **a** Explain why $x > 0$ in the inequality $|3x - 6| < x$.

b Solve $|3x - 6| < x$ graphically or by using graphing technology.

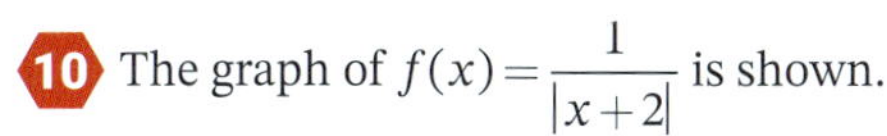

10 The graph of $f(x)=\dfrac{1}{|x+2|}$ is shown.

Use the graph to solve $\dfrac{x}{3}\leq\dfrac{1}{|x+2|}$.

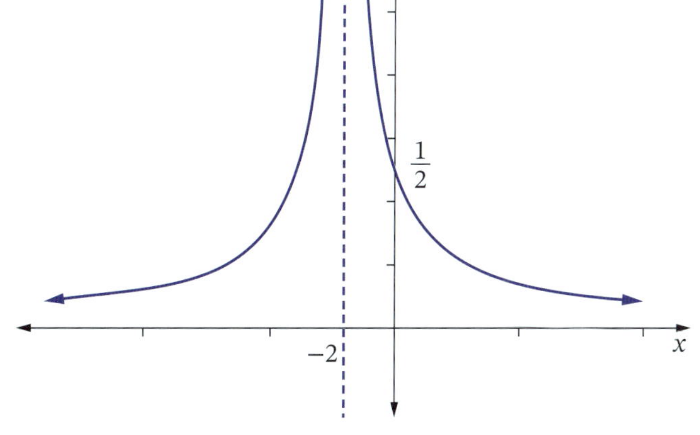

□ Foundation　○ Mastery　○ Complex

5.05 Circles and semicircles

Video Identifying graphs

Worksheets Equations of circles

Advanced graphs

Puzzles Circle equations

Matching graphs (Advanced)

The equation of a circle is not a function. It does not pass the vertical line test.

Circle with centre (0, 0)

We can use Pythagoras' theorem to find the equation of a circle, using a general point (x, y) on a circle with centre $(0, 0)$ and radius r.

$$c^2 = a^2 + b^2$$

$$\therefore r^2 = x^2 + y^2$$

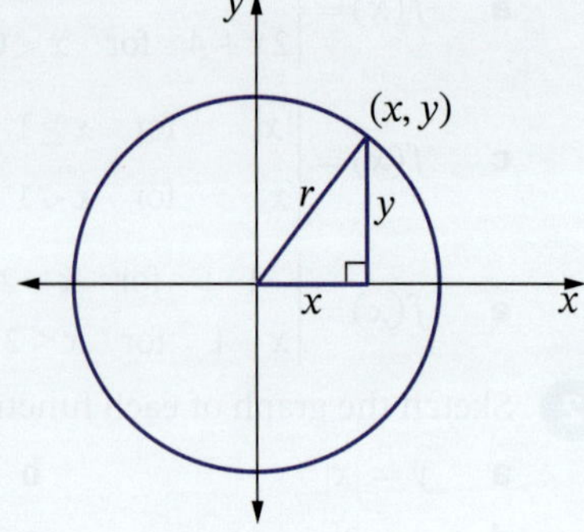

Equation of a circle with centre (0, 0)

The equation of a circle with centre $(0, 0)$ and radius r is $x^2 + y^2 = r^2$.

Example 10

a Sketch the graph of $x^2 + y^2 = 4$.

b Why is it not a function?

c State its domain and range.

Solution

a The equation is in the form $x^2 + y^2 = r^2$, where $r^2 = 4$.

Radius, $r = \sqrt{4} = 2$

This is a circle with radius 2 and centre (0, 0).

b The circle is not a function because a vertical line will cut the graph in more than one place. This means that an x value in the domain can be matched to 2 y values rather than one.

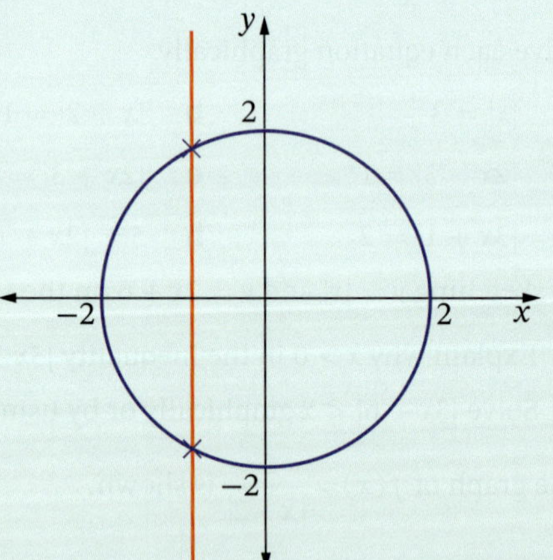

c The x values for this graph lie between -2 and 2 and the y values also lie between -2 and 2.

Domain: $[-2, 2]$ (or $-2 \leq x \leq 2$)

Range: $[-2, 2]$ (or $-2 \leq y \leq 2$)

Circle with centre (a, b)

We can use Pythagoras' theorem to find the equation of a circle using a general point (x, y) on a circle with centre (a, b) and radius r.

The smaller sides of the triangle are $x - a$ and $y - b$ and the hypotenuse is r, the radius.

$c^2 = a^2 + b^2$

$r^2 = (x - a)^2 + (y - b)^2$

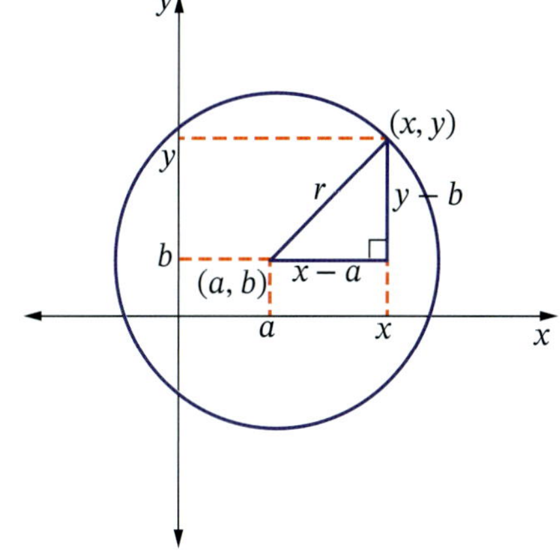

Equation of a circle with centre (a, b)

The equation of a circle with centre (a, b) and radius r is $(x - a)^2 + (y - b)^2 = r^2$.
Note that this is also the equation of the circle $x^2 + y^2 = r^2$ translated a units to the right and b units up.

Example 11

Video
Equation of a circle

a **i** Sketch the graph of the circle $(x - 1)^2 + (y + 2)^2 = 4$.

ii State its domain and range.

b Find the equation of a circle with radius 3 and centre $(-2, 1)$ in expanded form.

c Find the centre and radius of the circle with equation $x^2 + 2x + y^2 - 6y - 6 = 0$.

Solution

a **i** The equation is in the form $(x - a)^2 + (y - b)^2 = r^2$.

$$(x - 1)^2 + (y + 2)^2 = 4$$

$$(x - 1)^2 + (y - (-2))^2 = 2^2$$

So $a = 1$, $b = -2$ and $r = 2$.

This is a circle with centre $(1, -2)$ and radius 2.

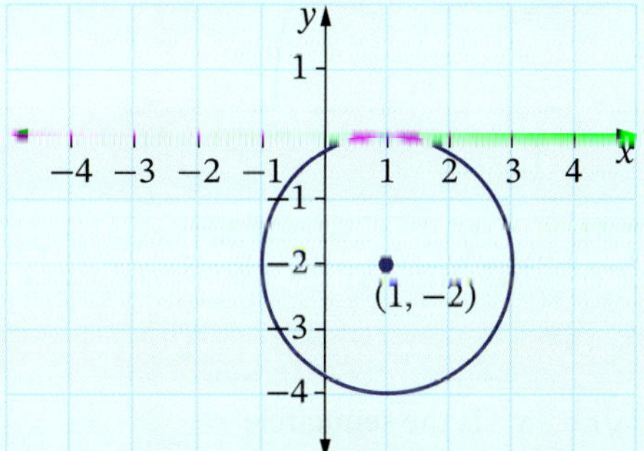

ii From the graph, we can see that all x values lie between -1 and 3 and all y values lie between -4 and 0.

Domain: $-1 \le x \le 3$ or $[-1, 3]$

Range: $-4 \le y \le 0$ or $[-4, 0]$

b Centre is $(-2, 1)$ so $a = -2$ and $b = 1$.

Radius is 3 so $r = 3$.

$$(x - a)^2 + (y - b)^2 = r^2$$

$$(x - (-2))^2 + (y - 1)^2 = 3^2$$

$$(x + 2)^2 + (y - 1)^2 = 9$$

Expanding:

$$x^2 + 4x + 4 + y^2 - 2y + 1 = 9$$

$$x^2 + 4x + y^2 - 2y - 4 = 0$$

c The equation of a circle is $(x-a)^2+(y-b)^2=r^2$.

We need to complete the square to put the equation into this form.

To complete the square on x^2+2x, we add $\left(\frac{2}{2}\right)^2=1$.

To complete the square on y^2-6y, we add $\left(\frac{6}{2}\right)^2=9$.

$$x^2+2x+y^2-6y-6=0$$
$$x^2+2x+y^2-6y=6$$
$$x^2+2x+1+y^2-6y+9=6+1+9$$
$$(x+1)^2+(y-3)^2=16$$
$$(x-(-1))^2+(y-3)^2=4^2$$

This is in the form $(x-a)^2+(y-b)^2=r^2$, where $a=-1$, $b=3$ and $r=4$.

So it is a circle with centre $(-1, 3)$ and radius 4 units.

Semicircles

By rearranging the equation of a circle, we can find the equations of 2 semicircles.

$$x^2+y^2=r^2$$
$$y^2=r^2-x^2$$
$$y=\pm\sqrt{r^2-x^2}$$

A circle is not a function, but we can split it into 2 semicircles that are functions.

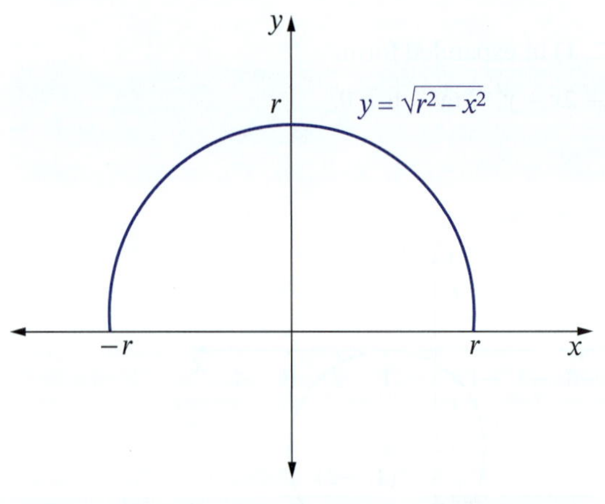

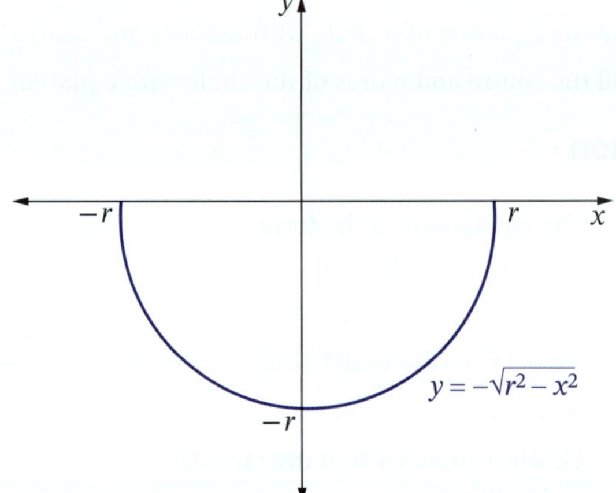

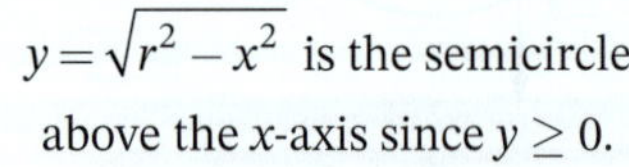

$y=\sqrt{r^2-x^2}$ is the semicircle above the x-axis since $y \geq 0$.

$y=-\sqrt{r^2-x^2}$ is the semicircle below the x-axis since $y \leq 0$.

Equations of a semicircle with centre (0, 0)

The equation of a semicircle above the x-axis with centre $(0,0)$ and radius r is

$$y=\sqrt{r^2-x^2}$$

The equation of a semicircle below the x-axis with centre $(0,0)$ and radius r is

$$y=-\sqrt{r^2-x^2}$$

Example 12

Sketch the graph of each function and state the domain and range.

a $f(x) = \sqrt{9 - x^2}$ **b** $y = -\sqrt{4 - x^2}$

Solution

a This is in the form $f(x) = \sqrt{r^2 - x^2}$ where $r^2 = 9$, so $r = 3$.

It is a semicircle above the x-axis with centre $(0, 0)$ and radius 3.

Domain: $[-3, 3]$

Range: $[0, 3]$

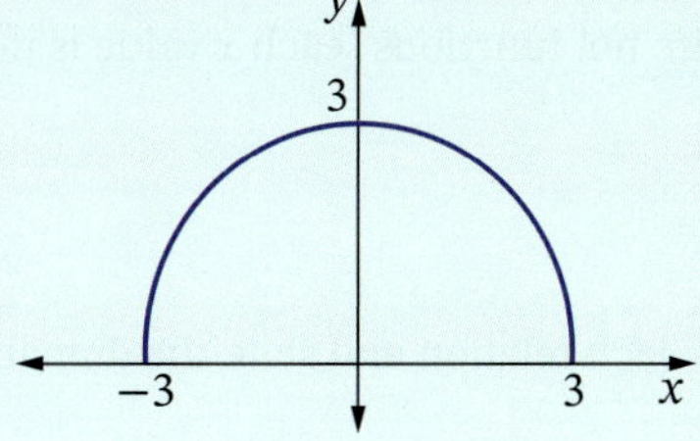

b This is in the form $y = -\sqrt{r^2 - x^2}$ where $r^2 = 4$, so $r = 2$.

It is a semicircle below the x-axis with centre $(0, 0)$ and radius 2.

Domain: $[-2, 2]$

Range: $[-2, 0]$

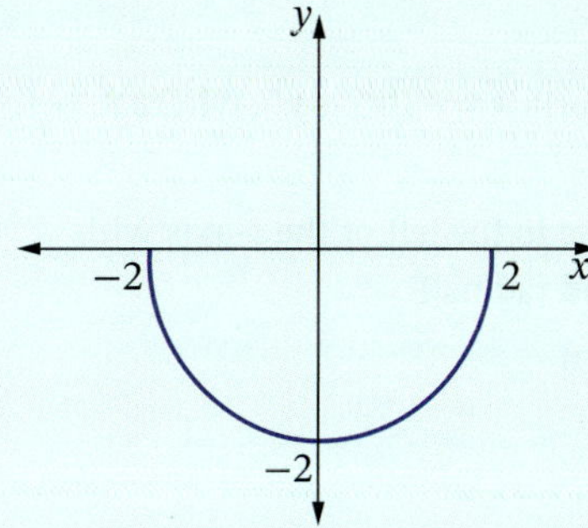

We can also halve the circle vertically to create 2 semicircles about the y-axis. This involves making x the subject of the equation.

$$x^2 + y^2 = r^2$$

$$x^2 = r^2 - y^2$$

$$x = \pm\sqrt{r^2 - y^2}$$

These 2 semicircles are not functions.

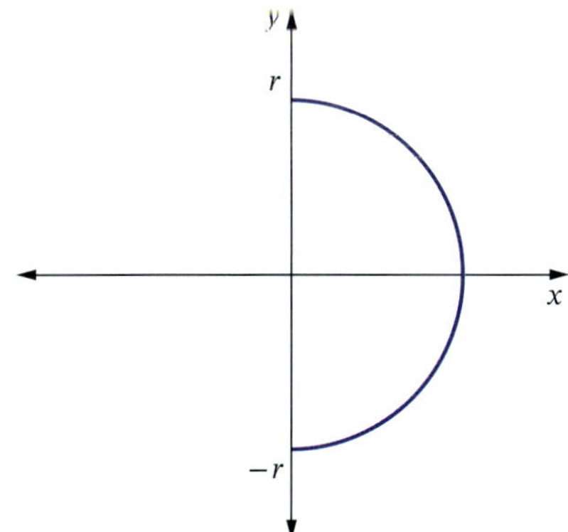

$x = \sqrt{r^2 - y^2}$ is the semicircle to the right of the y-axis since $x \geq 0$.

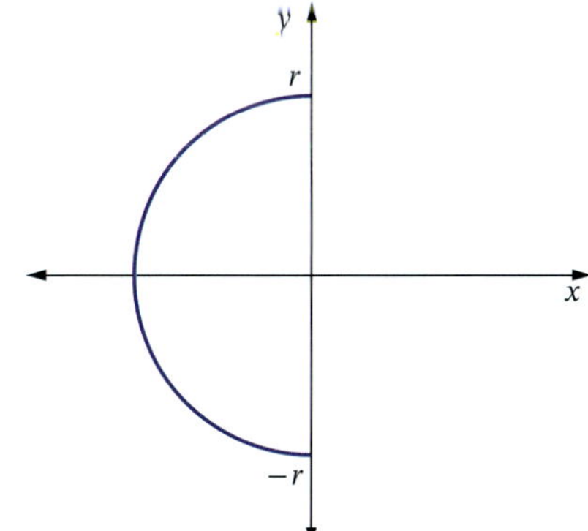

$x = -\sqrt{r^2 - y^2}$ is the semicircle to the left of the y-axis since $x \leq 0$.

Equations of a semicircle with centre (0, 0) about the y-axis

The equation of a semicircle to the right of the y-axis with centre $(0,0)$ and radius r is

$$x=\sqrt{r^2-y^2}$$

The equation of a semicircle to the left of the y-axis with centre $(0,0)$ and radius r is

$$x=-\sqrt{r^2-y^2}$$

These are not functions (each x value is matched with 2 non-zero y values).

Example 13

Sketch the graph of each relation and state the domain and range.

a $x=-\sqrt{1-y^2}$

b $x=\sqrt{49-y^2}$

Solution

a This is in the form $x=-\sqrt{r^2-y^2}$, where $r^2=1$, so $r=1$

It is a semicircle to the left of the y-axis with centre $(0,0)$ and radius 1.

Domain: $[-1,0]$

Range: $[-1,1]$

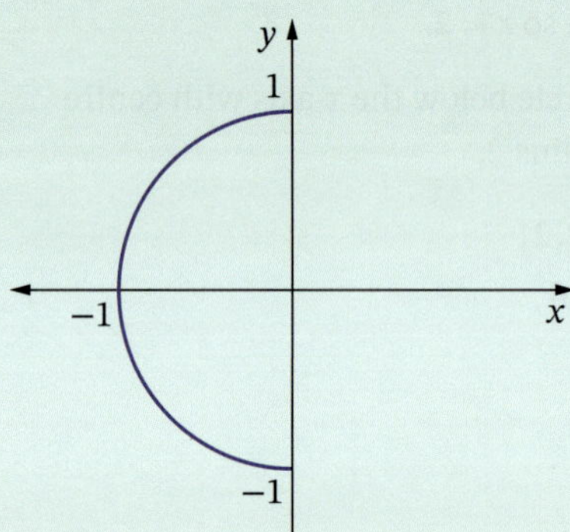

b This is in the form $x=\sqrt{r^2-y^2}$, where $r^2=49$, so $r=7$.

It is a semicircle to the right of the y-axis with centre $(0,0)$ and radius 7.

Domain: $[0,7]$

Range: $[-7,7]$

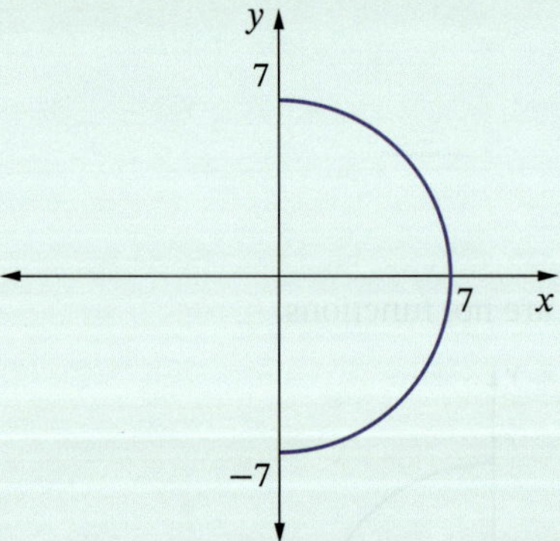

EXERCISE 5.05 Answers on p. 484

Circles and semicircles

1 For each equation:

i sketch the graph **ii** state the domain and range.

a $x^2+y^2=9$

b $x^2+y^2-16=0$

c $(x-2)^2+(y-1)^2=4$

d $(x+1)^2+y^2=9$

e $(x+2)^2+(y-1)^2=1$

□ Foundation ○ Mastery ⬡ Complex

EXAMPLES 12,13

2 For each semicircle:

- **i** state whether it is above or below the x-axis, or left or right of the y-axis
- **ii** sketch the graph
- **iii** state the domain and range.

a $y=-\sqrt{25-x^2}$ **b** $y=\sqrt{1-x^2}$ **c** $x=\sqrt{36-y^2}$

d $y=-\sqrt{64-x^2}$ **e** $y=-\sqrt{7-x^2}$ **f** $x=-\sqrt{81-y^2}$

3 Find the radius and the centre of each circle.

a $x^2+y^2=100$ **b** $x^2+y^2=5$ **c** $(x-4)^2+(y-5)^2=16$

d $(x-5)^2+(y+6)^2=49$ **e** $x^2+(y-3)^2=81$

4 Find the equation of each circle in expanded form.

a centre $(0, 0)$ and radius 4 **b** centre $(3, 2)$ and radius 5

c centre $(-1, 5)$ and radius 3 **d** centre $(2, 3)$ and radius 6

e centre $(-4, 2)$ and radius 5 **f** centre $(0, -2)$ and radius 1

g centre $(4, 2)$ and radius 7 **h** centre $(-3, -4)$ and radius 9

i centre $(-3, 0)$ and radius $\sqrt{5}$ **j** centre $(-4, -7)$ and radius $\sqrt{3}$

5 Find the radius and the centre of each circle.

a $x^2-4x+y^2-2y-4=0$ **b** $x^2+8x+y^2-4y-5=0$

c $x^2+y^2-2y=0$ **d** $x^2-10x+y^2+6y-2=0$

e $x^2+2x+y^2-2y+1=0$ **f** $x^2-12x+y^2=0$

g $x^2+6x+y^2-8y=0$ **h** $x^2+20x+y^2-4y+40=0$

i $x^2-14x+y^2+2y+25=0$ **j** $x^2+2x+y^2+4y-5=0$

6 **a** Find the centre and radius of the circle with equation $x^2+4x+y^2-2y+1=0$.

b Sketch its graph.

c Solve simultaneous equations to find any points where the circle and the line $y=1-x$ intersect.

7 **a** In how many ways can a circle and a line intersect?

b Show by solving simultaneous equations that there is only one point of intersection between x^2+y^2-1 and $4x-3y+5=0$.

8 Prove that $y=\sqrt{r^2-x^2}$ is an even function.

9 Hamish designs a circular disc with equations $x^2+y^2=25$ for the outer edge and x^2+y^2-4 for the inner edge.

a If the measurements are in mm, find the area of the disc. Round your answer to the nearest whole number.

b Hamish designs a second disc with circles $x^2+y^2=16$ and $x^2+y^2=9$ for its outer and inner edges. Find the area of this disc. Round your answer to the nearest whole number.

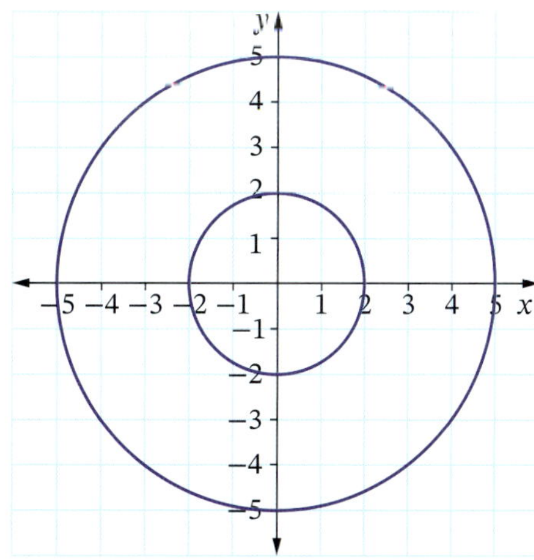

□ Foundation ○ Mastery ○ Complex

5.06 Composite functions

Worksheet
Circles and composite functions

A **composite function** $f(g(x))$ is a combination of 2 or more functions (a 'function of a function') where the output of one function, $g(x)$, becomes the input of a second function, $f(x)$. The input values of a composite function go through a 'double function'.

Alternatively, $g(f(x))$ is a different composite function where the output of $f(x)$ becomes the input of $g(x)$.

Example 14

a Find the composite function $f(g(x))$ given:

i $f(x) = x^2$ and $g(x) = 2x - 5$

ii $f(x) = x^3$ and $g(x) = x^2 + 3$

iii $f(x) = 5x - 3$ and $g(x) = x^3 + 2$

b Given $f(x) = 5x + 2$ and $g(x) = \frac{1}{x}$, find:

i $f(g(x))$ and $f(g(10))$ **ii** $g(f(x))$ and $g(f(-4))$

c Find the domain and range of $f(g(x))$ given $f(x) = \sqrt{x}$ and $g(x) = 9 - x^2$.

Solution

a **i**

$$\begin{aligned} f(g(x)) &= (2x-5)^2 \\ &= 4x^2 - 20x + 25 \end{aligned}$$

ii

$$\begin{aligned} f(g(x)) &= (x^2+3)^3 \\ &= (x^2+3)(x^2+3)^2 \\ &= (x^2+3)(x^4+6x^2+9) \\ &= x^6 + 9x^4 + 27x^2 + 27 \end{aligned}$$

iii

$$\begin{aligned} f(g(x)) &= 5(x^3+2) - 3 \\ &= 5x^3 + 10 - 3 \\ &= 5x^3 + 7 \end{aligned}$$

b **i**

$$\begin{aligned} f(g(x)) &= 5\left(\frac{1}{x}\right) + 2 \\ &= \frac{5}{x} + 2 \end{aligned}$$

$$f(g(10)) = \frac{5}{10} + 2 = 2\frac{1}{2}$$

ii

$$\begin{aligned} g(f(x)) &= \frac{1}{5x+2} \\ gf(-4)) &= \frac{1}{5(-4)+2} \\ &= -\frac{1}{18} \end{aligned}$$

c $f(g(x)) = \sqrt{9 - x^2}$

This is a semicircle above the x-axis with centre $(0, 0)$ and radius 3.

Domain: $[-3, 3]$, range: $[0, 3]$

EXERCISE 5.06 Answers on p. 486

Composite functions

1 Find the composite function $f(g(x))$ given:

- **a** $f(x) = x^2$ and $g(x) = x^2 + 1$
- **b** $f(x) = x^3$ and $g(x) = 5x - 3$
- **c** $f(x) = x^7$ and $g(x) = x^2 - 3x + 2$
- **d** $f(x) = \sqrt{x}$ and $g(x) = 2x - 1$
- **e** $f(x) = \sqrt[3]{x}$ and $g(x) = x^4 + 7x^2 - 4$
- **f** $f(x) = 3x$ and $g(x) = 2x + 1$
- **g** $f(x) = 2x - 7$ and $g(x) = x^3$
- **h** $f(x) = 6x - 5$ and $g(x) = x^2$
- **i** $f(x) = 2x^2$ and $g(x) = 3x$
- **j** $f(x) = 4x^2 + 1$ and $g(x) = x^2 + 3$

2 Find the domain and range of the composite function $f(g(x))$ given that:

- **a** $f(x) = x^2$ and $g(x) = x - 1$
- **b** $f(x) = x^3$ and $g(x) = x + 5$
- **c** $f(x) = \sqrt{x}$ and $g(x) = x - 2$
- **d** $f(x) = -\sqrt{x}$ and $g(x) = 3x + 9$
- **e** $f(x) = \sqrt{x}$ and $g(x) = 4 - x^2$
- **f** $f(x) = -\sqrt{x}$ and $g(x) = 1 - x^2$

3 If $f(x) = \sqrt{x}$ and $g(x) = x^3$, find:

- **a** $f(g(x))$ and $f(g(4))$
- **b** $g(f(x))$ and $g(f(4))$

4 If $f(x) = \frac{1}{x}$ and $g(x) = x^2 + 3$, find:

- **a** $g(f(x))$ and $g(f(-2))$
- **b** $y = f(g(x))$ and $f(g(-2))$

5 If $f(x) = x^3$ and $g(x) = x - 2$, then find:

- **a** $f(g(x))$ and state whether it is even, odd or neither
- **b** $g(f(x))$ and state whether it is even, odd or neither
- **c** $f(g(1))$
- **d** $g(f(-2))$

6 If $f(x) = (x + 1)^2$ and $g(x) = \sqrt{x}$, sketch the graph of:

- **a** $y = f(g(x))$
- **b** $y = g(f(x))$

7 If $y = f(x)$ is even, show that $y = g(f(x))$ is even.

8 If $y = f(x)$ is odd and $g(x)$ is even, show that $y = g(f(x))$ is even

Foundation | Mastery | Complex

9 Given the graphs of $y = f(x)$ and $y = g(x)$ below, choose which graph could represent $y = g(f(x))$.

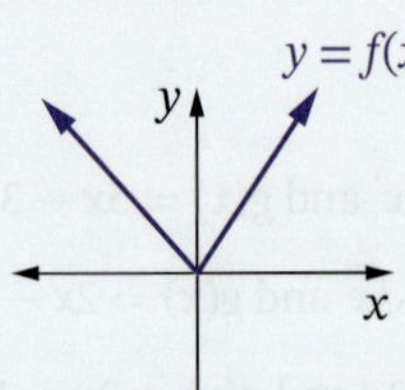

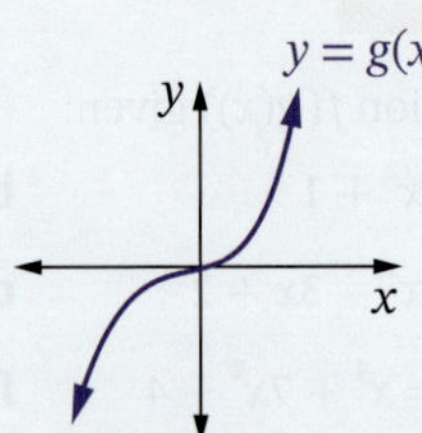

A

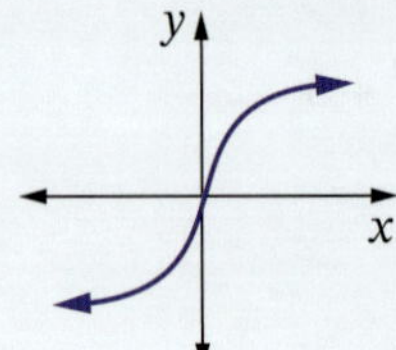

B

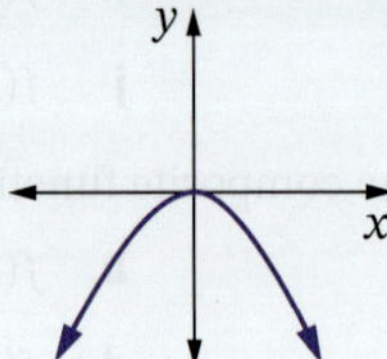

C

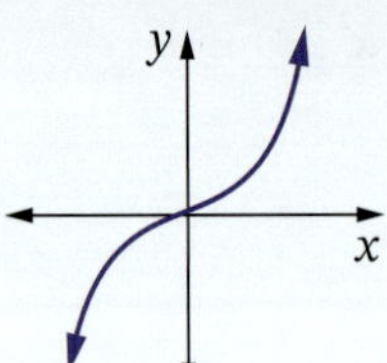

D

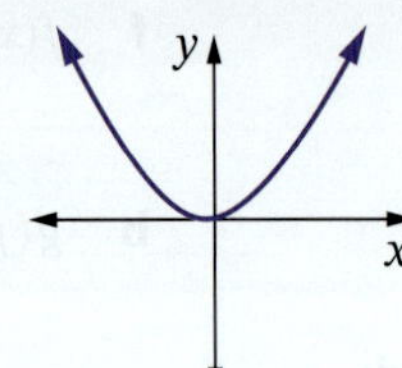

10 $f(x) = |x|$ and $g(x) = -\dfrac{1}{x}$.

a Find $y = g(f(x))$ and state its domain and range.

b Find $y = f(g(x))$ and state its domain and range.

Sample HSC problem Answers on p. 486

(7 marks)

a Sketch the graph of $f(x) = \begin{cases} x & \text{for } x < -2 \\ x^2 & \text{for } -2 \le x \le 0 \\ 2 & \text{for } x > 0 \end{cases}$ 3 marks

b Find any x values for which the function is discontinuous. 2 marks

c Find the domain and range of the function. 2 marks

Foundation Mastery Complex

CHAPTER SUMMARY

This chapter, *Further functions and graphs*, examined more advanced functions and composite functions.

Before you move on, consider what you learned in this chapter and revisit any sections that may have been unclear.

What you learned in this chapter...	Section
Identify values of a function where it is continuous and discontinuous	5.01 The hyperbola
Identify characteristics of a hyperbola, piecewise and absolute value functions, including domain and range	5.01 The hyperbola 5.04 Absolute value and piecewise functions
Solve absolute value equations graphically	5.04 Absolute value and piecewise functions
Understand direct and inverse variation and solve problems involving them	5.02 Direct variation 5.03 Inverse variation
Sketch graphs of circles and semicircles and find their equations	5.05 Circles and semicircles
Work with composite functions	5.06 Composite functions

To help master these techniques, make a summary of this topic. Use the chapter outline and the mind map below as a guide. Add your own words, symbols, diagrams, boxes and reminders. The summary should give you a 'whole picture' view of the topic and allow you to identify any weak areas to revisit in your revision.

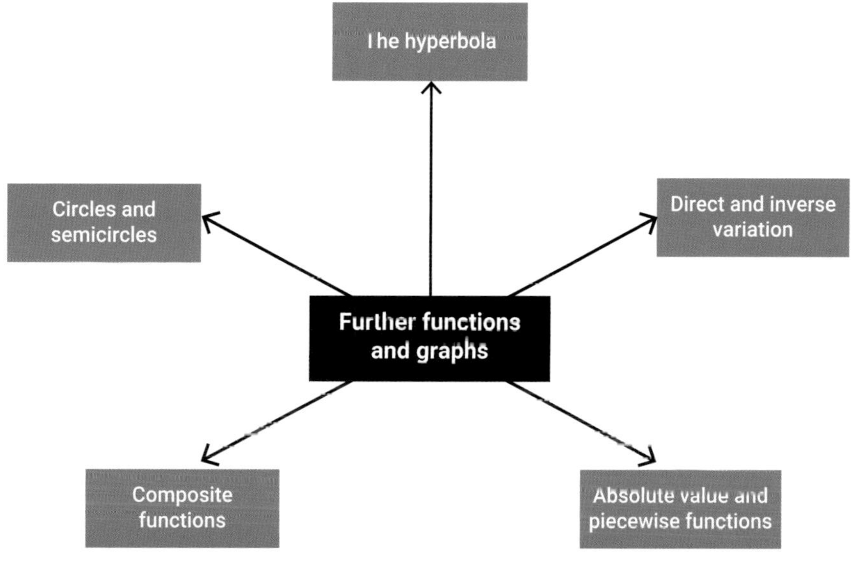

5 Test yourself

Answers on p. 486

For Questions **1** to **3**, select the correct answer **A**, **B**, **C** or **D**.

5.01 **1** The domain of $y = -\frac{3}{x-4}$ is:

A $x > 4$ **B** All real $x, x \neq 4$

C All real $x, x \neq -4$ **D** $x < 4$

5.05 **2** The equation of a circle with radius 3 and centre $(1, -2)$ is:

A $(x-1)^2 + (y+2)^2 = 9$ **B** $(x+1)^2 + (y-2)^2 = 9$

C $(x-1)^2 + (y+2)^2 = 3$ **D** $(x+1)^2 + (y-2)^2 = 3$

5.03 **3** A variable N varies inversely as the square of x. Its equation is:

A $N = kx$ **B** $N = \frac{k}{x}$ **C** $N = \frac{k}{x^2}$ **D** $N = kx^2$

5.03 **4** The area of a pizza slice varies inversely with the number of people sharing the pizza. When 5 people share the pizza, the area of each slice is 30 cm^2.

a Find the equation of the area, A, of a pizza slice in terms of the number of people sharing, n.

b What is the area of one pizza slice when:

i 10 people share? **ii** 8 people share?

c How many people are sharing the pizza when each slice has an area of:

i $16.67\,\text{cm}^2$? **ii** $25\,\text{cm}^2$?

5.01, 5.04 **5** Sketch the graph of each function or relation.

a $x^2 + y^2 = 1$ **b** $y = \frac{2}{x}$

5.05

c $y = |x+2|$ **d** $y = -\sqrt{4-x^2}$

5.05 **6** Find the radius and centre of the circle $x^2 - 6x + y^2 - 2y - 6 = 0$.

5.06 **7** If $f(x) = x^3$ and $g(x) = 3x - 1$, find the equation of:

a $y = f(g(x))$ **b** $y = g(f(x))$

5.05 **8** **a** Is the circle $x^2 + y^2 = 1$ a function?

b Change the subject of the equation to y in terms of x.

c Sketch the graphs of 2 separate functions that together make up the circle $x^2 + y^2 = 1$.

5.01, 5.04 **9** Find the domain and range of each relation.

5.05

a $x^2 + y^2 = 16$ **b** $y = \frac{1}{x+2}$

c $f(x) = |x| + 3$ **d** $y = \sqrt{9-x^2}$

□ Foundation ○ Mastery ⬡ Complex

10 The area (A) of a hexagon is proportional to the square of its length x. If the area is $448\,\text{cm}^2$ when $x = 8$, find: 5.02

a the equation for the area, A

b the area when $x = 10$

c x when the area is $1093.75\,\text{cm}^2$.

11 A variable a is inversely proportional to the square of b. When $b = 3$, $a = 2$. 

Evaluate b when $a = 10$, correct to 2 decimal places, if $b > 0$.

12 **a** Write down the domain and range of the curve $y = \dfrac{2}{x-3}$. 

b Sketch the graph of $y = \dfrac{2}{x-3}$.

13 Find the centre and radius of the circle with equation: 5.05

a $x^2 + y^2 = 100$

b $(x - 3)^2 + (y - 2)^2 = 121$

c $x^2 + 6x + y^2 + 2y + 1 = 0$

14 Find the x- and y-intercepts (where they exist) of:  5.05

a $P(x) = x^3 - 4x$

b $y = -\dfrac{2}{x+1}$

c $x^2 + y^2 = 9$

d $y = \sqrt{25 - x^2}$

e $f(x) = |x - 2| + 3$

15 **a** Sketch the graph of $y = |x + 1|$. 5.04

b From the graph, solve $|x + 1| = 3$.

Foundation Mastery Complex

5 Challenge exercise

Answers on p. 487

1 Solve $|2x+1| = 3x - 2$ graphically.

2 Given $f(x) = |x| + 3x - 4$, sketch the graph of:

a $y = f(x)$ **b** $y = -f(x)$

3 Find the centre and radius of the circle with equation given by $x^2 + 3x + y^2 - 2y - 3 = 0$.

4 Find the equation of the straight line through the centres of the circles with equations $x^2 + 4x + y^2 - 8y - 5 = 0$ and $x^2 - 2x + y^2 + 10y + 10 = 0$.

5 Sketch the graph of $y = \dfrac{|x|}{x^2}$.

6 **a** Show that $\dfrac{2x+7}{x+3} = 2 + \dfrac{1}{x+3}$.

b Find the domain and range of $y = \dfrac{2x+7}{x+3}$.

c Hence sketch the graph of $y = \dfrac{2x+7}{x+3}$.

7 Show that $x^2 - 2x + y^2 + 4y + 1 = 0$ and $x^2 - 2x + y^2 + 4y - 4 = 0$ are concentric.

8 Sketch the graph of $f(x) = 1 - \dfrac{1}{x^2}$.

☐ Foundation ◯ Mastery ◯ Complex

PRACTICE SET 2

Answers on p. 488

In Questions **1** to **10**, select the correct answer **A**, **B**, **C** or **D**.

1 Find an expression involving θ for this triangle (there may be more than one answer).

A $\cos\theta = \frac{5^2 + 4^2 - 7^2}{2 \times 5 \times 4}$

B $\frac{\sin\theta}{4} = \frac{\sin\alpha}{5}$

C $\frac{\sin\theta}{5} = \frac{\sin\alpha}{4}$

D $\cos\theta = \frac{5^2 + 7^2 - 4^2}{2 \times 5 \times 7}$

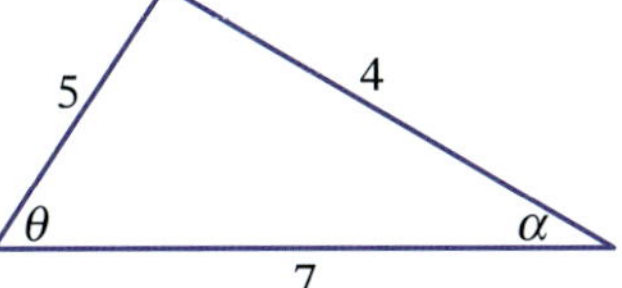

2 If $f(x) = \begin{cases} 8x^3 & \text{if } x > 3 \\ 3x^2 - 2 & \text{if } 0 \leq x \leq 3 \\ 9 & \text{if } x < 0 \end{cases}$

evaluate $f(3) + f(1) + f(-1)$.

A 35 **B** 53 **C** 226 **D** 233

3 The linear function with equation $4x - 2y + 3 = 0$ has:

A gradient -2, y-intercept $-1\frac{1}{2}$

B gradient $\frac{1}{2}$, y-intercept $\frac{3}{4}$

C gradient 2, y-intercept $1\frac{1}{2}$

D gradient 4, y-intercept 3

4 If $f(x) = x^2$ and $g(x) = 2x + 1$, the composite function $g(f(x))$ is:

A $(2x + 1)^2$ **B** $(2x)^2 + 1$ **C** $2x + 12$ **D** $2x^2 + 1$

5 Which of the following does not represent a function?

A

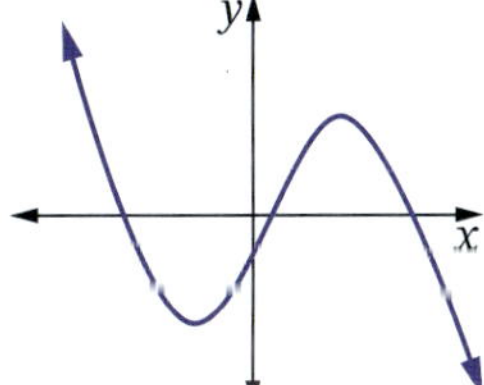

B

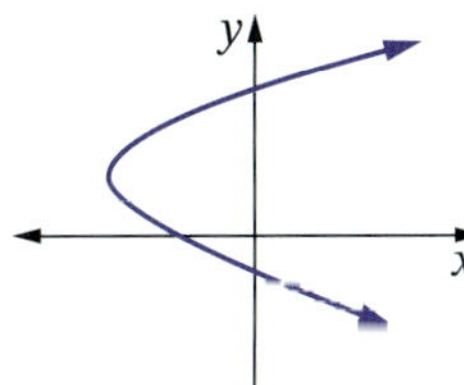

C

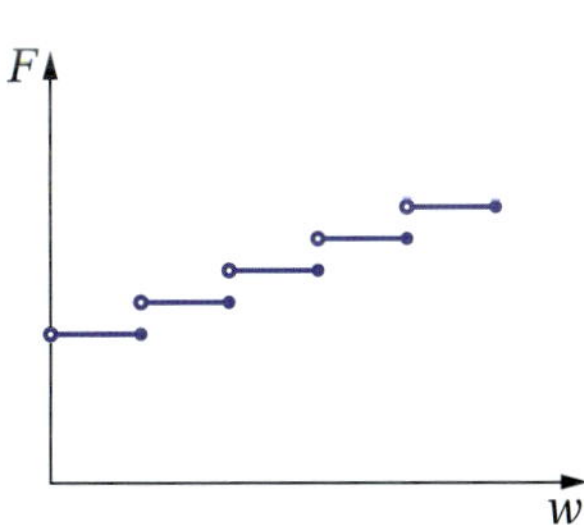

D (0, 3), (1, 3), (2, 5), (3, 1)

Foundation Mastery Complex

3.11 **6** The quadratic equation $x^2 + (k - 3)x + k = 0$ has real roots. Evaluate k.

A $k \le 1, k \ge 9$ **B** $k = 1, 9$

C $1 \le k \le 9$ **D** $k < 1, k > 9$

5.03 **7** At a department store, the number of employees required to do a stocktake, N, is inversely proportional to the time, t, it takes. What is the equation showing this information?

A $N = kt$ **B** $N = t + k$ **C** $N = \frac{k}{t}$ **D** $N = \frac{t}{k}$

5.05 **8** The equation of a circle with radius 3 and centre $(-1, 4)$ is:

A $(x - 1)^2 + (y + 4)^2 = 3$ **B** $(x - 1)^2 + (y + 4)^2 = 9$

C $(x + 1)^2 + (y - 4)^2 = 9$ **D** $(x + 1)^2 + (y - 4)^2 = 3$

4.03 **9** Find the exact value of $\sin 135° + \cos 120°$.

A $\frac{\sqrt{2} - \sqrt{3}}{2}$ **B** $\frac{\sqrt{2} + 1}{2}$ **C** $\frac{\sqrt{2} + \sqrt{3}}{2}$ **D** $\frac{\sqrt{2} - 1}{2}$

5.01 **10** Find the domain of $f(x) = \frac{2}{x + 7}$.

A all real x **B** all real $x, x \ne -7$

C all real $x, x \ne 7$ **D** all real $x, x \ne 2$

5.01 **11** Sketch the graph of each equation and find its domain and range.

5.04, 5.05

a $y = \frac{4}{2x - 4}$ **b** $y = x^3 + x^2 - 2x$ **c** $y = |x - 1|$

d $x^2 + y^2 = 25$ **e** $y = -\sqrt{1 - x^2}$ **f** $y = 1 - x^2$

5.05 **12** **a** Find the centre and radius of the circle $x^2 + 2x + y^2 - 6y - 6 = 0$.

b Find its domain and range.

4.05, 4.06 **13** A triangle has sides of length 5.1 m, 6.5 m and 8.2 m.

a Find the size of the angle opposite the 6.5 m side, correct to the nearest minute.

b Find the area of the triangle, correct to one decimal place.

3.06, 3.07 **14** Find the equation of the straight line:

a with gradient -2 and y-intercept 3

b with x-intercept 5 and y-intercept -1

c passing through $(2, 0)$ and $(-3, -4)$

d through $(5, -4)$ and parallel to the line through $(7, 4)$ and $(3, -1)$

e through $(3, -1)$ perpendicular to the line $3x - 2y - 7 = 0$

f through $(1, 2)$ and parallel to the line through $(-3, 4)$ and $(5, 5)$

g through $(1, 3)$ and making an angle of $135°$ with the positive x-axis.

3.08 **15** The area of a community garden in square metres is given by $A = 7x - x^2$, where x is the side length of the garden.

a Find the area of the garden when the side length is 4.5 m.

b Find, correct to one decimal place, the side length of the garden when the area is 8 m^2.

c Sketch the graph of the area function.

d Find the maximum possible area.

4.08 **16** Convert these angles into radians in terms of π:

a $60°$ **b** $150°$ **c** $90°$

d $10°$ **e** $315°$

3.04 **17** Sketch the graph of:

3.10, 3.13

a $5x - 2y - 10 = 0$ **b** $x = 2$ **c** $f(x) = (x - 3)^2$

d $y = x^2 - 5x + 4$ **e** $y = (x - 1)^3 + 2$

☐ Foundation ○ Mastery ○ Complex

18 Convert each value in radians into degrees and minutes: 4.08

a 1.7 **b** 0.36 **c** 2.54

19 The lines AB and AC have equations $3x - 4y + 9 = 0$ and $8x + 6y - 1 = 0$ respectively. 

a Show that the lines are perpendicular.

b Find the coordinates of the point of intersection.

20 If $f(x) = |x| - 2$, find: 

a $f(-2)$ **b** $f(0)$ **c** $f(m+1)$

21 If $g(x) = \begin{cases} 3 - x & \text{if } x > 1 \\ 2x & \text{if } x \le 1 \end{cases}$: 5.04

a find $g(2)$ and $g(-3)$

b sketch the graph of $y = g(x)$.

22 Find the value of x if $f(x) = 7$, where $f(x) = 2^x - 1$. 

23 If $f(x) = 9 - 2x^2$, find the value of $f(-1)$. 

24 Show that the line $3x - 4y + 10 = 0$ intersects with the circle $x^2 + y^2 = 4$ at only one point. 

25 Change each value in radians into degrees: 4.08

a $\frac{\pi}{4}$ **b** $\frac{3\pi}{2}$ **c** $\frac{\pi}{5}$ **d** $\frac{7\pi}{8}$ **e** 6π

26 Find the exact value of each expression. 

a $\tan 150°$ **b** $\cos(-45°)$ **c** $\sin 240°$

d $\cos\frac{7\pi}{4}$ **e** $\sin\frac{4\pi}{3}$ **f** $\tan\frac{5\pi}{6}$

27 Show that: 3.12

a $-x^2 + x - 9 < 0$ for all x **b** $x^2 - x + 3 > 0$ for all x.

28 Solve each equation or inequality. 3.09

a $x^2 - 2x - 3 = 0$ **b** $1 < 2x - 3 \le 7$

c $|3x + 1| - 4$ **d** $x^2 \ge 4$

29 The distance travelled by a runner is directly proportional to the time she takes. 
If Vesna runs 12 km in 2 hours 30 minutes, find:

a an equation for distance d in terms of time t 5.02

b how far Vesna runs in:

i 2 hours **ii** 5 hours

c how long it takes Vesna to run:

i 30 km **ii** 19 km

d her average speed.

30 An arc subtends an angle of 40° at the centre of a circle with radius 4 cm. Find in exact form: 

a the length of the arc **b** the area of the sector bordered by the arc.

31 Find the equation of the parabola with x-intercepts 3 and -1 and y-intercept -3. 3.11

32 Show that the quadratic equation $6x^2 + x - 15 = 0$ has 2 real, rational roots. 3.12

☐ Foundation ○ Mastery ⬡ Complex

4.09, 4.10 **33** The area of a circle is 5π and an arc 3 cm long cuts off a sector with an angle of θ subtended at the centre. Find θ in degrees and minutes.

4.05 **34** A soccer goal is 8 m wide. Tim shoots for goal when he is 9 m from one post and 11 m from the other. Within what angle must a shot be made in order to score a goal?

5.05 **35** Find the equation of the circle with centre $(-2, -3)$ and radius 5 units.

4.01 **36** Evaluate θ in degrees and minutes, to the nearest minute.

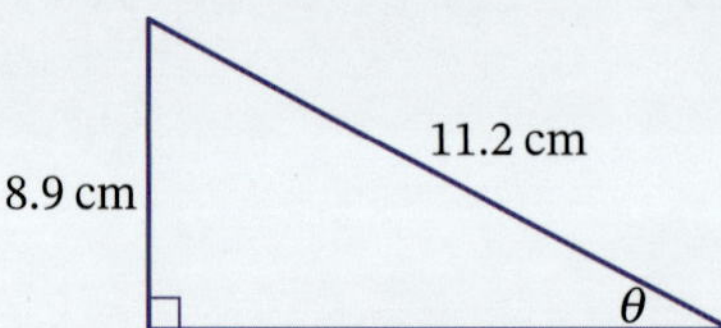

3.04, 3.07 **37** **a** Find the equation of the straight line l through $(-1, 2)$ that is perpendicular to the line $3x + 6y - 7 = 0$.

b Line l cuts the x-axis at P and the y-axis at Q. Find the coordinates of P and Q.

3.03 **38** Show that $f(x) = x^6 - x^2 - 3$ is an even function.

4.02 **39** Find the angle of depression from the top of a 5.6 m tall cliff down to a boat that is 150 m out from the base of the cliff. (Assume the cliff is vertical.)

5.06 **40** If $f(x) = x^3$ and $g(x) = 2x + 5$, find:

a $f(g(x))$ **b** $g(f(x))$ **c** $f(g(-1))$ **d** $g(f(3))$

4.03, 4.08 **41** Simplify each expression.

a $\tan(180° - \theta)$ **b** $\sin(\pi + \theta)$ **c** $\cos(2\pi - \theta)$

4.02 **42** Write each direction shown as:

i a compass bearing **ii** a true bearing.

a

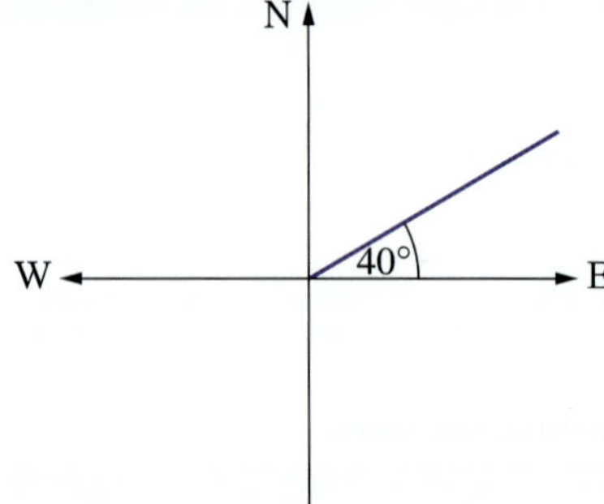

b

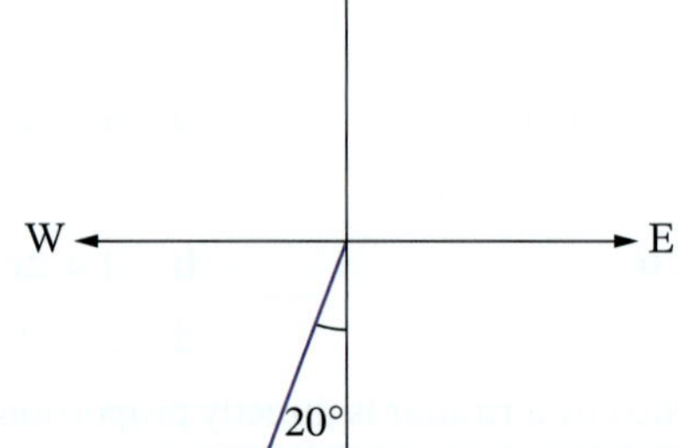

c

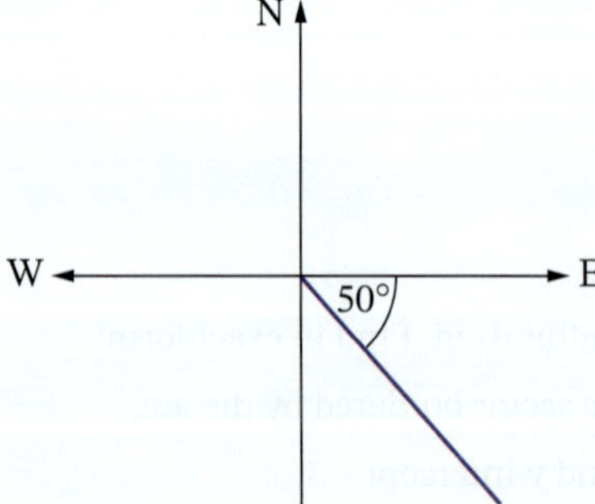

d

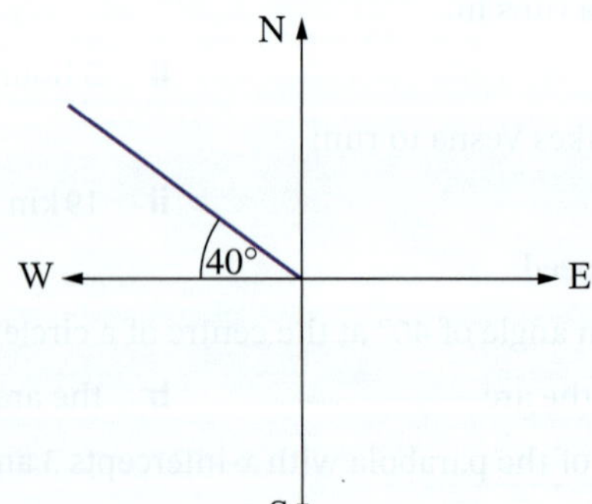

☐ Foundation ○ Mastery ○ Complex

43 An arc of a circle with radius 5 cm subtends an angle of 30° at the centre. Find the exact: 4.09–4.11

a arc length **b** area of the sector

c area of the minor segment inside the sector

44 Find α in degrees and minutes. 4.04

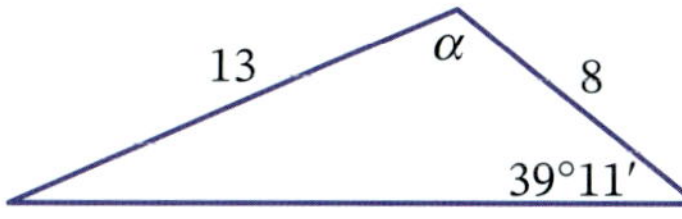

45 Find the value of y, correct to 3 significant figures. 4.01

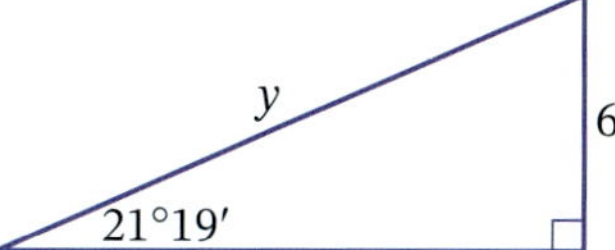

46 Find the intersection of the graphs: 3.04, 3.08

a $x + 3y - 1 = 0$ and $x - 2y - 6 = 0$

b $y = x^2$ and $x - 2y + 15 = 0$

47 For each quadratic function: 3.10

i find the equation of the axis of symmetry

ii state whether it has a maximum or minimum turning point and find its coordinates.

a $y = x^2 - 6x + 1$ **b** $y = -2x^2 - 4x - 3$

48 A hawk at the top of a 10 m tree sees a mouse on the ground. If the angle of depression is 34°51′, how far, correct to one decimal place, does the bird need to fly to reach the mouse? 4.02

49 **a** Find the length of AB to the nearest metre. 4.05, 4.06

b Find the area of $\triangle ABC$, correct to 3 significant figures.

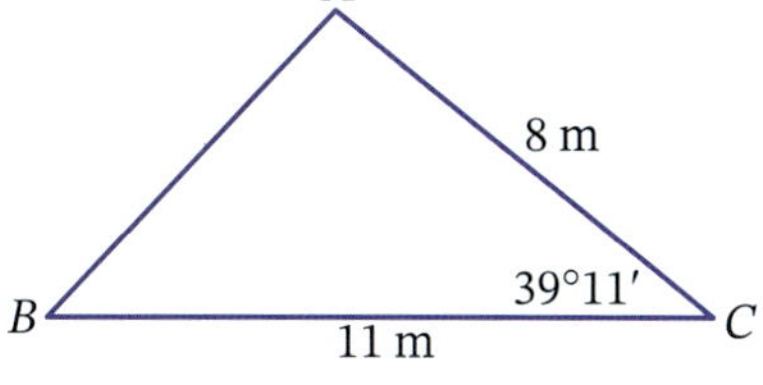

50 Two points A and B are 100 m apart on the same side of a tower. The angle of elevation of A to the top of the tower is 20° and the angle of elevation from B is 27°. Find the height of the tower, to the nearest metre. 4.02

51 The length of an arc in a circle of radius 6 cm is 7π cm. Find the area of the sector cut off by this arc. 4.09, 4.10

52 Jordan walks for 3.1 km due west, then turns and walks for 2.7 km on a bearing of 205°. How far is he from his starting point? 4.02

53 Find the domain and range of: 5.01

a $f(x) = \dfrac{3}{x+4}$ **b** $y = |x| + 2$ **c** $y = -\sqrt{4 - x^2}$ 5.04, 5.05

d $y = 4$ **e** $y = x^2 - 3$

54 Nalini leaves home and cycles west for 12.5 km then turns and rides south for 11.3 km. 4.02

a How far is Nalini from home?

b Find the bearing of Nalini from home.

55 Show that $f(x) = x^3 - 5x$ is an odd function. 3.03

56 Sketch the graph of: 3.10, 3.13

a $3x - 2y + 6 = 0$ **b** $y = x^2 - x - 2$

c $y = x^3 - 1$ **d** $y = x(x + 2)(x - 3)$

☐ Foundation ○ Mastery ○ Complex

4.09, 4.10 **57** The length of an arc in a circle of radius 2 cm is 1.6 cm. Find the area of the sector.

4.08 **58** Convert into degrees.

a 2π **b** $\dfrac{\pi}{6}$ **c** $\dfrac{9\pi}{4}$

4.02 **59** A plane flies on a bearing of 034° from Sydney for 875 km. How far due east of Sydney is the plane? Give your answer correct to one decimal place.

3.09 5.04 **60** Solve:

a $5b - 3 \geq 7$ **b** $x^2 - 3x = 0$ **c** $|2n + 5| = 9$ **d** $n^2 - 9 < 0$

PRACTICE SET 2

☐ Foundation ◯ Mastery ⬡ Complex

INTRODUCTION TO DIFFERENTIATION

Calculus is a very important branch of mathematics that involves the measurement of change. It can be applied to many areas such as science, economics, engineering, astronomy, sociology and medicine. Differentiation, a part of calculus, has many applications involving rates of change; for example, the spread of infectious diseases, population growth, inflation, unemployment, filling of our water reservoirs.

Chapter outline

In this chapter you will:

- understand the derivative of a function as the gradient of the tangent to the curve and a measure of a rate of change
- draw graphs of gradient functions
- identify functions that are continuous and discontinuous, and their differentiability
- differentiate from first principles
- differentiate functions involving powers of x, including terms with negative and fractional indices
- use derivatives to find rates of change, including displacement and velocity, and the behaviour of graphs of functions
- use derivatives to find gradients and equations of tangents and normals to curves
- find the derivative of composite functions, products and quotients of functions

Videos (8):

6.01 Sketching gradient functions
6.02 Differentiation from first principles
6.04 Displacement and velocity
6.05 Equation of a tangent • Equations of tangents and normals
6.06 The chain rule
6.07 The product rule
6.08 The quotient rule 1

Worksheets (20):

6.01 Gradient functions 1 • Gradient functions 2 • Rates of change 1 • Instantaneous rates of change • Graphs of rates of change
6.02 Differentiation from first principles • Limits • Finding derivatives from first principles
6.03 Derivatives of linear products • Derivatives of polynomials • Derivatives of a sum of terms • Basic differentiation
6.05 Tangents and normals • Slopes of curves • Tangents to a curve
6.06 The chain rule
6.07 The product rule
6.08 The quotient rule • Rules of differentiation • Mixed differentiation problems

Puzzles (2):

6.02 Rates of change - Gradients of secants
Test yourself Derivatives find-a-word

To access resources above, visit **cengage.com.au/nelsonmindtap**

Terminology

average rate of change	chain rule	derivative
differentiation from first principles	displacement	instantaneous rate of change
limit	normal	product rule
quotient rule	secant	stationary point
tangent	turning point	velocity

Did you know?

Newton and Leibniz

'Calculus' comes from the Latin word meaning 'pebble' or 'small stone', and the ancient Romans used pebbles on abacuses for counting and calculation. In many ancient civilisations stones were used for counting, but the mathematics they practised was quite sophisticated.

It was not until the 17th century that there was a breakthrough in calculus when scientists were searching for ways of measuring the motion of objects such as planets, pendulums and projectiles.

Georgios Kollidas/Shutterstock.com

Sir Isaac Newton

Sir Isaac Newton (1642–1727), an Englishman, discovered the main principles of calculus when he was 23 years old. At this time an epidemic of bubonic plague had closed Cambridge University where he was studying, so many of his discoveries were made at home. He first wrote about his calculus methods, which he called fluxions, in 1671, but his *Method of fluxions* was not published until 1704.

Gottfried Leibniz (1646–1716), a German, was studying the same methods and there was intense rivalry between the 2 countries over who was first to discover calculus.

Rates of change 6.01

Worksheets
Gradient functions 1
Gradient functions 2
Rates of change 1
Instantaneous rates of change
Graphs of rates of change

The **gradient** of a straight line measures the constant **rate of change** of y (the dependent variable) with respect to the change in x (the independent variable).

$$m = \frac{\text{rise}}{\text{run}}$$
$$= \frac{y_2 - y_1}{x_2 - x_1}$$

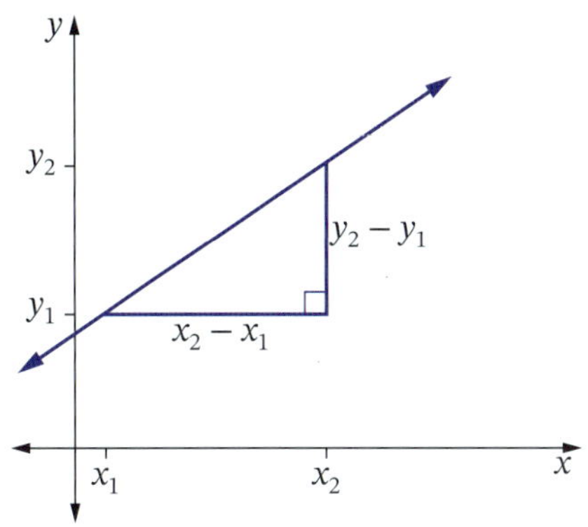

Notice that when the gradient of a straight line is positive the line is increasing, and when the gradient is negative the line is decreasing. Straight lines increase or decrease at a constant rate and the gradient is the same everywhere along the line.

This graph shows the distance travelled by a car over time.

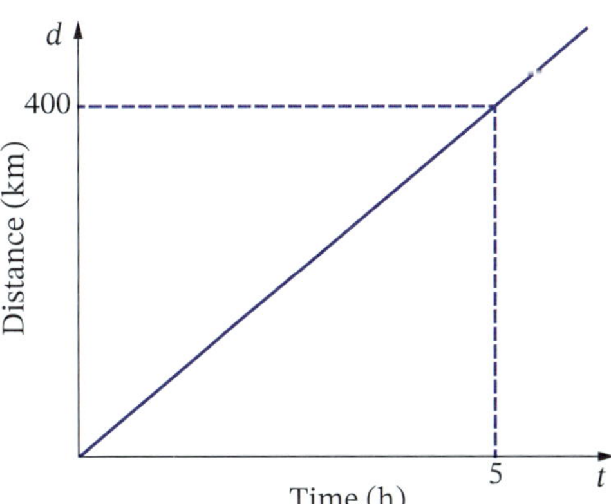

$$m = \frac{\text{rise}}{\text{run}}$$
$$= \frac{400}{5}$$
$$= 80$$

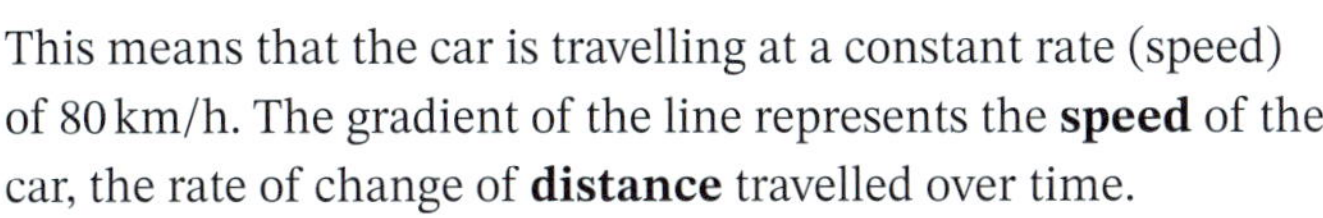

This means that the car is travelling at a constant rate (speed) of 80 km/h. The gradient of the line represents the **speed** of the car, the rate of change of **distance** travelled over time.

This graph shows the number of cases of flu reported in a town over several weeks.

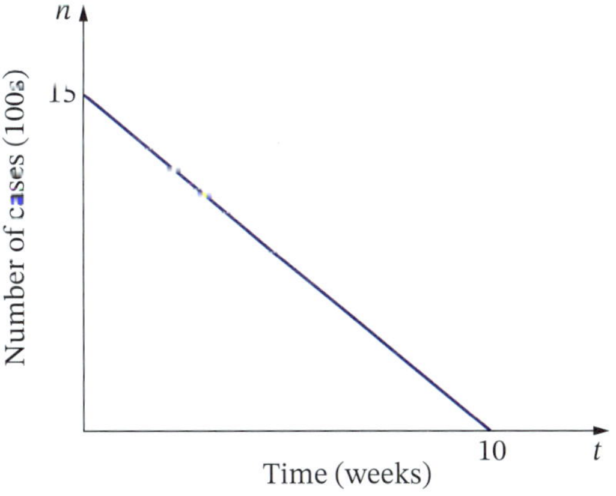

$$m = \frac{\text{rise}}{\text{run}}$$
$$= \frac{-1500}{10}$$
$$= -150$$

The line is decreasing, so it will have a negative gradient.

The 'rise' is a drop so it is negative.

This means that the rate is -150 cases/week, or the number of cases reported is decreasing by 150 cases/week.

Average rate of change

The **average rate of change** between 2 points on the graph of a function $y = f(x)$ is the gradient of the line joining those points. The line going through the 2 points $A(a, f(a))$ and $B(b, f(b))$ on the graph is called a **secant**.

$$m = \frac{y_2 - y_1}{x_2 - x_1}$$

$$= \frac{f(b) - f(a)}{b - a}$$

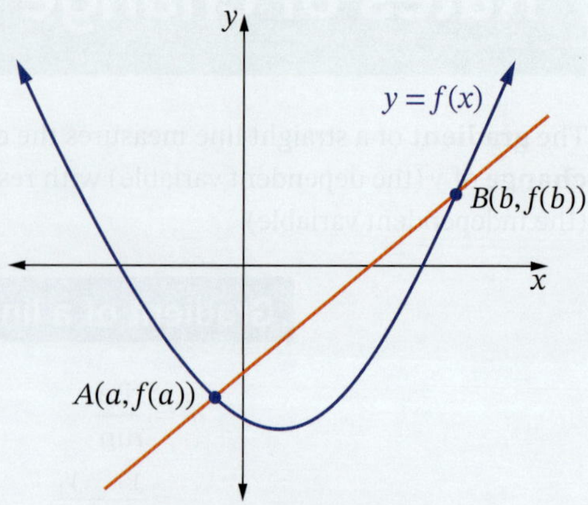

Average rate of change of $y = f(x)$ between points A and B.

The average rate of change of y with respect to x can be written using delta notation, where the Greek letter, Δ (delta), stands for 'change in' or 'difference in'.

$$\frac{\Delta y}{\Delta x} = \frac{\text{change in } y}{\text{change in } x}$$

Average rate of change

The average rate of change of y with respect to x for a function $y = f(x)$ over the domain $[a, b]$ is

$$\frac{\Delta y}{\Delta x} = \frac{f(b) - f(a)}{b - a}$$

which is the gradient of the secant through $(a, f(a))$ and $(b, f(b))$ on the graph of $y = f(x)$.

Note that if $f(x)$ is a linear function, then its gradient and rate of change are constant. Any secant drawn on its graph will be on the line, so the average rate of change is the gradient of the line.

Investigation

These 2 graphs show the distance that a bicycle travels over time. One is a straight line (constant speed) and the other is a curve (variable speed), but both travel the same distance in 4 hours.

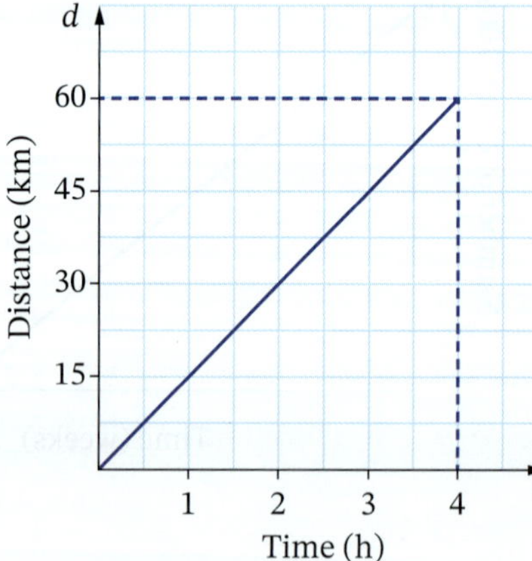

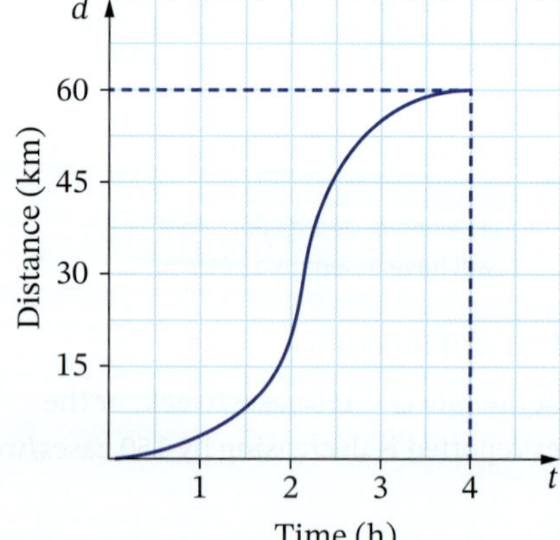

Is the average speed of the bicycle over the 4 hours the same in both cases?

What is different about the speed in the 2 graphs?

How could you measure the speed in the second graph at any one time during the journey?

Does it change? If so, how does it change?

Example 1

a This graph shows the distance d in km that a car travels over time t in hours. After one hour the car has travelled 50 km and after 3 hours the car has travelled 200 km. Find the average speed of the car.

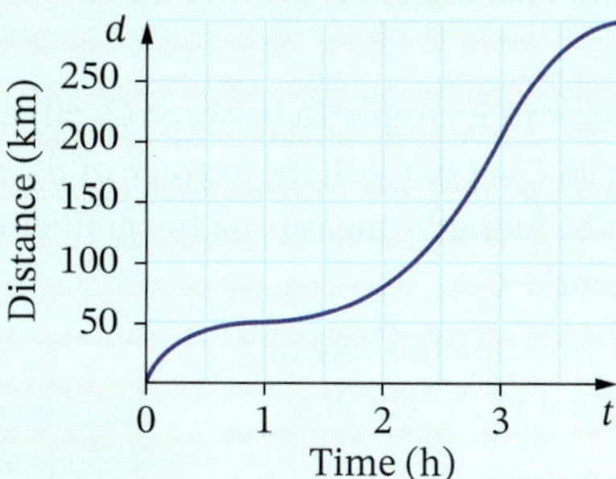

b Given the function $f(x) = x^2$, find the average rate of change between $x = 1$ and $x = 1.1$.

Solution

a Speed is the change in distance over time.

The gradient of the secant will give the average speed.

Average rate of change:

$$m = \frac{y_2 - y_1}{x_2 - x_1}$$

$$= \frac{200 - 50}{3 - 1}$$

$$= \frac{150}{2}$$

$$= 75$$

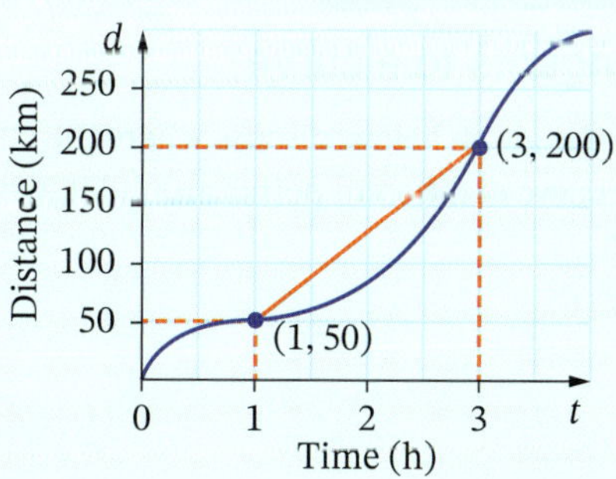

So the average speed is 75 km/h.

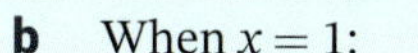

b When $x = 1$: When $x = 1.1$:

$f(1) = 1^2$ $\quad f(1.1) = 1.1^2$

$= 1$ $\quad = 1.21$

So the points are (1, 1) and (1.1, 1.21).

Average rate of change:

$$m = \frac{y_2 - y_1}{x_2 - x_1}$$

$$= \frac{1.21 - 1}{1.1 - 1}$$

$$= 2.1$$

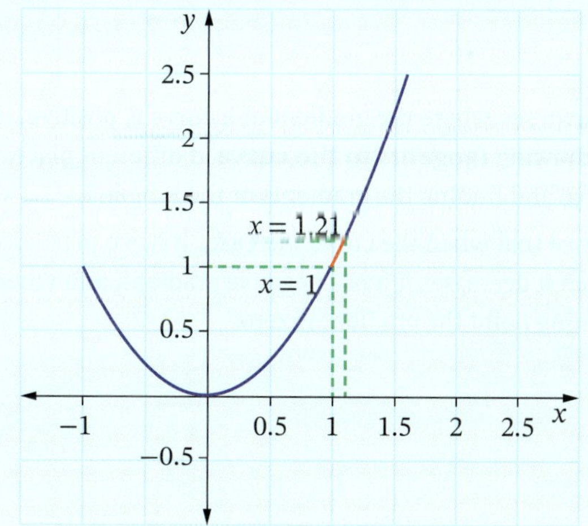

So the average rate of change is 2.1.

Notice that the secant (orange line) is very close to the shape of the curve itself. This is because the 2 points chosen are close together.

Instantaneous rate of change

While the average rate of change is the average of the rates of change between 2 points on $y = f(x)$, the **instantaneous rate of change** at one point on $y = f(x)$ is the actual rate of change at that point, measured by the gradient of the **tangent** at that point. The tangent is the line that touches the graph at that point (without crossing it).

For example, the **average speed** of the car in Example **1a** between (1, 50) and (3, 200) was 75 km/h, which means that during the 2-hour period from $t = 1$ to $t = 3$, the average of the speeds was 75 km/h. However, at $t = 1$, the **instantaneous speed** can be found by measuring the gradient of the graph at the point (1, 50), which means the actual speed at that time.

For any function $f(x)$, we can find the instantaneous rate of change at any point by finding the gradient of the graph at that point.

We can draw tangents to the curve to see how the gradient changes along the graph.

Note that if $f(x)$ is a linear function, then its rate of change is constant, so the average rate of change and instantaneous rate of change are the same.

Investigation

Tangents to a curve

For the graph below, discuss where the curve is increasing, decreasing or flat. If you draw more tangents to the curve at different points around it, where is the gradient positive, negative or zero?

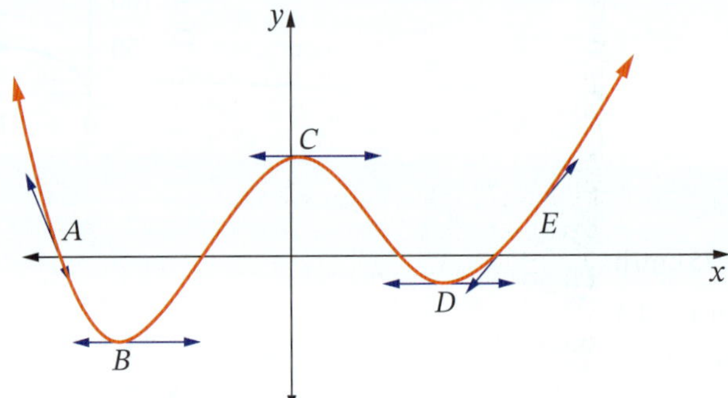

We can see where the gradient of a curve is positive, negative or zero by drawing **tangents to the curve** at different places around the curve and finding the gradients of the tangents.

Notice that when the curve increases it has a positive gradient, when it decreases it has a negative gradient, and when it is a turning point the gradient is zero.

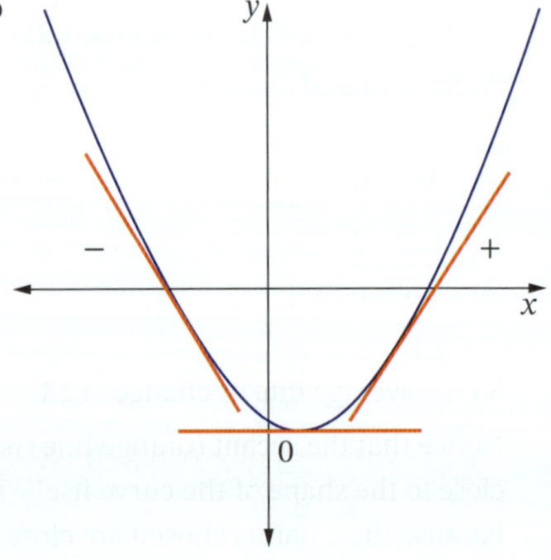

Example 2

Copy each curve and write the sign of its gradient along the curve.

a

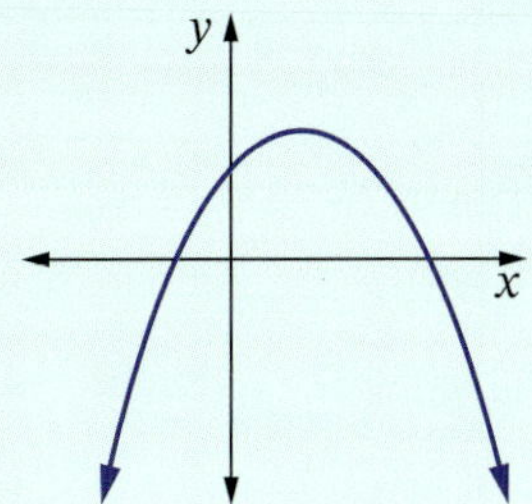

b

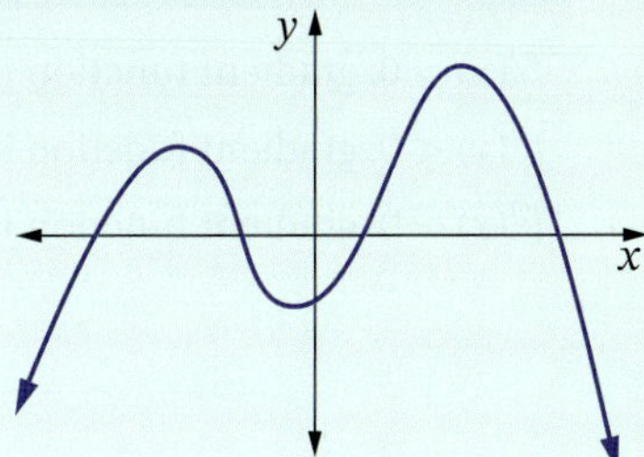

Solution

Where the curve increases, the gradient is positive. Where it decreases, it is negative. Where it is a turning point, it has a zero gradient.

a

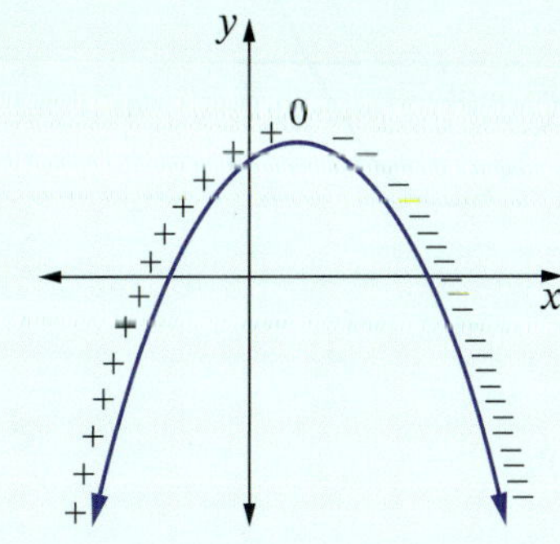

b

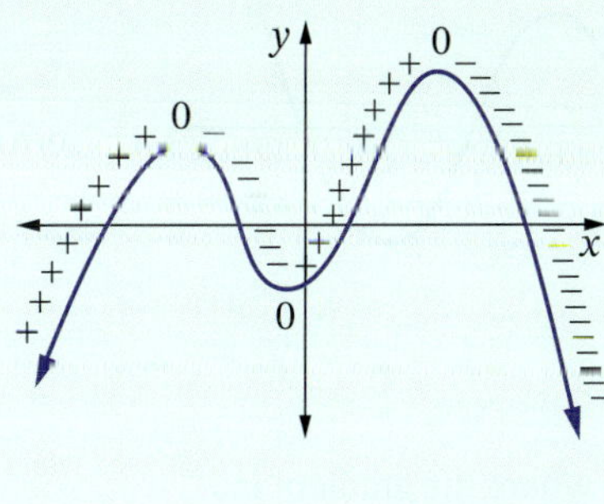

We find the gradient of a curve by measuring the **gradient of a tangent** to the curve at different points around the curve.

We can then sketch the graph of these gradient values, which we call $y = f'(x)$, the **gradient function** or the **derivative function**. $f'(x)$ is read 'f dash of x'.

Consider this graph of $f(x) = x^2$, with tangents drawn at values of x from -3 to 3. We can measure the gradient of the tangent at each point, and graph the gradient function $y = f'(x)$.

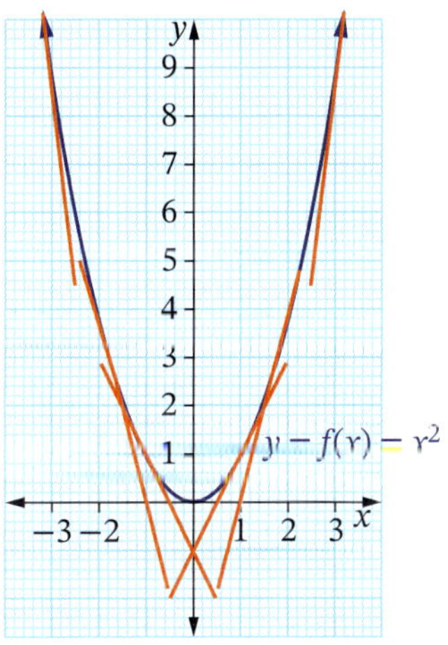

x	−3	−2	−1	0	1	2	3
Gradient, m	−6	−4	−2	0	2	4	6

Graphing the values of m, we find that $y = f'(x)$ is a linear function.

The graph of the quadratic function $f(x)$ is a parabola.

The graph of its derivative function $f'(x)$ is a line.

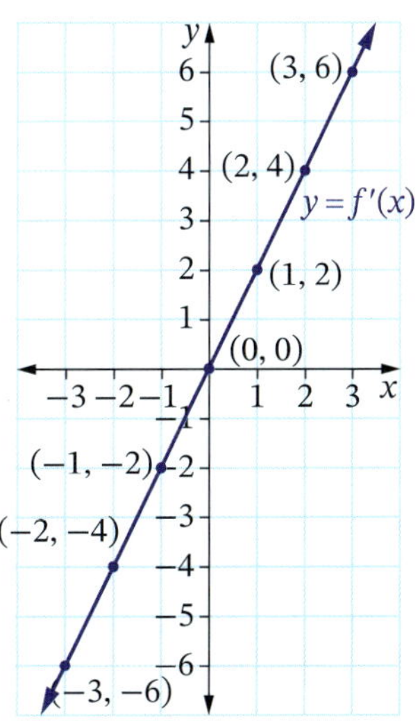

Because the gradient function comes from the original function, we say that we **differentiate** or 'derive' the original function to find the 'derivative' function, or just the **derivative**. This process is called **differentiation**. Notice in the graph of $f(x)$ shown above that where $m > 0$, the graph of the gradient function $y = f'(x)$ shown at right is above the x-axis; where $m = 0$, the gradient function is on the x-axis; and where $m < 0$, the gradient function is below the x-axis.

Since $m = f'(x)$, we can write the following:

Sketching gradient (derivative) functions

$f'(x) > 0$: gradient function is above the x-axis

$f'(x) < 0$: gradient function is below the x-axis

$f'(x) = 0$: gradient function is on the x-axis

Video
Sketching gradient functions

Example 3

Sketch a gradient function for each curve.

a

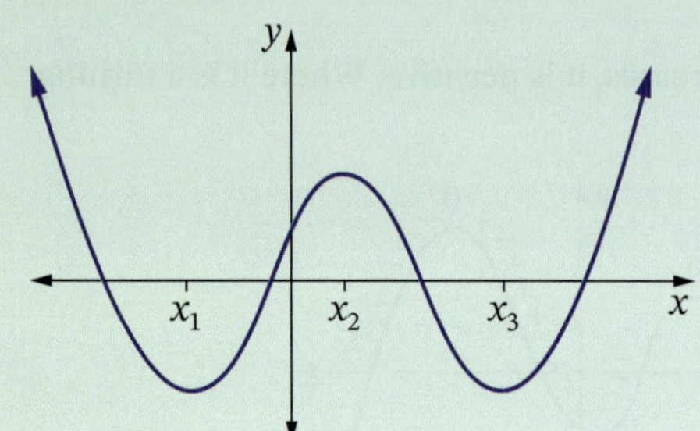

b

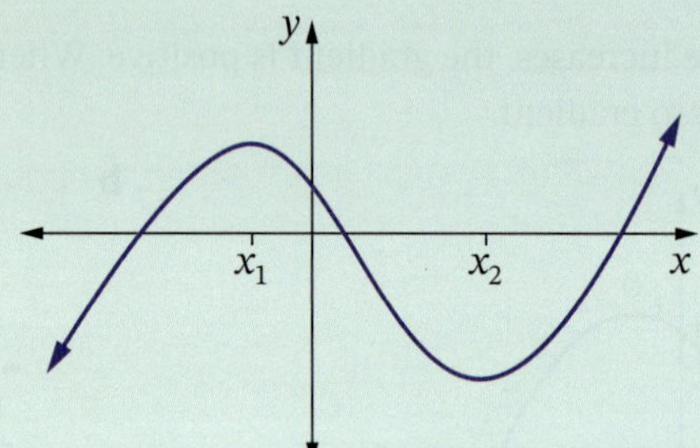

Solution

a First we mark in where the gradient is positive, negative and zero.

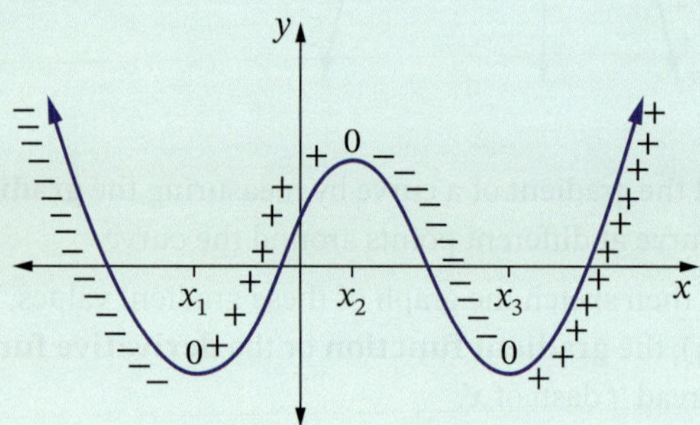

$f'(x) = 0$ at x_1, x_2 and x_3, so on the gradient graph these points will be on the x-axis (the x-intercepts of the gradient graph).

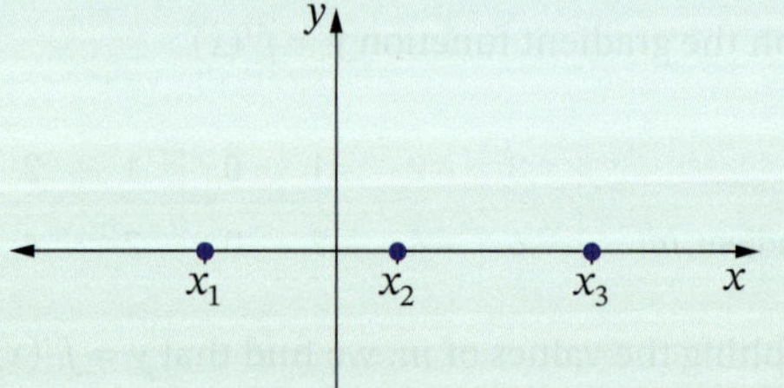

$f'(x) < 0$ to the left of x_1, so this part of the gradient graph will be below the x-axis.

$f'(x) > 0$ between x_1 and x_2, so the graph will be above the x-axis here.

$f'(x) < 0$ between x_2 and x_3, so the graph will be below the x-axis here.

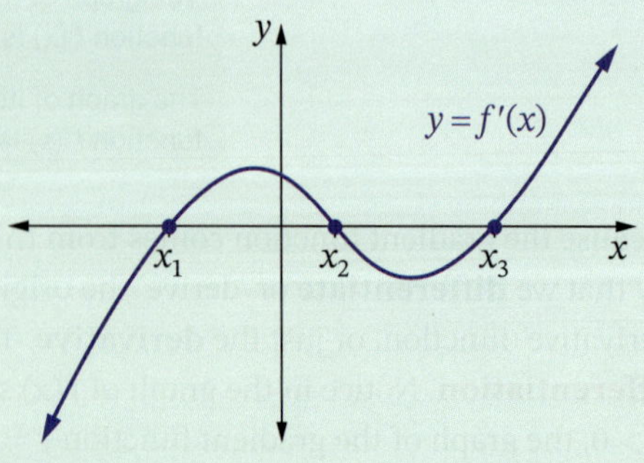

$f'(x) > 0$ to the right of x_3, so this part of the graph will be above the x-axis.

Sketching this information gives the graph of the gradient function $y = f'(x)$. Note that this is only a rough graph that shows the shape and sign rather than precise values.

b First, mark in where the gradient is positive, negative and zero.

$f'(x) = 0$ at x_1 and x_2. These points will be the x-intercepts of the gradient function graph.

$f'(x) > 0$ to the left of x_1, so the graph will be above the x-axis here.

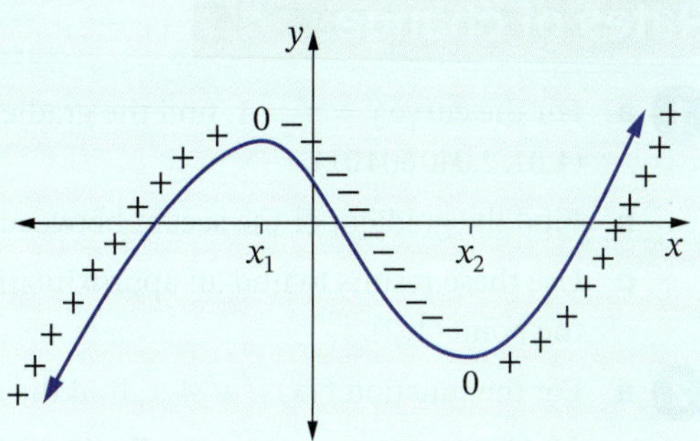

$f'(x) < 0$ between x_1 and x_2, so the graph will be below the x-axis here.

$f'(x) > 0$ to the right of x_2, so the graph will be above the x-axis here.

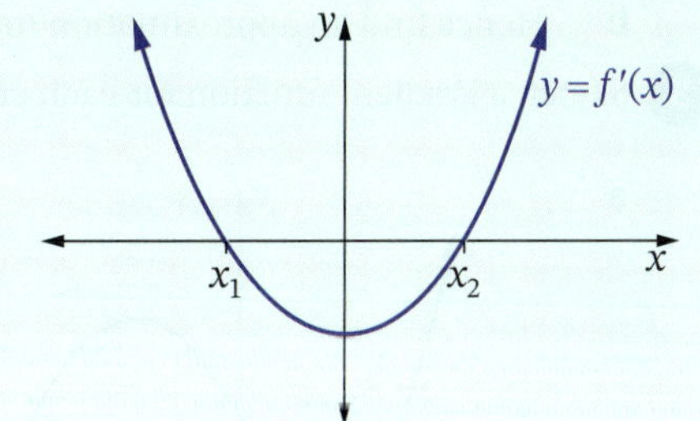

Technology

Tangents to a curve

There are some excellent graphing software, apps and websites that will draw tangents to a curve and sketch the gradient function.

Explore how to sketch gradient functions from the previous examples.

Stationary points

The points on a curve where the gradient $f'(x) = 0$ are called **stationary points** because the gradient there is flat, neither increasing nor decreasing.

For example, the curve shown decreases to a **minimum turning point**, which is a type of **stationary point**. It then increases to a **maximum turning point** (also a stationary point) and then decreases again.

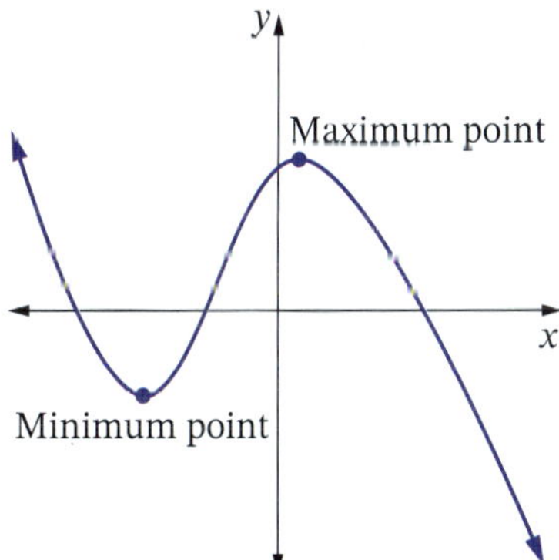

Stationary points

The points on the graph of $y = f(x)$ where the gradient of the tangent is 0, $f'(x) = 0$, are called **stationary points**. There are 3 types of stationary points.

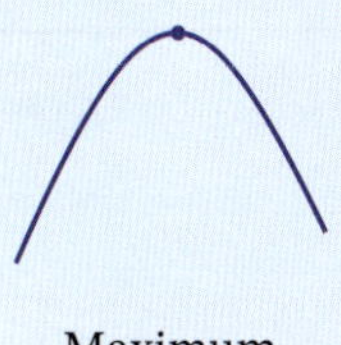

Maximum turning point

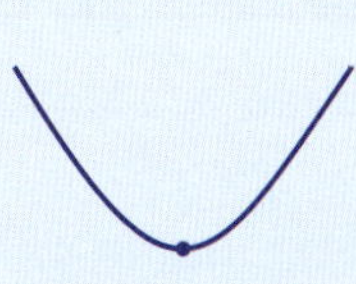

Minimum turning point

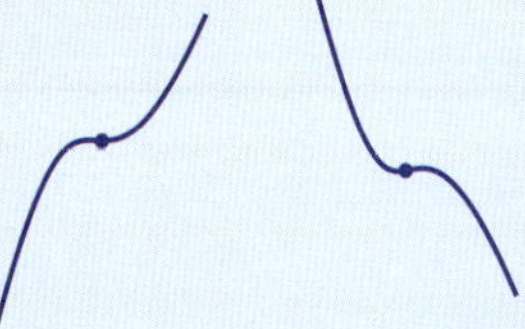

Horizontal point of inflection (such as at the centre of $y = x^3$)

EXERCISE 6.01 Answers on p. 490

Answers on p. 490

Rates of change

1 **a** For the curve $y = x^4 + 1$, find the gradient of the secant between the point $(1, 2)$ and the point $(1.01, 2.040\,604\,01)$.

b Find the gradient of the secant between $(1, 2)$ and the point $(0.999, 1.996\,005\,996)$.

c Use these results to find an approximation to the gradient of the tangent to the curve $y = x^4 + 1$ at the point $(1, 2)$.

2 **a** For the function $f(x) = x^3 + x$, find the average rate of change between the points $(2, 10)$ and:

i $(2.1, 11.361)$ **ii** $(2.01, 10.130\,601)$ **iii** $(1.99, 9.870\,599)$

b Hence find an approximation to the gradient of the tangent at the point $(2, 10)$.

EXAMPLES 2, 3

3 Sketch a gradient function for each curve.

a

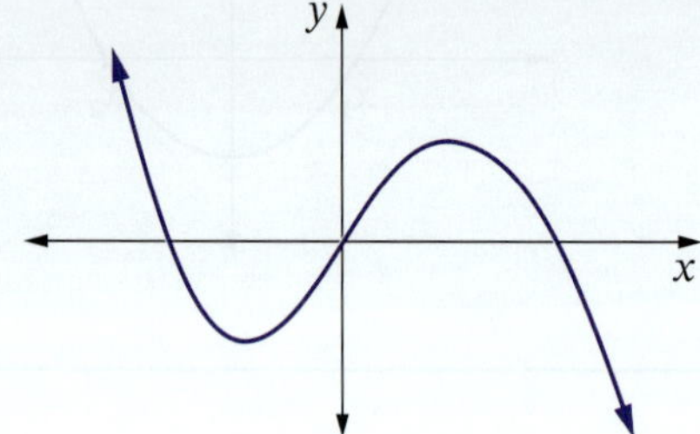

b

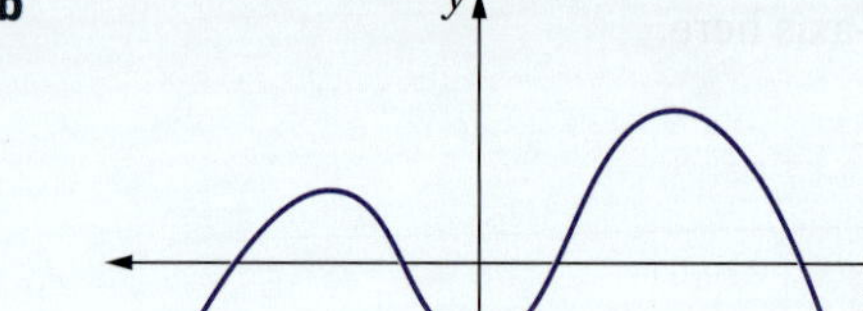

c

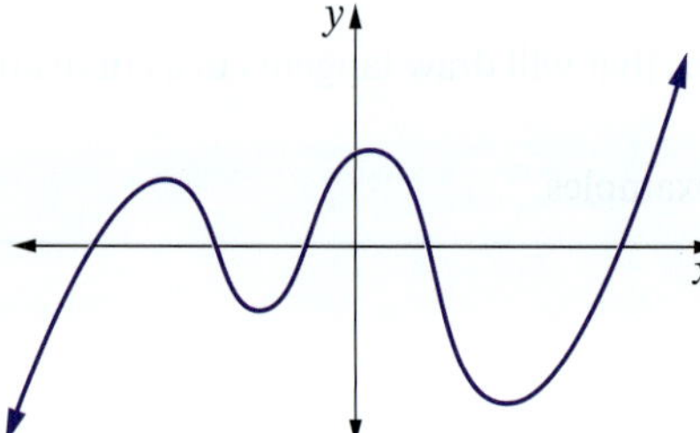

d

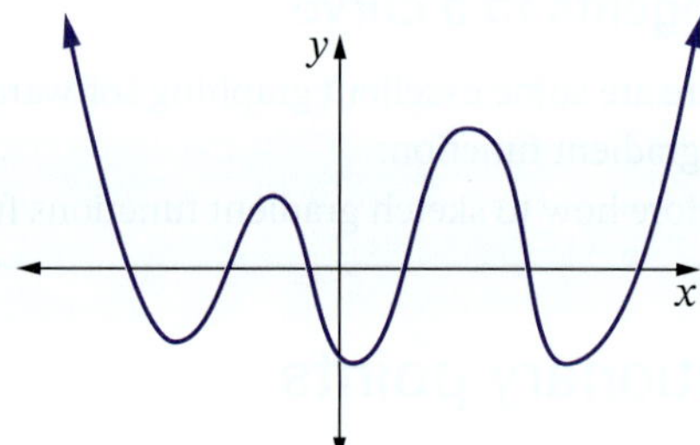

e

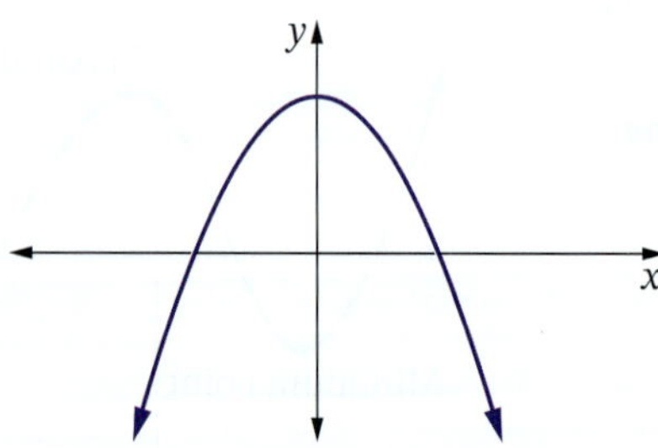

f

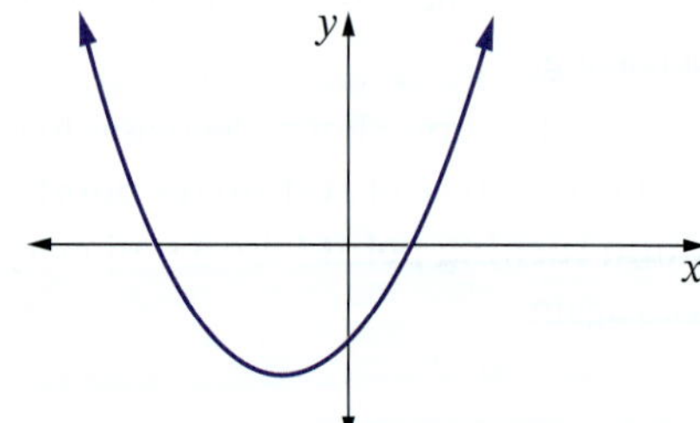

g

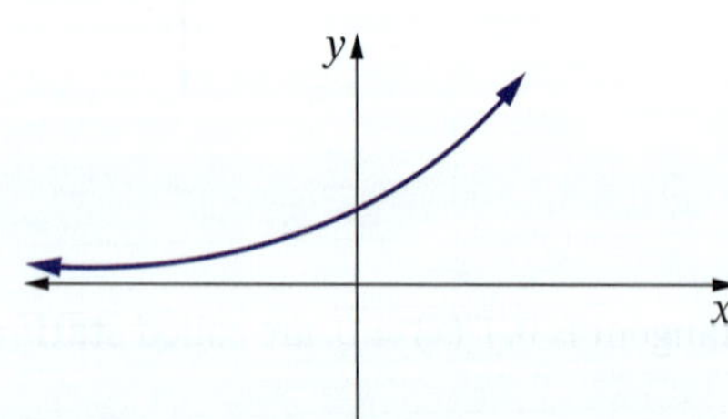

h

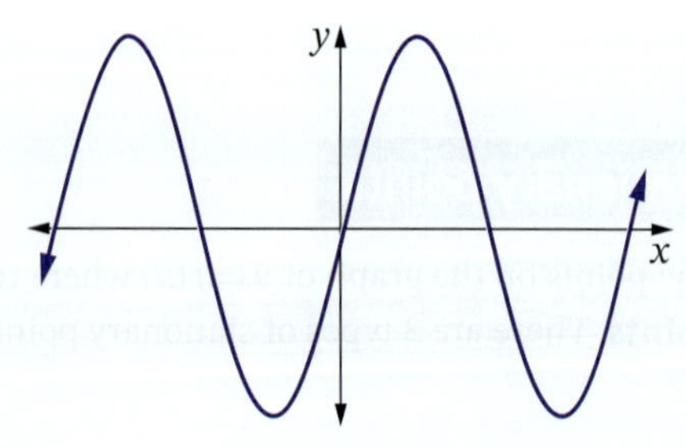

☐ Foundation ○ Mastery ⬡ Complex

i

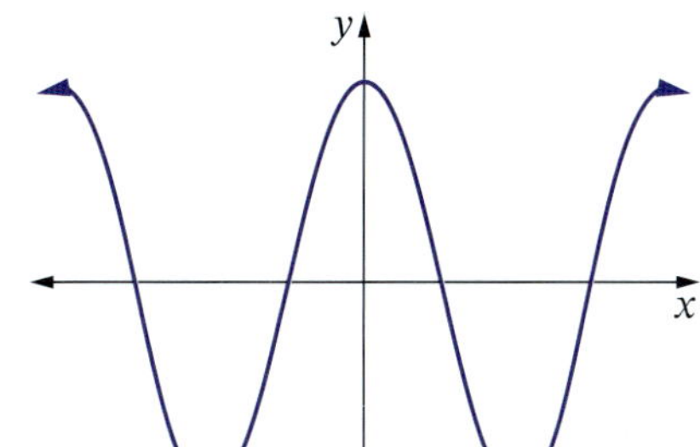

j

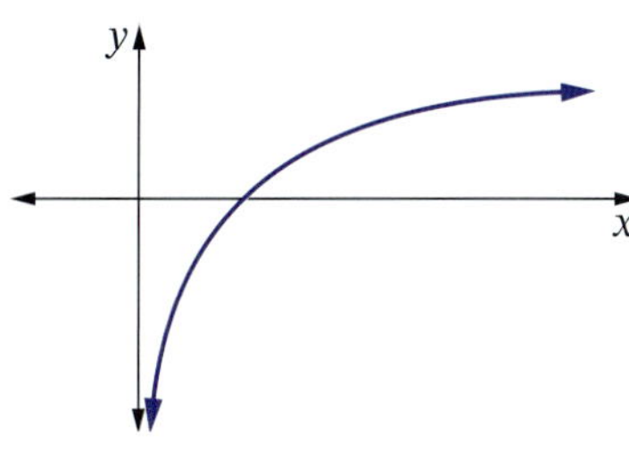

4 This graph below shows the number of people working at *Omega Arts* over time.

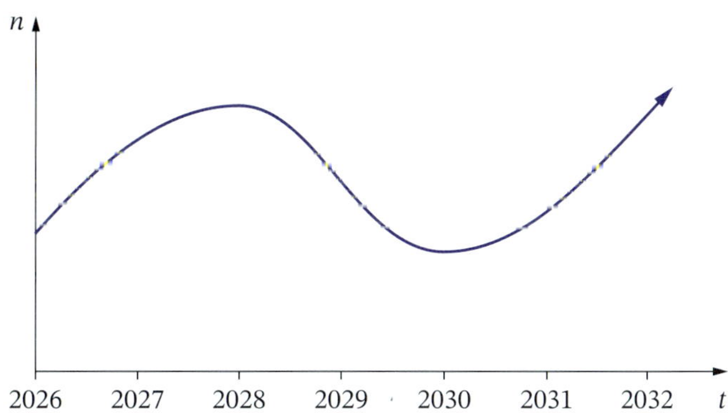

When is the number of workers:

a increasing?

b decreasing?

5 **a** For the graph $y = f(x)$ shown, find the values of x for which the derivative $f'(x)$ is:

i positive **ii** negative **iii** zero

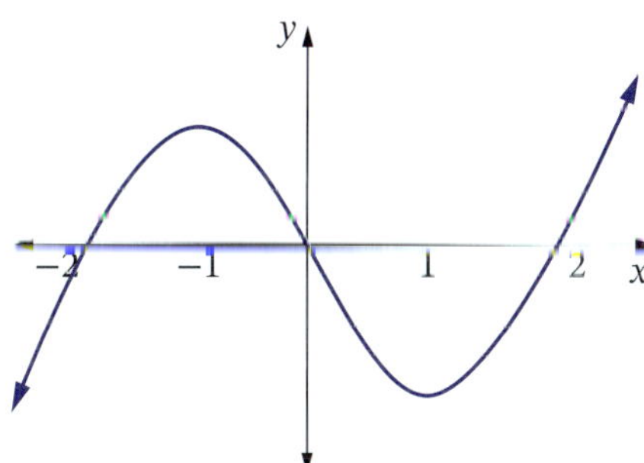

b Sketch the graph of $y = f'(x)$.

6 **a** For the quadratic function $y = ax^2$, where $a > 0$, find values of x for which the function is:

i increasing **ii** decreasing **iii** stationary

b Sketch its derivative function.

7 For the cubic function $f(x) = bx^3$, where $b < 0$, find the x values for which the function is:

a increasing **b** decreasing **c** stationary

☐ Foundation ◯ Mastery ⬡ Complex

8 If the graph below is that of a gradient function $y = f'(x)$, sketch a graph of the original function $y = f(x)$. Is there only one answer? Explain.

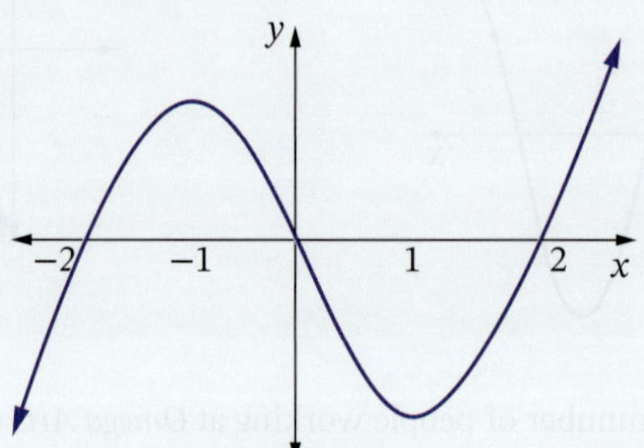

9 Describe the type of derivative function for each function.

a $f(x) = ax^2 + bx + c$

b $f(x) = (x - a)(x - b)(x - c)$

c $f(x) = mx + c$

10 For the graph of $y = f(x)$ below, describe the derivative $f'(x)$. What happens to $f'(x)$ as x becomes large?

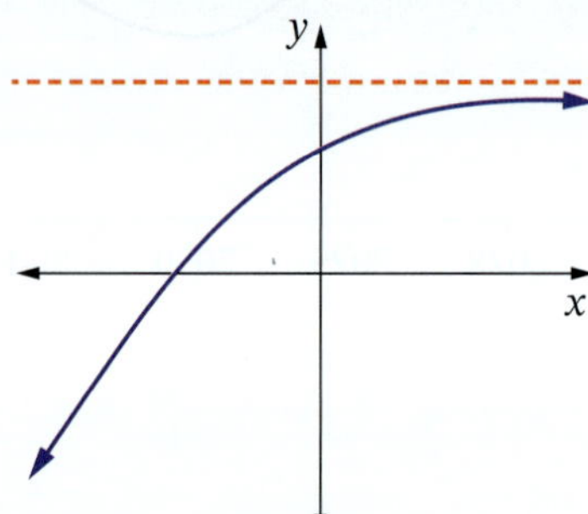

11 Find x values for which each function is decreasing.

a $f(x) = -\dfrac{3}{x-1}$

b $y = 2x^{-2}$

6.02 Differentiation from first principles

Gradient of a secant

Worksheets
Differentiation from first principles

Limits

Finding derivatives from first principles

Puzzle
Rates of change – Gradients of secants

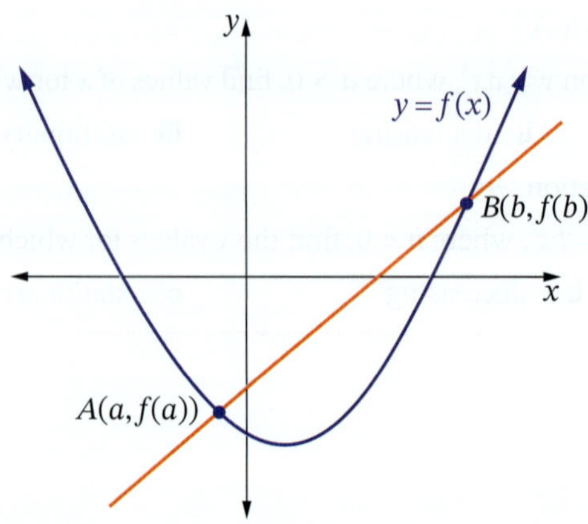

The line passing through the 2 points (x_1, y_1) and (x_2, y_2) on the graph of a function $y = f(x)$ is called a **secant**.

☐ Foundation ○ Mastery ○ Complex

Gradient of a secant

$$m = \frac{y_2 - y_1}{x_2 - x_1}$$

Video Differentiation from first principles

The **gradient of a secant** gives the average rate of change between the 2 points.

A good way to estimate the gradient of a tangent (instantaneous rate of change) is by finding the gradient of a secant between 2 points *very close together* (average rate of change).

If you look at a close-up of a graph, you can get some idea of this concept. When the curve is magnified, any 2 points close together appear to be joined by a straight line. We say the curve is **locally straight**.

Technology

Locally straight curves

Use graphing technology to sketch a curve and then zoom in on a section of the curve to see that it is locally straight.

For example, here is the parabola $y = x^2$.

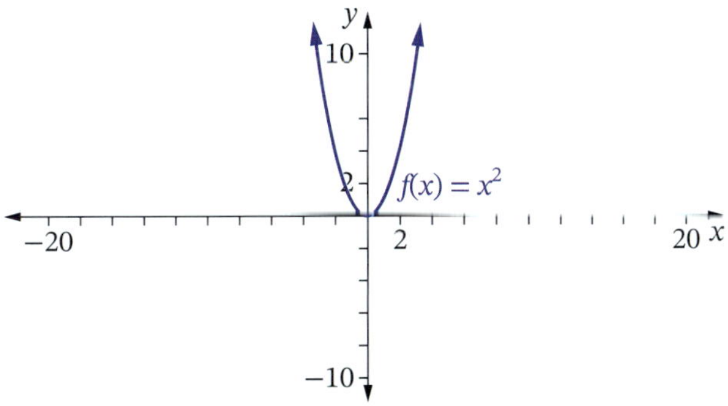

Notice how it looks straight when we zoom in on a point on the parabola. This gives a good idea of what the instantaneous rate of change of the curve is at that point.

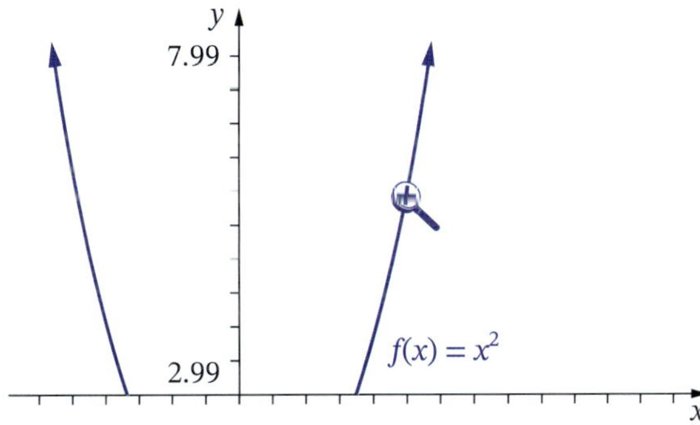

Use technology to sketch other curves and zoom in to show that they are locally straight.

We can calculate an approximate value for the gradient of the tangent at a point on a curve by taking another point close by, then calculating the gradient of the secant joining those 2 points.

Example 4

a For $f(x) = x^3$, find the gradient of the secant PQ where P is the point on the curve where $x = 2$ and Q is another point on the curve where $x = 2.1$. Then choose different points for Q closer to P and use these results to estimate $f'(2)$, the gradient of the tangent to the curve at P.

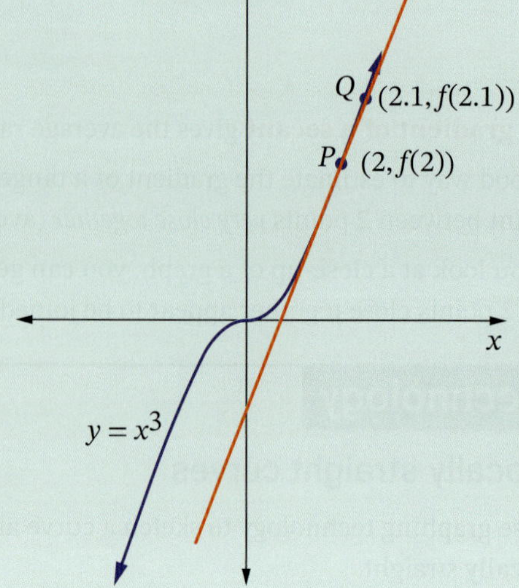

b For the curve $y = x^2$, find the gradient of the secant AB, where A is the point on the curve at which $x = 5$ and point B is close to A. Find an estimate of the gradient of the tangent to the curve at A by using 3 different points for B close to A.

Solution

a P is $(2, f(2)) = (2, 8)$. Take close values of x for point Q, starting with $x = 2.1$, and find the gradient of the secant using $m = \frac{f(b) - f(a)}{b - a}$.

Point *Q*	Gradient of secant *PQ*	Point *Q*	Gradient of secant *PQ*
$(2.1, f(2.1))$	$m = \frac{f(2.1) - f(2)}{2.1 - 2}$ $= \frac{2.1^3 - 2^3}{0.1}$ $= 12.61$	$(1.9, f(1.9))$	$m = \frac{f(1.9) - f(2)}{1.9 - 2}$ $= \frac{1.9^3 - 2^3}{-0.1}$ $= 11.41$
$(2.01, f(2.01))$	$m = \frac{f(2.01) - f(2)}{2.01 - 2}$ $= \frac{2.01^3 - 2^3}{0.01}$ $= 12.0601$	$(1.99, f(1.99))$	$m = \frac{f(1.99) - f(2)}{1.99 - 2}$ $= \frac{1.99^3 - 2^3}{-0.01}$ $= 11.9401$
$(2.001, f(2.001))$	$m = \frac{f(2.001) - f(2)}{2.001 - 2}$ $= \frac{2.001^3 - 2^3}{0.001}$ $= 12.006\,001$	$(1.999, f(1.999))$	$m = \frac{f(1.999) - f(2)}{1.999 - 2}$ $= \frac{1.999^3 - 2^3}{-0.001}$ $= 11.994\,001$

From these results, we can see that a good estimate for $f'(2)$, the gradient at P, is 12.

As $x \to 2, f'(2) \to 12$.

We use a special notation for **limits** to show this.

$$f'(2) = \lim_{x \to 2} \frac{f(x) - f(2)}{x - 2}$$
$$= 12$$

$\lim\limits_{x \to 2}$ is read 'the limit as x approaches 2'.

b $A = (5, f(5))$

Take 3 different values of x for point B; for example, $x = 4.9$, $x = 5.1$ and $x = 5.01$.

$B = (4.9, f(4.9))$

$$m = \frac{f(4.9) - f(5)}{4.9 - 5}$$
$$= \frac{4.9^2 - 5^2}{-0.1}$$
$$= 9.9$$

$B = (5.1, f(5.1))$

$$m = \frac{f(5.1) - f(5)}{5.1 - 5}$$
$$= \frac{5.1^2 - 5^2}{0.1}$$
$$= 10.1$$

$B = (5.01, f(5.01))$

$$m = \frac{f(5.01) - f(5)}{5.01 - 5}$$
$$= \frac{5.01^2 - 5^2}{0.01}$$
$$= 10.01$$

As $x \to 5$, $f'(5) \to 10$.

$$f'(5) = \lim_{x \to 5} \frac{f(x) - f(5)}{x - 5}$$
$$= 10$$

Gradient of a tangent

By choosing points closer together, the average rate of change (from the secant) becomes closer to the instantaneous rate of change (from the tangent). We measure the instantaneous rate of change at any point on the graph of a function by using limits to find the gradient of the tangent to the curve at that point. This is called **differentiation from first principles**. Using the method from the above example, we can find a general formula for the derivative function $y = f'(x)$.

We want to find the instantaneous rate of change or gradient of the tangent to a curve $y = f(x)$ at point $P(x, f(x))$.

Choose a point Q close to P with coordinates $(x + h, f(x + h))$, where h is small.

Now find the gradient of the secant PQ.

$$m = \frac{y_2 - y_1}{x_2 - x_1}$$
$$= \frac{f(x+h) - f(x)}{h}$$

To find the gradient of the tangent at P, we make h smaller as shown, so that Q becomes closer and closer to P.

As h approaches 0, the gradient of the tangent becomes $\lim_{h \to 0} \frac{f(x+h) - f(x)}{h}$.

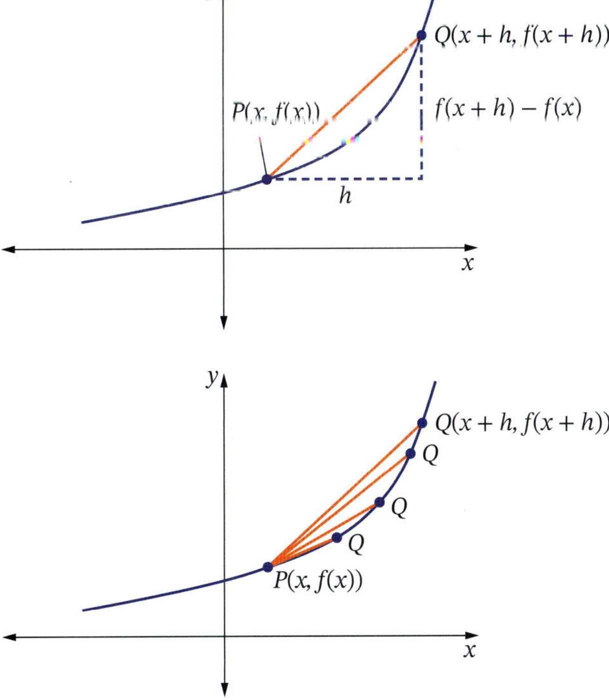

Differentiation from first principles

$$f'(x)=\lim_{h\to 0}\frac{f(x+h)-f(x)}{h}$$

Investigation

Two mathematicians, two notations

At the beginning of this chapter, we learned about the mathematicians Isaac Newton and Gottfried Leibniz. Newton used the notation $f'(x)$ for the derivative function, while Leibniz used the notation $\frac{dy}{dx}$ where d stood for 'difference'. Can you see why he would have used this? Use the internet to explore the different notations used in calculus and where they came from.

Differentiation notation

Newton wrote the derivative of $f(x)$ as $f'(x)$, but there are other notations.
Leibniz wrote the derivative as $\frac{dy}{dx}$ (which comes from $\frac{\Delta y}{\Delta x}$). This means the derivative of y with respect to x (or difference in y per difference in x).

Other notations are y', $\frac{d}{dx}(y)$ and $\frac{d}{dx}(f(x))$.

Example 5

a Differentiate from first principles to find the gradient of the tangent to the curve $y = x^2 + 3$ at the point where $x = 1$.

b Differentiate $f(x) = 2x^2 + 7x - 3$ from first principles.

Solution

a

$$f'(x)=\lim_{h\to 0}\frac{f(x+h)-f(x)}{h}$$

$$f(x) = x^2 + 3$$

$$f(x+h) = (x+h)^2 + 3$$

$$= x^2 + 2xh + h^2 + 3$$

$$f'(x)=\lim_{h\to 0}\frac{x^2+2xh+h^2+3-\left(x^2+3\right)}{h}$$

$$=\lim_{h\to 0}\frac{2xh+h^2}{h}$$

$$=\lim_{h\to 0}\frac{h(2x+h)}{h}$$

$$=\lim_{h\to 0}2x+h$$

$$=2x+0$$

$$=2x$$

$$f'(1) = 2\times 1$$

$$= 2$$

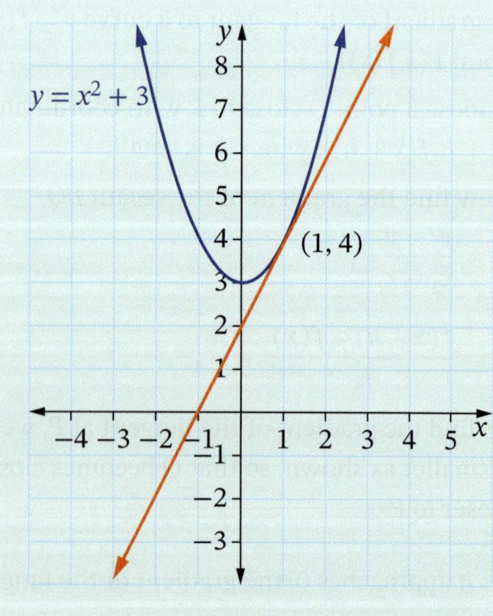

So the gradient of the tangent to the curve $y = x^2 + 3$ at the point $(1, 4)$ is 2.

b $f(x) = 2x^2 + 7x - 3$

$$f(x+h) = 2(x+h)^2 + 7(x+h) - 3$$
$$= 2(x^2 + 2xh + h^2) + 7x + 7h - 3$$
$$= 2x^2 + 4xh + 2h^2 + 7x + 7h - 3$$
$$f(x+h) - f(x) = 2x^2 + 4xh + 2h^2 + 7x + 7h - 3 - (2x^2 + 7x - 3)$$
$$= 2x^2 + 4xh + 2h^2 + 7x + 7h - 3 - 2x^2 - 7x + 3$$
$$= 4xh + 2h^2 + 7h$$
$$f'(x) = \lim_{h \to 0} \frac{f(x+h) - f(x)}{h}$$
$$= \lim_{h \to 0} \frac{4xh + 2h^2 + 7h}{h}$$
$$= \lim_{h \to 0} \frac{h(4x + 2h + 7)}{h}$$
$$= \lim_{h \to 0} (4x + 2h + 7)$$
$$= 4x + 0 + 7$$
$$= 4x + 7$$

So the derivative of $f(x) = 2x^2 + 7x - 3$ is $f'(x) = 4x + 7$.

Displacement and velocity

We can apply rates of change and differentiation to situations involving motion in a straight line, particularly displacement and velocity.

The **displacement** of an object can be positive, negative or 0. Displacement is a 'signed distance' from a fixed point (origin) that has direction as well as magnitude. A displacement of 20 metres means that it is 20 m in a positive direction from a point, while a displacement of -8 metres means that it is 8 m in the (opposite) negative direction from the point.

Similarly, the **velocity** of an object is a 'signed speed'. A velocity of -15 metres per second means that it is travelling at a speed of 15 m/s in the (opposite) negative direction.

Velocity is the rate of change of displacement over time, so it is the derivative of displacement with respect to time.

Displacement and velocity

If x is the displacement of an object at time t, then $v = \frac{dx}{dt}$ is its velocity at that time.

$\frac{dx}{dt}$ can also be written as $\dot{x}$.

$$v = \frac{dx}{dt} = \dot{x}$$

The derivative of velocity with respect to time is acceleration.

$$a = \frac{dv}{dt} = \dot{v}$$

Example 6

This displacement–time graph shows the displacement of a particle in centimetres moving along a straight line from a point O over time t seconds.

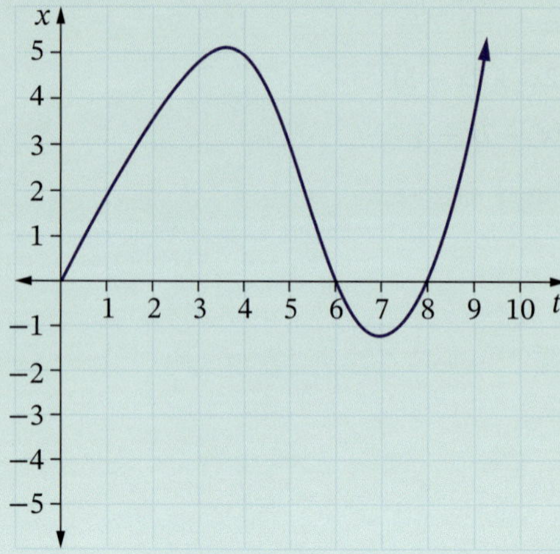

Note that x is on the vertical axis and stands for displacement, and x is a function of time, t.

a When is the particle at point O?

b By drawing tangents to the curve, estimate the velocity of the particle when it is at point O.

Solution

a The particle is at O when $x = 0$ (displacement is 0), which corresponds to the t-intercepts of the graph (along the x-axis).

So the particle is at O when $t = 0, 6, 8$ s.

b

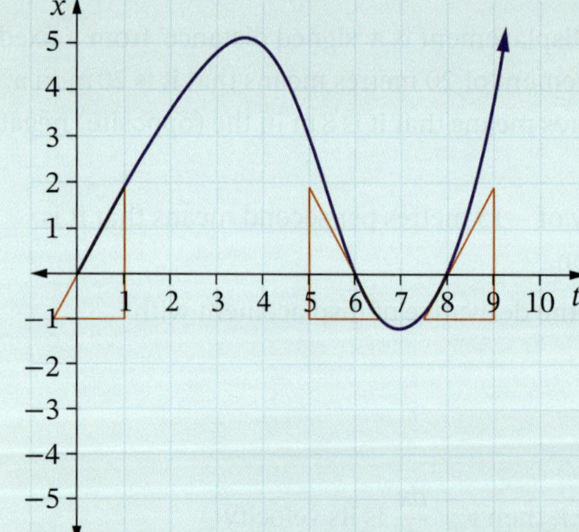

From the graph, the gradient at $t = 0$ and 8 is approximately 2, and the gradient at $t = 6$ is approximately -2.

So the velocity of the particle when it is at O is about 2 cm/s at $t = 0, 8$ and about -2 cm/s at $t = 6$.

EXERCISE 6.02 Answers on p. 491

Differentiation from first principles

1 For the function $f(x) = x^2 - 4$, find the gradient of the tangent at point P where $x = 3$ by selecting points near P and finding the gradient of the secant.

2 A function is given by $f(x) = x^2 + x + 5$.

a Find $f(x + h)$.

b Find $f(x + h) - f(x)$.

c Show that $\dfrac{f(x+h)-f(x)}{h} = 2x + h + 1$.

d Find $f'(2)$.

3 Given the curve $f(x) = 4x^2 - 3$, find:

a $f(x + h)$

b $f(x + h) - f(x)$

c the gradient of the tangent to the curve at the point where $x = -1$.

4 For the parabola $y = x^2 - 1$, find:

a $f(x + h)$ **b** $f(x + h) - f(x)$ **c** $f'(3)$.

5 For the function $f(x) = 4 - 3x - 5x^2$, find:

a $f'(x)$ **b** the gradient of the tangent at the point $(-2, -10)$.

6 A function is given by $f(x) = 2x^2 - 7x + 3$.

a Show that $f(x + h) = 2x^2 + 4xh + 2h^2 - 7x - 7h + 3$.

b Show that $f(x + h) - f(x) = 4xh + 2h^2 - 7h$.

c Show that $\dfrac{f(x+h)-f(x)}{h} = 4x + 2h - 7$.

d Find $f'(x)$.

7 A particle starts at a fixed point O and moves along a straight line so that its displacement from O over time t is as shown in this graph.

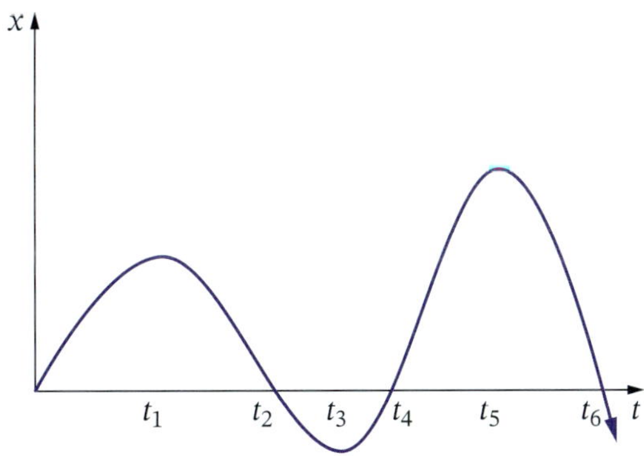

a At what times is the particle at O?

b When is the velocity zero?

c When is the particle at its furthest distance from O?

d Describe the distance from O of the particle between $t = 0$ and $t = t_4$.

e Describe the velocity of the particle between $t = 0$ and $t = t_4$.

8. Estimate the velocity at $t = 6$ on each displacement–time graph, where x is measured in centimetres and time is measured in seconds.

a

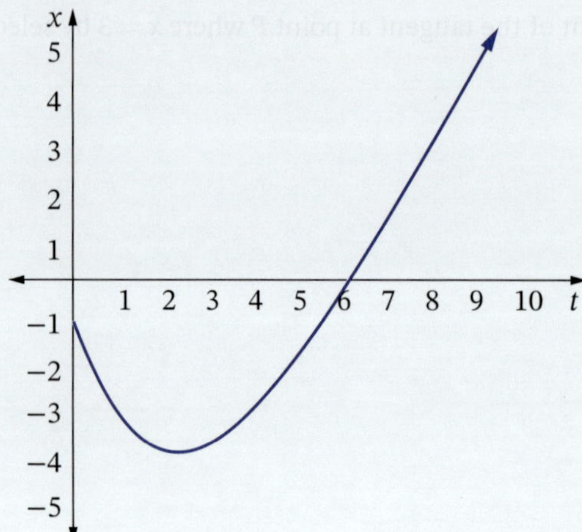

b

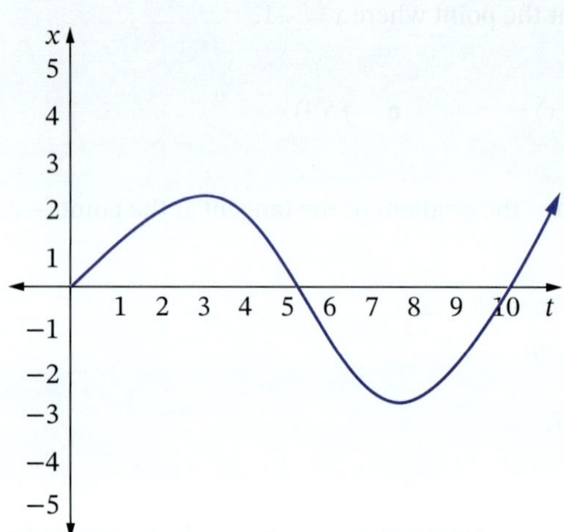

c

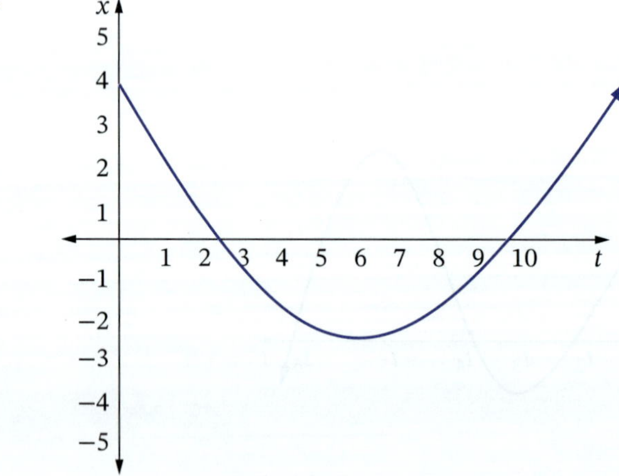

9. Differentiate from first principles to find the gradient of the tangent to the curve:

a $f(x) = x^2$ at the point where $x = 1$

b $y = x^2 + x$ at the point $(2, 6)$

c $f(x) = 2x^2 - 5$ at the point where $x = -3$

d $y = 3x^2 + 3x + 1$ at the point where $x = 2$

e $f(x) = x^2 - 7x - 4$ at the point $(-1, 4)$.

□ Foundation ○ Mastery ○ Complex

10 Find the derivative function for each function from first principles.

a $f(x) = x^2$ **b** $y = x^2 + 5x$

c $f(x) = 4x^2 - 4x - 3$ **d** $y = 5x^2 - x - 1$

11 **a** Find the average speed of an object between 2 and 3 seconds if distance d cm is given by $d = 8 + t^2$, where t is in seconds.

b Find the average speed of the object between 2 and 2.5 seconds.

c Explain how to get a good approximation to the instantaneous speed at 2 seconds.

d Differentiate by first principles to find the instantaneous speed at 2 seconds.

12 **a** Differentiate $f(x) = x^3 - x^2 - 2x$ from first principles.

b Find:

i $f'(2)$ **ii** $f'(-1)$

c Find exact x values for which $f'(x) = 0$.

d Describe the behaviour of the function when $f'(x) = 0$.

13 Differentiate $f(x) = \frac{1}{x}$ from first principles.

The derivative of x^n 6.03

Remember that the gradient of a straight line $y = mx + c$ is m. The tangent to the line is the line itself, so the gradient of the tangent is m everywhere along the line.

So if $y = mx$, $\frac{dy}{dx} = m$.

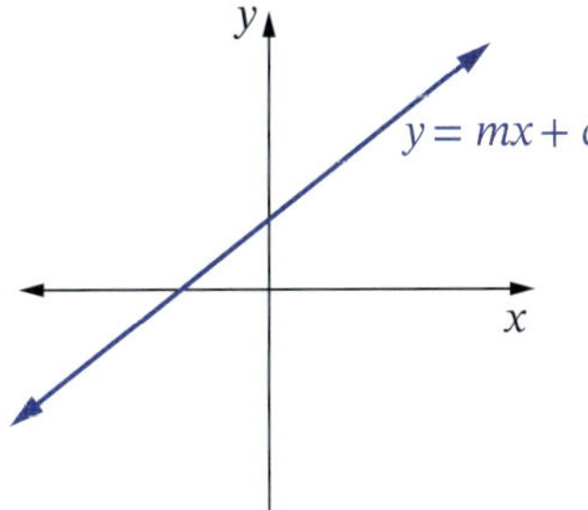

Worksheets
Derivatives of linear products
Derivatives of polynomials
Derivatives of a sum of terms
Basic differentiation

The derivative of kx

$$\frac{d}{dx}(kx) = k$$

A horizontal line $y = k$ has a gradient of zero.

So if $y = k$, $\frac{dy}{dx} = 0$.

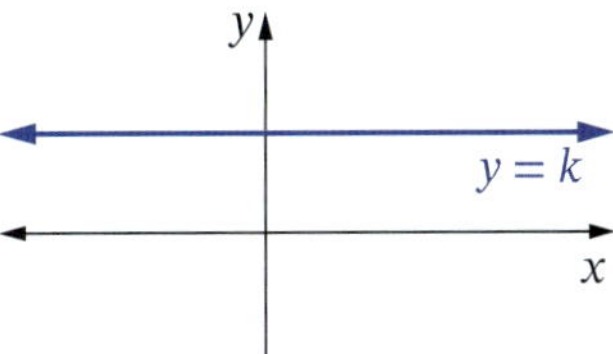

The derivative of k

$$\frac{d}{dx}(k) = 0$$

Both properties above can be easily proved by differentiation from first principles.

Foundation Mastery Complex

Technology

The derivative of powers of *x*

Find an approximation to the derivative of power functions such as $y = x^2, y = x^3, y = x^4$, $y = x^5$ by drawing the graph of $y = \frac{f(x+0.01) - f(x)}{0.01}$. You could use graphing technology to sketch the derivative for these functions and find its equation.

Can you find a pattern? Could you predict what the result would be for x^n?

When differentiating $y = x^n$ from first principles, a simple pattern appears:

- For $y = x$, $\quad f'(x) = 1x^0 = 1$
- For $y = x^2$, $\quad f'(x) = 2x^1 = 2x$
- For $y = x^3$, $\quad f'(x) = 3x^2$
- For $y = x^4$, $\quad f'(x) = 4x^3$
- For $y = x^5$, $\quad f'(x) = 5x^4$

The derivative of x^n

$$\frac{d}{dx}\left(x^n\right) = nx^{n-1}$$

There are some more properties of differentiation.

The derivative of kx^n

$$\frac{d}{dx}(kx^n) = knx^{n-1}$$

More generally:

The derivative of a constant multiple of a function

$$\frac{d}{dx}(kf(x)) = kf'(x)$$

Example 7

a Find the derivative of $3x^8$.

b Differentiate $f(x) = 7x^3$.

Solution

a

$$\frac{d}{dx}\left(x^n\right) = nx^{n-1}$$

$$\frac{d}{dx}(3x^8) = 3 \times 8x^{8-1}$$

$$= 24x^7$$

b

$$f'(x) = knx^{n-1}$$

$$f'(x) = 7 \times 3x^{3-1}$$

$$= 21x^2$$

If there are several terms in an expression, we differentiate each one separately.

The derivative of a sum of functions

$$\frac{d}{dx}(f(x) + g(x)) = f'(x) + g'(x)$$

Example 8

a Differentiate $x^3 + x^4$.

b Find the derivative of $7x$.

c Differentiate $f(x) = x^4 - x^3 + 5$.

d Find the derivative of $y = 4x^7$.

e If $f(x) = 2x^5 - 7x^3 + 5x - 4$, evaluate $f'(-1)$.

f Find the derivative of $f(x) = 2x^2(3x - 7)$.

g Find the derivative of $\frac{3x^2 + 5x}{2x}$.

h Differentiate $S = 6r^2 - 12r$ with respect to r.

Solution

a $\frac{d}{dx}(x^3 + x^4) = 3x^2 + 4x^3$

b $\frac{d}{dx}(7x) = 7$

c
$$\begin{aligned} f'(x) &= 4x^3 - 3x^2 + 0 \\ &= 4x^3 - 3x^2 \end{aligned}$$

d
$$\begin{aligned} \frac{dy}{dx} &= 4 \times 7x^6 \\ &= 28x^6 \end{aligned}$$

e
$$\begin{aligned} f'(x) &= 10x^4 - 21x^2 + 5 \\ f'(-1) &= 10(-1)^4 - 21(-1)^2 + 5 \\ &= -6 \end{aligned}$$

f Expand first.
$$\begin{aligned} f(x) &= 2x^2(3x - 7) \\ &= 6x^3 - 14x^2 \\ f'(x) &= 18x^2 - 28x \end{aligned}$$

g Simplify first.
$$\begin{aligned} \frac{3x^2 + 5x}{2x} &= \frac{x(3x + 5)}{2x} \\ &= \frac{3x + 5}{2} \\ \frac{d}{dx}\left(\frac{3x + 5}{2}\right) &= \frac{3 + 0}{2} \\ &= 1\frac{1}{2} \end{aligned}$$

h Differentiating with respect to r rather than x:
$$\begin{aligned} S &= 6r^2 - 12r \\ \frac{dS}{dr} &= 12r - 12 \end{aligned}$$

This can also be written as $S'(r)$.

Investigation

Families of curves

1 Differentiate:

x^2+1 $\quad x^2-3$ $\quad x^2+7$

x^2 $\quad x^2+20$ $\quad x^2-100$

What do you notice?

2 Differentiate:

x^3+5 $\quad x^3+11$ $\quad x^3-1$

x^3-6 $\quad x^3$ $\quad x^3+15$

What do you notice?

These groups of functions are families because they have the same derivatives. Can you find others?

Investigation

Derivative of x^{-1} and $x^{\frac{1}{2}}$

1 **a** Show that $\dfrac{1}{x+h}-\dfrac{1}{x}=\dfrac{-h}{x(x+h)}$.

b Hence differentiate $y=\dfrac{1}{x}$ from first principles.

c Differentiate $y=x^{-1}$ using the formula. Do you get the same answer as in part **b**?

2 **a** Show that $\left(\sqrt{x+h}-\sqrt{x}\right)\left(\sqrt{x+h}+\sqrt{x}\right)=h$.

b Hence differentiate $y=\sqrt{x}$ from first principles.

c Differentiate $y=x^{\frac{1}{2}}$ and show that this gives the same answer as in part **b**.

The nx^{n-1} rule of differentiation also applies to negative and fractional powers.

The derivative of $\frac{1}{x}$

$$\frac{d}{dx}\left(\frac{1}{x}\right)=-\frac{1}{x^2}$$

This can also be written as $\dfrac{d}{dx}(x^{-1})=-x^{-2}$

The derivative of $\sqrt{x}$

$$\frac{d}{dx}\left(\sqrt{x}\right)=\frac{1}{2\sqrt{x}}$$

This can also be written as $\dfrac{d}{dx}\left(x^{\frac{1}{2}}\right)=\dfrac{1}{2}x^{-\frac{1}{2}}$

Example 9

a Differentiate $f(x) = 7\sqrt[3]{x}$.

b Find the derivative of $y = \dfrac{4}{x^2}$ at the point where $x = 2$.

Solution

a $f(x) = 7\sqrt[3]{x} = 7x^{\frac{1}{3}}$

Convert the function to a power of x first.

$$f'(x) = 7 \times \frac{1}{3}x^{\frac{1}{3}-1}$$
$$= \frac{7}{3}x^{-\frac{2}{3}}$$
$$= \frac{7}{3} \times \frac{1}{x^{\frac{2}{3}}}$$
$$= \frac{7}{3} \times \frac{1}{\sqrt[3]{x^2}}$$
$$= \frac{7}{3\sqrt[3]{x^2}}$$

b $y = \dfrac{4}{x^2}$

$$= 4x^{-2}$$
$$\frac{dy}{dx} = -8x^{-3}$$
$$= -\frac{8}{x^3}$$

When $x = 2$:

$$\frac{dy}{dx} = -\frac{8}{2^3}$$
$$= -1$$

EXERCISE 6.03 Answers on p. 491

The derivative of x^n

1 Differentiate:

a $x + 2$ **b** $5x - 9$ **c** $x^2 + 3x + 4$

d $5x^2 - x - 8$ **e** $x^3 + 2x^2 - 7x - 3$ **f** $2x^3 - 7x^2 + 7x - 1$

g $3x^4 - 2x^2 + 5x$ **h** $x^6 - 5x^3 - 2x^4$ **i** $2x^5 - 4x^3 + x^2 - 2x + 4$

2 Find the derivative of:

a $x(2x + 1)$ **b** $(2x - 3)^2$ **c** $(x + 4)(x - 4)$

d $(2x^2 - 3)^2$ **e** $(2x + 5)(x^2 - x + 1)$

3 Find the derivative of:

a $\dfrac{x^2}{6} - x$ **b** $\dfrac{x^4}{2} - \dfrac{x^3}{3} + 4$ **c** $\dfrac{1}{3}x^6(x^2 - 3)$

d $\dfrac{2x^3 + 5x}{x}$ **e** $\dfrac{x^2 + 2x}{4x}$ **f** $\dfrac{2x^5 - 3x^4 + 6x^3 - 2x^2}{3x^2}$

4 Find $f'(x)$ when $f(x) = 8x^2 - 7x + 4$.

5 If $y = x^4 - 2x^3 + 5$, find $\dfrac{dy}{dx}$ when $x = -2$.

6 Find $\dfrac{dy}{dx}$ if $y = 6x^{10} - 5x^8 + 7x^5 - 3x + 8$.

7 If $s = 5t^2 - 20t$, find $\dfrac{ds}{dt}$.

8 Find $g'(x)$ given $g(x) = 5x^4$.

☐ Foundation ◯ Mastery ◯ Complex

9 Find $\dfrac{dv}{dt}$ when $v = 15t^2 - 9$.

10 If $h = 40t - 2t^2$, find $\dfrac{dh}{dt}$.

11 Given $V = \dfrac{4}{3}\pi r^3$, find $\dfrac{dV}{dr}$.

12 If $f(x) = 2x^3 - 3x + 4$, evaluate $f'(1)$.

13 Given $f(x) = x^2 - x + 5$, evaluate:

a $f'(3)$ **b** $f'(-2)$ **c** x when $f'(x) = 7$

14 If $y = x^3 - 7$, evaluate:

a the derivative when $x = 2$ **b** x when $\dfrac{dy}{dx} = 12$

15 Evaluate $g'(2)$ when $g(t) = 3t^3 - 4t^2 - 2t + 1$.

EXAMPLE 9

16 Differentiate:

a x^{-3} **b** $x^{1.4}$ **c** $6x^{0.2}$ **d** $x^{\frac{1}{2}}$

e $2x^{\frac{1}{2}} - 3x^{-1}$ **f** $3x^{\frac{1}{3}}$ **g** $8x^{\frac{3}{4}}$ **h** $-2x^{-\frac{1}{2}}$

17 Find the derivative function.

a $\dfrac{1}{x}$ **b** $5\sqrt{x}$ **c** $\sqrt[6]{x}$ **d** $\dfrac{2}{x^5}$

e $-\dfrac{5}{x^3}$ **f** $\dfrac{1}{\sqrt{x}}$ **g** $\dfrac{1}{2x^6}$ **h** $x\sqrt{x}$

i $\dfrac{2}{3x}$ **j** $\dfrac{1}{4x^2} + \dfrac{3}{x^4}$

18 Find the derivative of $y = \sqrt[3]{x}$ at the point where $x = 27$.

19 If $x = \dfrac{12}{t}$, find $\dfrac{dx}{dt}$ when $t = 2$.

20 A function is given by $f(x) = \sqrt[4]{x}$. Evaluate $f'(16)$.

21 Find the derivative of $y = \dfrac{3}{2x^2}$ at the point $\left(1, 1\dfrac{1}{2}\right)$.

22 Find $\dfrac{dy}{dx}$ if $y = \left(x + \sqrt{x}\right)^2$.

23 A function $f(x) = \dfrac{\sqrt{x}}{2}$ has a tangent at $(4, 1)$. Find its gradient.

24 **a** Differentiate $\dfrac{\sqrt{x}}{x}$.

b Hence find the derivative of $y = \dfrac{\sqrt{x}}{x}$ at the point where $x = 4$.

25 The function $f(x) = 3\sqrt{x}$ has $f'(x) = \dfrac{3}{4}$ at $x = a$. Find a.

26 The hyperbola $y = \dfrac{2}{x}$ has 2 tangents with gradient $-\dfrac{2}{25}$.

Find the points where these tangents touch the hyperbola.

☐ Foundation ◯ Mastery ⬡ Complex

Did you know?

Motion and calculus

Galileo (1564–1642) was very interested in the behaviour of bodies in motion. He dropped stones from the Leaning Tower of Pisa to try to prove that they would fall with equal speed. He rolled balls down slopes to prove that they move with uniform speed until friction slows them down. He showed that a body moving through the air follows a curved path at a fairly constant speed.

Galileo

John Wallis (1616–1703) continued this study with his publication *Mechanica, sive Tractatus de Motu Geometricus*. He applied mathematical principles to the laws of motion and stimulated interest in the subject of mechanics.

Soon after Wallis' publication, **Christiaan Huygens** (1629–1695) wrote *Horologium Oscillatorium sive de Motu Pendulorum*, in which he described various mechanical principles. He invented the pendulum clock, improved the telescope and investigated circular motion and the descent of heavy bodies.

These 3 mathematicians provided the foundations of mechanics. **Sir Isaac Newton** (1642–1727) used calculus to increase the understanding of the laws of motion. He also used these concepts as a basis for his theories on gravity and inertia.

Applications of differentiation 6.04

Differentiability

A function is differentiable at any point if it is continuous and smooth, because we can draw a tangent and find its gradient at any point.

A function is not differentiable at any point where it is discontinuous, has a gap or a sharp vertex (different gradients on either side).

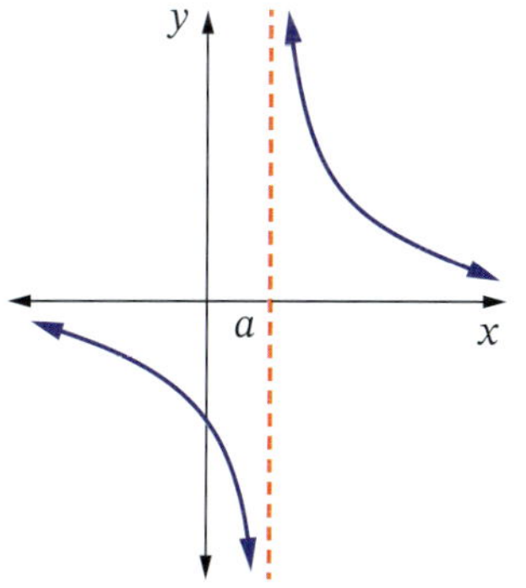

Not differentiable at $x = a$

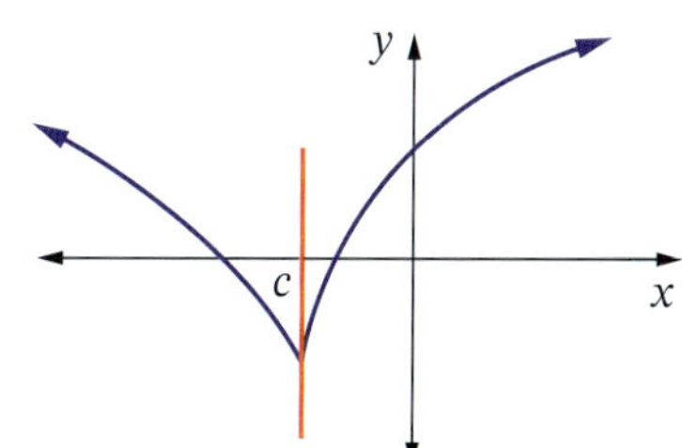

Not differentiable at $x = b$

y

c

x

Not differentiable at $x = c$

Graphing applications

Example 10

For the graph of $f(x) = 2x^3 + 9x^2 - 60x + 5$:

a find $f'(1)$ and $f'(3)$ and explain whether the function is increasing or decreasing at these points

b find any points where the tangent to the graph is parallel to the x-axis and explain the behaviour of the function at those points.

Solution

a $f'(x) = 6x^2 + 18x - 60$

$f'(1) = 6(1)^2 + 18(1) - 60$

$= -36$

$f'(1) < 0$, so the function is decreasing at $x = 1$.

$f'(3) = 6(3)^2 + 18(3) - 60$

$= 48$

$f'(3) > 0$, so the function is increasing at $x = 3$.

b If the tangent is parallel to the x-axis, then it is flat and $f'(x) = 0$ at that point, a stationary point.

$6x^2 + 18x - 60 = 0$

$x^2 + 3x - 10 = 0$

$(x + 5)(x - 2) = 0$

$x = -5, x = 2$

When $x = -5$:

$f(-5) = 2(-5)^3 + 9(-5)^2 - 60(-5) + 5$

$= 280$

When $x = 2$:

$f(2) = 2(2)^3 + 9(2)^2 - 60(2) + 5$

$= -63$

This means that there is a stationary point at $(-5, 280)$ and $(2, -63)$.

Video
Displacement and velocity

Applications of rates of change

Example 11

a The number of bacteria in a culture increases according to the function $B = 2t^4 - t^2 + 2000$, where t is time in hours. Find:

i the number of bacteria initially

ii the average rate of change in the number of bacteria between 2 and 3 hours

iii the number of bacteria after 5 hours

iv the rate at which the number of bacteria is increasing after 5 hours.

b An object travels a distance according to the function $D = t^2 + t + 5$, where D is in metres and t is in seconds. Find the speed at which it is travelling at:

i 4 s　　**ii** 10 s

Solution

a **i** $B = 2t^4 - t^2 + 2000$

Initially, $t = 0$:

$B = 2(0)^4 - (0)^2 + 2000$

$= 2000$

So there are 2000 bacteria initially.

ii When $t = 2$, $B = 2(2)^4 - (2)^2 + 2000$

$= 2028$

When $t = 3$, $B = 2(3)^4 - (3)^2 + 2000$

$= 2153$

Average rate of change $= \dfrac{B_2 - B_1}{t_2 - t_1}$

$= \dfrac{2153 - 2028}{3 - 2}$

$= 125$

So the average rate of change is 125 bacteria per hour.

iii When $t = 5$, $B = 2(5)^4 - (5)^2 + 2000$

$= 3225$

So there will be 3225 bacteria after 5 hours.

iv The instantaneous rate of change is given by the derivative $\dfrac{dB}{dt} = 8t^3 - 2t$.

When $t = 5$, $\dfrac{dB}{dt} = 8(5)^3 - 2(5)$

$= 990$

So the rate of increase after 5 hours will be 990 bacteria per hour.

b Speed is the rate of change of distance over time: $\dfrac{dD}{dt} = 2t + 1$.

i When $t = 4$, $\dfrac{dD}{dt} = 2(4) + 1$

$= 9$

So speed after 4 s is 9 m/s.

ii When $t = 10$, $\dfrac{dD}{dt} = 2(10) + 1$

$= 21$

So speed after 10 s is 21 m/s.

Example 12

a This graph shows the amount of bacteria in a sample of cheese over a period of time.

Describe how the amount of bacteria changes, including the rate of change.

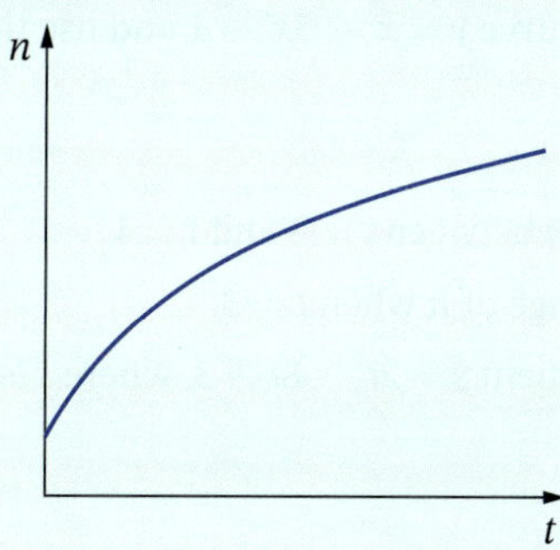

b This distance–time graph shows a car journey over a period of time, with t_1, t_2, t_3, t_4 and t_5 being specific values of time.

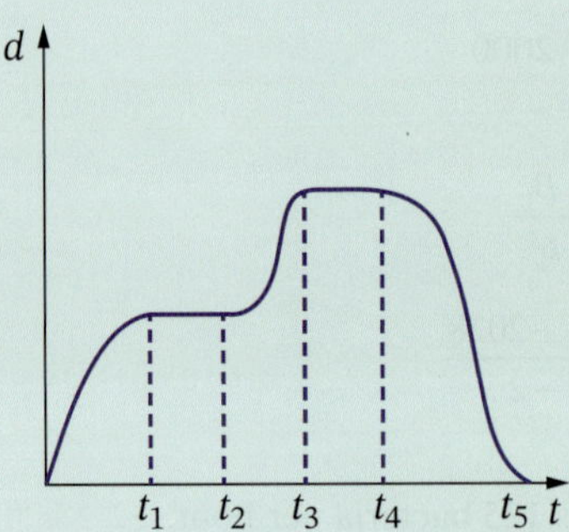

Describe the car's journey in terms of distance and speed.

Solution

a The graph increases quickly with a steep gradient but then it starts to flatten out. While it is still increasing, its gradient (or the gradient of its tangents) is decreasing so the rate of change is decreasing. This means that the amount of bacteria is increasing, but the rate at which it is increasing is slowing down or decreasing.

b The car sets out and travels a long distance quickly between time 0 and t_1, with its steep gradient suggesting a high speed.

It flattens out at t_1, meaning that the car slows down (lower speed) and the flat graph between t_1 and t_2 means that it stops then (no change in distance, and gradient and speed are 0).

After t_2, the car starts off again slowly (low gradients) but then starts to speed up until it stops between t_3 and t_4. Then the graph goes down which means that the car changes direction and returns towards its starting position. The graph starts off flat and then the gradient becomes steep again, meaning an increase in speed in the reverse direction.

The car slows down to a stop at t_5, when it has returned to its starting position.

EXERCISE 6.04 Answers on p. 492

Applications of differentiation

1 For the graph of $g(x) = 8x^3 - 12x^2 - 480x + 5$:

- **a** determine whether $g(x)$ is increasing or decreasing at $x = -10$ and the origin
- **b** find the coordinates of all stationary points.

2 For the parabola with equation $y = x^2 - 4x + 1$, show that its vertex is a stationary point and describe its nature.

3 Find any stationary points on the curve $y = x^3 - 3x^2 + 1$ and use them to sketch the graph of the function.

4 If $h = t^3 - 7t + 5$, find:

- **a** the average rate of change of h between $t = 3$ and $t = 4$
- **b** the instantaneous rate of change of h when $t = 3$.

5 A particle is moving with displacement $s = 2t^2 - 8t + 3$, where s is in metres and t is in seconds.

- **a** Find its initial velocity.
- **b** Find its displacement at 5 s.
- **c** Find when the particle's velocity is zero.
- **d** What will the particle's displacement be at that time?

6 **a** Find the equation of the axis of symmetry and the minimum point of the graph of $y = x^2 - 4x$.

- **b** Find the derivative of the function at this point and explain this result.

Foundation Mastery Complex

7 The volume of water V in litres flowing through a pipe after t seconds is given by $V = t^2 + 3t$. Find the rate at which the water is flowing when $t = 5$.

8 For the graph of $f(x) = -6x^3 - 9x^2 + 216x - 14$:

a find the values of x for which $f(x)$ is increasing

b show that $f(x)$ has a minimum point at $x = -4$.

9 The mass in grams of a melting ice block is given by the formula $M = t - 2t^2 + 100$, where t is time in minutes.

a Find the average rate of change at which the ice block is melting between:

i 1 and 3 minutes ii 2 and 5 minutes.

b Find the rate at which the ice block is melting at 5 minutes.

10 The graph on the left shows the displacement of an object while the graph on the right shows the velocity of the same object. Describe what is happening with the movement of the object.

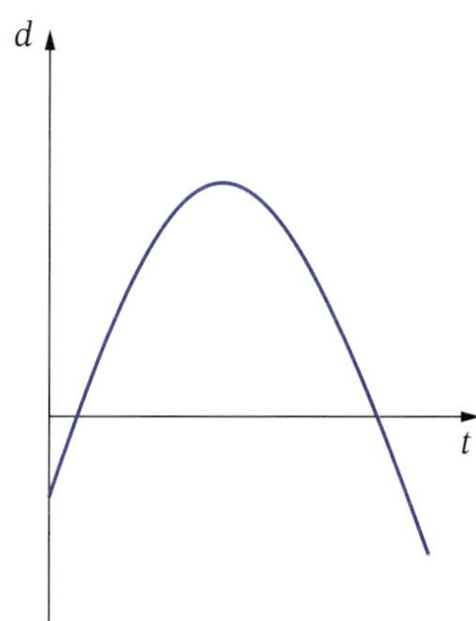

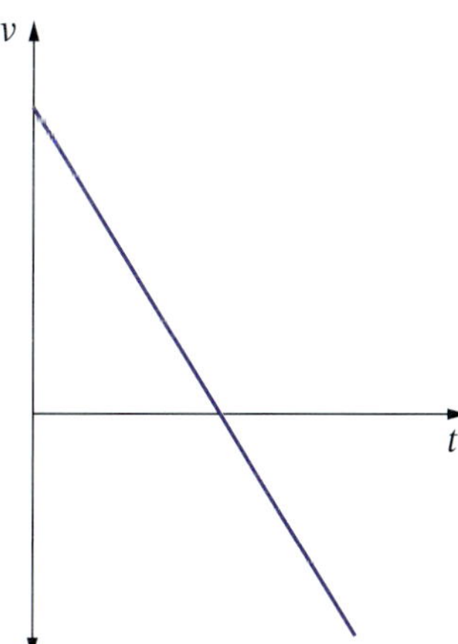

11 Each graph shows the change in temperature at a beach over a 4-hour period. Describe how the temperature is changing and its rate of change.

a

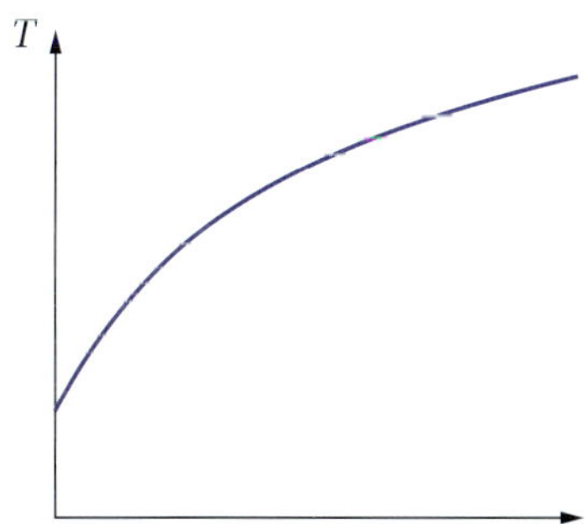

b

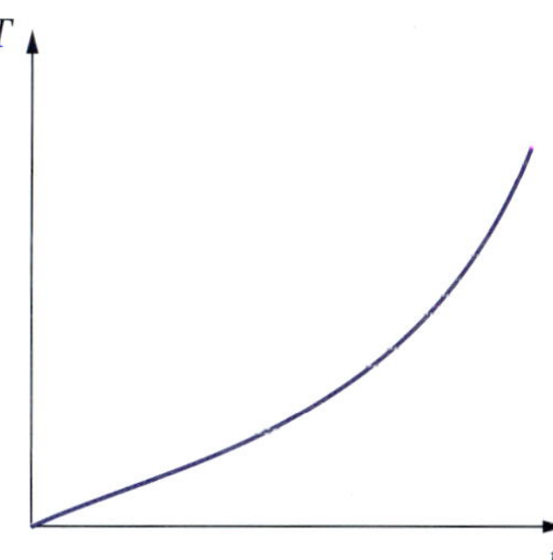

c

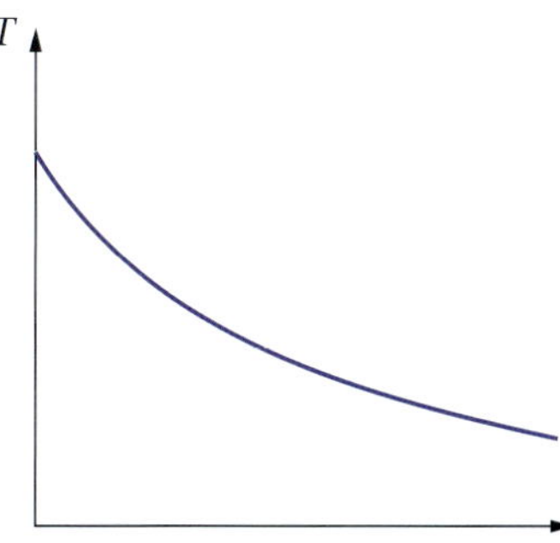

d

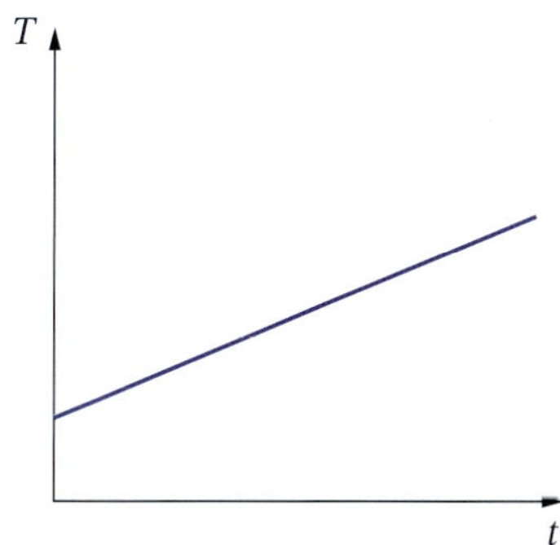

Foundation Mastery Complex

e

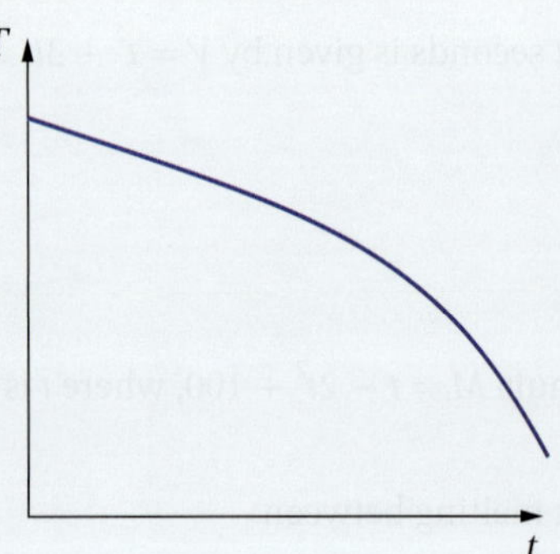

12 **a** Describe the movement of a bird shown by this displacement–time graph.

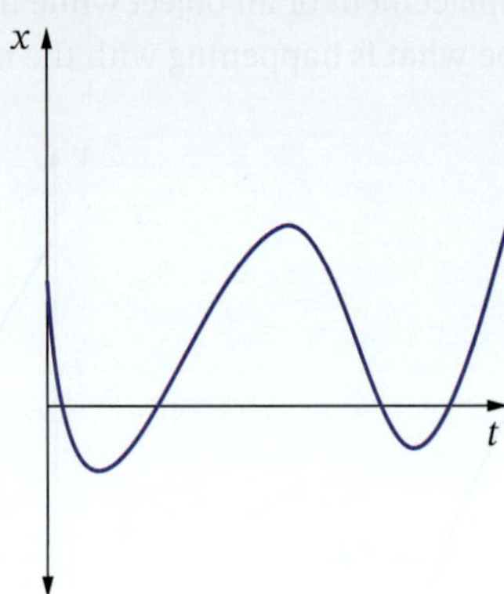

b Sketch a possible velocity–time graph.

13 Find a function that has each derivative.

a $f'(x) = 3x^2 + 2x - 3$

b $\frac{dy}{dx} = 5x^4 + 4x^3 - 6x$

c $f'(x) = x^3 - x^2 + x - 7$

14 Find any points on the curve $y = \sqrt{x}$ where the angle of inclination to the tangent is 30°.

15 The surface area in cm^2 of a balloon being inflated is given by $S = t^3 - 2t^2 + 5t + 2$, where t is time in seconds. Find the rate of increase in the balloon's surface area at 8 seconds.

16 The displacement of a particle is $x = t^3 - 9t$ cm, where t is time in seconds.

a Find the velocity of the particle at 3 s.

b Show that the particle is initially at the origin and find any other times that the particle will be at the origin.

c Find the exact time when the particle has zero velocity (where the particle is at rest).

17 Use the derivative to prove that the parabola $y = ax^2 + bx + c$ has a stationary point at $x = -\frac{b}{2a}$.

18 A ball is thrown into the air and its height h metres after t seconds is given by $h = 3 + 2t - t^2$.

a How high is the ball when it is thrown?

b When does the ball reach its highest point?

c What is the highest point it reaches?

d What is the velocity of the ball after 2 s?

e What is the velocity of the ball when it reaches the ground?

19 The volume V of a cylinder is given by $V = \frac{200\pi r^3}{3}$. Find the rate at which the volume is changing, to the nearest integer, as the radius changes, when the radius is:

a 5 cm

b 10 cm

Foundation Mastery Complex

20 The displacement (s mm) of a particle from point P over time t seconds is given by $s = t^3 - 16t^2 + 63t$.

- **a** Find any times when the particle is at P.
- **b** When is the velocity of the particle 18 mm/s?
- **c** Find the displacement of the particle at these times.

21 Show that the function $f(x) = ax^3 + x$, where $a > 0$, never decreases.

22 Show that the tangents to the function $f(x) = \dfrac{3}{x}$ at the points $A\,(3, 1)$ and $B\,(-3, -1)$ are parallel.

23 Given $y = px^n$, where p is a constant, show that $x\dfrac{dy}{dx} = ny$.

Tangents and normals 6.05

Tangents to a curve

A tangent is a line, so we can use the formula $y = mx + c$ or $y - y_1 = m(x - x_1)$ to find its equation.

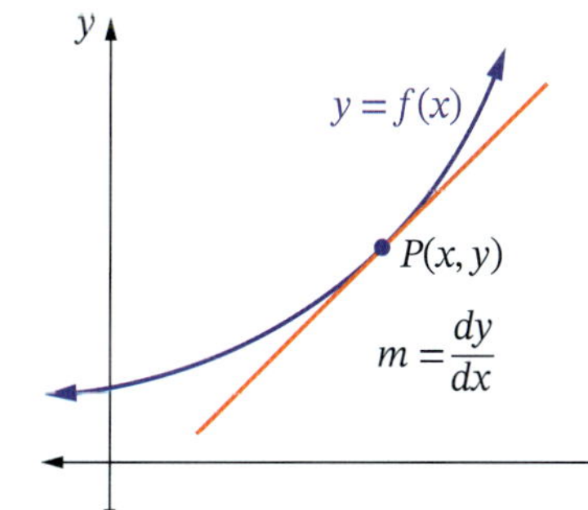

Videos
Equation of a tangent
Equations of tangents and normals

Worksheets
Tangents and normals
Slopes of curves
Tangents to a curve

Example 13

- **a** Find values of x for which the gradient of the tangent to the curve $y = 2x^3 - 6x^2 + 1$ is equal to 18.
- **b** Find the equation of the tangent to the curve $y = x^4 - 3x^3 + 7x - 2$ at the point $(2, 4)$.

Solution

a $\dfrac{dy}{dx} = 6x^2 - 12x$

Gradient is 18 so $\dfrac{dy}{dx} = 18$.

$$18 = 6x^2 - 12x$$
$$0 = 6x^2 - 12x - 18$$
$$x^2 - 2x - 3 = 0$$
$$(x - 3)(x + 1) = 0$$
$$\therefore x = 3, -1$$

b $\dfrac{dy}{dx} = 4x^3 - 9x^2 + 7$

At $(2, 4)$, $\dfrac{dy}{dx} = 4(2)^3 - 9(2)^2 + 7$

$$= 3$$

So the gradient of the tangent at $(2, 4)$ is 3.

Equation of the tangent:

$$y - y_1 = m(x - x_1)$$
$$y - 4 = 3(x - 2)$$
$$= 3x - 6$$
$$y = 3x - 2 \text{ (or } 3x - y - 2 = 0)$$

☐ Foundation ○ Mastery ⬡ Complex

Normals to a curve

The **normal** is a straight line **perpendicular** to the tangent at the same point of contact with the curve.

Remember the rule for perpendicular lines from Chapter 3, *Functions and graphs*:

Gradients of perpendicular lines

If 2 lines with gradients m_1 and m_2 are perpendicular, then $m_1m_2 = -1$ or $m_2 = -\frac{1}{m_1}$.

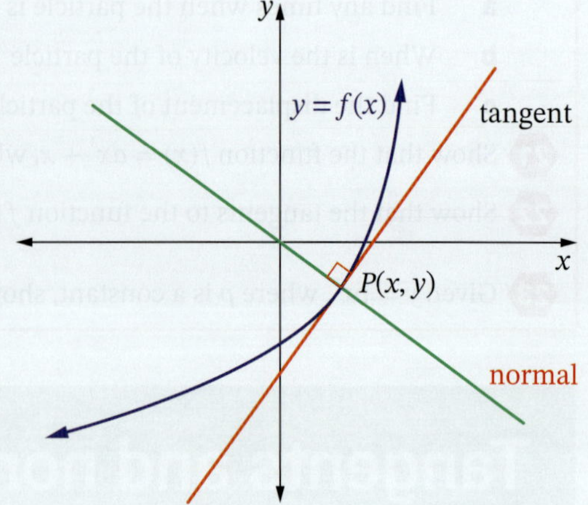

Example 14

a Find the gradient of the normal to the curve $y = 2x^2 - 3x + 5$ at the point where $x = 4$.

b Find the equation of the normal to the curve $y = x^3 + 3x^2 - 2x - 1$ at $(-1, 3)$.

Solution

a $\frac{dy}{dx} = 4x - 3$

When $x = 4$:

$$\frac{dy}{dx} = 4 \times 4 - 3$$
$$= 13$$

So $m_1 = 13$.

The normal is perpendicular to the tangent, so $m_1m_2 = -1$.

$$13m_2 = -1$$
$$m_2 = -\frac{1}{13}$$

So the gradient of the normal is $-\frac{1}{13}$.

b $\frac{dy}{dx} = 3x^2 + 6x - 2$

When $x = -1$:

$$\frac{dy}{dx} = 3(-1)^2 + 6(-1) - 2$$
$$= -5$$

So $m_1 = -5$.

The normal is perpendicular to the tangent, so $m_1m_2 = -1$.

$$-5m_2 = -1$$
$$m_2 = \frac{1}{5}$$

So the gradient of the normal is $\frac{1}{5}$.

Equation of the normal:

$$y - y_1 = m(x - x_1)$$
$$y - 3 = \frac{1}{5}(x - (-1))$$
$$5y - 15 = x + 1$$
$$x - 5y + 16 = 0$$

EXERCISE 6.05 Answers on p. 492

Tangents and normals

1 Find the gradient of the tangent to the curve:

- **a** $y = x^3 - 3x$ at the point where $x = 5$
- **b** $f(x) = x^2 + x - 4$ at the point $(-7, 38)$
- **c** $v = 2t^2 + 3t - 5$ at the point where $t = 2$
- **d** $Q = 3r^3 - 2r^2 + 8r - 4$ at the point where $r = 4$
- **e** $h = t^4 - 4t$ where $t = 0$
- **f** $f(t) = 3t^5 - 8t^3 + 5t$ at the point where $t = 2$.

2 Find the gradient of the normal to the curve:

- **a** $f(x) = 2x^3 + 2x - 1$ at the point where $x = -2$
- **b** $y = 3x^2 + 5x - 2$ at $(-5, 48)$
- **c** $f(x) = x^2 - 2x - 7$ at the point where $x = -9$
- **d** $y = x^3 + x^2 + 3x - 2$ at $(-4, -62)$
- **e** $V = h^3 - 4h + 9$ at $(2, 9)$
- **f** $g(x) = x^4 - 2x^2 + 5x - 3$ at the point where $x = -1$.

3 A function $f(x) = x^2 + 4x - 12$ has a tangent with a gradient of -6 at point P on the curve. Find the coordinates of P.

4 The tangent at point P on the curve $y = 4x^2 + 1$ is parallel to the x-axis. Find the coordinates of P.

5 Find the coordinates of point Q where the tangent to the curve $y = 5x^2 - 3x$ is parallel to the line $7x - y + 3 = 0$.

6 Find the coordinates of point S where the tangent to the curve $y = x^2 + 4x - 1$ is perpendicular to the line $4x + 2y + 7 = 0$.

7 Find the equation of the tangent to the curve:

- **a** $y = x^4 - 5x + 1$ at $(2, 7)$
- **b** $f(x) = 5x^3 - 3x^2 - 2x + 6$ at $(1, 6)$
- **c** $y = x^2 + 2x - 8$ at $(-3, -5)$
- **d** $y = 3x^3 + 1$ where $x = 2$
- **e** $v = 4t^4 - 7t^3 - 2$ where $t = 2$

8 The curve $y = 3x^2 - 4$ has a gradient of 6 at point A.

- **a** Find the coordinates of A.
- **b** Find the equation of the tangent to the curve at A.

9 Find the equation of the tangent to the curve $y = \dfrac{1}{x^3}$ at $\left(2, \dfrac{1}{8}\right)$.

10 Find the equation of the normal to the curve:

- **a** $f(x) = x^3 - 3x + 5$ at $(3, 23)$
- **b** $y = x^2 - 4x - 5$ at $(-2, 7)$
- **c** $f(x) = 7x - 2x^2$ where $x = 6$
- **d** $y = 7x^2 - 3x - 3$ at $(-3, 69)$
- **e** $y = x^4 - 2x^3 + 4x + 1$ where $x = 1$.

▷

☐ Foundation ○ Mastery ○ Complex

11 Find the equation of the tangent to $f(x) = 6\sqrt{x}$ at the point where $x = 9$.

12 Find the equation of the tangent to the curve $y = \frac{4}{x}$ at $\left(8, \frac{1}{2}\right)$.

13 If the gradient of the tangent to $y = \sqrt{x}$ is $\frac{1}{6}$ at point A, find the coordinates of A.

14 A tangent to the function $y = 2x^2 + 3x - 1$ is inclined at an angle of 135° to the x-axis in the positive direction. Find the equation of the tangent.

15 For the function $f(x) = x^2 - 3x$, find:

a the angle of inclination to the nearest minute that the tangent at each x-intercept makes with the x-axis

b the point where the normal to the curve at $(2, -2)$ cuts the function again.

16 For the function $y = x^2 - 2x$, find the point of intersection of the tangent to the curve at (3, 3) and the normal at $(-2, 8)$.

17 **a** Find the equation of the tangent to $y = x^3$ at the point $(2, 8)$.

b Find the area of the triangle between $(2, 8)$, the origin and the x-intercept of the tangent.

18 The tangent and normal to the curve $y = x^2 + 3x - 1$ at point $P(-1, -3)$ have x-intercepts at A and B respectively.

a Find the length of AB. **b** Find the area of $\triangle ABP$.

19 The tangent to the curve $y = x^2 + x - 3$ at $P(n, n^2 + n - 3)$ passes through the point $Q(-1, -12)$. Find the equations of the tangent to the curve at P for the 2 values of n that fit this information.

6.06 The chain rule

The **chain rule** is a method for differentiating composite functions. It is also called the **composite function rule** or the '**function of a function**' **rule**.

Worksheet
The chain rule

The chain rule

If a function y can be written as a composite function where $y = f(u(x))$, then:

$$\frac{dy}{dx} = \frac{dy}{du} \times \frac{du}{dx}$$

Video
The chain rule

Example 15

Differentiate:

a $y = (5x + 4)^7$ **b** $y = (3x^2 + 2x - 1)^9$ **c** $y = \sqrt{3 - x}$

Solution

a Let $u = 5x + 4$.

Then $\frac{du}{dx} = 5$.

$y = u^7$

$\therefore \frac{dy}{du} = 7u^6$

$$\frac{dy}{dx} = \frac{dy}{du} \times \frac{du}{dx}$$
$$= 7u^6 \times 5$$
$$= 35u^6$$
$$= 35(5x + 4)^6$$

Foundation Mastery Complex

b Let $u = 3x^2 + 2x - 1$.

Then $\frac{du}{dx} = 6x + 2$.

$y = u^9$

$\therefore \frac{dy}{du} = 9u^8$

$$\frac{dy}{dx} = \frac{dy}{du} \times \frac{du}{dx}$$
$$= 9u^8 \times (6x + 2)$$
$$= 9(3x^2 + 2x - 1)^8(6x + 2)$$
$$= 9(6x + 2)(3x^2 + 2x - 1)^8$$

c $y = \sqrt{3-x} = (3-x)^{\frac{1}{2}}$

Let $u = 3 - x$.

Then $\frac{du}{dx} = -1$.

$y = u^{\frac{1}{2}}$

$\therefore \frac{dy}{du} = \frac{1}{2}u^{-\frac{1}{2}}$

$$\frac{dy}{dx} = \frac{dy}{du} \times \frac{du}{dx}$$
$$= \frac{1}{2}u^{-\frac{1}{2}} \times (-1)$$
$$= -\frac{1}{2}(3-x)^{-\frac{1}{2}}$$
$$= -\frac{1}{2\sqrt{3-x}}$$

You might see a pattern when using the chain rule. The derivative of a composite function is the product of the derivatives of 2 functions. The chain rule can also be written this way:

The chain rule

If $h(x) = f(g(x))$, then $h'(x) = f'(g(x))\, g'(x)$.

We can use a special version of this when differentiating a power of a function $[f(x)]^n$:

The derivative of $[f(x)]^n$

$$\frac{d}{dx}[f(x)]^n = f'(x) \times n[f(x)]^{n-1}$$

Example 16

Differentiate:

a $y = (8x^3 - 1)^5$

b $y = \frac{1}{(6x+1)^2}$

Solution

a
$$\frac{dy}{dx} = f'(x) \times n[f(x)]^{n-1}$$
$$= 24x^2 \times 5(8x^3 - 1)^4$$
$$= 120x^2(8x^3 - 1)^4$$

b $y = \frac{1}{(6x+1)^2} = (6x+1)^{-2}$

$$\frac{dy}{dx} = f'(x) \times n[f(x)]^{n-1}$$
$$= 6 \times (-2)(6x + 1)^{-3}$$
$$= -12(6x + 1)^{-3}$$
$$= -\frac{12}{(6x+1)^3}$$

EXERCISE 6.06 Answers on p. 492

The chain rule

1 Differentiate:

a $y = (x + 3)^4$ **b** $y = (2x - 1)^3$ **c** $y = (5x^2 - 4)^7$

d $y = (8x + 3)^6$ **e** $y = (1 - x)^5$ **f** $y = 3(5x + 9)^9$

g $y = 2(x - 4)^2$ **h** $y = (x^6 - 2x^2 + 3)^6$ **i** $y = (3x - 1)^{\frac{1}{2}}$

j $y = (4 - x)^{-2}$ **k** $y = (x^2 - 9)^{-3}$ **l** $y = (5x + 4)^{\frac{1}{3}}$

m $y = \sqrt{3x + 4}$ **n** $y = \dfrac{1}{5x - 2}$ **o** $y = \dfrac{1}{(x^2 + 1)^4}$

p $y = \sqrt[3]{(7 - 3x)^2}$ **q** $y = \dfrac{5}{\sqrt{4 + x}}$ **r** $y = \dfrac{1}{2\sqrt{3x - 1}}$

s $y = \dfrac{1}{x^4 - 3x^3 + 3x}$ **t** $y = \sqrt[3]{(4x + 1)^4}$ **u** $y = \dfrac{1}{\sqrt[4]{(7 - x)^5}}$

2 Find the gradient of the tangent to the curve $y = (3x - 2)^3$ at the point $(1, 1)$.

3 If $f(x) = 2(x^2 - 3)^5$, evaluate $f'(2)$.

4 The curve $y = \sqrt{x - 3}$ has a tangent with gradient $\frac{1}{2}$ at point N. Find the coordinates of N.

5 For what values of x does the function $f(x) = \dfrac{1}{4x - 1}$ have $f'(x) = -\dfrac{4}{49}$?

6 Find the equation of the tangent to the curve $y = (2x - 1)^8$ at the point where $x = 1$.

7 Find the equation of the normal to the curve $y = (x^2 + 1)^4$ at $(1, 16)$.

8 Find the equation of **a** the tangent and **b** the normal to the curve $f(x) = \dfrac{1}{2x + 3}$ at the point where $x = -1$.

9 The displacement D mm of a particle over time t s is given by $D = (2t^2 - 5)^3$.

a When is the displacement negative?

b Explain why the velocity is never negative.

c Find the velocity when $t = 2$.

10 **a** Find the equation of the semicircle above the x-axis with centre $(0, 0)$ and radius 5.

b Differentiate the function of the semicircle.

c Show that the normal at $(4, 3)$ passes through the centre of the circle.

11 **a** Find the equation of the tangent to the curve $f(x) = (x - 1)^5$ at $A\ (2, 1)$.

b Find the distance from A to B, the x-intercept of the tangent.

c Find the area of the triangle ABC where C is the point on the x-axis directly below A.

12 Find the value of n if the tangent to $y = (2x + n)^2$ has a gradient of 4 at the point where $x = 2$.

13 Find any values of p if the tangent to the function $y = \dfrac{1}{x + p}$ has an angle of inclination of 135° at the point where $x = 3$.

14 If $y = f(x)$ is even, show that $f'(x)$ is an odd function.

15 If $y = g(x)$ is odd, show that $g'(x)$ is an even function.

16 Given $p(x) = f(g(x))$, where $f(x)$ is even and $g(x)$ is odd, show that $p'(x)$ is odd.

☐ Foundation ○ Mastery ○ Complex

The product rule 6.07

The **product rule** is a method for differentiating the product of 2 functions.

The product rule

If $y = uv$, where u and v are functions, then:

$$\frac{dy}{dx} = v\frac{du}{dx} + u\frac{dv}{dx}$$

We can also write this as $y' = u'v + v'u$.

Also if $h(x) = f(x)g(x)$, then $h'(x) = f'(x)g(x) + g'(x)f(x)$.

We can also write the product rule the other way round (differentiating v first), but differentiating u first helps us to remember the quotient rule, in the next section.

Example 17

Differentiate $y = (9x^3 - x + 4)(2x - 7)$.

Solution

$y = uv$, where $u = 9x^3 - x + 4$ and $v = 2x - 7$

$u' = 27x^2 - 1 \qquad v' = 2$

$$\begin{aligned} y' &= u'v + v'u \\ &= (27x^2 - 1)(2x - 7) + 2(9x^3 - x + 4) \\ &= 54x^3 - 189x^2 - 2x + 7 + 18x^3 - 2x + 8 \\ &= 72x^3 - 189x^2 - 4x + 15 \end{aligned}$$

Check this answer by expanding $y = (9x^3 - x + 4)(2x - 7)$ first, then differentiating to get the same answer.

We can use the product rule together with the chain rule.

Example 18

Differentiate:

a $y = 2x^5(5x + 3)^3$ **b** $y = (3x - 4)\sqrt{5 - 2x}$

Solution

a $y = uv$, where $u = 2x^5$ and $v = (5x + 3)^3$

$u' = 10x^4 \qquad v' = 5 \times 3(5x + 3)^2$ using chain rule

$\qquad\qquad\qquad\; = 15(5x + 3)^2$

$$\begin{aligned} y' &= u'v + v'u \\ &= 10x^4(5x + 3)^3 + 15(5x + 3)^2\, 2x^5 \\ &= 10x^4(5x + 3)^3 + 30x^5(5x + 3)^2 \\ &= 10x^4(5x + 3)^2[(5x + 3) + 3x] \\ &= 10x^4(5x + 3)^2(8x + 3) \end{aligned}$$

Factorising out $10x^4(5x + 3)^2$

b $y = uv$, where $u = 3x - 4$ and $v = \sqrt{5-2x} = (5-2x)^{\frac{1}{2}}$

$u' = 3$

$$v' = -2 \times \frac{1}{2}(5-2x)^{-\frac{1}{2}}$$ using chain rule

$$= -(5-2x)^{-\frac{1}{2}}$$

$$= -\frac{1}{(5-2x)^{\frac{1}{2}}}$$

$$= -\frac{1}{\sqrt{5-2x}}$$

$$y' = u'v + v'u$$

$$= 3 \times \sqrt{5-2x} + -\frac{1}{\sqrt{5-2x}}(3x-4)$$

$$= 3\sqrt{5-2x} - \frac{3x-4}{\sqrt{5-2x}}$$

$$= \frac{3\sqrt{5-2x} \times \sqrt{5-2x}}{\sqrt{5-2x}} - \frac{3x-4}{\sqrt{5-2x}}$$

$$= \frac{3(5-2x)}{\sqrt{5-2x}} - \frac{3x-4}{\sqrt{5-2x}}$$

$$= \frac{15-6x-3x+4}{\sqrt{5-2x}}$$

$$= \frac{19-9x}{\sqrt{5-2x}}$$

EXERCISE 6.07 Answers on p. 493

The product rule

1 Differentiate each function using the product rule.

a $y = x^3(2x+3)$ **b** $y = (3x-2)(2x+1)$ **c** $y = 3x(5x+7)$

d $y = 4x^4(3x^2-1)$ **e** $y = 2x(3x^4-x)$ **f** $y = x^2(x+1)^3$

g $y = 4x(3x-2)^5$ **h** $y = 3x^4(4-x)^3$ **i** $y = (x+1)(2x+5)^4$

2 Find the gradient of the tangent to the curve $y = 2x(3x-2)^4$ at (1, 2).

3 If $f(x) = (2x+3)(3x-1)^5$, evaluate $f'(1)$.

4 Find the exact gradient of the tangent to the curve $y = x\sqrt{2x+5}$ at the point where $x = 1$.

5 Find the gradient of the tangent where $t = 3$ given $x = (2t-5)(t+1)^3$.

6 Find the equation of the tangent to the curve $y = x^2(2x-1)^4$ at (1, 1).

7 Find the equation of the tangent to $h = (t+1)^2(t-1)^7$ at (2, 9).

8 Find exact values of x for which the gradient of the tangent to the curve $y = 2x(x+3)^2$ is 14.

9 **a** Given $f(x) = (4x-1)(3x+2)^2$, find the equation of the tangent at the point where $x = -1$.

b Find other points on the graph of $y = f(x)$ that are intersected by this tangent.

10 The volume V litres of water in a large aquarium in terms of its depth x m is given by $V = 3x(2x+1)^3$.

a What is the initial volume in the aquarium?

b What is the average rate of change in volume between a depth of 5 m and 10 m?

c Find the instantaneous rate of change in volume when the depth is:

i 3 m **ii** 12 m

☐ Foundation ○ Mastery ⬡ Complex

11 Find the equation of the normal to the curve $y = (2x+1)^2(x-3)^3$ at the point where $x = -1$.

12 Find the equation of the normal to the function $f(x) = (3x-2)\sqrt{2x+3}$ at the point $(11, 155)$.

13 Given the function $f(x) = x(x+2)^2$, evaluate $f'(x)$ for each x value and describe the behaviour of the graph of $f(x)$ at that value.

a $f'(-4)$ **b** $f'(3)$ **c** $f'(-2)$

14 **a** Differentiate $y = p(x-a)(x-b)$, where p, a and b are constants.

b Show that the tangents to the curve at the x-intercepts have gradients that are equal but opposite in sign.

c Show that the tangents intersect at the point where $x = \dfrac{a+b}{2}$.

The quotient rule 6.08

Video
The quotient rule 1

The **quotient rule** is a method for differentiating the ratio of 2 functions.

The quotient rule

If $y = \dfrac{u}{v}$ where u and v are functions, then:

$$\frac{dy}{dx} = \frac{v\dfrac{du}{dx} - u\dfrac{dv}{dx}}{v^2}$$

We can also write this as $y' = \dfrac{u'v - v'u}{v^2}$.

Also if $h(x) = \dfrac{f(x)}{g(x)}$, then $h'(x) = \dfrac{g(x)f'(x) - f(x)g'(x)}{[g(x)]^2}$.

Remember to differentiate u first in the numerator.

Worksheets
The quotient rule

Rules of differentiation

Mixed differentiation problems

Example 19

Differentiate:

a $y = \dfrac{3x-5}{5x+2}$ **b** $y = \dfrac{4x^3 - 5x + 2}{x^3 - 1}$

Solution

a $y = \dfrac{u}{v}$, where $u = 3x - 5$ and $v = 5x + 2$

$u' = 3$ $v' = 5$

$$y' = \frac{u'v - v'u}{v^2}$$

$$= \frac{3(5x+2) - 5(3x-5)}{(5x+2)^2}$$

$$= \frac{15x+6 \quad 15x+25}{(5x+2)^2}$$

$$= \frac{31}{(5x+2)^2}$$

Foundation Mastery Complex

b $y = \frac{u}{v}$, where $u = 4x^3 - 5x + 2$ and $v = x^3 - 1$

$u' = 12x^2 - 5$ $\quad v' = 3x^2$

$$y' = \frac{u'v - v'u}{v^2}$$

$$= \frac{(12x^2 - 5)(x^3 - 1) - 3x^2(4x^3 - 5x + 2)}{(x^3 - 1)^2}$$

$$= \frac{12x^5 - 12x^2 - 5x^3 + 5 - 12x^5 + 15x^3 - 6x^2}{(x^3 - 1)^2}$$

$$= \frac{10x^3 - 18x^2 + 5}{(x^3 - 1)^2}$$

EXERCISE 6.08 Answers on p. 493

The quotient rule

EXAMPLE 19

1 Differentiate each function using the quotient rule.

a $y = \frac{1}{2x - 1}$ **b** $y = \frac{3x}{x + 5}$ **c** $y = \frac{x^3}{x^2 - 4}$ **d** $y = \frac{x - 3}{5x + 1}$

e $y = \frac{x - 7}{x^2}$ **f** $y = \frac{5x + 4}{x + 3}$ **g** $y = \frac{x}{2x^2 - 1}$ **h** $y = \frac{x + 1}{3x^2 - 7}$

i $y = \frac{2x^2}{2x - 3}$ **j** $y = \frac{x^2 + 4}{x^2 - 5}$ **k** $y = \frac{x^3}{x + 4}$ **l** $y = \frac{x^3 + 2x - 1}{x + 3}$

m $y = \frac{2x}{(x + 5)^{\frac{1}{2}}}$ **n** $y = \frac{x - 1}{(7x + 2)^4}$ **o** $y = \frac{3x + 1}{\sqrt{x + 1}}$ **p** $y = \frac{\sqrt{x - 1}}{2x - 3}$

2 Find the gradient of the tangent to the curve $y = \frac{2x}{3x + 1}$ at $\left(1, \frac{1}{2}\right)$.

3 If $f(x) = \frac{4x + 5}{2x - 1}$, evaluate $f'(2)$.

4 Find values of x for which the gradient of the tangent to $y = \frac{4x - 1}{2x - 1}$ is -2.

5 Given $f(x) = \frac{2x}{x + 3}$, find x if $f'(x) = \frac{1}{6}$.

6 Find the equation of the tangent to the curve $y = \frac{x}{x + 2}$ at $\left(4, \frac{2}{3}\right)$.

7 Find the equation of the tangent to the curve $y = \frac{x^2 - 1}{x + 3}$ at $x = 2$.

8 Find the exact value of the x-coordinate of the stationary point on the curve $y = \frac{x^2 - 1}{x + 2}$.

□ Foundation ○ Mastery ⬠ Complex

9 Use an appropriate rule to differentiate each function.

a $y = 3(8x - 1)^5$

b $y = \dfrac{5x+3}{2x-1}$

c $y = x(2x + 9)^3$

d $f(t) = (t - 2)(3t + 5)^{-1}$

e $A(h) = \dfrac{4h-1}{\sqrt{h}}$

10 **a** Find any x values on the curve $y = \dfrac{x^3}{x-1}$ where the function is not differentiable.

b Find the equation of the normal to the curve at the point where $x = 2$.

Sample HSC problem Answers on p. 493

(7 marks) The displacement x cm of an object over time t s is given by $x = 5t^3 - 7t^2 + 3t - 1$.

a Find the average velocity of the object in the first second. 2 marks

b Find the instantaneous velocity:

i initially 2 marks

ii after 1 second. 1 mark

c Find when the velocity is 11 cm/s. 2 marks

☐ Foundation ○ Mastery ⬡ Complex

CHAPTER SUMMARY

This chapter, *Introduction to differentiation*, is the first calculus topic, a branch of mathematics that will dominate the rest of this HSC course. Calculus, and differentiation in particular, is the mathematics of measuring change.

Before you move on, consider what you learned in this chapter and revisit any sections that may have been unclear.

What you learned in this chapter...	Section	
Understand the derivative of a function as the gradient of the tangent to the curve and a measure of a rate of change	6.01	Rates of change
Draw graphs of gradient functions	6.01	Rates of change
Identify functions that are continuous and discontinuous, and their differentiability	6.04	Applications of differentiation
Differentiate from first principles	6.02	Differentiation from first principles
Differentiate functions involving powers of x, including terms with negative and fractional indices	6.03	The derivative of x^n
Use derivatives to find rates of change, including displacement and velocity, and the behaviour of graphs of functions	6.03	The derivative of x^n
Use derivatives to find gradients and equations of tangents and normals to curves	6.05	Tangents and normals
Find the derivative of composite functions, products and quotients of functions	6.06 6.07 6.08	The chain rule The product rule The quotient rule

To help master these techniques, make a summary of this topic. Use the chapter outline and the mind map below as a guide. Add your own words, symbols, diagrams, boxes and reminders. The summary should give you a 'whole picture' view of the topic and allow you to identify any weak areas to revisit in your revision.

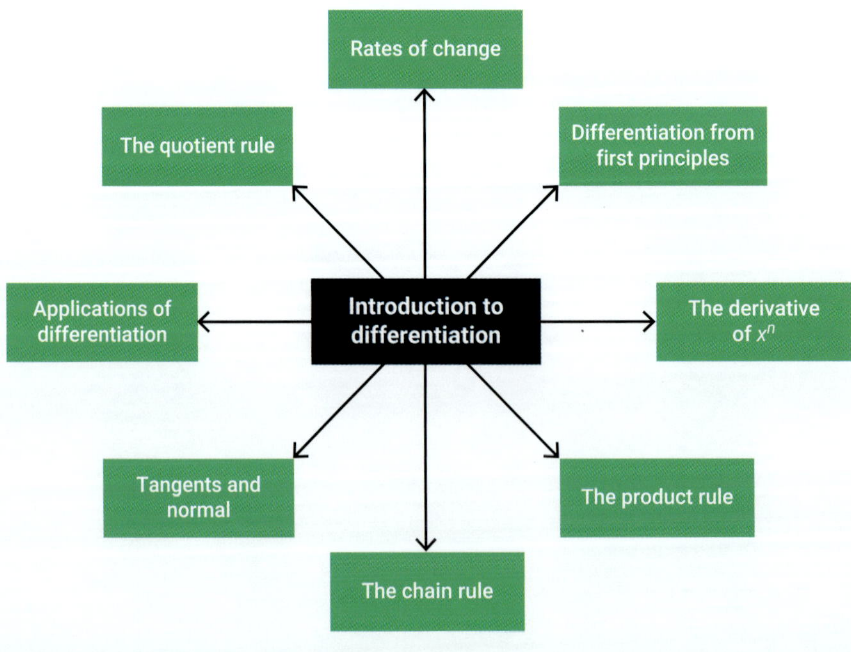

Test yourself Answers on p. 493 6

Puzzle
Derivatives find-a-word

For Questions **1** to **4**, select the correct answer **A**, **B**, **C** or **D**.

1 Find the derivative of $\frac{2}{3x^4}$. 6.03

A $\frac{8}{3x^5}$ **B** $-\frac{8}{3x^3}$ **C** $-\frac{8}{3x^5}$ **D** $\frac{8}{3x^3}$

2 Differentiate $3x(x^3 - 5)$. 6.07

A $4x^3$ **B** $12x^3 - 15$ **C** $9x^2$ **D** $3x^4 - 15x$

3 The derivative of $y = f(x)$ is given by: 6.02

A $\lim\limits_{h \to 0} \frac{f(x+h) - f(x)}{x - h}$ **B** $\lim\limits_{x \to 0} \frac{f(x+h) - f(x)}{x}$

C $\lim\limits_{h \to 0} \frac{f(x) - f(x+h)}{h}$ **D** $\lim\limits_{h \to 0} \frac{f(x+h) - f(x)}{h}$

4 Given $n(x) = f(x)\,g(x)$, where $f(3) = 2$, $f'(3) = -1$, $g(3) = 5$ and $g'(3) = -2$, find the gradient of the tangent to $n(x)$ at the point where $x = 3$. 6.07

A 10 **B** -12 **C** 1 **D** -9

5 Sketch the derivative function of each graph. 6.01

a

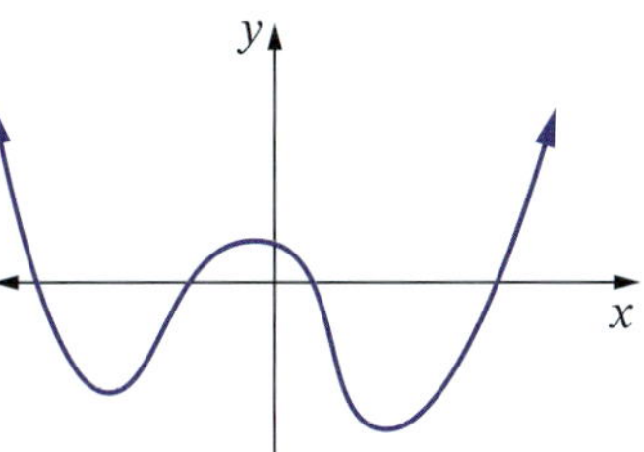

b

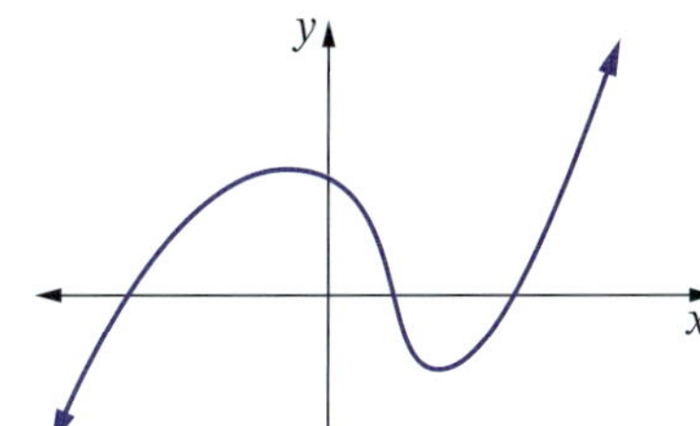

6 Differentiate $y = 5x^2 - 3x + 2$ from first principles. 6.02

7 Differentiate. 6.03, 6.06

a $y = 7x^6 - 3x^3 + x^2 - 8x - 4$

b $y = \frac{2}{(x+1)^4}$

c $y = x^2\sqrt{x}$ 6.08

d $y = (x^2 + 4x - 2)^9$

e $y = \frac{3x-2}{2x+1}$

f $y = x^3(3x + 1)^6$

8 Find $\frac{dv}{dt}$ if $v = 2t^2 - 3t - 4$. 6.02

9 Find the gradient of the tangent to the curve $y = x^3 + 3x^2 + x - 5$ at $(1, 0)$. 6.05

10 If $h = 60t - 3t^2$, find $\frac{dh}{dt}$ when $t = 3$. 6.03

Foundation Mastery Complex

6.01 **11** **a** Find the gradient of the secant on the graph of the function $f(x) = 3^x$ between:

i $x = 2$ and $x = 2.1$

ii $x = 2$ and $x = 2.01$

iii $x = 2$ and $x = 2.001$

b Estimate the gradient of the tangent at $x = 2$.

6.06-6.08 **12** Differentiate:

a $y = \frac{4}{x}$ **b** $f(x) = \sqrt[5]{x}$ **c** $f(x) = 2(4x + 9)^4$

d $y = (3x + 2)(x - 1)^3$ **e** $f(x) = \frac{x^3 - 3}{2x + 5}$

6.01 **13** Sketch the derivative function of this curve.

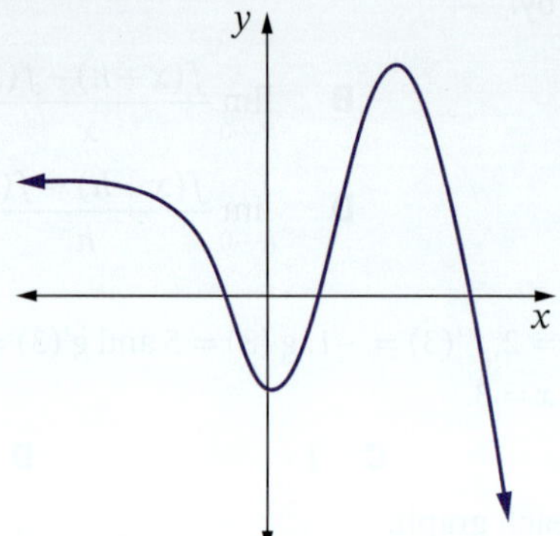

6.05 **14** Find the equation of the tangent to the curve $y = x^2 + 5x - 3$ at $(2, 11)$.

6.05 **15** Find the point on the curve $y = x^2 - x + 1$ at which the tangent has a gradient of 3.

6.03 **16** Find $\frac{dS}{dr}$ if $S = 4\pi r^2$.

6.01 **17** Find the gradient of the secant on the curve $f(x) = x^2 - 3x + 1$ between the points where $x = 1$ and $x = 1.1$.

6.04 **18** At which points on the curve $y = 2x^3 - 9x^2 - 60x + 3$ are the tangents horizontal?

6.05 **19** Find the equation of the tangent to the curve $y = x^2 + 2x - 5$ that is parallel to the line $y = 4x - 1$.

6.03 **20** **a** Differentiate $s = ut + \frac{1}{2}at^2$ with respect to t.

b Find the value of t for which $\frac{ds}{dt} = 5$, $u = 7$ and $a = -10$.

6.05 **21** Find the equation of the tangent to the curve $y = \frac{1}{3x}$ at the point where $x = \frac{1}{6}$.

□ Foundation ○ Mastery ⬡ Complex

22 A ball is thrown into the air and its height h metres over t seconds is given by $h = 4t - t^2$. 6.04

a Find the height of the ball:

i initially **ii** at 2 s

iii at 3 s **iv** at 3.5 s.

b Find the average rate of change of the height between:

i 1 and 2 seconds **ii** 2 and 3 seconds.

c Find the rate at which the ball is moving:

i initially **ii** at 2 s **iii** at 3 s.

23 If $f(x) = x^2 - 3x + 5$, find: 6.02

a $f(x + h)$ **b** $f(x + h) - f(x)$ **c** $f'(x)$

24 Given $f(x) = (4x - 3)^5$, find the value of: 6.06

a $f(1)$ **b** $f'(1)$

25 Find $f'(4)$ when $f(x) = (x - 3)^9$ 6.06

26 For what values of x is the graph of the function $y = x^3 - 9x^2 + 24x - 7$: 6.04

a increasing **b** decreasing **c** stationary?

27 Find any stationary points on the graph of the function $f(x) = 2x^3 - 3x^2 - 36x + 1$. 6.04

28 The displacement x in cm of a particle over time t seconds is given by $x = 5 + 6t - 3t^2$. 6.04

a Find the initial:

i displacement **ii** velocity.

b What is the velocity at 1 s?

c When is the particle at rest ($v = 0$)?

d What is the maximum displacement?

29 This graph shows the displacement of a particle. 6.04

a When is the particle at the origin?

b When is it at rest?

c When is it travelling at its greatest speed?

d Sketch the graph of its velocity.

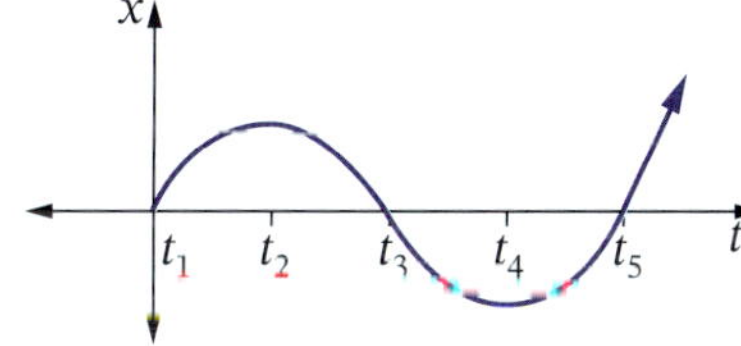

6 Challenge exercise

Answers on p. 494

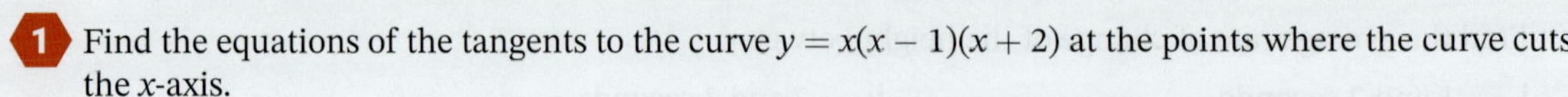

1 Find the equations of the tangents to the curve $y = x(x-1)(x+2)$ at the points where the curve cuts the x-axis.

2 **a** Find the points on the curve $y = x^3 - 6$ where the tangents are parallel to the line $y = 12x - 1$.

b Hence find the equations of the normals to the curve at those points.

3 The normal to the curve $y = x^2 + 1$ at the point where $x = 2$ cuts the curve again at point P. Find the coordinates of P.

4 The equation of the tangent to the curve $y = x^4 - nx^2 + 3x - 2$ at the point where $x = -2$ is given by $3x - y - 2 = 0$. Evaluate n.

5 **a** Find any points at which the graphed function is not differentiable.

b Sketch the derivative function for the graph.

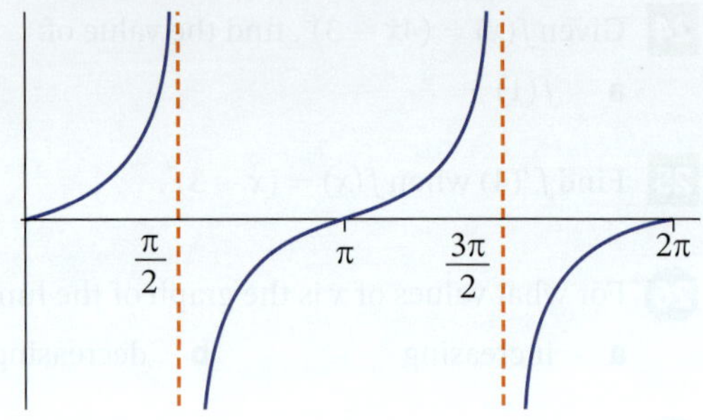

6 Find the exact gradient of the tangent to the curve $y = \sqrt{x^2 - 3}$ at the point where $x = 5$.

7 Find the equation of the normal to the curve $y = 3\sqrt{x+1}$ at the point where $x = 8$.

8 **a** Find the equations of the tangents to the parabola $y = 2x^2$ at the points where the line $6x - 8y + 1 = 0$ intersects with the parabola.

b Show that the tangents are perpendicular.

9 Find any x values of the function $f(x) = \dfrac{2}{x^3 - 8x^2 + 12x}$ where it is not differentiable.

10 For the function $f(x) = ax^2 + bx + c$, $f(2) = 4$, $f'(1) = 0$ and $f'(-3) = 8$. Evaluate a, b and c.

11 The displacement of a particle is given by $x = (t^3 + 1)^6$, where x is in metres and t is in seconds.

a Find the initial displacement and velocity of the particle.

b Show that the particle is never at the origin.

☐ Foundation ○ Mastery ⬡ Complex

EXPONENTIAL AND LOGARITHMIC FUNCTIONS

In this chapter you will study the definition and laws of logarithms and their relationship with the exponential and logarithmic functions.

You will meet a new irrational number, e, that has special properties, differentiate exponential functions, use logarithms and their laws solve exponential and logarithmic equations, and examine applications of exponential and logarithmic functions.

The photo above shows the algae in a salt lake in Melbourne's Westlake Park, turning the lake pink every summer.

Chapter outline

In this chapter you will:

- graph exponential and logarithmic functions
- understand and use Euler's number, e
- differentiate exponential functions
- convert between exponential and logarithmic forms using the definition of a logarithm
- identify and apply logarithm laws, including solving logarithmic equations
- solve exponential equations using logarithms

Videos (9):

7.01 The exponential curve

7.03 Differentiating exponential functions

7.04 Logarithms

7.05 Logarithm laws 1 • Logarithm laws 2

7.07 Exponential equations 3 • Exponential equations 1 • Exponential equations 2 • Exponential functions

Worksheets (11):

7.01 Graphing exponentials • Exponential functions • Translating exponential graphs

7.03 Differentiating exponential functions • Derivatives of exponential functions

7.04 Logarithms

7.05 Logarithms review

7.06 Plotting log functions

7.07 Logarithmic and exponential equations • Solving exponential equations • Using exponential models

Puzzles (7):

7.04 Logarithms 1 • Logarithms 2

7.05 Logarithm laws

7.06 Logarithmic graphs match-up • Exponential and log graphs match-up

7.07 Logarithms - Solving equations 1 • Logarithms - Solving equations 2

Nelson MindTap

To access resources above, visit **cengage.com.au/nelsonmindtap**

Terminology

base	change of base	Euler's number	exponential function
logarithm	logarithm laws	logarithmic function	

Exponential functions

7.01

An **exponential function** is in the form $y = a^x$, where $a > 0$ and $a \neq 1$.

Video
The exponential curve

Worksheets
Graphing exponentials
Exponential functions
Translating exponential graphs

Example 1

Sketch the graph of the function $y = 5^x$ and state its domain and range.

Solution

Complete a table of values for $y = 5^x$.

x	−3	−2	−1	0	1	2	3
y	$\frac{1}{125}$	$\frac{1}{25}$	$\frac{1}{5}$	1	5	25	125

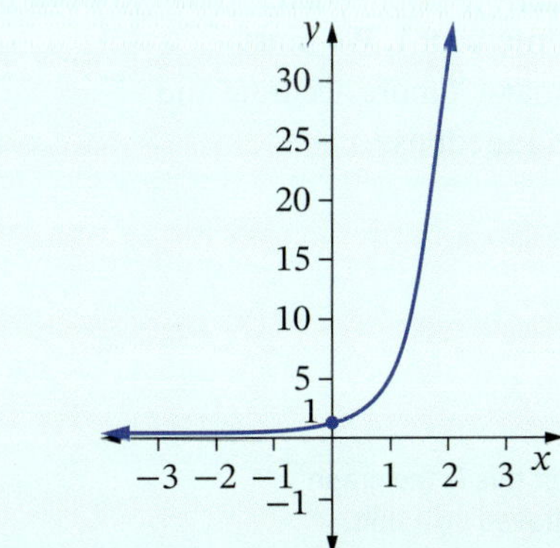

The values of y are increasing by a factor of 5 each time.

Notice that a^x is always positive. So there is no x-intercept and $y > 0$.

For the y-intercept, when $x = 0$, $y = 5^0 = 1$.

The y-intercept is 1.

From the graph, the domain is all real x and the range is $y > 0$.

Investigation

The value of a in $y = a^x$

Notice that the exponential function $y = a^x$ is only defined for $a > 0$ and $a \neq 1$.

1 Suppose $a = 0$. What would the function $y = 0^x$ look like? Try completing a table of values or use technology to sketch the graph. Is the function defined for positive values of x, negative values of x or when $x = 0$? What if x is a fraction?

2 Suppose $a = 1$. What would the function $y = 1^x$ look like? Try completing a table of values or use technology to sketch the graph. Is the function defined for positive values of x, negative values of x or when $x = 0$? What if x is a fraction?

3 Suppose $0 < a < 1$. Graph $y = \left(\frac{1}{3}\right)^x$. How is this graph related to the graph of $y = 3^x$?
(Hint: Rewrite $\left(\frac{1}{3}\right)^x$ as a power of 3 and see Example **4b** on page 291).

4 Suppose $a < 0$. What would the function $y = (-2)^x$ look like?

5 For $y = 0^x$, $y = 1^x$ and $y = (-2)^x$:

a is it possible to graph these functions at all?

b are there any discontinuities on the graphs?

c do they have a domain and range?

The exponential function $y = a^x$

- Domain $(-\infty, \infty)$, range $(0, \infty)$.
- The y-intercept $(x = 0)$ is always 1 because $a^0 = 1$.
- The graph is always above the x-axis and there is no x-intercept $(y = 0)$ because $a^x > 0$ for all values of x.
- The x-axis is an **asymptote**.
- $y = a^x$ is a monotonically increasing function (unless a is a fraction, $0 < a < 1$, when it is monotonically decreasing).

Example 2

Sketch the graph of:

a $f(x) = 3^x$

b $y = 2^x + 1$

Solution

a The curve is above the x-axis with y-intercept 1. We must show another point on the curve to make it more accurate and distinguish it from other exponential functions.

$f(1) = 3^1 = 3$

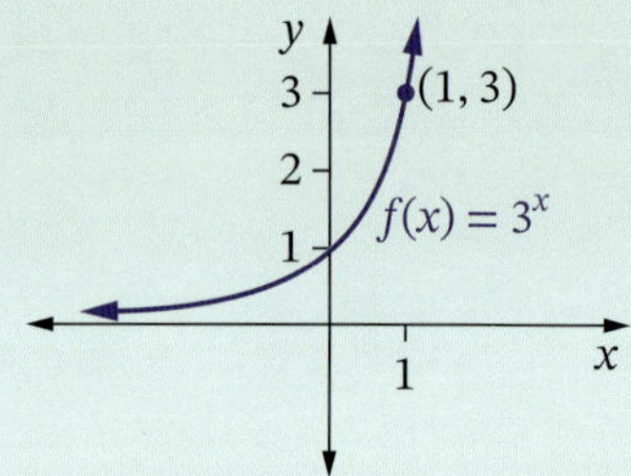

b For y-intercept, $x = 0$:

$$y = 2^0 + 1$$
$$= 1 + 1$$
$$= 2$$

Notice that this is the graph of $y = 2^x$ moved up 1 unit.

Find another point, for example, when $x = -1$:

$$y = 2^{-1} + 1$$
$$= \frac{1}{2} + 1$$
$$= 1\frac{1}{2}$$

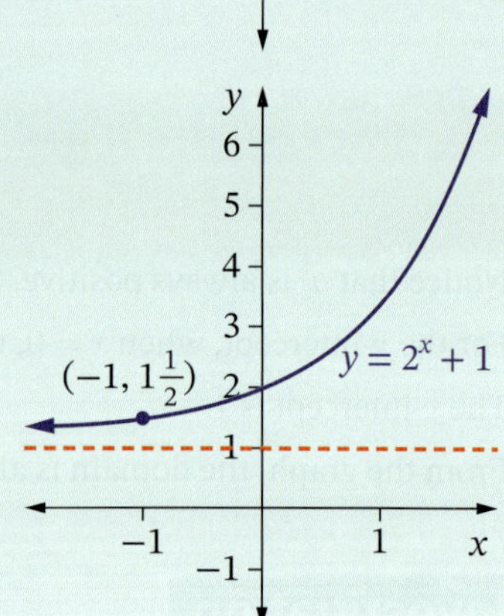

Example 3

Sketch the graph of:

a $f(x) = 3(4^x)$

b $y = 2^{x+1}$

Solution

a The values of $f(x) = 3(4^x)$ will be 3 times greater/higher than 4^x so its curve will be steeper.

$$f(-1) = 3(4^{-1}) = 0.75$$
$$f(0) = 3(4^0) = 3$$
$$f(1) = 3(4^1) = 12$$
$$f(2) = 3(4^2) = 48$$

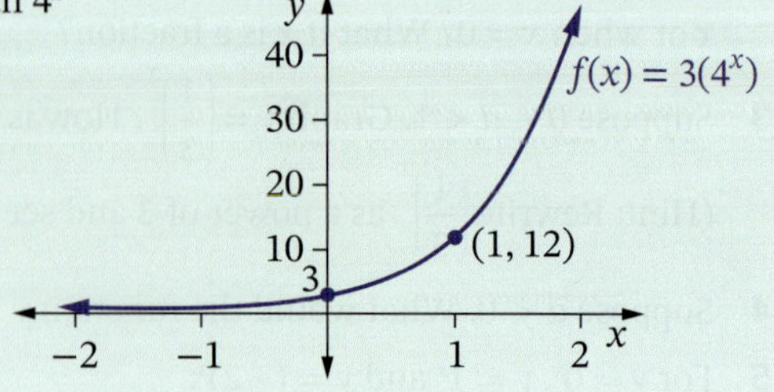

b $x = -1, y = 2^{-1+1} = 1$

$x = 0, \quad y = 2^{0+1} = 2$

$x = 1, \quad y = 2^{1+1} = 4$

$x = 1, \quad y = 2^{2+1} = 8$

This is the graph of $y = 2^x$ shifted 1 unit left.

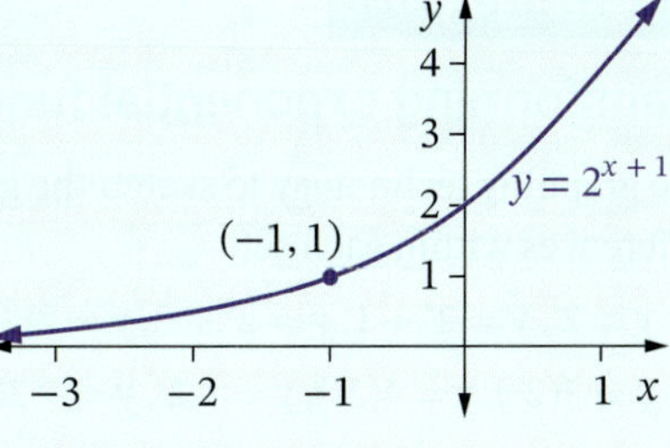

When we know the shape and properties of the exponential function, we can draw the graph using just one or 2 points to make it more accurate.

The exponential function $y = ka^x$

- The k value affects the height of the curve compared with $y = a^x$.
- It is either monotonically increasing or decreasing depending upon the sign of k and the value of a.
- The y-intercept ($x = 0$) is always k because $a^0 = 1$.
- The x-axis is an asymptote.

For $k > 0$:

- Domain $(-\infty, \infty)$, range $(0, \infty)$.
- The graph is always above the x-axis and there is no x-intercept because $a^x > 0$ for all x.
- As $x \to \infty, y \to \infty$ if $a > 1$.
- As $x \to -\infty, y \to 0$ if $a > 1$.

For $k < 0$:

- Domain $(-\infty, \infty)$, range $(-\infty, 0)$.
- The graph is always below the x-axis and there is no x-intercept because $a^x > 0$ for all x.
- As $x \to \infty, y \to -\infty$ if $a > 1$.
- As $x \to -\infty, y \to 0$ if $a > 1$.

Example 4

Given $f(x) = 3^x$, sketch the graph of:

a $y = -3^x$

b $y = 3^{-x}$

Solution

a Given $f(x) = 3^x$, then $y = -f(x) = -3^x$. This means that the y values in $f(x) = 3^x$ change from positive to negative, which is a reflection of $f(x)$ in the x-axis.

Note: -3^x means $-(3^x)$, *not* $(-3)^x$.

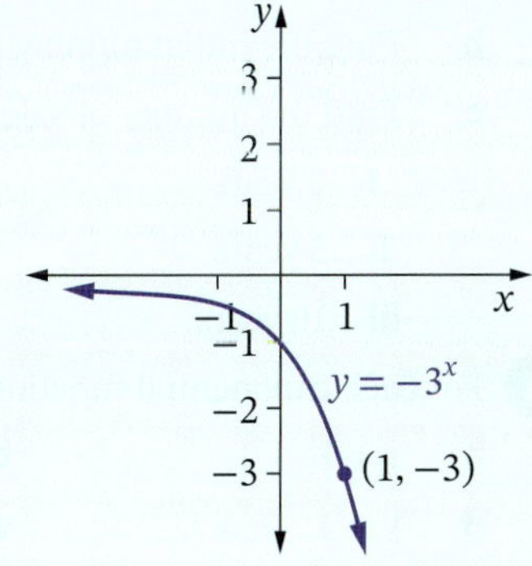

b Given $f(x) = 3^x$, then $y = f(-x) = 3^{-x}$. This means that the x values in $f(x) = 3^x$ change sign, from negative to positive and the other way round, which is a reflection of $f(x)$ in the y-axis.

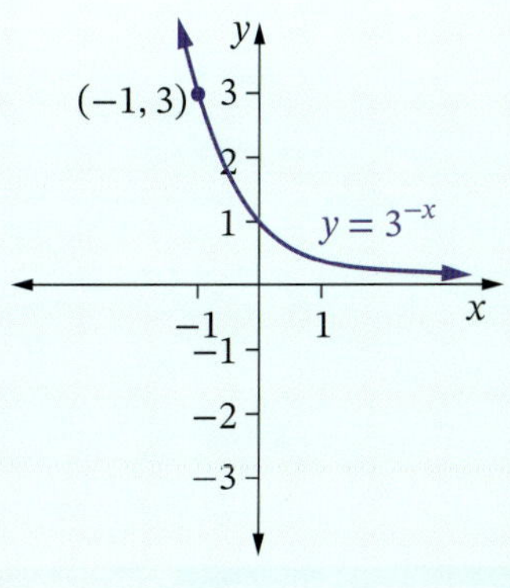

Investigation

Transforming exponential functions

Use graphing technology to sketch the graphs of the exponential functions below. Look for similarities and differences within each set.

1 $y = 2^x, y = 2^x + 1, y = 2^x + 3, y = 2^x - 5$

2 $y = 3(2^x), y = 4(2^x), y = -2^x, y = -3(2^x)$

3 $y = 3(2^x) + 1, y = 4(2^x) + 3, y = -2^x + 1, y = -3(2^x) - 3$

4 $y = 2^{x+1}, y = 2^{x+2}, y = 2^{x-1}, y = 2^{x-3}, y = 2^{-x}$

5 $y = \left(\frac{1}{2}\right)^{-x}, y = 2\left(\frac{1}{2}\right)^{-x}, y = \left(\frac{1}{2}\right)^{-x}, y = -3\left(\frac{1}{2}\right)^{-x}, y = \left(\frac{1}{2}\right)^{-x-1}$

Can you predict what the graph of $y = 3^{2x} - 5$ would look like?

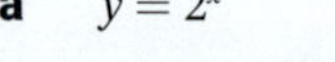

EXERCISE 7.01 Answers on p. 494

Exponential functions

EXAMPLES 1–3

1 Sketch each exponential function, with or without using graphing technology.

a $y = 2^x$ **b** $y = 4^x$ **c** $f(x) = 3^x + 2$ **d** $y = 2^x - 1$

e $f(x) = 3(2^x)$ **f** $y = 4^{x+1}$ **g** $y = 3(4^{2x}) - 1$ **h** $f(x) = -2^x$

i $y = 2(4^{-x})$ **j** $f(x) = -3(5^{-x}) + 4$

2 State the domain and range of each function.

a $f(x) = 2^x$ **b** $y = 3^x + 5$ **c** $f(x) = 10^{-x}$ **d** $f(x) = -5^x + 1$

3 Given $f(x) = 2^x$ and $g(x) = 3x - 4$, find:

a $f(g(x))$ **b** $g(f(x))$.

EXAMPLE 4

4 **a** Sketch the graph of $f(x) = 4(3^x) + 1$.

b Sketch the graph of:

i $y = 4(3^{-x}) + 1$ **ii** $y = -[4(3^x) + 1]$ **iii** $y = -[4(3^{-x}) + 1]$.

5 Sales numbers, N, of a new solar battery are growing over t years according to the formula $N = 450(3^{0.9t})$.

a Draw a graph of this function.

b Find the initial number of sales when $t = 0$.

c Find the number of sales after:

i 3 years

ii 5 years

iii 10 years.

6 For each exponential function, find the domain, range, y-intercept and equation of the asymptote.

a $y = 7^x - 3$ **b** $f(x) = -3^{2x}$ **c** $y = 5(2^x) + 1$

d $y = 4^{-x} - 3$ **e** $f(x) = -2^{2x} + 3$

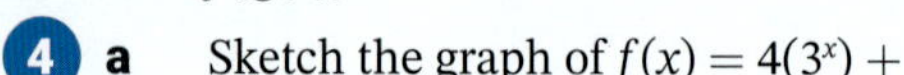

Euler's number, e

The gradient (derivative) function of an exponential function is interesting. Notice that the gradient of an exponential function is always increasing, and increases at an increasing rate.

If you sketch the derivative function of an exponential function, then it too is an exponential function. Drawn below are the graphs of the derivative functions (in blue) of $y = 2^x$ and $y = 3^x$ (in red) together with their equations.

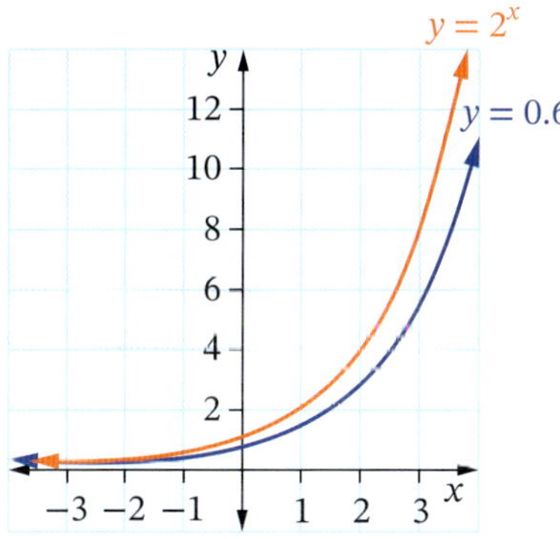

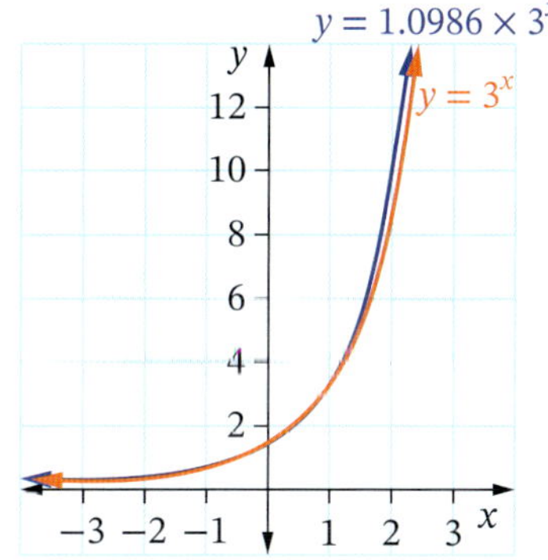

Notice that the graph of the derivative function is a constant multiple of itself, that is, the derivative of $y = a^x$ is $y = ka^x$, where k is a constant. Also notice that the graph of the derivative of $y = 3^x$ is very close to the graph of $y = 3^x$, that is, the value of k is close to 1.

Technology

When the derivative of $y = a^x$ is itself

1. Use graphing technology to graph $y = a^x$ and its derivative function $y = ka^x$ for different values of a between 2 and 3. The 2 graphs should be similar.
2. Find a decimal value of a where the derivative function is the same as the original function, that is, where $k = 1$ and the derivative of $y = a^x$ is itself $y = a^x$.
3. Can you find the equation of the derivative function? What do you notice?
4. If the derivative of $y = a^x$ is itself, then when $x = 0$, not only is $y = 1$ (the y-intercept), but the gradient of the function at $(0, 1)$ should also be 1. Check that the gradient of the tangent at $(0, 1)$ is 1.

We can find a number close to 3 that gives exactly the same derivative function as the original graph. This number is approximately 2.71828, and is called **Euler's number**, e.

Euler's number

$$e \approx 2.71828$$

Like π, the number e is irrational.

Did you know?

Leonhard Euler

Euler's number, e, is a **transcendental** number, which is an irrational number that is not a surd. This was proven by a French mathematician, **Charles Hermite**, in 1873. The Swiss mathematician **Leonhard Euler** (1707–1783) gave e its symbol, and he gave an approximation of e to 23 decimal places. Now e has been calculated to over a trillion decimal places.

Euler gave mathematics much of its important notation. He popularised π as the standard notation for pi and used i for the square root of -1. He also introduced the symbol Σ for sums and $f(x)$ notation for functions.

Example 5

Sketch the graph of the exponential function $y = e^x$.

Solution

Use e^x on your calculator to draw up a table of values. For example, to calculate e^{-3}:

Casio scientific	Sharp scientific
SHIFT e^x (−) 3 =	2ndF e^x +/− 3 =

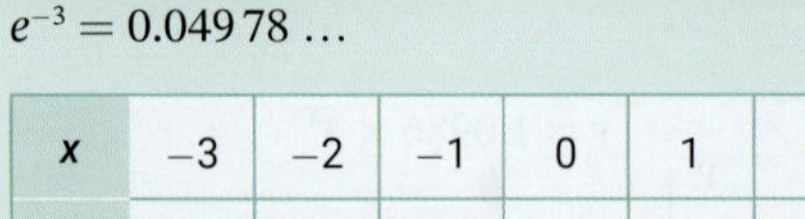

The value of e, 2.718..., can be found by evaluating e^1 on your calculator.

$e^{-3} = 0.049\,78\ldots$

x	−3	−2	−1	0	1	2	3
y	0.05	0.1	0.4	1	2.7	7.4	20.1

(rounded figures)

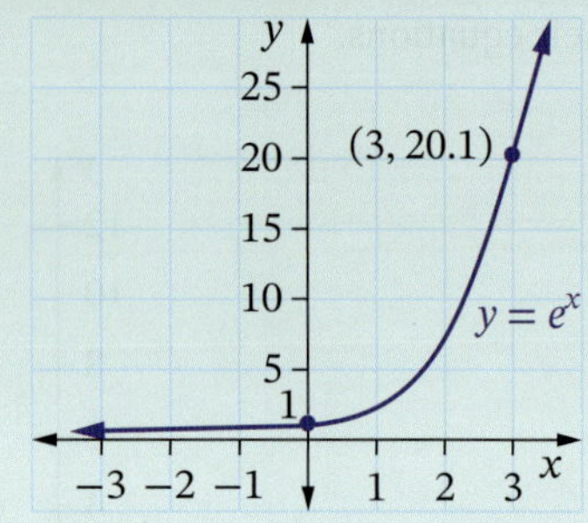

Example 6

The salmon population in a river over time can be described by the exponential function $P = 200e^{0.3t}$, where t is time in years.

a Find the population after 3 years.

b Draw the graph of the population.

Solution

a $P = 200e^{0.3t}$

When $t = 3$:

$P = 200e^{0.3 \times 3}$

$= 491.9206\ldots$

≈ 492

So after 3 years there are 492 salmon.

b The graph is an exponential curve. Finding some points will help us graph it accurately.

When $t = 0$: $P = 200e^{0.3 \times 0}$

$= 200$ This is also the P-intercept.

When $t = 1$: $P = 200e^{0.3 \times 1}$

$= 269.9717\ldots$

≈ 270

When $t = 2$: $P = 200e^{0.3 \times 2}$

$= 364.4237\ldots$

≈ 364

We already know $P \approx 492$ when $t = 3$.

Time $t \geq 0$, so don't sketch the curve for negative values of t.

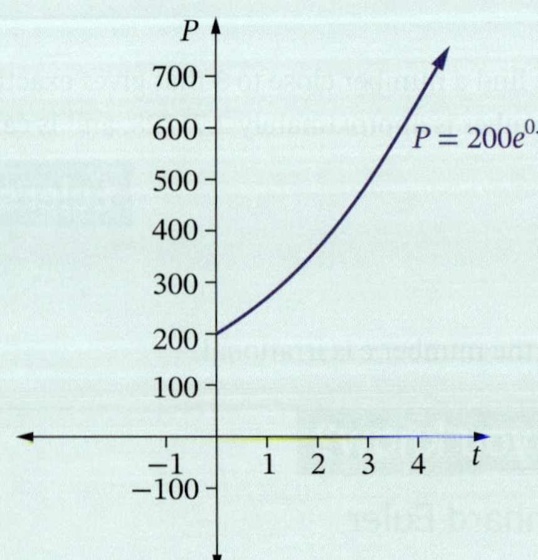

EXERCISE 7.02 Answers on p. 496

Euler's number, e

EXAMPLE 5

1. Sketch the graph of $f(x) = e^x$.
2. Evaluate, correct to 2 decimal places:

   a $e^{1.5}$ **b** e^{-2} **c** $2e^{0.3}$ **d** $\frac{1}{e^3}$ **e** $-3e^{-3.1}$

3. Sketch each exponential function.

   a $y = 2e^x$ **b** $f(x) = e^x + 1$ **c** $y = -e^x$

   d $y = e^{-x}$ **e** $y = 2e^{x-2}$

4. State the domain and range of $f(x) = e^x - 2$.
5. If $f(x) = e^x$ and $g(x) = x^3 + 3$, find:

   a $f(g(x))$ **b** $g(f(x))$.

6. The volume V of a metal in mm³ expands as it is heated over time according to the formula $V = 25e^{0.7t}$, where t is in minutes.

   a Sketch the graph of $V = 25e^{0.7t}$.

   b Find the volume of the metal at:

   i 3 minutes **ii** 8 minutes.

   c Is this formula a good model for the rise in volume? Why?

7. The mass of a radioactive substance in g is given by $M = 150e^{-0.014t}$, where t is in years. Find the mass after:

   a 10 years **b** 50 years **c** 250 years.

8. The number of koalas in a forest is declining according to the formula $N = 873e^{-0.078t}$, where t is the time in years.

   a Sketch a graph showing this decline in numbers of koalas for the first 10 years.

   b Find the number of koalas:

   i initially

   ii after 5 years

   iii after 10 years.

Photo courtesy of Margaret Grove

9. An object is cooling down according to the exponential function $T = 23 + 125e^{-0.06t}$, where T is the temperature in °C and t is time in minutes.

   a Find the initial temperature.

   b Find the temperature at:

   i 2 minutes **ii** 5 minutes **iii** 10 minutes **iv** 2 hours.

   c What temperature is the object tending towards? Can you explain why?

10. A population is growing exponentially. If the initial population is 20 000 and after 5 years the population is 80 000, draw a graph showing this information.
11. The temperature of a piece of iron in a smelter is 1000°C and it is cooling down exponentially. After 10 minutes, the temperature is 650°C. Draw a graph showing this information.

□ Foundation ○ Mastery ○ Complex

7.03 The derivative of exponential functions

Worksheets
Differentiating exponential functions

Derivatives of exponential functions

Euler's number, e, is the special number such that the derivative of $y = e^x$ is itself. To distinguish it from other exponential functions, $y = e^x$ is sometimes called the **natural exponential function**.

The derivative of e^x

$$\frac{d}{dx}(e^x) = e^x$$

Example 7

a Differentiate $y = e^x - 5x^2$.

b Find the equation of the tangent to the curve $y = e^x$ at the point $(1, e)$.

Solution

a $\frac{dy}{dx} = e^x - 10x$

b Gradient of the tangent:

$$\frac{dy}{dx} = e^x$$

At $(1, e)$:

$$\begin{aligned}\frac{dy}{dx} &= e^1\\ &= e\end{aligned}$$

So, $m = e$.

Remember, e = 2.718... is a number.

Equation:

$$\begin{aligned}y - y_1 &= m(x - x_1)\\ y - e &= e(x - 1)\\ &= ex - e\\ y &= ex\end{aligned}$$

The rule for differentiating $kf(x)$ works with the rule for e^x as well.

The derivative of ke^x

$$\frac{d}{dx}(ke^x) = ke^x$$

Example 8

a Differentiate $y = 5e^x$.

b Find the gradient of the normal to the curve $y = 3e^x$ at the point $(0, 3)$.

Solution

a $\frac{dy}{dx} = 5e^x$

b Gradient of tangent:

$$\frac{dy}{dx} = 3e^x$$

At $(0, 3)$:

$$\begin{aligned}\frac{dy}{dx} &= 3e^0\\ &= 3 \text{ since } e^0 = 1\end{aligned}$$

So $m_1 = 3$.

For normal:

$$\begin{aligned}m_1 m_2 &= -1\\ 3m_2 &= -1\\ m_2 &= -\frac{1}{3}\end{aligned}$$

So the gradient of the normal at $(0, 3)$ is $-\frac{1}{3}$.

We can also use the chain rule, product rule and quotient rule with exponential functions.

Example 9

Differentiate:

a $y = e^{9x}$

b $y = e^{-5x}$

Solution

a Let $u = 9x$.

Then $\frac{du}{dx} = 9$.

$y = e^u$

$\frac{dy}{du} = e^u$

$\frac{dy}{dx} = \frac{dy}{du} \times \frac{du}{dx}$

$= e^u \times 9$

$= 9e^u$

$= 9e^{9x}$

b Let $u = -5x$.

Then $\frac{du}{dx} = -5$.

$y = e^u$

$\frac{dy}{du} = e^u$

$\frac{dy}{dx} = \frac{dy}{du} \times \frac{du}{dx}$

$= e^u \times (-5)$

$= -5e^u$

$= -5e^{-5x}$

The derivative of e^{ax}

$$\frac{d}{dx}(e^{ax}) = ae^{ax}$$

Example 10

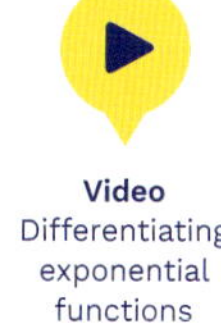

Video
Differentiating exponential functions

Differentiate:

a $y = (1 + e^x)^3$

b $y = \frac{2x+3}{e^x}$

Solution

a $\frac{dy}{dx} = 3(1 + e^x)^2 \times e^x$

$= 3e^x(1 + e^x)^2$

b $\frac{dy}{dx} = \frac{u'v - v'u}{v^2}$

$= \frac{2e^x - e^x(2x+3)}{(e^x)^2}$

$= \frac{2e^x - 2xe^x - 3e^x}{e^{2x}}$

$= \frac{-e^x - 2xe^x}{e^{2x}}$

$= \frac{-e^x(1+2x)}{e^{2x}}$

$= \frac{-(1+2x)}{e^x}$

EXERCISE 7.03 Answers on p. 498

The derivative of exponential functions

EXAMPLES 7–10

1 Differentiate:

a $y = 9e^x$ **b** $y = -e^x$ **c** $y = e^x + x^2$

d $y = 2x^3 - 3x^2 + 5x - e^x$ **e** $y = (e^x + 1)^3$ **f** $y = (e^x + 5)^7$

g $y = (2e^x - 3)^2$ **h** $y = xe^x$ **i** $y = \dfrac{e^x}{x}$

j $y = x^2e^x$ **k** $y = e^x(2x + 1)$ **l** $y = \dfrac{e^x}{7x-3}$

m $y = \dfrac{5x}{e^x}$ **n** $y = e^{2x}$ **o** $y = e^{-x}$

p $y = 2e^{3x}$ **q** $y = -e^{7x}$ **r** $y = -3e^{2x} + x^2$

s $y = e^{2x} - e^{-2x}$ **t** $y = 5e^{-x} - 3x + 2$ **u** $y = xe^{4x}$

v $y = \dfrac{2e^{3x}-3}{x+1}$ **w** $y = (9e^{3x} + 2)^5$

2 If $f(x) = x^3 + 3x - e^x$, find $f'(1)$ in terms of e.

3 Find the exact gradient of the tangent to the curve $y = e^x$ at the point $(1, e)$.

'Exact' here means to leave your answers in terms of e (not as a rounded decimal).

4 Find the exact gradient of the normal to the curve $y = e^{2x}$ at the point where $x = 5$.

5 Find the gradient of the tangent to the curve $y = 4e^x$, correct to 2 decimal places, at the point where $x = 1.6$.

6 Find the equation of the tangent to the curve $y = -e^x$ at the point $(1, -e)$.

7 Find the equation of the normal to the curve $y = e^{-x}$ at the point where $x = 3$.

8 A population P of insects over time t weeks is given by $P = 3e^{1.4t} + 12\,569$.

a What is the initial population?

b Find the rate of change in the number of insects after:

i 3 weeks **ii** 7 weeks.

9 The displacement of a particle over time t seconds is given by $x = 2e^{4t}$ metres.

a What is the initial displacement?

b What is the exact velocity after 10 s?

10 The displacement of an object in cm over time t seconds is given by $x = 6e^{-0.34t} - 5$. Find:

a the initial displacement

b the initial velocity

c the displacement after 4 s

d the velocity after 9 s.

11 The volume V of a balloon in mm³ as it expands over time t seconds is given by $V = 3e^{0.8t}$.

a Find the volume of the balloon at:

i 3 s **ii** 5 s.

b Find the rate at which the volume is increasing at:

i 3 s **ii** 5 s.

□ Foundation ○ Mastery ⬡ Complex

12 The population of a city is changing over t years according to the formula $P = 34\,500e^{0.025t}$.

a Find (to the nearest whole number) the population after:

i 5 years **ii** 10 years.

b Find the rate at which the population is changing after:

i 5 years **ii** 10 years.

13 The depth of water (in metres) in a dam is decreasing over t months according to the formula $D = 3e^{-0.017t}$.

a Find, correct to 2 decimal places, the depth after:

i 1 month **ii** 2 months **iii** 3 months.

b Find, correct to 3 decimal places, the rate at which the depth is changing after:

i 1 month **ii** 2 months **iii** 3 months.

Logarithms 7.04

The **logarithm** of a positive number is the **power** to which a **base**, a, must be raised in order to produce that number. For example, $\log_2 8 = 3$ because $2^3 = 8$.

We say the logarithm of 8 to the base 2 is 3 (the logarithm is the power of the base number).

Just as the exponential function $y = a^x$ is defined for positive bases only ($a > 0, a \neq 1$), logarithms are also defined for $a > 0, a \neq 1$.

Video
Logarithms

Worksheet
Logarithms

Puzzles
Logarithms 1

Logarithms 2

Logarithms

If $y = a^x$ then $x = \log_a y$ $\quad (a > 0, a \neq 1, y > 0)$

Logarithms are related to exponentials (powers) in the same way square roots are related to squares. One undoes the other.

Example 11

a Write $\log_4 x = 3$ in index form and solve for x.

b Write $5^2 = 25$ in logarithm form.

c Solve $\log_x 36 = 2$.

d Evaluate $\log_3 81$.

e Find the value of $\log_2 \frac{1}{4}$.

Solution

a $\log_4 x = 3$ means $x = 4^3$

So $x = 64$.

b $y = a^x$ means $\log_a y = x$

So $25 = 5^2$ means $\log_5 25 = 2$.

c $\log_x 36 = 2$ means $36 = x^2$, where $x > 0$

$x = \sqrt{36}$

$= 6$

Foundation Mastery Complex

d $\log_3 81 = x$ means $81 = 3^x$

Solving $3^x = 81$:

$$3^x = 3^4$$

$\therefore x = 4$

So $\log_3 81 = 4$.

e Let $\log_2\left(\frac{1}{4}\right) = x$.

Then $2^x = \frac{1}{4}$

$$= \frac{1}{2^2}$$

$$= 2^{-2}$$

$\therefore x = -2$

So $\log_2\left(\frac{1}{4}\right) = -2$.

Properties of logarithms

Example 12

Simplify:

a $\log_8 1$ **b** $\log_8 8$ **c** $\log_8 8^3$

d $\log_a a^x$ **e** $3^{\log_3 7}$ **f** $a^{\log_a x}$

Solution

a $\log_8 1 = 0$ because $8^0 = 1$.

b $\log_8 8 = 1$ because $8^1 = 8$.

c $\log_8 8^3 = 3$ because $8^3 = 8^3$.

d $\log_a a^x = x$ because $a^x = a^x$.

e Let $\log_3 7 = y$.

Then $3^y = 7$.

So substituting for y:

$3^{\log_3 7} = 7$

f Let $\log_a x = y$.

Then $a^y = x$.

So substituting for y:

$a^{\log_a x} = x$

Because logarithms and exponentials are inverse operations, we can use one to undo the other.

Properties of logarithms

$$\log_a a = 1$$

$$\log_a 1 = 0$$

$$\log_a a^x = x$$

$$a^{\log_a x} = x$$

Did you know?

The origins of logarithms

John Napier (1550–1617), a Scottish theologian and an amateur mathematician, was the first to invent logarithms. These 'natural', or 'Naperian', logarithms were based on e. Napier originally used the compound interest formula to find the value of e.

Napier was also one of the first mathematicians to use decimals rather than fractions. He invented decimal notation, using either a comma or a point. The point was used in England, but some European countries use a comma.

Henry Briggs (1561–1630), an Englishman who was a professor at Oxford, decided that logarithms would be more useful if they were based on 10 (our decimal system). Briggs painstakingly produced a table of common logarithms correct to 14 decimal places.

The work on logarithms was greatly appreciated by **Kepler**, **Galileo** and other astronomers at the time, since they allowed the computation of very large numbers.

Common and natural logarithms

There are 2 types of logarithms that you can find on your calculator.

- **Common logarithms (base 10)**: $\log_{10} x$, indicated by the log key on a calculator
- **Natural (Naperian) logarithms (base *e*)**: $\log_e x$ or $\ln x$, indicated by the ln key on a calculator

Example 13

a Find $\log_{10} 5.3$, correct to one decimal place.

b Evaluate $\log_e 80$, correct to 3 significant figures.

c Loudness in decibels is given by the formula $L = 10 \log_{10}\left(\frac{I}{I_0}\right)$, where I_0 is threshold sound, or sound that can barely be heard. Sound louder than 85 decibels can cause hearing damage.

i The loudness of a vacuum cleaner is 10 000 000 times the threshold level, or $10\,000\,000 I_0$. How many decibels is this?

ii If the loudness of the sound of rustling leaves is 20 dB, find its loudness in terms of I_0.

Solution

a $\log_{10} 5.3 = 0.7242\ldots$ log 5.3 =

≈ 0.7

b $\log_e 80 = 4.3820\ldots$ ln 80 =

≈ 4.38

c **i** $L = 10 \log_{10}\left(\frac{10\,000\,000 I_0}{I_0}\right)$

$= 10 \log_{10}(10\,000\,000)$

$= 10 \times 7$

$= 70$

So the loudness of the vacuum cleaner is 70 dB.

ii $L = 10 \log_{10}\left(\frac{I}{I_0}\right)$

$20 = 10 \log_{10}\left(\frac{I}{I_0}\right)$

$2 = \log_{10}\left(\frac{I}{I_0}\right)$

Using the definition of a logarithm:

$10^2 = \frac{I}{I_0}$

$100 = \frac{I}{I_0}$

$100 I_0 = I$

So the loudness of rustling leaves is 100 times threshold sound.

EXERCISE 7.04 Answers on p. 498

Logarithms

EXAMPLE 11

1 Write in logarithmic form:

a $3^x = y$ **b** $5^x = z$ **c** $x^2 = y$ **d** $2^b = a$

e $b^3 = d$ **f** $y = 8^x$ **g** $y = a^x$ **h** $Q = e^x$

2 Write in index form:

a $\log_3 5 = x$ **b** $\log_a 7 = x$ **c** $\log_3 a = b$

d $\log_x y = 9$ **e** $\log_a b = y$ **f** $y = \log_2 6$

g $y = \log_3 x$ **h** $y = \log_{10} 9$ **i** $y = \ln 4$

3 Solve for x, correct to one decimal place where necessary:

a $\log_{10} x = 6$ **b** $\log_3 x = 5$ **c** $\log_x 343 = 3$

d $\log_x 64 = 6$ **e** $\log_5 \left(\frac{1}{5}\right) = x$ **f** $\log_x \sqrt{3} = \frac{1}{2}$

g $\ln x = 3.8$ **h** $3 \log_{10} x - 2 = 10$ **i** $\log_4 x = \frac{3}{2}$

4 Evaluate:

a $\log_2 16$ **b** $\log_4 16$ **c** $\log_5 125$

d $\log_3 3$ **e** $\log_7 49$ **f** $\log_7 7$

g $\log_5 1$ **h** $\log_2 128$ **i** $\log_8 8$

5 Evaluate:

a $3 \log_2 8$ **b** $\log_5 25 + 1$ **c** $3 - \log_3 81$

d $4 \log_3 27$ **e** $2 \log_{10} 10\,000$ **f** $1 + \log_4 64$

g $3 \log_4 64 + 5$ **h** $\frac{\log_3 9}{2}$ **i** $\frac{\log_8 64 + 4}{\log_2 8}$

6 Evaluate:

a $\log_2 \left(\frac{1}{2}\right)$ **b** $\log_3 \sqrt{3}$ **c** $\log_4 2$

d $\log_5 \left(\frac{1}{25}\right)$ **e** $\log_7 \sqrt[4]{7}$ **f** $\log_3 \left(\frac{1}{\sqrt[3]{3}}\right)$

g $\log_4 \left(\frac{1}{2}\right)$ **h** $\log_8 2$ **i** $\log_6 6\sqrt{6}$

j $\log_2 \left(\frac{\sqrt{2}}{4}\right)$

EXAMPLE 12

7 Evaluate:

a $2^{\log_2 3}$ **b** $7^{\log_7 4}$ **c** $3^{\log_3 29}$

EXAMPLE 13

8 Use a calculator to evaluate correct to 2 decimal places:

a $\log_{10} 1200$ **b** $\log_{10} 875$ **c** $\log_e 25$

d $\ln 140$ **e** $5 \ln 8$ **f** $\log_{10} 350 + 4.5$

g $\frac{\log_{10} 15}{2}$ **h** $\ln 9.8 + \log_{10} 17$ **i** $\frac{\log_{10} 30}{\log_e 30}$

9 Solve $\log_y 125 = 3$.

10 If $\log_{10} x = 1.65$, evaluate x correct to one decimal place.

11 Evaluate b to 3 significant figures if $\log_e b = 0.894$.

Foundation Mastery Complex

12 Find the value of $\log_2 1$. What is the value of $\log_a 1$?

13 Evaluate $\log_5 5$. What is the value of $\log_a a$?

14 **a** Simplify $\ln e$.

b Simplify each expression, then check your answer using a calculator.

i $\log_e e^3$ **ii** $\log_e e^2$ **iii** $\ln e^5$ **iv** $\log_e \sqrt{e}$

v $\ln \frac{1}{e}$ **vi** $e^{\ln 2}$ **vii** $e^{\ln 3}$ **viii** $e^{\ln 5}$

ix $e^{\ln 7}$ **x** $e^{\ln 1}$ **xi** $e^{\ln e}$

15 A class was given musical facts to learn. The students were then tested on these facts and each week they were given similar tests to find out how much they were able to remember. The formula $A = 85 - 55\log_{10}(t + 2)$ seemed to model the average score after t weeks.

a What was the initial average score?

b What was the average score after:

i 1 week **ii** 3 weeks?

c After how many weeks was the average score 30?

16 The pH of a solution is defined as $\text{pH} = -\log_{10}[\text{H}^+]$, where $[\text{H}^+]$ is the hydrogen ion concentration. A solution is acidic if its pH is less than 7, alkaline if pH is greater than 7 and neutral if pH is 7. For each question find the pH and state whether the solution is acidic, alkaline or neutral.

a Fruit juice whose hydrogen ion concentration is 0.0035

b Water with $[\text{H}^+] = 10^{-7}$

c Baking soda with $[\text{H}^+] = 10^{-9}$

d Coca-Cola whose hydrogen ion concentration is 0.01

e Bleach with $[\text{H}^+] = 1.2 \times 10^{-12}$

f Coffee with $[\text{H}^+] = 0.00001$

17 If $f(x) = \log x$ and $g(x) = 2x - 7$, find:

a $f(g(x))$ **b** $g(f(x))$.

Investigation

History of bases and number systems

Common logarithms use base 10 like our decimal number system. We might have developed a different system if we had a different number of fingers! The Mayans, in ancient times, used base 20 for their number system since they counted with both their fingers and toes.

1 Research the history and types of other number systems, including those of Aboriginal and Torres Strait Islander peoples. Did any cultures use systems other than base 10? Why?

2 Explore computer-based systems. Digital technologies use binary (base 2), octal (base 8) and hexadecimal (base 16). Find out why these bases are used.

☐ Foundation ○ Mastery ○ Complex

7.05 Logarithm laws

Videos
Logarithm laws 1
Logarithm laws 2

Worksheet
Logarithms review

Puzzle
Logarithm laws

Because logarithms are just another way of writing indices (powers), there are logarithm laws that correspond to the index laws.

Logarithm of a product

$$\log_a(xy) = \log_a x + \log_a y$$

Proof

Let $x = a^m$ and $y = a^n$.

Then $m = \log_a x$ and $n = \log_a y$.

$$xy = a^m \times a^n$$
$$= a^{m+n}$$
$$\therefore \log_a(xy) = m + n \quad \text{(by definition)}$$
$$= \log_a x + \log_a y$$

Logarithm of a quotient

$$\log_a\left(\frac{x}{y}\right) = \log_a x - \log_a y$$

Proof

Let $x = a^m$ and $y = a^n$.

Then $m = \log_a x$ and $n = \log_a y$.

$$\frac{x}{y} = a^m \div a^n$$
$$= a^{m-n}$$
$$\therefore \log_a\left(\frac{x}{y}\right) = m - n \quad \text{(by definition)}$$
$$= \log_a x - \log_a y$$

Logarithm of a power

$$\log_a x^n = n\log_a x$$

Proof

Let $x = a^m$.

Then $m = \log_a x$.

$$x^n = (a^m)^n$$
$$= a^{mn}$$
$$\therefore \log_a x^n = mn \quad \text{(by definition)}$$
$$= n\log_a x$$

Logarithm of a reciprocal

$$\log_a\left(\frac{1}{x}\right) = -\log_a x$$

Proof

$$\begin{aligned}\log_a\left(\frac{1}{x}\right) &= \log_a 1 - \log_a x\\ &= 0 - \log_a x\\ &= -\log_a x\end{aligned}$$

Example 14

a If $\log_5 3 = 0.68$ and $\log_5 4 = 0.86$, evaluate:

i $\log_5 12$ **ii** $\log_5 0.75$ **iii** $\log_5 9$ **iv** $\log_5 20$.

b Solve $\log_2 12 = \log_2 3 + \log_2 x$.

c Simplify $\log_a 21$ if $\log_a 3 = p$ and $\log_a 7 = q$.

d The formula for measuring R, the strength of an earthquake on the Richter scale, is $R = \log\left(\frac{I}{S}\right)$, where I is the maximum seismograph signal of the earthquake being measured and S is the signal of a standard earthquake. Show that:

i $\log I = R + \log S$ **ii** $I = S(10^R)$.

Solution

a **i**
$$\begin{aligned}\log_5 12 &= \log_5(3\times 4)\\ &= \log_5 3 + \log_5 4\\ &= 0.68 + 0.86\\ &= 1.54\end{aligned}$$

ii
$$\begin{aligned}\log_5 0.75 &= \log_5\left(\frac{3}{4}\right)\\ &= \log_5 3 - \log_5 4\\ &= 0.68 - 0.86\\ &= -0.18\end{aligned}$$

iii
$$\begin{aligned}\log_5 9 &= \log_5 3^2\\ &= 2\log_5 3\\ &= 2\times 0.68\\ &= 1.36\end{aligned}$$

iv
$$\begin{aligned}\log_5 20 &= \log_5(5\times 4)\\ &= \log_5 5 + \log_5 4\\ &= 1 + 0.86\\ &= 1.86\end{aligned}$$

b
$$\begin{aligned}\log_2 12 &= \log_2 3 + \log_2 x\\ &= \log_2 3x\end{aligned}$$

So $12 = 3x$

$4 = x$

c
$$\begin{aligned}\log_a 21 &= \log_a(3\times 7)\\ &= \log_a 3 + \log_a 7\\ &= p + q\end{aligned}$$

d **i**
$$\begin{aligned}R &= \log\left(\frac{I}{S}\right)\\ &= \log I - \log S\end{aligned}$$

$$R + \log S = \log I$$

ii
$$R = \log\left(\frac{I}{S}\right)$$

$$\frac{I}{S} = 10^R$$

$$I = S(10^R)$$

Change of base law

If we need to evaluate logarithms such as $\log_5 2$ using base 10 or base e logarithms, we use the change of base law.

Change of base law

$$\log_a x = \frac{\log_b x}{\log_b a}$$

Proof

Let $y = \log_a x$.

Then $x = a^y$.

Take logarithms to the base b of both sides of the equation:

$$\log_b x = \log_b a^y$$
$$= y \log_b a$$
$$\therefore \frac{\log_b x}{\log_b a} = y$$
$$= \log_a x$$

To find the logarithm of any number, such as $\log_5 2$, you can change it to either $\log_{10} x$ or $\log_e x$.

Example 15

a Evaluate $\log_5 2$ correct to 2 decimal places.

b Find the value of $\log_2 3$ correct to one decimal place.

Some calculators have a general log key that calculates the logarithm for *any base*.

Solution

a $\log_5 2 = \dfrac{\log 2}{\log 5}$

≈ 0.43

b $\log_2 3 = \dfrac{\log 3}{\log 2}$

≈ 1.6

EXERCISE 7.05 Answers on p. 498

Logarithm laws

EXAMPLE 14

1 Simplify each expression.

a $\log_a 4 + \log_a y$

b $\log_a 4 + \log_a 5$

c $\log_a 12 - \log_a 3$

d $\log_a b - \log_a 5$

e $3 \log_x y + \log_x z$

f $2 \log_k 3 + 3 \log_k y$

g $5 \log_a x - 2 \log_a y$

h $\log_a x + \log_a y - \log_a z$

i $\log_{10} a + 4 \log_{10} b + 3 \log_{10} c$

j $3 \log_3 p + \log_3 q - 2 \log_3 r$

k $\log_4 \left(\dfrac{1}{n}\right)$

l $\log_x \left(\dfrac{1}{6}\right)$

2 Evaluate:

a $\log_5 5^2$

b $\log_7 7^6$

3 If $\log_7 2 = 0.36$ and $\log_7 5 = 0.83$, evaluate:

a $\log_7 10$

b $\log_7 0.4$

c $\log_7 20$

d $\log_7 25$

e $\log_7 8$

f $\log_7 14$

g $\log_7 50$

h $\log_7 35$

i $\log_7 98$

Foundation | Mastery | Complex

4 Use the logarithm laws to evaluate:

a $\log_5 50 - \log_5 2$
b $\log_2 16 + \log_2 4$
c $\log_4 2 + \log_4 8$
d $\log_5 500 - \log_5 4$
e $\log_9 117 - \log_9 13$
f $\log_8 32 + \log_8 16$
g $3\log_2 2 + 2\log_2 4$
h $2\log_4 6 - (2\log_4 3 + \log_4 2)$
i $\log_6 4 - 2\log_6 12$
j $2\log_3 6 + \log_3 18 - 3\log_3 2$

5 Evaluate to 2 decimal places: EXAMPLE 15

a $\log_4 9$
b $\log_6 25$
c $\log_9 200$
d $\log_2 12$
e $\log_3 23$
f $\log_8 250$
g $\log_5 9.5$
h $2\log_4 23.4$
i $7 - \log_7 108$
j $3\log_{11} 340$

6 If $\log_a 3 = x$ and $\log_a 5 = y$, find an expression in terms of x and y for:

a $\log_a 15$
b $\log_a 0.6$
c $\log_a 27$
d $\log_a 25$
e $\log_a 9$
f $\log_a 75$
g $\log_a 3a$
h $\log_a\left(\frac{a}{5}\right)$
i $\log_a 9a$

7 If $\log_a b = 3.4$ and $\log_a c = 4.7$, evaluate:

a $\log_a\left(\frac{c}{b}\right)$
b $\log_a bc^2$
c $\log_a (bc)^2$
d $\log_a abc$
e $\log_a a^2c$
f $\log_a b^7$
g $\log_a\left(\frac{a}{c}\right)$
h $\log_a a^3$
i $\log_a bc^4$

8 If $\log_a x = p$ and $\log_a y = q$, find, in terms of p and q:

a $\log_a xy$
b $\log_a y^3$
c $\log_a \frac{y}{x}$
d $\log_a x^2$
e $\log_a xy^5$
f $\log_a\left(\frac{x^2}{y}\right)$
g $\log_a ax$
h $\log_a\left(\frac{a}{y^2}\right)$
i $\log_a a^3y$
j $\log_a\left(\frac{x}{ay}\right)$

9 Solve:

a $\log_4 12 = \log_4 x + \log_4 3$
b $\log_3 4 = \log_3 y - \log_3 7$
c $\log_a 6 = \log_a x - 3\log_a 2$
d $\log_2 81 = 4\log_2 x$
e $\log_x 54 = \log_x k + 2\log_x 3$
f $\log_3(x+1) + \log_3(2x-1) = 2$
g $\log_4(x+1) + \log_4(x-2) = 1$
h $\log_5(x+7) + \log_5(x+3) - \log_5(2x+6) = 1$

10 a Change the subject of $L = 10\log_{10}\left(\frac{I}{I_0}\right)$ to I.

b Find the value of I in terms of I_0 when $L = 45$.

11 a Show that the formula $A = 100 - 50\log_{10}(t+1)$ can be written as:

i $\log(t+1) = \frac{100-A}{50}$
ii $t = 10^{\frac{100-A}{50}} - 1$

b Hence find:

i A when $t = 3$
ii t when $A = 75$

12 Make x the subject of each equation.

a $\log_b x^a = 3$
b $\log_a x^2 = b$

☐ Foundation ○ Mastery ⬡ Complex

7.06 Logarithmic functions

Worksheet
Plotting log functions

Puzzles
Logarithmic graphs match-up

Exponential and log graphs match-up

A **logarithmic function** is a function of the form $y = \log_a x$.

Because logarithms and exponentials are inverse operations, the logarithmic function is the inverse function of the exponential function. It 'undoes' the exponential function and its graph is the graph of $y = a^x$ reflected in the line $y = x$.

Example 16

Sketch the graph of $y = \log_2 x$.

Solution

y-intercept ($x = 0$): No y-intercept because $x > 0$.

x-intercept ($y = 0$): $0 = \log_2 x$

$x = 2^0 = 1$, so x-intercept is 1.

Complete a table of values.

$y = \log_2 x$ means $x = 2^y$.

x	$\frac{1}{4}$	$\frac{1}{2}$	1	2	4	6	8
y	−2	−1	0	1	2	2.58	3

For $x = 6$, use the change of base formula, $\log_2 6 = \frac{\log 6}{\log 2}$.

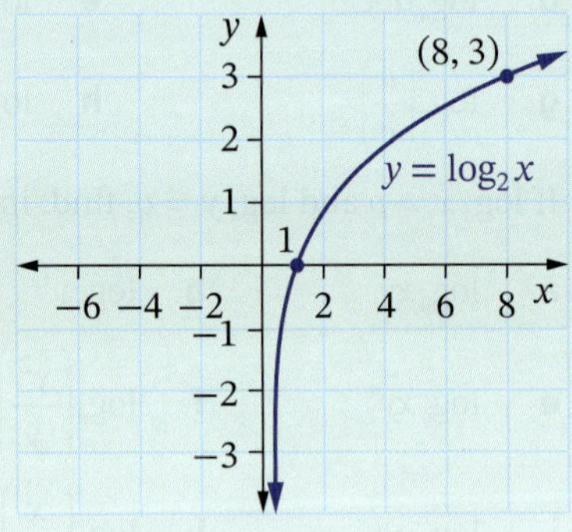

Logarithmic functions

- The logarithmic function $y = \log_a x$ is the inverse function of the exponential function $y = a^x$.
- Domain $(0, \infty)$, range $(-\infty, \infty)$.
- $x > 0$ so the curve is always to the right of the y-axis (no y-intercept).
- The y-axis is an **asymptote**.
- The x-intercept is always 1 because $\log_a 1 = 0$.
- As $x \to \infty, y \to \infty$, for $a > 1$.
- As $x \to 0, y \to -\infty$, for $a > 1$.

Notice that its domain is the same as the range of $y = a^x$ and its range is the same as the domain of $y = a^x$.

Technology

Graphs of logarithmic functions

1 Substitute different values of x into the logarithmic function $y = \log x$: positive, negative and zero. What do you notice?

2 Use graphing technology to sketch the graphs of different logarithmic functions such as

a $y = \log_2 x, y = \log_3 x, y = \log_4 x, y = \log_5 x, y = \log_6 x$

b $y = \log_2 x + 1, y = \log_2 x + 2, y = \log_2 x + 3, y = \log_2 x - 1, y = \log_2 x - 2$

c $y = 2\log_2 x, y = 3\log_2 x, y = -\log_2 x, y = -2\log_2 x, y = -3\log_2 x$

d $y = 2\log_2 x + 1, y = 2\log_2 x + 2, y = 2\log_2 x + 3, y = 2\log_2 x - 1, y = 2\log_2 x - 2$

e $y = 3\log_4 x + 1, y = 5\log_3 x + 2, y = -\log_5 x + 3, y = -2\log_2 x - 1, y = 4\log_7 x - 2$

3 Try sketching the graph of $y = \log_{\frac{1}{2}} x$. What does the table of values look like?

Could you find the domain and range? Use graphing technology to sketch this graph. What do you find?

4 Try sketching the graph of $y = \log_{-2} x$. What does the table of values look like? Are there any discontinuities on the graph? Why? Could you find the domain and range?

Use graphing technology to sketch this graph. What do you find?

$y = \log_e x$ with base e is called the **natural logarithmic function**.

Example 17

Sketch the graph of:

a $y = \log_e x - 1$

b $y = 3\log_{10} x + 4$

Solution

a Complete a table of values for this graph using the ln key on the calculator.

x	1	2	3	4
y	−1	−0.3	0.1	0.4

No y-intercept ($x = 0$) because $\log_e 0$ is undefined. The y-axis is an asymptote.

For x-intercept, $y = 0$:

$0 = \log_e x - 1$

$1 = \log_e x$

$x = e^1$

≈ 2.7

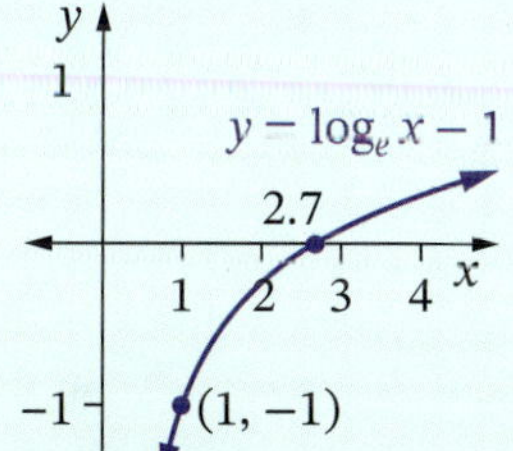

Notice that this is the graph of $y = \log_e x$ moved down 1 unit.

b Complete a table of values using the log key on the calculator.

$y = 3\log_{10} x + 4$

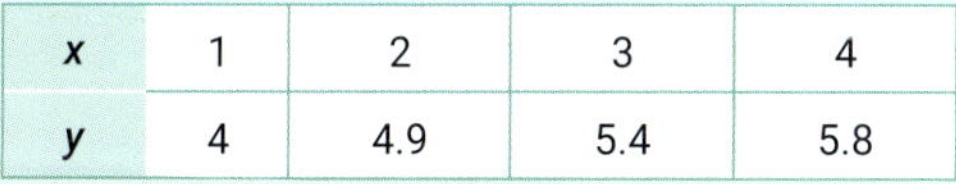

x	1	2	3	4
y	4	4.9	5.4	5.8

No y-intercept.

For x-intercept, $y = 0$:

$0 = 3\log_{10} x + 4$

$-4 = 3\log_{10} x$

$-\frac{4}{3} = \log_{10} x$

$10^{-\frac{4}{3}} = x$

$x = 0.04641\ldots$

≈ 0.046

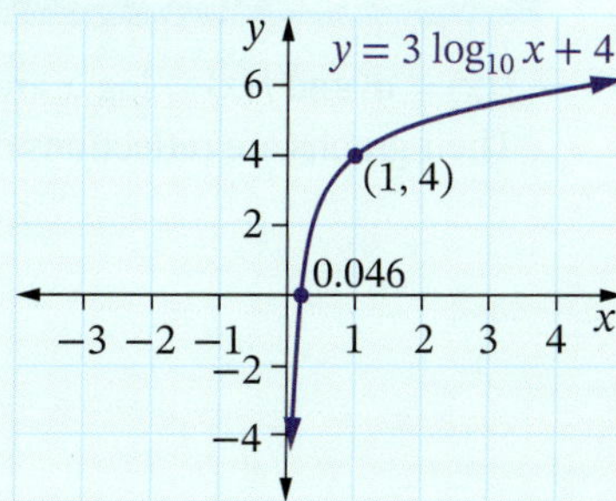

Example 18

a Sketch the graphs of $y = e^x$, $y = \log_e x$ and $y = x$ on the same set of axes.

b What relationship do these graphs have?

c If $f(x) = \log_a x$, sketch the graph of $y = -f(x)$ and state its domain and range.

Solution

a Drawing $y = e^x$ gives an exponential curve with y-intercept 1.

Find another point, say $x = 2$:

$y = e^2$

$= 7.3890\ldots$

≈ 7.4

Drawing $y = \log_e x$ gives a logarithmic curve with x-intercept 1.

Find another point, say $x = 2$:

$y = \ln 2$

$= 0.6931\ldots$

≈ 0.7

$y = x$ is a linear function with gradient 1 and y-intercept 0.

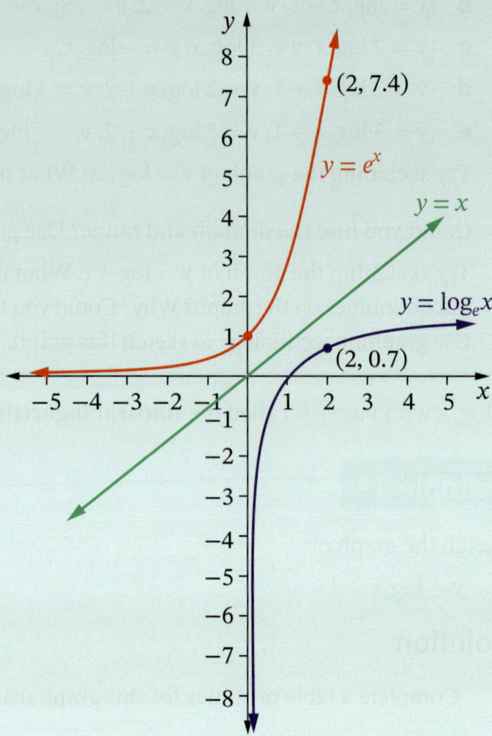

b The graphs of $y = e^x$ and $y = \log_e x$ are reflections of each other in the line $y = x$. They are **inverse functions.**

c Given $f(x) = \log_a x$,

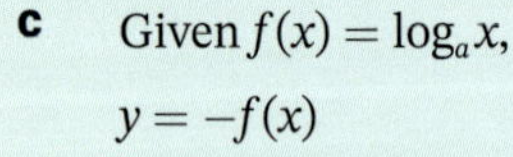

$y = -f(x)$

$= -\log_a x$

This is a reflection of $f(x)$ in the x-axis.

Domain $(0, \infty)$, range $(-\infty, \infty)$

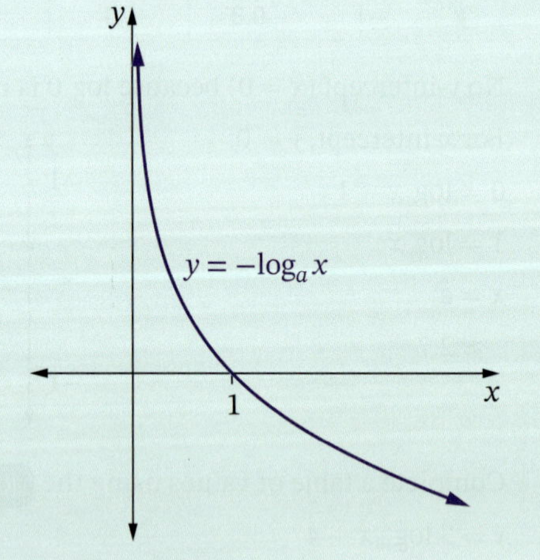

The exponential and logarithmic functions

$f(x) = a^x$ and $f(x) = \log_a x$ are inverse functions.
Their graphs are reflections of each other in the line $y = x$.

Did you know?

Logarithmic scales

It is difficult to describe and graph exponential functions because their y values increase so quickly. We can use logarithms and **logarithmic scales** on the y-axis to solve this problem. Graphs that showed the number of cases rising during the worldwide COVID-19 pandemic in 2020 used logarithmic scales to represent large numbers.

On a base 10 logarithmic scale, an axis or number line has units that don't increase by 1, but by powers of 10.

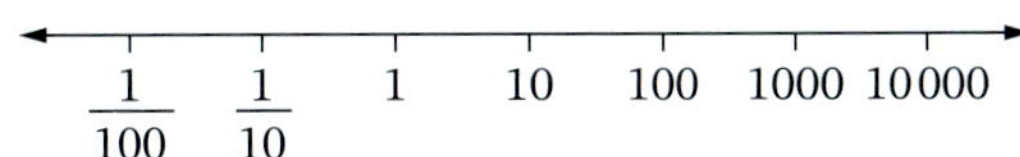

Examples of base 10 logarithmic scales are:

- the Richter scale for measuring earthquake magnitude
- the pH scale for measuring acidity in chemistry.

You will learn more about logarithmic scales in Year 12.

EXERCISE 7.06 Answers on p. 498

Logarithmic functions

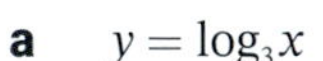

1 Sketch the graph of each logarithmic function and state its domain and range.  EXAMPLES 16,17

a $y = \log_3 x$ **b** $f(x) = 2\log_4 x$ **c** $y = \log_2 x + 1$

d $y = \log_5 x - 1$ **e** $y = 5\ln x + 3$ **f** $f(x) = -3\log_{10} x + 2$

2 Sketch the graphs of $y = 10^x$, $y = \log_{10} x$ and $y = x$ on the same number plane.  EXAMPLE 18
What do you notice about the relationship of the curves to the line?

3 Sketch the graphs of $y = \log_2 x$, $y = 2^x$ and $y = x$ on the same set of axes.
Find the inverse function of $y = \log_2 x$.

4 Sketch the graph of $y = \log_2 x$ and $y = \log_2(-x)$ on the same set of axes and describe their relationship.

5 Find the inverse function of each function.

a $y = \log_7 x$ **b** $y = \log_9 x$ **c** $y = \log_e x$

d $y = 2^x$ **e** $y = 6^x$ **f** $y = e^x$

6 Find the domain of each logarithmic function.

a $y = \ln(x + 1)$

b $f(x) = \log_2(3x - 1)$

c $y = \ln(2 - x)$

7.07 Exponential equations

Videos
Exponential equations 3
Exponential equations 1
Exponential equations 2
Exponential functions

Worksheets
Logarithmic and exponential equations
Solving exponential equations
Using exponential models

Puzzles
Logarithms - Solving equations 1
Logarithms - Solving equations 2

You can solve **exponential equations** using logarithms or the change of base formula.

Example 19

Solve $5^x = 7$, correct to one decimal place.

Solution

Method 1: Logarithms

Take logarithms of both sides:

$$\log 5^x = \log 7$$
$$x \log 5 = \log 7$$
$$x = \frac{\log 7}{\log 5}$$
$$= 1.2090\ldots$$
$$\approx 1.2$$

Method 2: Change of base formula

Convert to logarithm form:

$5^x = 7$ means $\log_5 7 = x$

Using the change of base formula to evaluate x:

$$x = \log_5 7$$
$$= \frac{\log 7}{\log 5}$$
$$= 1.2090\ldots$$
$$\approx 1.2$$

Exponential equations have many practical applications.

Example 20

a Solve $e^{3.4x} = 100$, correct to 2 decimal places.

b The temperature T in °C of a metal as it cools down over t minutes is given by $T = 27 + 219e^{-0.032t}$. Find, correct to one decimal place, the time it takes for the metal to cool down to 100°C.

Solution

a With an equation involving e we use $\ln x$, which is $\log_e x$.

Take natural logs of both sides:

$$\ln e^{3.4x} = \ln 100$$
$$3.4x = \ln 100$$

$\ln x$ and e^x are inverses

$$x = \frac{\ln 100}{3.4}$$
$$= 1.3544\ldots$$
$$\approx 1.35$$

b When $T = 100$:

$$100 = 27 + 219e^{-0.032t}$$
$$73 = 219e^{-0.032t}$$
$$\frac{73}{219} = e^{-0.032t}$$
$$\log_e\left(\frac{73}{219}\right) = \log_e(e^{-0.032t})$$
$$= -0.032t$$
$$t = \frac{\log_e\left(\frac{73}{219}\right)}{-0.032}$$
$$= 34.3316\ldots$$
$$\approx 34.3$$

So it takes 34.3 minutes to cool down to 100°C.

EXERCISE 7.07 Answers on p. 499

Exponential equations

1 Solve each equation, correct to 2 significant figures: (EXAMPLE 19)

a $4^x = 9$ **b** $3^x = 5$ **c** $7^x = 14$ **d** $2^x = 15$

e $5^x = 34$ **f** $6^x = 60$ **g** $2^x = 76$ **h** $4^x = 50$

i $3^x = 23$ **j** $9^x = 210$

2 Solve, correct to 2 decimal places:

a $2^x = 6$ **b** $5^y = 15$ **c** $3^x = 20$ **d** $7^m = 32$

e $4^k = 50$ **f** $3^t = 4$ **g** $8^x = 11$ **h** $2^p = 57$

i $4^x = 81.3$ **j** $6^n = 102.6$

3 Solve, to one decimal place:

a $3^{x+1} = 8$ **b** $5^{3n} = 71$ **c** $2^{x-3} = 12$

d $4^{2n-1} = 7$ **e** $7^{5x+2} = 11$ **f** $8^{3-n} = 5.7$

g $2^{x+4} = 18.3$ **h** $3^{7k-3} = 32.9$ **i** $9^{\frac{x}{2}} = 50$

4 Solve each equation, correct to 3 significant figures: (EXAMPLE 20)

a $e^x = 200$ **b** $e^{3t} = 5$ **c** $2e^t = 75$

d $45 = e^x$ **e** $3000 = 100e^n$ **f** $100 = 20e^{3t}$

g $2000 = 50e^{0.15t}$ **h** $15\,000 = 2000e^{0.03k}$ **i** $3Q = Qe^{0.02t}$

5 The amount A of money in a bank account after n years grows with compound interest according to the formula $A = 850(1.025)^n$.

a Find:

i the initial amount in the bank **ii** the amount after 7 years.

b Find how many years it will take for the amount in the bank to be \$1000.

6 The population of a city is given by $P = 35\,000e^{0.024t}$, where t is time in years.

a Find the population:

i initially **ii** after 10 years **iii** after 50 years.

b Find when the population will reach:

i 80 000 **ii** 200 000.

7 A species of wattle is gradually dying out in a Blue Mountains region. The number of wattle trees over time t years is given by $N = 8900e^{-0.048t}$.

a Find the number of wattle trees:

i initially

ii after 5 years

iii after 70 years.

b After how many years will there be:

i 5000 wattle trees?

ii 200 wattle trees?

Photo courtesy of Margaret Grove

8 A formula for the mass M g of plutonium after t years is given by $M = 100e^{-0.000\,03t}$. Find:

a initial mass **b** mass after 50 years **c** mass after 500 years

d its half-life (the time taken to decay to half of its initial mass).

Foundation Mastery Complex

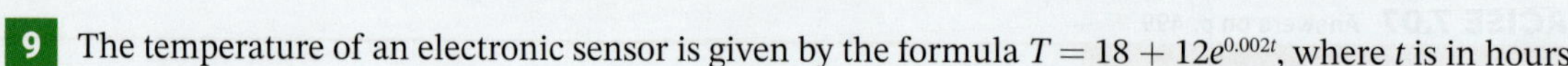

9 The temperature of an electronic sensor is given by the formula $T = 18 + 12e^{0.002t}$, where t is in hours.

a What is the temperature of the sensor after 5 hours?

b When the temperature reaches 50°C the sensor needs to be shut down to cool. After how many hours does this happen?

10 A particle is moving along a straight line with displacement x cm over time t s according to the formula $x = 5e^t + 23$. Find:

a the initial displacement

b the exact velocity after 20 s

c the displacement after 6 s

d the time when displacement is 85 cm

e the time when the velocity is 1000 cm s^{-1}.

11 **a** Find how many powers of 3 lie between 100 and 1000.

b Solve $3^x < 10\,000$, where x is an integer.

12 Solve each equation.

a $2^{2x} - 3(2^x) + 2 = 0$ **b** $3^{2x} - 12(3^x) + 27 = 0$ **c** $4^x - 12(2^x) + 32 = 0$

Sample HSC problem

Answers on p. 499

(3 marks) The depth of water (in metres) in a dam is decreasing over t months according to the formula

$$D = 4e^{-0.012t}.$$

a Find, correct to 2 significant figures, the depth after 6 months. 1 mark

b Find, correct to 2 significant figures, the rate at which the depth is changing after 6 months. 2 marks

Foundation Mastery Complex

CHAPTER SUMMARY

This chapter, *Exponential and logarithmic functions*, introduced exponential functions, Euler's number, e, the derivative of e^x, logarithms, logarithmic functions and exponential equations. Most of the content builds upon your knowledge of functions and calculus.

Before you move on, consider what you learned in this chapter and revisit any sections that may have been unclear.

What you learned in this chapter...	Section	
Graph exponential and logarithmic functions	7.01	Exponential functions
	7.06	Logarithmic functions
Understand and use Euler's number, e	7.02	Euler's number, e
Differentiate exponential functions	7.03	The derivative of exponential functions
Convert between exponential and logarithmic forms using the definition of a logarithm	7.04	Logarithms
Identify and apply logarithm laws, including solving logarithmic equations	7.05	Logarithm laws
Solve exponential equations using logarithms	7.07	Exponential equations

To help master the theory, formulas and terminology associated with this topic, including the logarithm laws and the graphs of the functions, make a summary. Use the chapter outline and the mind map below as a guide. Add your own words, symbols, diagrams, boxes and reminders. The summary should give you a 'whole picture' view of the topic and allow you to identify any weak areas to revisit in your revision.

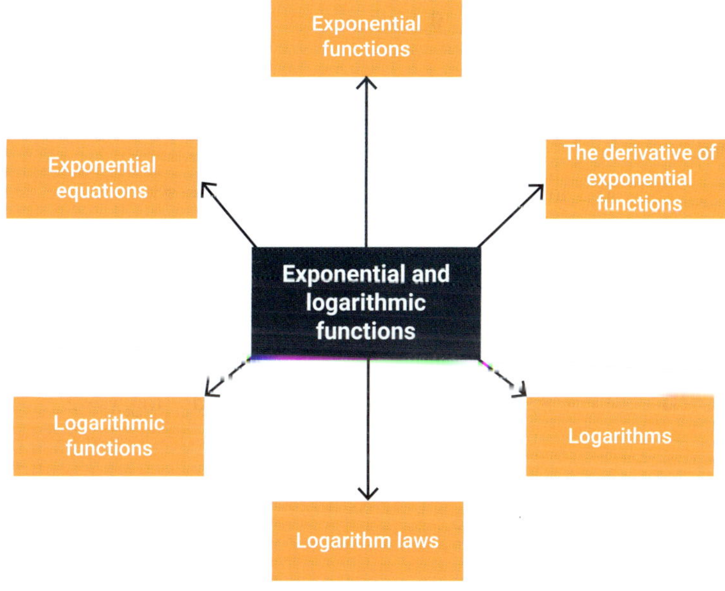

7 Test yourself

Answers on p. 499

For Questions **1** to **3**, select the correct answer **A**, **B**, **C** or **D**.

7.05 **1** Simplify $\log_a 15 - \log_a 3$:

A $\log_a 45$ **B** $\dfrac{\log_a 15}{\log_a 3}$ **C** $\log_a 15 \times \log_a 3$ **D** $\log_a 5$

7.04 **2** Write $a^x = y$ as a logarithm.

A $\log_y x = a$ **B** $\log_a y = x$ **C** $\log_a x = y$ **D** $\log_x a = y$

7.05 **3** Solve $5^x = 4$ (there is more than one answer).

A $x = \dfrac{\log 4}{\log 5}$ **B** $x = \dfrac{\log 5}{\log 4}$ **C** $x = \dfrac{\ln 4}{\ln 5}$ **D** $x = \dfrac{\ln 5}{\ln 4}$

7.04 **4** Evaluate:

a $\log_2 8$ **b** $\log_7 7$ **c** $\log_{10} 1000$ **d** $\log_9 81$

e $\log_e e$ **f** $\log_4 64$ **g** $\log_9 3$ **h** $\log_2 \dfrac{1}{2}$

i $\log_5 \dfrac{1}{25}$ **j** $\ln e^3$

7.02, 7.05 **5** Evaluate correct to 3 significant figures:

a $e^2 - 1$ **b** $\log_{10} 95$ **c** $\log_e 26$ **d** $\log_4 7$

e $\log_4 3$ **f** $\ln 50$ **g** $e + 3$ **h** $\dfrac{5e^3}{\ln 4}$

7.04 **6** Evaluate:

a $e^{\ln 6}$ **b** $e^{\ln 2}$

7.04 **7** Write in index form:

a $\log_3 a = x$ **b** $\ln b = y$ **c** $\log c = z$

7.05 **8** If $\log_7 2 = 0.36$ and $\log_7 3 = 0.56$, find the value of:

a $\log_7 6$ **b** $\log_7 8$ **c** $\log_7 1.5$

d $\log_7 14$ **e** $\log_7 3.5$

7.07, 7.05 **9** Solve each equation.

a $3^x = 8$ **b** $2^{3x-4} = 3$ **c** $\log_x 81 = 4$ **d** $\log_6 x = 2$

7.07 **10** Solve $12 = 10e^{0.01t}$.

7.05 **11** Evaluate $\log_9 8$, correct to one decimal place.

7.05 **12** Evaluate:

a $\log_6 12 + \log_6 3$ **b** $\log_{10} 25 + \log_{10} 4$ **c** $2 \log_4 8$

d $\log_8 72 - \log_8 9$ **e** $\log_{10} 53\,000 - \log_{10} 53$

7.05 **13** Simplify:

a $5 \log_a x + 3 \log_a y$ **b** $2 \log_x k - \log_x 3 + \log_x p$

7.04 **14** Evaluate, correct to 2 significant figures:

a $\log_{10} 4.5$ **b** $\ln 3.7$

7.01 **15** Sketch the graph of $y = 2^x + 1$ and state its domain and range.

□ Foundation ○ Mastery ⬡ Complex

16 Solve each equation. 7.07, 7.05

a $2^x = 9$ b $3^x = 7$ c $5^{x+1} = 6$ d $4^{2y} = 11$

e $8^{3n-2} = 5$ f $\log_x 16 = 4$ g $\log_3 y = 3$ h $\log_7 n = 2$

i $\log_x 64 = \frac{1}{2}$ j $\log_8 m = \frac{1}{3}$

17 Write as a logarithm: 7.04

a $2^x = y$ b $5^a = b$ c $10^x = y$

d $e^x = z$ e $3^{x+1} = y$

18 Solve correct to one decimal place: 7.07

a $e^x = 15$ b $2.7^x = 21.8$ c $10^x = 128.7$

19 Sketch the graph of each exponential equation. 7.01

a $y = 5(3^{x+2})$ b $y = 2(3^x) - 5$ c $f(x) = -3^x$ d $y = 3(2^{-x})$

20 Sketch the graph of each logarithmic equation. 7.06

a $f(x) = \log_3 x$ b $y = 3 \ln x - 4$

21 If $\log_x 2 = a$ and $\log_x 3 = b$, find in terms of a and b: 7.05

a $\log_x 6$ b $\log_x 1.5$ c $\log_x 8$

d $\log_x 18$ e $\log_x 27$

22 Solve each logarithmic function. 7.05

a $\log_2(3x - 1) + \log_2 x = 2$

b $\log_3(8x + 1) - \log_3(x - 4) = 2$

c $\log_2(x + 5) - \log_2(x + 2) + \log_2 x = 1$

23 Simplify: 7.05

a $\log_a\left(\frac{1}{x}\right)$ b $\log_e\left(\frac{1}{y}\right)$

24 If $f(x) = \log_e x$, $g(x) = e^x$ and $h(x) = 6x^2 - 1$, find: 7.06

a $f(h(x))$ b $g(h(x))$ c $h(g(x))$

d $f(g(x))$ e $g(f(x))$

25 The amount of money in dollars Dion invested in the bank after n years is given by $A = 5280(1.019)^n$. 

a Find the amount in the bank:

i initially   ii after 3 years   iii after 4 years.

b Find how long it will take for the amount of money in the bank to reach:

i \$6000   ii \$10 000

26 Differentiate each function. 7.03

a $y = e^{-2x}$ b $y = 5e^{4x}$ c $y = -2e^{8x} + 5x^3 - 1$

d $y = x^2e^{2x}$ e $y = (4e^{3x} - 1)^9$ f $y = \frac{x}{e^{2x}}$

27 The formula for the number of wombats in Teragram Grove after t years is $N = 1118 - 37e^{0.032t}$. 

a Find the initial number of wombats in this region.

b How many wombats are there after 5 years?

c How long will it take until the number of wombats in the region is:

i 500?   ii 100?

28 Differentiate each function. 7.03

a $y = e^x + x$ b $y = -4e^x$ c $y = 3e^{-x}$

d $y = (3 + e^x)^9$ e $y = 3x^5e^x$ f $y = \frac{e^x}{7x-2}$

□ Foundation ○ Mastery ⬡ Complex

7 Challenge exercise

Answers on p. 501

1. If $\log_b 2 = 0.6$ and $\log_b 3 = 1.1$, find:
   a $\log_b 6b$ **b** $\log_b 8b$ **c** $\log_b 1.5b^2$
2. Find the point of intersection of the curves $y = \log_e x$ and $y = \log_{10} x$.
3. Sketch the graph of $y = \log_2(x - 1)$ and state its domain and range.
4. By substituting $u = 3^x$, solve $3^{2x} - 3^x - 2 = 0$, correct to 2 decimal places.
5. Draw a sketch of the function $y = 2a^x$, where $0 < a < 1$.
6. If $y = 8 + \log_2(x + 2)$:
   a show that $x = 2(2^{y-9} - 1)$
   b find, correct to 2 decimal places:
   i y when $x = 5$ **ii** x when $y = 1$
7. Find the equation of **a** the tangent and **b** the normal to the curve $y = 3e^x - 5$ at the point $(2, 3e^2 - 5)$.

□ Foundation ○ Mastery ○ Complex

PRACTICE SET 3

Answers on p. 501

For Questions **1** to **6**, select the correct answer **A**, **B**, **C** or **D**.

1 The quotient rule for differentiating $y = \dfrac{u}{v}$ is: 6.08

A $y' = \dfrac{uv' - vu'}{v^2}$ **B** $\dfrac{u'v - v'u}{v^2}$

C $y' = u'v + v'u$ **D** $y' = uv' + vu'$

2 Find the derivative of $(3x - 2)^8$. 6.06

A $(3x - 2)^7$ **B** $8(3x - 2)^7$

C $8x^7(3x - 2)$ **D** $24(3x - 2)^7$

3 Which statement is the same as $3^x = 7$? There is more than one answer. 7.04, 7.05

A $x = \log\left(\dfrac{7}{3}\right)$ **B** $\log_3 x = 7$ **C** $\log_3 7 = x$ **D** $x = \dfrac{\log 7}{\log 3}$

4 The derivative of $x^2(2x + 9)^2$ is: 6.06

A $4x(2x + 9)$ **B** $2x(2x + 9)^2 + 2x^2(2x + 9)$

C $2x(2x + 9)$ **D** $2x(2x + 9)^2 + 4x^2(2x + 9)$

5 If the displacement of a particle is given by $x = 2t^3 + 6t^2 - 4t + 10$, the initial velocity is: 6.04

A -4 **B** 10 **C** 12 **D** 14

6 Given $y = \dfrac{f(x)}{g(x)}$, where $f(-2) = 2$, $f'(-2) = 4$, $g(-2) = -1$, $g'(-2) = 3$, find y' at $x = -2$. 6.08

A 10 **B** 2.5 **C** -10 **D** -2.5

7 Differentiate each function. 6.03, 6.08

a $y = x^9 - 4x^2 + 7x + 3$ **b** $y = 2x(x^2 - 1)$ **c** $y = 3x^{-4}$

d $y = \dfrac{5}{2x^5}$ **e** $y = \sqrt{x^3}$ **f** $y = (2x + 3)^7$

g $y = \dfrac{1}{(x^2 - 7)^4}$ **h** $y = \sqrt[3]{5x + 1}$ **i** $y = \dfrac{5x^2 - 1}{2x + 3}$

8 The volume in litres of a rectangular container that is leaking over time t minutes is given by $V = -t^2 + 4t + 100$. Find: 6.04

a the initial volume

b the volume after 10 minutes

c the rate of change in volume after 10 minutes

d how long it will take, correct to one decimal place, until the container is empty.

9 **a** Find the equation of the tangent to the curve $y = x^3 - 3x$ at the point $P = (-2, -2)$. 6.05

b Find the equation of the normal to $y = x^3 - 3x$ at P.

c Find the point Q where this normal cuts the x-axis.

10 Differentiate each exponential function. 7.03

a $y = e^x - x$ **b** $y = 3e^x + 1$ **c** $y = (e^x - 2)^4$

d $y = e^x(4x + 1)^3$ **e** $y = \dfrac{e^x}{5x - 2}$ **f** $y = 5e^{7x}$

☐ Foundation ○ Mastery ○ Complex

6.05 **11** The function $f(x) = ax^2 + bx + c$ has a tangent at $(1, -3)$ with a gradient of -1. It also passes through $(4, 3)$. Find the values of a, b and c.

7.04 **12** Find $\log_5\left(\frac{1}{25}\right)$.

7.04 **13** Solve $\log_x\left(\frac{1}{16}\right) = 4$.

6.02 **14** Consider the function $f(x) = 3x^2 - 4x + 9$.

a Find $f(x + h) - f(x)$.

b Show by differentiating from first principles that $f'(x) = 6x - 4$.

6.05 **15** **a** Find the equation of the tangent to the curve $y = x^3 - 2$ at the point $P(1, -1)$.

b The curve $y = x^3 - 2$ meets the y-axis at Q. Find the equation of PQ.

c Find the equation of the normal to $y = x^3 - 2$ at the point $(-1, -3)$.

d Find the point R where this normal cuts the x-axis.

7.06 **16** Sketch the graph of each logarithmic function.

a $y = \log_3 x$ **b** $y = 3\log_2 x - 1$

7.04 **17** **a** Write $y = \log_e x$ as an equation with x in terms of y.

b Hence find the value of x, to 3 significant figures, when $y = 1.23$.

7.07 **18** Solve $72^x = 3$.

7.06 **19** Sketch the graph of the inverse function of $y = \log_2 x$.

6.03 **20** If $f(x) = 2x^3 - 5x^2 + 4x - 1$, find $f(-2)$ and $f'(-2)$.

6.02 **21** **a** Find the gradient of the secant to the curve $f(x) = 2x^3 - 7$ between the point $(2, 9)$ and the point where:

i $x = 2.01$ **ii** $x = 1.99$

b Hence estimate the gradient of the tangent to the curve at $(2, 9)$.

6.01 **22** Sketch the gradient function for each curve.

a

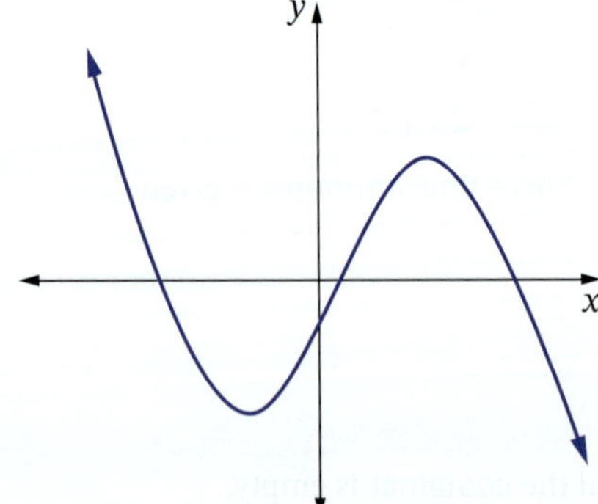

b

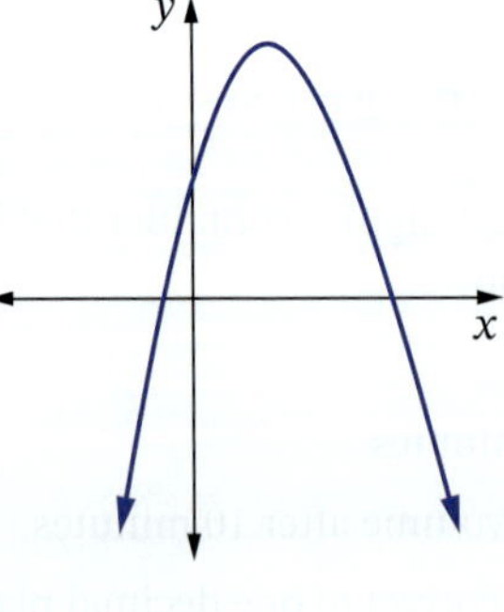

7.01 **23** Sketch the graph of each exponential function.

a $y = e^{-x}$ **b** $y = 2e^{3x} + 1$

Foundation Mastery Complex

24 The displacement x cm of an object moving along a straight line over time t seconds is given by $x = 2t^3 - 13t^2 + 17t + 12$. 6.04

a Find the initial displacement and velocity.

b Find the displacement and velocity after 2 seconds.

25 Evaluate each expression, correct to 2 decimal places where appropriate. 

a $\log_2 16$ **b** $\log_3 3$ **c** $\log_4 2$

d $\log_{10} 109.7$ **e** $\ln 43.1$ **f** $\log_3 11$

26 Find the equation of the tangent to the curve $y = 3x^2 - 6x + 7$ at the point $(2, 7)$. 6.05

27 Solve each equation. 7.05, 7.07

a $\log_2 64 = x$

b $\log_x 81 = 4$

c $\log_3 x + \log_3 (x + 3) - \log_3 (x - 1) = 2$

d $\log_{10} 10^x + e^{\ln x^2} + 10 = 0$

e $3^{x+1} = 7$, correct to 2 decimal places.

28 $f(x) = x^3 - 6x^2 + 3x - 1$. 

a Find $f'(2)$ and explain the behaviour of the function when $x = 2$.

b Find the exact values of x where the function is stationary.

29 Find the equation of the normal to the curve $y = x^2 - 4x + 1$ at the point $(3, -2)$. 6.05

30 Differentiate each function. 6.03, 6.08

a $y = 2x^4 - 5x^3 + 3x^2 - x - 4$ **b** $y = \dfrac{1}{2x^5}$

c $y = \sqrt{x}$ **d** $y = (2x - 3)^7$

e $y = 3x^4(2x - 5)^7$ **f** $y = \dfrac{5x + 7}{3x - 2}$

31 Find the equation of the tangent to the curve $y = 5e^x$ at the point $(2, 5e^2)$. 7.03

□ Foundation ○ Mastery ○ Complex

GRAPH TRANSFORMATIONS

8.

TRANSFORMATIONS OF FUNCTIONS

In this chapter you will explore transformations on the graph of the function $y = f(x)$ that move or stretch the function. We have already met some transformations of functions informally, such as how the graph of $y = f(x) + c$ is the graph of $y = f(x)$ translated upwards or downwards, how $y = -f(x)$ is $y = f(x)$ reflected in the x-axis, and how $y = kf(x)$ is $y = f(x)$ stretched (dilated) vertically. In this chapter, we will examine horizontal and vertical translations, reflections and dilations more formally.

Chapter outline

Sergey Nivens/Adobe Stock Photos

In this chapter you will:

- understand and apply translations of functions and relations
- understand and apply reflections of functions and relations in the x- and y-axes
- understand and apply dilations of functions and relations from the x- and y-axes
- apply combinations of transformations to functions and relations
- use transformations to sketch the graphs of different types of functions and relations

Videos (3):

8.02 Translations of functions • Horizontal translations of functions

8.05 Horizontal dilations of functions

Worksheets (7):

8.02 Translations of functions • Graphing translations of functions

8.03 Advanced graphs

8.05 Dilations of functions • Advanced graphs

8.06 Combinations of transformations

8.07 Graphing transformed functions • Solving inequalities graphically

Puzzle (1)

8.03 Matching graphs (Advanced)

Nelson MindTap

To access resources above, visit **cengage.com.au/nelsonmindtap**

Terminology

dilation	dilation factor	enlargement	horizontal
image	parameter	reduction	reflection
factor	shift	shrink	stretch
transformation	translation	vertical	

Vertical translations of functions

8.01

A **transformation** of a function or graph is the general name for the process of changing it by translating (shifting), reflecting (flipping) or dilating (stretching/shrinking) it.

A **vertical translation** of a function shifts the graph of the function up or down without changing its size or shape.

Technology

Vertical translations

Some graphing technologies have a dynamic feature to show how a constant c (a **parameter**) changes the graph of a function.

Use dynamic geometry software to explore the effect of c on each graph below. If you don't have dynamic software, substitute different values for c into the equation. Use positive and negative values, integers and fractions.

$f(x) = x + c$ $\quad$ $f(x) = x^2 + c$ $\quad$ $f(x) = x^3 + c$ $\quad$ $f(x) = x^4 + c$

$f(x) = e^x + c$ $\quad$ $f(x) = \ln x + c$ $\quad$ $f(x) = \frac{1}{x} + c$ $\quad$ $f(x) = |x| + c$

How does the value of c transform the graph?

What is the difference between positive and negative values of c?

The value and sign of c determines the amount and direction of the vertical translation of $f(x)$.

Vertical translation

For the function $y = f(x)$:

$y = f(x) + c$ translates the graph vertically (along the y-axis).

If $c > 0$, the graph is translated upwards by c units.

If $c < 0$, the graph is translated downwards.

A vertical translation changes the y values of the function.

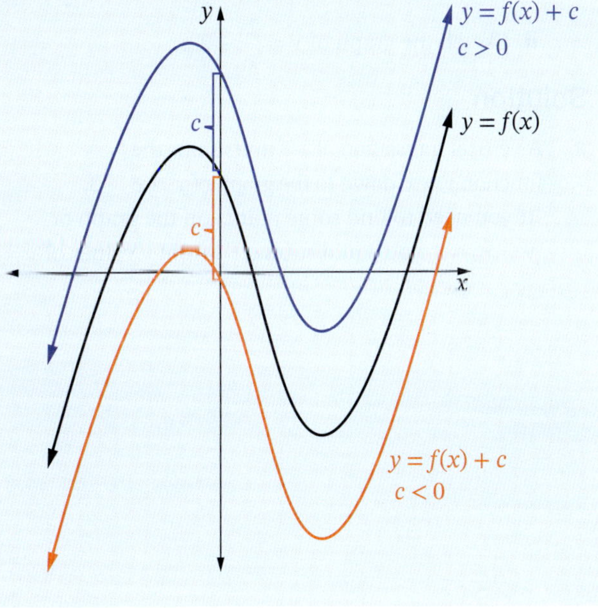

Example 1

a Explain how the graph of $y = x^2 + 2$ is related to the graph of $y = x^2$.

b If the graph of the function $y = x^2 + 7x + 1$ is translated 4 units down, find the equation of the transformed function.

c The point $P(3, -2)$ lies on the function $y = f(x)$. Find the transformed point (the image of P) if the function is translated:

i 6 units down **ii** 8 units up

Solution

a The graph of $y = x^2 + 2$ is a vertical translation 2 units up from the original (parent) function $y = x^2$.

b For a vertical translation 4 units down:

$y = f(x) + c$ where $c = -4$

$y = x^2 + 7x + 1 - 4$

$= x^2 + 7x - 3$

The equation of the transformed function is $y = x^2 + 7x - 3$.

c **i** $P(3, -2)$ is translated 6 units down, so subtract 6 from the y value.

The transformed point is $(3, -2 - 6) = (3, -8)$.

ii $P(3, -2)$ is translated 8 units up, so add 8 to the y value.

The transformed point is $(3, -2 + 8) = (3, 6)$.

Example 2

a Sketch the graph of $y = x^3 - 3$.

b **i** State the relationship of $y = \frac{1}{x} - 2$ to $y = \frac{1}{x}$.

ii Sketch the graph of $y = \frac{1}{x} - 2$.

Solution

a A vertical translation of -3 units shifts the function $y = x^3$ down to the graph of $y = x^3 - 3$.

If you need to find some points on the graph of $y = x^3 - 3$, you could subtract 3 from y values of $y = x^3$.

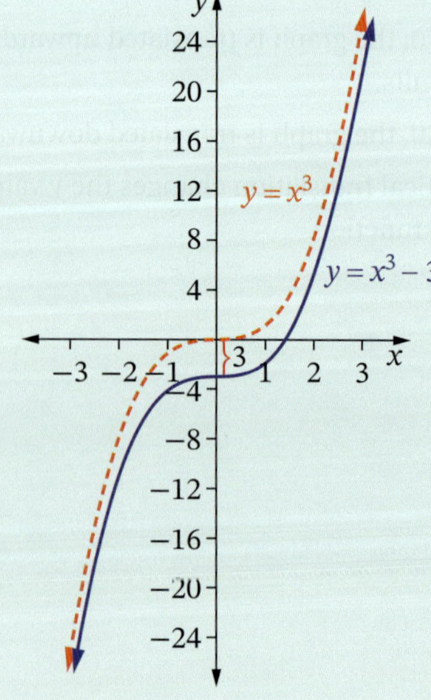

b **i** $y = \frac{1}{x} - 2$ is a vertical translation 2 units down of $y = \frac{1}{x}$.

ii Shift the graph of $y = \frac{1}{x}$, including the horizontal asymptote $y = 0$, down 2 units. The new horizontal asymptote is at $y = -2$.

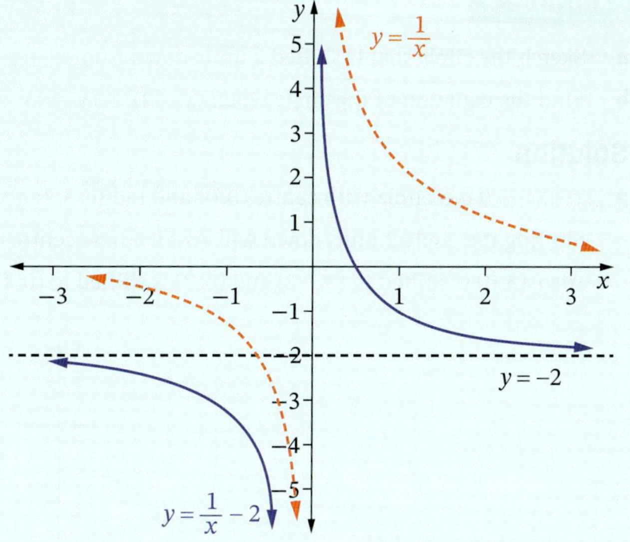

Example 3

What is the equation of $y = g(x)$ in relation to $y = f(x)$?

Solution

$y = g(x)$ is 5 units up from $y = f(x)$.

Its equation is $y = f(x) + c$ where $c = 5$.

$y = f(x) + 5$

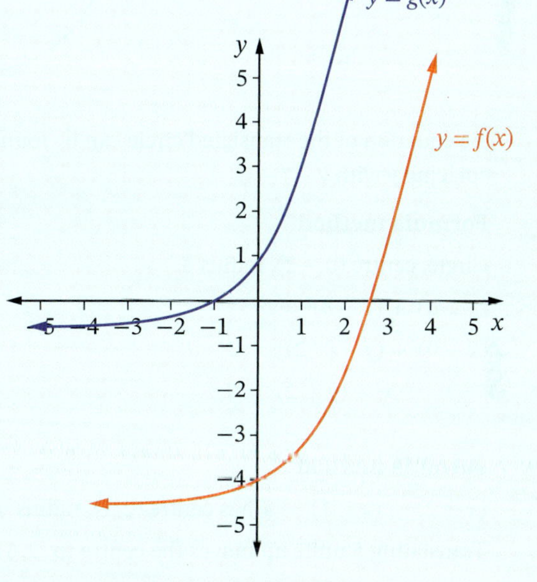

The function $y = f(x) + c$ translates the graph of $y = f(x)$ vertically, changing its y values.

Another way of saying this is that to translate $y = f(x)$ vertically by c units, replace y with $y - c$.

$y - c = f(x)$ gives $y = f(x) + c$.

Vertical translation

Replacing y with $y - c$ in the equation of a function translates its graph vertically.

If $c > 0$, the graph is translated upwards.

If $c < 0$, the graph is translated downwards.

This method also works for translating relations that are not functions.

Example 4

a Sketch the circle that is shifted 2 units down from $x^2 + y^2 = 9$ and find its equation.

b Find the equation of the circle when $(x - 2)^2 + (y - 1)^2 = 4$ is translated 5 units upwards.

Solution

a $x^2 + y^2 = 9$ is a circle with centre $(0, 0)$ and radius 3. It also goes through $(3, 0)$ and $(0, 3)$.

Shifting the circle 2 units down will give it a new centre of $(0, -2)$.

Also, $(3, 0)$ is shifted to $(3, -2)$ and $(0, 3)$ is shifted to $(0, 1)$.

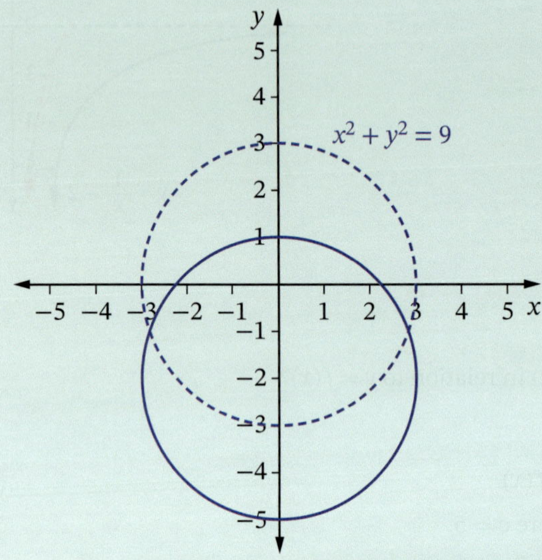

The equation of the translated circle can be found using the formula $(x - a)^2 + (y - b)^2 = r^2$ or by replacing y with $y - (-2)$.

Formula method

Circle, centre $(0, -2)$, radius 3.

$(x - a)^2 + (y - b)^2 = r^2$

$(x - 0)^2 + (y - (-2))^2 = 3^2$

$x^2 + (y + 2)^2 = 9$

Replace y by $y - (-2)$ method

$x^2 + y^2 = 9$

$x^2 + (y - (-2))^2 = 9$

$x^2 + (y + 2)^2 = 9$

b **Formula method**

$(x - 2)^2 + (y - 1)^2 = 4$ has centre $(2, 1)$, radius $\sqrt{4} = 2$.

Translating 5 units up moves the centre to $(2, 6)$.

$(x - 2)^2 + (y - 6)^2 = 4$

Replace y by $y - 5$ method

$(x - 2)^2 + (y - 1)^2 = 4$

$(x - 2)^2 + (y - 5 - 1)^2 = 4$

$(x - 2)^2 + (y - 6)^2 = 4$

EXERCISE 8.01 Answers on p. 502

Vertical translations of functions

1 Describe how the graph of each function is related to the graph of $y = x^2$.

a $y = x^2 + 3$ **b** $y = x^2 - 7$

c $y = x^2 - 1$ **d** $y = x^2 + 5$

2 Describe how the graph of each function is related to the graph of $y = x^3$.

a $y = x^3 + 1$ **b** $y = x^3 - 4$ **c** $y = x^3 + 8$

3 Describe how the graph of $y = \frac{1}{x}$ transforms to the graph of $y = \frac{1}{x} + 9$.

4 Find the equation of each translated function.

a $y = x^2$ is translated 3 units downwards

b $f(x) = 2^x$ is translated 8 units upwards

c $y = |x|$ is translated 1 unit upwards

d $y = x^3$ is translated 4 units downwards

e $f(x) = \log_{10} x$ is translated 3 units upwards

f $y = \frac{2}{x}$ is translated 7 units downwards

5 Describe the relationship between the graph of $f(x) = x^4$ and:

a $f(x) = x^4 - 1$ **b** $f(x) = x^4 + 6$

6 Find the equation of the transformed function if:

a $y = 2x^3 + 3$ is translated:

i 5 units down **ii** 3 units up

b $y = |x| - 4$ is translated:

i 1 unit up **ii** 2 units down

c $y = e^x + 2$ is translated:

i 1 unit down **ii** 3 units up

d $y = \log_e x - 1$ is translated:

i 11 units up **ii** 7 units down.

7 If $P(1, -3)$ lies on the function $y = f(x)$, find the transformed (image) point of P if the function is translated.

a 2 units up **b** 6 units down **c** m units up for $m > 0$.

8 Find the original point P on the function $y = f(x)$ if the coordinates of its transformed image are $P'(-1, 2)$ when the function is translated:

a 1 unit up **b** 3 units down.

EXAMPLE 2

9 Sketch each set of functions on the same number plane.

a $y = x^2$, $y = x^2 + 2$ and $y = x^2 - 3$

b $y = 3^x$ and $y = 3^x - 4$

c $y = |x|$ and $y = |x| - 3$

Foundation Mastery Complex

EXAMPLE 3

10 For each pair of graphs, find the equation of $y = g(x)$ in relation to $y = f(x)$.

a

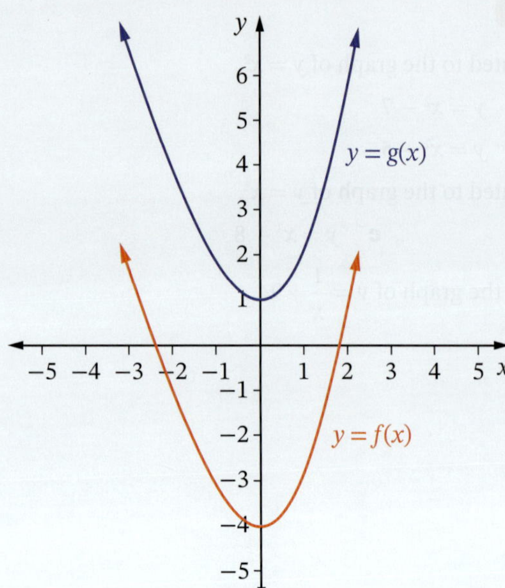

b

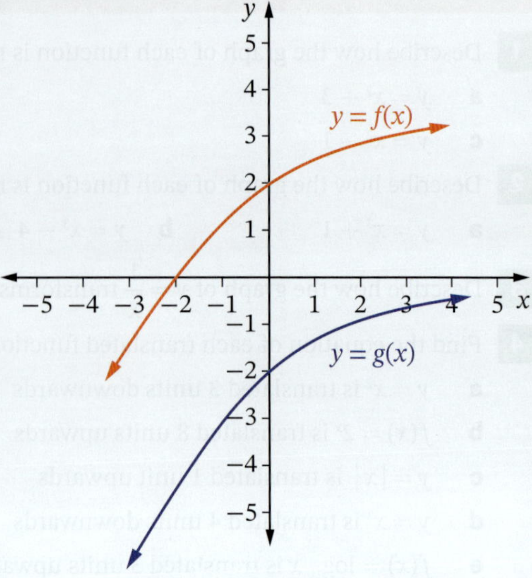

c

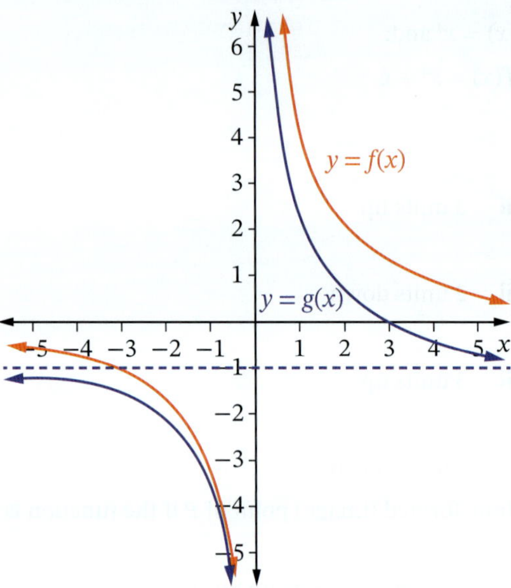

11 Given the graph of $y = f(x)$, sketch the graph of:

a $y = f(x) - 2$

b $y = f(x) + 1$

c $y = f(x) - 3$

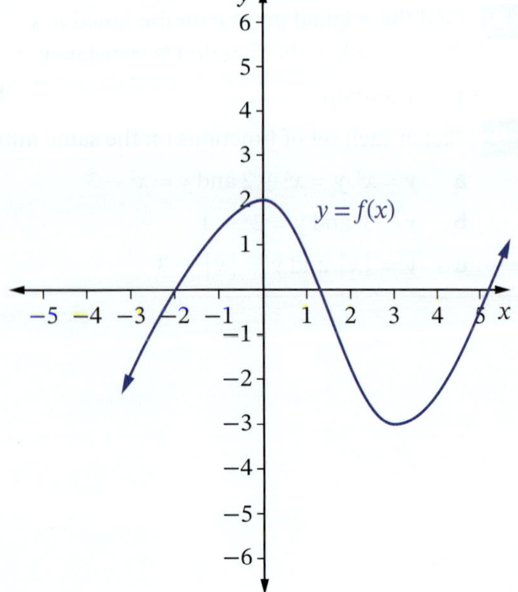

☐ Foundation ○ Mastery ⬡ Complex

12 The graph shows $y = f(x)$. Sketch the graph of:

a $y = f(x) - 1$

b $y = f(x) + 2$

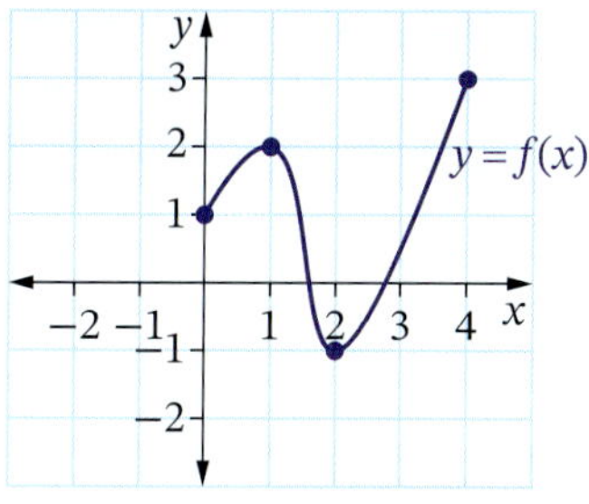

13 The circle $x^2 + y^2 = 1$ is translated 4 units up. Find the equation of the resulting circle.

EXAMPLE 4

14 Find the equation of the resulting circle if $(x + 1)^2 + (y + 2)^2 = 16$ is shifted:

a 9 units downwards

b 9 units upwards.

15 The function $y = f(x)$ is a quadratic function $y = ax^2 + bx + c$ with $a = -1$. It has x-intercepts -1 and 4. Find the equation of the resulting function if $y = f(x)$ is shifted 6 units downwards.

16 The function $y = g(x)$ is the function $y = f(x)$ shifted 3 units up. Find:

a the equation of $y = g(x)$ in terms of $y = f(x)$

b the coordinates of the point Q on $y = g(x)$ corresponding to the point $P(-1, -3)$ on $y = f(x)$

c the equation of the tangent line at Q if the gradient of the tangent to $y = f(x)$ at P is -7.

17 The function $y = h(x)$ is the function $y = f(x)$ translated 4 units down. Find:

a the equation of $y = h(x)$ in terms of $y = f(x)$

b the coordinates of the point B on $y = h(x)$ corresponding to the point $A(x, y)$ on $y = f(x)$

c the gradient of the normal at B if the gradient of the tangent to $y = f(x)$ at A is 5.

18 a Show that $\frac{3x+1}{x} = \frac{1}{x} + 3$ for $x \neq 0$.

b Hence or otherwise, sketch the graph of $y = \frac{3x+1}{x}$.

19 This is the resultant graph when an exponential function has been translated 3 units upwards. Sketch the original function.

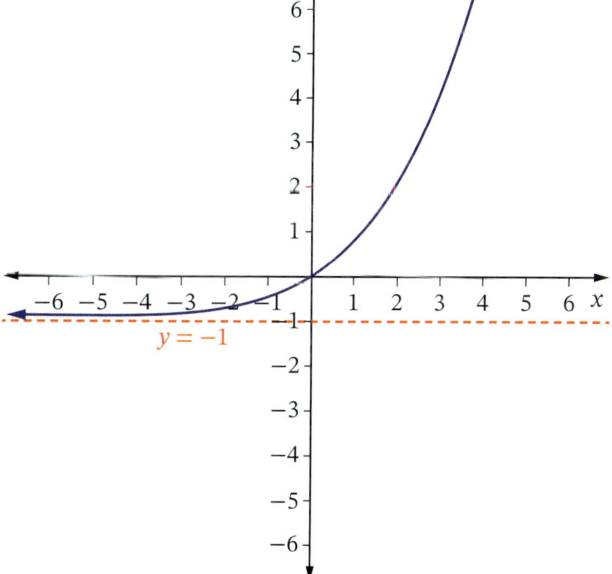

20 The circle $x^2 + 8x + y^2 - 2y + 1 = 0$ is translated 2 units downwards. Find the radius and centre of the transformed circle.

21 Find the equation of the circle and state its radius and centre if $(x - 4)^2 + (y - 5)^2 = 81$ is shifted 7 units up.

☐ Foundation ○ Mastery ○ Complex

8.02 Horizontal translations of functions

A **horizontal translation** of a function shifts the graph of the function left or right without changing its size or shape.

Video
Translations of functions

Worksheets
Translations of functions

Graphing translations of functions

Technology

Horizontal translations

Use graphing technology to explore the effect of parameter a on each graph below. If you don't have dynamic software, substitute different values for a into the equation. Use positive and negative values, integers and fractions for a.

$f(x) = (x - a)^2$ $\quad f(x) = (x - a)^3$ $\quad f(x) = (x - a)^4$ $\quad f(x) = e^{(x - a)}$

$f(x) = \ln(x - a)$ $\quad f(x) = \dfrac{1}{x - a}$ $\quad f(x) = |x - a|$

How does the graph change as the value of a changes?

What is the difference between positive and negative values of a?

The value and sign of a determines the amount and direction of the horizontal translation of $f(x)$.

Horizontal translations

For the function $y = f(x)$:

$y = f(x - a)$ translates the graph horizontally (along the x-axis).

Replacing x with $x - a$ in the equation of a function translates its graph horizontally.

If $a > 0$, the graph is translated to the right by a units.

If $a < 0$, the graph is translated to the left.

A horizontal translation changes the x values of the function.

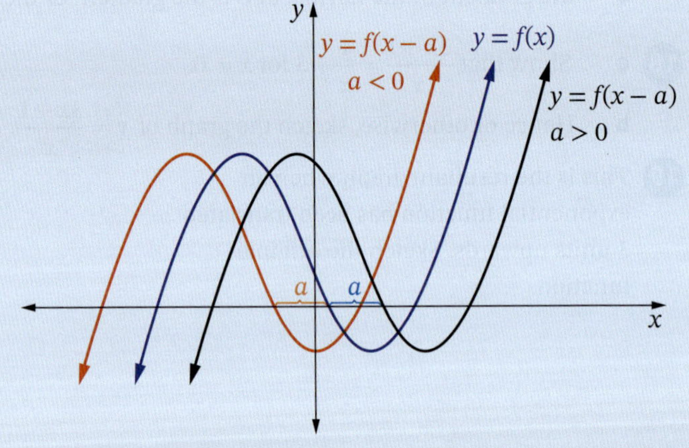

Example 5

a What is the relationship of $f(x) = \log_2(x + 3)$ to $f(x) = \log_2 x$?

b If the graph $y = (x - 4)^3$ is translated 7 units to the right, find the equation of the transformed function.

c The point $P(2, 5)$ lies on the function $y = f(x)$. Find the image (corresponding point) of P on the function $y = f(x + 1)$.

d The point $Q(3, -4)$ on the graph of $y = f(x - 2)$ is the image of point $P(x, y)$ on $y = f(x)$. Find the coordinates of P.

Solution

a $f(x) = \log_2(x + 3) = \log_2(x - (-3))$ is a horizontal translation 3 units to the left from the function $f(x) = \log_2 x$.

b If $y = (x - 4)^3$ is translated 7 units to the right, replace x with $x - 7$.

$y = f(x - a)$ where $a = 7$ and $f(x) = (x - 4)^3$

$y = (x - 7 - 4)^3 = (x - 11)^3$

So the equation of the transformed function is $y = (x - 11)^3$.

c $y = f(x + 1) = f(x - (-1))$ means a translation 1 unit left, so x values shift 1 unit to the left.

Image of P is $(2 - 1, 5) = (1, 5)$

d $y = f(x - 2)$ means a translation 2 units right so, x values shift 2 units to the right.

But $Q(3, -4)$ is the image of $P(x, y)$.

So P is 2 units *left* of Q. Thus, the x value of P will be $3 - 2 = 1$.

Therefore, the coordinates of P are $(1, -4)$.

Example 6

a The graph of $y = f(x)$ shown is transformed into $y = f(x - 3)$. Sketch the transformed graph.

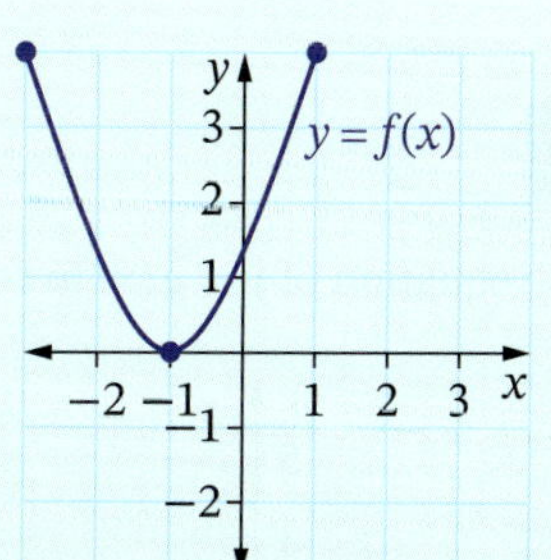

Video
Horizontal translations of functions

b Sketch the graph of:

i $y = |x + 3|$ **ii** $y = \dfrac{1}{x - 2}$

Solution

a The graph $y = f(x - 3)$ describes a horizontal translation of $y = f(x)$ 3 units to the right.

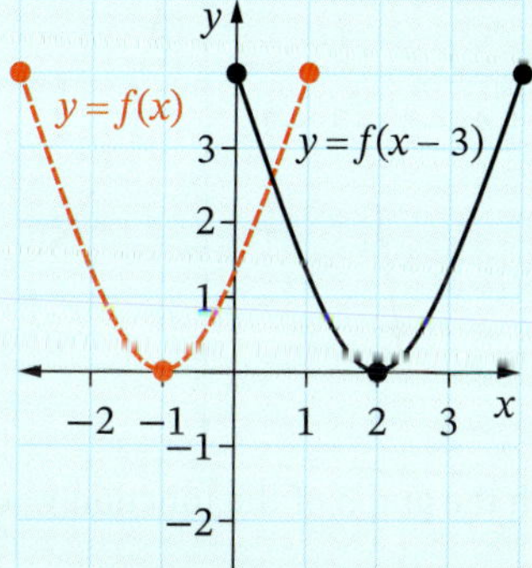

b **i** The function $y = |x + 3|$ is in the form $y = f(x - a)$ where $a = -3$.

Since $a < 0$, $y = |x|$ is shifted 3 units to the left.

If you need to find some points on the graph of $y = |x + 3|$ you could subtract 3 from x values of $y = |x|$.

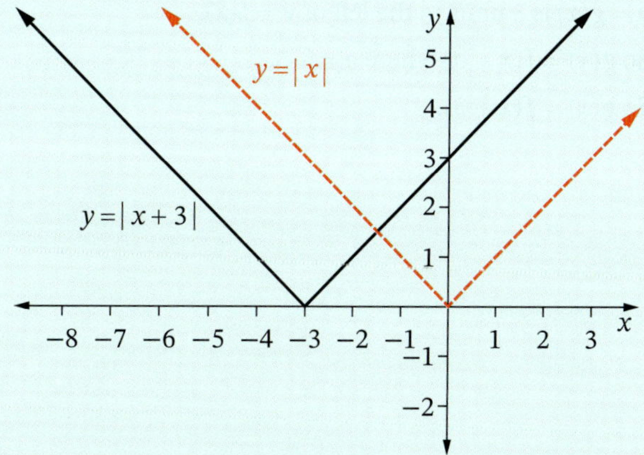

ii $y = \dfrac{1}{x-2}$ is in the form $y = f(x - a)$, where $a = 2$.

$y = \dfrac{1}{x}$ is shifted 2 units to the right.

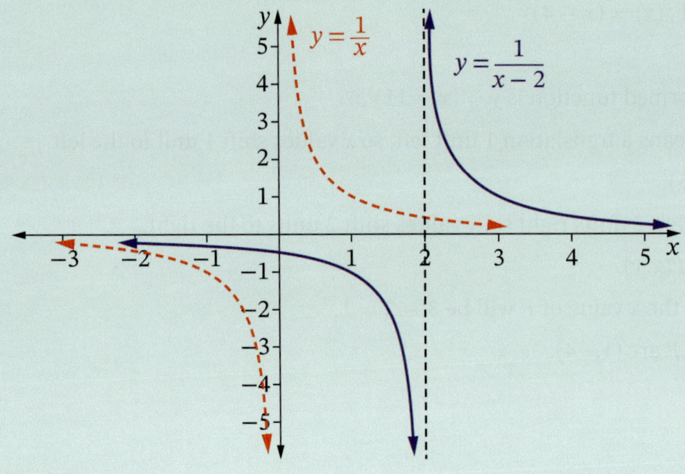

Example 7

a What is the equation of $y = g(x)$ in relation to $y = f(x)$?

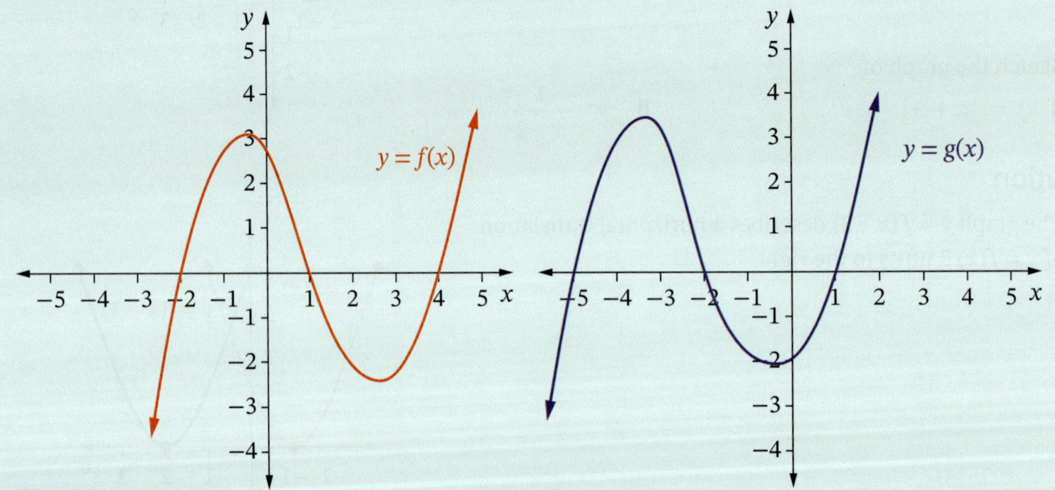

b Sketch the graph of the function $y = x^2 - x$ translated 2 units to the right and find its equation.

Solution

a $y = g(x)$ is 3 units to the left of $y = f(x)$.

So $g(x) = f(x - a)$ where $a = -3$.

So $g(x) = f(x + 3)$.

b First, graph $y = x^2 - x$.

For x-intercepts, $y = 0$:

$x^2 - x = 0$

$x(x - 1) = 0$

$x = 0, x = 1$

For y-intercept, $x = 0$:

$y = (0)^2 - 0$

$= 0$

This is a concave up parabola.

Shift this graph 2 units to the right for the transformed graph.

So x-intercepts of the new graph are $0 + 2 = 2$ and $1 + 2 = 3$.

To find its equation and translate $y = f(x)$ 2 units to the right, replace x with $x - 2$, that is, find $f(x - 2)$.

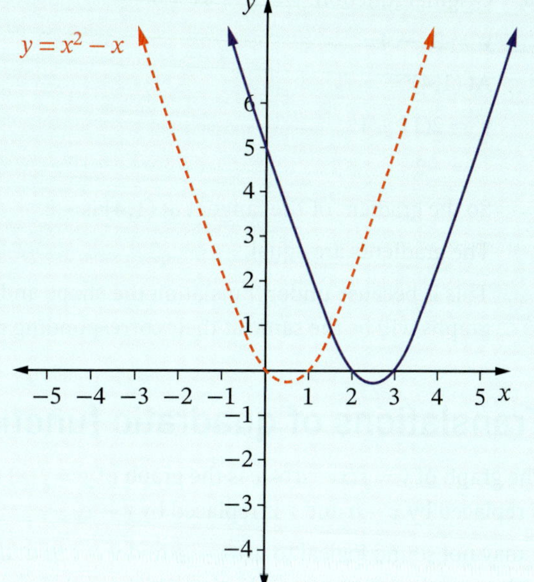

$f(x) = x^2 - x$

$f(x - 2) = (x - 2)^2 - (x - 2)$

$= x^2 - 4x + 4 \quad x + 2$

$= x^2 - 5x + 6$

It can be shown algebraically that this graph has x-intercepts 2 and 3, and y-intercept 6, as shown on the graph.

Example 8

The point $P(1, 4)$ lies on the function $y = x^2 - 4x + 7$. The function is shifted 3 units to the left.

a Find the gradient of the tangent at the the image of P on the resultant function. Is the function increasing, decreasing or stationary at this point?

b Show that this gradient is the same as the gradient of the original function at $(1, 4)$ and explain why this is the case.

Solution

a If the point $(1, 4)$ on $y = x^2 - 4x + 7$ is translated 3 units to the left, the corresponding point is $(1 - 3, 4) \equiv (-2, 4)$.

Find the resultant function:

$y = f(x - (-3)) = f(x + 3)$.

$y = (x + 3)^2 - 4(x + 3) + 7$

$= x^2 + 6x + 9 - 4x - 12 + 7$

$= x^2 + 2x + 4$

$y' = 2x + 2$

At $(-2, 4)$:

$y' = 2(-2) + 2$

$= -2$

So the gradient of the tangent at $(-2, 4)$ is -2.

Since $y' < 0$, the curve is decreasing at $(-2, 4)$.

b Original function: $y = x^2 - 4x + 7$

$y' = 2x - 4$

At $(1, 4)$:

$y' = 2(1) - 4$

$= -2$

So the gradient of the tangent at $(1, 4)$ is -2.

The gradients are equal.

This is because under translation the shape and size of the graph don't change. The gradient of both graphs will be the same at their corresponding points.

Translations of quadratic functions, cubic functions and circles

The graph of $y = f(x - a) + c$ is the graph of $y = f(x)$ translated a units to the right and c units upwards, where x is replaced by $x - a$ and y is replaced by $y - c$.

It may not sound logical to use $x - a$ and $y - c$ to *add* to x and y values respectively, but this is consistent with some equations of graphs that we learned about in Chapter 3, *Functions and graphs* and Chapter 5, *Further functions and graphs*.

- The vertex form of a quadratic **function** is $y = a(x - b)^2 + c$, where the vertex has been translated from $(0, 0)$ to (b, c).
- The **cubic function** is $y = k(x - b)^3 + c$, where the point of inflection has been translated from $(0, 0)$ to (b, c).
- The **equation of a circle** is $(x - a)^2 + (y - b)^2 = r^2$, where the centre has been translated from $(0, 0)$ to (a, b).

So why is it $x - a$ and not $x + a$? Because for $y = f(x - a)$, if we make x the subject, we get $g(y) = x - a$, and $x = g(y) + a$ so the value of x has increased by a.

For example, for $y = (x - 2)^3$ is the translation of $y = x^3$ by 2 units to the right because if we make x the subject, we can see that the x value has increased by 2:

$\sqrt[3]{y} = x - 2$

$x = \sqrt[3]{y} + 2$

EXERCISE 8.02 Answers on p. 503

Horizontal translations of functions

1 Describe how the graph of each function is related to the graph of $y = x^2$.

a $y = (x - 4)^2$ **b** $y = (x + 2)^2$

2 Describe how the graph of each function is related to the graph of $y = x^3$.

a $y = (x - 5)^3$ **b** $y = (x + 3)^3$

3 Find the equation of each translated graph.

a $y = x^2$ translated 3 units to the left **b** $f(x) = 2^x$ translated 8 units to the right

c $y = |x|$ translated 1 unit to the left **d** $y = x^3$ translated 4 units to the right

e $f(x) = \log_{10} x$ translated 3 units left

4 Describe how $y = \frac{1}{x}$ transforms to $y = \frac{1}{x - 3}$.

5 Describe the relationship between $f(x) = x^4$ and:

a $f(x) = (x + 2)^4$ **b** $f(x) = (x - 5)^4$

☐ Foundation ○ Mastery ○ Complex

6 Find the equation if:

a $y = -x^2$ is translated

i 4 units to the left ii 8 units to the right

b $y = |x|$ is translated

i 3 units to the right ii 4 units to the left

c $y = e^{x+2}$ is translated

i 4 units to the left ii 7 units to the right

d $y = \log_2(x - 3)$ is translated

i 2 units to the right ii 3 units to the left.

7 If $P = (1, -3)$ lies on the function $y = f(x)$, find the coordinates of the image point of P if the function is transformed to $y = f(x - a)$ where:

a $a = 4$ b $a = -9$ c $a = -t$

8 Find the original point P on the function $y = f(x)$ if the coordinates of its image are $P'(-1, 2)$ when the function is translated:

a 4 units to the left b 8 units to the right.

9 Sketch on the same number plane:

a $y = x^3$ and $y = (x + 1)^3$ b $f(x) = \ln x$ and $f(x) = \ln(x + 2)$

EXAMPLE 6

10 The graph shown is $y = f(x)$. Sketch the graph of:

a $y = f(x - 1)$ b $y = f(x + 3)$

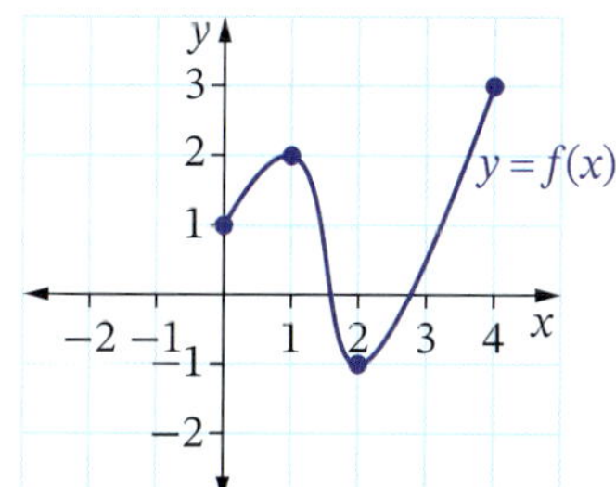

11 Find the equation of the transformed function if $f(x) = x^5$ is translated:

a 5 units down b 3 units to the right

c 2 units up d 7 units to the left.

12 The point $P'(3, -2)$ is the image of a point P on $y = f(x)$ after it has been translated 4 units to the left. Find the original point.

13 For each graph of $y = f(x)$ and $y = g(x)$, find the equation of $y = g(x)$ in terms of $f(x)$.

EXAMPLE 7

a

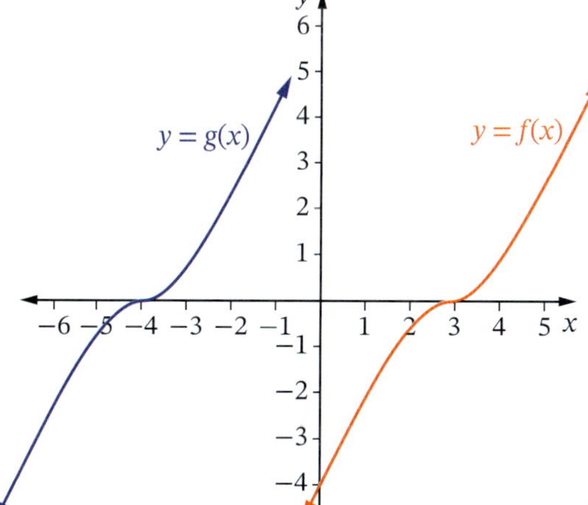

b

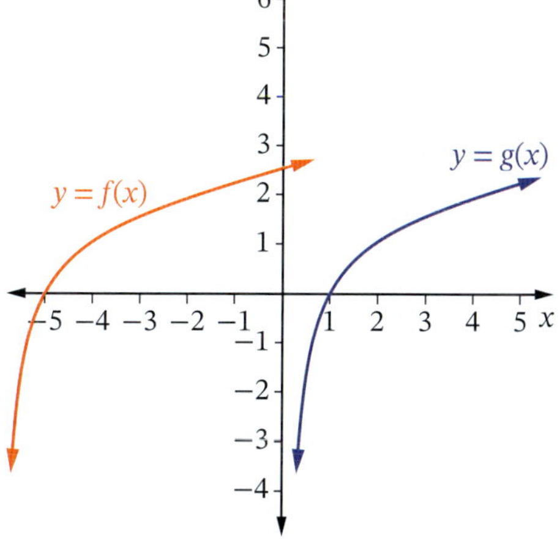

14 The circle $x^2 + y^2 = 9$ is translated 7 units to the left. Find the equation of the resulting circle.

Foundation Mastery Complex

15 Describe the translations that transformed $f(x) = x^2$ into $g(x) = (x + 8)^2 - 3$.

16 The function $y = f(x)$ is a cubic function $y = ax^3 + bx^2 + cx + d$, where $a = -2$. It has x-intercepts $-3, 1, 4$. Find the equation of the function $y = f(x)$ translated 4 units to the left.

17 The graph shows the resultant graph when a circle has been translated 2 units to the left. Sketch the original function and find its equation.

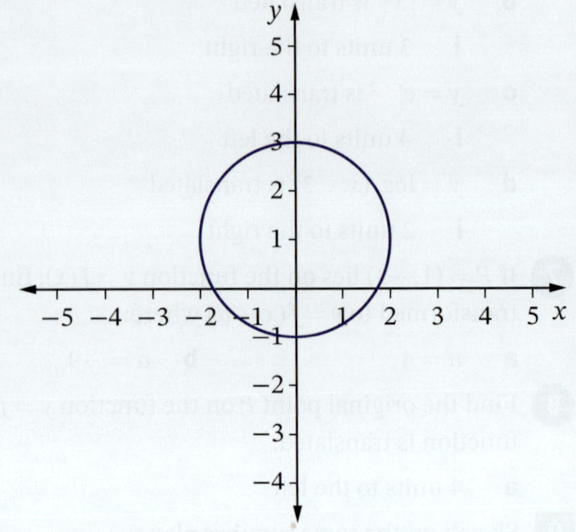

18 Find the equation of the circle $(x + 3)^2 + (y - 1)^2 = 9$ shifted 6 units to the left.

19 The function $y = g(x)$ is $y = f(x)$ translated 7 units to the left. Find:

a the equation of $y = g(x)$ in terms of $y = f(x)$

b the coordinates of the point N on $y = g(x)$ corresponding to the point $M(3, 2)$ on $y = f(x)$

c the equation of the tangent line at N if the gradient of the tangent to $y = f(x)$ at M is 3.

20 Find the equation of the circle if $x^2 + y^2 = 9$ is translated:

a 3 units to the left and 2 units up **b** 3 units to the right and 2 units down.

21 The function $y = V(x)$ is $y = U(x)$ translated 6 units to the right. Find:

a the equation of $y = V(x)$ in terms of $y = U(x)$

b the coordinates of the point Q on $y = V(x)$ corresponding to the point $P(x, y)$ on $y = U(x)$

c the gradient of the normal at Q if the gradient of the tangent to $y = U(x)$ at P is -2.

22 The circle $x^2 + 2x + y^2 - 6y - 6 = 0$ is translated 5 units to the right. Find the radius and centre of the resulting circle.

☐ Foundation ○ Mastery ○ Complex

Reflections of functions

8.03

Reflections in the x-axis

A **reflection** of a function flips the graph of the function across a line to create a (back-to-front) mirror image on the other side. In this course, we only study reflections in the x- and y-axes.

A reflection in the x-axis is a **vertical reflection**.

Worksheets
Advanced graphs

Matching graphs (Advanced)

Example 9

For each function, sketch the graph of $y = f(x)$ and $y = -f(x)$ on the same number plane.

a $f(x) = x^2 - 2x$ **b** $f(x) = x^3$

Solution

a $f(x) = x^2 - 2x$ is a concave up parabola.

For x-intercepts: $f(x) = 0$

$$x^2 - 2x = 0$$
$$x(x - 2) = 0$$
$$x = 0, 2$$

For y-intercept, $x = 0$:

$f(0) = 0^2 - 2(0) = 0$

Axis of symmetry at $x = 1$ (halfway between 0 and 2):

$f(1) = 1^2 - 2(1) = -1$

Minimum turning point at $(1, -1)$.

$$y = -f(x)$$
$$= -(x^2 - 2x)$$
$$= -x^2 + 2x$$

$y = -x^2 + 2x$ is a concave down parabola also with x-intercepts 0, 2, y-intercept 0 and axis of symmetry at $x = 1$.

$f(1) = -1^2 + 2(1) = 1$

Maximum turning point at $(1, 1)$.

Draw both graphs on the same set of axes.

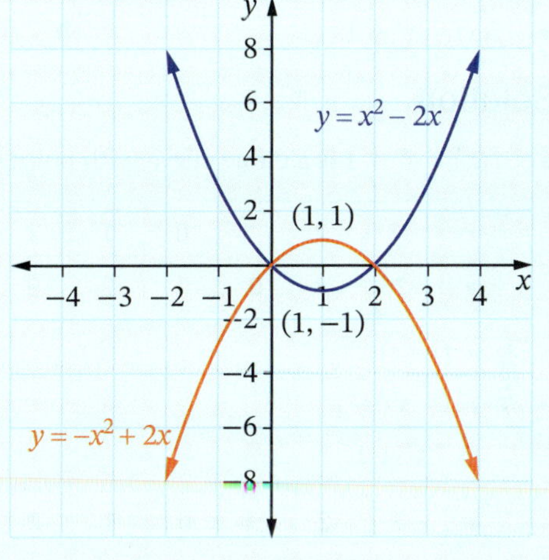

b $f(x) = x^3$ is a cubic curve with a point of inflection at $(0, 0)$.

x	−3	−2	−1	0	1	2	3
y	−27	−8	−1	0	1	8	27

$$y = -f(x)$$
$$= -x^3$$

x	−3	−2	−1	0	1	2	3
y	27	8	1	0	−1	−8	−27

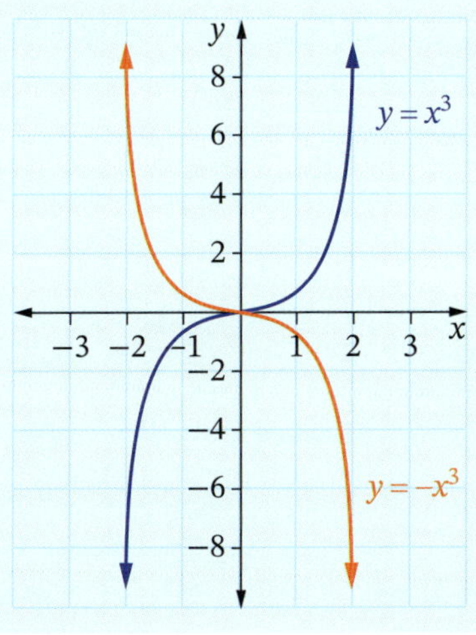

$y = -f(x)$ changes the sign of the y values of the original function: from positive to negative, or negative to positive. On the number plane, this means reflecting the graph in the x-axis.

The graph of $y = -f(x)$

The graph of $y = -f(x)$ is a reflection of the graph of $y = f(x)$ in the x-axis.

Replacing y with $-y$ in the equation of a function reflects its graph vertically in the x-axis.

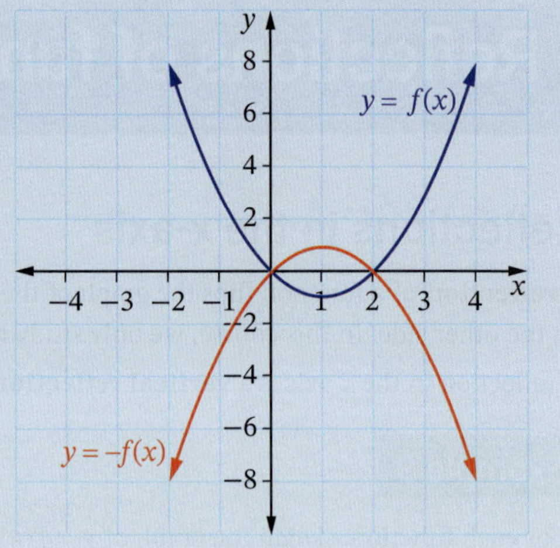

Reflections in the y-axis

We have already seen that some functions are even or odd by finding $f(-x)$. We can see the relationship between $f(x)$ and $f(-x)$ by drawing their graphs.

Example 10

For each function, sketch the graph of $y = f(x)$ and $y = f(-x)$ on a number plane.

a $f(x) = x^3 + 1$

b $f(x) = \dfrac{1}{x-2}$

Solution

a $f(x) = x^3 + 1$ is a cubic curve with point of inflection at (0, 1).

x	−3	−2	−1	0	1	2	3
y	−26	−7	0	1	2	9	28

$y = f(-x)$
$= (-x)^3 + 1$
$= -x^3 + 1$

x	−3	−2	−1	0	1	2	3
y	28	9	2	1	0	−7	−26

Draw $y = x^3 + 1$ and $y = -x^3 + 1$ on the same set of axes.

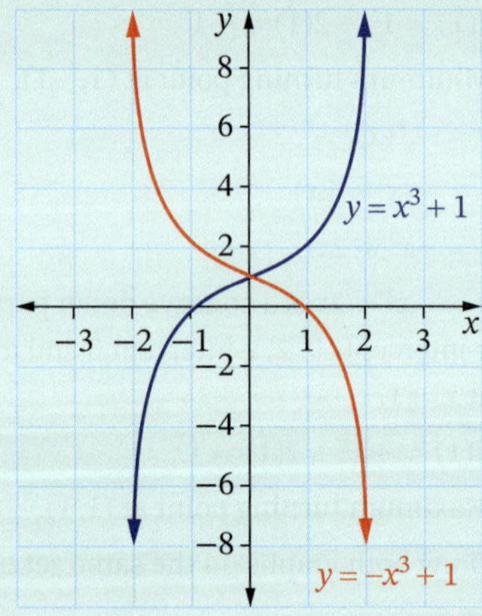

b $f(x) = \dfrac{1}{x-2}$ is a hyperbola with asymptotes at $x = 2$ and $y = 0$.

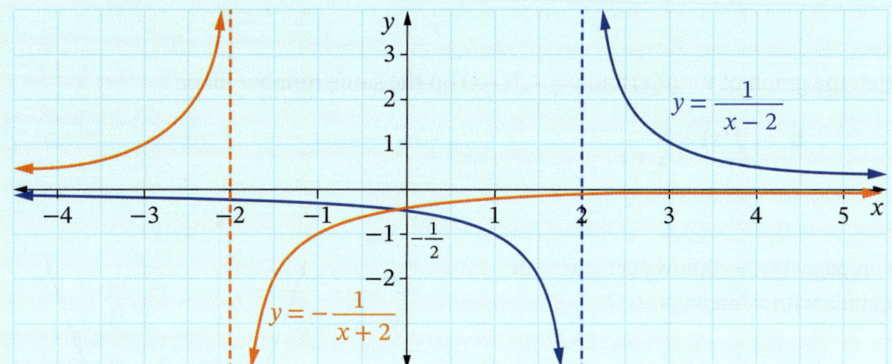

y-intercept, $x = 0$:

$$f(0) = \frac{1}{0-2} = -\frac{1}{2}$$

$$y = f(-x)$$

$$= \frac{1}{-x-2}$$

$$= -\frac{1}{x+2}$$

Asymptotes at $x = -2$ and $y = 0$.

y-intercept, $x = 0$:

$$f(0) = -\frac{1}{0+2} = -\frac{1}{2}$$

Draw $y = \dfrac{1}{x-2}$ and $y = -\dfrac{1}{x+2}$ on the same set of axes.

$y = f(-x)$ changes the sign of the x value of the original function: from positive to negative, or negative to positive. On the number plane, this means reflecting the graph in the y-axis.

Notice that for even functions, $y = f(x) = f(-x)$. Even functions are already symmetrical about the y-axis so its reflected graph is the same as the original graph.

The graph of $y = f(-x)$

The graph of $y = f(-x)$ is a reflection of the graph of $y = f(x)$ in the y-axis.

Replacing x with $-x$ in the equation of a function reflects its graph horizontally in the y-axis.

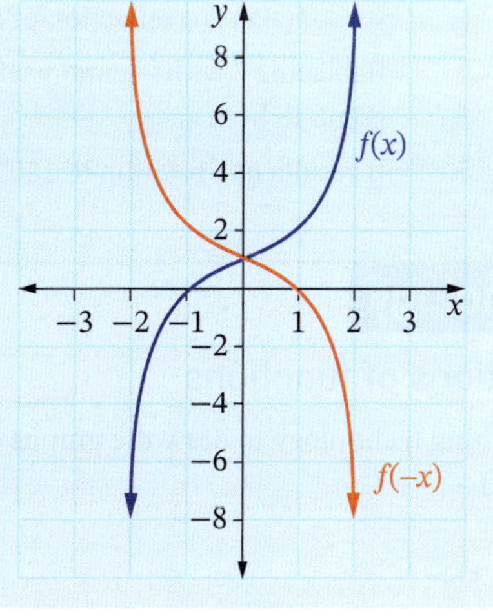

Combined reflections in both axes

Example 11

For each function, sketch the graph of $y = f(x)$ and $y = -f(-x)$ on the same number plane.

a $f(x) = x^2 - 2x$ **b** $f(x) = \frac{2}{x+1}$

Solution

a From Example **9a** on page 339, we know $f(x) = x^2 - 2x$ is a concave up parabola with x-intercepts 0, 2.

$$y = -f(-x)$$
$$= -[(-x)^2 - 2(-x)]$$
$$= -(x^2 + 2x)$$
$$= -x^2 - 2x$$

This is a concave down parabola with x-intercepts 0, −2.

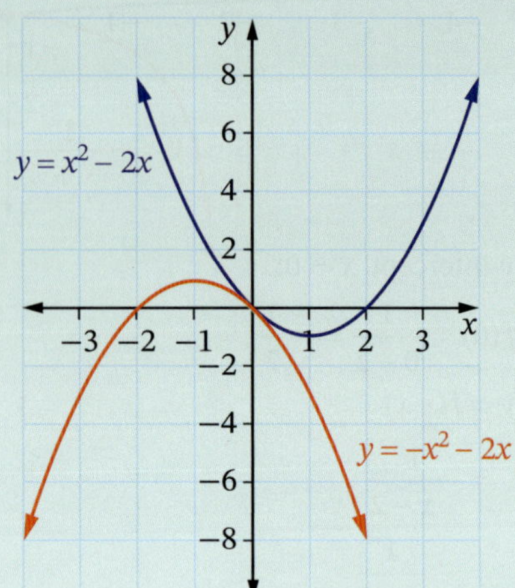

b $f(x) = \frac{2}{x+1}$

This is a hyperbola with asymptotes at $x = -1$, $y = 0$ and y-intercept $f(0) = 2$.

$$y = -f(-x)$$
$$= -\frac{2}{-x+1}$$
$$= \frac{2}{x-1}$$

This is a hyperbola with asymptotes at $x = 1$, $y = 0$ and y-intercept $f(0) = -2$.

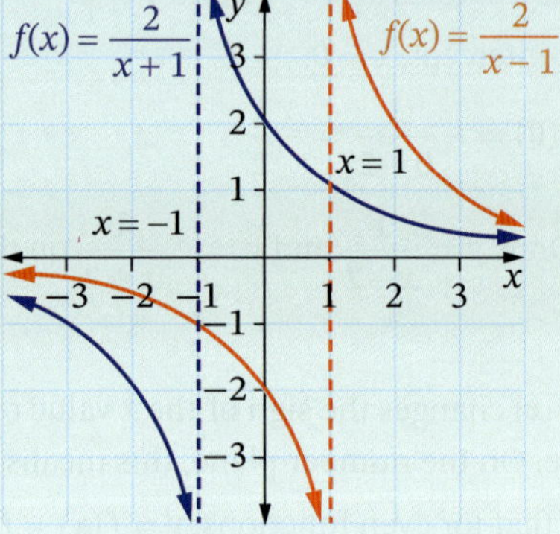

The graph of $y = -f(-x)$

$y = -f(-x)$ is a reflection of the graph of $y = f(x)$ in both the x- and y-axes.

Replacing x with $-x$ and y with $-y$ in the equation of a function reflects its graph in both axes.

It results in a rotation of 180° about the origin.

Technology

Reflections of functions

Use graphing technology to draw the graphs of different functions $y = f(x)$ together with:

$y = -f(x)$

$y = f(-x)$

$y = -f(-x)$.

Some functions you could use are $y = x^2$, $y = x^3$, $y = \frac{1}{x}$, $y = |x|$, $y = a^x$, $y = \log_a x$, $y = \sqrt{x}$.

Are any of these functions the same as $y = f(x)$ if $f(x)$ is an even or odd function? Why?

Example 12

The graph of $y = f(x)$ is shown below.

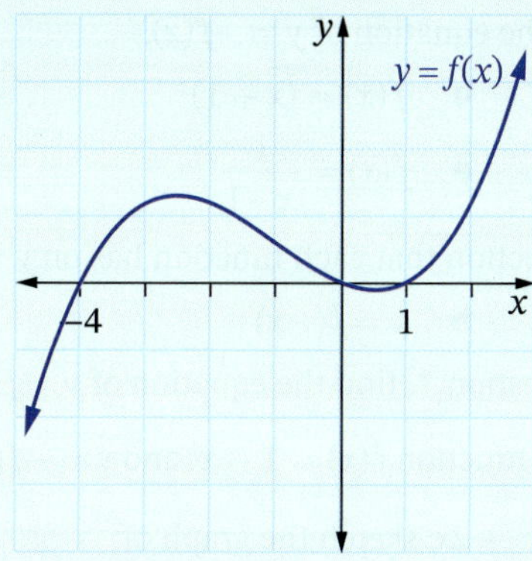

Sketch the graph of:

a $y = -f(x)$ **b** $y = f(-x)$ **c** $y = -f(-x)$

Solution

a $y = -f(x)$ is a reflection in the x-axis.

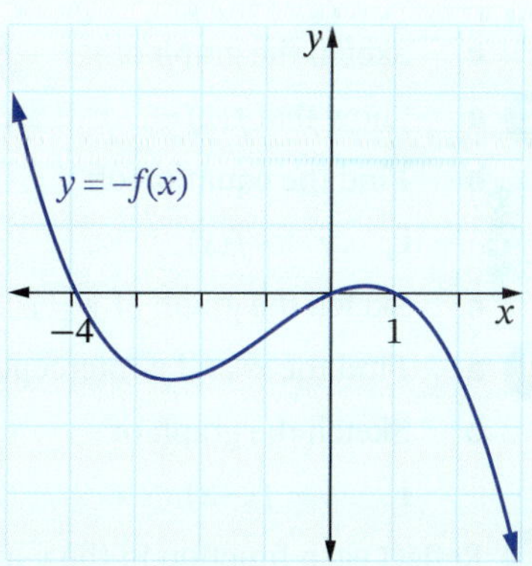

b $y = f(-x)$ is a reflection in the y-axis.

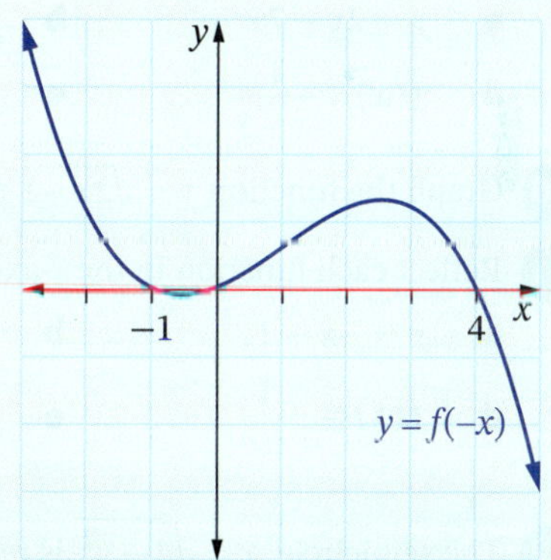

c $y = -f(-x)$ is a reflection in both the x-and y-axes.
Using the graph from part **a** that has been reflected in the x-axis and reflecting it in the y-axis gives the graph on the right.

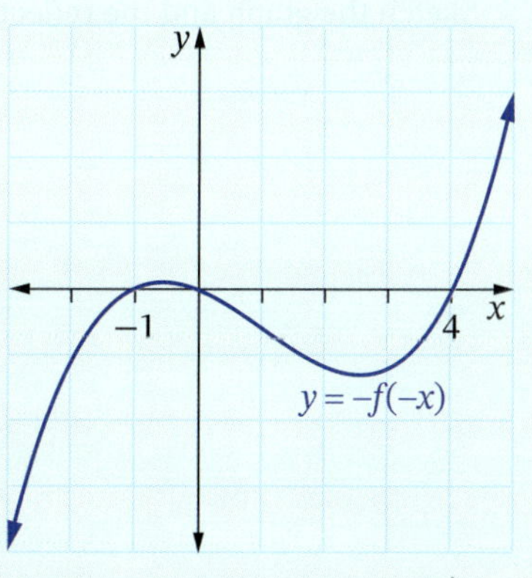

EXERCISE 8.03 Answers on p. 504

Reflections of functions

EXAMPLE 9

1 For each function, find the equation of $y = -f(x)$.

a $f(x) = x^2 - 2$ **b** $f(x) = (x + 1)^3$ **c** $f(x) = 5x - 3$

d $f(x) = |2x + 5|$ **e** $f(x) = \dfrac{1}{x-1}$ **f** $f(x) = 3^x$

2 Describe the type of reflection that each function has on $y = f(x)$.

a $y = -f(x)$ **b** $y = f(-x)$ **c** $y = -f(-x)$

EXAMPLE 10

3 For each function in Question **1**, find the equation of $y = f(-x)$.

4 Sketch the graphs of the function $f(x) = 1 - x^3$ and $y = -f(x)$ on the same number plane.

5 For the function $f(x) = x^2 + 2x$, sketch the graph of:

a $y = f(x)$ **b** $y = -f(x)$ **c** $y = f(-x)$ **d** $y = -f(-x)$

6 **a** Show that $f(x) = 2x^2$ is an even function.

b Find the equation of:

i $y = f(-x)$ **ii** $y = -f(x)$

c Sketch the graph of $y = -f(-x)$.

7 **a** Show that $f(x) = -x^3$ is an odd function.

b Find the equation of:

i $y = -f(x)$ **ii** $y = -f(-x)$

c Sketch the graph of $y = f(-x)$.

8 **a** Find the x- and y-intercepts of the graph of $f(x) = x^3 - 7x^2 + 12x$ and sketch the graph.

b Sketch the graph of:

i $y = f(-x)$ **ii** $y = -f(x)$ **iii** $y = -f(-x)$

9 Reflect each function in the y-axis and find its new equation.

a $y = 2x - 7$ **b** $y = x^2 - 2x + 5$ **c** $f(x) = x^3 + x^2 - 4x + 1$

d $f(x) = \dfrac{1}{x+3}$ **e** $y = 2^x$ **f** $f(x) = \log_2 x$

10 Graph the function $y = \sqrt{2x-1}$ and its reflection in the y-axis.

11 Reflect each function in the x-axis and find its new equation.

a $y = -x + 1$ **b** $y = x^2 + 3$ **c** $f(x) = x^3 + 4x - 3$

d $f(x) = -\dfrac{1}{2x}$ **e** $y = 5^{3x}$ **f** $f(x) = \ln x$

12 The semicircle $x = \sqrt{4 - y^2}$ is reflected in the y-axis.

Sketch the graph and the reflected graph on the same number plane.

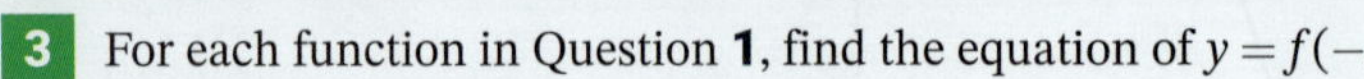

Foundation Mastery Complex

13 The function $y = \log_2 x$ is reflected in both the x- and y-axes. EXAMPLE 11

a Find the equation of the reflected graph.

b Sketch both graphs on the number plane.

14 For each graph of $y = f(x)$, sketch the graph of $y = f(-x)$.

a

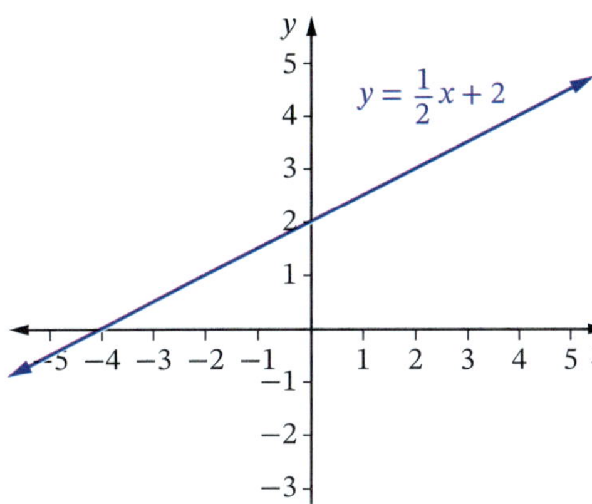

b

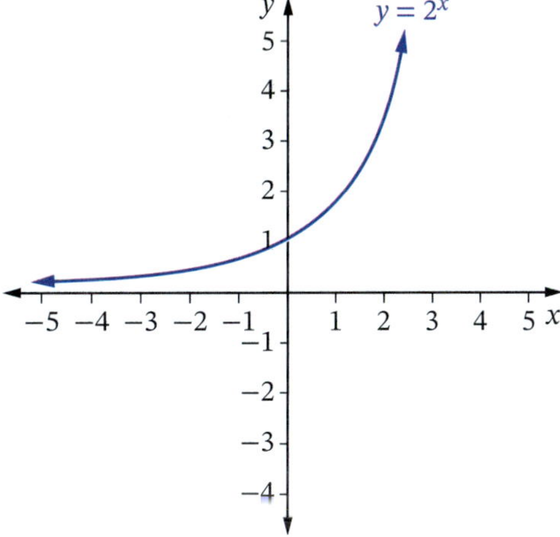

c

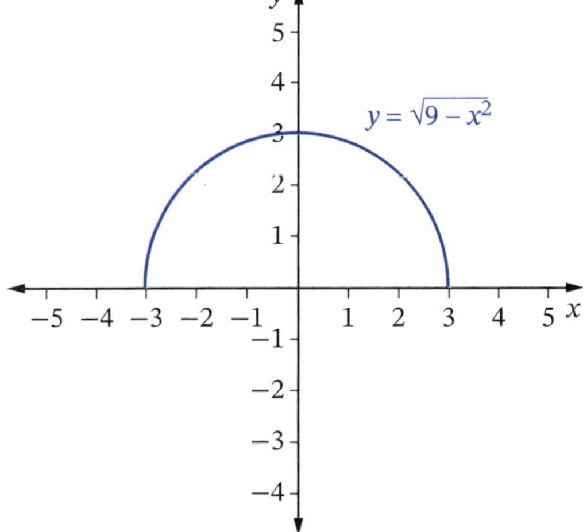

15 For the function $y = e^x$, sketch the graph of: EXAMPLE 12

a $y = -f(x)$ **b** $y = f(-x)$ **c** $y = -f(-x)$

16 **a** Find the equation of the function $y = f(x)$ that is shifted 5 units up and then reflected in the y-axis. Would this be the same equation if the reflection came first?

b Find the equation of the function $y = f(x)$ that is shifted 5 units down and then reflected in the x-axis. Would this be the same equation if the reflection came first?

☐ Foundation ○ Mastery ○ Complex

17 For each graph of $y = f(x)$, sketch the graph of $y = -f(-x)$.

a

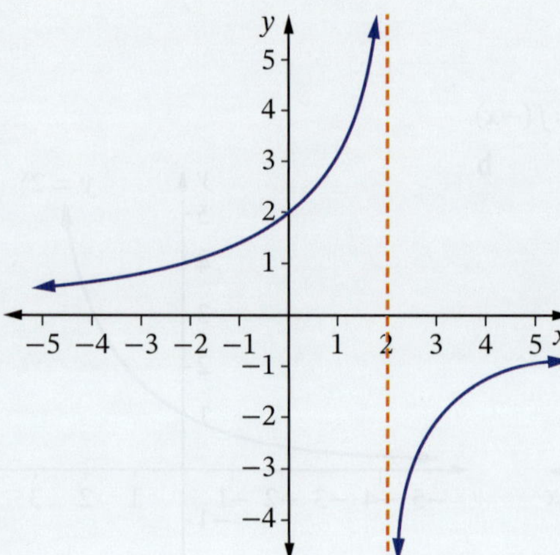

b

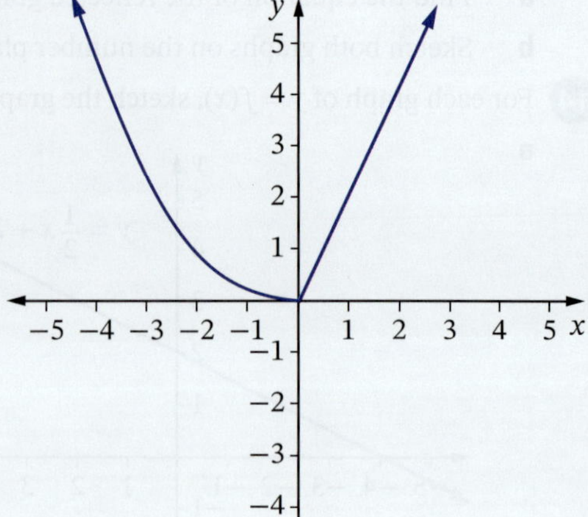

c

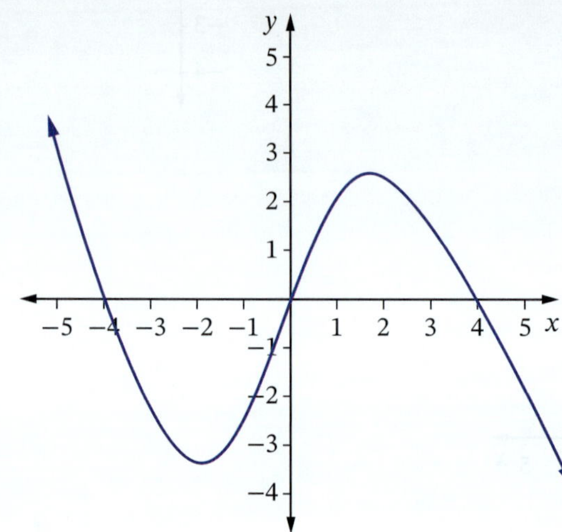

18 Find the equation of the resultant function if $f(x) = x^3 - 2x$ is:

a translated 4 units to the left then reflected in the x-axis

b translated 4 units to the right then reflected in the y-axis

c reflected in the x-axis then translated 4 units to the left

d reflected in the y-axis then translated 4 units to the right.

19 The circle $x^2 + 4x + y^2 - 6y + 9 = 0$ is reflected in the y-axis. Sketch the graph of the reflected circle, showing its radius and centre.

20 The function $y = 2|x - 1| + 3$ is reflected in both the x- and y-axis. Find the equation of the transformed function.

☐ Foundation ○ Mastery ○ Complex

Vertical dilations of functions 8.04

A **dilation** stretches or shrinks a function, changing its size and shape.

In this course, we only study dilations from the x- and y-axes, in the vertical and horizontal direction respectively.

A **vertical dilation** stretches (enlarges) or shrinks (reduces) the graph up or down, away from or towards the x-axis.

Technology

Vertical dilation

Use graphing technology to explore the effect of parameter k on each graph below. If you don't have dynamic geometry software, substitute different values for k into the equation. Use positive and negative values, integers and fractions for k.

$f(x) = kx$	$f(x) = kx^2$	$f(x) = kx^3$	$f(x) = kx^4$
$f(x) = ke^x$	$f(x) = k\ln x$	$f(x) = \dfrac{k}{x}$	$f(x) = k\lvert x \rvert$

How does the graph change as the value of k changes?

What is the difference between positive and negative values of k?

Notice that k stretches or shrinks the graph vertically, up and down from the x-axis (or along the y-axis) and changes its shape. The value of the parameter k controls the amount of stretching or shrinking.

We call k the **dilation factor** or **factor**.

Vertical dilations

For the curve $y = f(x)$:

$y = kf(x)$ dilates the curve vertically (from the x-axis) by a dilation factor of k.

This is the same as replacing y with $\dfrac{y}{k}$ in $y = f(x)$.

If $k > 1$, the graph is stretched, or expanded, away from the x-axis.

If $0 < k < 1$, the graph is shrunk, or reduced, towards the x-axis.

A vertical dilation changes the y values of the function.

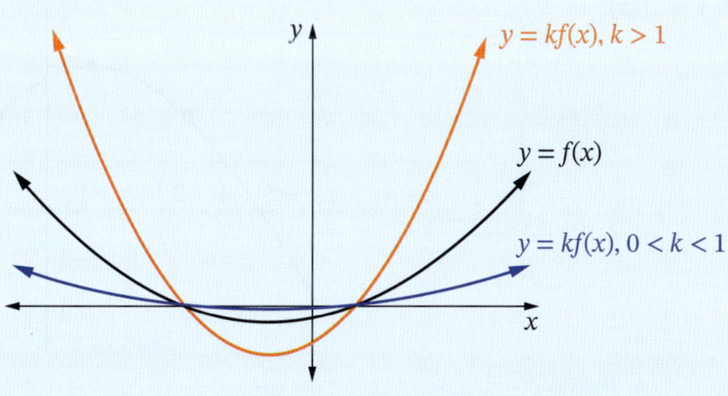

Example 13

a The function $y = x^7$ is dilated vertically by a factor of 3. Find the equation of the transformed function.

b Describe how the function $f(x) = \dfrac{\log_2 x}{2}$ is related to the function $f(x) = \log_2 x$.

c Find the factor of each dilation of a basic function and state whether the dilation stretches or shrinks the graph.

i $y = 7x^2$ **ii** $y = \dfrac{e^x}{5}$

Solution

a If a function $y = f(x)$ has a vertical dilation with factor k, the equation of its transformed function is $y = kf(x)$.

So if the function $y = x^7$ has a vertical dilation with factor 3, the equation of the transformed function is $y = 3x^7$.

b $f(x) = \dfrac{\log_2 x}{2}$

$= \dfrac{1}{2}\log_2 x$

So the function is in the form $y = kf(x)$, where $k = \dfrac{1}{2}$.

Since $0 < k < 1$, the function is shrunk vertically towards the x-axis.

So $f(x) = \dfrac{\log_2 x}{2}$ is the result of $f(x) = \log_2 x$ being shrunk vertically by a dilation factor of $\dfrac{1}{2}$.

c **i** $y = 7x^2$ has factor 7 (stretched).

ii $y = \dfrac{e^x}{5} = \dfrac{1}{5}e^x$

Factor is $\dfrac{1}{5}$ (shrunk).

Example 14

a The point $N(-1, 8)$ lies on the graph of the function $y = f(x)$. Find the image of N on the graph of the function $y = kf(x)$ when:

i $k = 5$ **ii** $k = \dfrac{1}{2}$

b The graph shown is $y = f(x)$.
Sketch the graph of $y = 2f(x)$.

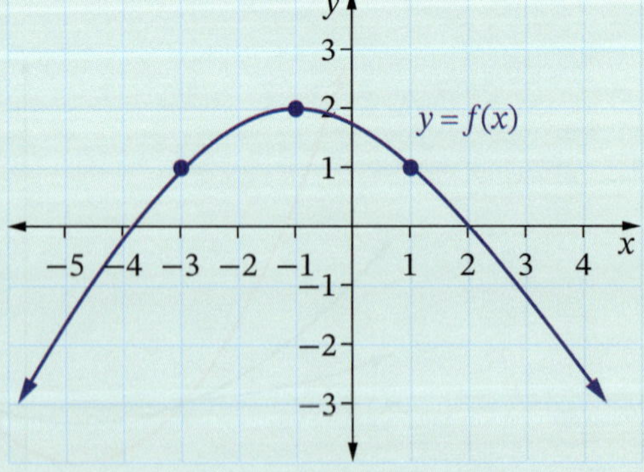

Solution

a With $y = kf(x)$, the y values of $f(x)$ will change.

i When $k = 5$: y values are multiplied by a factor of 5.

Image of N is $(-1, 8 \times 5) = (-1, 40)$

ii When $k = \frac{1}{2}$:

Image of N is $\left(-1, 8 \times \frac{1}{2}\right) = (-1, 4)$

b The graph of $y = 2f(x)$ is a vertical dilation of $y = f(x)$ with factor 2.

So each y value is doubled and the graph is twice as high as the original graph.

For example:

$y = 1$ becomes $y = 2$

$y = 2$ becomes $y = 4$

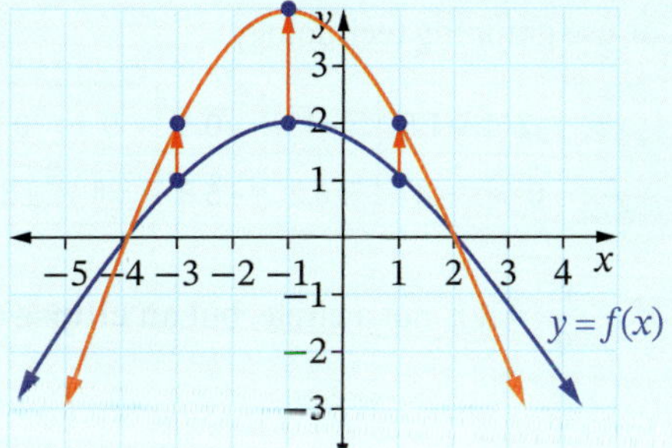

The transformed graph is still a parabola. However it is higher (stretched) and narrower than the original graph.

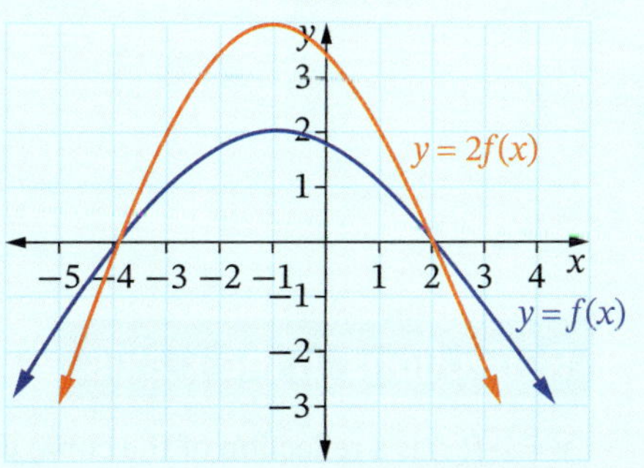

Example 15

a Find the equation of the circle $x^2 + y^2 = 4$ if it is vertically dilated by a factor of 3.

b Find the domain and range of the transformed relation.

c Sketch the graph of the transformed relation.

Solution

a $x^2 + y^2 = 4$

For a vertical dilation with factor 3, replace y with $\frac{y}{3}$.

$$x^2 + \left(\frac{y}{3}\right)^2 = 4$$

$$x^2 + \frac{y^2}{9} = 4$$

OR

$$9x^2 + y^2 = 36$$

b $x^2 + y^2 = 4$ is a circle with centre $(0, 0)$ and radius 2, so its domain and range are both $[-2, 2]$.

It is vertically dilated by a factor of 3, so the y values are multiplied by 3, so the circle is 3 times taller, with new y-intercepts 6 and -6.

So its domain is $-2 \le x \le 2$ and its range is $-6 \le y \le 6$.

c The graph of $x^2 + \frac{y^2}{9} = 4$ can be drawn by multiplying the y values of the circle $x^2 + y^2 = 4$ by 3. You could also make y the subject $\left(y = \pm\sqrt{36 - 9x^2}\right)$ and plot points from a table or use graphing technology.

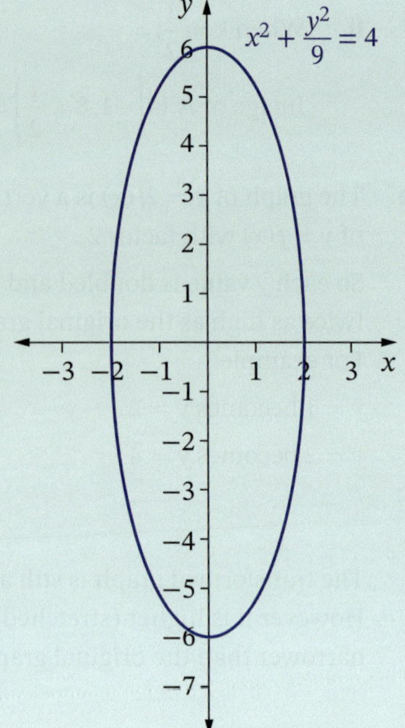

x	−2	−1.5	−1	−0.5	0	0.5	1	1.5	2
y	0	±4	±5.2	±5.8	±6	±5.8	±5.2	±4	0

$x^2 + \frac{y^2}{9} = 4$ is not a circle, but an **ellipse** (oval or egg shape).

Reflections in the x-axis

$y = -f(x)$ is a reflection of the curve $y = f(x)$ in the x-axis.

This is also a vertical dilation with factor $k = -1$.

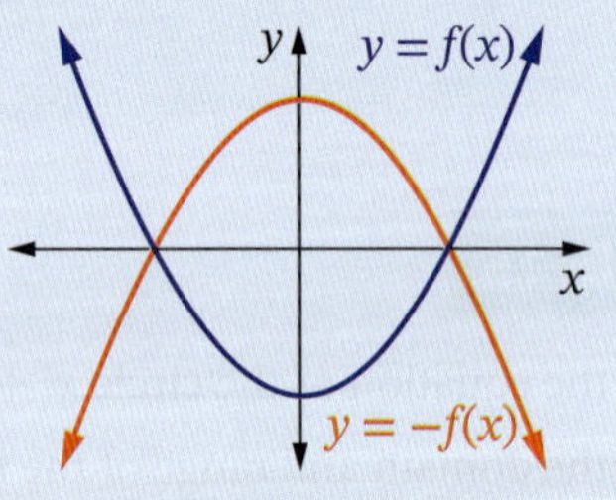

EXERCISE 8.04 Answers on p. 508

Vertical dilations of functions

1 For each basic function, describe how the graph of each function in parts **i** to **iii** is related to it.

a $y = x$

i $y = 6x$ **ii** $y = \frac{x}{2}$ **iii** $y = -x$

b $y = x^2$

i $y = 2x^2$ **ii** $y = \frac{x^2}{6}$ **iii** $y = -x^2$

c $y = x^3$

i $y = 4x^3$ **ii** $y = \frac{x^3}{7}$ **iii** $y = \frac{4x^3}{3}$

d $y = x^4$

i $y = 9x^4$ **ii** $y = \frac{x^4}{3}$ **iii** $y = \frac{3x^4}{8}$

e $y = |x|$

i $y = 5|x|$ **ii** $y = \frac{|x|}{8}$ **iii** $y = -|x|$

f $f(x) = \log x$

i $f(x) = 9 \log x$ **ii** $f(x) = -\log x$ **iii** $f(x) = \frac{2 \log x}{5}$

2 Find the equation of each dilated graph and state its domain and range.

a $y = x^2$ dilated vertically with a factor of 6

b $y = \ln x$ dilated vertically with a factor of $\frac{1}{4}$

c $f(x) = |x|$ reflected in the x-axis

d $f(x) = e^x$ dilated vertically with a factor of 4

e $y = \frac{1}{x}$ dilated vertically with a factor of 7

3 Find the equation of each transformed function after the vertical dilation given.

a $y = 3^x$ with factor 5 **b** $f(x) = x^2$ with factor $\frac{1}{3}$

c $y = \frac{1}{x}$ with factor $\frac{1}{2}$ **d** $y = |x|$ with factor $\frac{2}{3}$

4 Point $M = (3, 6)$ lies on the graph of $y = f(x)$. Find the coordinates of the image of M when $f(x)$ is:

a dilated vertically with a factor of 4 **b** reflected in the x-axis

c dilated vertically with a factor of 12 **d** dilated vertically with a factor of $\frac{5}{6}$.

5 The coordinates of the image of $X(x, y)$ are $(4, 12)$ when $y = f(x)$ is vertically dilated. Find the coordinates of X if the dilation factor is:

a 3 **b** 2 **c** $\frac{1}{3}$ **d** $\frac{3}{4}$ **e** -1

6 Sketch each pair of functions on the same set of axes.

a $f(x) = \log_2 x$ and $f(x) = 2 \log_2 x$ **b** $y = 3^x$ and $y = 2(3^x)$

c $y = |x|$ and $y = 2|x|$ **d** $y = x^3$ and $y = -x^3$

7 Points on a function $y = f(x)$ are shown on the graph. Sketch the graph of the transformed function showing the image points, given a vertical dilation with factor:

a 3 **b** $\frac{1}{2}$ **c** -1

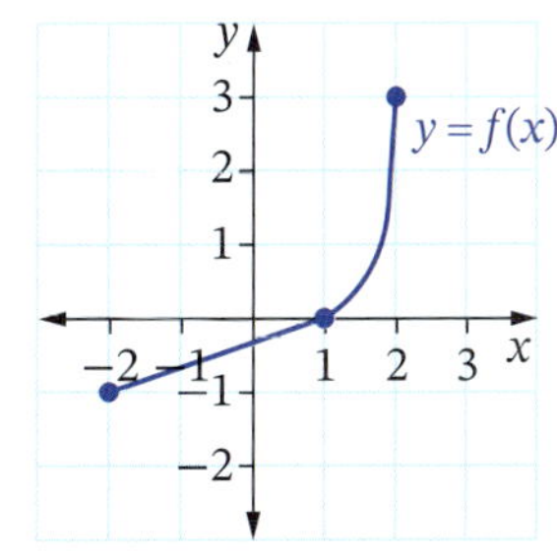

8 Sketch the graph of $y = 2\sqrt{1 - x^2}$ and find its domain and range.

□ Foundation ○ Mastery ⬡ Complex

9 **a** $y = g(x)$ is the function $f(x) = x^2 - 2x + 5$ dilated vertically with a factor 4. Find the image point Q on $y = g(x)$ of point $P(-2, 13)$ on $y = f(x)$.

b Find the gradient of the tangent to $y = f(x)$ at P.

c Find the gradient of the tangent to $y = g(x)$ at Q, and compare with and explain the results in part **b**.

EXAMPLE 15

10 This graph shows the vertical dilation of a circle. Find its equation.

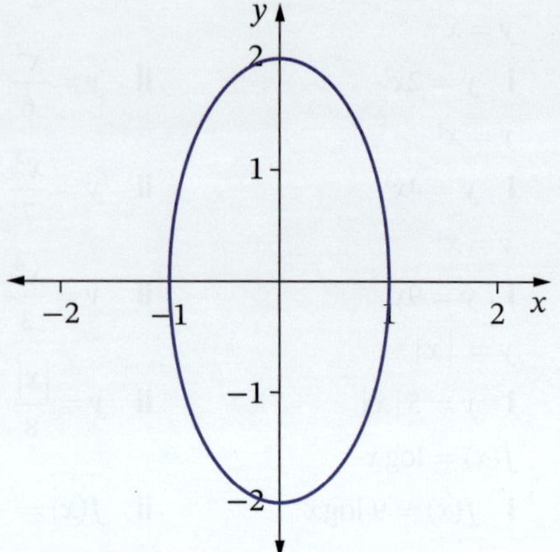

11 Find the equation of $f(x)$ if this graph shows $y = 3f(x)$.

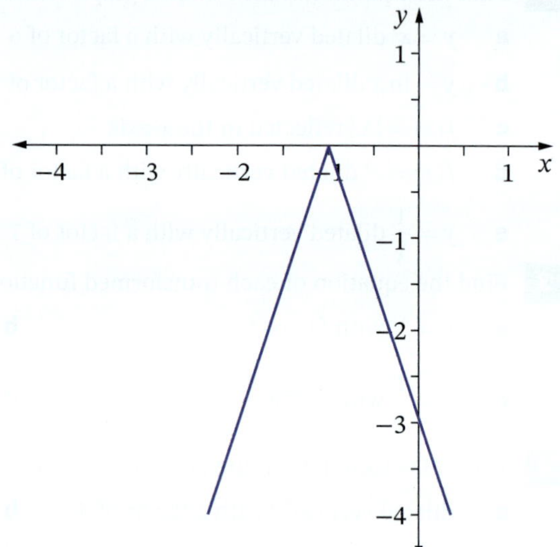

12 Find the equation of the circle $x^2 + y^2 = 16$ if it is vertically dilated with factor 2, then translated 5 units to the right.

13 The function $y = x^2 - 4$ is translated 2 units up then vertically dilated with factor 3. Find the equation of the resulting function.

14 Find the equation of the circle $x^2 + y^2 = 1$ if it is translated 2 units to the left, 3 units down and then vertically dilated with factor $\frac{1}{3}$.

□ Foundation ○ Mastery ○ Complex

Horizontal dilations of functions

8.05

A **horizontal dilation** stretches or shrinks the graph left or right, away from or towards the *y*-axis.

Worksheets Dilations of functions

Advanced graphs

Technology

Horizontal dilations

Use dynamic geometry software to explore the effect of parameter *a* on each graph below. If you don't have dynamic software, substitute different values for *a* into the equation. Use positive and negative values, integers and fractions for *a*.

$f(x) = ax$ $f(x) = (ax)^2$ $f(x) = (ax)^3$ $f(x) = (ax)^4$

$f(x) = e^{ax}$ $f(x) = \ln ax$ $f(x) = \dfrac{1}{ax}$ $f(x) = |ax|$

How does *a* transform the graph as the value of *a* changes?

What is the difference between positive and negative values of *a*?

This diagram compares the graphs of $y = x^2$ and $y = \left(\frac{1}{2}x\right)^2$.

x	−3	−2	−1	0	1	2	3
$y = x^2$	9	4	1	0	1	4	9
$y = \left(\frac{1}{2}x\right)^2$	2.25	1	0.25	0	0.25	1	2.25

The transformed graph $y = \left(\frac{1}{2}x\right)^2$ stretches away from the *y*-axis, with a dilation factor of 2.

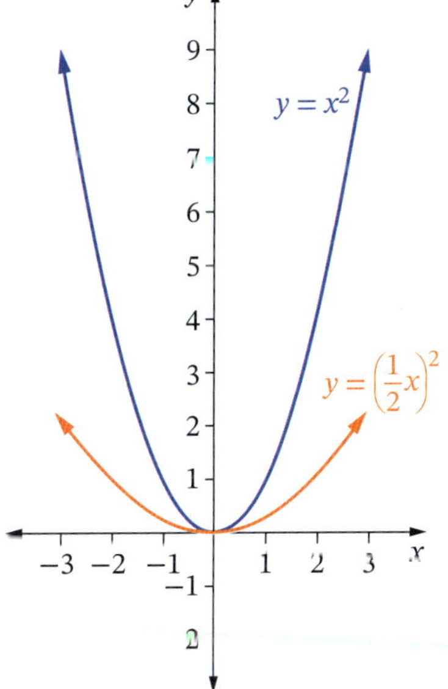

This diagram compares the graphs of $y = x^2$ and $y = (3x)^2$.

x	−3	−2	−1	0	1	2	3
$y = x^2$	9	4	1	0	1	4	9
$y = (3x)^2$	81	36	9	0	9	36	81

The transformed graph $y = (3x)^2$ shrinks towards the *y*-axis, with a dilation factor of $\frac{1}{3}$.

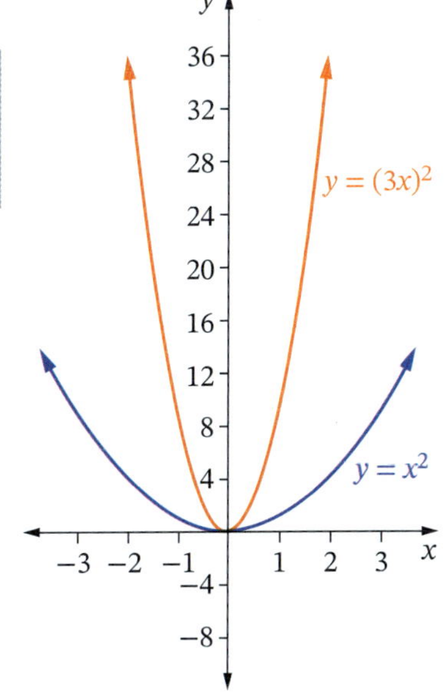

Notice that with **horizontal dilations**, the higher the value of a, the more the graph is shrunk along the x-axis from left and right. This is inverse variation and the dilation factor for horizontal dilations is $\frac{1}{a}$.

Again, it may not sound logical that $f(ax)$ *divides* the x-value by a. To see why, we change the subject of the function to x. For example:

$$y = (3x)^2$$
$$\sqrt{y} = 3x$$
$$\frac{\sqrt{y}}{3} = x$$
$$x = \frac{1}{3}\sqrt{y}$$

This shows a dilation factor of $\frac{1}{3}$.

Like horizontal translations, a horizontal stretch works the opposite way to what you would expect, because the equation is in the form $y = f(x)$ rather than $x = f(y)$.

If we let $k = \frac{1}{a}$, then we can have simpler formulas for a horizontal dilation factor of k.

Then $a = \frac{1}{k}$, and $f(ax)$ becomes $f\left(\frac{1}{k}x\right) = f\left(\frac{x}{k}\right)$. This means we can use $k > 1$ for a stretched curve and $0 < k < 1$ for a shrunk curve, like with vertical dilations.

Horizontal dilations

For the curve $y = f(x)$:

$y = f\left(\frac{x}{k}\right)$ dilates the curve horizontally (from the y-axis) by a factor of k.

This is the same as replacing x with $\frac{x}{k}$ in $y = f(x)$.

If $k > 1$, the graph is stretched away from the y-axis.

If $0 < k < 1$, the graph is shrunk towards the y-axis.

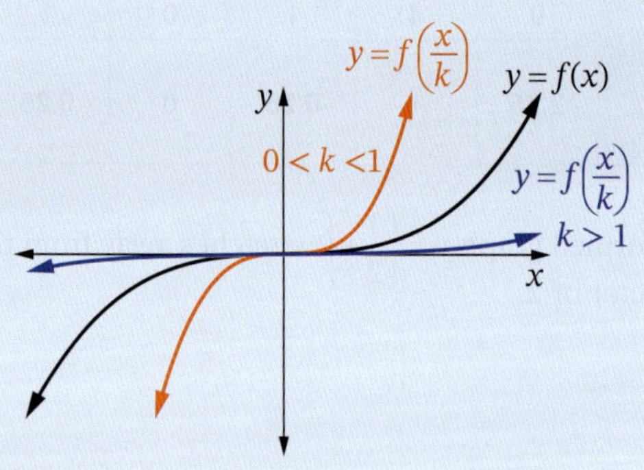

Video Horizontal dilations of functions

Example 16

a Describe how the function $f(x) = x^3$ is transformed to the function $f(x) = (4x)^3$.

b The function $y = \ln x$ is dilated horizontally by a factor of 2. Find the equation of the transformed function.

c Find the dilation factor of each function and state whether it stretches or shrinks the graph.

i $y = e^{3x}$ **ii** $f(x) = \left|\frac{x}{4}\right|$

Solution

a The function $y = f\left(\frac{x}{k}\right)$ is a horizontal dilation of $y = f(x)$ with factor k.

So for $f(x) = (4x)^3$, $\frac{1}{k} = 4$ and $k = \frac{1}{4}$.

So the function $f(x) = (4x)^3$ is a horizontal dilation of $f(x) = x^3$ with factor $\frac{1}{4}$.

b If $y = \ln x$ is dilated horizontally by a factor of 2, replace x by $\frac{x}{2}$ in $y = \ln x$.

So $y = \ln\left(\frac{x}{2}\right)$.

c **i** $y = e^{3x}$ is in the form $y = f\left(\frac{x}{k}\right)$, where $f(x) = e^x$ and $\frac{1}{k} = 3$.

Dilation factor is $k = \frac{1}{3}$ (shrunk towards the y-axis)

ii $f(x) = \left|\frac{x}{4}\right|$ is in the form $y = f\left(\frac{x}{k}\right)$, where $k = 4$.

Dilation factor is $k = 4$ (stretched from the y-axis)

Example 17

a The graph of $y = b^x$ shown is transformed to $y = b^{2x}$.
Sketch the graph of the transformed function.

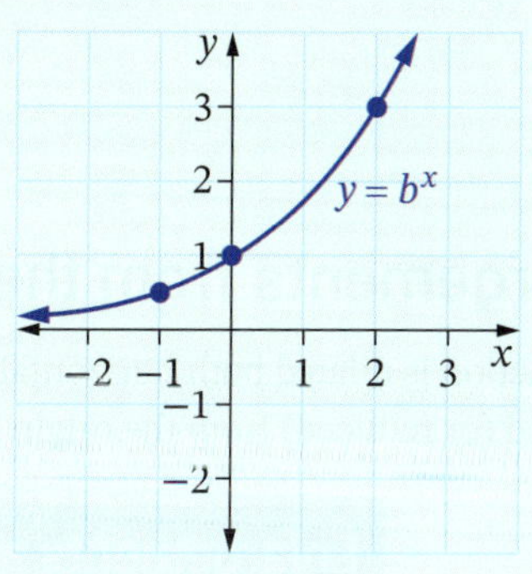

b State the factor if the graph $y = |x|$ is transformed to $y = \left|\frac{x}{2}\right|$ and sketch both graphs on the same set of axes.

Solution

a The graph of $y = b^{2x}$ describes a horizontal dilation of $y = b^x$ with factor $\frac{1}{2}$.

So we halve the x values.

$x = -1$ becomes $x = -\frac{1}{2}$

$x = 0$ becomes $x = 0$

$x = 2$ becomes $x = 1$

The y-intercepts stay the same since a horizontal dilation doesn't affect y values.

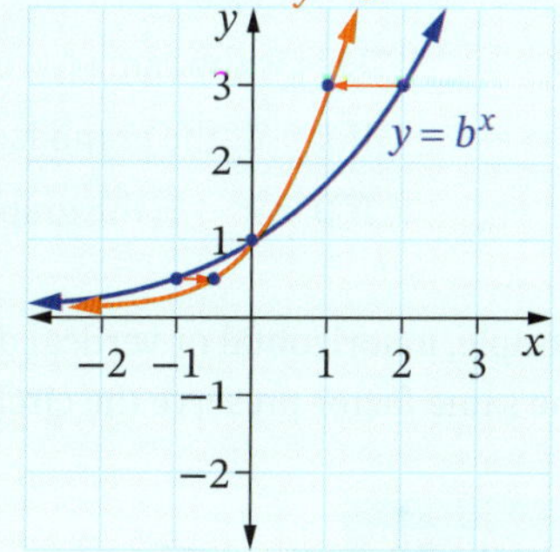

b 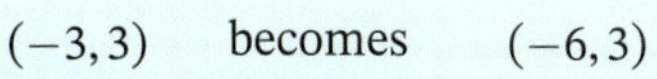The graph $y = \left|\frac{x}{2}\right|$ is a horizontal dilation of $y = |x|$ with a factor 2.

We double the x values.

$(-3, 3)$	becomes	$(-6, 3)$
$(-2, 2)$	becomes	$(-4, 2)$
$(-1, 1)$	becomes	$(-2, 1)$
$(0, 0)$	becomes	$(0, 0)$
$(1, 1)$	becomes	$(2, 1)$
$(2, 2)$	becomes	$(4, 2)$
$(3, 3)$	becomes	$(6, 3)$

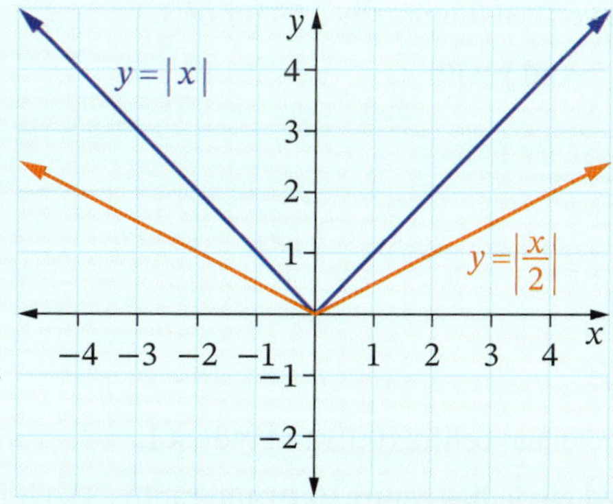

Reflections in the y-axis

$y = f(-x)$ is a reflection of the curve $y = f(x)$ in the y-axis.

This is a horizontal dilation with factor $k = \frac{1}{-1} = -1$.

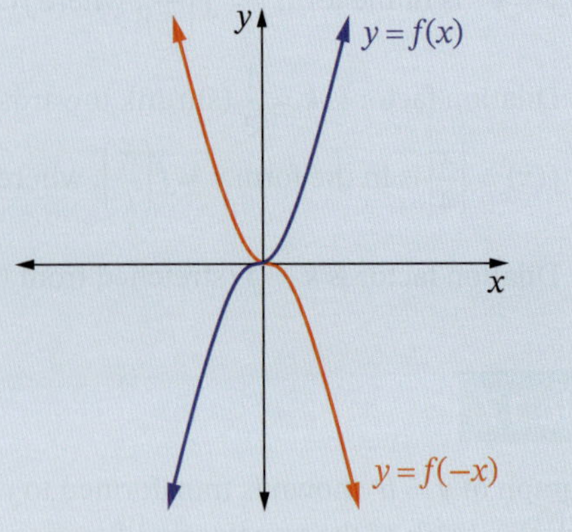

Enlargements from the origin

If a function is dilated both horizontally and vertically by the same dilation factor k, then the function is enlarged (or reduced) from the origin by a factor of k.

Enlargements and reductions

For the curve $y = f(x)$:

$y = kf\left(\frac{x}{k}\right)$ dilates the curve from the origin, in all directions, by a factor of k.

This is the same as replacing x with $\frac{x}{k}$ and y with $\frac{y}{k}$.

If $k > 1$, the graph is enlarged.

If $0 < k < 1$, the graph is reduced.

For example, a horizontal or vertical dilation of a circle changes it into an ellipse, but both dilations together with the same factor preserve the circle shape.

Example 18

Find the equation of the circle $x^2 + y^2 = 1$ if it is enlarged from the origin by a dilation factor of 3.

Solution

Horizontal and vertical dilation by a factor of 3.

Replace x with $\frac{x}{3}$ and y with $\frac{y}{3}$.

$$\left(\frac{x}{3}\right)^2 + \left(\frac{y}{3}\right)^2 = 1$$

$$\frac{x^2}{9} + \frac{y^2}{9} = 1$$

$$x^2 + y^2 = 9$$

The original circle had centre $(0, 0)$ and radius 1.

The transformed circle has centre $(0, 0)$ and radius 3 so it is an enlargement of the circle.

Notice that the radius is 3 times the radius of the original circle.

EXERCISE 8.05 Answers on p. 509

Horizontal dilations of functions

EXAMPLE 16

1 Describe the transformation each function has undergone from $f(x) = x^4$ and state the factor.

a $f(x) = (8x)^4$ **b** $f(x) = \left(\frac{x}{5}\right)^4$

c $f(x) = \left(\frac{3x}{7}\right)^4$ **d** $f(x) = (-x)^4$

2 For each basic function, describe how the graph of each function in parts **i** to **iii** has undergone a horizontal or vertical dilation and state the factor.

a $y = x^2$

i $y = (2x)^2$ **ii** $y = (5x)^2$ **iii** $y = \left(\frac{x}{3}\right)^2$

b $y = x^3$

i $y = 4x^3$ **ii** $y = \left(\frac{x}{2}\right)^3$ **iii** $y = (-x)^3$

c $y = x^4$

i $y = (7x)^4$ **ii** $y = \frac{x^4}{8}$ **iii** $y = \left(\frac{3x}{4}\right)^4$

d $y = |x|$

i $y = |5x|$ **ii** $y = \left|\frac{x}{2}\right|$ **iii** $y = \left|\frac{3x}{5}\right|$

e $y = 5^x$

i $y = 5^{3x}$ **ii** $y = -5^x$ **iii** $y = 5^{\frac{x}{2}}$

f $f(x) = \log x$

i $f(x) = 8 \log x$ **ii** $f(x) = \log(-x)$ **iii** $f(x) = \log\left(\frac{x}{7}\right)$

3 Find the equation of each transformed graph and state its domain and range.

a $f(x) = |x|$ is dilated horizontally with a factor of $\frac{1}{5}$

b $y = x^2$ is dilated horizontally with a factor of 3

c $y = x^3$ is reflected in the y-axis.

d $y = e^x$ is dilated vertically with a factor $\frac{1}{9}$

e $y = \log_4 x$ is reflected in the x-axis

4 Point $X(-2, 7)$ lies on $y = f(x)$. Find the coordinates of the image of X on $y = f\left(\frac{x}{k}\right)$ given:

a $k = \frac{1}{2}$ **b** $k = -1$ **c** $k = 3$

5 The function $y = f(x)$ is transformed into the function $y = f\left(\frac{x}{k}\right)$. The coordinates of the image point of $P(x, y)$ on the original function are $(-24, 1)$ on the transformed function. Find the coordinates of P if:

a $k = \frac{1}{3}$ **b** $k = \frac{1}{2}$ **c** $k = 4$

EXAMPLE 17

6 Sketch each pair of functions on the same set of axes:

a $f(x) = \ln x$ and $f(x) = \ln(2x)$ **b** $y = 2^x$ and $y = 2^{\frac{x}{3}}$

c $y = \frac{1}{x}$ and $y = \frac{1}{3x}$ **d** $y = |x|$ and $y = |2x|$

e $f(x) = x^2$ and $f(x) = (3x)^2$ **f** $y = \ln x$ and $y = \ln(-x)$

7 Sketch the graphs of $y = e^x$, $y = e^{2x}$ and $y = 2e^x$ on the same set of axes.

8 Explain why a reflection in the y-axis does not change the graph of:

a $y = x^2$ **b** $f(x) = |x|$

☐ Foundation ○ Mastery ○ Complex

9 Sketch the graph of $y = f(ax)$ given the graph of $y = f(x)$ shown, when:

a $a = \frac{1}{2}$ **b** $a = 2$.

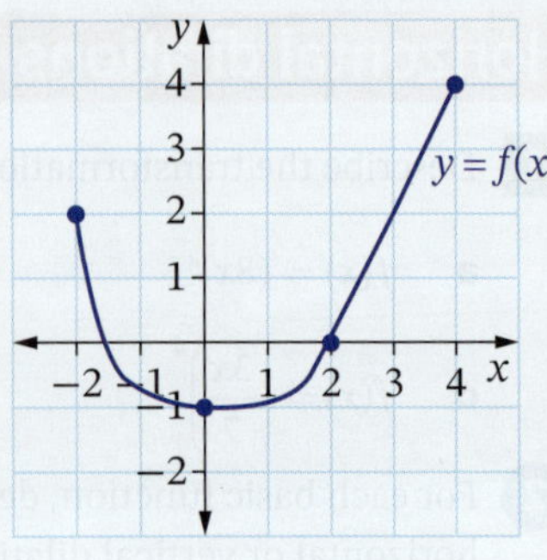

10 $f(x) = x^2 - 3x + 1$ is horizontally dilated with a factor of $\frac{1}{4}$.

a Find the equation of the transformed function.

b Find the point on the transformed function that corresponds with $(2, -1)$ on the original function.

11 The circle $x^2 + y^2 = 4$ is horizontally dilated with a factor of 2.

a Find the equation of the transformed relation.

b Sketch its graph and state its domain and range.

EXAMPLE 18

12 **a** The circle $x^2 + y^2 = 9$ is vertically dilated with factor 2 then horizontally dilated with factor 3. Find the equation of the resultant relation and its domain and range.

b The circle $x^2 + y^2 = 9$ is vertically and horizontally dilated with factor 2. Find the equation of the resultant relation and explain this result.

c The circle $x^2 + y^2 = 9$ is vertically and horizontally dilated with factor $\frac{1}{2}$. Find the equation of the resultant relation and explain this result.

13 This ellipse was horizontally dilated from a circle.

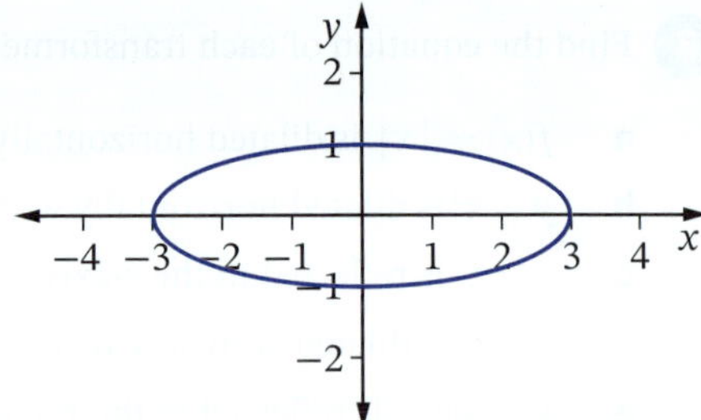

Find the equation of the circle that was transformed.

14 The point $P(2, 2)$ lies on the function $y = \sqrt{x + 2}$. If the function is dilated horizontally with a factor of $\frac{1}{2}$, find:

a the point Q on the new function that is the image of P

b the gradient of the tangent to the new function at Q.

15 Find the equation of the transformed function if $y = |x|$ is:

a horizontally dilated with factor $\frac{1}{5}$, then horizontally translated 3 units to the right

b vertically translated 4 units down, then horizontally dilated with factor 2

c reflected in the x-axis, then horizontally dilated with factor 3

d horizontally dilated with factor $\frac{1}{2}$, then vertically dilated with factor 6.

☐ Foundation ○ Mastery ○ Complex

Combined transformations

8.06

We can perform multiple transformations on a function $y = f(x)$.

We also can combine all the transformation formulas into one single formula.

Worksheet
Combinations of transformations

Combined transformations

For the curve $y = f(x)$:

$y = lf\left(\frac{1}{k}[x - a]\right) + b$ is a combination of the following transformations:

Transformation	Function formula	Replacement formula
Horizontal dilation of factor k stretches if $k > 1$, shrinks if $0 < k < 1$	$y = f\left(\frac{x}{k}\right)$	replace x with $\frac{x}{k}$
Vertical dilation of factor l stretches if $l > 1$, shrinks if $0 < l < 1$	$y = lf(x)$	replace y with $\frac{y}{l}$
Horizontal reflection in the y-axis, $k = -1$	$y = f(-x)$	replace x with $-x$
Vertical reflection in the x-axis, $l = -1$	$y = -f(x)$	replace y with $-y$
Horizontal translation of a units right if $a > 0$, left if $a < 0$	$y = f(x - a)$	replace x with $x - a$
Vertical translation of b units up if $b > 0$, down if $b < 0$	$y = f(x) + b$	replace y with $y - b$

Note: The combined formula $y = lf\left(\frac{1}{k}[x - a]\right) + b$ works only if dilations and reflections are done before translation. Otherwise, the replacement formulas must be applied separately in order.

Example 19

a Find the equation of the transformed function if $y = x^4$ is shifted 2 units down and 5 units to the left.

b Find the equation of the transformed function if $y = e^x$ is dilated vertically by a factor of 3 and translated horizontally 2 units to the right.

Solution

a

Transformation	Function	Replacement
	$y = x^4$	
Translation 2 units down	$y = x^4 - 2$	replace y with $y - (-2) = y + 2$
Translation 5 units left	$y = (x + 5)^4 - 2$	replace x with $x - (-5) = x + 5$

New equation is $y = (x + 5)^4 - 2$.

b

Transformation	Function	Replacement
	$y = e^x$	
Vertical dilation of factor 3	$y = 3e^x$	replace y with $\frac{y}{3}$
Translation 2 units right	$y = 3e^{x-2}$	replace x with $x - 2$

New equation is $y = 3e^{x-2}$.

Technology

Order of transformations

Choose some functions and relations upon which to perform multiple transformations, for example, $y = x^2, y = x^3, y = \frac{1}{x}, y = |x|, y = a^x, y = \log_a x, y = \sqrt{x}, x^2 + y^2 = 1$.

Use graphing technology to sketch the combined transformations.

Is the order of combined transformations important? Try the following combinations on different functions and relations and notice any patterns.

1. A reflection in the x-axis and y-axis in different orders.
2. A horizontal and vertical translation in different orders.
3. A horizontal and vertical dilation in different orders.
4. A horizontal dilation and horizontal translation in different orders.
5. A vertical dilation and vertical translation in different orders.
6. A horizontal translation and vertical dilation in different orders.
7. A vertical translation and horizontal dilation in different orders.

What do you discover? Which combinations give the same result when done in a different order and which give a different result?

When the transformations are *both* vertical or *both* horizontal, then the order is important.

Example 20

a When the function $y = x^2$ is translated 3 units up (vertically) and vertically dilated by factor 4, the equation of the transformed function is $y = 4x^2 + 3$. Find the order in which the transformations were done.

b The equation of the transformed function is $y = (2x + 5)^3$ when the function $y = x^3$ is horizontally dilated by factor $\frac{1}{2}$ and translated 5 units to the left. In which order were the transformations done?

c The equation of the transformed function is $y = \ln\left(\frac{x-2}{3}\right)$ when the function $y = \ln x$ is horizontally dilated by factor 3 and translated 2 units to the right. In which order were the transformations done?

Solution

a

Transformation	Function	Replacement
	$y = x^2$	
Translation 3 units up	$y = x^2 + 3$	replace y with $y - 3$
Vertical dilation of factor 4	$y = 4(x^2 + 3)$ Not the correct function.	replace y with $\frac{y}{4}$
Try reversing the order:		
	$y = x^2$	
Vertical dilation of factor 4	$y = 4x^2$	replace y with $\frac{y}{4}$
Translation 3 units up	$y = 4x^2 + 3$ The correct function.	replace y with $y - 3$

So, the order was the vertical dilation, then the vertical translation.

b

Transformation	Function	Replacement
	$y = x^3$	
Horizontal dilation of factor $\frac{1}{2}$	$y = (2x)^3$	replace x with $\frac{x}{\frac{1}{2}} = 2x$
Translation 5 units left	$y = [2(x+5)]^3 = (2x+10)^3$ Not the correct function.	replace x with $x - (-5) = x + 5$
Try reversing the order:		
	$y = x^3$	
Translation 5 units left	$y = (x+5)^3$	replace x with $x + 5$
Horizontal dilation of factor $\frac{1}{2}$	$y = (2x+5)^3$ The correct function.	replace x with $2x$

So the order was the horizontal translation, then the horizontal dilation.

c

Transformation	Function	Replacement
	$y = \ln x$	
Horizontal dilation of factor 3	$y = \ln\left(\frac{x}{3}\right)$	replace x with $\frac{x}{3}$
Translation 2 units right	$y = \ln\left(\frac{x-2}{3}\right)$ The correct function.	replace x with $x - 2$

So the order was the horizontal dilation, then the horizontal translation.

Note that doing the horizontal dilation first in part **c** gives $y = f\left(\frac{1}{k}[x-a]\right)$, while doing the horizontal translation first in part **b** gives $y = f\left(\frac{1}{k}x - a\right)$ (no brackets around $x - a$).

Example 21

Find the equation of the function if $y = x^2$ is horizontally dilated with factor 4, then translated 3 units down.

Solution

Transformation	Function	Replacement
	$y = x^2$	
Horizontal dilation of factor 4	$y = \left(\frac{x}{4}\right)^2$	replace x with $\frac{x}{4}$
Translation 3 units down	$y = \left(\frac{x}{4}\right)^2 - 3$	replace y with $y - (-3) = y + 3$

The equation is $y = \left(\frac{x}{4}\right)^2 - 3$.

Order of combined transformations

To use the formula $y = lf\left(\frac{1}{k}[x-a]\right)+b$:

1 Do dilations (including reflections) first: k and l in any order

2 Do translations last: a and b in any order

The order of transformations is similar to the order of operations with numbers where multiplying and dividing are done before adding and subtracting.

Note that the horizontal dilation and translation parameters k and a are inside the brackets (they change x values) and the vertical dilation and translation parameters l and b are outside the brackets (they change the y values).

Example 22

a The point $P(3,4)$ lies on the function $y = f(x)$. Find the coordinates of the image (corresponding) point on the transformed function $y = 7f(3[x+2])+5$.

b Find the equation of the transformed function if $y = g(x)$ undergoes a vertical dilation of factor 5, a reflection in the y-axis and a translation 4 units to the right and 9 units down.

Solution

a Dilations first, translations last. Comparing $y = 7f(3[x+2])+5$ with $y = lf\left(\frac{1}{k}[x-a]\right)+b$:
$l = 7, k = \frac{1}{3}, a = -2, b = 5$.

Transformation	Operation	Point
		$(3, 4)$
$l = 7$: vertical dilation of factor 7	y value $\times$ 7	$(3, 28)$
$k = \frac{1}{3}$: horizontal dilation of factor $\frac{1}{3}$	x value $\times \frac{1}{3}$	$(1, 28)$
$a = -2$: translation 2 units left	x value $-$ 2	$(-1, 28)$
$b = 5$: translation 5 units up	y value $+$ 5	$(-1, 33)$

Image point is $(-1, 33)$.

b

Transformation	Function	Replacement
	$y = g(x)$	
Vertical dilation of factor 5	$y = 5g(x)$	replace y with $\frac{y}{5}$
Reflection in the y-axis	$y = 5g(-x)$	replace x with $-x$
Translation 4 units right	$y = 5g(-(x-4)) = 5g(4-x)$	replace x with $x - 4$
Translation 9 units down	$y = 5g(5-x) - 9$	replace y with $y - (-9) = y + 9$

Domain and range

We can use transformations to help us find the domain and range of functions.

Horizontal transformations change x values so they affect the domain.

Vertical transformations change y values so they affect the range.

Example 23

Find the domain and range of each function.

a $f(x) = -3(x-2)^2 + 5$ **b** $y = 5\sqrt{2x+1}$

Solution

a The basic function $y = x^2$ has domain all real x and range $y \geq 0$.

Horizontal transformations affect the domain:

No horizontal dilation.

Horizontal translation 2 units right: domain of $x - 2$ is all real x so domain of $f(x)$ is unchanged.

Vertical transformations affect the range:

Vertical dilation, factor -3: range of y is $y \geq 0$, so range of $-3y$ is $-3y \leq 0$, which simplifies to $y \leq 0$.

Vertical translation 5 units up: add 5 to y values so range of $-3y + 5$ is $y \leq 5$.

So $y = -3(x-2)^2 + 5$ has domain all real x and range $y \leq 5$.

b The basic function $y = \sqrt{x}$ has domain $x \geq 0$ and range $y \geq 0$.

Horizontal transformations affect the domain:

$$
\begin{aligned}
2x + 1 &\geq 0 \\
2x &\geq -1 \\
x &\geq -\frac{1}{2}
\end{aligned}
$$

Vertical transformations affect the range:

Vertical dilation, factor 5: Range of $\sqrt{2x+1}$ is $y \geq 0$ so range of $5\sqrt{2x+1}$ is $5y \geq 0$, which simplifies to $y \geq 0$.

No vertical translation.

So $y = 5\sqrt{2x+1}$ has domain $x \geq -\frac{1}{2}$ and range $y \geq 0$.

EXERCISE 8.06 Answers on p. 510

Combined transformations

1 The point $(2, -6)$ lies on the function $y = f(x)$. Find the coordinates of its image if the function is:

- **a** horizontally translated 3 units to the right and vertically translated 5 units down
- **b** translated 4 units up and 3 units to the left
- **c** translated 7 units to the right and 9 units up
- **d** translated 11 units down and 4 units to the left.

2 Find the equation of the transformed function $g(x)$ where $f(x) = x^5$ is reflected:

- **a** in the x-axis and vertically dilated with factor 4
- **b** in the y-axis and horizontally dilated with factor 3.

3 Describe the transformations to $y = x^3$ in the correct order if the transformed function has equation:

a $y = (x-1)^3 + 7$	**b** $y = 4x^3 - 1$	**c** $y = -5x^3 - 3$
d $y = 2(x+7)^3$	**e** $y = 6(2x-4)^3 + 5$	**f** $y = 2(3x+9)^3 - 10$

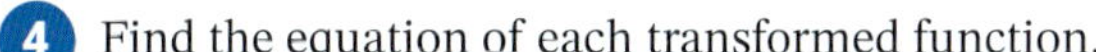

4 Find the equation of each transformed function.

- **a** $y = x^3$ is translated 3 units down and 4 units to the left
- **b** $f(x) = |x|$ is translated 9 units up and 1 unit to the right
- **c** $f(x) = x$ is dilated vertically with a factor of 3 and translated down 6 units
- **d** $y = e^x$ is reflected in the x-axis and translated up 2 units

Foundation | Mastery | Complex

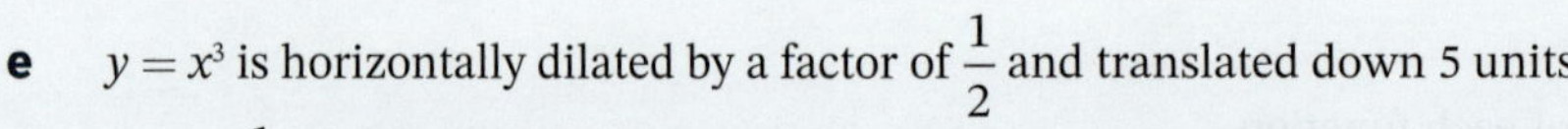

e $y = x^3$ is horizontally dilated by a factor of $\frac{1}{2}$ and translated down 5 units

f $f(x) = \frac{1}{x}$ is vertically dilated by a factor of 2 and horizontally dilated by a factor of 3

g $f(x) = \sqrt{x}$ is reflected in the y-axis, vertically dilated by a factor of 3 and horizontally dilated by a factor of $\frac{1}{2}$

h $y = \ln x$ is horizontally dilated by a factor of 3 and translated upwards by 2 units

i $f(x) = \log_2 x$ is horizontally dilated by a factor of $\frac{1}{4}$ and vertically dilated by a factor of 3

j $y = x^2$ is horizontally dilated by a factor of 2 and translated down 3 units

5 Describe the transformations in their correct order for each function from:

a $y = \log x$ to $y = 2\log(x + 3) - 1$ **b** $f(x) = x^2$ to $g(x) = -(3x)^2 + 9$

c $y = e^x$ to $y = 2e^{5x} - 3$ **d** $f(x) = \sqrt{x}$ to $g(x) = 4\sqrt{x - 7} + 1$

e $y = |x|$ to $y = |-2(x + 1)| - 1$ **f** $y = \frac{1}{x}$ to $y = \frac{1}{2x} + 8$

EXAMPLE 22

6 The point $(8, -12)$ lies on the function $y = f(x)$. Find the coordinates of the image point when the function is transformed into:

a $y = 3f(x - 1) + 5$ **b** $y = -f(2x) - 7$ **c** $y = 2f(x + 3) - 1$

d $y = 6f(-x) + 5$ **e** $y = -2f[2(x - 2)] - 3$

7 Given the function $y = f(x)$, find the coordinates of the image of $P(x, y)$ if the function is:

a translated 6 units down and 3 units to the right

b reflected in the y-axis and translated 6 units up

c vertically dilated with factor 2 and translated 5 units to the left

d horizontally dilated with factor 3 and translated 5 units up

e reflected in the x-axis, vertically dilated with factor 8, horizontally dilated with factor 5, translated 1 unit down and translated 6 units to the left

8 Find the equation of the transformed function if $y = f(x)$ is:

a translated 2 units down and 1 unit to the left

b translated 5 units to the right and 3 units up

c reflected in the x-axis and translated 4 units to the right

d reflected in the y-axis and translated 2 units up

e reflected in the x-axis and horizontally dilated with a factor of 4

f vertically dilated by a factor of 2 and translated 2 units down.

EXAMPLE 23

9 Find the domain and range of each function.

a $f(x) = (x + 3)^2 + 5$ **b** $y = 5|-2x| - 2$ **c** $f(x) = \frac{1}{2x - 4} + 1$

d $y = 4^{3x} + 2$ **e** $f(x) = 3\log(3x - 6) - 5$

10 **a** By completing the square, write the equation for the parabola $y = x^2 + 2x - 7$ in the form $y = (x + a)^2 + b$.

b Describe the transformations on $y = x^2$ that result in the function $y = x^2 + 2x - 7$.

11 Describe the transformations that change $y = x^2$ into the function $y = x^2 - 10x - 3$.

Foundation Mastery Complex

12 The gradient of the tangent to $y = f(x)$ is -4 at the point $(-2, 1)$.
Find the gradient of the tangent to the curve $y = 4f(2x - 6) + 1$ at the point $(2, 5)$.

13 The tangent line to the function $y = f(x)$ at the point $(6, 2)$ has equation $4x - y + 1 = 0$.
Find the equation of the tangent line to the curve $y = 5f(2x + 8) - 4$ at the point $(-1, 6)$.

14 The function $f(x) = x^2$ is dilated vertically with factor a, vertically translated b units, where $b > 0$, then dilated horizontally with factor $\frac{1}{2}$.
If the equation of the transformed function is $y = 12x^2 + 5$, find the values of a and b.

Graphing combined transformations 8.07

We can find points and sketch the graphs of functions that are changed by a combination of transformations. Translations are the easiest transformations to use since they shift the graph while keeping it the same size and shape.

Worksheets
Graphing transformed functions

Solving inequalities graphically

Example 24

a Sketch the graph of $y = (x - 2)^2 - 5$.

b Use the graph to solve the inequality $(x - 2)^2 - 5 > 4$.

Solution

a $y = (x - 2)^2 - 5$ is transformed from $y = x^2$ by a horizontal translation of 2 units to the right and a vertical translation of 5 units down.

The vertex of $y = x^2$ is $(0, 0)$.

So the vertex of $y = (x - 2)^2 - 5$ is $(0 + 2, 0 - 5) \equiv (2, -5)$.

We can find the intercepts for a more accurate graph.

For x-intercepts, $y = 0$:

$$0 = (x - 2)^2 - 5$$
$$5 = (x - 2)^2$$
$$\pm\sqrt{5} = x - 2$$
$$2 \pm \sqrt{5} = x$$

For y-intercepts, $x = 0$:

$$y = (0 - 2)^2 - 5$$
$$= 4 - 5$$
$$= -1$$

So the x-intercepts are approximately 4.2, -0.2.

b To solve $(x - 2)^2 - 5 > 4$, draw the horizontal line $y = 4$ and look for the x values where the parabola is above the line.

Solutions are $x < -1, x > 5$.

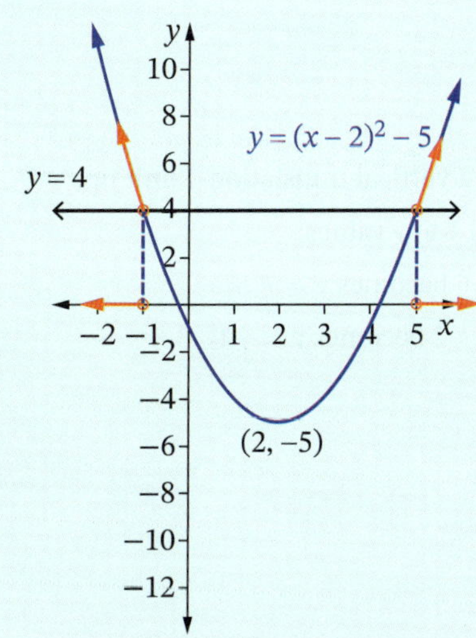

Foundation Mastery Complex

Example 25

The graph $y = f(x)$ shown is reflected in the y-axis, dilated vertically with a factor of 2 and translated 1 unit up.

Sketch the graph of the transformed function.

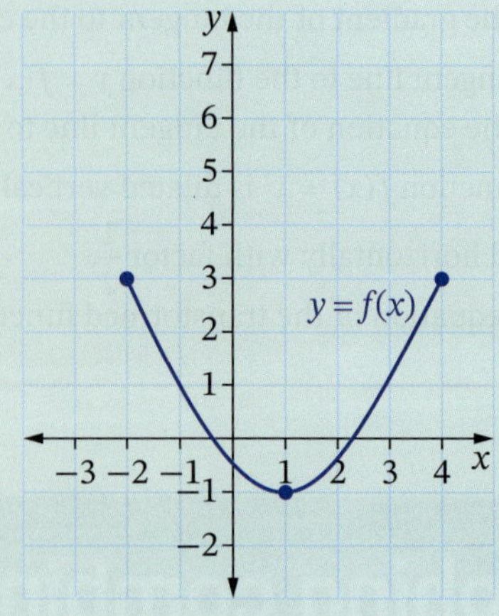

Solution

A reflection in the y-axis is a horizontal dilation with factor -1.

Multiply each x value by -1.

$x = 1$ becomes $x = -1$

$x = 4$ becomes $x = -4$

$x = -2$ becomes $x = 2$

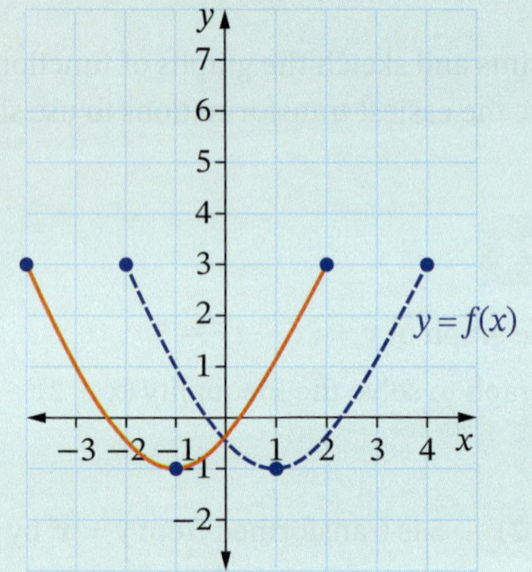

For a vertical dilation with factor 2:

Multiply each y value by 2.

$y = -1$ becomes $y = -2$

$y = 3$ becomes $y = 6$

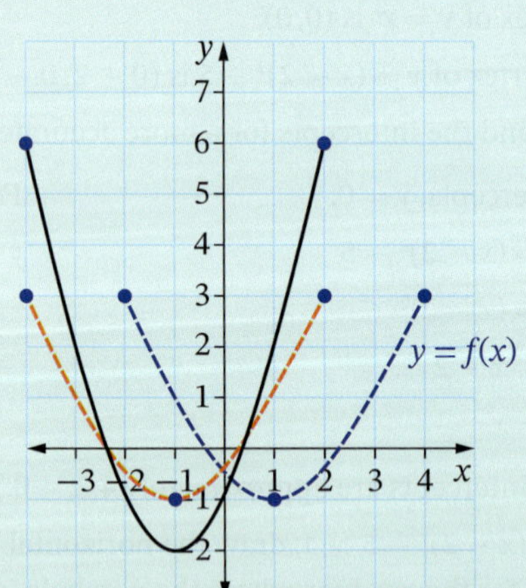

For a vertical translation 1 unit up:

Add 1 to y values.

$y = 6$ becomes $y = 7$

$y = -2$ becomes $y = -1$

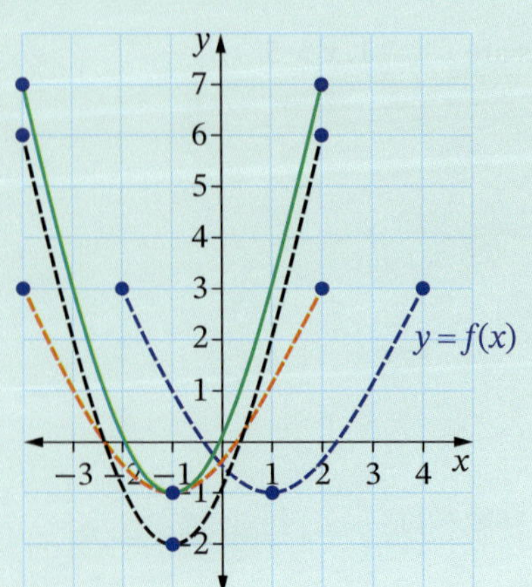

Example 26

a Describe the transformations if $y = |x|$ is transformed to $y = -\left|\frac{x}{2}\right| + 3$ and the image of point (x, y) on the original function.

b Sketch the transformed function.

Solution

a

Function	Replacement	Transformation
$y = \|x\|$		(x, y)
$y = \left\|\frac{x}{2}\right\|$	replace x with $\frac{x}{2}$	Horizontal dilation of factor 2: $(2x, y)$
$y = -\left\|\frac{x}{2}\right\|$	replace y with $-y$	Reflection in the x-axis: $(2x, -y)$
$y = -\left\|\frac{x}{2}\right\| + 3$	replace y with $y - 3$: $-(y - 3) = (-y + 3)$	Translation 3 units up: $(2x, -y + 3)$

Transformations are horizontal dilation of factor 2, reflection in the x-axis and translation 3 units up.

The image of (x, y) is $(2x, -y + 3)$.

b The vertex of $y = |x|$ is $(0, 0)$, which is transformed to $(2(0), -0 + 3) = (0, 3)$.

$(1, 1)$ is transformed to $(2(1), -1 + 3) = (2, 2)$.

$(-1, 1)$ is transformed to $(2(-1), -1 + 3) = (-2, 2)$.

x-intercepts on $y = -\left|\frac{x}{2}\right| + 3$:

$$0 = -\left|\frac{x}{2}\right| + 3$$

$$\left|\frac{x}{2}\right| = 3$$

$$\frac{x}{2} = \pm 3$$

$$x = \pm 6$$

Sketching this information using an appropriate scale gives the graph.

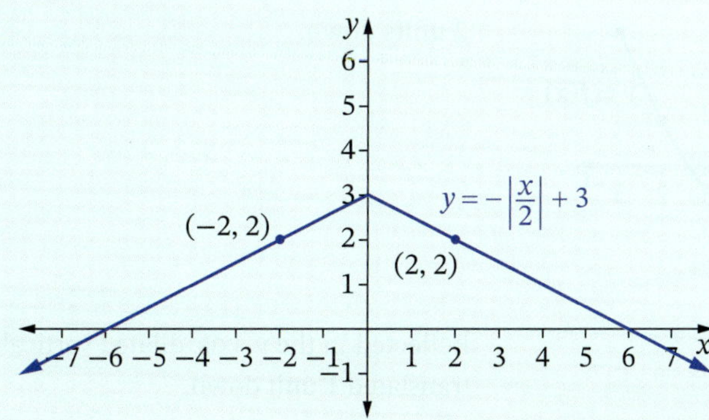

EXERCISE 8.07 Answers on p. 511

Graphing combined transformations

EXAMPLE 24

1 The function $y = x^2$ is transformed into $y = 3(x+1)^2 - 2$.

a Describe the transformations that it underwent.

b Sketch the graph of $y = 3(x+1)^2 - 2$

c Solve:

i $3(x+1)^2 - 2 = 10$

ii $3(x+1)^2 - 2 \le 1$

2 Given this graph of $y = f(x)$, sketch the graph of $y = 2f(x+1)$.

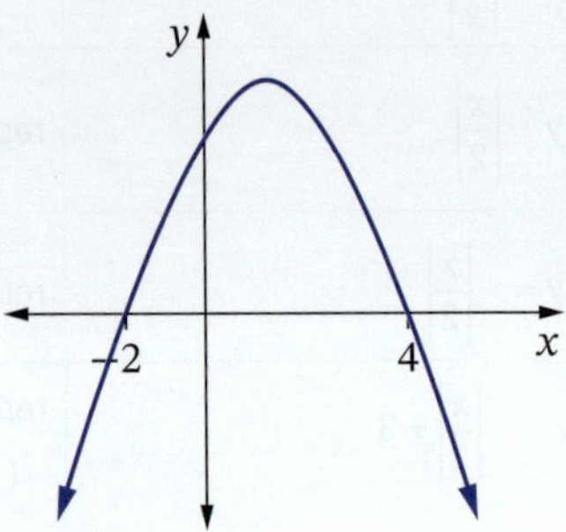

3 Sketch the graph of the transformed function if the cubic function $y = x^3$ is transformed into:

a $y = (x-1)^3 + 2$ **b** $y = (x-2)^3 - 3$ **c** $y = -(x+1)^3 + 4$

d $y = 2(x+3)^3 - 5$ **e** $y = 3(x-1)^3 - 2$

4 Sketch the graph of:

a $y = -3(x-2)^3 + 1$ **b** $y = 2e^{x+1} - 4$

c $f(x) = 3\sqrt{x-2} - 1$ **d** $y = 2|3x| + 4$

EXAMPLE 25

5 Given each function $y = f(x)$, sketch the graph of the transformed function.

a

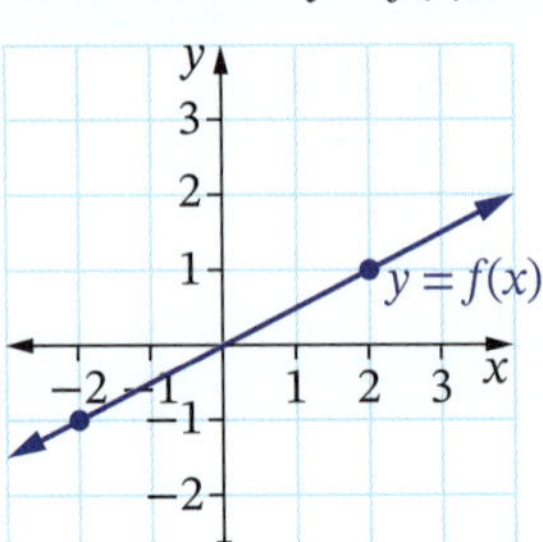
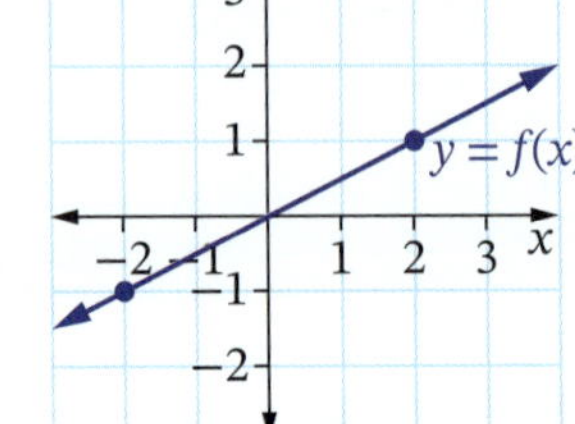

Dilated horizontally by factor $\frac{1}{2}$, reflected in the x-axis and translated 3 units up.

b

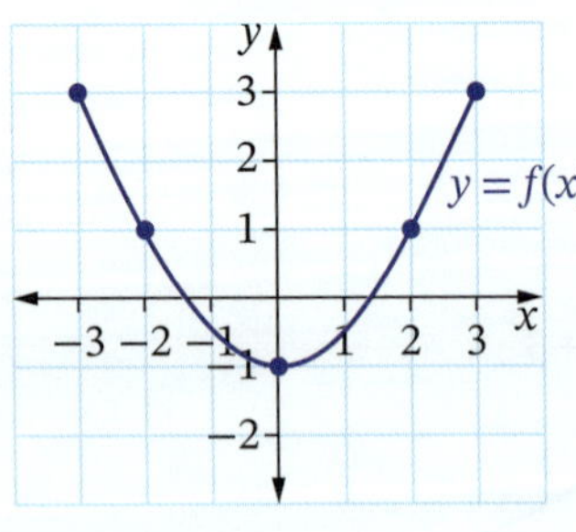

Dilated vertically by factor 3, translated 3 units left and 2 units down.

c

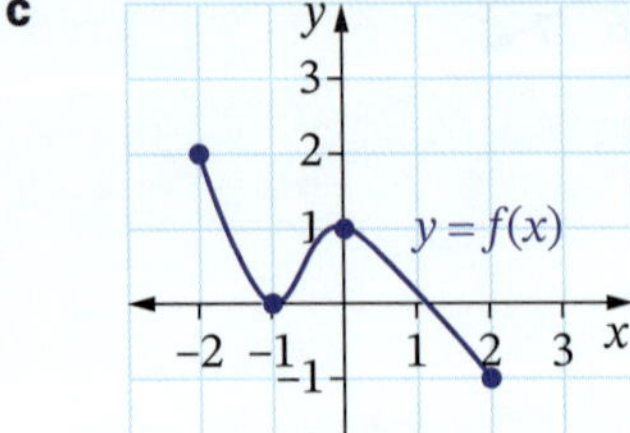

Reflected in the y-axis, dilated vertically by factor 2, translated 1 unit down.

☐ Foundation ◯ Mastery ◯ Complex

6 For the function $y = f(x)$, with turning points as shown, sketch the transformed function if it is vertically dilated with factor 3, translated 4 units down, and horizontally translated 2 units to the left.

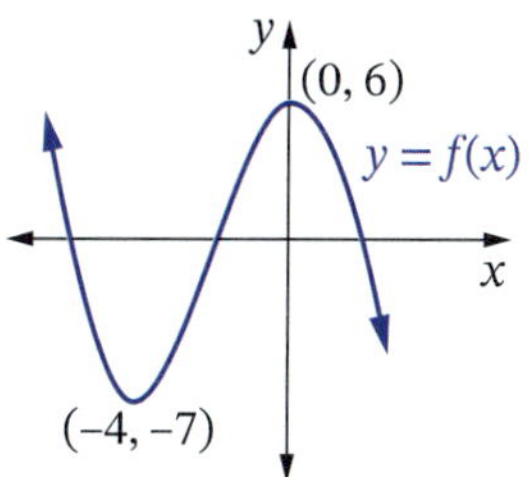

7 Sketch the graph of:

a $y = -(3x)^2 + 1$ **b** $f(x) = -2e^x + 1$

c $y = 1 - (x + 1)^3$ **d** $y = \frac{2}{x-1} + 3$

8 **a** Describe the transformations if $y = x$ is transformed to $y = -2(x - 3) + 1$ and the image of point (x, y) on the original function. EXAMPLE 26

b Sketch the transformed function.

9 **a** Describe the transformations if $y = \ln x$ is transformed to $y = 3 - 2\ln x$ and the image of point (x, y) on the original function.

b Sketch the transformed function.

10 **a** Sketch the graph of $y = 3|x - 2| + 4$.

b Use the graph to determine how many solutions the equation $y = 3|x - 2| + 4 = 1$ has.

c Solve $3|x - 2| + 4 > 10$ graphically.

11 **a** Show that $y = \frac{x+2}{x+5}$ can be expressed as $1 - \frac{3}{x+5}$.

b Describe what transformations on $y = \frac{1}{x}$ give the function $y = 1 - \frac{3}{x+5}$.

c Hence sketch the graph of $y = 1 - \frac{3}{x+5}$, showing any asymptotes and intercepts, and state its domain and range.

d Sketch on the same number plane the graph of $y = \frac{1}{2}x$ and solve the inequality $\frac{x+2}{x+5} = \frac{1}{2}x$ graphically.

e Hence solve the inequality $\frac{x+2}{x+5} \geq \frac{1}{2}x$ graphically.

12 **a** The coordinates of the image of $P(x, y)$ when $y = f(x)$ is transformed to $y = 3f(x - 2) + 1$ are $P'(-3, 2)$. Find the original point $P(x, y)$.

b Sketch the graph of the original function $y = f(x)$ if $y = 3f(x - 2) + 1$ is a cubic function with turning points $(-3, 2)$ and $(2, -4)$.

13 The coordinates of the image point of the vertex $P(x, y)$ of a parabola are $(-24, 18)$ when $y = f(x)$ is transformed as shown below.

Find the coordinates of the original point $P(x, y)$ and sketch the graph of the original quadratic function.

a $y = 3f(x - 2) - 5$ **b** $y = -5f[3(x + 1)]$ **c** $y = 2f(2x - 6) - 3$

14 This graph shows a transformation of $y = |x|$.

a Find the equation of this graph.

b If the line $y = ax$ is drawn on the graph, find the values of a for which the line crosses the graph at 2 distinct points.

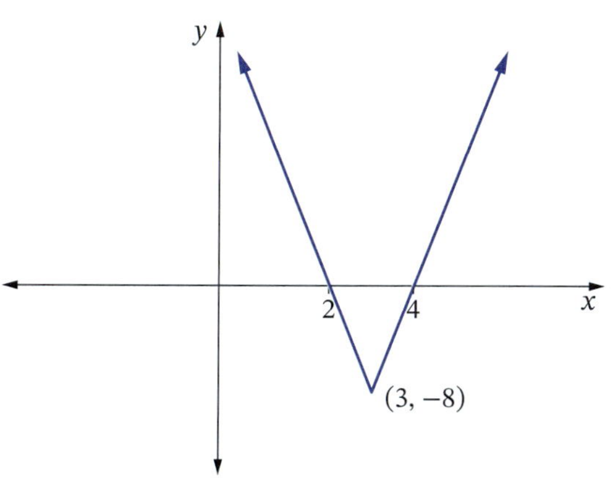

Foundation Mastery Complex

Sample HSC problem

Answers on p. 515

(10 marks)

a Show that $\dfrac{x+2}{x-3}$ can be expressed as $1+\dfrac{5}{x-3}$. **1 mark**

b Describe what transformations on $y=\dfrac{1}{x}$ give the function $y=1+\dfrac{5}{x-3}$. **3 marks**

c Hence sketch the graph of $y=\dfrac{x+2}{x-3}$, showing any asymptotes and intercepts, and state its domain and range. **3 marks**

d Sketch on the same number plane the graph of $y=-x-2$ and solve the equation $\dfrac{x+2}{x-3}=-x-2$ graphically. **2 marks**

e Hence solve the inequality $\dfrac{x+2}{x-3}<-x-2$ graphically. **1 mark**

□ Foundation ○ Mastery ○ Complex

CHAPTER SUMMARY

This chapter, *Transformations of functions*, examined 3 types of transformations of functions and their graphs: translation, reflection in the x- and y-axes, and dilation from the x- and y-axes.

Before you move on, consider what you learned in this chapter and revisit any sections that may have been unclear.

What you learned in this chapter...	Section	
Understand and apply translations of functions and relations	8.01	Vertical translations of functions
	8.02	Horizontal translations of functions
Understand and apply reflections of functions and relations in the x- and y-axes	8.03	Reflections of functions
Understand and apply dilations of functions and relations from the x- and y-axes	8.04	Vertical dilations of functions
	8.05	Horizontal dilations of functions
Apply combinations of transformations to functions and relations	8.06	Combined transformations
Use transformations to sketch the graphs of different types of functions and relations	8.07	Graphing combined transformations

To help master these techniques, make a summary of this topic. Use the chapter outline and the mind map below as a guide. Add your own words, symbols, diagrams, boxes and reminders. The summary should give you a 'whole picture' view of the topic and allow you to identify any weak areas to revisit in your revision.

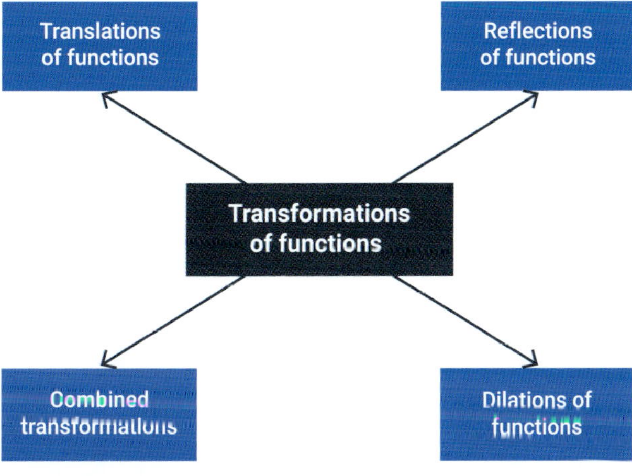

8 Test yourself

Answers on p. 515

For Questions **1** to **6**, choose the correct answer **A**, **B**, **C** or **D**.

1 The function $y = f(x)$ transformed to $y = f(x - 8)$ is:

A a vertical translation 8 units up
B a horizontal translation 8 units to the right
C a vertical translation 8 units down
D a horizontal translation 8 units to the left

2 The graph below is a transformation of $y = x^2$. Find its equation.

A $y = (-x + 3)^2$
B $y = (-x - 3)^2$
C $y = -(x + 3)^2$
D $y = -(x - 3)^2$

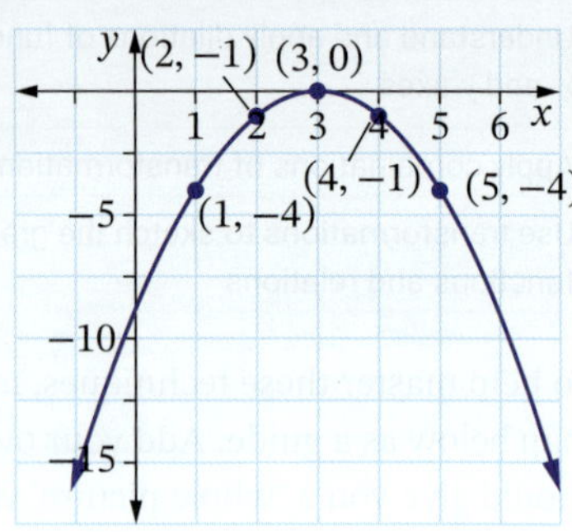

3 Find the coordinates of the image of (x, y) when the function $y = f(x)$ is transformed to $y = -2f(x + 1) + 4$.

A $(x + 1, -2y - 4)$
B $(x + 1, -2y + 4)$
C $(x - 1, -2y + 4)$
D $(-x + 1, 2y + 4)$

4 Name the transformation on the graph of $y = \sqrt{x}$ to create this graph.

A vertical dilation
B reflection in y-axis
C reflection in x-axis
D horizontal dilation

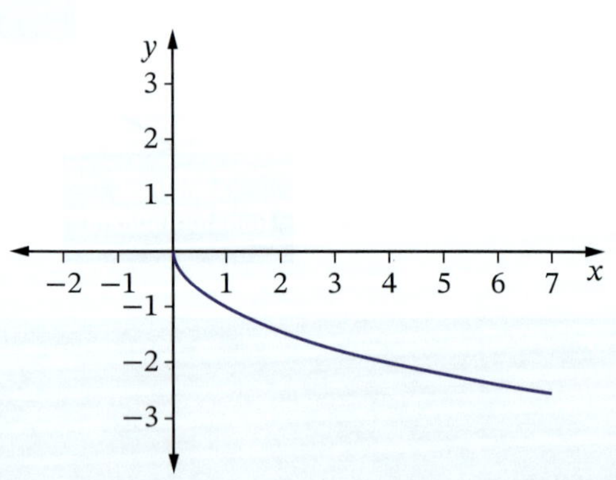

8.06

5 Find the domain and range of the relation $x^2 + y^2 = 9$ if it is dilated vertically with factor 2, translated 3 units to the right and translated 2 units down.

A domain $[0, 6]$, range $[-8, 4]$
B domain $[0, 3]$, range $[-5, 1]$
C domain $[0, 6]$, range $[-5, 3]$
D domain $[0, 3]$, range $[-8, 4]$

□ Foundation ○ Mastery ○ Complex

6 This is the graph of $y = f(x)$. 8.06

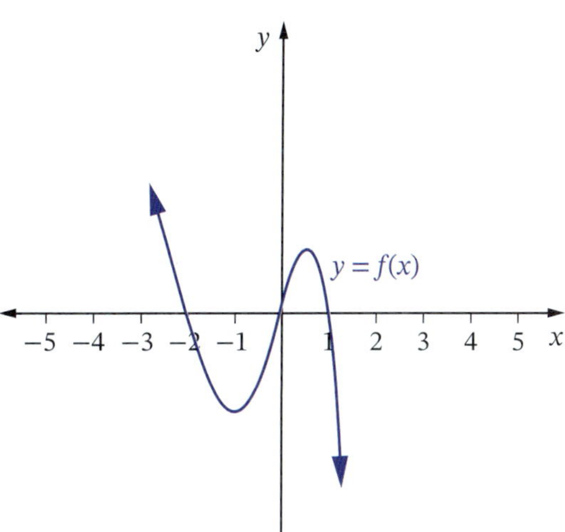

Which of the following is the graph of $y = f(2x - 6)$?

A

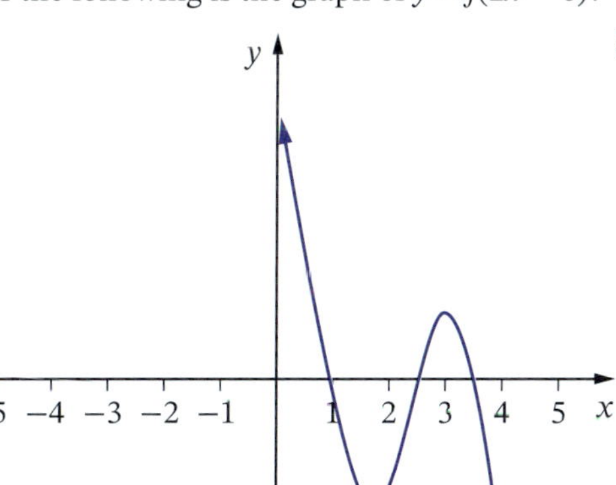

B

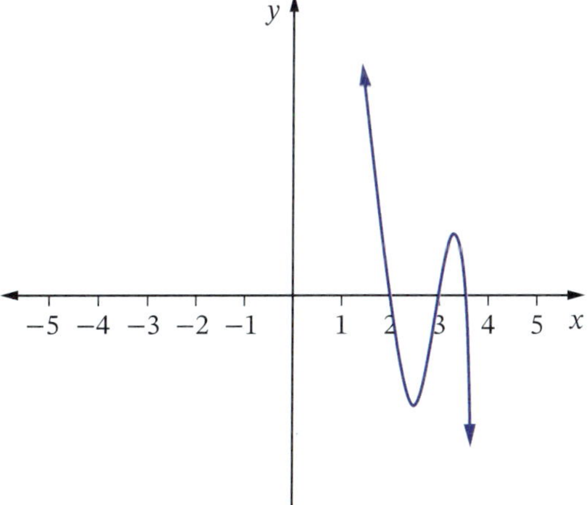

C

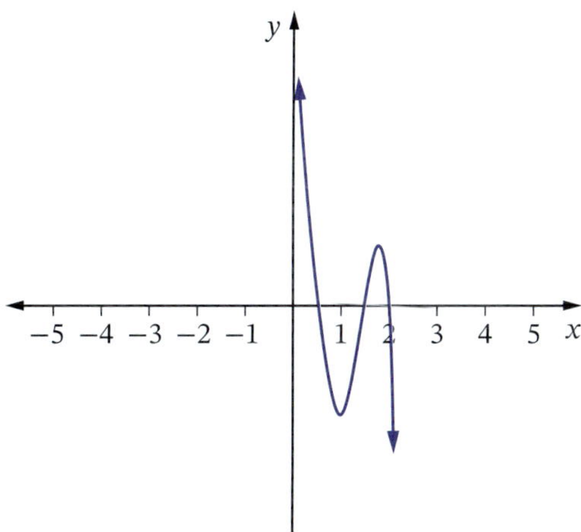

D

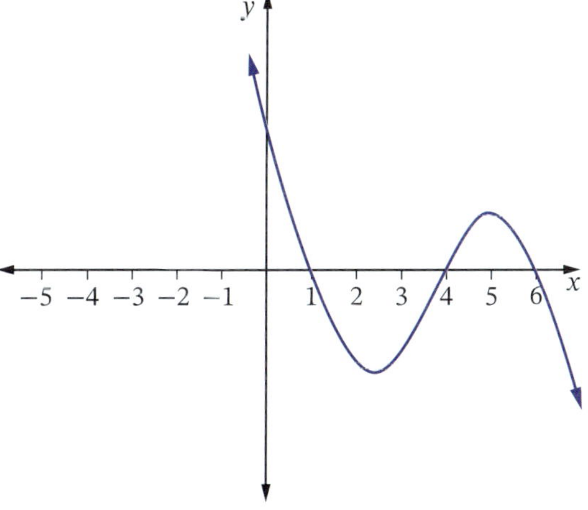

7 **a** State the meaning of the constants a, b, k and l in the function $y = l\,f\left(\frac{1}{k}[x - a]\right) + b$ and describe the effect they have on the graph of the function $y = f(x)$. 8.06

b Describe the effect on the graph of the function if:

i $k = -1$ **ii** $l = -1$

8 Show that if $y = x^2$ is dilated vertically with factor 3, reflected in the x-axis and translated 1 unit up, the transformed function is even. 8.06

9 **a** Draw the graph of $y = e^{x-1} - 2$. 8.07

b Use the graph to solve $e^{x-1} - 2 = 8$.

c Solve $e^{x-1} - 2 = 20$ algebraically.

□ Foundation ○ Mastery ○ Complex

8.06 **10** Find the equation of each transformed function.

a $y = x^3$ is translated:

i 3 units up **ii** 7 units to the left.

b $y = |x|$ is dilated:

i vertically with factor 3 **ii** horizontally with factor 2.

c $f(x) = \ln x$ is dilated vertically with factor 5 and reflected in the y-axis.

d $f(x) = \frac{1}{x}$ is reflected in the x-axis and translated 4 units to the right.

e $f(x) = 3^x$ is dilated vertically with factor 9, dilated horizontally with factor $\frac{1}{3}$ and translated 6 units down and 2 units to the right.

8.02 **11** Find the equation of the transformed function if $f(x) = x^4$ is horizontally translated 4 units to the left.

8.06 **12** If $(8, 2)$ lies on the graph of $y = f(x)$, find the coordinates of the image of this point when the function is transformed to $y = -4f[2(x + 1)] - 3$.

8.07 **13** **a** Show that $\frac{2x-7}{x-3} = -\frac{1}{x-3} + 2$.

b Sketch the graph of $y = \frac{2x-7}{x-3}$ and state its domain and range.

c Solve:

i $\frac{2x-7}{x-3} \geq 0$ **ii** $\frac{2x-7}{x-3} < 2$

8.06 **14** The function $y = f(x)$ is transformed to $y = -7f(x - 3) - 4$.

a Find the coordinates of the image of (x, y).

b If the image point is $(-3, 3)$, find the value of x and y.

8.07 **15** From the graph of $y = f(x)$ shown, draw the graph of:

a $y = 2f(x - 1)$

b $y = -f(x) - 2$

8.07 **16** By drawing the graph of $y = 2(x + 1)^2 - 8$, solve:

a $2(x + 1)^2 - 8 \leq 0$

b $2(x + 1)^2 - 8 > 0$

8.07 **17** Sketch on the same set of axes:

a $y = x^2$ and $y = -4x^2 + 3$

b $y = |x|$ and $y = -|x - 1| + 2$

c $f(x) = e^x$ and $f(x) = \frac{e^{x+2}}{2} - 1$ **d** $y = \frac{1}{x}$ and $y = \frac{1}{x+2} + 1$

e $y = x^3$ and $y = 2(x - 3)^3 + 1$ **f** $f(x) = \ln x$ and $f(x) = \ln(-x) + 5$

g $y = \sqrt{x}$ and $y = 2\sqrt{x+4} - 1$

8.04, 8.05 **18** State whether the function $y = f(x)$ is stretched or shrunk if it is dilated:

a horizontally with factor $\frac{1}{6}$ **b** horizontally with factor 3

c vertically with factor $\frac{1}{4}$ **d** horizontally with factor $\frac{7}{6}$

8.06 **19** Find the domain and range of:

a $y = 3(x - 7)^2 - 10$ **b** $y = -|x + 1| + 2$ **c** $y = -\frac{2}{x-3} - 5$

☐ Foundation ○ Mastery ○ Complex

Challenge exercise 8

Answers on p. 517

1 The point (x, y) lies on the function $y = f(x)$. The image of (x, y) is the point $(12, 6)$ when the function is transformed to $y = -6f(2x + 8)$. Find the coordinates of (x, y).

2 **a** If $(4, -3)$ lies on the function $y = f(x)$, find the coordinates of its image point.

- **i** P on $y = 3f(x + 3) + 1$
- **ii** Q on $y = -f(2x) - 3$
- **iii** R on $y = f(2x - 2) + 1$

b Find the equation of the linear function passing through P that is perpendicular to QR.

c If $y = x$ is transformed into the linear function from part **b**, describe the transformations.

3 The equation of the normal line to the function $y = f(x)$ at $(3, -2)$ is $3x + y - 2 = 0$.
Find the equation of the tangent line to the transformed function $y = -f(x - 7) + 1$ at $(10, 3)$.

4 **a** If $y = \frac{1}{x}$ is dilated horizontally with factor 2, explain why the equation of the transformed function is the same as if it was dilated vertically with factor 2.

b Is this the same result for the function $y = \frac{1}{x^2}$? Why?

5 **a** What is the equation of the axis of symmetry of the quadratic function $f(x) = ax^2 + bx + c$?

b What types of transformations on this function will change the axis of symmetry?

c Find the equation of the axis of symmetry of the quadratic function:

- **i** $f(x) = 2(x + 1)^2 - 2$
- **ii** $y = -(x - 3)^2 + 7$
- **iii** $y = k(x + b)^2 + c$
- **iv** $y = k(ax + b)^2 + c$

6 The circle $x^2 + 4x + y^2 - 6y + 12 = 0$ is transformed by a vertical translation 3 units down and a horizontal translation 5 units right. Find the equation of the transformed circle.

7 The function $y = 2^x$ is transformed to $y = 3(2^{-3x-6}) - 5$. Describe the transformations applied to the function.

8 $P(x) = x^3 - 3x - 2$ is translated up 2 units and then reflected in the y-axis.Find the equation of the transformed function.

Foundation | Mastery | Complex

9.

TRIGONOMETRIC IDENTITIES AND FUNCTIONS

In this chapter, you will learn about trigonometric functions and their graphs, trigonometric identities and solving trigonometric equations.

Some physical changes such as tides, annual temperatures and phases of the Moon are described as cyclic or periodic because they repeat regularly. Trigonometric functions are also periodic and we can use them to model real-world situations.

Chapter outline

In this chapter you will:

- examine and apply the reciprocal trigonometric ratios cosecant, secant and cotangent
- examine and apply the complementary angle identities such as $\sin\theta = \cos(90° - \theta)$
- examine and apply the trigonometric identities $\tan\theta = \dfrac{\sin\theta}{\cos\theta}$, $\cot\theta = \dfrac{\cos\theta}{\sin\theta}$, $\cos^2x + \sin^2x = 1$, $1 + \tan^2x = \sec^2x$, $1 + \cot^2x = \operatorname{cosec}^2x$
- examine the trigonometric functions, including the reciprocal trigonometric functions, their features, properties and graphs
- solve trigonometric equations in degrees and radians for any domain

Videos (5):

9.01 The reciprocal trigonometric ratios

9.02 Amplitude and period
• Graphing the cosine function

9.03 Trigonometric equations 1
• Trigonometric equations 2

Worksheets (6):

9.01 Trigonometric identities

9.02 Sine and cosine curves • Trigonometric graphs • Sketching periodic functions: amplitude and period • Sketching periodic functions: phase and vertical shift

9.03 Trigonometric equations

Puzzles (2):

9.01 Simplifying trigonometric functions

9.02 Trigonometric graphs match-up

Nelson MindTap

To access resources above, visit **cengage.com.au/nelsonmindtap**

Terminology

amplitude	centre	cosecant (cosec)	cotangent (cot)
identity	period	periodic function	phase
reciprocal trigonometric ratio	secant (sec)		

Trigonometric identities 9.01

The reciprocal trigonometric ratios

The **reciprocal trigonometric ratios** are the reciprocals of the sine, cosine and tangent ratios.

Worksheet Trigonometric identities

Puzzle Simplifying trigonometric functions

The reciprocal trigonometric ratios

Cosecant $\operatorname{cosec}\theta = \dfrac{1}{\sin\theta} = \dfrac{\text{hypotenuse}}{\text{opposite}}$

Secant $\sec\theta = \dfrac{1}{\cos\theta} = \dfrac{\text{hypotenuse}}{\text{adjacent}}$

Cotangent $\cot\theta = \dfrac{1}{\tan\theta} = \dfrac{\text{adjacent}}{\text{opposite}}$

Hint: The 3rd letter of each ratio gives the 1st letter of its reciprocal ratio!

The reciprocal ratios have the same signs as their related ratios in the different quadrants. For example, in the 3rd and 4th quadrants, $\sin\theta < 0$, so $\operatorname{cosec}\theta < 0$.

Example 1

a Find $\operatorname{cosec}\alpha$, $\sec\alpha$ and $\cot\alpha$ for this triangle.

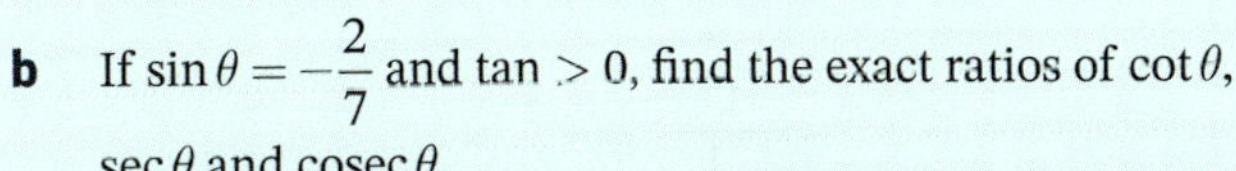

b If $\sin\theta = -\dfrac{2}{7}$ and $\tan > 0$, find the exact ratios of $\cot\theta$, $\sec\theta$ and $\operatorname{cosec}\theta$.

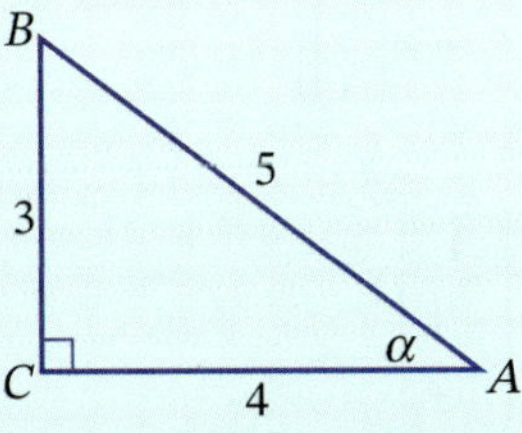

c State the quadrants where $\operatorname{cosec}\theta$ is negative.

Solution

a $\operatorname{cosec}\alpha = \dfrac{1}{\sin\alpha} = \dfrac{\text{hypotenuse}}{\text{opposite}} = \dfrac{5}{3}$

$\sec\alpha = \dfrac{1}{\cos\alpha} = \dfrac{\text{hypotenuse}}{\text{adjacent}} = \dfrac{5}{4}$

$\cot\alpha = \dfrac{1}{\tan\alpha} = \dfrac{\text{adjacent}}{\text{opposite}} = \dfrac{4}{3}$

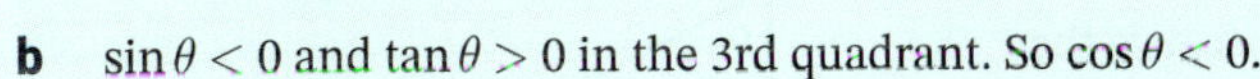

b $\sin\theta < 0$ and $\tan\theta > 0$ in the 3rd quadrant. So $\cos\theta < 0$.

By Pythagoras' theorem:

$$7^2 = a^2 + 2^2$$
$$a^2 + 4 = 49$$
$$a^2 = 45$$
$$a = \sqrt{45}$$
$$= 3\sqrt{5}$$

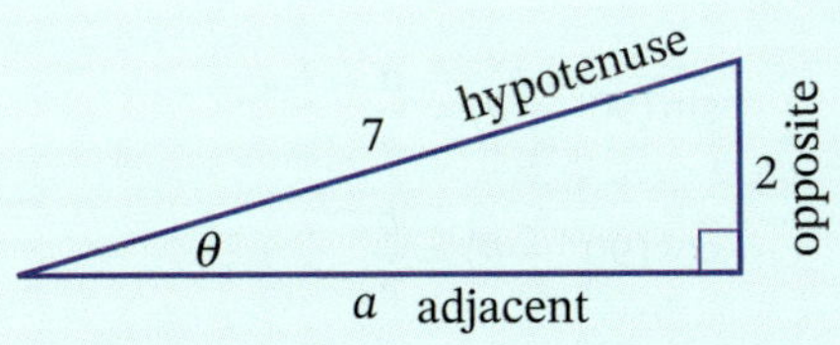

$\cot\theta = \dfrac{1}{\tan\theta} = \dfrac{\text{adjacent}}{\text{opposite}} = \dfrac{3\sqrt{5}}{2}$

$\sec\theta = \dfrac{1}{\cos\theta} = \dfrac{\text{hypotenuse}}{\text{adjacent}} = -\dfrac{7}{3\sqrt{5}} = -\dfrac{7\sqrt{5}}{15}$

$\operatorname{cosec}\theta = \dfrac{1}{\sin\theta} = \dfrac{\text{hypotenuse}}{\text{opposite}} = -\dfrac{7}{2}$

c $\sin\theta < 0$ in the 3rd and 4th quadrants.

So $\operatorname{cosec}\theta < 0$ in the 3rd and 4th quadrants.

Video
The reciprocal trigonometric ratios

Example 2

Find the exact value of each expression.

a $\cot 45°$ **b** $\sec^2\left(\frac{2\pi}{3}\right)$ **c** $\operatorname{cosec}\left(\frac{7\pi}{4}\right)$

Solution

a
$$\cot 45° = \frac{1}{\tan 45°} = \frac{1}{1} = 1$$

b
$$\sec^2\left(\frac{2\pi}{3}\right) = \left[\frac{1}{\cos\left(\frac{2\pi}{3}\right)}\right]^2 = \left[\frac{1}{\cos\left(\pi - \frac{\pi}{3}\right)}\right]^2 = \left[\frac{1}{-\cos\left(\frac{\pi}{3}\right)}\right]^2 = \left(\frac{1}{-\frac{1}{2}}\right)^2 = (-2)^2 = 4$$

c
$$\operatorname{cosec}\left(\frac{7\pi}{4}\right) = \frac{1}{\sin\left(\frac{7\pi}{4}\right)} = \left[\frac{1}{\sin\left(2\pi - \frac{\pi}{4}\right)}\right]^2 = \left[\frac{1}{-\sin\left(\frac{\pi}{4}\right)}\right]^2 = \frac{1}{-\frac{1}{\sqrt{2}}} = -\sqrt{2}$$

The complementary angle identities

In ABC if $\angle B = \theta$ then $\angle A = 90° - \theta$ (by the angle sum of a triangle). $\angle B$ and $\angle A$ are complementary angles because they add up to 90°.

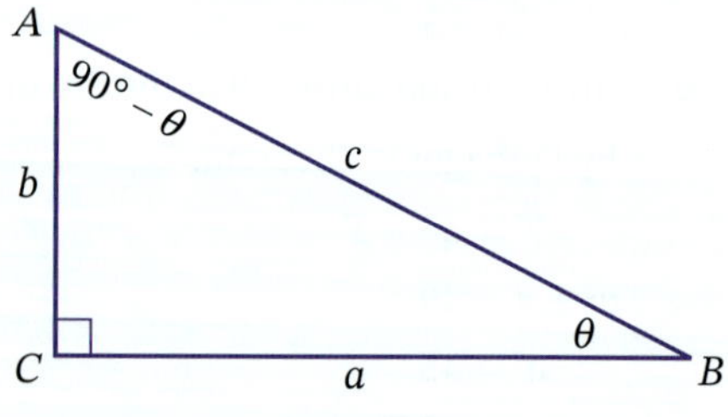

$$\sin\theta = \frac{b}{c} \qquad \sin(90° - \theta) = \frac{a}{c}$$

$$\cos\theta = \frac{a}{c} \qquad \cos(90° - \theta) = \frac{b}{c}$$

$$\tan\theta = \frac{b}{a} \qquad \tan(90° - \theta) = \frac{a}{b}$$

$$\sec\theta = \frac{c}{a} \qquad \sec(90° - \theta) = \frac{c}{b}$$

$$\operatorname{cosec}\theta = \frac{c}{b} \qquad \operatorname{cosec}(90° - \theta) = \frac{c}{a}$$

$$\cot\theta = \frac{a}{b} \qquad \cot(90° - \theta) = \frac{b}{a}$$

Notice the pairs of trigonometric ratios that are equal:

The complementary angle identities

$\sin\theta = \cos(90° - \theta)$	$\tan\theta = \cot(90° - \theta)$	$\sec\theta = \text{cosec}(90° - \theta)$
$\cos\theta = \sin(90° - \theta)$	$\cot\theta = \tan(90° - \theta)$	$\text{cosec}\,\theta = \sec(90° - \theta)$

That is why cosine, cotangent and cosecant have those names: cos = co-sine, cot = co-tan, cosec = co-sec.

Example 3

a Simplify $\tan 50° - \cot 40°$.

b Find the value of m if $\sec 55° = \text{cosec}(2m - 15)°$.

Solution

a
$$\begin{aligned}\tan 50° - \cot 40° &= \tan 50° - \cot(90° - 50°)\\ &= \tan 50° - \tan 50°\\ &= 0\end{aligned}$$

b
$$\begin{aligned}\sec 55° &= \text{cosec}(90° - 55°)\\ &= \text{cosec}\,35°\end{aligned}$$

$$\begin{aligned}\text{So } 2m - 15 &= 35\\ 2m &= 50\\ m &= 25\end{aligned}$$

The tangent identity

In the work on angles of any magnitude in Chapter 4, *Trigonometry*, we saw that $\sin\theta = y$, $\cos\theta = x$ and $\tan\theta = \frac{y}{x}$.

Now we can also write $\text{cosec}\,\theta = \frac{1}{y}$, $\sec\theta = \frac{1}{x}$ and $\cot\theta = \frac{x}{y}$.

From this we get the following trigonometric identities:

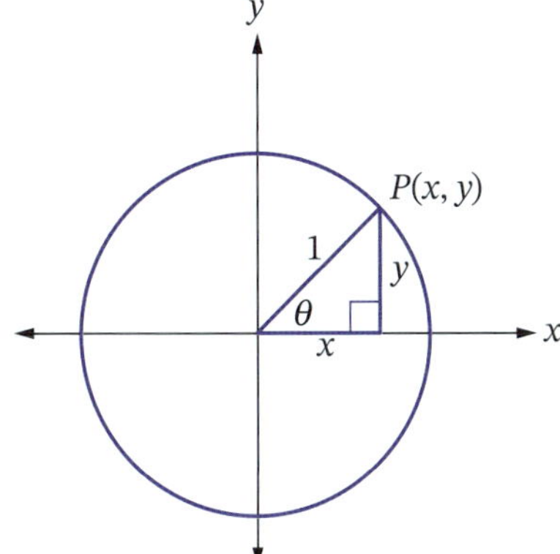

The tangent identity

For any value of θ:

$$\tan\theta = \frac{\sin\theta}{\cos\theta} \qquad \cot\theta = \frac{\cos\theta}{\sin\theta}$$

This is called an **identity** because the formulas are identical statements that are true for every value of θ.

Example 4

Simplify $\sin\theta\cot\theta$.

Solution

$$\begin{aligned}\sin\theta\cot\theta &= \sin\theta \times \frac{\cos\theta}{\sin\theta}\\ &= \cos\theta\end{aligned}$$

The Pythagorean identities

The unit circle above has equation $x^2 + y^2 = 1$, because of Pythagoras' theorem.

But $\sin\theta = y$ and $\cos\theta = x$, so

$(\cos\theta)^2 + (\sin\theta)^2 = 1$

A shorter way of writing this is:

$\cos^2\theta + \sin^2\theta = 1$

This formula is called a Pythagorean identity because it is based on Pythagoras' theorem in the unit circle.

There are 2 other identities that can be derived from this identity.

Dividing each term by $\cos^2\theta$:

$$\frac{\cos^2\theta}{\cos^2\theta} + \frac{\sin^2\theta}{\cos^2\theta} = \frac{1}{\cos^2\theta}$$

$$1 + \tan^2\theta = \sec^2\theta$$

Dividing each term by $\sin^2\theta$:

$$\frac{\cos^2\theta}{\sin^2\theta} + \frac{\sin^2\theta}{\sin^2\theta} = \frac{1}{\sin^2\theta}$$

$$\cot^2\theta + 1 = \operatorname{cosec}^2\theta$$

The Pythagorean identities

For any value of x:

$$\cos^2 x + \sin^2 x = 1$$
$$1 + \tan^2 x = \sec^2 x$$
$$1 + \cot^2 x = \operatorname{cosec}^2 x$$

$\cos^2 x + \sin^2 x = 1$ can also be rearranged to give:

$$\cos^2 x = 1 - \sin^2 x$$

or

$$\sin^2 x = 1 - \cos^2 x$$

Example 5

Prove that:

a $\cot x + \tan x = \operatorname{cosec} x \sec x$

b $\dfrac{1-\cos x}{\sin^2 x} = \dfrac{1}{1+\cos x}$

Solution

a
$$\begin{aligned}
\text{LHS} &= \cot x + \tan x \\
&= \frac{\cos x}{\sin x} + \frac{\sin x}{\cos x} \\
&= \frac{\cos^2 x + \sin^2 x}{\sin x \cos x} \\
&= \frac{1}{\sin x \cos x} \\
&= \frac{1}{\sin x} \times \frac{1}{\cos x} \\
&= \operatorname{cosec} x \sec x \\
&= \text{RHS}
\end{aligned}$$

$\therefore \cot x + \tan x = \operatorname{cosec} x \sec x$

b
$$\begin{aligned}
\text{LHS} &= \frac{1-\cos x}{\sin^2 x} \\
&= \frac{1-\cos x}{1-\cos^2 x} \\
&= \frac{1-\cos x}{(1+\cos x)(1-\cos x)} \\
&= \frac{1}{1+\cos x} \\
&= \text{RHS}
\end{aligned}$$

$\therefore \dfrac{1-\cos x}{\sin^2 x} = \dfrac{1}{1+\cos x}$

EXERCISE 9.01 Answers on p. 517

Trigonometric identities

1 For this triangle, find the exact ratios of $\sec x$, $\cot x$ and $\operatorname{cosec} x$.

2 If $\sin\theta = \frac{5}{13}$, find $\operatorname{cosec}\theta$, $\sec\theta$ and $\cot\theta$.

3 If $\cos\theta = \frac{4}{7}$, find exact values of $\operatorname{cosec}\theta$, $\sec\theta$ and $\cot\theta$.

4 If $\sec\theta = -\frac{6}{5}$ and $\sin\theta > 0$, find exact values of $\tan\theta$, $\operatorname{cosec}\theta$ and $\cot\theta$.

5 If $\cot\theta = 0.6$ and $\operatorname{cosec}\theta < 0$, find the exact values of $\sin\theta$, $\operatorname{cosec}\theta$, $\tan\theta$ and $\sec\theta$.

6 Find the exact value of each expression, with a rational denominator if needed.

a $\operatorname{cosec} 45°$ **b** $\sec 60°$ **c** $\cot\left(\frac{\pi}{6}\right)$

d $\sin^2 60°$ **e** $\sin^2 45°$ **f** $\cos^2 30° + \sin^2 30°$

g $\sec\left(\frac{\pi}{4}\right)$ **h** $\operatorname{cosec}\left(\frac{5\pi}{6}\right)$ **i** $1 + \tan^2 30°$

j $\sec^2\left(\frac{\pi}{3}\right)$ **k** $\frac{\cot 30°}{\sec 60°}$ **l** $1 - \sin^2\left(\frac{5\pi}{4}\right)$

m $\operatorname{cosec}^2\left(\frac{4\pi}{3}\right)$ **n** $\cot^2\left(\frac{7\pi}{6}\right)$ **o** $\frac{2 - \tan 60°}{\sec^2 45°}$

7 By looking at the sides of a right-angled triangle and using a calculator, find the limit of the value of $\sin x$, $\cos x$, $\tan x$, $\operatorname{cosec} x$, $\sec x$ and $\cot x$ as x becomes close to:

a $0°$ **b** $90°$

8 Show that $\sin 67° = \cos 23°$.

9 Show that $\sec 82° = \operatorname{cosec} 8°$.

10 Show that $\tan 48° = \cot 42°$.

11 Simplify:

a $\cos 61° + \sin 29°$ **b** $\sec\theta - \operatorname{cosec}(90° - \theta)$

c $\tan 70° + \cot 20° - 2\tan 70°$ **d** $\frac{\sin 55°}{\cos 35°}$

e $\frac{\cot 25° + \tan 65°}{\cot 25°}$

12 Find the value of x if $\sin 80° = \cos(90 - x)°$.

13 Find the value of y if $\tan 22° = \cot(90 - y)°$.

14 Find the value of p if $\cos 49° = \sin(p + 10)°$.

15 Find the value of b if $\sin 35° = \cos(b + 30)°$.

16 Find the value of t if $\cot(2t + 5)° = \tan(3t - 15)°$.

17 Find the value of k if $\tan(15 - k)° = \cot(2k + 60)°$.

18 Simplify:

a $\tan\theta\cos\theta$ **b** $\tan\theta\operatorname{cosec}\theta$ **c** $\sec x\cot x$

d $1 - \sin^2 x$ **e** $\sqrt{1 - \cos^2\alpha}$ **f** $\cot^2 x + 1$

g $1 + \tan^2 x$ **h** $\sec^2\theta - 1$ **i** $5\cot^2\theta + 5$

j $\frac{1}{\operatorname{cosec}^2 x}$ **k** $\sin^2\alpha\operatorname{cosec}^2\alpha$ **l** $\cot\theta - \cot\theta\cos^2\theta$

Foundation Mastery Complex

EXAMPLE 5

19 Simplify each expression.

a $\cos x \cot x$ **b** $\dfrac{\sin x}{\tan x}$

c $\sin^2 3x - 1$ **d** $\dfrac{\sin 5a}{\cos 5a}$

e $\operatorname{cosec}^2 2x - 1$ **f** $\dfrac{\cos(3\theta - a)}{\sin(3\theta - a)}$

g $\sec^2(x+2) - \tan^2(x+2)$ **h** $\sqrt{9 - 9\sin^2\theta}$

i $(\sin x - \cos x)^2$ **j** $\sin^2 35° + \cos^2 35°$

k $\dfrac{\sin a + 1}{\cos a}$ **l** $\tan^2\left(\dfrac{\pi}{5}\right) - \sec^2\left(\dfrac{\pi}{5}\right)$

20 Prove that:

a $\cos^2 x - 1 = -\sin^2 x$ **b** $\sec\theta + \tan\theta = \dfrac{1 + \sin\theta}{\cos\theta}$

c $3 + 3\tan^2\alpha = \dfrac{3}{1 - \sin^2\alpha}$ **d** $\sec^2 x - \tan^2 x = \operatorname{cosec}^2 x - \cot^2 x$

e $(\sin x - \cos x)^3 = \sin x - \cos x - 2\sin^2 x\cos x + 2\sin x\cos^2 x$

f $\cot\theta + 2\sec\theta = \dfrac{1 - \sin^2\theta + 2\sin\theta}{\sin\theta\cos\theta}$ **g** $\cos^2(90° - \theta)\cot\theta = \sin\theta\cos\theta$

h $(\operatorname{cosec} x + \cot x)(\operatorname{cosec} x - \cot x) = 1$ **i** $\dfrac{1 - \sin^2\theta\cos^2\theta}{\cos^2\theta} = \tan^2\theta + \cos^2\theta$

21 Show that $\operatorname{cosec}\theta - \sin\theta = \cos\theta\cot\theta$.

22 Prove that $\dfrac{1 - \sin^2\theta}{\operatorname{cosec}^2\theta - 1} = \sin^2\theta$.

Andrii Chagovets/Shutterstock.com

Foundation Mastery Complex

Trigonometric functions

9.02

Investigation

Trigonometric ratios of 0°, 90°, 180°, 270° and 360°

Remember the results from the unit circle:

$\sin\theta = y$

$\cos\theta = x$

$\tan\theta = \dfrac{y}{x}$

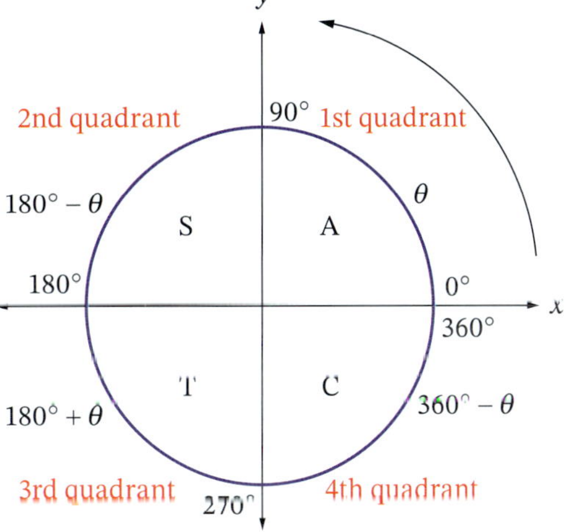

Angle 0° is at the point $(1, 0)$ on the unit circle. Use the circle results to find $\sin 0°$, $\cos 0°$ and $\tan 0°$.

Angle 90° is at the point $(0, 1)$. Use the circle results to find $\sin 90°$, $\cos 90°$ and $\tan 90°$.
Discuss the result for $\tan 90°$ and why this happens.

Angle 180° is at the point $(-1, 0)$. Find $\sin 180°$, $\cos 180°$ and $\tan 180°$.

Angle 270° is at the point $(0, -1)$. Find $\sin 270°$, $\cos 270°$ and $\tan 270°$. Discuss the result for $\tan 270°$ and why this happens.

What are the results for $\sin 360°$, $\cos 360°$ and $\tan 360°$? Why?

Check these results on your calculator.

Explore the ratios of these angles for the reciprocal trigonometric ratios cosec, sec and cot.
Where they are not defined, explore nearby angles.

Video
Amplitude and period

Worksheets
Sine and cosine curves

Trigonometric graphs

Sketching periodic functions: amplitude and period

Sketching periodic functions: phase and vertical shift

Puzzle
Trigonometric graphs match-up

The sine function

Using all the results from the investigation, we can draw up a table of values for $y = \sin x$.

x	0°	90°	180°	270°	360°
y	0	1	0	−1	0

We could draw up a table of values for $\sin x$ using other angles for a more accurate graph. The table shows the exact values of y but you could use decimal values to help you graph. Remember that $\sin x$ is positive in the 1st and 2nd quadrants and negative in the 3rd and 4th quadrants.

x	0°	30°	45°	60°	90°	120°	135°	150°	180°	210°	225°	240°	270°	300°	315°	330°	360°
y	0	$\frac{1}{2}$	$\frac{1}{\sqrt{2}}$	$\frac{\sqrt{3}}{2}$	1	$\frac{\sqrt{3}}{2}$	$\frac{1}{\sqrt{2}}$	$\frac{1}{2}$	0	$-\frac{1}{2}$	$-\frac{1}{\sqrt{2}}$	$-\frac{\sqrt{3}}{2}$	−1	$-\frac{\sqrt{3}}{2}$	$-\frac{1}{\sqrt{2}}$	$-\frac{1}{2}$	0

Drawing the graph gives a smooth 'wave' curve.

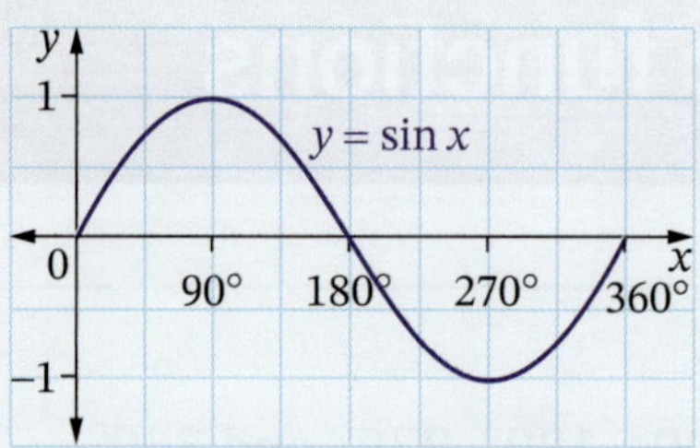

As we go around the unit circle and graph the y values of the points on the circle, the graph should repeat itself every 360°.

$y = \sin x$ has domain $(-\infty, \infty)$ and range $[-1, 1]$. It is an odd function.

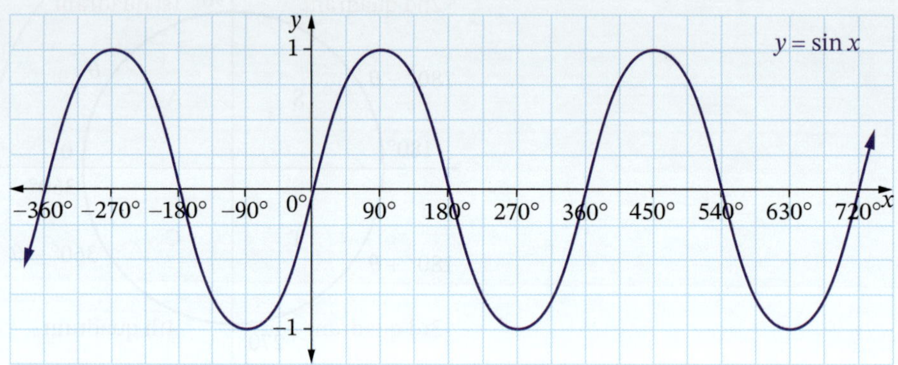

The cosine function

We could also draw up a table of values to graph for $y = \cos x$, which is positive in the 1st and 4th quadrants and negative in the 2nd and 3rd quadrants. Its graph has the same shape as the graph of the sine function.

x	0°	30°	45°	60°	90°	120°	135°	150°	180°	210°	225°	240°	270°	300°	315°	330°	360°
y	1	$\frac{\sqrt{3}}{2}$	$\frac{1}{\sqrt{2}}$	$\frac{1}{2}$	0	$-\frac{1}{2}$	$-\frac{1}{\sqrt{2}}$	$-\frac{\sqrt{3}}{2}$	−1	$-\frac{\sqrt{3}}{2}$	$-\frac{1}{\sqrt{2}}$	$-\frac{1}{2}$	0	$\frac{1}{2}$	$\frac{1}{\sqrt{2}}$	$\frac{\sqrt{3}}{2}$	1

If you look at the graph and the table of values, you may notice that the cosine function is actually the sine function translated 90° to the left. So $\cos x = \sin(x + 90)$.

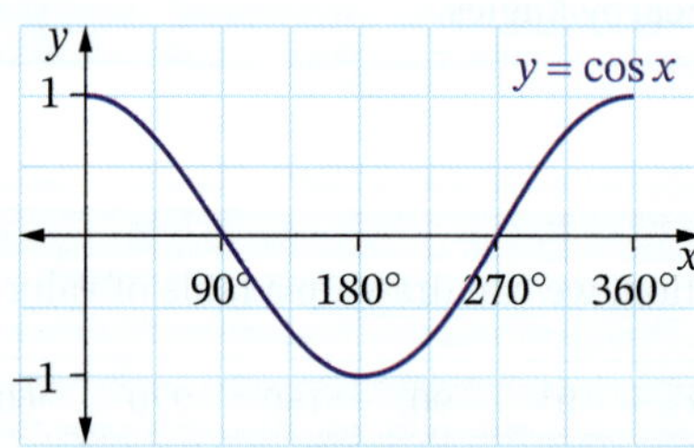

As we go around the unit circle and graph the x values of the points on the circle, the graph should repeat itself every 360°.

$y = \cos x$ has domain $(-\infty, \infty)$ and range $[-1, 1]$. It is an even function.

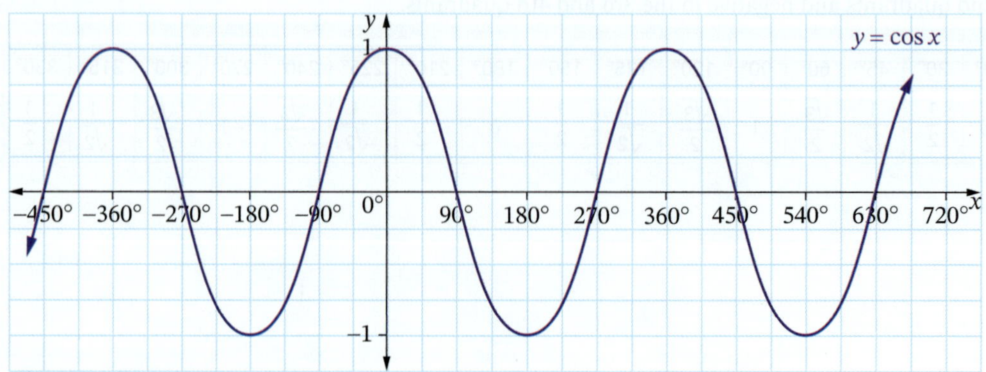

The tangent function

$y = \tan x$ is positive in the 1st and 3rd quadrants and negative in the 2nd and 4th quadrants.

It is also undefined for $x = 90°$ and $x = 270°$ so there are vertical asymptotes at those x values, where the function is discontinuous.

Find $\tan x$ for values close to these angles, for example, 89°59′, 90°01′.

x	0°	30°	45°	60°	90°	120°	135°	150°	180°	210°	225°	240°	270°	300°	315°	330°	360°
y	0	$\frac{1}{\sqrt{3}}$	1	$\sqrt{3}$	–	$-\sqrt{3}$	−1	$-\frac{1}{\sqrt{3}}$	0	$\frac{1}{\sqrt{3}}$	1	$\sqrt{3}$	–	$-\sqrt{3}$	−1	$-\frac{1}{\sqrt{3}}$	0

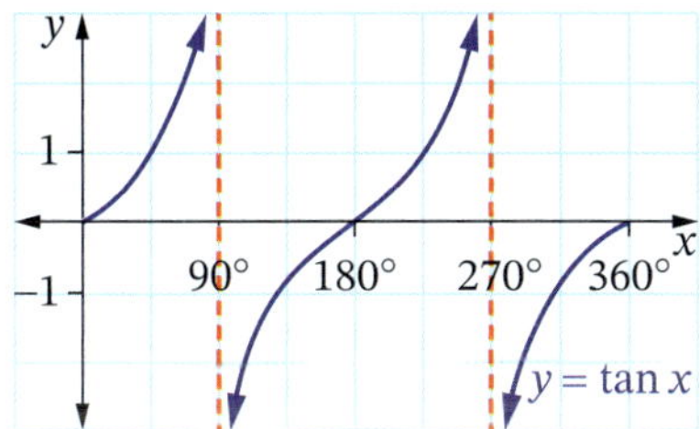

As we go around the unit circle and graph the values of $\frac{y}{x}$, the graph repeats itself every 180°.

$y = \tan x$ has domain $(-\infty, \infty)$ (all real x) except for when $x = 90°, 270°, 540°, \ldots$ (odd multiples of 90°) and range $(-\infty, \infty)$ (all real y). It is an odd function.

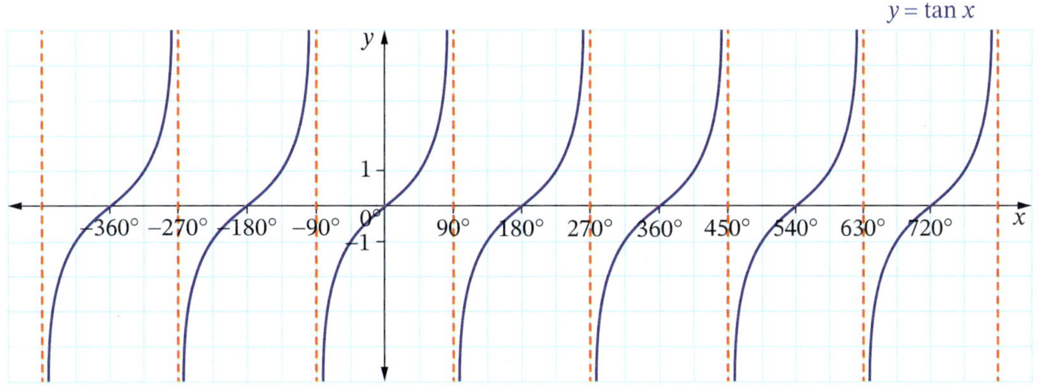

Radians

It is more practical to express the trigonometric functions in terms of radians (not degrees), so here are the graphs in radians:

$y = \sin x$

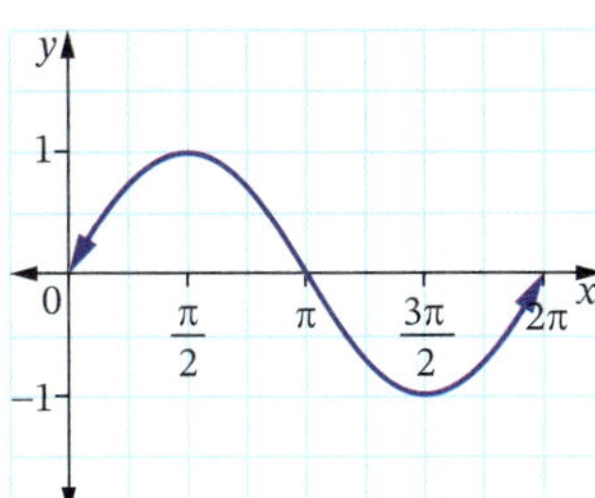

domain $(-\infty, \infty)$,
range $[-1, 1]$ odd function

$y = \cos x$

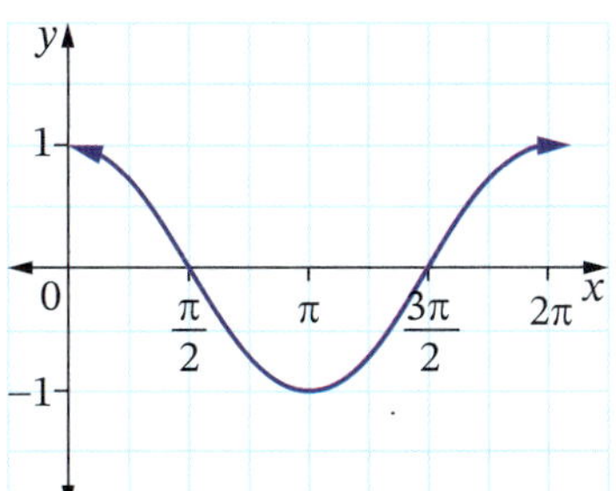

domain $(-\infty, \infty)$,
range $[-1, 1]$ even function

$y = \tan x$

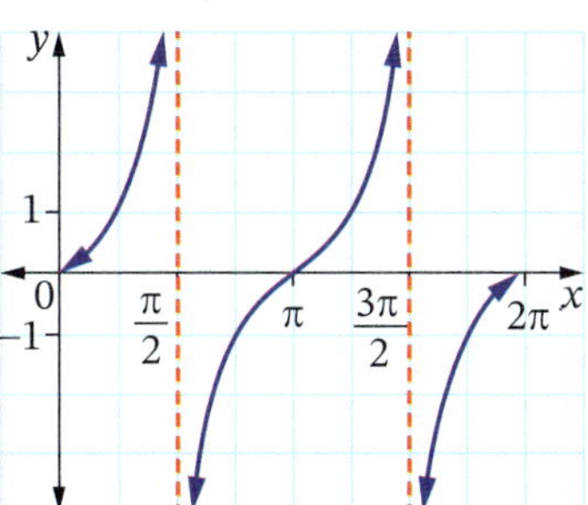

domain $(-\infty, \infty)$ except for $\frac{\pi}{2}, \frac{3\pi}{2}, \frac{5\pi}{2}, \ldots$
(odd multiples of $\frac{\pi}{2}$),
range $(-\infty, \infty)$ odd function

Remember that π is a number, so $90° = \frac{\pi}{2} = 1.5707\ldots$ radians, $180° = \pi = 3.1415\ldots$ radians, and so on.

Properties of the trigonometric functions

All the trigonometric functions have graphs that repeat at regular intervals, so they are called **periodic functions**. The **period** is the length of 1 cycle of a periodic function on the x-axis, before the function repeats itself.

> You will learn about properties and transformations of trigonometric functions again in Year 12 Chapter 2, *Trigonometric functions*.

The **centre** of a periodic function is its mean value and is equidistant from the maximum and minimum values. The mean value of $y = \sin x$, $y = \cos x$ and $y = \tan x$ is 0, represented by the x-axis.

The **amplitude** is the height from the centre of a periodic function to the maximum or minimum values (peaks and troughs of its graph respectively). The range of $y = \sin x$ and $y = \cos x$ is $[-1, 1]$.

Period and amplitude of $y = \sin x$, $y = \cos x$ and $y = \tan x$

	Period	Amplitude	0°	90°	180°	270°	360°
$y = \sin x$	2π	1	0	1	0	−1	0
$y = \cos x$	2π	1	1	0	−1	0	1
$y = \tan x$	π	No amplitude	0	–	0	–	0

Investigation

Transforming trigonometric graphs

Use graphing technology to draw the graphs of trigonometric functions with different values.

1. Graphs in the form $y = k \sin x$, $y = k \cos x$ and $y = k \tan x$, where $k = ..., -3, -2, -1, 2, 3, ...$
2. Graphs in the form $y = \sin ax$, $y = \cos ax$ and $y = \tan ax$, where $a = ..., -3, -2, -1, 2, 3, ...$
3. Graphs in the form $y = \sin x + c$, $y = \cos x + c$ and $y = \tan x + c$, where $c = ..., -3, -2, -1, 2, 3, ...$
4. Graphs in the form $y = \sin(x + b)$, $y = \cos(x + b)$ and $y = \tan(x + b)$, where $b = ..., \pm\frac{\pi}{2}, \pm\pi, \pm\frac{\pi}{4}, ...$
5. Do the same with $\operatorname{cosec} x$, $\sec x$ and $\tan x$.

Can you see patterns? Could you predict what different graphs look like?

Now we shall examine more general trigonometric functions of the form $y = k \sin ax$, $y = k \cos ax$ and $y = k \tan ax$, where k and a are constants.

Period and amplitude of trigonometric functions

$y = k \sin ax$ has amplitude k and period $\frac{2\pi}{a}$.

$y = k \cos ax$ has amplitude k and period $\frac{2\pi}{a}$.

$y = k \tan ax$ has no amplitude and has period $\frac{\pi}{a}$.

> The amplitude is affected by the vertical dilation of factor k, while the period is affected by the horizontal dilation of factor $\frac{1}{a}$.

Example 6

a Sketch each function in the domain $[0, 2\pi]$.

i $y = 5\sin x$ **ii** $y = \sin 4x$ **iii** $y = 5\sin 4x$

b Sketch the graph of $y = 2\tan\frac{x}{2}$ for $[0, 2\pi]$.

Solution

a i The graph of $y = 5\sin x$ is the graph of $y = \sin x$ dilated vertically by factor 5 from the x-axis, so this function has amplitude 5 and period 2π. We draw 1 period of the sine 'shape' with y-values between 5 and -5.

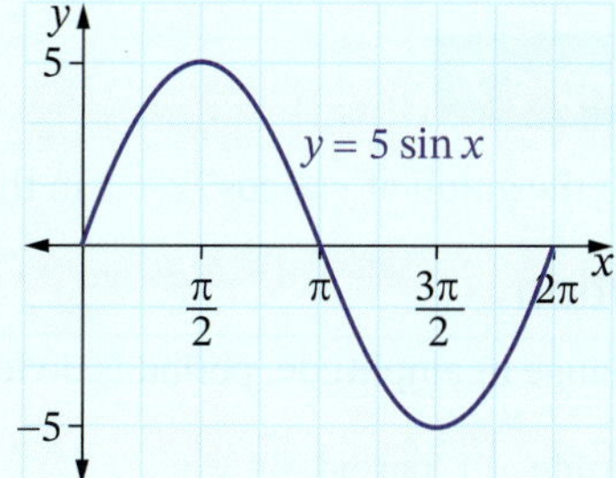

ii The graph $y = \sin 4x$ is the graph of $y = \sin x$ dilated horizontally by factor $\frac{1}{4}$ towards the y-axis, so its period is shortened. It has amplitude 1 and period $\frac{2\pi}{4} = \frac{\pi}{2}$.

The curve repeats every $\frac{\pi}{2}$, so in the domain $[0, 2\pi]$ there will be 4 repetitions or cycles. The '4' in $\sin 4x$ shrinks the graph of $y = \sin x$ horizontally.

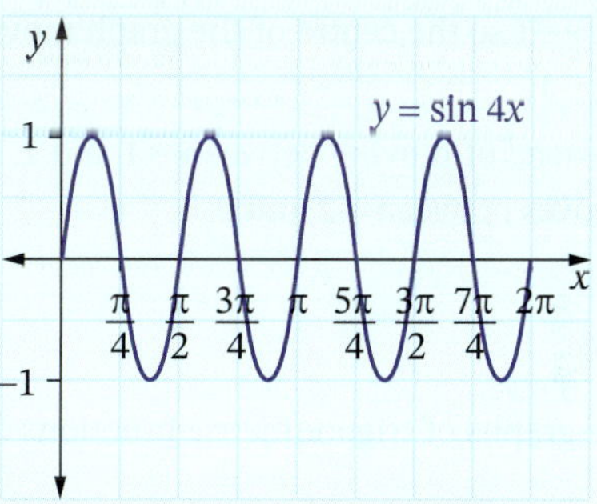

4 cycles between 0 and 2π.

iii The graph $y = 5\sin 4x$ has amplitude 5 and period $\frac{\pi}{2}$. It is a combination of graphs **i** and **ii**.

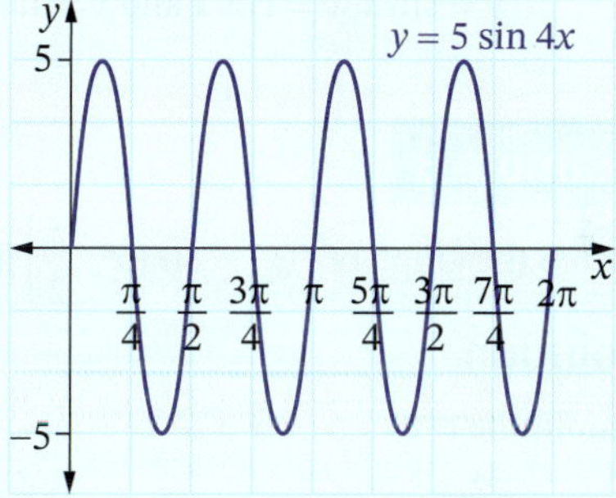

b $y = 2\tan\left(\frac{x}{2}\right)$ has no amplitude.

It is $y = \tan x$ dilated vertically by factor 2 and dilated horizontally by factor 2. It has been enlarged from the origin by dilation factor 2. Its period is doubled from π to 2π.

Period $= \frac{\pi}{\frac{1}{2}} = 2\pi$

So there will be 1 period in the domain $[0, 2\pi]$.

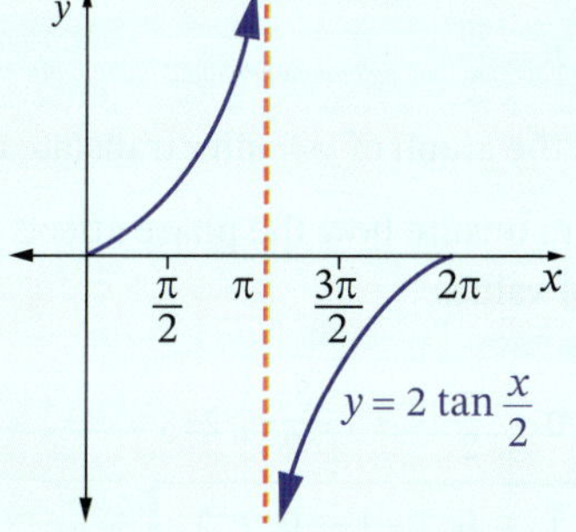

The graphs of trigonometric functions can change their centre, a shift up or down.

Centre of trigonometric functions

$y = \sin x + c$, $y = \cos x + c$ and $y = \tan x + c$ have centre c, which is a shift up from $y = \sin x$, $y = \cos x$ and $y = \tan x$ respectively if $c > 0$ and down if $c < 0$.

This shift of the centre or mean value is a vertical translation.

Video
Graphing the cosine function

Example 7

Sketch the graph of $y = \cos 2x - 1$ in the domain $[0, 2\pi]$.

Solution

No change in amplitude, period is divided by 2.

amplitude = 1, period $\frac{2\pi}{2} = \pi$

$c = -1$ so the centre of the graph moves down to -1.

Instead of moving between -1 and 1, the graph moves between -2 and 0.

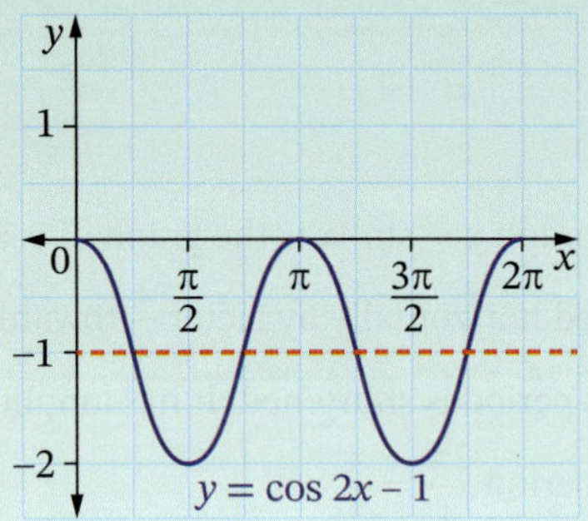

The graphs of trigonometric functions can change their **phase**, a shift to the left or right.

Phase shift of trigonometric functions

$y = \sin(x + b)$, $y = \cos(x + b)$ and $y = \tan(x + b)$ have phase b, which is a shift b units from $y = \sin x$, $y = \cos x$ and $y = \tan x$ respectively, to the left if $b > 0$ and to the right if $b < 0$.

Example 8

Sketch the graph of $f(x) = \sin\left(x + \frac{\pi}{2}\right)$ for $[0, 2\pi]$.

Solution

Amplitude = 1

Period $= \frac{2\pi}{1} = 2\pi$

Phase: $b = \frac{\pi}{2}$

This is the graph of $y = \sin x$ translated $\frac{\pi}{2}$ units to the left.

If you're unsure how the phase affects the graph, create a table of values.

x	0	$\frac{\pi}{2}$	π	$\frac{3\pi}{2}$	2π
y	1	0	-1	0	1

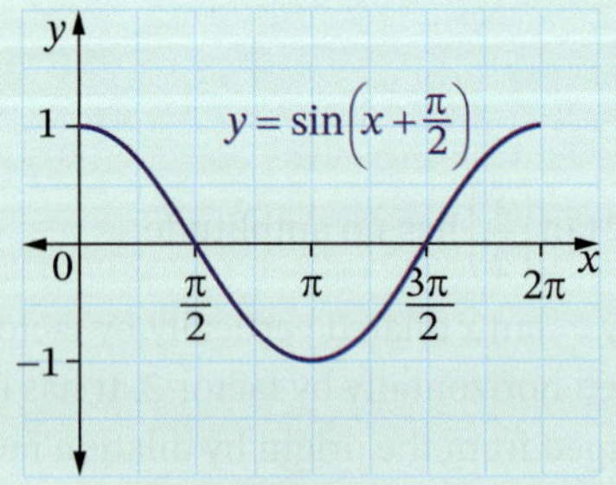

Notice that this function is the same as $y = \cos x$.

General trigonometric functions

	Amplitude	Period	Phase	Centre
$y = k\sin[a(x+b)] + c$	k	$\frac{2\pi}{a}$	b Shift left if $b > 0$ Shift right if $b < 0$	$y = c$ Shift up if $c > 0$ Shift down if $c < 0$
$y = k\cos[a(x+b)] + c$	k	$\frac{2\pi}{a}$		
$y = k\tan[a(x+b)] + c$	No amplitude	$\frac{\pi}{a}$		

Example 9

For the function $y = 3\cos(2x - \pi)$, find:

a the amplitude **b** the period **c** the phase.

Solution

$y = 3\cos(2x - \pi)$

$= 3\cos\left[2\left(x - \frac{\pi}{2}\right)\right]$

a amplitude $= 3$

b period $= \frac{2\pi}{2}$

$= \pi$

c phase $= \frac{\pi}{2}$ units to the right

EXERCISE 9.02 Answers on p. 518

Trigonometric functions

EXAMPLE 6

1 **a** Sketch the graph of $f(x) = \cos x$ in the domain $[0, 2\pi]$.

b Sketch the graph of $y = -f(x)$ in the same domain.

2 Sketch the graph of each function in the domain $[0, 2\pi]$.

a $y = \cos 2x$ **b** $y = \tan 2x$ **c** $y = \sin 3x$

d $f(x) = 3\cos 4x$ **e** $y = 6\cos 3x$ **f** $y = \tan\left(\frac{x}{2}\right)$

g $f(x) = 2\tan 3x$ **h** $y = 3\cos\left(\frac{x}{2}\right)$ **i** $y = 2\sin\left(\frac{x}{2}\right)$

3 Sketch the graph of each function in the domain $[0, 2\pi]$.

a $f(x) = 2\sin x$ **b** $y = \sin x + 1$ **c** $y = 2 - \sin x$

d $f(x) = -3\cos x$ **e** $y = 4\sin x$ **f** $f(x) = \cos x + 3$

g $y = 5\tan x$ **h** $f(x) = \tan x + 3$ **i** $y = 1 - 2\tan x$

4 Sketch the graph of each function in the domain $[-\pi, \pi]$:

a $y = -\sin 2x$ **b** $y = 7\cos 4x$ **c** $f(x) = -\tan 4x$

d $y = 5\sin 4x$ **e** $f(x) = 2\cos 2x$ **f** $f(x) = 3\tan x - 1$

5 Sketch the graph of $y = 8\sin\left(\frac{x}{2}\right)$ in the domain $[0, 4\pi]$.

Foundation Mastery Complex

EXAMPLES 7, 8

6 Sketch over the interval $[0, 2\pi]$ the graph of:

a $y = \sin(x + \pi)$ **b** $y = \tan\left(x + \frac{\pi}{2}\right)$ **c** $f(x) = \cos(x - \pi)$

d $y = 3\sin\left(x - \frac{\pi}{2}\right)$ **e** $f(x) = 2\cos\left(x + \frac{\pi}{2}\right)$ **f** $y = 4\sin\left(2x + \frac{\pi}{2}\right)$

g $y = \cos\left(x - \frac{\pi}{4}\right)$ **h** $y = \tan\left(x + \frac{\pi}{4}\right)$

7 **a** Describe the transformations that change $y = \sin x$ to $y = 3\sin\left(\frac{x}{2} - 4\right) - 1$.

b Sketch the graph of $y = 3\sin\left(\frac{x}{2} - 4\right) - 1$ in the domain $-4\pi \le x \le 4\pi$.

c State its range.

8 Sketch over the interval $[-2, 2]$ the graph of:

a $y = \sin \pi x$ **b** $y = 3\cos 2\pi x$

EXAMPLE 9

9 For each function, find:

i the amplitude **ii** the period **iii** the centre **iv** the phase

a $y = 5\sin 2x$ **b** $f(x) = -\cos(x - \pi)$ **c** $y = 2\tan(4x) - 2$

d $y = 3\sin\left(x + \frac{\pi}{4}\right) + 1$ **e** $y = 8\cos(\pi x - 2) - 3$ **f** $f(x) = 3\tan\left(5x + \frac{\pi}{2}\right) + 2$

10 Find the domain and range of each function.

a $y = 4\sin x - 1$ **b** $f(x) = -3\cos 5x + 7$

c $y = 3\sin 2x - 1$ **d** $y = 2\cos x + 3$

11 Find the values of a, b and c for this graph of $y = a\cos bx + c$.

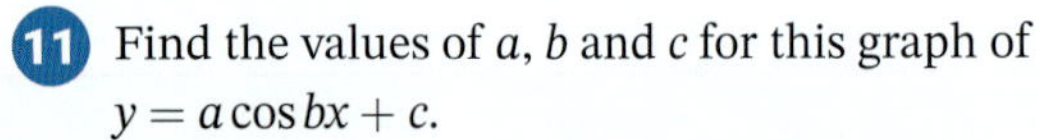

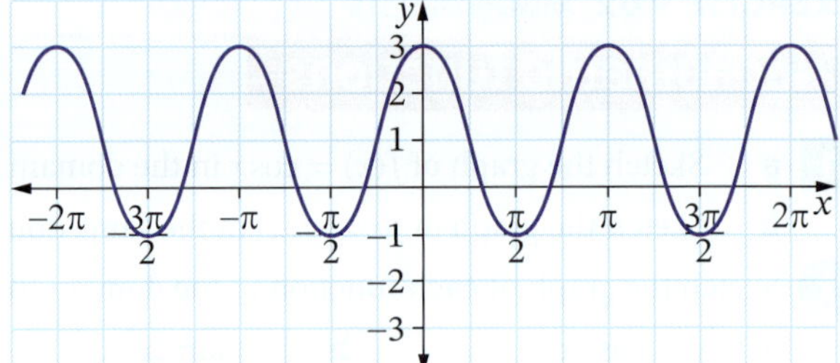

☐ Foundation ○ Mastery ⬡ Complex

Trigonometric equations

9.03

Now that we know how to find the trigonometric ratios of angles of any magnitude, we can learn that there can be more than one solution to a trigonometric equation if we consider all 4 quadrants and larger domains.

Video Trigonometric equations 1

Worksheet Trigonometric equations

Example 10

Solve each equation for the interval [0°, 360°].

a $\sin x = 0.34$ **b** $\cos x = \frac{\sqrt{3}}{2}$ **c** $\tan \theta = -1$

Solution

a 0.34 is positive and $\sin x > 0$ in 1st and 2nd quadrants, so this equation has 2 possible solutions.

$\sin x = 0.34$

$x \approx 19°53', 180° - 19°53'$

$= 19°53', 160°7'$

19°53′ is the **principal solution** but there is another solution in the 2nd quadrant.

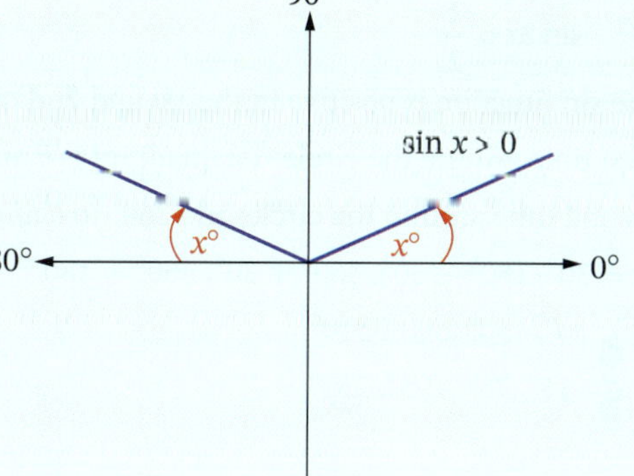

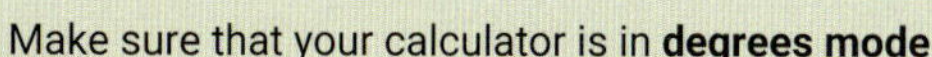

Make sure that your calculator is in **degrees mode**.

Check your solution by substituting back into the equation.

b $\cos x > 0$ in the 1st and 4th quadrants.

$\cos x = \frac{\sqrt{3}}{2}$

$x = 30°, 360° - 30°$

$= 30°, 330°$

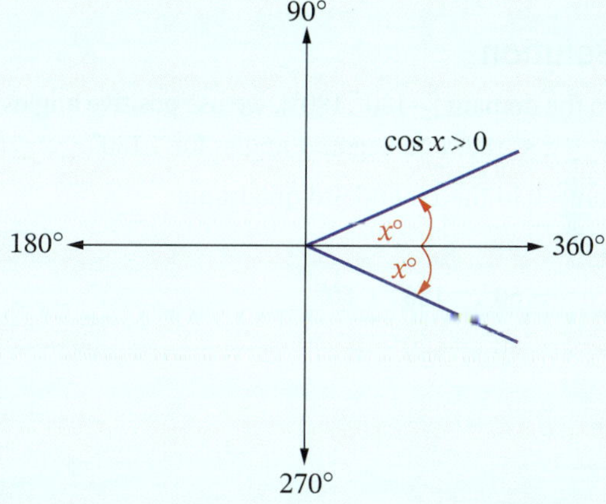

c $\tan \theta < 0$ in the 2nd and 4th quadrants.

For $\tan \theta = -1$:

$\theta = 180° - 45°, 360° - 45°$

$= 135°, 315°$

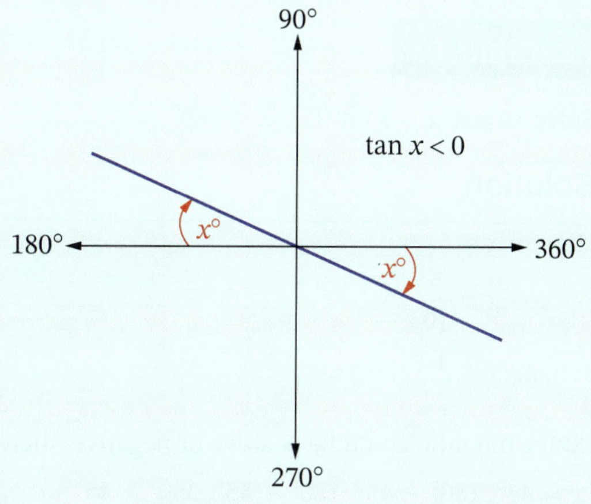

If we are solving an equation involving $2x$ or $3x$, for example, we need to change the domain to find all possible solutions.

Example 11

Solve $2\sin 2x - 1 = 0$ for $[0°, 360°]$.

Solution

Notice that the angle is $2x$ but the domain is for x.

If $\quad 0° \le x \le 360°$

then $0° \le 2x \le 720°$.

This means that we can find the solutions by going around the circle twice.

$$2\sin 2x - 1 = 0$$
$$2\sin 2x = 1$$
$$\sin 2x = \frac{1}{2}$$

The sin function is positive in the 1st and 2nd quadrants.

First time around the circle: 1st quadrant is θ and the 2nd quadrant is $180° - \theta$.

Second time around the circle: add $360°$ to θ and $180° - \theta$.

$$2x = 30°, 180° - 30°, 360° + 30°, 360° + 180° - 30°$$
$$= 30°, 150°, 390°, 510°$$
$$\therefore x = 15°, 75°, 195°, 255°$$

Example 12

Solve $\tan x = \sqrt{3}$ for $[-180°, 180°]$.

Solution

In the domain $[-180°, 180°]$, we use positive angles for $0° \le x \le 180°$ and negative angles for $-180° \le x \le 0°$.

$\tan > 0$ in the 1st and 3rd quadrants.

$$\tan x = \sqrt{3}$$
$$x = 60°, -180° + 60°$$
$$= 60°, -120°$$

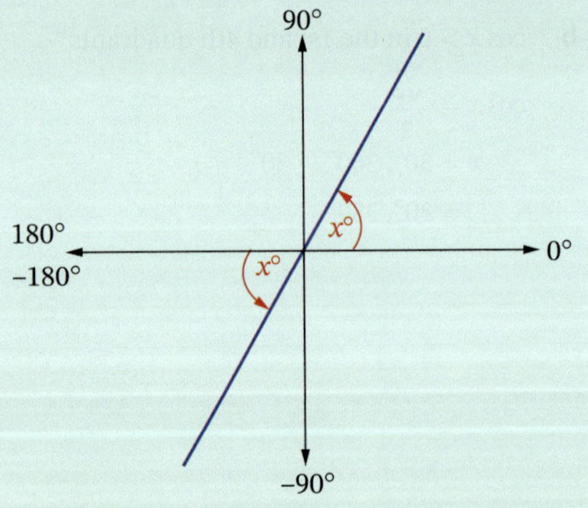

Example 13

Solve $10\cos^2 x = 5$ for $0° \le x \le 360°$.

Solution

$$10\cos^2 x = 5$$
$$\cos^2 x = \frac{5}{10}$$
$$\cos^2 x = \frac{1}{2}$$
$$\cos x = \pm\frac{\sqrt{1}}{\sqrt{2}}$$
$$= \pm\frac{1}{\sqrt{2}}$$

Since the ratio could be positive or negative, there are solutions in all 4 quadrants.

$$x = 45°, 180° - 45°, 180° + 45°, 360° - 45°$$
$$= 45°, 135°, 225°, 315°$$

Equations involving radians

You can solve trigonometric equations involving radians. You can recognise these because the domain is in radians.

Example 14

Video Trigonometric equations 2

Solve each equation for $[0, 2\pi]$.

a $\cos x = 0.34$

b $\sin \alpha = -\frac{1}{\sqrt{2}}$

c $\sin^2 x - \sin x - 2$

d $\tan\left(x - \frac{\pi}{4}\right) = \sqrt{3}$

The domain $[0, 2\pi]$ tells us that the solutions will be in **radians**. Make sure that your calculator is in radian mode here.

Solution

a $\cos x > 0$ in the 1st and 4th quadrants.

$\cos x = 0.34$

$x \approx 1.224, 2\pi - 1.224$

$= 1.224, 5.059$

b $\sin \alpha$ is negative in the 3rd and 4th quadrants.

$\sin \alpha = -\frac{1}{\sqrt{2}}$

$\alpha = \pi + \frac{\pi}{1}, 2\pi - \frac{\pi}{1}$

$= \frac{5\pi}{4}, \frac{7\pi}{4}$

c $\sin^2 x - \sin x = 2$

$\sin^2 x - \sin x - 2 = 0$

This is a quadratic equation in terms of $\sin x$.

Let $u = \sin x$:

$u^2 - u - 2 = 0$

$(u - 2)(u + 1) = 0$ factorising

$u = 2, u = -1$

Change u back to $\sin x$:

$(\sin x - 2)(\sin x + 1) = 0$

$\sin x = 2 \qquad \sin x = -1$

$\sin x = 2$ has no solutions since $-1 \leq \sin x \leq 1$

$\sin x = -1$ has solution $x = \frac{3\pi}{2}$ This comes from the graph of $y = \sin x$ or the y-coordinate of $(0, -1)$ on the unit circle for angle $\frac{3\pi}{2}$.

So $x = \frac{3\pi}{2}$.

d $\tan(\theta) > 0$ in the 1st and 3rd quadrants.

For the domain $[0, 2\pi]$, the equivalent interval for $\left(x - \frac{\pi}{4}\right)$ is

$\left[0 - \frac{\pi}{4}, 2\pi - \frac{\pi}{4}\right] = \left[-\frac{\pi}{4}, \frac{7\pi}{4}\right]$ (one revolution of the circle starting at $-\frac{\pi}{4}$). Thus, $\tan\left(x - \frac{\pi}{4}\right) > 0$ on the two intervals, $\left(-\frac{\pi}{4}, \frac{\pi}{4}\right)$ and $\left(\frac{3\pi}{4}, \frac{5\pi}{4}\right)$

$\tan\left(x - \frac{\pi}{4}\right) = \sqrt{3}$

$x - \frac{\pi}{4} = \frac{\pi}{3}, \pi + \frac{\pi}{3}$

$= \frac{\pi}{3}, \frac{4\pi}{3}$

$x = \frac{\pi}{3} + \frac{\pi}{4}, \frac{4\pi}{3} + \frac{\pi}{4}$

$= \frac{7\pi}{12}, \frac{19\pi}{12}$

Equations involving trigonometric identities

Example 15

Solve $2\sin^2 x - 3\cos x = 0$ for $0° \leq x \leq 360°$.

Solution

Write $\sin^2 x$ as $1 - \cos^2 x$ to create an equation involving $\cos x$ only.

$$2\sin^2 x - 3\cos x = 0$$
$$2(1 - \cos^2 x) - 3\cos x = 0$$
$$2 - 2\cos^2 x - 3\cos x = 0$$
$$-2\cos^2 x - 3\cos x + 2 = 0$$
$$2\cos^2 x + 3\cos x - 2 = 0$$

This is a quadratic equation in terms of $\cos x$.

Let $u = \cos x$:

$$2u^2 + 3u - 2 = 0$$
$$(2u - 1)(u + 2) = 0 \quad \text{factorising}$$
$$2u - 1 = 0, u + 2 = 0$$
$$u = \frac{1}{2} \text{ or } -2$$

Change u back to $\cos x$:

$$\cos x = \frac{1}{2}, \cos x = -2$$

$\cos x = -2$ has no solution

$\cos x = \frac{1}{2}$ has solutions in the 1st and 4th quadrants.

$$x = 60°, 360° - 60°$$
$$= 60°, 300°$$

EXERCISE 9.03 Answers on p. 521

Trigonometric equations

1 Solve each equation for $[0°, 360°]$:

a $\sin\theta = 0.35$ **b** $\cos\theta = -\frac{1}{2}$ **c** $\tan\theta = -1$ **d** $\sin\theta = \frac{\sqrt{3}}{2}$

e $\tan\theta = -\frac{1}{\sqrt{3}}$ **f** $2\cos\theta = \sqrt{3}$ **g** $\tan 2\theta = \sqrt{3}$ **h** $2\cos 2\theta - 1 = 0$

i $2\sin 3\theta = -1$ **j** $\tan^2 3\theta = 1$ **k** $\sin^2 x = 1$ **l** $2\cos^2 x - \cos x = 0$

2 Solve for $[-180°, 180°]$:

a $\cos\theta = 0.187$ **b** $\sin\theta = \frac{1}{2}$ **c** $\tan\theta = 1$ **d** $\sin\theta = -\frac{\sqrt{3}}{2}$

e $\tan\theta = -\frac{1}{\sqrt{3}}$ **f** $3\tan^2\theta = 1$ **g** $\tan\theta + 1 = 0$ **h** $\tan 2\theta = 1$

3 Solve for $0° \leq x \leq 360°$:

a $\cos x = 1$ **b** $\sin x + 1 = 0$ **c** $\cos^2 x = 1$

d $\sin x = 1$ **e** $\tan x = 0$ **f** $\sin^2 x + \sin x = 0$

g $\cos^2 x - \cos x = 0$ **h** $\tan^2 x = \tan x$ **i** $\tan^2 x = 3$

▷

Foundation Mastery Complex

EXAMPLE 14

4 Solve for $[0, 2\pi]$:

a $\sin x = 0$ b $\tan 2x = 0$ c $\sin x = -1$

d $\cos x - 1 = 0$ e $\cos x = -1$

5 Solve for $0 \le x \le 2\pi$:

a $\cos x = \frac{1}{2}$ b $\sin x = -\frac{1}{\sqrt{2}}$ c $\tan x = 1$

d $\tan x = \sqrt{3}$ e $\cos x = -\frac{\sqrt{3}}{2}$

6 Solve for $-\pi \le x \le \pi$:

a $2\sin x = \sqrt{3}$ b $2\cos x = 0$ c $3\tan^2 x = 1$

7 Solve $2\cos x = -1$ in the domain $[-2\pi, 2\pi]$.

EXAMPLE 15

8 Solve $1 - 2\cos^2 x + \sin x = 0$ for $-180° \le x \le 180°$.

9 Solve each equation for $0 \le x \le 2\pi$.

a $2\sin x = 3\cos x$ b $\sec x = 2$ c $\cot x = -1$

d $\operatorname{cosec} x = -2$ e $\cot^2 x - \operatorname{cosec} x = 1$ f $3\sec x + 4\operatorname{cosec} x = 0$

10 Solve for $-\pi \le \theta \le \pi$:

a $2\cos 2\theta = -1$ b $\sin 3\theta = -1$

11 Solve for $0 \le x \le 2\pi$:

a $\tan\left(x - \frac{\pi}{3}\right) = 1$ b $2\sin\left(x + \frac{\pi}{6}\right) = 1$ c $\cos\left(x - \frac{\pi}{6}\right) = -\frac{1}{\sqrt{2}}$

12 Solve for $[0, 2\pi]$:

a $\tan^2 x + \tan x = 0$ b $\sin^2 x - \sin x = 0$ c $2\cos^2 x - \cos x - 1 = 0$

d $4\sin^2 x = 1$ e $\tan x \cos x + \tan x = 0$ f $\sin^2 x + 2\cos x - 2 = 0$

g $\cos^2 x + \sin x + 1 = 0$ h $\sec^2 x + 2\tan x - 4 = 0$ i $4\cos^2 3x = 3$

13 a Write $\cot x \operatorname{cosec}^2 x - \cot^2 x - 4\cot x + 3$ in terms of $\cot x$.

b Solve $\cot x \operatorname{cosec}^2 x - \cot^2 x - 4\cot x + 3 = 0$ for $0° \le x \le 360°$.

14 Solve for $0° \le x \le 360°$:

a $2\sin^2 x - \sin x - 1 = 0$ b $\sec^2 x - \sec x - 2 = 0$

c $\tan^2 x + \tan x - \sqrt{3}\tan x - \sqrt{3} = 0$ d $4\cos^2 x - 2\cos x - 2\sqrt{3}\cos x + \sqrt{3} = 0$

15 Solve $2\cos^3 x - 2\cos^2 x - \cos x + 1 = 0$ for $0 \le x \le 2\pi$.

Sample HSC problem

Answers on p. 521

(7 marks)

a Sketch the graph of $y = 2\sin\left(\frac{x}{2}\right) - 1$ for $[0, 2\pi]$. 3 marks

b Solve the equation $2\sin\left(\frac{x}{2}\right) - 1 = 0$ graphically. 2 marks

c Solve the equation $2\sin\left(\frac{x}{2}\right) - 1 = 0$ algebraically. 2 marks

Foundation Mastery Complex

CHAPTER SUMMARY

This chapter, *Trigonometric identities and functions*, examined the relationships and properties between the different trigonometric ratios, the features of trigonometric functions and their graphs, and the different methods of solving trigonometric equations.

Before you move on, consider what you learned in this chapter and revisit any sections that may have been unclear.

What you learned in this chapter...	Section
Examine and apply the reciprocal trigonometric ratios cosecant, secant and cotangent	9.01 Trigonometric identities
Examine and apply the complementary angle results such as $\sin\theta = \cos(90° - \theta)$	9.01 Trigonometric identities
Examine and apply the trigonometric identities $\tan\theta = \frac{\sin\theta}{\cos\theta}$, $\cot\theta = \frac{\cos\theta}{\sin\theta}$, $\cos^2 x + \sin^2 x = 1$, $1 + \tan^2 x = \sec^2 x$, $1 + \cot^2 x = \operatorname{cosec}^2 x$	9.01 Trigonometric identities
Examine the trigonometric functions, including the reciprocal trigonometric functions, their features, properties and graphs	9.02 Trigonometric functions
Solve trigonometric equations in degrees and radians for any domain	9.03 Trigonometric equations

To help master this topic, make a summary mind map. Use the chapter outline and the mind map below as a guide. Add your own words, symbols, diagrams, boxes and reminders. The summary should give you a 'whole picture' view of the topic and allow you to identify any weak areas to revisit in your revision.

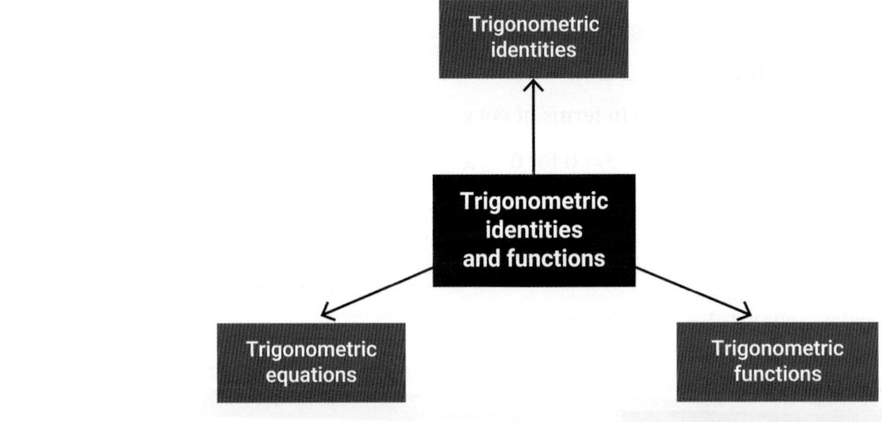

Test yourself 9

Answers on p. 522

For Questions **1** to **5** select the correct answer **A**, **B**, **C** or **D**.

1 The function $y = 2\cos 3x - 7$ has: 9.02

A amplitude 2, period 3 and centre -7

B amplitude 7, period $\frac{1}{3}$ and centre 2

C amplitude 2, period $\frac{2\pi}{3}$ and centre 7

D amplitude 2, period $\frac{2\pi}{3}$ and centre -7

2 The equation of a function with phase π units to the left is: 9.02

A $y = \tan(x + \pi)$ **B** $y = \tan(\pi x)$ **C** $y = \tan(x) + \pi$ **D** $y = \tan(x - \pi)$

3 The equation of the graph below is: 

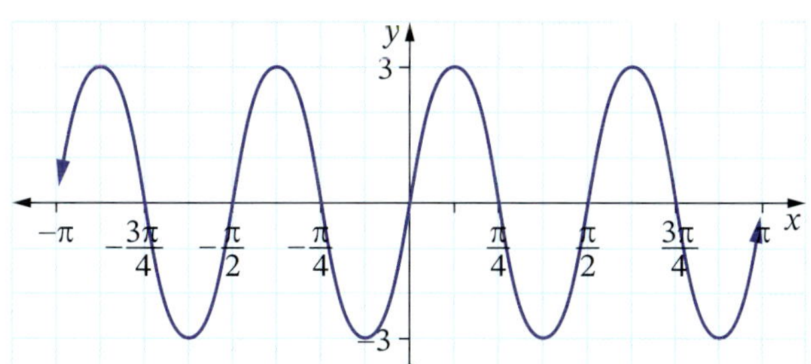

A $y = 3\cos 4x$ **B** $y = 3\sin 4x$ **C** $y = 4\sin 3x$ **D** $y = 4\cos 3x$

4 $\text{cosec}^2 x$ is equal to: 9.01

A $\cot^2 x - 1$ **B** $1 - \cos^2 x$ **C** $1 + \cot^2 x$ **D** $\tan^2 x + 1$

5 The solution of $\cos 2x = 1$ in the domain $[0, 2\pi]$ is: 9.03

A $x = 0, 2\pi$ **B** $x = 0, \pi, 2\pi$ **C** $x = \frac{\pi}{2}, \frac{3\pi}{2}$ **D** $x = \pi$

6 Solve for $0° \leq x \leq 360°$: 9.03

a $\sin x = \frac{\sqrt{3}}{2}$ **b** $\tan x = 1$ **c** $2\cos x + 1 = 0$

d $\sin^2 x = \frac{3}{4}$ **e** $\tan 2x = \frac{1}{\sqrt{3}}$

7 Solve for $0 \leq x \leq 2\pi$: 9.03

a $\tan x = -1$ **b** $2\sin x = 1$ **c** $\tan^2 x = 3$

d $\cos x = 1$ **e** $\sin x = -1$

8 For $0 \leq x \leq 2\pi$, sketch the graph of: 9.02

a $y = 3\cos 2x$ **b** $y = 7\sin\left(\frac{x}{2}\right)$ **c** $y = \sin x - 2$

d $y = -\sin x$ **e** $y = 3\cos 2x - 1$

9 Simplify: 9.01

a $\tan x \cos x$ **b** $\sqrt{4 - 4\sin^2(A)}$

c $\cos(90° - x)$ **d** $\cot\beta \tan\beta$

Foundation Mastery Complex

9.03 **10** Solve for $[0, 2\pi]$:

a $4\cos^2 x = 3$ **b** $2\sin 2x = 1$

c $\cos\left(x - \frac{\pi}{2}\right) = -1$ **d** $\tan^2\left(x + \frac{\pi}{6}\right) = 3$

9.03 **11** Solve for $[-\pi, \pi]$:

a $2\cos 2x = 1$ **b** $\tan\left(x - \frac{\pi}{4}\right) = -\frac{1}{\sqrt{3}}$ **c** $\sin\left(x + \frac{\pi}{2}\right) = \frac{1}{\sqrt{2}}$

9.03 **12** Find the equation of each transformed function of $y = \cos x$ that has:

a period π **b** amplitude 5 **c** a reflection in the x-axis

d a phase of $\frac{\pi}{6}$ units to the right **e** centre 4

9.03 **13** Solve for $[0°, 360°]$:

a $\tan^2 x = \tan x$ **b** $2\cos^2 x + \cos x = 1$

c $2\cos x \tan x = 1$ **d** $2\sin^2 x + 3\cos x - 3 = 0$

9.01 **14** Prove that $\frac{2\cos^2\theta}{1 - \sin\theta} = 2 + 2\sin\theta$.

9.01 **15** Find the value of b if $\sin b = \cos(2b - 30)°$.

9.03 **16** Solve for $[0°, 360°]$:

a $\tan(x + 45°) = 1$ **b** $\sqrt{2}\cos(x - 20°) + 1 = 0$ **c** $3\sin[2(x + 10°)] - 2 = 0$

9.03 **17** Solve for $[-180°, 180°]$:

a $5\tan 2x = -5$ **b** $\cos[3(x - 30°)] + 1 = 0$

9.03 **18** Find the exact value of:

a $\sec\left(\frac{\pi}{4}\right)$ **b** $\cot\left(\frac{\pi}{6}\right)$ **c** $\operatorname{cosec}\left(\frac{\pi}{3}\right)$ **d** $\frac{\cos\left(\frac{\pi}{6}\right)}{\sin\left(\frac{\pi}{6}\right)}$

9.02 **19** Find the domain and range of each function.

a $y = -6\sin(2x) + 5$ **b** $f(x) = 4\cos x - 3$

9.02 **20** State the amplitude, period, centre and phase of each function.

a $y = 2\sin 3x - 1$ **b** $y = \cos\left(\frac{x}{2} + \pi\right)$ **c** $y = -3\tan\left(5x - \frac{\pi}{4}\right)$

9.03 **21** Solve for $[0, 2\pi]$:

a $\sin^3 x = \sin x$ **b** $\tan^2 x + \tan x - 2 = 0$ **c** $15\sin^3 x - \sin x - 6 = 0$

☐ Foundation ○ Mastery ⬡ Complex

Challenge exercise 9

Answers on p. 523

1 Find the exact value of:

a $\sin 600°$ **b** $\tan(-405°)$

2 Solve $2\cos(\theta + 10°) = -1$ for $0° \leq \theta \leq 360°$.

3 If $f(x) = 3\cos \pi x$:

a find the period and amplitude of the function

b sketch the graph of $f(x)$ for $0 \leq x \leq 4$.

4 For $0 \leq x \leq 2\pi$, sketch the graph of:

a $f(x) = 2\cos\left(x + \frac{\pi}{2}\right) + 1$ **b** $y = 2 - 3\sin\left(x - \frac{\pi}{2}\right)$

c $y = \sin 2x - \sin x$ **d** $y = \sin x + 2\cos 2x$

e $y = 3\cos x - \cos 2x$ **f** $y = \sin x - \sin\left(\frac{x}{2}\right)$

5 **a** Find the equation of the transformation of $y = \cos x$ if it has a vertical translation 5 units down, a horizontal translation $\frac{\pi}{6}$ units to the right, then a vertical stretch, factor 4 and a horizontal stretch, factor 3.

b Find the amplitude, period, centre and phase of this transformed function.

6 **a** Find the amplitude, period and phase of the function $y = 2\cos\left(2x - \frac{\pi}{2}\right)$.

b Solve $2\cos\left(2x - \frac{\pi}{2}\right) = \sqrt{3}$ for $[0, 2\pi]$.

7 Sketch the graph of $y = 3\sec 2x$ for $[0, 2\pi]$.

8 Solve $\text{cosec}\, 2x = \sqrt{2}$ for $[0, 2\pi]$.

9 Show that $\dfrac{\cos\theta(\sin\theta + \cos\theta)}{(1+\sin\theta)(1-\sin\theta)} = 1 + \tan\theta$.

10 Write the equation of a sine function with period π, centre -2 and range $[-5, 1]$.

Foundation Mastery Complex

PROBABILITY AND DATA

Probability is the study of how likely it is that something will happen. It is used to make predictions and decisions in areas such as business, investment, weather forecasting and insurance. Statistics is also about analysing data to make decisions. Probability and data are closely related.

In this chapter we will look at how to find the probability of something happening, using both experimental data and theoretical probability.

Chapter outline

In this chapter you will:

- use the language and notation of sets, Venn diagrams and probability theory
- understand and use the addition rule for probability $P(A \cup B) = P(A) + P(B) - P(A \cap B)$
- understand and use the product rule for probability for independent events $P(A \cap B) = P(A) \times P(B)$
- apply the addition and product rules to tree diagrams to solve multi-stage probability problems
- solve problems involving conditional probability, including using the formula $P(A \mid B) = \dfrac{P(A \cap B)}{P(B)}$
- identify discrete and continuous random variables
- organise data in frequency tables using cumulative frequency, relative frequency and cumulative relative frequency
- display data using histograms and polygons, including cumulative frequency histograms and polygons (ogives)
- find the mode and median of a data set from tables, histograms and polygons
- use relative frequency to estimate probability

Videos (7):

10.01 Venn diagrams 1 • Venn diagrams 2
10.04 Dependent events
10.05 Tree diagrams 1 • Tree diagrams 2
10.06 Conditional probability
10.08 Experimental probability

Worksheets (11):

10.01 Venn diagrams 1 • Venn diagrams 2 • Set operations
10.02 Theoretical probability
10.04 Multi-stage problems
10.05 Tree diagrams • Probability trees
10.06 Conditional probability
10.07 Graphing data • Cumulative frequency graphs
10.08 Experimental probability

Puzzles (5):

10.01 Venn diagrams matching activity
10.02 Matching probabilities
10.06 Conditional probability: Two-way tables
Chapter summary Probability crossword • Probability find-a-word

Nelson MindTap

To access resources above, visit **cengage.com.au/nelsonmindtap**

Terminology

addition rule	complement	conditional probability	cumulative frequency
disjoint	element	frequency polygon	histogram
independent events	intersection	median	mode
mutually exclusive events	ogive	probability tree	product rule
random variable	relative frequency	sample space	set
subset	tree diagram	union	Venn diagram

Set notation and Venn diagrams

10.01

A **set** is a collection of items called **elements**. The items are distinct (different) and can be numbers or things. We used sets when introducing functions in Chapter 3, *Functions and graphs*, and learned that the domain of a function is the set of all of the x values of a function.

For example, $A = \{1, 2, 3, 4, 5\}$ is the set of integers from 1 to 5 and $B = \{1, 3, 5, 7, 9, 11, ...\}$ is the set of all odd, positive integers.

The number of elements in set A is written $n(A)$ or $|A|$, and is called the **cardinality** of set A.

In the above example, $n(A) = 5$.

If a set has no elements, then it is called the **empty set**, written as $\varnothing$ or { }.

In the above example, the set A is called a **finite set** because it has a finite number of elements, while the set B is called an **infinite set** because it has an infinite number of elements.

A part of a set, containing some (or all) of the elements of the set, is called a **subset** of the original set. Set X is a subset of set Y if all the elements of X are also elements of Y.
For example, for $P = \{3, 8, 10\}$ and $Q = \{1, 2, 3, 4, 5, 6, 7, 8, 9, 10\}$, set P is a subset of set Q.

Videos
Venn diagrams 1
Venn diagrams 2

Worksheets
Venn diagrams 1
Venn diagrams 2
Set operations

Puzzle
Venn diagrams matching activity

Union and intersection

$A \cup B$ means A **union** B and is the set of all elements in set A or set B (or both).

$A \cap B$ means A **intersection** B and is the set of all elements that are in **both** sets A and B.

If 2 sets have no elements in common, then $A \cap B = \varnothing$, and A and B are called **disjoint sets**.

Example 1

Set A contains the numbers 3, 7, 12 and 15.

Set B contains the numbers 2, 9, 12, 13 and 17.

a Write sets A and B in set notation.

b Find $A \cup B$ and $A \cap B$.

c Let $C = \{2, 8, 10, 17, 20, 21\}$. Given sets A, B and C, name 2 disjoint sets.

Solution

a $A = \{3, 7, 12, 15\}$.

$B = \{2, 9, 12, 13, 17\}$.

b $A \cup B = \{2, 3, 7, 9, 12, 13, 15, 17\}$. It includes all the numbers in either set A or set B.

$A \cap B = \{12\}$. It includes any numbers that are in both set A and set B.

c Sets A and C are disjoint because they have no elements in common. $A \cap C = \varnothing$

The universal set U and the complementary set $\bar{A}$

The **universal set** U is the set of all possible elements of sets.

The **complement** of set A is the set of all elements (from U) that are *not* in set A. This complementary set is written as $\bar{A}$, A' or A^c.

Example 2

Let set U contain the integers from 1 to 10. Set A contains the odd numbers that are a subset of U and set B contains the prime numbers that are a subset of U. Find:

a $A \cup B$ **b** $A \cap B$ **c** $\overline{A}$ **d** $\overline{B}$ **e** $A \cap \overline{B}$

Solution

$U = \{1, 2, 3, 4, 5, 6, 7, 8, 9, 10\}$

$A = \{1, 3, 5, 7, 9\}$

$B = \{2, 3, 5, 7\}$

a $A \cup B = \{1, 2, 3, 5, 7, 9\}$

b $A \cap B = \{3, 5, 7\}$

c $\overline{A}$ is the complement of A, all elements in U that are not in A.

$\overline{A} = \{2, 4, 6, 8, 10\}$

d $\overline{B} = \{1, 4, 6, 8, 9, 10\}$

e $A \cap \overline{B} = \{1, 9\}$

Venn diagrams

A **Venn diagram** is a visual way to show the relationships between 2 or more sets. A rectangle represents the whole group, the **universal set**, while the circles represent individual sets. Items common to 2 or more sets are placed in the intersection (overlapping region) of the circles.

Venn diagram

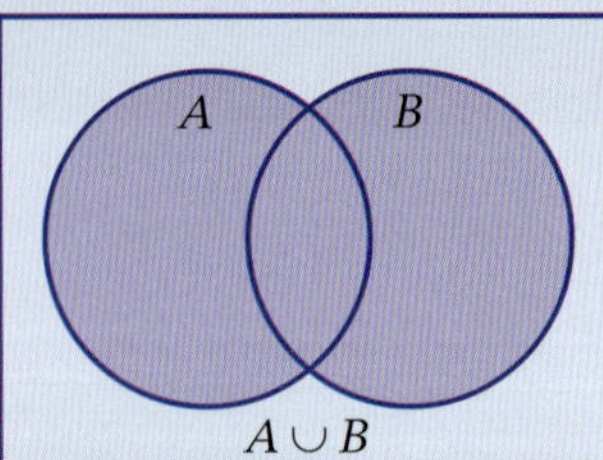

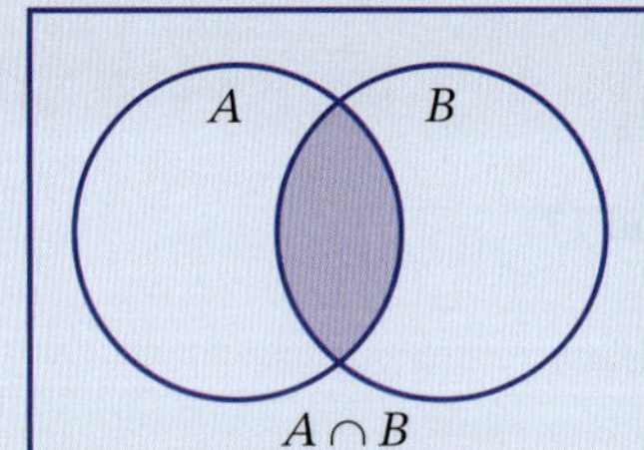

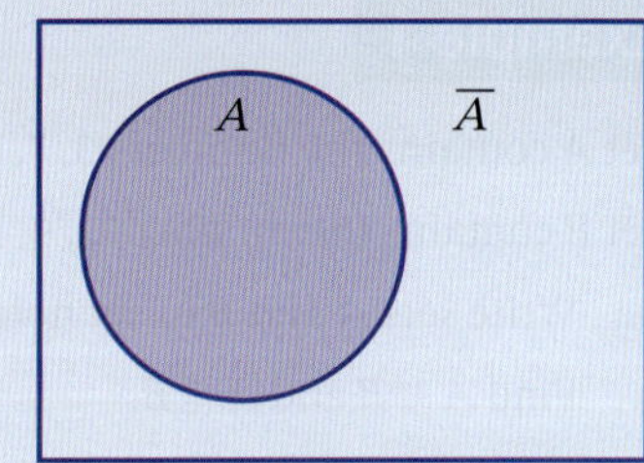

Example 3

Draw a Venn diagram for the integers from 1 to 10, given $A = \{1, 3, 4, 5, 8\}$ and $B = \{3, 6, 8, 9, 10\}$.

Solution

Draw 2 overlapping circles inside a rectangle and name them A and B.

$A \cap B = \{3, 8\}$, so place these numbers in the overlapping part.

Place the remaining elements of A in the other part of circle A and the remaining elements of B in the other part of circle B.

The numbers 2 and 7 are not in A or B, so place them outside the circles.

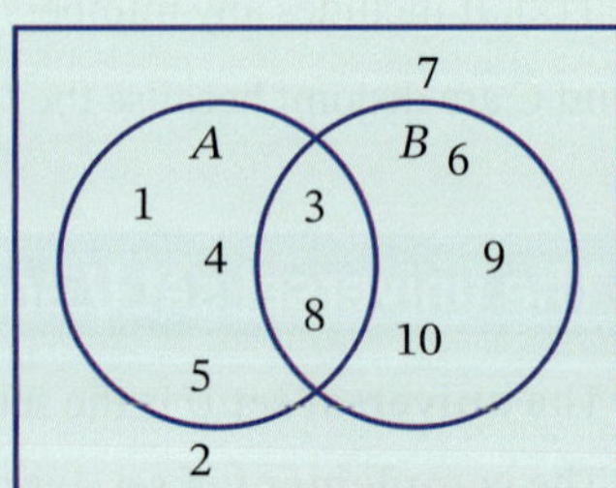

Example 4

In a class of 30 students, all students study Modern History or Drama.

Of these, 21 students do Modern History and 22 students do Drama.

Draw a Venn diagram to represent these numbers.

Solution

a $21 + 22 = 43 > 30$.

There are only 30 students, so there is some overlap of students doing both subjects (they have been counted twice).

$43 - 30 = 13$, so 13 students do both Modern History and Drama.

Draw a Venn diagram with overlapping circles for Modern History and Drama and write 13 in the overlapping region:

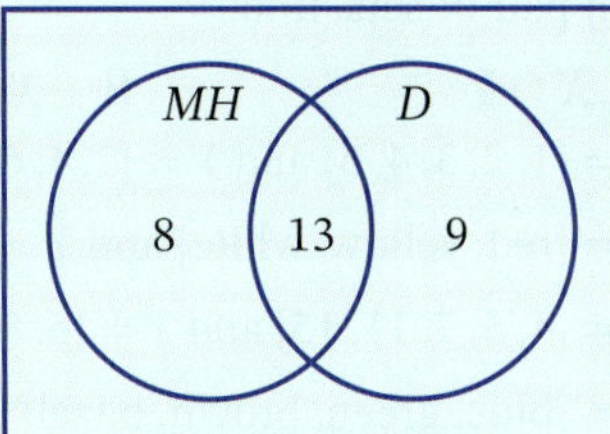

To work out what values go in the other parts of the circles:

For Modern History only, $21 - 13 = 8$, so 8 students only do Modern History and not Drama.

For Drama only, $22 - 13 = 9$, so 9 students only do Drama and not Modern History.

Check that the numbers add up to 30 and that the individual totals add up to 22 and 21

Number of elements in $A \cup B$

$$|A \cup B| = |A| + |B| - |A \cap B|$$

To avoid 'double counting' for the total, we must subtract $|A \cap B|$, the overlapping section of the Venn diagram.

Example 5

For all the integers from 1 to 100, find $|A \cup B|$ if set A contains all the multiples of 5 and set B contains all the even numbers.

Solution

$A = \{5, 10, 15, 20, \ldots 100\}$

$B = \{2, 4, 6, 8, 10, \ldots 100\}$

Dividing 100 by 5, we can see that A has 20 elements: $|A| = 20$

B has 50 elements (half the numbers from 1 to 100): $|B| = 50$

$A \cap B = \{10, 20, 30, \ldots 100\}$ as these elements are both even and multiples of 5.

So $|A \cap B| = 10$.

$$\begin{aligned} |A \cup B| &= |A| + |B| - |A \cap B| \\ &= 20 + 50 - 10 \\ &= 60 \end{aligned}$$

EXERCISE 10.01 Answers on p. 523

Set notation and Venn diagrams

1 Given sets $P = \{5, 6, 7, 8, 9\}$ and $Q = \{1, 2, 3, 4, 5, 6, 7, 8, 9\}$:

a find $n(P)$ and $n(Q)$

b describe the relationship between P and Q.

EXAMPLE 1

2 For each pair of sets, find:

i $X \cap Y$ **ii** $X \cup Y$.

a $X = \{1, 2, 3, 4, 5\}$ and $Y = \{2, 4, 6\}$

b $X =$ {red, yellow, white} and $Y =$ {red, white}

c $X = \{4, 5, 7, 11, 15\}$ and $Y = \{6, 8, 9, 10, 12\}$

d $X =$ {blue, green, brown, hazel} and $Y =$ {brown, grey, blue}

e $X = \{1, 3, 5, 7, 9\}$ and $Y = \{2, 4, 6, 8, 10\}$

3 If $A = \{2, 5, 6, 9\}$ and $B = \{1, 3, 4, 8, 10\}$, find $A \cap B$ and describe the relationship between sets A and B.

EXAMPLE 2

4 Find the complement of each set.

a A is the set of all odd numbers.

b $B =$ {all Toyota cars}

c $C =$ {all multiples of 3}

d D is the set of tossed coins that come up heads.

e $E =$ {red, white} when red, green, white, blue and yellow lollies are selected from a bag.

f F is the set of all winning teams in a soccer round.

5 Let set U be the integers from 1 to 20, set A be the factors of 20 and set B be the multiples of 4 that are a subset of U. Find:

a $A \cup B$ **b** $A \cap B$ **c** $\overline{A}$ **d** $\overline{B}$ **e** $A \cap \overline{B}$

EXAMPLE 3

6 Draw a Venn diagram for each pair of sets.

a $A = \{10, 12, 13, 14, 15\}$ and $B = \{12, 14, 15, 16\}$

b $P =$ {red, yellow, white} and $Q =$ {red, green, white}

c $X = \{2, 3, 5, 7, 8\}$ and $Y = \{1, 2, 5, 7, 9, 10\}$

d $A =$ {Toyota, Mazda, Kia, Nissan, MG} and $B =$ {Mazda, Nissan, Hyundai, Ford, MG}

e $X =$ {rectangle, square, trapezium} and $Y =$ {square, parallelogram, trapezium, kite}

7 Draw a Venn diagram for each pair of sets.

a Event $K =$ {Monday, Thursday, Friday} and Event $L =$ {Tuesday, Thursday, Saturday} out of days of the week.

b Event $A = \{3, 5, 6, 8\}$ and Event $B = \{4, 5, 7, 9\}$ with cards each with a number from 1 to 10 drawn out of a hat.

8 For the sets in Question **7**, find the value of:

a $|K \cup L|$ **b** $|K'|$ **c** $|A \cap B|$ **d** $|\overline{B}|$ **e** $|\overline{A \cup B}|$

EXAMPLE 4

9 Draw a Venn diagram for the numbers of elements in each situation.

a All students in a class study Music or Economics. 11 students study only music, 9 study only Economics and 3 study both subjects.

b A street cart sells coffee and sandwiches. Today, 27 people only bought coffee, 12 only bought a sandwich and 31 people bought both.

c In a class of 25 students, 14 play basketball and 18 play football.

d In a group of 8 friends, 5 like to watch NetTix and 7 like to watch YouTock.

e There are 108 students doing the HSC at Belvetia High School. 83 study Maths, 59 study History and 10 study neither subject.

☐ Foundation ○ Mastery ○ Complex

10 A sample of 50 people were surveyed to see if they liked pop, jazz or country music. All liked at least one of these music forms. Draw a Venn diagram if 31 liked jazz, 10 liked only pop, 4 liked pop and country but not jazz, 8 liked pop and jazz but not country, 12 liked country and jazz but not pop, and 4 liked all 3 forms of music.

11 Find $|A \cup B|$ for each description.

a A is the set of multiples of 10 from 1 to 200 and B is the set of integers from 1 to 50

b $A = \{2, 4, 6, 8, \ldots 60\}$ and $B = \{3, 6, 9, 12, \ldots 60\}$

Did you know?

John Venn

Venn diagrams are named after **John Venn** (1834–1923), an English probabilist and logician.

Probability 10.02

We can usually predict the probability or likelihood of something happening, an event, by using surveys, experiments or trials.

Worksheet Theoretical probability

Puzzle Matching probabilities

Investigation

Tossing coins

Toss a coin 20 times and count the number of heads and tails. What would you expect to happen when tossing a coin this many times? Did your results surprise you?

Combine your results with others in your classroom into a table with frequencies of heads and tails.

Do the combined results differ from your own?

From the table, find the probability of tossing

a heads **b** tails.

If a coin came up tails every time when it was tossed 20 times, do you think it would it be more likely to come up heads the next time? Why?

Now toss 2 coins 20 or more times and record your individual and combined results for 2 heads, 2 tails, one head and one tail.

Are you surprised by your results? Why?

Sample space

Experiments and surveys can give a good prediction of the probability of future events, but they are not very accurate. For example, it is reasonable to assume that if you toss a coin many times you would get similar numbers of heads and tails. Yet in an experiment a coin may come up heads every time, purely by chance.

- An **outcome** is a possible result of a random chance experiment that can be repeated.
- Each running of the chance experiment is called a **trial**.
- The **sample space**, S, is the set of all possible outcomes for a chance experiment.
- An **event** is a set of one or more outcomes, a **subset** of S.

The sample space for rolling a die is $S = \{1, 2, 3, 4, 5, 6\}$. $|S| = 6$.

The sample space for rolling 2 dice is shown in the table. There are 36 possible outcomes. $|S| = 36$.

		2nd die					
		1	2	3	4	5	6
1st die	1	1, 1	1, 2	1, 3	1, 4	1, 5	1, 6
	2	2, 1	2, 2	2, 3	2, 4	2, 5	2, 6
	3	3, 1	3, 2	3, 3	3, 4	3, 5	3, 6
	4	4, 1	4, 2	4, 3	4, 4	4, 5	4, 6
	5	5, 1	5, 2	5, 3	5, 4	5, 5	5, 6
	6	6, 1	6, 2	6, 3	6, 4	6, 5	6, 6

If you were surprised by the outcomes for tossing 2 coins in the above investigation, you may have thought there were only 3 outcomes: 2 heads, 2 tails, one of each.

But there are really 4 possible outcomes: $S = \{\text{HH, TT, HT or TH}\}$

'1 head, 1 tail' can occur in 2 different ways, so the event of tossing a head and a tail is made up of 2 different outcomes.

The probability of an event

To find the probability of an event happening, we compare the number of ways the event can occur with the total number of possible outcomes (the sample space).

$$\text{Probability of an event} = \frac{\text{number of favourable outcomes}}{\text{total number of outcomes}}$$

If we call the event A and the sample space S we can write this as:

Probability formula

$$P(A) = \frac{n(A)}{n(S)} = \frac{|A|}{|S|}$$

where $P(A)$ means the probability of event A, $|A|$ is the number of outcomes in the event, and $|S|$ is the number of outcomes in the sample space, where each outcome is **equally likely**.

The probability of each outcome in the sample space is $\frac{1}{|S|}$.

We usually write a probability as a fraction, but we could also write it as a decimal or percentage.

Example 6

30 people were surveyed on their favourite sport: 11 preferred football, 4 preferred basketball, 7 preferred tennis, 2 preferred golf and 6 preferred swimming.

Find the probability that any one of these people selected at random will prefer:

a swimming **b** football **c** golf.

Solution

The size of the sample space $|S| = 30$ since 30 people were surveyed.

a $P(\text{swimming}) = \frac{6}{30}$

$= \frac{1}{5}$

b $P(\text{football}) = \frac{11}{30}$

c $P(\text{golf}) = \frac{2}{30}$

$= \frac{1}{15}$

Mutually exclusive events

Mutually exclusive events are events that cannot occur at the same time. For example, when throwing a die you cannot throw a number that is both a 5 and a 6. On a Venn diagram, mutually exclusive events are represented by separate circles that do not overlap.

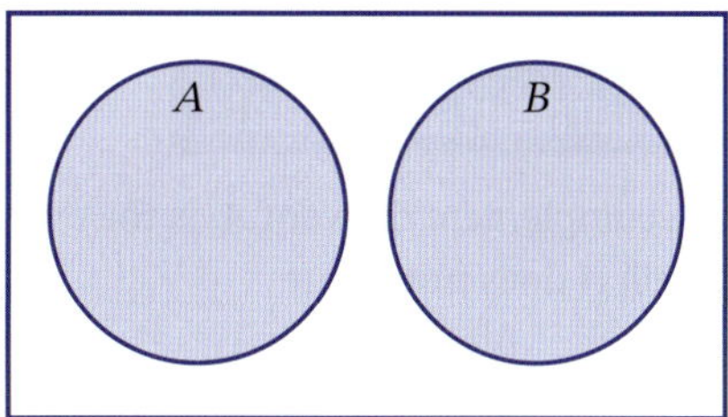

Mutually exclusive events:
'A or B' means A or B but not both

Mutually exclusive events are also disjoint subsets of the sample space as they have no elements in common.

The addition rule for mutually exclusive events

When events A and B are mutually exclusive:

$$P(A \cup B) = P(A) + P(B)$$

Example 7

A clothes basket holds 5 blue, 3 white and 7 yellow T-shirts. If one T-shirt is selected at random, find the probability of selecting:

a a white T-shirt

b a white or blue T-shirt

c a yellow, white or blue T-shirt

d a red T-shirt.

Solution

Blue, white and yellow are mutually exclusive events.

$|S| = 5 + 3 + 7 = 15$

a $P(\text{W}) = \frac{3}{15}$

$= \frac{1}{5}$

b $P(\text{W} \cup \text{B}) = P(\text{W}) + P(\text{B})$

$= \frac{3}{15} + \frac{5}{15}$

$= \frac{8}{15}$

$P(W \cup B)$ means the probability of W or B.

c $P(\text{Y} \cup \text{W} \cup \text{B}) = P(\text{Y}) + P(\text{W}) + P(\text{B})$

$= \frac{7}{15} + \frac{3}{15} + \frac{5}{15}$

$= \frac{15}{15}$

$= 1$

d $P(\text{R}) = \frac{0}{15}$

$= 0$

The range of probabilities

If $P(A) = 0$, the event is impossible.

If $P(A) = 1$, the event is certain (it has to happen).

$0 \leq P(A) \leq 1$

The sum of all (mutually exclusive) probabilities is 1.

As a percentage, this is from 0% to 100%.

Complementary events

We know that set $\overline{A}$ is the complement of set A, the set of all elements that are not in A.

In probability, event $\overline{A}$ is the complement of event A, all the possible outcomes in the sample space that are not A.

So, $\overline{A}$ is the event that something other than A happens, that is, that A does not occur.

Complementary events

$\overline{A}$, the complement of event A, is the event in which A does **not** occur.

$$P(\overline{A}) = 1 - P(A)$$

$$P(A) + P(\overline{A}) = 1$$

Example 8

a The probability of winning a raffle is $\frac{1}{350}$. What is the probability of not winning?

b The probability of a tree surviving a fire is 72%. Find the probability of the tree being destroyed by a fire.

Solution

a $P(\text{not win}) = 1 - P(\text{win})$

$= 1 - \frac{1}{350}$

$= \frac{349}{350}$

b $P(\text{failing to survive}) = 1 - P(\text{surviving})$

$= 100\% - 72\%$

$= 28\%$

We can use probability to make predictions or decisions based on expected frequencies.

Example 9

The probability that a traffic light will turn green as a car approaches it is $\frac{5}{12}$.

A taxi goes through 192 intersections where there are traffic lights.
How many of these would be expected to show green as the taxi approached?

Solution

It is expected that $\frac{5}{12}$ of the traffic lights would show green.

$\frac{5}{12} \times 192 = 80$

So it would be expected that 80 traffic lights would show green as the taxi approached.

EXERCISE 10.02 Answers on p. 525

Probability

1 Write the sample space in set notation for each chance situation.

- **a** Tossing a coin.
- **b** Randomly selecting a card from five cards with numbers 1 to 5 written.
- **c** Rolling a die.
- **d** Selecting a jelly snake from a packet containing red, green, yellow and blue jelly snakes.
- **e** Rolling an 8-sided die numbered 1 to 8.

2 A standard pack of 52 cards, has one card which is the ace of diamonds. If one card is chosen at random, find the probability that it:

- **a** will be the ace of diamonds
- **b** will not be the ace of diamonds.

3 Alannah is in a class of 30 students. If one student is chosen at random to make a speech, find the probability that the student chosen:

- **a** will be Alannah
- **b** will not be Alannah.

4 A shoe shop orders in 20 pairs of black, 14 pairs of navy and 3 pairs of brown school shoes. If the boxes are all mixed up, find the probability that one box selected at random will contain brown shoes. 

5 In a lottery, 200 000 tickets are sold. If Lucia buys 10 tickets, what is the probability of her winning first prize?

6 A bag contains 6 red balls and 8 white balls. If Peter draws one ball out of the bag at random, find the probability that it will be:

- **a** white
- **b** red.

7 A bag contains 5 black marbles, 4 yellow marbles and 11 green marbles. Find the probability of drawing one marble out at random and getting: 

- **a** a green marble
- **b** a yellow or a green marble.

8 What is the probability of each possible outcome when:

- **a** choosing a particular coloured cup from a box containing a blue, a white, a green, a pink and a grey cup?
- **b** rolling 2 dice?
- **c** tossing 2 coins?
- **d** tossing 3 coins?

9 A raffle is held in which 200 tickets are sold. If I buy 5 tickets, what is the probability of: 

- **a** my winning?
- **b** my not winning the prize in the raffle?

10 The probability of a certain seed producing a plant with a pink flower is $\frac{7}{9}$. 

- **a** Find the probability of the seed producing a flower of a different colour.
- **b** If 189 of these plants are grown, how many of them would be expected to have a pink flower?

11 If a baby has a 0.2% chance of being born with a disability, find the probability of the baby being born without a disability.

12 A die is rolled. Calculate the probability of rolling:

- **a** a 6
- **b** an even number
- **c** a number less than 3.

13 A book has 124 pages. If any page is selected at random, find the probability of the page number being:

- **a** either 80 or 90
- **b** a multiple of 10
- **c** an odd number
- **d** less than 100.

14 A machine has a 1.5% chance of breaking down at any given time.

- **a** What is the probability of the machine not breaking down?
- **b** If 2600 of these machines are manufactured, how many of them would be expected to:
  - **i** break down?
  - **ii** not break down?

15 The probabilities when 3 coins are tossed are as follows:

$P(3 \text{ heads}) = \frac{1}{8}$ $P(2 \text{ heads}) = \frac{3}{8}$

$P(1 \text{ head}) = \frac{3}{8}$ $P(3 \text{ tails}) = \frac{1}{8}$

Find the probability of tossing at least one head.

16 In the game of pool, there are 15 balls, each with the number 1 to 15 on it. In Kelly pool, each person chooses a number at random to determine which ball to sink.
If Tiana chooses a number, find the probability that her ball will be:

a an odd number **b** a number less than 8 **c** the 8 ball.

17 **a** Find the probability of a coin coming up heads when tossed.

b If the coin is double-headed, find the probability of tossing a head.

18 The probability of a bus arriving on time is estimated at $\frac{18}{33}$.

a What is the probability that the bus will not arrive on time?

b If there are 352 buses each day, how many would be expected to arrive on time?

19 There are 29 red, 17 blue, 21 yellow and 19 green chocolate beans in a packet.
If Kate chooses one at random, find the probability that it will be red or yellow.

20 The probability of breeding a white budgerigar is $\frac{2}{9}$.
If Mr Seed breeds 153 budgerigars over the year, how many would be expected to be white?

21 A student is chosen at random to write about his or her favourite sport. If 12 students like tennis best, 7 prefer soccer, 3 prefer squash, 5 prefer basketball and 4 prefer swimming, find the probability that the student chosen will write about:

a soccer **b** squash or swimming **c** tennis.

22 A ball is chosen at random from a bag containing red, white and blue balls. The probability of getting a red ball is $\frac{3}{17}$ and the probability of getting a blue ball is $\frac{7}{17}$. Find the probability of getting a white ball.

23 Discuss whether each probability statement is true.

a The probability of one particular horse winning the Melbourne Cup is $\frac{1}{20}$ if there are 20 horses in the race.

b The probability of a player winning a masters golf tournament is $\frac{1}{15}$ if there are 15 players competing in the tournament.

c A coin came up tails 8 times in a row. So, the next toss must be a head.

d A family has 3 sons and is expecting a fourth child. There is more chance of the new baby being a daughter.

e The probability of a Ducati motorcycle winning the MotoGP (international motorcycle grand prix) this year is $\frac{6}{47}$ because there are 47 motorcycles competing and 6 of them are Ducatis.

24 A biased coin is weighted so that heads comes up twice as often as tails.
Find the probability of tossing a tail.

25 A die has the centre dot painted white on the 5 so that it appears as a 4.
Find the probability of throwing:

a a 2 **b** a 4 **c** a number less than 5.

☐ Foundation ◯ Mastery ⬡ Complex

26 The probabilities of a certain number of seeds germinating when 4 seeds are planted are:

Number of seeds	0	1	2	3	4
Probability	$\frac{3}{49}$	$\frac{18}{49}$	$\frac{16}{49}$	$\frac{8}{49}$	$\frac{4}{49}$

Find the probability of at least one seed germinating.

27 The probabilities of 4 friends being chosen for a soccer team are:

$P(4 \text{ chosen}) = \frac{1}{15}$ $\qquad P(3 \text{ chosen}) = \frac{4}{15}$

$P(2 \text{ chosen}) = \frac{6}{15}$ $\qquad P(1 \text{ chosen}) = \frac{2}{15}$

Find the probability of:

a none of the friends being chosen

b at least 1 of the friends being chosen.

28 If 2 events are mutually exclusive, what could you say about $P(A \cap B)$?

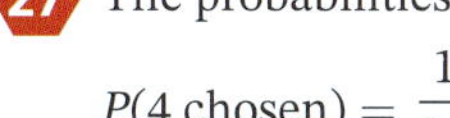

Did you know?

The origins of probability

Girolamo Cardano (1501–76) was a doctor and mathematician who developed the first theory of probability. He was a great gambler, and he wrote *De Ludo Aleae* ('On Games of Chance'). This work was largely ignored, and it is said that the first book on probability was written by **Christiaan Huygens** (1629–95).

The main study of probability was done by **Blaise Pascal** (1623–62), and **Pierre de Fermat** (1601–65). Pascal developed the 'arithmetical triangle' that has properties which are applicable to probability as well.

The addition rule of probability 10.03

Sometimes, there is an overlap where more than one event can occur at the same time.
We call these **non mutually exclusive events**. As we saw when counting elements in Venn diagrams in Example 5 on page 407, we need to be careful not to count the overlapping outcomes $A \cap B$ twice.
We know that $|A \cup B| = |A| + |B| - |A \cap B|$.

The addition rule of probability

For events A and B:

$P(A \cup B) = P(A) + P(B) - P(A \cap B)$

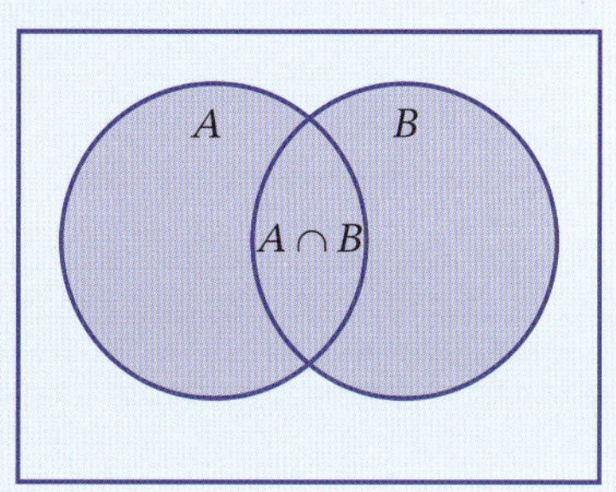

If A and B are mutually exclusive, then $P(A \cap B) = 0$ so $P(A \cup B) = P(A) + P(B)$, as we saw on page 411.

Foundation Mastery Complex

Example 10

One card is selected at random from a pack of 100 cards numbered from 1 to 100.
Find the probability that the number on this card is even or less than 20.

Solution

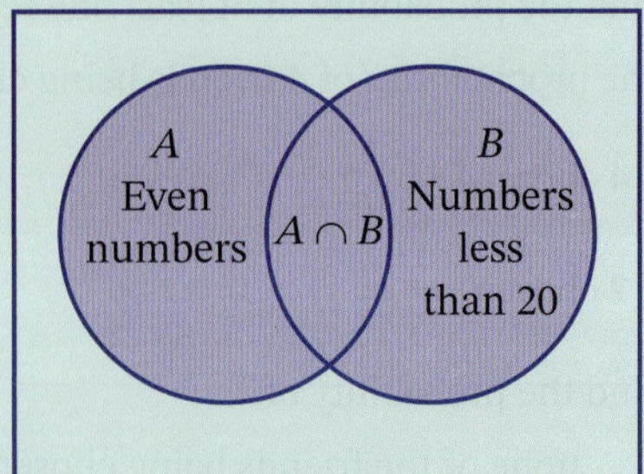

Even: $A = \{2, 4, 6, ..., 100\}$

There are 50 even numbers between 1 and 100.

Less than 20: $B = \{1, 2, 3, ..., 19\}$

There are 19 numbers less than 20.

Even and less than 20: $A \cap B = \{2, 4, 6, 8, 10, 12, 14, 16, 18\}$

There are 9 numbers that are both even and less than 20.

P(even or less than 20):

$$
\begin{aligned}
P(A \cup B) &= P(A) + P(B) - P(A \cap B) \\
&= \frac{50}{100} + \frac{19}{100} - \frac{9}{100} \\
&= \frac{60}{100} \\
&= \frac{3}{5}
\end{aligned}
$$

This is to avoid counting the 9 'overlapping' numbers twice.

Sometimes for more complex problems, a Venn diagram is useful.

Example 11

In Year 7 at Mt Random High School, every student must do Art or Music. In a group of 100 students surveyed, 47 do Music and 59 do Art. If one student is chosen at random from Year 7, find the probability that this student does:

a both Art and Music **b** only Art **c** only Music.

Solution

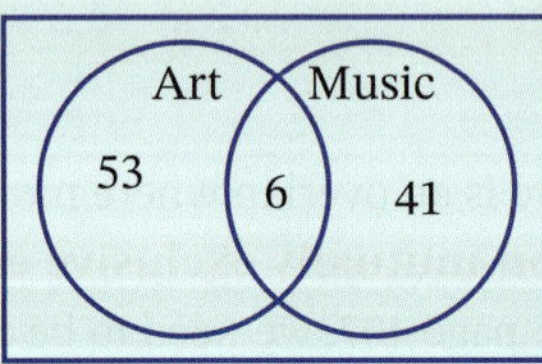

Number of students: $47 + 59 = 106$

But there are only 100 students.

This means 6 students do both Art and Music.

Students doing Art only: $59 - 6 = 53$

Students doing Music only: $47 - 6 = 41$

a $P(\text{both}) = \frac{6}{100} = \frac{3}{50}$

b $P(\text{Art only}) = \frac{53}{100}$

c $P(\text{Music only}) = \frac{41}{100}$

EXERCISE 10.03 Answers on p. 525

The addition rule of probability

1 A number is chosen at random from the numbers 1 to 20. Find the probability that the number chosen will be:

- **a** divisible by 3
- **b** less than 10 or divisible by 3
- **c** a composite number
- **d** a composite number or a number greater than 12.

2 A set of 50 cards is labelled from 1 to 50. One card is drawn out at random. Find the probability that the card will be:

- **a** a multiple of 5
- **b** an odd number
- **c** a multiple of 5 or an odd number
- **d** a number greater than 40 or an even number
- **e** less than 20.

3 A set of 26 cards, each with a different letter of the alphabet on it, is placed in a box and one card is drawn out at random. Find the probability that the letter on the card is:

- **a** a vowel
- **b** a vowel or one of the letters in the word 'random'
- **c** a consonant or one of the letters in the word 'movies'.

4 A set of discs is numbered 1 to 100 and one is chosen at random. Find the probability that the number on the disc will be:

- **a** less than 30
- **b** an odd number or a number greater than 70
- **c** divisible by 5 or less than 20.

5 In Lotto, a machine holds 45 balls, each with a different number between 1 and 45 on it. The machine draws out one ball at a time at random. Find the probability that the first ball drawn out will be:

- **a** less than 10 or an even number
- **b** between 1 and 15 inclusive, or divisible by 6
- **c** greater than 30 or an odd number.

6 A class of 28 students puts on a concert with all class members performing. If 15 dance and 19 sing in the performance, find the probability that any one student chosen at random from the class will:

- **a** both sing and dance
- **b** only sing
- **c** only dance.

7 A survey of 80 people with dark hair or brown eyes showed that 63 had dark hair and 59 had brown eyes. Find the probability that one of the people surveyed chosen at random has:

- **a** dark hair but not brown eyes
- **b** brown eyes but not dark hair
- **c** both brown eyes and dark hair.

8 A list is made up of 20 people with experience in coding or graphic design. On the list, 13 have coding experience while 9 have graphic design experience. Find the probability that a person chosen at random from the list will have experience in:

- **a** both coding and graphic design
- **b** coding only
- **c** graphic design only.

9 Of a group of 75 students, all study either History or Geography. Altogether 54 take History and 31 take Geography. Find the probability that a student selected at random studies:

- **a** only Geography
- **b** both History and Geography
- **c** History but not Geography.

10 In a group of 20 dogs at dog training school, 14 dogs will walk to heel and 12 will stay when told. All dogs will do one or the other, or both. If one dog is chosen at random, find the probability that it will:

a both walk to heel and stay

b walk to heel but not stay

c stay but not walk to heel.

11 A survey of 25 people found that 12 liked to drink hot chocolate, 12 liked to drink tea and 15 liked to drink coffee. Of this group, 3 liked to drink all 3 hot beverages, 2 liked only tea and coffee, 2 liked only coffee and hot chocolate, and 4 liked only hot chocolate and tea. Find the probability that if one of these people was selected at random, they liked to drink:

a only coffee **b** only coffee and hot chocolate **c** only tea.

12 Of 112 tourists surveyed, 19 had holidayed only in Europe and Asia, 3 had holidayed in Europe, Asia and the USA, 27 had holidayed only in Asia, 13 had holidayed only in the USA, 7 had holidayed only in the USA and Asia, 12 had holidayed only in Europe and 12 had holidayed elsewhere. If a tourist is selected at random, find the probability that they had holidayed only in:

a the USA and Europe **b** Europe.

10.04 The product rule of probability

Worksheet
Multi-stage problems

Tossing 2 coins and rolling 2 dice are examples of **multi-stage experiments**, where 2 or more outcomes happen together. The sample space becomes more complicated, so to list all possible outcomes we use tables or **tree diagrams**.

Example 12

Find the sample space and the probability of each outcome for:

a tossing 2 coins

b rolling 2 dice and calculating their sum.

Solution

a Using a table:

		2nd coin	
		H	T
1st coin	H	HH	HT
	T	TH	TT

Using a tree diagram:

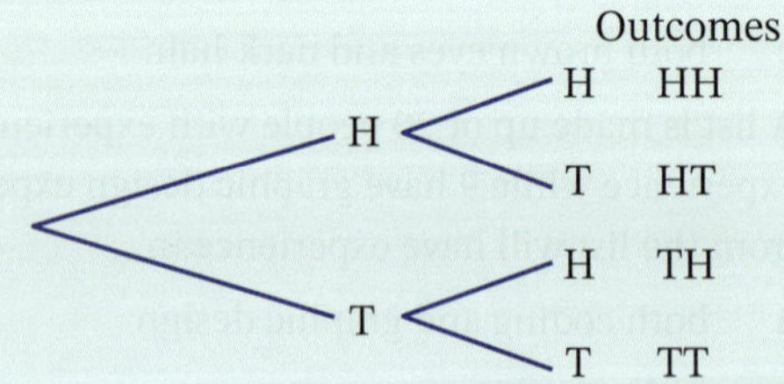

Since there are 4 possible outcomes (HH, HT, TH, TT), each outcome has a probability of $\frac{1}{4}$.

Remember that each outcome when tossing one coin is $\frac{1}{2}$.

Notice that $\frac{1}{2} \times \frac{1}{2} = \frac{1}{4}$.

☐ Foundation ○ Mastery ⬡ Complex

b A tree diagram would be too big to draw for rolling 2 dice.

Using a table:

		2nd die					
		1	2	3	4	5	6
1st die	1	2	3	4	5	6	7
	2	3	4	5	6	7	8
	3	4	5	6	7	8	9
	4	5	6	7	8	9	10
	5	6	7	8	9	10	11
	6	7	8	9	10	11	12

Since there are 36 outcomes, each has a probability of $\frac{1}{36}$.

Remember that each outcome when rolling one die is $\frac{1}{6}$.

Notice that $\frac{1}{6} \times \frac{1}{6} = \frac{1}{36}$.

If A and B are **independent events**, then A occurring does not affect the probability of B occurring. The probability of both occurring is the product of their probabilities.

Product rule for independent events

$$P(A \cap B) = P(A)P(B)$$

$A \cap B$ means both events A and B occur together.

Example 13

a Find the probability of rolling a double 6 on 2 dice.

b The probability that an archer will hit a target is $\frac{7}{8}$. Find the probability that the archer will:

i hit the target twice

ii miss the target twice.

Solution

a $P(A \cap B) = P(A)P(B)$

$$\begin{aligned} P(6 \cap 6) &= P(6)P(6) \\ &= \frac{1}{6} \times \frac{1}{6} \\ &= \frac{1}{36} \end{aligned}$$

b **i** $P(A \cap B) = P(A)P(B)$

Let H = hit, M = miss

We could also write $\overline{\text{H}}$ = miss (not hit).

$$P(\text{H} \cap \text{H}) = P(\text{H})P(\text{H})$$
$$= \frac{7}{8} \times \frac{7}{8}$$
$$= \frac{49}{64}$$

ii $$P(\text{M}) = P(\overline{\text{H}})$$
$$= 1 - \frac{7}{8}$$
$$= \frac{1}{8}$$

$$P(\text{M} \cap \text{M}) = P(\text{M})P(\text{M})$$
$$= \frac{1}{8} \times \frac{1}{8}$$
$$= \frac{1}{64}$$

The sample space changes when events are not independent. The second event is dependent or **conditional** on the first event.

Video
Dependent events

Example 14

Maryam buys 5 tickets in a raffle in which 95 tickets are sold altogether. There are 2 prizes in the raffle. What is the probability of her winning:

a both first and second prizes

b neither prize

c at least one of the prizes?

Solution

a Probability of winning first prize: $P(\text{W}_1) = \frac{5}{95}$

After winning first prize, she has 4 tickets left in the raffle out of a total of 94 tickets left.

Probability of winning second prize: $P(\text{W}_2) = \frac{4}{94}$

Probability of winning both prizes:

$$P(\text{W}_1 \cap \text{W}_2) = \frac{5}{95} \times \frac{4}{94}$$
$$= \frac{20}{8930}$$
$$= \frac{2}{893}$$

b Probability of not winning first prize:

$$P(\overline{\text{W}_1}) = 1 - \frac{5}{95}$$
$$= \frac{90}{95}$$
$$= \frac{18}{19}$$

After not winning first prize, Maryam's 5 tickets are all left in the draw, but the winning ticket is taken out, leaving 94 tickets in the raffle.

Probability of winning second prize: $P(\text{W}_2) = \frac{5}{94}$

Probability of not winning second prize:

$$P(\overline{W}_2) = 1 - \frac{5}{94}$$
$$= \frac{89}{94}$$

Probability of winning neither prize:

$$P(\overline{W}_1 \cap \overline{W}_2) = P(\overline{W}_1)P(\overline{W}_2)$$
$$= \frac{90}{95} \times \frac{89}{94}$$
$$= \frac{8010}{8930}$$
$$= \frac{801}{893}$$

c Probability of at least one win:

$$P(\geq 1 \text{ win}) = 1 - P(0 \text{ wins})$$
$$= 1 - \frac{801}{893} \quad \text{from part } \mathbf{b}$$
$$= \frac{92}{893}$$

'at least one' is the same as 'not none'. In this example, it means 1 or 2.

EXERCISE 10.04 Answers on p. 525

The product rule of probability

1. Find the probability of getting 2 heads if a coin is tossed twice.
2. A coin is tossed 3 times. Find the probability of tossing 3 tails.
3. A family has 2 children. What is the probability that they are both girls?
4. A box contains 2 black balls, 5 red balls and 4 green balls. If I draw out 2 balls at random, replacing the first before drawing out the second, find the probability that they will both be red.

5. The probability of a conveyor belt in a factory breaking down at any one time is 0.21. If the factory has 2 conveyor belts, find the probability that at any one time:
   - **a** both conveyor belts will break down
   - **b** neither conveyor belt will break down.
6. The probability of a certain plant flowering is 93%. If a nursery has 3 of these plants, find the probability that they will all flower.
7. An archery student has a 69% chance of hitting a target. If she fires 3 arrows at a target, find the probability that she will hit the target each time.
8. The probability of a pair of small parrots breeding an albino bird is $\frac{2}{33}$. If they lay 3 eggs, find the probability of the pair:
   - **a** not breeding any albinos
   - **b** having all 3 albinos
   - **c** breeding at least one albino.
9. A photocopier has a paper jam on average around once every 2400 sheets of paper.
   - **a** What is the probability that a particular sheet of paper will jam?
   - **b** What is the probability that 2 particular sheets of paper will jam?
   - **c** What is the probability that 2 particular sheets of paper will both not jam?

10. In the game Yahtzee, 5 dice are rolled. Find the probability of rolling:
   - **a** five 6s
   - **b** no 6s
   - **c** at least one 6.

Foundation Mastery Complex

11 The probability of a faulty computer part being manufactured at Omikron Computer Factory is $\frac{3}{5000}$. If 2 computer parts are examined, find the probability that:

a both are faulty **b** neither is faulty **c** at least one is faulty.

12 A set of 10 cards is numbered 1 to 10 and 2 cards are drawn out at random with replacement. Find the probability that the numbers on both cards are:

a odd numbers **b** divisible by 3 **c** less than 4

13 The probability of an arrow hitting a target is 85%. If 3 arrows are shot, find the probability as a percentage, correct to 2 decimal places, of:

a all arrows hitting the target

b no arrows hitting the target

c at least one arrow hitting the target.

14 A coin is tossed n times. Find the probability in terms of n of tossing:

a no tails **b** at least one tail.

15 A bag contains 8 yellow and 6 green lollies. If I choose 2 lollies at random, find the probability that they will both be green:

a if I replace the first lolly before selecting the second

b if I don't replace the first lolly.

16 Mala buys 10 tickets in a raffle in which 250 tickets are sold. Find the probability that she wins both first and second prizes.

17 Two cards are drawn from a deck of 20 red and 25 blue cards (without replacement). Find the probability that they will both be red.

18 A bag contains 100 cards numbered 1 to 100. Scott draws 2 cards out of the bag without replacement. Find the probability that:

a both cards are less than 10

b both cards are even

c neither card is a multiple of 5

19 A box of pegs contains 23 green pegs and 19 red pegs. If 2 pegs are taken out of the box at random, find the probability that both will be:

a green **b** red.

20 Find the probability of selecting 2 apples at random from a fruit bowl that contains 8 apples, 9 oranges and 3 peaches.

Foundation Mastery Complex

Probability trees

10.05

A **probability tree** is a tree diagram that shows the probabilities on the branches. We use a probability tree when the probability of each outcome is not equally likely. It is useful in multi-stage problems where the probabilities change with each stage depending upon the outcomes of the previous stage. A probability tree would make Example 14 on page 420 much simpler to do. After Example 15, go back and do Example 14 using a probability tree.

Videos
Tree diagrams 1

Tree diagrams 2

Worksheets
Tree diagrams

Probability trees

Probability trees

Use the product rule along the branches to find $P(A \cap B)$, the probability of 'A and B'.

Use the addition rule for different branches to find $P(A \cup B)$, the probability of 'A or B'.

Example 15

a Robert has a 0.2 chance of winning a prize in a Taekwondo competition.
If he enters 3 competitions, find the probability of him winning:

i 2 competitions **ii** at least one competition.

b A bag contains 3 red, 4 white and 7 blue tubes of paint. Two tubes are drawn at random from the bag without replacement. Find the probability that the paint colours are red and white.

Solution

a $P(\text{W}) = 0.2$, $P(\text{L}) = 1 - 0.2 = 0.8$ W = win, L = lose

Draw a probability tree with 3 levels of branches as shown.

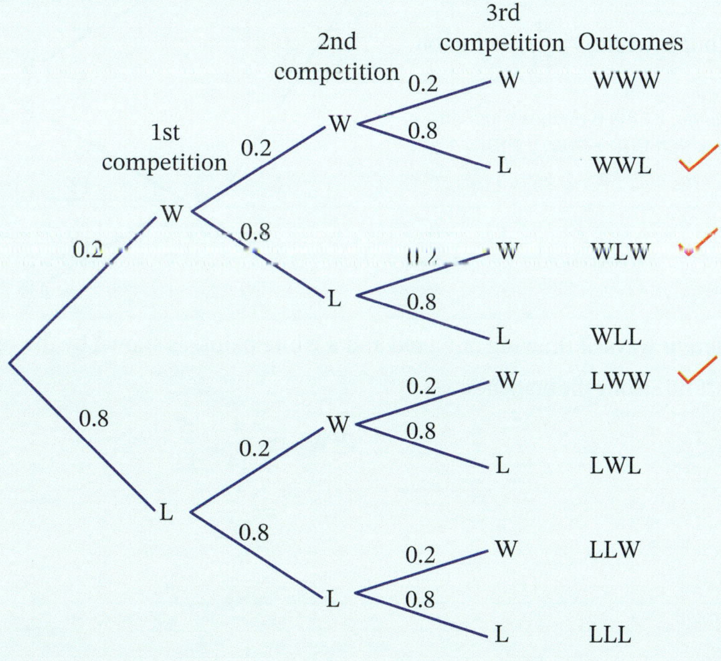

9780170498197

i There are 3 different ways of winning 2 competitions (WWL, WLW and LWW, shown by the red ticks).

Using the product rule along the branches:

$P(\text{WWL}) = 0.2 \times 0.2 \times 0.8 = 0.032$ 'P(win and win and lose)'

$P(\text{WLW}) = 0.2 \times 0.8 \times 0.2 = 0.032$

$P(\text{LWW}) = 0.8 \times 0.2 \times 0.2 = 0.032$

Using the addition rule for the different results:

$P(2 \text{ wins}) = P(\text{WWL}) + P(\text{WLW}) + P(\text{LWW})$ 'P(WWL or WLW or LWW)'

$= 0.032 + 0.032 + 0.032$

$= 0.096$

ii $P(\geq 1\text{W}) = 1 - P(\text{LLL})$

$= 1 - 0.8 \times 0.8 \times 0.8$

$= 0.488$

b First paint colour:

$P(\text{R}) = \frac{3}{14}$ $P(\text{W}) = \frac{4}{14}$

$P(\text{B}) = \frac{7}{14}$

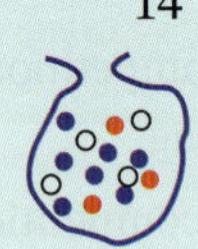

The probabilities for the second paint colour are dependent on the outcome of the first draw. Also, there are now 13 paint tubes left to choose from. If the first paint colour chosen was red, then there are 2 reds remaining,

so P(2nd paint colour is red) $= \frac{2}{13}$,

P(2nd paint colour is white) $= \frac{4}{13}$, and so on.

The probabilities on the branches must add up to 1. Check that P(R) + P(W) + P(B) = 1 at every point.

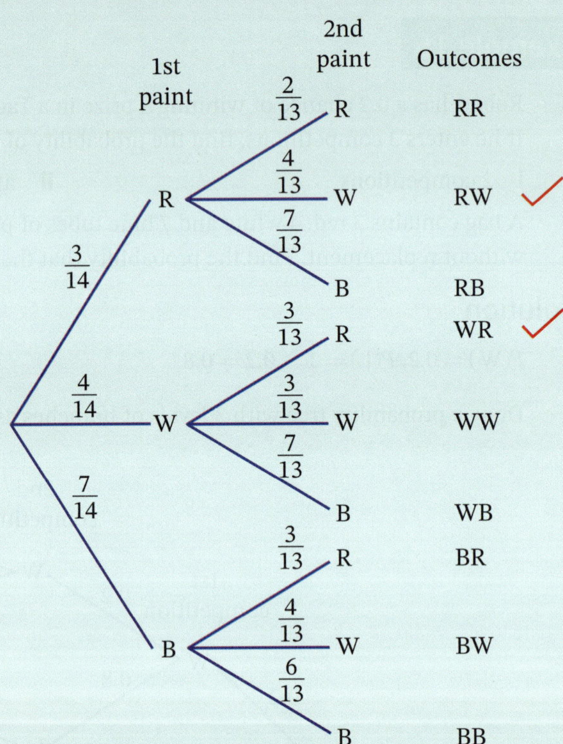

There are 2 different ways of drawing out a red and a white paint, as shown by the red ticks: RW, WR.

Using the product rule along the branches:

$P(\text{RW}) = \frac{3}{14} \times \frac{4}{13}$

$= \frac{12}{182}$

$= \frac{6}{91}$

$P(\text{WR}) = \frac{4}{14} \times \frac{3}{13}$

$= \frac{12}{182}$

$= \frac{6}{91}$

Using the addition rule for the different results:

$P(\text{RW or WR}) = \frac{6}{91} + \frac{6}{91}$

$= \frac{12}{91}$

EXERCISE 10.05 Answers on p. 525

Probability trees

1 Three coins are tossed. Find the probability of getting:

a 3 tails **b** 2 heads and 1 tail **c** at least 1 head.

2 In a set of 30 cards, each card has a number on it from 1 to 30. If one card is drawn out, then replaced and another drawn out, find the probability of getting:

EXAMPLE 15

a 2 8s

b a 3 on the first card and an 18 on the second card

c a 3 on one of the cards and an 18 on the other card.

3 A bag contains 5 red marbles and 8 blue marbles. If 2 marbles are chosen at random, with the first replaced before the second is drawn out, find the probability of getting:

a 2 red marbles **b** a red marble and a blue marble.

4 A certain breed of cat has a 35% probability of producing a white kitten. If a cat has 3 kittens, find the probability that she will produce:

a no white kittens **b** 2 white kittens **c** at least 1 white kitten.

5 The probability of rain on any day in May each year is $\frac{3}{10}$. A school holds a fete on a Sunday in May for 3 years running. Find the probability that it will rain:

a during 2 of the fetes **b** during 1 fete **c** during at least 1 fete.

6 A certain type of plant has a probability of 0.85 of producing a variegated leaf. If I grow 3 of these plants, find the probability of getting a variegated leaf in:

a 2 of the plants **b** none of the plants **c** at least 1 plant.

7 A bag contains 3 yellow balls, 4 pink balls and 2 black balls. If 2 balls are chosen at random, find the probability of getting a yellow ball and a black ball:

a with replacement **b** without replacement.

8 Anh buys 4 tickets in a raffle in which 100 tickets are sold altogether. There are 2 prizes in the raffle. Find the probability that Anh will win:

a first prize **b** both prizes **c** 1 prize

d no prizes **e** at least 1 prize.

9 Two singers are selected at random to compete against each other in a TV singing contest. One person is chosen from Team A, which has 8 females and 7 males, and the other is chosen from Team B, which has 6 females and 9 males. Find the probability of choosing:

a 2 females **b** 1 female and 1 male.

10 Two tennis players are said to have a probability of $\frac{2}{5}$ and $\frac{3}{4}$ respectively of winning a tournament. Find the probability that:

a 1 of them will win **b** neither one will win.

11 In a batch of 100 cars, past experience would suggest that 3 could be faulty. If 3 cars are selected at random, find the probability that:

a 1 is faulty **b** none is faulty **c** all 3 cars are faulty.

12 In a certain poll, 46% of people surveyed liked the current government, 42% liked the Opposition and 12% had no preference. If 2 people from the survey are selected at random, find the probability that:

a both will prefer the Opposition

b one will prefer the government and the other will have no preference

c both will prefer the government.

13 A manufacturer of X energy drink surveyed a group of people and found that 31 people liked X drinks best, 19 liked another brand better and 5 did not drink energy drinks. If any 2 people are selected at random from that group, find the probability that:

a one person likes the X brand of energy drink

b both people do not drink energy drinks.

14 In a group of people, 32 are Australian-born, 12 were born in Asia and 7 were born in Europe. If 2 of the people are selected at random, find the probability that:

a they were both born in Asia

b at least one of them will be Australian-born

c both were born in Europe.

15 There are 34 men and 32 women at a party. Of these, 13 men and 19 women are married. If 2 people are chosen at random, find the probability that:

a both will be men

b one will be a married woman and the other an unmarried man

c both will be married.

16 Frankie rolls 3 dice. Find the probability she rolls:

a three 6s **b** two 6s **c** at least one 6.

17 A set of 5 cards, each labelled with one of the letters A, B, C, D and E, is placed in a hat and 2 cards are selected at random without replacement. Find the probability of getting:

a D and E

b neither D nor E on either card

c at least one card has D.

18 The ratio of girls to boys at a school is 4 : 5. Two students are surveyed at random from the school. Find the probability that the students are:

a both boys **b** a girl and a boy **c** at least one girl.

□ Foundation ○ Mastery ○ Complex

Conditional probability

10.06

Conditional probability is the probability that an event A occurs when it is known that another event B has already occurred. You have already used conditional probability in multi-stage events when the outcome of the second event was dependent on the outcome of the first event. Examples include selections **without replacement**.

Worksheet
Conditional probability

Puzzle
Conditional probability: Two-way tables

We write the probability of event A happening given that event B has happened as $P(A|B)$.
This is called 'the probability of A *given* B (has ocurred).'

Example 16

All 30 students in a class study either Health or Graphic Design. If 18 only do Graphic Design and 8 do both subjects, find the probability that a student does Graphic Design, given that the student does Health.

Solution

Draw a Venn diagram using H = Health and G = Graphic Design.

8 students do both Health and Graphic Design.

18 students only do Graphic Design.

So $|G| = 18 + 8 = 26$.

So $|H \text{ only}| = 30 - 26 = 4$.

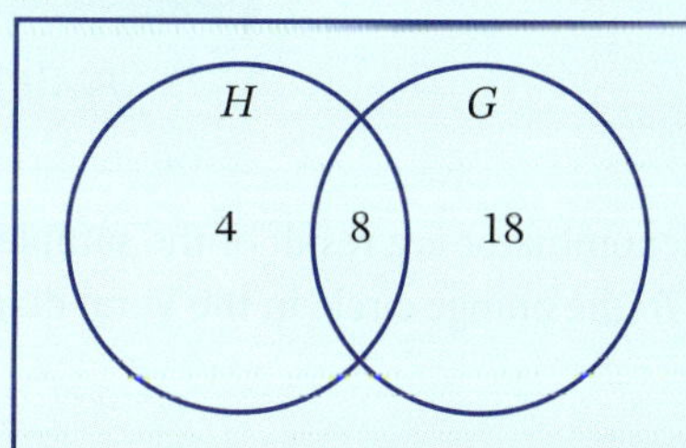

There are $8 + 4 = 12$ students doing Health, of whom 8 also do Graphic Design.

$$\text{So } P(G|H) = \frac{8}{12} = \frac{2}{3}$$

With conditional probability, knowing that an event has already occurred reduces the sample space. In the example above, the sample space changed from 30 to 12.

Example 17

The table shows the results of a survey into vaccinations against a new virus.

	Vaccinated	Not vaccinated	Totals
Infected	13	159	172
Not infected	227	38	265
Totals	240	197	437

Find the probability that a person selected at random is:

a not vaccinated

b infected given that the person is vaccinated

c not infected given that the person is not vaccinated

d vaccinated given that the person is infected.

Solution

a $n(S) = 437$, $n(\text{not vaccinated}) = 197$

$P(\text{not vaccinated}) = \frac{197}{437}$

b $n(\text{vaccinated}) = 240$

$n(\text{infected} \mid \text{vaccinated}) = 13$

$P(\text{infected} \mid \text{vaccinated}) = \frac{13}{240}$

So the sample space has reduced to 240.

c $n(\text{not vaccinated}) = 197$

$n(\text{not infected} \mid \text{not vaccinated}) = 38$

$P(\text{not infected} \mid \text{not vaccinated}) = \frac{38}{197}$

New sample space is 197.

d $n(\text{infected}) = 172$

$n(\text{vaccinated} \mid \text{infected}) = 13$

$P(\text{vaccinated} \mid \text{infected}) = \frac{13}{172}$

Notice that $P(\text{infected} \mid \text{vaccinated}) \neq P(\text{vaccinated} \mid \text{infected})$.

Conditional probability

$$P(A|B) = \frac{P(A \cap B)}{P(B)}$$

The $P(B)$ in the denominator is a result of the sample space being reduced to B (the orange circle in the Venn diagram).

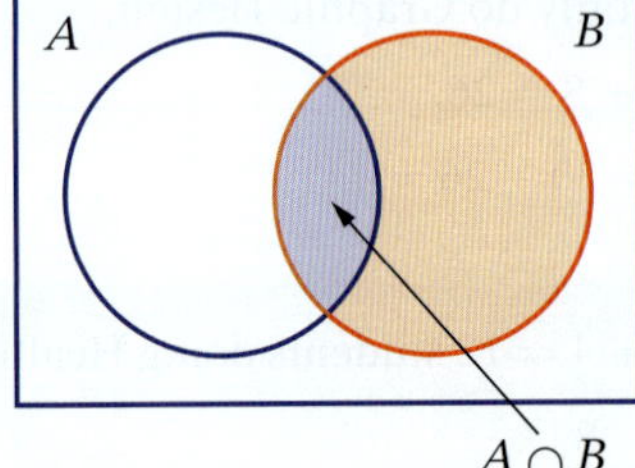

Proof

$$P(A|B) = \frac{|A \cap B|}{|B|}$$

$$= \frac{\frac{|A \cap B|}{|S|}}{\frac{|B|}{|S|}}$$

$$= \frac{P(A \cap B)}{P(B)}$$

Example 18

Lara is an athlete who enters a swimming race and a running race. She has a 44% chance of winning the swimming race and a 37% chance of winning both races. Find to the nearest whole percentage the probability that she wins the running race if she has won the swimming race.

Solution

If Lara has already won the swimming race (S), then the probability of her winning the running race (R) is conditional.

$P(S) = 44\%$
$= 0.44$

$P(R \text{ and } S) = P(R \cap S)$
$= 37\%$
$= 0.37$

$$P(R|S) = \frac{P(R \cap S)}{P(S)}$$

$P(A|B) = \frac{P(A \cap B)}{P(B)}$

$$= \frac{0.37}{0.44}$$

$\approx 0.84\,09\ldots$

$\approx 84\%$

So the probability of Lara winning the running race given that she has won the swimming race is 84%.

Video
Conditional probability

Example 19

A zoo has a probability of 80% of having an article published in the newspaper when there is a birth of a baby animal. When there is no birth, the zoo has a probability of only 30% of having an article published. The probability of the zoo having an animal born at any one time is 40%.

Find the percentage probability that a baby animal was born given that an article was published.

Solution

We can draw up a probability tree showing the probabilities of having an article published (A) and a baby animal being born (B).

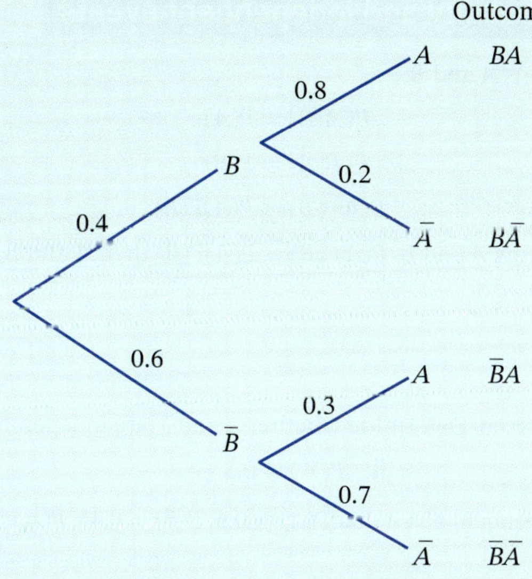

We want $P(B|A)$. According to the formula:

$$P(B|A) = \frac{P(B \cap A)}{P(A)}$$

From the probability tree:

$$\begin{aligned} P(B \cap A) &= P(BA) \\ &= 0.4 \times 0.8 \\ &= 0.32 \end{aligned}$$ The event (numerator)

$$\begin{aligned} P(A) &= 0.4 \times 0.8 + 0.6 \times 0.3 \\ &= 0.5 \end{aligned}$$ The sample space (denominator)

$$\begin{aligned} P(B|A) &= \frac{P(B \cap A)}{P(A)} \\ &= \frac{0.32}{0.5} \\ &= 0.64 \\ &= 64\% \end{aligned}$$

So the probability that an animal was born given that an article was published is 64%.

Conditional probability and independent events

We can rearrange the formula:

$$P(A|B) = \frac{P(A \cap B)}{P(B)}$$

to create a product rule for conditional probability $P(A \cap B) = P(A|B)P(B)$.

But if A and B are **independent events**, $P(A \cap B) = P(A)P(B)$ (the product rule), which means:

$P(A|B) = P(A)$.

Similarly, $P(B|A) = P(B)$.

Conditional probability and independent events

For independent events A and B:

$$P(A|B) = P(A)$$
$$P(B|A) = P(B)$$
$$P(A \cap B) = P(A)P(B)$$

Conversely, if, for events A and B, $P(A \cap B) = P(A)P(B)$ then A and B are independent.

Example 20

a $P(X) = 0.2$ and $P(X \cap Y) = 0.06$. Determine whether X and Y are independent if:

i $P(Y) = 0.6$ **ii** $P(Y) = 0.3$

b Show that A and B are independent given that $P(A) = 0.6$, $P(B) = 0.45$, $P(A \cup B) = 0.78$.

Solution

a For independent events, the product rule is $P(X \cap Y) = P(X)P(Y)$.

i $P(X \cap Y) = 0.06$

$P(X)P(Y) = 0.2 \times 0.6$

$= 0.12$

$\neq P(X \cap Y)$

$\therefore X$ and Y are not independent.

ii $P(X \cap Y) = 0.06$

$P(X)P(Y) = 0.2 \times 0.3$

$= 0.06$

$= P(X \cap Y)$

$\therefore X$ and Y are independent.

b Using the addition rule:

$P(A \cup B) = P(A) + P(B) - P(A \cap B)$

$0.78 = 0.6 + 0.45 - P(A \cap B)$

$0.78 = 1.05 - P(A \cap B)$

$P(A \cap B) = 1.05 - 0.78$

$= 0.27$

Using the product rule for independent events:

$P(A \cap B) = P(A)\,P(B)$

$= 0.6 \times 0.45$

$= 0.27$, as above

$\therefore A$ and B are independent.

EXERCISE 10.06 Answers on p. 526

Conditional probability

EXAMPLE 16

1 A bag contains 9 black and 8 white balls. I draw out 2 at random. If the first ball is white, find the probability that the next ball is:

a black **b** white.

2 A class has 13 boys and 15 girls. Two students are chosen at random to carry a box of equipment. Find the probability that the second person chosen is a boy given that the first student chosen was a girl.

3 Two dice are rolled. Find the probability of rolling:

a a double 6 if the first die was a 6.

b a total of 8 or more if the first die was a 3.

4 A team has a probability of 52% of winning its first season and a 39% chance of winning both seasons 1 and 2. What is the probability of the team winning the second season given that it wins the first season?

5 A missile has a probability of 0.75 of hitting a target. It has a probability of 0.65 of hitting 2 targets in a row. What is the probability that the missile will hit the second target given that it has hit the first target?

6 Danuta has an 80% probability of passing her first English assessment and she has a 45% probability of passing both the first and second assessments. Find the probability that Danuta will pass the second assessment given that she passes the first one.

7 A group of 10 friends all prepared to go out in the sun by putting on either sunscreen or a hat. If 5 put on only sunscreen and 3 put on both sunscreen and a hat, find the probability that a friend who:

a put on sunscreen also put on a hat

b put on a hat didn't put on sunscreen.

8 A container holds 20 cards numbered 1 to 20. Two cards are selected at random. Find the probability that the second card is:

a an odd number given that the first card was a 7

b a number less than 5 given that the first card was a 12

c a number divisible by 3 if the first number was 6.

9 A group of 12 people met at a café for lunch. If 9 people had a pie and 7 had chips, find the probability that one of the people:

a had chips given that this person had a pie

b did not have a pie given that this person had chips.

10 All except for 3 people out of 25 on a European tour had studied either French or Spanish. Nine people studied only French and 5 studied both French and Spanish. Find the probability that one of these people:

a studied Spanish if that person studied French

b did not study French given that the person studied Spanish.

11 The two-way table shows the numbers of students who own smartphones and tablets.

	Smartphone	No smartphone	Totals
Tablet	23	8	31
No tablet	65	3	68
Totals	88	11	99

Find the probability that a person selected at random:

a owns a smartphone given that the person:

i owns a tablet **ii** doesn't own a tablet.

b owns a tablet given that the person:

i owns a smartphone **ii** doesn't own a smartphone.

12 The table below shows the number of local people with casual and permanent jobs.

	Women	Men
Permanent	23	38
Casual	79	64

Find the probability that a person chosen at random:

a has a permanent job given that she is a woman

b has a casual job given that he is a man

c is a man given that the person has a casual job

d is a man if the person has a permanent job.

13 In a group of 35 friends, all play either sport or a musical instrument. If 14 play both and 8 only play sport, find the probability that a friend chosen at random will:

a play a musical instrument given that the friend plays sport

b not play sport given that the friend plays a musical instrument.

14 The 2-way table shows the results of a survey into attendance at a local TAFE college.

	Under 25	Between 25 and 50	Over 50	Totals
At TAFE	53	68	34	155
Not at TAFE	85	105	88	278
Totals	138	173	122	433

Find the probability that a person:

a attends TAFE given that this person is over 50

b is between 25 and 50 if that person does not attend TAFE

c is not at TAFE given that this person is under 25

d is over 50 if the person is at TAFE

e is at TAFE given the person is aged 25 or over.

EXAMPLE 19

15 A tennis team has a probability of 76% of winning a match when they are at home and 45% of winning a match when they are away. If the team plays 58% of their matches away, find the probability that the team:

a wins their match given that they are away

b are at home given that they win a match

c are away given that they lose a match.

16 A factory produces solar batteries. The probability of a new battery being defective is 3%. However, if the manager is on duty, the probability of a new battery being defective changes to 2%. The manager is on duty 39% of the time. Find the probability that the manager is on duty if a new battery is defective.

17 The chance of a bushfire is 85% after a period of no rain and 21% after rain. The chance of rain is 46%. Find the probability that:

a there is not a bushfire given that it has rained

b it has rained given that there is a bushfire

c it has not rained given there is a bushfire

d it has rained given there is not a bushfire.

☐ Foundation ○ Mastery ○ Complex

18 If $P(A|B) = 0.67$ and $P(B) = 0.31$, find the value of $P(A \cap B)$.

19 If $P(L) = 0.17$, $P(L \cap M) = 0.0204$ and $P(M) = 0.12$, show that L and M are independent.

EXAMPLE 20

20 Given $P(X) = 0.3$, $P(Y) = 0.42$ and $P(X \cup Y) = 0.594$, show that X and Y are independent.

21 A bag contains 7 black and 9 white marbles. Two marbles are selected at random.

a Find the probability of drawing out:

i 2 white marbles

ii a white marble given that the first marble was white.

b If the first marble is put back in the bag before the second one is selected, show that selecting a white marble given that the first marble is white is an independent event.

22 The probability of customers buying a cup of coffee (C) at City Café is $P(C) = \frac{9}{11}$.

Some of these customers also buy a muffin (M), with the probabilities $P(M|C) = \frac{3}{8}$ and $P(C|M) = \frac{5}{6}$.

One customer is surveyed at random going into the café.

a Is the probability of this person buying a muffin independent of the probability of buying a coffee?

b What is the probability that this person buys a muffin?

Random variables and data 10.07

A **random variable** is a variable whose value is the outcome of a random experiment. It can take on different values depending on the outcome of a random process, such as the number of vehicles in a car park. Random variables can be **discrete** or **continuous**. Discrete variables such as year of birth or number of children take on specific finite values and can be counted, while continuous variables such as length or temperature are measured along a continuous scale.

Example 21

Is each random variable discrete or continuous?

a The number of goals scored by a netball team.

b The height of a student.

c The shoe size of a Year 11 student.

Solution

a The number of goals scored is a specific whole number so it is a discrete random variable.

b Height is measured on a continuous scale so the height of a student is a continuous random variable.

c Shoe sizes are specific values that can be listed so it is a discrete random variable.

Foundation Mastery Complex

We use a capital letter such as X for a random variable and a lower-case letter such as x for the values of X.

Example 22

Find the set of possible values for each random variable.

a The number rolled on a die.

b The number of girls in a family of 3 children.

c The number of heads when tossing a coin 8 times.

Solution

a Any number from 1 to 6 can be rolled on a die.

So $X = \{1, 2, 3, 4, 5, 6\}$.

b It is possible to have no girls, 1 girl, 2 girls or 3 girls.

So $X = \{0, 1, 2, 3\}$.

c The coin could come up heads 0, 1, 2, ... 8 times.

So $X = \{0, 1, 2, 3, 4, 5, 6, 7, 8\}$.

Worksheets
Graphing data

Cumulative frequency graphs

Data and frequency

Frequency tables, **histograms**, **frequency polygons** and other graphs can be used to display the data of random variables.

Example 23

a For the following Year 12 English essay marks (out of 10):

8, 4, 5, 4, 8, 6, 7, 8, 9, 5, 6, 7, 7, 5, 4, 6, 7, 9, 3, 5, 5

- **i** draw a frequency distribution table
- **ii** draw a histogram for this data
- **iii** draw a frequency polygon on the same set of axes as the histogram
- **iv** how many marks are less than 5?
- **v** what percentage of marks are over 6?

b The assessment scores for a Year 12 Mathematics class are below.

75, 53, 58, 71, 68, 51, 60, 87, 62, 62, 89, 65, 69, 47, 70, 72, 75, 68, 76, 83, 62, 88, 94, 53, 85

- **i** Draw a frequency table that shows the results of the class test using groups of 40–49, 50–59 and so on.
- **ii** Add a column for class centre and cumulative frequency.
- **iii** Draw a **cumulative frequency histogram** and a cumulative frequency polygon (ogive).
- **iv** What percentage of students scored less than 60?

Solution

a **i** The scores range from 3 to 9.
We arrange them in a table as shown.

Score	Tally	Frequency
3	|	1
4	|||	3
5	卌	5
6	|||	3
7	||||	4
8	|||	3
9	||	2

ii The histogram is a column graph where the centre of the column is lined up with the score and the columns join together.

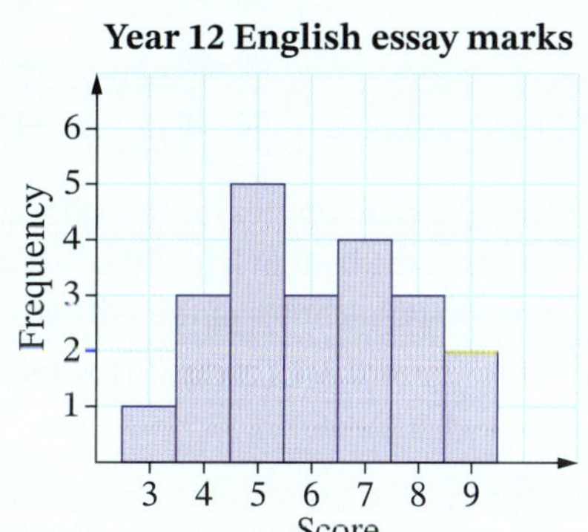

iii The frequency polygon is a line graph as shown. It starts and ends on the horizontal axis.

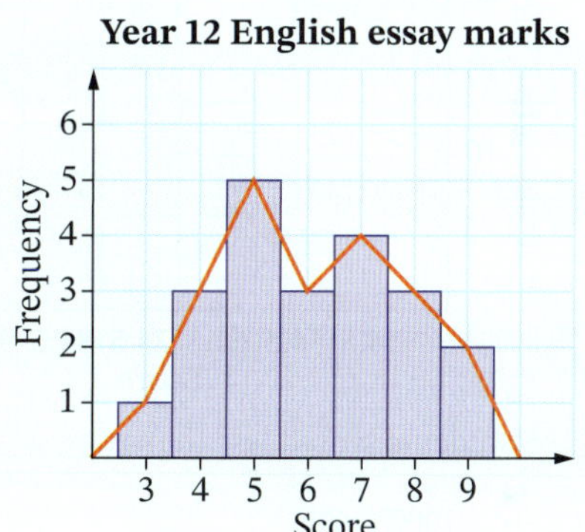

iv Less than 5 means a data value of 3 or 4. Reading from either the table or the graphs, there is one value of 3 and 3 values of 4.

So there are $1 + 3 = 4$ values less than 5.

v There are $4 + 3 + 2 = 9$ values over 6 out of a total of 21 data values.

$\frac{9}{21} \times 100\% \approx 42.9\%$

b **i**

Scores	Tally	Frequency
40–49	|	1
50–59	||||	4
60–69	卌 |||	8
70–79	卌 |	6
80–89	卌	5
90–99	|	1

ii The class centre is the average of the highest and lowest possible value in each group.

For example, $\frac{40+49}{2} = 44.5$.

The **cumulative frequency** is a running total of frequencies. Add each value to the previous total for cumulative frequencies.

Scores	Class centre	Frequency	Cumulative frequency
40–49	44.5	1	1
50–59	54.5	4	5
60–69	64.5	8	13
70–79	74.5	6	19
80–89	84.5	5	24
90–99	94.5	1	25

iii Use the class centres for the scores on the graph. The **cumulative frequency polygon** or **ogive** starts at the bottom left of the first column, joins the top right corners of each column, and ends at the top right corner of the last column.

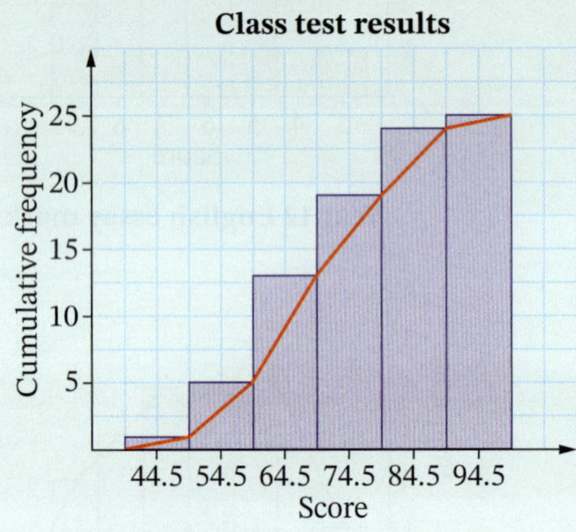

Notice that the cumulative frequency polygon is drawn differently from the frequency polygon. Unlike the frequency polygon, it is completely inside the histogram, joining top right corners instead of midpoints of columns.

iv 5 students scored less than 60.

$\frac{5}{25} \times 100\% = 20\%$

The mode and median

When we analyse data, we try to find a 'typical', 'normal' or 'average' value to represent the data set.

You would usually expect to find this value somewhere in the centre of the data.

The **mode** and the **median** are **measures of centre**.

The **mode** is the most frequent value(s) in the data set.

There could be 2 or more modes, or no mode if all the values are different.

The **median** is the middle value when all the data values are arranged in order.

If there are an even number of data values, then there are 2 middle values, and the median is the average of these 2 values.

The median of n values is the $\left(\frac{n+1}{2}\right)$th value if n is odd; otherwise find the average of the values on either side of the $\left(\frac{n+1}{2}\right)$th position.

For example, the ages of the students in a school band are:

18, 15, 20, 18, 17, 16, 11, 13

Arranging them in order from youngest to oldest:

11, 13, 15, 16, 17, 18, 18, 20

The mode is 18, the most common value (frequency 2).

There are 8 values, so the median is the average of the 2 middle values 16 and 17 (the 4th and 5th values).
[Using the formula, $\frac{8+1}{2}$ = 4.5, so the middle values are the 4th and 5th values.]

11, 13, 15, 16, 17, 18, 18, 20

median = $\frac{16+17}{2}$ = 16.5

Example 24

Find the mode and median of this set of data.

Score	Frequency
4	4
5	2
6	3
7	6
8	2
9	2
10	1

Solution

The mode is 7, the score with the highest frequency (6).

To find the median, complete a cumulative frequency column.

Score	Frequency	Cumulative frequency
4	4	4
5	2	6
6	3	9
7	6	15
8	2	17
9	2	19
10	1	20

There are 20 data values so the 2 middle values are the 10th and 11th values.
[Using the formula, $\frac{20+1}{2}$ = 10.5, so the middle values are the 10th and 11th values.]
From the cumulative frequency column, the 9th value is 6, the 15th value is 7, so the 10th and 11th values are both 7.

The median is $\frac{7+7}{2}$ = 7.

Another way to find the median is to draw an ogive (cumulative frequency polygon) and read off the halfway mark on the cumulative frequency axis.

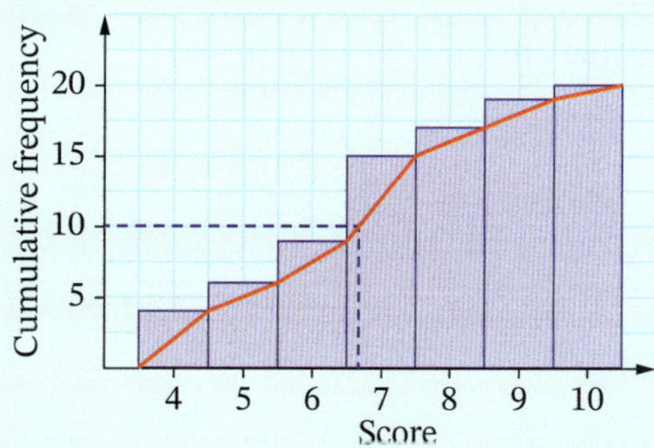

The ogive goes up to 20 for cumulative frequency, so the halfway point is at the 10th value as shown on the cumulative frequency axis. The dotted line meets the ogive inside the 7 column.

The median is 7.

EXERCISE 10.07 Answers on p. 526

Random variables and data

EXAMPLE 21

1 For each random variable, state whether it is discrete or continuous.

- **a** A film critic's rating of a film, from 0 to 4 stars.
- **b** The speed of a car.
- **c** The sum rolled on a pair of dice.
- **d** The winning ticket number drawn from a raffle.
- **e** The weight of parcels at a post office.
- **f** The size of jeans in a shop.
- **g** The temperature of a metal as it cools.
- **h** The amount of water in different types of fruit drink.
- **i** The number of cars passing the school over a 10-minute period.
- **j** The number of cities in each country in Europe.
- **k** The number of heads when tossing a coin 50 times.
- **l** The number of correct answers in a 10-question test.

EXAMPLE 22

2 Write the set of possible values for each discrete random variable.

- **a** Number of daughters in a one-child family.
- **b** Number of 6s on 10 rolls of a die.
- **c** Number of people aged over 50 in a group of 20 people.
- **d** The number of days it rains in March.
- **e** The sum of the 2 numbers rolled on a pair of dice.

EXAMPLE 23

3 For each set of data below:

- **i** draw a frequency distribution table
- **ii** draw a histogram and frequency polygon
- **iii** find the highest and lowest scores (groups for part **d**)
- **iv** find the mode (group for part **d**).

- **a** Results of a class quiz:

  8, 6, 5, 7, 6, 8, 3, 2, 6, 5, 8, 4, 7, 3, 8, 7, 5, 6, 5, 8, 6, 4, 9, 6, 5

- **b** The number of people ordering pizzas each night:

  15, 12, 17, 18, 18 ,15, 16, 13, 15, 17, 18, 12, 17, 14, 16, 15, 17, 18, 19, 15, 15, 12

- **c** The number of people attending a gym:

  110, 112, 114, 109, 112, 113, 108, 110, 113, 112, 113, 110, 109, 110, 110, 112, 114, 114, 112, 114, 113

- **d** The heights of students (in cm) in a Year 12 class:

  159, 173, 182, 166, 172, 179, 181, 163, 178, 169, 183, 158, 162, 167, 174, 175, 180, 174, 176, 159, 161, 171, 174, 179, 180, 159, 157

  (Use groups of 155–159, 160–164, 165–169 and so on.)

☐ Foundation ○ Mastery ⬡ Complex

4 For each data set:

i add a cumulative frequency column and class centre where necessary

ii sketch a cumulative frequency histogram and ogive (cumulative frequency polygon).

a Number of cars in a school car park:

Number of cars	Frequency
10	4
11	8
12	11
13	9
14	5

b Results of a science experiment:

Score	Frequency
1	7
2	1
3	3
4	0
5	2
6	5

5 The table shows the number of daily rescues at a beach over a period of time.

a Draw a histogram showing this data.

b How many times were more than 6 rescues made?

c What was the mode number of daily rescues during the survey?

Number of rescues	Frequency
4	1
5	3
6	6
7	4
8	5
9	0
10	2

6 For each data set, find:

i the mode **ii** the median.

a Number of people auditioning for parts in a play: 5, 5, 7, 6, 5, 6, 1

b Number of minutes for an ambulance to respond to a call: 1, 4, 6, 8, 7, 4, 6, 4, 5

c Ages of students on a basketball team: 15, 18, 14, 19, 18, 17, 11

d Scores on a class quiz: 4, 6, 5, 4, 7, 8

7 For each data set, find:

EXAMPLE 24

i the median **ii** the mode.

a Judge's scores on a dance contest:

Score	Frequency
3	3
4	4
5	2
6	7
7	6
8	2
9	3

b Number of matches in each match box surveyed:

Number of matches	Frequency
50	1
51	6
52	5
53	3
54	4
55	2

☐ Foundation ◯ Mastery ◯ Complex

c Results in a History assignment:

Score	Frequency
14	4
15	2
16	1
17	4
18	3
19	5
20	6

d Attendances at hockey matches:

Attendance	Frequency
100	3
101	0
102	2
103	1
104	6
105	5

8 Find the median from each ogive:

a

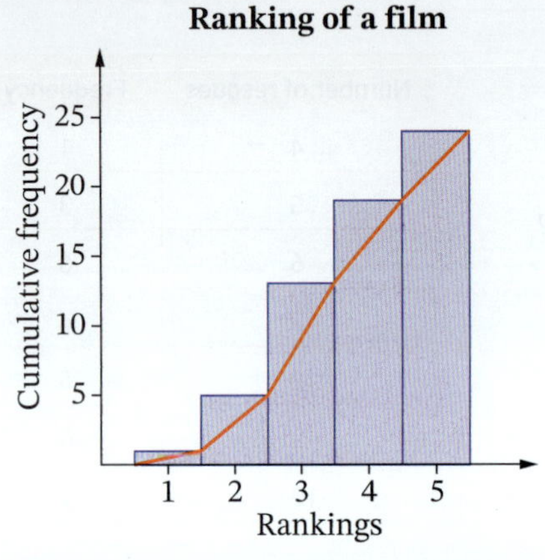

b

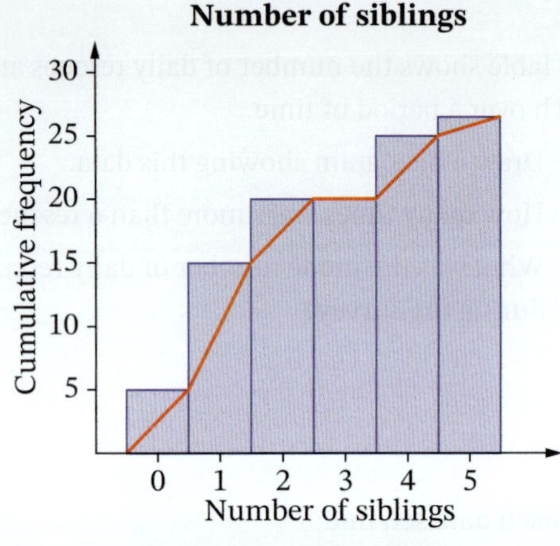

c

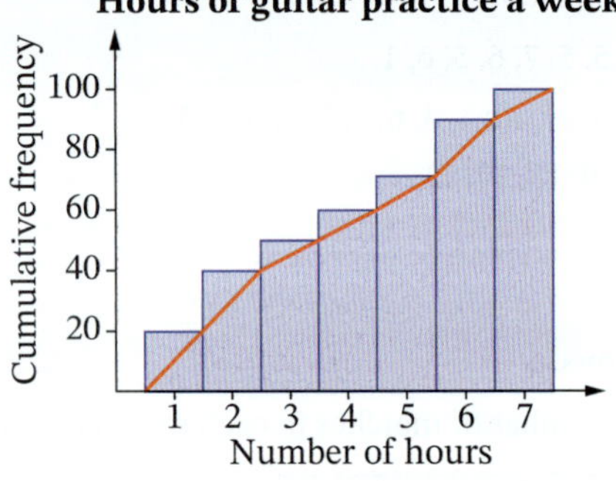

d

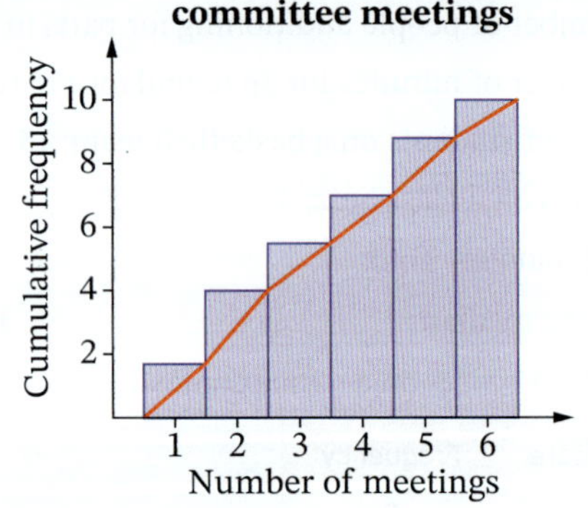

9 Find the mode of each data set represented by the ogives in Question **8**.

☐ Foundation ○ Mastery ⬡ Complex

10 For each data set:

i add a cumulative frequency column

ii draw a cumulative frequency polygon

iii find the median from the graph (estimate for parts **c** and **d**).

a Number of athletes representing their school over a 30-year period:

Number of athletes	Frequency
1	5
2	6
3	4
4	8
5	5
6	2

b Number of lollies in a bag:

Number of lollies	Frequency
45	3
46	5
47	1
48	7
49	3
50	1

c Hours a week worked by employees in a cafe:

Number of hours	Frequency
1–5	7
6–10	5
11–15	3
16–20	6
21–25	7
26–30	2

d Time to complete a race:

Time (min)	Frequency
2.5–2.8	3
2.9–3.2	2
3.3–3.6	0
3.7–4.0	6
4.1–4.4	1
4.5–4.8	4
4.9–5.2	4

11 For each data set:

i draw a frequency distribution table including cumulative frequency

ii find the mode

iii draw an ogive

iv find the median from the ogive

a Home runs scored over a baseball season:

4, 6, 5, 8, 8, 6, 5, 3, 4, 9, 6, 3, 5, 6, 5, 4, 7, 5, 8, 5, 6, 2, 3

b Number of movies seen in a year:

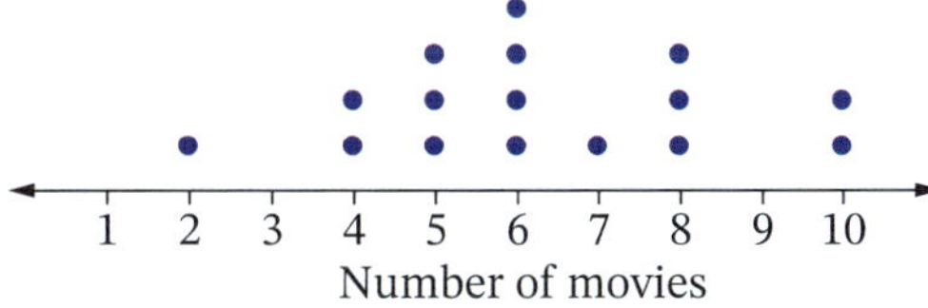

Foundation Mastery Complex

12 The frequency histogram shows the results on a maths quiz.

a Find the mode.

b Draw a cumulative frequency histogram and polygon and use them to find the median.

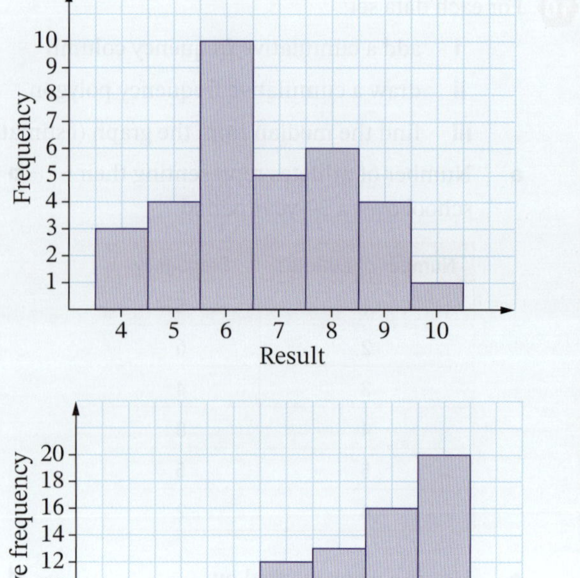

13 The cumulative frequency histogram shows data collected in a survey on the number of junk mail items that people received daily in their inbox.

a Draw a frequency distribution table to show the number of junk mail items people receive daily.

b Construct a frequency histogram to show this data.

c Find the mode and the median of this data set.

Cumulative frequency
20
18
16
14
12
10
8
6
4
2
1 2 3 4 5 6 7
Number of junk mail items

10.08 Relative frequency and probability

Video Experimental probability

Worksheet Experimental probability

We can use frequency distribution tables to find the probability of an event using **relative frequency**, the frequency of the event as a fraction of the total frequency.

Example 25

This table shows the number of items bought by a group of people surveyed in a shopping centre.

Number of items	Frequency
0	6
1	4
2	5
3	3
4	7

a Find the relative frequency for each number of items as a fraction.

b Find the probability that a person surveyed at random would buy:

i no items

ii at least 3 items.

Foundation Mastery Complex

Solution

a The sum of the frequencies is 25. This means that 25 people were surveyed.

From the table, 0 has a frequency of 6. The relative frequency is 6 out of 25 $= \frac{6}{25}$.

Similarly, other relative frequencies are:

1 item: $\frac{4}{25}$ 2 items: $\frac{5}{25} = \frac{1}{5}$ 3 items: $\frac{3}{25}$ 4 items: $\frac{7}{25}$

b **i** $P(0) = \frac{6}{25}$

ii At least 3 items means 3 or 4 items. Relative frequency of 3 or 4 is $3 + 7 = 10$.

$P(\geq 3) = \frac{10}{25} = \frac{2}{5}$

Example 26

This table shows students' scores on a reaction test.

Score	Frequency
4	1
5	3
6	8
7	7
8	5
9	4
10	2

a Add columns for relative frequency and cumulative relative frequency.

b Draw a relative frequency histogram.

c Find the probability of choosing a student from this class who scored:

i less than 8 **ii** 6 or less **iii** more than 7.

Solution

a The total number of students (frequency) is 30.

Relative frequencies are the frequencies divided by 30.

The cumulative relative frequency column is a running total of the relative frequencies.

Score	Frequency	Relative frequency	Cumulative relative frequency
4	1	$\frac{1}{30}$	$\frac{1}{30}$
5	3	$\frac{3}{30} = \frac{1}{10}$	$\frac{4}{30} = \frac{2}{15}$
6	8	$\frac{8}{30} = \frac{4}{15}$	$\frac{12}{30} = \frac{2}{5}$
7	7	$\frac{7}{30}$	$\frac{19}{30}$
8	5	$\frac{5}{30} = \frac{1}{6}$	$\frac{24}{30} = \frac{4}{5}$
9	4	$\frac{4}{30} = \frac{2}{15}$	$\frac{28}{30} = \frac{14}{15}$
10	2	$\frac{2}{30} = \frac{1}{15}$	$\frac{30}{30} = 1$
Total	**30**		

Notice that the cumulative relative frequencies always end with 1, since the total of all probabilities is 1. The cumulative frequency column gives the probability of the score and all the scores below it.

b

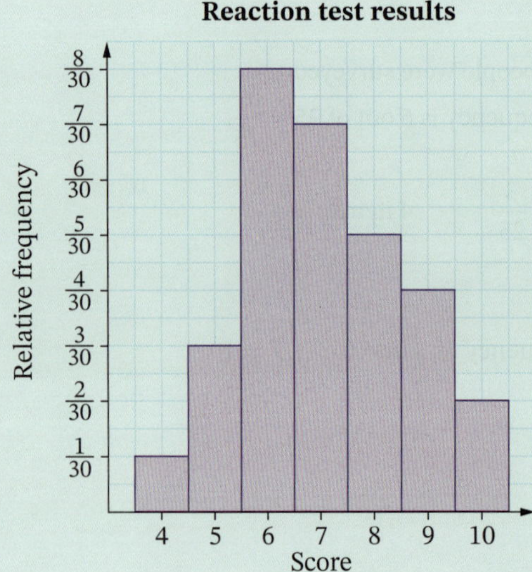

c **i** $P(<8) = P(\leq 7)$

$= \frac{19}{30}$

ii $P(\leq 6) = \frac{2}{5}$

iii Scores more than 7 are the scores 8, 9 and 10

However, this is the complement to '≤ 7', scores 7 or less.

$P(>7) = 1 - P(\leq 7)$

$= 1 - \frac{19}{30}$

$= \frac{11}{30}$

EXERCISE 10.08 Answers on p. 529

Relative frequency and probability

EXAMPLE 25

1 This table shows the scores that a class earned on a music performance.

Score	Frequency
4	6
5	4
6	1
7	7
8	2
9	3

a Find the relative frequency for each score in the table, in fraction form.

b If a student is chosen at random from this class, find the probability that this student:

i scored 8

ii scored less than 7

iii scored 5 or more.

c What score is:

i most likely?

ii least likely?

2 The table shows the results of a survey into the number of days students study each week.

Number of days	Frequency
1	3
2	6
3	1
4	7
5	2
6	1

a Find the relative frequency as a percentage for each number of days.

b If a student was selected at random, find the most likely number of days this student studies.

c Find the probability that this student would study for:

i 1 day

ii 5 days

iii 3 or 4 days

iv at least 4 days

v fewer than 3 days.

☐ Foundation ○ Mastery ○ Complex

3 The table shows the results of a trial HSC exam.

a Calculate the relative frequency as a decimal for each class.

b Find the probability that a student chosen at random from these students scored:

i between 20 and 39

ii between 60 and 99

iii less than 40.

Class	Frequency
0–19	9
20–39	12
40–59	18
60–79	7
80–99	4

4 This table shows the results of a science experiment to find the velocity of an object when it is rolled down a ramp.

a Write the relative frequency of each velocity as a fraction.

b Find the probability that an object selected at random rolls down the ramp with a velocity between:

i 5 and 7 m/s **ii** 11 and 13 m/s **iii** 8 and 10 m/s

iv 11 and 16 m/s **v** 2 and 10 m/s.

c Find the probability that the object has a velocity:

i less than 8 m/s **ii** 5 m/s or more **iii** more than 7 m/s.

Velocity (m/s)	Frequency
2–4	2
5–7	7
8–10	4
11–13	1
14–16	6

5 A telemarketing company records the number of sales it makes per minute over a half-hour period. The results are in the table.

a What percentage of the time were there 3 sales per minute?

b Write the relative frequencies as percentages.

c What is the most likely number of sales/minute?

d Find the probability of making:

i 2 sales/minute

ii 5 sales/minute

iii more than 2 sales per minute.

Sales/min	Frequency
0	4
1	12
2	6
3	3
4	0
5	5

EXAMPLE 26

6 **a** The waiting times in minutes between trains arriving at a station are shown below. Organise the scores below in a frequency distribution table, including relative frequency and cumulative relative frequency columns.

9, 5, 4, 7, 7, 9, 4, 6, 5, 8, 9, 6, 7, 4, 4, 3, 8, 5, 6, 9

b Draw a relative frequency histogram for this data.

c Find the probability that the waiting time between trains is:

i 7 minutes **ii** at least 8 minutes **iii** less than 5 minutes **iv** 7 minutes or less.

7 **a** From the dot plot of ages of children at a fun park, draw up a frequency distribution table.

b Draw a relative frequency polygon for this data.

c Find the probability (as a decimal) that a child chosen at random from the fun park is aged:

i 8 **ii** at least 6

iii older than 7 **iv** 5 or older

v 8 or younger.

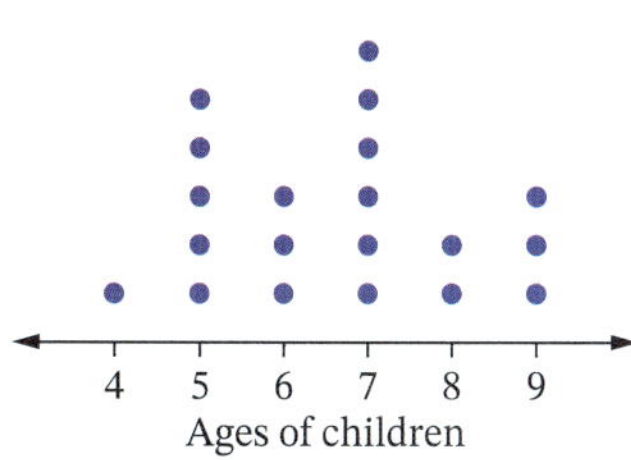

Foundation Mastery Complex

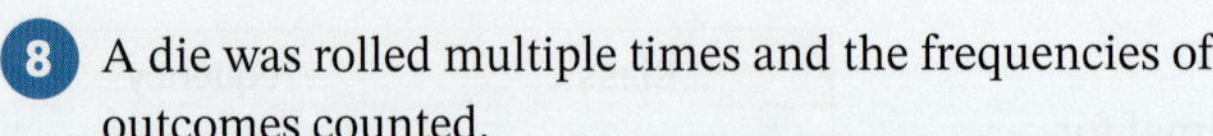

8 A die was rolled multiple times and the frequencies of outcomes counted.

Outcome	Frequency
1	21
2	18
3	15
4	24
5	16
6	12

a Add columns for relative frequency and cumulative relative frequency

b How many rolls of the die were made?

c Draw a cumulative relative frequency histogram and polygon.

d Based on these results, find the probability of rolling:

i a 6 **ii** less than 3 **iii** more than 4.

9 In a chance experiment, a coin and a die were tossed together. H = heads, T = tails.

Result	Frequency
H, 1	8
H, 2	3
H, 3	5
H, 4	7
H, 5	1
H, 6	6
T, 1	7
T, 2	12
T, 3	4
T, 4	9
T, 5	8
T, 6	6

Find the probabilities based on this experiment of getting:

a a head and 6

b an odd number and a tail

c a tail and 5 or more

d less than 4 and a head.

10 In an experiment where a coin was tossed 250 times, the results were 97 heads and 153 tails. Explain this result.

Sample HSC problem

Answers on p. 531

(4 marks) A bag contains 3 red and 4 blue pens. Two pens are taken out of the bag at random.

a Find the probability of drawing at most one blue pen. **2 marks**

b Find the probability that the second pen taken out is blue, given that the first pen taken out is blue. **2 marks**

□ Foundation ○ Mastery ⬡ Complex

CHAPTER SUMMARY

Puzzles
Probability crossword

Probability find-a-word

This chapter, *Probability and data*, revised and extended probability concepts, skills, terminology, notations and formulas from Years 9–10 before introducing the idea of random variables and relative frequency, which will be examined more deeply in Year 12.

Before you move on, consider what you learned in this chapter and revisit any sections that may have been unclear.

What you learned in this chapter...	Section	
Use the language and notation of sets, Venn diagrams and probability theory	10.01 10.02	Set notation and Venn diagrams Probability
Understand and use the addition rule for probability $P(A \cup B) = P(A) + P(B) - P(A \cup B)$	10.03	The addition rule of probability
Understand and use the product rule for probability for independent events $P(A \cap B) = P(A) \times P(B)$	10.04	The product rule of probability
Apply the addition and product rules to tree diagrams to solve multi-stage probability problems	10.05	Probability trees
Solve problems involving conditional probability, including using the formula $P(A \mid B) = \dfrac{P(A \cap B)}{P(B)}$	10.06	Conditional probability
Identify discrete and continuous random variables	10.07	Random variables and data
Organise data in frequency tables using relative frequency and cumulative relative frequency	10.07	Random variables and data
Display data using histograms and polygons, including cumulative frequency histograms and polygons (ogives)	10.07	Random variables and data
Find the mode and median of a data set from tables, histograms and polygons	10.07	Random variables and data
Use relative frequency to estimate probability	10.08	Relative frequency and probability

To help master these techniques, make a summary mind map. Use the chapter outline and the mind map below as a guide. Add your own words, symbols, diagrams, boxes and reminders. The summary should give you a 'whole picture' view of the topic and allow you to identify any weak areas to revisit in your revision.

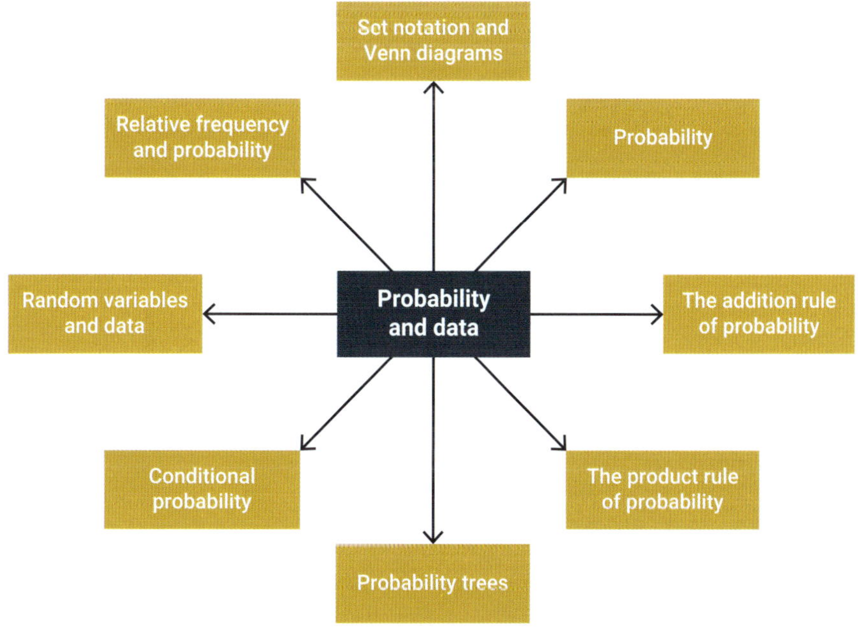

10 Test yourself

Answers on p. 531

For Questions **1 – 6**, select the correct answer **A, B, C** or **D**.

10.05 **1** The probability of getting at least one 1 when rolling 2 dice is:

A $\frac{1}{3}$ **B** $\frac{1}{6}$ **C** $\frac{11}{36}$ **D** $\frac{5}{18}$

10.05 **2** A bag contains 7 white and 5 blue balls. Two balls are selected at random without replacement. The probability of selecting a white and a blue ball is:

A $\frac{35}{132}$ **B** $\frac{35}{72}$ **C** $\frac{35}{144}$ **D** $\frac{35}{66}$

10.01 **3** If $A = \{5, 7, 8\}$ and $B = \{3, 7, 9\}$ then the set $\{7\}$ represents:

A $A - B$ **B** $A \cap B$ **C** $A + B$ **D** $A \cup B$

10.08 **4** For the table, the relative frequency of a score of 11 is: (there may be more than one answer)

A 32% **B** $\frac{8}{11}$

C 0.032 **D** $\frac{8}{25}$

Score	Frequency
8	5
9	2
10	9
11	8
12	1

10.01 **5** Which of these formulas is true (there may be more than one answer)?

A $|A \cap B| = |A| + |B| - |A \cup B|$

B $|A \cap B| = |A| + |B| + |A \cup B|$

C $|A \cup B| = |A| + |B| - |A \cap B|$

D $|A \cup B| = |A \cap B| - |A| + |B|$

10.06 **6** A bag contains 5 blue and 3 white balls. The probability tree shows the outcomes for drawing out 2 balls without replacement.

If one of the balls that is selected is blue, find the probability that the second ball is also blue.

A $\frac{5}{14}$

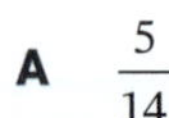

B $\frac{4}{7}$

C $\frac{5}{8}$

D $\frac{7}{10}$

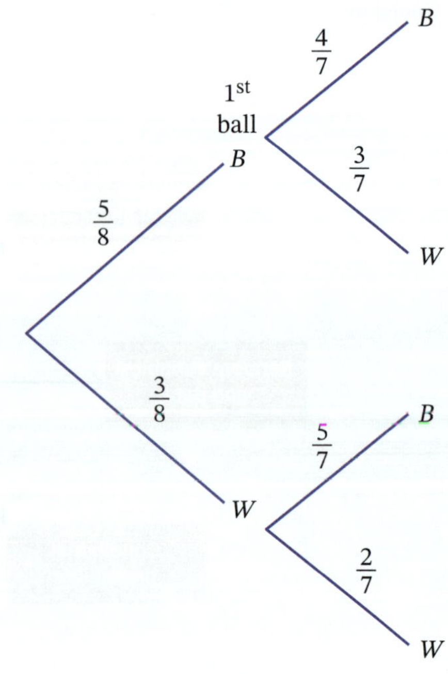

7 The probability that a certain type of seed will germinate is 93%. 10.05
If 3 of these seeds are planted, find the probability that:

a all will germinate **b** just one will germinate **c** at least one will germinate.

8 The table shows the results of an experiment when throwing a die. 10.07

Outcome	Frequency
1	17
2	21
3	14
4	20
5	18
6	10

a Add a column for relative frequencies and cumulative relative frequencies (as fractions).

b From the table, find the probability of throwing:

i 3 **ii** more than 4
iii 6 **iv** 1 or 2
v less than 4

9 A game is played where the differences of the numbers rolled on 2 dice are taken. 10.05

a Draw a table showing the sample space (all possibilities).

b Find the probability of rolling a difference of:

i 3 **ii** 0 **iii** 1 or 2

10 Mark buys 5 tickets in a raffle in which 200 are sold altogether. 10.05

a What is the probability that he will:

i win the raffle? **ii** not win the raffle?

b If the raffle has 2 prizes, find the probability that Mark will win just one prize.

11 In a group of 30 tourists, 17 have been to France, 11 have been to Germany and 5 haven't been to either country. 10.03
Find the probability that a tourist chosen at random from this group has been to:

a Germany but not France

b both countries

c Germany, given that they've been to France

d France given that they've been to Germany.

12 'In the casino, when tossing 2 coins, 2 tails came up 10 times in a row. So there is less chance that 2 tails will come up next time.' Is this statement true? Why or why not? 10.06

13 A set of 100 cards numbered 1 to 100 is placed in a box and one is drawn at random. 10.02
Find the probability that the card chosen is:

a odd **b** less than 30 **c** a multiple of 5
d less than 30 or a multiple of 5 **e** odd or less than 30.

14 Jenny has a probability of $\frac{3}{5}$ of winning a game of chess and a probability of $\frac{2}{3}$ of winning a card game. If she plays one of each game, find the probability that she wins: 10.05

a both games **b** 1 game **c** neither game.

15 **a** Given event $A = \{3, 5, 6, 8, 10\}$ and event $B = \{5, 7, 8, 9, 11, 12\}$, find: 10.01

i $A \cup B$ **ii** $A \cap B$

b Draw a Venn diagram showing this information.

16 Find the sample space for each situation. 10.01

a Tossing 2 coins **b** Choosing a colour from the Australian National flag

17 Each machine in a factory has a probability of 4.5% of breaking down at any time. If the factory has 3 of these machines, find the probability that: 10.05

a all will be broken down **b** at least one will be broken down.

18 A bag contains 4 yellow, 3 red and 6 blue balls. Two are chosen at random. 10.05

a Find the probability of choosing:

i 2 yellow balls **ii** a red and a blue ball **iii** 2 blue balls.

b Find the probability that the second ball is:

i yellow, given that the first ball is blue **ii** red, given the first ball is yellow.

☐ Foundation ○ Mastery ○ Complex

10.03 **19** In a group of 12 friends, 8 have seen the movie *Star Wars 20* and 9 have seen the movie *Mission Impossible 9*. Everyone in the group has seen at least one of these movies. If one of the friends is chosen at random, find the probability that this person has seen:

a both movies **b** only *Mission Impossible 9*.

10.05 **20** A game of chance offers a $\frac{2}{5}$ probability of a win or a $\frac{3}{8}$ probability of a draw.

a If Billal plays one of these games, find the probability that he loses.

b If Sonya plays 2 of these games, find the probability of:

i a win and a draw **ii** a loss and a draw **iii** 2 wins.

10.04 **21** Yukari plays 2 games of Super Maria Sisters and the probability that she completes a level in at least one of the games is 75%. Find the probability that she completes a level in both games.

10.05 **22** A loaded die has a $\frac{2}{3}$ probability of coming up 6. The other numbers have an equal probability of coming up. If the die is rolled, find the probability that it comes up:

a 2 **b** even.

10.05 **23** Amie buys 3 raffle tickets. If 150 tickets are sold altogether, find the probability that Amie wins:

a 1st prize **b** only 2nd prize **c** 1st and 2nd prizes **d** neither prize.

10.05 **24** A bag contains 6 white, 8 red and 5 blue balls. If 2 balls are selected at random, find the probability of choosing a red and a blue ball:

a with replacement **b** without replacement.

10.03 **25** A group of 9 friends go to the movies. All buy popcorn or an ice-cream. If 5 buy popcorn and 7 buy ice-creams, find the probability that one friend chosen at random will have:

a popcorn but not ice-cream **b** both popcorn and ice-cream

c popcorn given they also have an ice-cream.

10.05 **26** Ed's probability of winning at tennis is $\frac{3}{5}$ and his probability of winning at squash is $\frac{7}{10}$.
Find the probability of Ed winning:

a both games **b** neither game **c** one game.

10.07 **27** For each random variable, write the set of possible values.

a The number of 6s when rolling a die 5 times.

b The number of heads when tossing a coin 10 times.

c The first day the temperature rises above 28°C in November.

d The number of doubles when rolling 2 dice twice.

e The number of red cards selected in 9 trials when pulling a card from a hat that contains 20 red and 20 blue cards.

10.06 **28** Given $P(A \cap B) = 0.162$ and $P(B) = 0.36$, show that A and B are independent events if $P(A) = 0.45$.

29 Two cards are selected randomly from a box containing 30 cards, each labelled with a different number from 1 to 30. Find the probability of selecting a 2nd even card given that the first card selected was even.

10.07 **30** Nikola conducted a survey to find the number of people in each car that travelled across the Anzac Bridge one morning. Her data is shown in the table.

Number of occupants	Frequency
1	43
2	32
3	12
4	8
5	5

a Draw a cumulative frequency polygon to illustrate the results.

b From the polygon, find the median number of people in a car.

c Find the mode number of people in a car.

□ Foundation ○ Mastery ○ Complex

Challenge exercise Answers on p. 531 10

1 In a group of 35 students, 25 go to the movies and 15 go to the league game.
If all the students like at least one of these activities, and 2 students are chosen from this group at random, find the probability that:

a both only go to the movies

b one only goes to the league game and the other goes to both the game and movies.

2 A certain soccer team has a probability of 0.5 of winning a match and a probability of 0.2 of a draw.
If the team plays 2 matches, find the probability that it will:

a draw both matches **b** win at least one match **c** not win either match.

3 A game of poker uses a deck of 52 cards with 4 suits (hearts, diamonds, spades and clubs).
Each suit has 13 cards, consisting of an ace, cards numbered from 2 to 10, a jack, queen and king.
If a person is dealt 5 cards, find the probability of getting 4 aces.

4 There are 4 coins, one of which has 2 heads. One coin is randomly selected and tossed.

a Find the probability that this coin shows heads.

b Given that the coin shows heads, find the probability that it is the double headed coin.

5 Brayden does not select the numbers 1, 2, 3, 4, 5 and 6 for Lotto because he says this combination would never win. Is he correct?

6 Out of a class of 30 students, 19 play a musical instrument and 7 play both a musical instrument and a sport. Two students play neither.

a One student is selected from the class at random.
Find the probability that this person plays a sport but not a musical instrument.

b Two people are selected at random from the class.
Find the probability that both these people only play a sport.

7 A game involves tossing 2 coins and rolling 2 dice.
The scoring is shown in the table.

Result	Score (points)
2 heads and double 6	5
2 heads and double (not 6)	3
2 tails and double 6	4
2 tails and double (not 6)	2

a Find the probability of getting 2 heads and a double 6.

b Find the probability of getting 2 tails and a double that is not 6.

c What is the probability of obtaining a score of 13 in 3 moves?

8 Silvana has a 3.8% probability of passing on a defective gene to a daughter and a 0.6% probability of passing on a defective gene to a son. The probability of Silvana having a son is 52%.
Find the probability that she has a son given that Silvana passes on a defective gene.

Foundation Mastery Complex

PRACTICE SET 4

Answers on p. 532

For Questions **1** to **7**, select the correct answer **A**, **B**, **C** or **D**.

9.02 **1** Find the amplitude and period of $y = 5\sin 3x$.

A amplitude 3, period 5 **B** amplitude 5, period 3

C amplitude 5, period $\frac{2\pi}{3}$ **D** amplitude 3, period $\frac{2\pi}{5}$

8.05 **2** The transformation of $y = f(x)$ to $y = 3f(2x)$ is:

A vertical dilation of factor 3, horizontal dilation of factor 2

B horizontal dilation of factor 3, vertical dilation of factor 2

C vertical dilation of factor 3, horizontal dilation of factor $\frac{1}{2}$

D horizontal dilation of factor 3, vertical dilation of factor $\frac{1}{2}$

9.01 **3** Simplify $\frac{\sin\theta}{\cos^2\theta\sec\theta}$.

A $\cot\theta$ **B** $\tan^2\theta$ **C** $\tan\theta$ **D** $\cot^2\theta$

10.05 **4** Find the probability of drawing out a blue and a white ball from a bag containing 7 blue and 5 white balls if the first ball is not replaced before taking out the second.

A $\frac{70}{121}$ **B** $\frac{70}{144}$ **C** $\frac{1225}{17424}$ **D** $\frac{35}{66}$

10.02 **5** In a group of 25 students, 19 catch a train to school and 21 catch a bus. If one of these students is chosen at random, find the probability that the student only catches a bus to school.

A $\frac{6}{25}$ **B** $\frac{21}{25}$ **C** $\frac{3}{5}$ **D** $\frac{3}{20}$

9.02 **6** $y = \cos(x + \pi) + 3$ has a phase shift of:

A π units to the right **B** 3 units up

C 3 units down **D** π units to the left.

10.06 **7** Conditional probability $P(A|B)$ is given by:

A $\frac{P(A\cup B)}{P(B)}$ **B** $\frac{P(A\cap B)}{P(A)}$ **C** $\frac{P(A\cup B)}{P(A)}$ **D** $\frac{P(A\cap B)}{P(B)}$

8.06 **8** **a** Sketch the graphs of $y = x^2$ and $y = -(x + 2)^2$ on the same set of axes.

b Describe the transformations that changed $y = x^2$ into the transformed function.

9.03 **9** Solve each equation for $[0°, 360°]$.

a $\tan 2x + 1 = 0$ **b** $2\cos 3x = 1$ **c** $2\sin(x - 90°) = \sqrt{3}$

d $\tan(x - 180°) = \sqrt{3}$ **e** $2\cos^2(x + 45°) = 1$

9.02 **10** Describe the amplitude, period, centre and phase shift of each function.

a $y = 4\cos 5x$ **b** $y = -2\sin\left(x - \frac{\pi}{6}\right) + 1$ **c** $y = \tan\left(\frac{x}{4} + 2\right)$

10.02 **11** In a class of 25 students, 11 play guitar, 9 play the piano, while 8 don't play either instrument. If one student is selected at random from the class, find the probability that this student plays:

a both guitar and piano **b** neither guitar nor piano **c** only guitar.

9.02 **12** For $0 \le x \le 2\pi$, sketch the graph of:

a $y = 2\sin 4x$ **b** $y = \tan\left(\frac{x}{2}\right)$ **c** $y = -\cos x$

☐ Foundation ◯ Mastery ◯ Complex

13 Copy the graph $y = f(x)$ and sketch:

a $y = 2f(x)$

b $y = f(x) + 1$

c $y = f(x - 2)$

d $y = f(2x)$

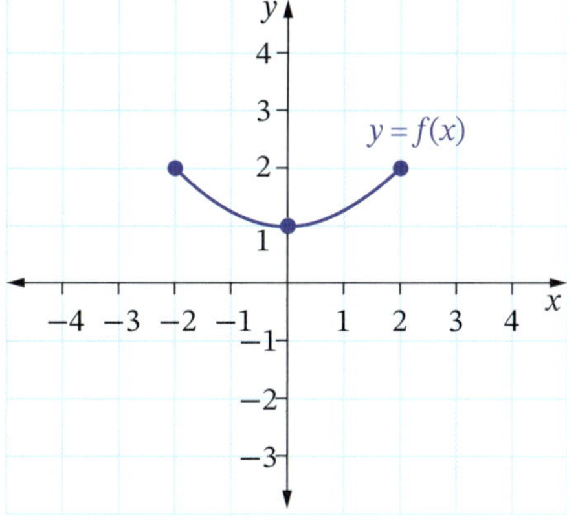

8.06

14 For each random variable X, write the set of possible values.

10.07

a The number of rolls of a die until a 6 turns up.

b The number of red cards selected when choosing 12 cards from a bag containing 15 red and 15 black cards.

c The first rainy day in January.

15 Two dice are rolled. Find the probability of rolling:

10.05

a double 1 **b** any double **c** at least one 3

d a total of 6 **e** a total of at least 8.

16 100 cards are numbered from 1 to 100. If one card is chosen at random, find the probability of selecting:

10.02

a an even number less than 30 **b** an odd number or a number divisible by 9.

17 
9.02

If $\tan x = -\frac{4}{3}$ and $\cos x > 0$, evaluate $\sin x$ and $\cos x$.

18 Solve for $0 \le x \le 2\pi$:

9.03

a $2\cos x + 1 = 0$ **b** $\tan^2 x = 1$

c $\cos x = 0$ **d** $\sin 2x = \frac{1}{2}$

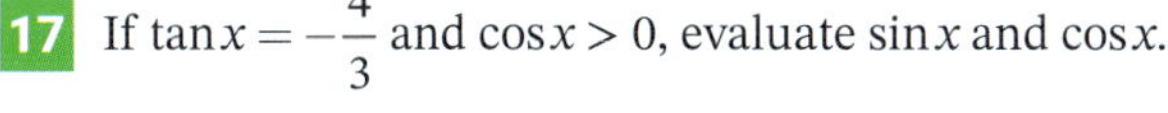

19 Solve each equation for $[0°, 360°]$.

9.03

a $6\sin^2 x - 7\sin x + 2 = 0$ **b** $2\sin[2(x - 30)] = 1$

20 Find the equation of the function if $y = \sqrt{x}$ is dilated vertically with factor 4, dilated horizontally with factor $\frac{1}{3}$, translated vertically 1 unit down and translated horizontally [illegible] units to the left.

8.06

21 Describe the transformations on $y = x^3$ if the equation of the transformed function is $y = 4(x - 1)^3 - 3$, and state whether any dilations stretch or shrink the graph of the function.

8.06

22 A bag contains 5 white, 6 yellow and 3 blue balls.
2 balls are chosen at random from the bag without replacement. Find the probability of choosing:

10.05

a 2 blue balls **b** a white ball and a yellow ball.

23 If Santigo buys 10 tickets, find the probability that he wins both first and second prizes in a raffle in which 100 tickets are sold.

10.05

24 Two dice are rolled. Find the probability of rolling a total:

10.05

a of 8 **b** less than 7 **c** greater than 9

d of 4 or 5 **e** that is an odd number.

☐ Foundation ○ Mastery ○ Complex

10.06 **25** For this Venn diagram, find:

a $P(A|B)$

b $P(B|A)$

10.06 **26** A bag contains 5 red, 7 blue and 9 yellow balls. Cherylanne chooses 2 balls at random from the bag. Find the probability that she chooses:

a blue given the first ball was yellow

b red given the first ball was blue.

10.07 **27** For this ogive, find:

a the median

b the mode

c the number of scores.

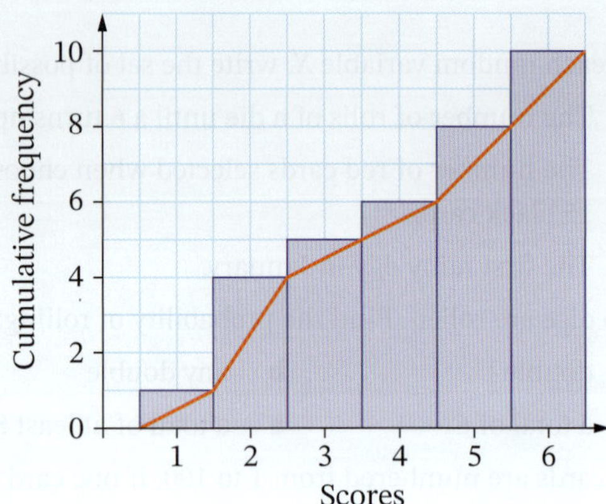

9.03 **28** Solve for $0° \leq x \leq 360°$:

a $\tan x = -1$ **b** $2\sin x = 1$ **c** $2\cos^2 x = 1$ **d** $\tan 2x = \sqrt{3}$

10.01 **29** $A = \{1, 3, 4, 5\}$ and $B = \{2, 3, 5, 6\}$.

a Find $A \cup B$.

b Find $A \cap B$.

c Draw a Venn diagram showing this information.

9.01 **30** Evaluate x if $\sec x = \text{cosec}(2x - 30°)$. Give values of x in the first quadrant only.

9.02 **31** Sketch the graph of each function in the domain $[0, 2\pi]$.

a $y = -7\cos x$ **b** $y = 2\sin x$ **c** $y = \cos x + 1$

d $y = \tan\left(x + \frac{\pi}{2}\right)$ **e** $y = 3\cos 2x$ **f** $y = -4\sin\left(\frac{x}{2}\right) + 3$

9.01 **32** Prove each identity.

a $\cot x \sec x = \text{cosec}\, x$ **b** $\sin^2 x \,\text{cosec}^2 x - \sin^2 x = \cos^2 x$

9.02 **33** Find the equation of the transformed function of $y = \sin x$ if the function has:

a amplitude 2

b period 4π

c centre -3

d a reflection in the x-axis and amplitude 5

e a phase shift $\frac{\pi}{2}$ units to the left

f amplitude 5, period 6π, centre 1 and a phase shift of π units to the right.

10.05 **34** A coin is tossed and a die thrown. Find the probability of getting:

a a head and a 6 **b** a tail and an odd number.

Foundation Mastery Complex

35 Given $P(X) = 0.26$, $P(Y) = 0.15$ and $P(X \cup Y) = 0.371$, show that X and Y are independent. 10.06

36 State whether events A and B are mutually exclusive if $P(A) = 0.18$, $P(A \cup B) = 0.5$ and $P(B) = 0.32$. 10.06

37 The probability that Despina passes her first maths test is 64% and the probability that she will pass both the first and second tests is 48%. Find the probability that Despina passes the second test given that she passes the first test.

38 If $P(L) = 45\%$, $P(L \cap M) = 5.4\%$ and $P(M) = 12\%$, show that L and M are independent.

39 Solve each equation for $[0, 2\pi]$.

a $\cos x = 0.62$ **b** $\tan^2 x = 1$ **c** $2\sin x - 1 = 0$

d $\cos x = 0$ **e** $4\sin^2 x = 3$ **f** $2\cos 2x + 1 = 0$

40 The function $y = f(x)$ is transformed into $y = -2f(2x - 6) + 5$.

a Describe the transformations applied.

b Find the point on the transformed function that corresponds to $(-2, 3)$ on the original function.

c The gradient of the tangent at $(-2, 3)$ on the original function is -4.
Find the gradient of the normal to the transformed function at the corresponding point.

41 Simplify:

a $5 + 5\tan^2 x$ **b** $\dfrac{(1+\sin x)(1-\sin x)}{\sin x \cos x}$

42 The function $y = f(x)$ is transformed to $y = 5f(3x + 6) - 7$. Describe the transformations in order. 8.06

☐ Foundation ○ Mastery ⬡ Complex

ANSWERS

Chapter 1

Exercise 1.01

1 **a** 500 **b** 145 **c** $\frac{1}{64}$
d 3 **e** 2

2 **a** 13.7 **b** 1.1 **c** 0.8
d 2.7 **e** −2.6 **f** 0.5

3 **a** a^{17} **b** $y^0 = 1$ **c** a^{-4}
d w **e** x^5 **f** x^{21}
g $4x^{10}$ **h** $81y^{-8}$ **i** a
j $\frac{x^{10}}{y^{45}}$ **k** w^{10} **l** p^5
m x^{-3} **n** $a^{-2}b^3$ or $\frac{b^3}{a^2}$ **o** $x^{-5}y^2$ or $\frac{y^2}{x^5}$

4 **a** x^{14} **b** a^{-7} **c** m^4
d k^{10} **e** a^{-8} **f** x
g mn^2 **h** p^{-1} **i** $9x^{22}$

5 **a** p^5q^{15} **b** $\frac{a^8}{b^8}$ **c** $\frac{64a^3}{b^{12}}$
d $49a^{10}b^2$ **e** $8m^{17}$ **f** x^4y^{10}
g $\frac{2k^{23}}{27}$ **h** $16y^{47}$ **i** a^3

6 $4\frac{1}{2}$ **7** 324 **8** $2\frac{10}{27}$

9 **a** 4^{x+1} **b** 3^{y-2} **c** 7^{12z}
d 5^{x+1} **e** 3^{4a-b}

10 **a** 17 033.83 **b** 10 067.77

11 $\frac{4}{9}$ **12** $\frac{1}{18}$

13 See worked solutions.

14 **a** 98^y **b** 24^a **c** 48^b
d $4^x = 2^{2x}$ **e** 3^a

15 **a** $\$2.866 \times 10^{11}$ **b** 2.26

16 **a** 9.35
b 3 112 676 or 3.1×10^6
c 333 054 or 3.3×10^5

Exercise 1.02

1 **a** $\frac{1}{27}$ **b** $\frac{1}{4}$ **c** $\frac{1}{343}$ **d** $\frac{1}{10000}$
e $\frac{1}{256}$ **f** 1 **g** $\frac{1}{32}$ **h** $\frac{1}{81}$
i $\frac{1}{7}$ **j** $\frac{1}{81}$ **k** $\frac{1}{64}$ **l** $\frac{1}{9}$
m 1 **n** $\frac{1}{36}$ **o** $\frac{1}{125}$

2 **a** 1 **b** 16 **c** $1\frac{1}{2}$ **d** $1\frac{11}{25}$
e 1 **f** $2\frac{1}{3}$ **g** 1 **h** $1\frac{13}{36}$
i $1\frac{19}{81}$ **j** 1 **k** 16 **l** $-15\frac{5}{8}$
m $-\frac{7}{23}$ **n** 1 **o** $\frac{16}{25}$

3 **a** m^{-3} **b** x^{-1} **c** p^{-7}
d d^{-9} **e** $2x^{-4}$ **f** $3y^{-2}$
g $\frac{1}{2}z^{-6}$ or $\frac{z^{-6}}{2}$ **h** $\frac{3t^{-8}}{5}$ **i** $\frac{2x^{-1}}{7}$
j $\frac{5m^{-6}}{2}$ **k** $(3x+4)^{-2}$ **l** $(a+b)^{-8}$
m $(x-2)^{-1}$ **n** $\frac{(x+1)^{-11}}{4}$ **o** $\frac{5(a+3b)^{-7}}{9}$

4 **a** $\frac{1}{t^5}$ **b** $\frac{1}{x^6}$ **c** $\frac{1}{y^3}$
d $\frac{1}{n^8}$ **e** $\frac{1}{w^{10}}$ **f** $\frac{2}{x}$
g $\frac{3}{m^4}$ **h** $\frac{5}{x^7}$ **i** $\frac{1}{8x^3}$
j $\frac{1}{4n}$ **k** $\frac{1}{(x+1)^6}$ **l** $\frac{1}{8y+z}$
m $\frac{1}{(k-3)^2}$ **n** $\frac{1}{(3x+2y)^9}$ **o** x^5
p y^{10} **q** $\frac{p}{2}$ **r** $(a+b)^2$
s $\frac{x-y}{x+y}$ **t** $\left(\frac{3x+y}{2w-z}\right)^7$

5 **a** $\frac{1}{36}$ **b** $3\frac{1}{8}$ **c** $\frac{17}{72}$
d $\frac{3}{4}$ **e** $-\frac{160}{189}$

6 **a** $\frac{1}{81}$ **b** $-1\frac{1}{2}$ **c** $3\frac{1}{4}$
d $-1\frac{1}{8}$ **e** $-\frac{2}{9}$

Exercise 1.03

1 **a** 9 **b** 3 **c** 4 **d** 2 **e** 7
f 10 **g** 2 **h** 8 **i** 4 **j** 1
k 3 **l** 2 **m** 0 **n** 5 **o** 7
p 2 **q** 4 **r** 25 **s** 32 **t** 4
u 27 **v** $\frac{1}{2}$ **w** $\frac{1}{3}$ **x** $\frac{1}{2}$ **y** $\frac{1}{16}$

2 **a** 2.19 **b** 2.60 **c** 1.53 **d** 0.60
e 0.90 **f** 0.29

3 **a** $\sqrt[3]{y}$ **b** $\sqrt[6]{x}$ **c** $\sqrt{a}$ **d** $\sqrt[9]{t}$
e $\sqrt[3]{y^2}$ or $\left(\sqrt[3]{y}\right)^2$ **f** $\sqrt[4]{x^3}$ **g** $\sqrt[5]{b^2}$ **h** $\sqrt[7]{a^4}$
i $\frac{1}{\sqrt{x}}$ **j** $\frac{1}{\sqrt[3]{d}}$ **k** $\frac{1}{\sqrt[8]{x}}$ **l** $\frac{1}{\sqrt[3]{y}}$
m $\frac{1}{\sqrt[4]{a}}$ **n** $\frac{1}{\sqrt[4]{z^3}}$ **o** $\frac{1}{\sqrt[5]{y^3}}$ **p** $\sqrt{2x+5}$
q $\sqrt[3]{6q+r}$ **r** $\sqrt[9]{a+b}$ **s** $\frac{1}{\sqrt{3x-1}}$
t $\frac{1}{\sqrt[5]{(x+7)^2}}$ or $\frac{1}{\left(\sqrt[5]{x+7}\right)^2}$

4 **a** $t^{\frac{1}{2}}$ **b** $y^{\frac{1}{5}}$ **c** $x^{\frac{3}{2}}$
d $(9-x)^{\frac{1}{3}}$ **e** $(4s+1)^{\frac{1}{2}}$ **f** $(3x+1)^{\frac{5}{2}}$
g $(2t+3)^{-\frac{1}{2}}$ **h** $(5x-y)^{-\frac{3}{2}}$ **i** $(x-2)^{-\frac{2}{3}}$
j $\frac{1}{2}(y+7)^{-\frac{1}{2}}$ **k** $5(x+4)^{-\frac{1}{3}}$ **l** $\frac{1}{3}(y^2-1)^{-\frac{1}{2}}$
m $\frac{3}{5}(x^2+2)^{-\frac{3}{4}}$

5 **a** $x^{\frac{3}{2}}$ **b** $x^{-\frac{1}{2}}$ **c** $x^{\frac{2}{3}}$
d $x^{\frac{5}{3}}$ **e** $x^{\frac{5}{4}}$

6 **a** $\frac{1}{\sqrt[3]{a-2b}}$ **b** $\frac{1}{\sqrt[3]{(y-3)^2}}$ **c** $\frac{4}{\sqrt[7]{(6a+1)^4}}$
d $\frac{1}{3\sqrt[4]{(x+y)^5}}$ **e** $\frac{6}{7\sqrt[9]{(3x+8)^2}}$

Exercise 1.04

1 **a** $6x-6y$ **b** $a-3b$
c $4xy+2y$ **d** $-6ab^2$
e $m^2-6m+12$ **f** p^2-2p-6
g $10b-2ab$ **h** $2bc-ac$
i $2a^5-9x^3+1$ **j** $x^3-2xy^2+3x^2y+2y^3$

2 **a** $6a^2b^3$ **b** $-10a^3b^2$ **c** $21p^3q^4$
d $5a^3b^3$ **e** $-8h^{10}$ **f** k^3p^3
g $81t^{12}$ **h** $-14m^{11}$ **i** $24x^6y^3$

3 **a** $8a$ **b** $4a$ **c** $\frac{y}{2}$
d $3p$ **e** $\frac{ab}{2}$ **f** $\frac{4}{3y}$
g $3a$ **h** $\frac{1}{3ab^2}$ **i** $\frac{-2}{qs}$
j $\frac{z^2}{2x^2}$ **k** $6p^4q$ **l** $\frac{a^4b^7}{4c}$
m $\frac{b^6}{2a}$ **n** $-\frac{x^3z^3}{3y}$ **o** $\frac{a^{13}}{2b^6}$

4 **a** $2x-8$ **b** $6h+9$ **c** $-5a+10$
d $2xy+3x$ **e** x^2-2x **f** $6a^2-16ab$
g $2a^2b+ab^2$ **h** $5n^2-20n$ **i** $3x^2y^2+6x^2y^3$
j $4k+7$ **k** $2t-17$ **l** $4y^2+11y$
m $-5b-6$ **n** $8-2x$ **o** $-3m+1$
p $8h-19$ **q** $d-6$ **r** a^2-2a+4
s $3x^2-9x-5$ **t** $2ab-2a^2b+b$ **u** $4x-1$
v $-7y+4$ **w** $2b$ **x** $5t-6$

Exercise 1.05

1 **a** $a^2+7a+10$ **b** x^2+2x-3
c $2y^2+7y-15$ **d** m^2-6m+8
e $x^2+7x+12$ **f** $y^2-3y-10$
g $2x^2+x-6$ **h** $h^2-10h+21$
i x^2-25 **j** $15a^2-17a+4$
k $8y^2+6y-9$ **l** $xy+7x-4y-28$
m x^3-2x^2+3x-6 **n** n^2-4
o a^2-4b^2 **p** $9x^2-16y^2$
q $x^2-2xy+11x-18y+18$
r $2ab+2b^2-7b-6a+3$
s x^3+8 **t** a^3-27
u $a^2+18a+81$ **v** $k^2-8k+16$
w $9a^2+24ab+16b^2$ **x** $x^2-10xy+25y^2$

2 **a–c** See worked solutions.

3 **a** $x^2-\frac{x}{3}-\frac{2}{9}$ **b** $\frac{y^2}{2}-\frac{5y}{6}-\frac{1}{3}$
c $m^3-2m-\frac{3}{m}$ **d** $x^2-\frac{1}{x^2}$
e $x^3+\frac{1}{x^3}-2$

4 **a** $t^2+8t+16$ **b** $z^2-12z+36$
c x^2-2x+1 **d** $y^2+16y+64$
e q^2+6q+9 **f** $k^2-14k+49$
g n^2+2n+1 **h** $4b^2+20b+25$
i $9-6x+x^2$ **j** $9y^2-6y+1$
k $x^2+2xy+y^2$ **l** $9a^2-6ab+b^2$
m $16d^2+40de+25e^2$ **n** t^2-16
o x^2-9 **p** p^2-1
q r^2-36 **r** x^2-100
s $4a^2-9$ **t** x^2-25y^2
u $16a^2-1$ **v** $49-9x^2$
w x^4-4 **x** x^4+10x^2+25

5 **a** $9a^2b^2-16c^2$ **b** $x^2+4+\frac{4}{x^2}$
c $a^2-\frac{1}{a^2}$ **d** x^2-y^2+4y-4
e $a^2+2ab+b^2+2ac+2bc+c^2$
f $x^2+2x+1-2xy-2y+y^2$
g $12a$ **h** $32-z^2$
i $9x^2+8x-3$ **j** $x^2+3xy+y^2-2x$
k $14n^2-4$ **l** $x^3-12x^2+48x-64$
m x^2 **n** $x^4-2x^2y^2+y^4$
o $8a^3+60a^2+150a+125$

6 See worked solutions.

7 See worked solutions.

8 **a** $11y^2+4y-13$ **b** $-2h^2+38h-143$
c $3x^2-20x+34$ **d** $10p^2+4p+26$
e $6x^2+y^2$

Exercise 1.06

1 **a** $m(m-3)$ **b** $2y(y+2)$
c $-3a(5+a)$ **d** $ab(b+1)$
e $2xy(2x-1)$ **f** $3mn(n^2+3)$
g $-2xz(4x+z)$ **h** $a(6b+3-2a)$
i $x(5x-2+y)$ **j** $q^2(3q^3-2)$
k $5b^2(b+3)$ **l** $-3a^2b^2(2b+a)$
m $(m+5)(r+7)$ **n** $(y-1)(2-y)$
o $(7+y)(4-3x)$ **p** $(a-2)(6x+5)$
q $-(2t+1)(x+y)$ **r** $5m^2n(7mn^3-5)$
s $8ab^2(3ab^3+2)$ **t** $2\pi r(r+h)$
u $(x-3)(x+2)$ **v** $(x+4)(y^2+2)$
w $-(a+1)(2a+1)$

2 **a** $(x+4)(2+b)$ **b** $(y-3)(a+b)$
c $(x+5)(x+2)$ **d** $(m-2)(m+3)$
e $(d-c)(a+b)$ **f** $(x+1)(x^2+3)$
g $(5a-3)(b+2)$ **h** $(2y-x)(x+y)$
i $(y+1)(a+1)$ **j** $(x+5)(x-1)$
k $(y+3)(1+a)$ **l** $(m-2)(1-2y)$
m $(x+5y)(2x-3y)$ **n** $(a+b^2)(ab-4)$
o $(5-x)(x+3)$ **p** $(x+7)(x^3-4)$
q $(x-3)(7-y)$ **r** $(d+3)(4-e)$
s $(y+7)(x-4)$ **t** $(x-4)(x^3-5)$
u $2(2x-3)(x^2+2)$
v $3(a+3)(a+2h)$ **w** $5(y-3)(1+2x)$
x $(r+2)(\pi r-3)$

3 **a** $10(3^x)$ **b** $6(5^n)$ **c** 2^{n-2} **d** 32.5

4 **a** $(3^x+1)(3^{x+1}+7)$ **b** $(5^b+4)(5^{b-1}+3)$
c $(7^{b+1}-2)(7^b+5)$ **d** $(3^{a+1}-2)(3^a-7)$
e $(2^n+1)(2^{n+1}-3)$

Exercise 1.07

1 a $(x+3)(x+1)$ b $(y+4)(y+3)$
c $(m+1)^2$ d $(t+4)^2$
e $(z+3)(z-2)$ f $(x+1)(x-6)$
g $(v-3)(v-5)$ h $(t-3)^2$
i $(x+10)(x-1)$ j $(y-7)(y-3)$
k $(m-6)(m-3)$ l $(y+12)(y-3)$
m $(x-8)(x+3)$ n $(a-2)^2$
o $(x-2)(x+16)$ p $(y+4)(y-9)$
q $(n-6)(n-4)$ r $(x-5)^2$

2 a $3(p+3)(p-4)$ b $2(x-1)(x-4)$
c $5(y-2)(y+5)$ d $11(a-3)(a-2)$
e $x(x+1)(x+5)$ f $x(x-5)(x+2)$

3 a $(2^x+3)(2^x+7)$ b $(7^k-3)(7^k-5)$
c $(3^m+4)(3^m-2)$ d $(5^h-1)(5^h+4)$
e $(3^n-7)(3^n-4)$

4 a $(2a+1)(a+5)$ b $(5y+2)(y+1)$
c $(3x+7)(x+1)$ d $(3x+2)(x+2)$
e $(2b-3)(b-1)$ f $(7x-2)(x-1)$
g $(3y-1)(y+2)$ h $(2x+3)(x+4)$
i $(5p-2)(p+3)$ j $(3x+4)(2x-3)$
k $(6y-1)(y+8)$ l $(4n-3)(n-2)$
m $(4t-1)(2t+5)$ n $(3q+2)(4q+5)$
o $(4r-1)(r+3)$ p $(2x-5)(2x+3)$
q $(6y-1)(y-2)$ r $(4y+3)^2$
s $(5k-2)^2$ t $(6a-1)^2$
u $(7m+6)^2$

5 a $3(2a+5)(a+4)$ b $5(3y-2)(y+3)$
c $2(5m-1)(2m-3)$ d $2(3x-2)(x+2)$
e $3(3x-2)(2x+5)$ f $x(2x+9)(x-4)$
g $y(3y-2)(y-1)$ h $n^2(3n+2)(2n+5)$
i $7b(3b-5)(3b-2)$

6 a $(3\times 2^n+4)(2^n+3)$
b $(5\times 3^x-1)(3^x+4)$
c $(2\times 7^k-3)(7^k-5)$
d $2(3\times 5^m+2)(5^m-2)$
e $(5\times 2^x-3)(3\times 2^x+7)$

Exercise 1.08

1 a $(y-1)^2$ b $(x+3)^2$ c $(m+5)^2$
d $(t-2)^2$ e $(x-6)^2$ f $(2x+3)^2$
g $(4b-1)^2$ h $(3a+2)^2$ i $(5x-4)^2$
j $(7y+1)^2$ k $(3y-5)^2$ l $(4k-3)^2$
m $3(x+1)^2$ n $5(a-3)^2$ o $7(y-5)^2$

2 a $(a+2)(a-2)$ b $(x+3)(x-3)$
c $(y+1)(y-1)$ d $(x+5)(x-5)$
e $(2x+7)(2x-7)$ f $(4y+3)(4y-3)$
g $(1+2z)(1-2z)$ h $(5t+1)(5t-1)$
i $(3t+2)(3t-2)$ j $(3+4x)(3-4x)$
k $(x+2y)(x-2y)$ l $(6x+y)(6x-y)$
m $(2a+3b)(2a-3b)$ n $(x+10y)(x-10y)$
o $(2a+9b)(2a-9b)$ p $(x+y+2)(x-y+2)$
q $(a+b-3)(a-b+1)$ r $(z+w+1)(z-w-1)$
s $\left(x+\frac{1}{2}\right)\left(x-\frac{1}{2}\right)$ t $\left(\frac{y}{3}+1\right)\left(\frac{y}{3}-1\right)$
u $(x+2y+3)(x-2y+1)$
v $(x^2+1)(x-1)(x+1)$
w $(3x^3+2y)(3x^3-2y)$
x $(x^2+4y^2)(x+2y)(x-2y)$

3 a $4a(a+3)(a-3)$ b $2(x+3)(x-3)$
c $3(p-4)(p+3)$ d $5(y+1)(y-1)$
e $5(a-1)^2$ f $3z(z+5)(z+4)$
g $ab(3+2ab)(3-2ab)$ h $x(x+1)(x-1)$
i $2(3x-2)(x+2)$ j $(y+5)(y+4)(y-4)$
k $x(x+1)(x-1)(x+8)$ l $(y^3+2)(y^3-2)$
m $x(x+2)(x-5)$ n $(x+3)(x-3)^2$
o $y(2xy+1)(2xy-1)$ p $6(2+b)(2-b)$
q $3(3x-2)(2x+5)$ r $3(x-1)^2$
s $(x+2)(x+5)(x-5)$ t $z(z+3)^2$
u $3(y+5)^2$ v $a(b+3)(b-3)$
w $4k(k+5)^2$ x $3(x+1)(x-1)(x+3)$
y $2ab(a+2b)(2a-1)$

Exercise 1.09

1 a $a+2$ b $2t-1$ c $\frac{4y+1}{3}$
d $\frac{4}{2d-1}$ e $\frac{x}{5x-2}$ f $\frac{1}{y-4}$
g $\frac{2(b-2a)}{a-3}$ h $\frac{s-1}{s+3}$ i b^2+1
j $\frac{p+5}{3}$ k $\frac{a+1}{a+3}$ l $\frac{3+y}{x+2}$
m $x-3$ n $\frac{p-2}{p}$ o $\frac{a+b}{2a-b}$

2 a $\frac{5x}{4}$ b $\frac{13y+3}{15}$ c $\frac{a+8}{12}$
d $\frac{4p+3}{6}$ e $\frac{x-13}{6}$

3 a 6 b $\frac{5b(a-2)}{3}$ c $\frac{2(t+5)}{5y}$
d $\frac{5(a-3)}{4}$ e $\frac{5-y}{35}$ f $\frac{b}{2a-1}$
g $\frac{b^2(x+2y)}{10(2b-1)}$ h $\frac{xy}{ab}$ i $\frac{(x-3)(x-1)}{(x-5)(x-2)}$
j $\frac{5(p-2)}{3(q+1)}$

4 a $\frac{5}{x}$ b $\frac{2-x}{x(x-1)}$
c $\frac{a+b+3}{a+b}$ d $\frac{2x}{x+2}$
e $\frac{(p+q)(p-q)+1}{p+q}$ f $\frac{2(x-1)}{(x+1)(x-3)}$
g $\frac{8-3x}{(x+2)(x-2)}$ h $\frac{a+2}{(a+1)^2}$

5 a $\frac{(y+2)(y+1)}{15y}$ b $\frac{(x-2)(x+12)}{2(x-3)(x-4)}$
c $\frac{3b^2-5b-10}{2b(b+1)}$ d x

6 a $\frac{3-5x}{(x+2)(x-2)}$ b $\frac{(3p-q)(p+2q)}{pq(p+q)(p-q)}$
c $\frac{a^2-2ab-b^2+1}{(a+b)(a-b)}$

Exercise 1.10

1 a $2\sqrt{3}$ b $3\sqrt{7}$ c $2\sqrt{6}$ d $5\sqrt{2}$ e $6\sqrt{2}$
f $10\sqrt{2}$ g $4\sqrt{3}$ h $5\sqrt{3}$ i $4\sqrt{2}$ j $3\sqrt{6}$
k $4\sqrt{7}$ l $10\sqrt{3}$ m $8\sqrt{2}$ n $9\sqrt{3}$ o $7\sqrt{5}$
p $6\sqrt{3}$ q $3\sqrt{11}$ r $5\sqrt{5}$

2 a $6\sqrt{3}$ b $20\sqrt{5}$ c $28\sqrt{2}$ d $4\sqrt{7}$ e $16\sqrt{5}$
f $8\sqrt{14}$ g $72\sqrt{5}$ h $30\sqrt{2}$ i $14\sqrt{10}$ j $24\sqrt{5}$

3 a $\sqrt{18}$ b $\sqrt{20}$ c $\sqrt{176}$ d $\sqrt{128}$ e $\sqrt{75}$
f $\sqrt{160}$ g $\sqrt{117}$ h $\sqrt{98}$ i $\sqrt{363}$ j $\sqrt{1008}$

4 **a** $8\sqrt{5}$ **b** $-4\sqrt{2}$ **c** $4\sqrt{5}$
d $\sqrt{2}$ **e** $5\sqrt{3}$ **f** $-\sqrt{3}$
g $\sqrt{2}$ **h** $5\sqrt{7}$ **i** $\sqrt{2}$
j $13\sqrt{6}$ **k** $-9\sqrt{10}$ **l** $47\sqrt{3}$
m $5\sqrt{2}-2\sqrt{3}$ **n** $\sqrt{7}-5\sqrt{2}$ **o** $-2\sqrt{3}-4\sqrt{5}$

5 **a** $\sqrt{21}$ **b** $\sqrt{15}$ **c** $3\sqrt{6}$
d $10\sqrt{14}$ **e** $-6\sqrt{6}$ **f** 30
g $-12\sqrt{55}$ **h** 14 **i** 60
j $2\sqrt{3}$ **k** 2 **l** 28
m $\sqrt{30}$ **n** $-2\sqrt{105}$ **o** 18

6 **a** $2\sqrt{6}$ **b** $4\sqrt{3}$ **c** 1
d $\frac{8}{\sqrt{6}}$ **e** $2\sqrt{3}$ **f** $\frac{1}{3\sqrt{10}}$
g $\frac{1}{2\sqrt{5}}$ **h** $\frac{1}{3\sqrt{5}}$ **i** $\frac{1}{2}$
j $\frac{\sqrt{3}}{2\sqrt{2}}$ **k** $\frac{2}{3}$ **l** $\frac{5}{7}$

7 **a** $\sqrt{10}+\sqrt{6}$ **b** $2\sqrt{6}-\sqrt{15}$
c $12+8\sqrt{15}$ **d** $5\sqrt{14}-2\sqrt{21}$
e $-\sqrt{6}+12\sqrt{2}$ **f** $5\sqrt{33}+3\sqrt{21}$
g $-6-12\sqrt{6}$ **h** $5-5\sqrt{15}$
i $6+\sqrt{30}$ **j** $6\sqrt{6}+6$
k $-8+24\sqrt{3}$ **l** $210-14\sqrt{15}$
m $10\sqrt{6}-120$ **n** $-\sqrt{10}-2\sqrt{2}$
o $4\sqrt{3}-12$

8 **a** $\sqrt{10}+3\sqrt{6}+3\sqrt{5}+9\sqrt{3}$
b $\sqrt{10}-\sqrt{35}-2+\sqrt{14}$
c $2\sqrt{10}-6+10\sqrt{15}-15\sqrt{6}$
d $24\sqrt{5}+36\sqrt{15}-8\sqrt{10}-12\sqrt{30}$
e $52-13\sqrt{10}$
f $15-\sqrt{15}+18\sqrt{10}-6\sqrt{6}$ **g** 4
h -1 **i** -12 **j** $11-4\sqrt{6}$
k $25+6\sqrt{14}$ **l** $57+12\sqrt{15}$ **m** $27-4\sqrt{35}$
n $77-24\sqrt{10}$ **o** $53+12\sqrt{10}$

9 **a** 18 **b** $108\sqrt{2}$ **c** $432\sqrt{2}$
d $19+6\sqrt{2}$ **e** 9

10 **a** $a-1$ **b** $2p-1-2\sqrt{p(p-1)}$

11 $2x-3y-5\sqrt{xy}$

12 $a=17, b=240$

Exercise 1.11

1 **a** $\frac{\sqrt{7}}{7}$ **b** $\frac{\sqrt{6}}{4}$
c $\frac{2\sqrt{15}}{5}$ **d** $\frac{3\sqrt{14}}{5}$
e $\frac{\sqrt{3}+\sqrt{6}}{3}$ **f** $\frac{2\sqrt{3}-5\sqrt{2}}{2}$
g $\frac{5+2\sqrt{10}}{5}$ **h** $\frac{3\sqrt{14}-4\sqrt{7}}{14}$
i $\frac{8\sqrt{5}+3\sqrt{10}}{20}$ **j** $\frac{4\sqrt{15}-2\sqrt{10}}{35}$

2 **a** $4\sqrt{3}-4\sqrt{2}$
b $\frac{-\left(\sqrt{6}+7\sqrt{3}\right)}{47}$
c $\frac{12\sqrt{2}-2\sqrt{15}}{19}$
d $\frac{8\sqrt{3}-19}{13}$
e $\sqrt{6}+2+5\sqrt{3}+5\sqrt{2}$
f $\frac{6\sqrt{15}-9\sqrt{6}+2\sqrt{10}-6}{2}$

3 **a** $2\sqrt{2}$ **b** $-2-\sqrt{6}+3\sqrt{2}-3\sqrt{3}$
c -4 **d** $4\sqrt{2}$
e $\frac{6+9\sqrt{2}+2\sqrt{3}}{6}$ **f** $\frac{4\sqrt{6}+9\sqrt{3}}{21}$
g $\frac{15\sqrt{30}-30\sqrt{5}-4\sqrt{3}}{30}$
h $\frac{28-2\sqrt{6}-7\sqrt{3}}{13}$
i $\frac{2\sqrt{15}+2\sqrt{10}-2\sqrt{6}-\sqrt{3}-5}{2}$

4 **a** $a=45, b=10$ **b** $a=1, b=8$
c $a=-\frac{1}{2}, b=\frac{1}{2}$ **d** $a=-1\frac{5}{9}, b=-\frac{8}{9}$
e $a=5, b=32$

5 3, so rational.

6 **a** 4 **b** 14 **c** 16

Sample HSC problem

$a=107$ and $b=-42$

Test yourself 1

1 B, C **2** D **3** D **4** C
5 C **6** B **7** A **8** D

9 **a** $\frac{1}{49}$ **b** $\frac{1}{5}$ **c** $\frac{1}{3}$

10 **a** x^9 **b** $25y^6$ **c** $a^{11}b^6$
d $\frac{8x^{18}}{27}$ **e** 1

11 **a** 6 **b** $\frac{1}{64}$ **c** 4
d $\frac{1}{7}$ **e** 2 **f** 1

12 **a** a^5 **b** $x^{30}y^{18}$ **c** p^9
d $16b^{36}$ **e** $8x^{11}y$

13 **a** $n^{\frac{1}{2}}$ **b** x^{-5} **c** $(x+y)^{-1}$
d $(x+1)^{\frac{1}{4}}$ **e** $(a+b)^{\frac{1}{7}}$ **f** $2x^{-1}$
g $\frac{1}{2}x^{-3}$ **h** $x^{\frac{4}{3}}$ **i** $(5x+3)^{\frac{9}{7}}$
j $m^{\frac{1}{4}}$

14 **a** $\frac{1}{a^5}$ **b** $\sqrt[4]{n}$ **c** $\sqrt{x+1}$
d $\frac{1}{x-y}$ **e** $\frac{1}{(4t-7)^4}$ **f** $\sqrt[5]{a+b}$
g $\frac{1}{\sqrt[3]{x}}$ **h** $\sqrt[4]{b^3}$ **i** $\sqrt[3]{(2x+3)^4}$
j $\frac{1}{\sqrt{x^3}}$

15 **a** $\frac{3\sqrt{7}}{7}$ **b** $\frac{\sqrt{6}}{15}$ **c** $\frac{\sqrt{5}+1}{2}$
d $\frac{12-2\sqrt{6}}{15}$ **e** $\frac{20+3\sqrt{15}+4\sqrt{10}+3\sqrt{6}}{53}$

16 **a** $\frac{x+10}{10}$ **b** $\frac{17a-15}{21}$
c $\frac{3-2x}{(x+1)(x-1)}$ **d** $\frac{1}{k-1}$
e $\frac{\sqrt{15}-\sqrt{6}-15\sqrt{3}-15\sqrt{2}}{3}$

17 1

18 $\frac{1}{192}$

19 a $x^{\frac{1}{2}}$ **b** y^{-1} **c** $(x+3)^{\frac{1}{6}}$
d $(2x-3)^{-11}$ **e** $y^{\frac{7}{3}}$

20 a $\frac{1}{x^3}$ **b** $\frac{1}{2a+5}$ **c** $\left(\frac{b}{a}\right)^5$

21 a $-2y$ **b** $a+4$ **c** $-6k^5$
d $\frac{5x+3y}{15}$ **e** $3a-8b$ **f** $6\sqrt{2}$
g $4\sqrt{5}$

22 a $(x+6)(x-6)$ **b** $(a+3)(a-1)$
c $4ab(b-2)$ **d** $(y-3)(5+x)$
e $2(2n-p+3)$

23 a $4b-6$ **b** $2x^2+5x-3$
c $4m+17$ **d** $16x^2-24x+9$
e p^2-25 **f** $-1-7a$
g $2\sqrt{6}-5\sqrt{3}$ **h** $3\sqrt{3}-6+\sqrt{21}-2\sqrt{7}$
i $11+4\sqrt{7}$

24 a $\frac{8}{b^2(a+3)}$ **b** $\frac{15}{(m-2)^2}$

25 $\frac{21\sqrt{5}-46-\sqrt{2}}{7}$

26 a 17 **b** $\frac{6\sqrt{15}-9}{17}$

27 $\frac{4x+5}{(x+3)(x-2)}$

28 a $6\sqrt{2}$ **b** $-8\sqrt{6}$ **c** $2\sqrt{3}$
d $\frac{4}{\sqrt{3}}$ **e** $30a^2b$ **f** $\frac{m}{3n^4}$
g $2x-3y$

29 a $\frac{1}{\sqrt{5}}$ **b** 8

30 $\frac{3\sqrt{3}+1}{2}$

31 a $2\sqrt{6}+4$
b $10\sqrt{14}-5\sqrt{21}-6\sqrt{10}+3\sqrt{15}$
c 7 **d** 43 **e** $65-6\sqrt{14}$

32 a $n=48$ **b** $n=175$ **c** $n=392$
d $n=5547$ **e** $n=1445$

33 a $3(x-3)(x+3)$ **b** $6(x-3)(x+1)$
c $5(y-3)^2$

34 a 99 **b** $24\sqrt{3}$

35 a a^2-b^2 **b** $a^2+2ab+b^2$

Challenge exercise 1

1 4
2 See worked solutions.
3 $-2^4\times 3^5$
4 a $2a^2b-8ab^2+6a^3$ **b** y^4-4
c $8x^3-60x^2+150x-125$

5 $\frac{11\sqrt{3}+2\sqrt{5}+14}{11}$ **6** $\frac{1}{2\sqrt{2}}$ or $\frac{\sqrt{2}}{4}$

7 a $(x+4)(x+9)$ **b** $(x^2-3y)(x^2+2y)$
c $(a-2)(a+2)(b-2)$

8 $\frac{y+1}{2(x-1)}$ **9** $\frac{(a+1)^2}{a-1}$ **10** $\left(\frac{2}{x}+\frac{a}{b}\right)\left(\frac{2}{x}-\frac{a}{b}\right)$

11 a $8x^3-12x^2+6x-1$ **b** $\frac{3x+4}{(2x-1)^2}$

12 a 3×2^x **b** $\frac{2}{9}$
c $\frac{x^3+x^2+3x+1}{x(x+1)(x-1)}$ **d** $22\times 3^{n-1}$
e $\frac{x-2}{(x+2)(x-1)}$

13 a $x+x^2+2x^{\frac{3}{2}}$ **b** $a^{\frac{2}{3}}-b^{\frac{2}{3}}$
c $p^2+p^{-1}+2p^{\frac{1}{2}}$ **d** $x+x^{-1}+2$

Chapter 2

Exercise 2.01

1 a $a=-7$ **b** $y=3$ **c** $x=3$
d $a=-1\frac{2}{3}$ **e** $t=-4$ **f** $x=1.2$
g $a=1.6$ **h** $b=\frac{1}{8}$ **i** $t=39$
j $p=5$ **k** $x\approx 4.41$ **l** $b=3\frac{1}{3}$
m $x=1\frac{9}{35}$ **n** $x=36$ **o** $x=-3$

2 a $y=-1.2$ **b** $x=69$ **c** $w=13$
d $t=30$ **e** $x=14$ **f** $x=-1$
g $x=-0.4$ **h** $p=3$ **i** $t=8.2$
j $x=-9.5$ **k** $q=22$ **l** $x=-3$

3 $t=8.5$ **4** $b=8$ **5** $a=41$
6 $y=4$ **7** $r=6.68$ **8** $x=6.44$
9 $y_1=3\frac{2}{3}$ **10** $h\approx 3.7$

11 a BMI $=25.39$ **b** $w=69.66$ **c** $h\approx 1.94$

12 a $x=\frac{y+1}{2}$ **b** $y=\frac{4-x}{5}$ **c** $h=\frac{S}{lb}$
d i $r^3=\frac{3V}{4\pi}$ **ii** $r=\sqrt[3]{\frac{3V}{4\pi}}$
e i $y_2=m(x_2-x_1)+y_1$
ii $x_2=\frac{y_2-y_1}{m}+x_1$
iii $x_1=\frac{mx_2-y_2+y_1}{m}$
f i $a=\frac{S}{n}-\frac{(n-1)d}{2}=\frac{2S-n(n-1)d}{2n}$
ii $d=\frac{2(S-an)}{n(n-1)}$

13 $x_1=-9$ **14** $x=5.5$ **15** $r\approx 3.3$

16 a $x=3.3$ **b** $n=3\frac{2}{3}$ **c** $y=2.7$

Exercise 2.02

1 a $x>3$

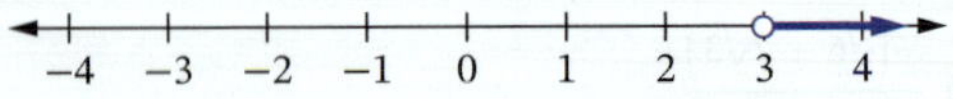

b $y\le 4$

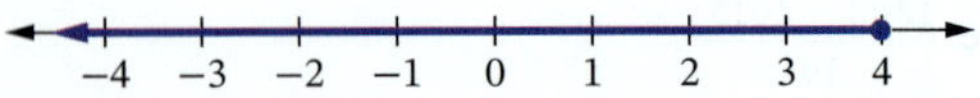

2 a $t>7$ **b** $x\ge 3$ **c** $p>-1$
d $x\ge -2$ **e** $y>-9$ **f** $a\ge -1$
g $y\ge -2\frac{1}{2}$ **h** $x<-2$ **i** $a\le -6$
j $y<12$ **k** $b<-18$ **l** $x>30$

3 **a** $x \le 3\frac{3}{4}$ **b** $m > 14\frac{2}{3}$ **c** $b \ge 16\frac{1}{4}$
d $r \le -9$ **e** $z > 8$ **f** $w < 2\frac{4}{5}$
g $x \ge 35$ **h** $t \ge -9$ **i** $q > -6\frac{2}{5}$
j $x > -1\frac{2}{3}$ **k** $b \le -11\frac{1}{4}$

4 **a** $1 < x < 7$

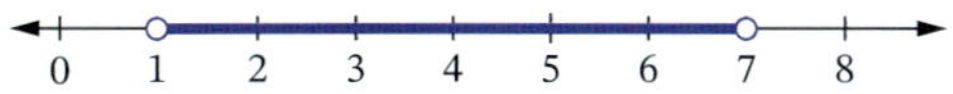

b $-2 \le p < 5$

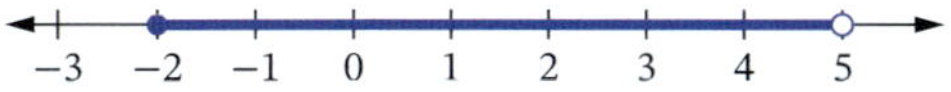

c $1 < x < 4$

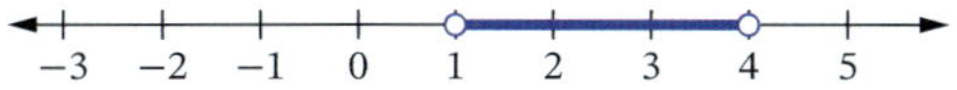

d $-3 \le y \le 5$

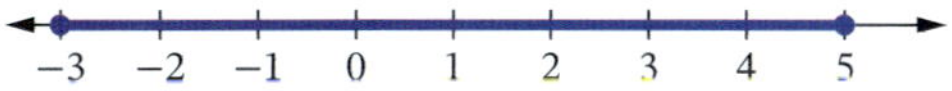

e $\frac{1}{6} < y < 1\frac{2}{3}$

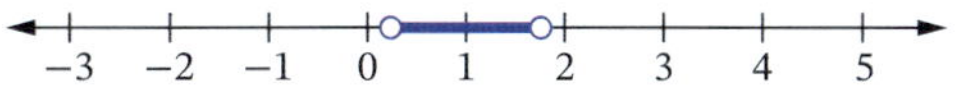

5 **a** $x > 50$ **b** $x \le 13$ **c** $x < 60$
d $x \ge 9$ **e** $x > 3\frac{1}{2}$ **f** $x < -20$
g $x \le \frac{5}{6}$

Exercise 2.03

1 **a** 7 **b** 5 **c** 6 **d** 0 **e** 2
f 11 **g** 6 **h** 24 **i** 25 **j** 125
2 **a** 5 **b** -1 **c** 2 **d** 14 **e** 4
f -67 **g** 7 **h** 12 **i** -6 **j** 10
3 **a** 3 **b** 3 **c** 1 **d** 3 **e** 1
4 **a** a **b** $-a$ **c** 0
d $3a$ **e** $-3a$ **f** 0
g $a + 1$ **h** $-a - 1$ **i** $x - 2$
5 **a** $6 \le 6$ **b** $3 \le 3$ **c** $1 \le 5$
d $1 \le 9$ **e** $10 \le 10$
6 See worked solutions.
7 ± 3 **8** ± 1 **9** $\pm 1, x \ne 2$

Exercise 2.04

1 **a** $x = \pm 5$ **b** $y = \pm 8$ **c** $x = 0$
2 **a** $x = 5, -9$ **b** $n = 4, -2$ **c** $x = 3, -6$
d $x = 5, -4\frac{5}{7}$ **e** $x = \pm 12$
3 **a** $x = 2, -0.75$ **b** $n = 1\frac{1}{3}, 2$ **c** $t = 2.4, -4$
d $y = -6, 15$ **e** $x = -3, 1\frac{2}{3}$
4 **a** $h = -2, 1\frac{1}{3}$ **b** $k = -17, 13$
c $t = 1\frac{3}{8}, 3\frac{5}{8}$ **d** $y = 0, -4$

Exercise 2.05

1 **a** $y = 0, -1$ **b** $b = 2, -1$ **c** $p = 3, -5$
d $t = 0, 5$ **e** $x = -2, -7$ **f** $q = \pm 3$
g $x = \pm 1$ **h** $a = 0, -3$ **i** $x = 0, -4$
j $x = \pm\frac{1}{2}$ **k** $x = -1, -1\frac{1}{3}$ **l** $y = 1, -1\frac{1}{2}$
m $b = \frac{3}{4}, \frac{1}{2}$ **n** $x = 5, -2$ **o** $x = 0, \frac{2}{3}$
p $x = 1, 2\frac{1}{2}$ **q** $x = 0, 5$ **r** $y = -1, 2$
s $n = 3, 5$ **t** $x = 3, 4$ **u** $m = -6, 1$
2 $-8, 1$ **3** $5, -3$
4 **a** $x = \pm 2$ **b** $x = -3, 1$ **c** $x = 2, 3$
d $x = 4$ **e** $x = -5, \frac{1}{3}$
5 **a** $n = 3, 8$ **b** $n = 2, 22$
6 9 cm, 12 cm

Exercise 2.06

1 **a** $x = -1 \pm \sqrt{7}$ **b** $y = -5 \pm \sqrt{5}$
c $a = 3 \pm \sqrt{6}$ **d** $x = 2 \pm \sqrt{13}$
e $y = \frac{-3 \pm \sqrt{2}}{2}$
2 **a** $h \approx 1.9, -5.9$ **b** $a \approx 3.8, -1.8$
c $x \approx 8.1, -0.1$ **d** $y \approx -2.4, -11.6$
e $x \approx 1.5, -0.8$
3 **a** $(y + 1)^2 - 8$ **b** $(a - 8)^2 - 61$
c $(n + 6)^2 - 37$ **d** $(a - 5)^2 - 20$
e $(g + 10)^2 - 119$ **f** $(x - 1)^2 - 12$
4 **a** $x = -2 \pm \sqrt{5}$ **b** $a = 3 \pm \sqrt{7}$
c $y = 4 \pm \sqrt{23}$ **d** $x = -1 \pm \sqrt{13}$
e $p = -7 \pm \sqrt{44} = -7 \pm 2\sqrt{11}$
f $x = 5 \pm \sqrt{28} = 5 \pm 2\sqrt{7}$
5 **a** $x \approx 3.45, -1.45$ **b** $x \approx -4.59, -7.41$
c $q \approx 0.0554, -18.1$ **d** $x \approx 4.45, -0.449$
e $b \approx -4.26, -11.7$ **f** $x \approx 17.7, 6.34$
6 **a** $x = 3 \pm \sqrt{10}$ **b** $y = \frac{-6 \pm \sqrt{33}}{3}$
c $a = \frac{8 \pm \sqrt{70}}{2}$ **d** $y = \frac{-20 \pm 2\sqrt{105}}{5}$
e $x = \frac{18 \pm \sqrt{330}}{3}$ **f** $x = \frac{-3 \pm \sqrt{17}}{2}$

Exercise 2.07

1 **a** $x = \frac{-1 \pm \sqrt{17}}{2}$ **b** $x = \frac{5 \pm \sqrt{13}}{6}$
c $q = 2 \pm \sqrt{7}$ **d** $h = \frac{-3 \pm 2\sqrt{2}}{2}$
e $a = \frac{4 \pm \sqrt{10}}{3}$ **f** $k = \frac{-11 \pm \sqrt{133}}{2}$
g $d = \frac{-5 \pm \sqrt{73}}{12}$ **h** $x = 1 \pm 2\sqrt{2}$
i $t = \frac{1 \pm \sqrt{5}}{2}$
2 **a** $y \approx -0.354, -5.65$ **b** $x = 1, 1.5$
c $b \approx 3.54, -2.54$ **d** $x = 1, -0.5$
e $x \approx -0.553, 0.678$ **f** $n \approx 0.243, -8.24$
g $m = -2, -5$ **h** $x = 0, 7$
i $x = 1, -6$
3 $1 \pm \sqrt{5}$ **4** 8.41, 0.59
5 **a** $x \approx 4.8, 0.2$ **b** $y \approx 1.1, -7.1$
c $n \approx 3.2, -2.2$ **d** $x \approx 0.2, -1.5$
e $b \approx 1.7, 0.3$

Exercise 2.08

1 $-3 < x < 0$ **2** $0 < y < 4$
3 $n \le 0, n \ge 1$ **4** $x \le -2, x \ge 2$
5 $n < -1, n > 1$ **6** $-5 \le n \le 3$
7 $c < -1, c > 2$ **8** $-4 \le x \le -2$
9 $4 < x < 5$ **10** $b \le -2, b \ge -\frac{1}{2}$
11 $a < -1, a > \frac{1}{3}$ **12** $y < -1\frac{1}{2}, y > 2$

ANSWERS

13 $x \le \frac{2}{3}, x \ge 1$

14 $b < -3, b > \frac{2}{5}$

15 $-1\frac{1}{2} \le x \le -\frac{1}{3}$

16 $-4 \le y \le 3$

17 $x < -4, x > 4$

18 $-1 \le a \le 1$

19 $-2 < x < 3$

20 $x \le -1, x \ge 3$

21 $0 < x < 2$

22 $1 \le a \le 1\frac{1}{2}$

23 $y \le -2, y \ge \frac{4}{5}$

24 $m < -1\frac{2}{3}, m > 1\frac{1}{2}$

Investigation

23 adults and 16 children

Exercise 2.09

1 **a** $a = 1, b = 3$ **b** $x = 2, y = 1$
c $p = 2, q = -1$ **d** $x = 6, y = 17$
e $x = -10, y = 2$ **f** $t = 3, v = 1$
g $x = -3, y = 2$ **h** $x = -64, y = -39$
i $x = 3, y = -4$ **j** $m = 2, n = 3$
k $a = -2, b = 0$ **l** $k = -4, h = 1$

2 4, 9

3 7 parents, 10 children

4 3 shirts, 6 dresses

5 Paola 10, Natalia 5

6 7, 14

7 $a = 2, b = -1, c = 4$

Exercise 2.10

1 **a** $x = 0, y = 0$ and $x = 1, y = 1$
b $x = 0, y = 0$ and $x = -2, y = 4$
c $x = 0, y = 3$ and $x = 3, y = 0$
d $x = 4, y = -3$ and $x = 3, y = -4$
e $x = -1, y = -3$
f $x = 3, y = 9$
g $x = 1, y = 2$ and $x = -1, y = -2$
h $x = 0, y = 0$ and $x = 1, y = 1$
i $x = 2, y = 1$ and $x = -1, y = -2$
j $x = 0, y = 1$
k $x = 1, y = 5$ and $x = 4, y = 11$
l $x = \frac{1}{4}, y = 4$ and $x = -1, y = -1$
m $t = -\frac{1}{2}, h = \frac{1}{4}$
n $x = \frac{1}{2}, y = 2\frac{3}{4}$

2 2 cm, 6 cm

Sample HSC problem

$x = 3, y = 9$ or $x = -1, y = 1$

Test yourself 2

1 C **2** A, D **3** B **4** C

5 **a** 15 **b** −2 **c** 12 **d** 11

6 **a** $a = -116$ **b** $x = -7$ **c** $p \le 4$

7 **a** $x = -2, y = 5$
b $x = 4, y = 1$ and $x = -\frac{1}{2}, y = -8$

8 $b = 2, -1\frac{1}{3}$

9 $A = 36$ **b** $b = 12$

10 $x = \frac{1}{2}, 1$

11 $-1 < y \le 3$

12 **a** $x \approx -0.298, -6.70$ **b** $y \approx 4.16, -2.16$
c $n \approx 0.869, -1.54$

13 **a** $A \approx 764.5$ **b** $r \approx 2.9$

14 $x > 71\frac{1}{4}$

15 $x < 2, x > 9$

16 $x = 2.4, y = 3.2$

17 **a** $y > 3$ **b** $-3 \le n \le 0$
c $x = 3, -1\frac{2}{5}$ **d** $-4 \le x \le 2$
e $y < -2, y > 2$ **f** $x \le -1, x \ge 1$
g $-1 < x < 3$ **h** $m \le -3, m \ge 2$

18 **a** B **b** A **c** A **d** C **e** B

Challenge exercise 2

1 $y = 1$

2 $x < -a, x > a$

3 The number is any value greater than 9.

4 $x \approx 2.56, -1.56$

5 **a** $(x + 3)(x - 3)(x^3 - 8)$ **b** $x = \pm 3, 2$

6 $x = 1, y = 2$ and $x = -1, y = 0$

7 $b = 16, x = 4 \pm \sqrt{17}$

8 $x = 1$

9 $-3 \le x \le 8$

10 $x = \frac{1}{4}$

11 $x = \pm\sqrt{b + a^2} + a$

12 $x = \frac{2\left(4 \pm \sqrt{10}\right)}{3}$

13 $x = 2, -4$

14 $x \approx 0.6, -1.3$

15 5 large, 6 small

16 **a** $n = 2$
b $n = \frac{3 \pm \sqrt{37}}{2}$
c $n = -0.12, -2.88$

Practice set 1

1 B **2** A **3** C **4** B

5 D **6** D

7 **a** $x = 10$ **b** $b = 6$ **c** $x = 3$
d $y = 16$ **e** $z = 6$ **f** $x = \pm 4\sqrt{2}$
g $x = 0, 3$ **h** $x = 3, -7$ **i** $a = 2, -1.2$

8 $p = 9$

9 $4\sqrt{3}$

10 $2(5 + y)(x - y)$

11 **a** x^{-1} **b** $x^{\frac{4}{3}}$

12 $6y - 10$

13 $\frac{25 + 5\sqrt{2}}{23}$

14 $x = 1.78, -0.281$

15 $\frac{2}{x - 3}$

16 $-\sqrt{3}$

17 $x^3 + 2x^2 - 16x + 3$

18 $3\sqrt{10} - 4$

19 $a > -3$

20 $a = 3, b = 2$

21 $x > 1$

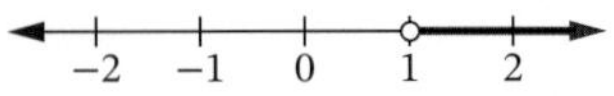

22 $x = \frac{4 \pm \sqrt{12}}{2} = 2 \pm \sqrt{3}$

23 $\frac{1}{49}$

24 $x = 4, y = 11$ or $x = -1, y = -4$

25 $x = 2, y = -1$

26 7

27 $8(x + 2)(x - 2)$

28 $\frac{6\sqrt{15} + 2\sqrt{6}}{43}$

29 7

30 $-2\sqrt{10} + 3\sqrt{5} - 2\sqrt{2} + 3$

31 $a^{-21}b^{10} = \dfrac{b^{10}}{a^{21}}$ **32** $\dfrac{1}{8}$

33 $-x - 7$ **34** $(x + 3)^{-1}$

35 $\dfrac{1}{\sqrt{3x+2}}$

36 **a** $12x - 8y$ **b** $2\sqrt{31}$

c $\dfrac{x-3}{2x-1}$ **d** $3\sqrt{2} + 1$

e $\dfrac{-(x+5)}{(x+1)(x-1)}$ **f** $x^{-14}y^{7}z^{-11}$ or $\dfrac{y^7}{x^{14}z^{11}}$

g $\dfrac{3}{5a(a+b)(1+2b)}$ **h** $8\sqrt{5}$

37 $r = \dfrac{2}{\sqrt[3]{\pi}}$ cm **38** $k = 20$

39 $9xy\sqrt{y}$

40 **a** $5(a-2)(a+6)$ **b** $(3a+4b)(a-6b+2c)$

41 $-1\frac{1}{8} \le x < 5\frac{3}{4}$ **42** $\dfrac{-2x-7}{15}$

43 $x = 0, 5$ **44** $x = 5.19, -0.19$

45 $9\sqrt{2}$ **46** $\dfrac{-2x-17}{x(x+5)}$

47 **a** $(x-4)(x+2)$ **b** $(a+3)(a-3)$

c $(y+3)^2$ **d** $(t+4)^2$

e $(3x-2)(x-3)$

48 **a** $-1 < a < 1$ **b** $y \le -3, y \ge 0$

c $-1 \le y \le 2$ **d** $x < -3, x > 3$

e $0 \le d \le 2$

49 **a** $y = 4.74, -2.74$ **b** $b = 2, -5$

c $-3 \le m \le 3$

Chapter 3

Exercise 3.01

1 **a** (Wade, black), (Scott, blond), (Geoff, grey), (Deng, black), (Mila, brown), (Stevie, blond); function

b (1, 1), (1, 4), (2, 3), (3, 1), (4, 4); not a function

c (1, A), (2, D), (3, A), (4, B), (5, C); function

d (3, 5), (5, −2), (5, 2), (8, −7), (9, 3), (5, 6), (8, 0); not a function

e (1, 9), (2, 15), (3, 27), (4, 33), (5, 45); function

2 **a** yes **b** no **c** no **d** yes

e yes **f** yes **g** no **h** yes

i yes **j** no **k** yes **l** no

m yes **n** no **o** yes

3 **a** $\{-3, -1, 0, 1, 6\}$ **b** $\{-2, 4, 5, 8\}$ **c** yes

4 **a** no

b $\{-2, -1, 0, 2, 3\}$

c $\{-6, -2, 0, 1, 5, 7\}$

Exercise 3.02

1 $f(1) = 4, f(-3) = 0$

2 $h(0) = -2, h(2) = 2, h(-4) = 14$

3 $f(5) = -25, f(-1) = -1, f(3) = -9, f(-2) = -4$

4 14 **5** −35 **6** $x = 9$

7 $x = \pm 5$ **8** $x = -3$ **9** $z = 1, -4$

10 $f(p) = 2p - 9, f(x+h) = 2x + 2h - 9$

11 $g(x-1) = x^2 + 2$ **12** $f(k) = (k-1)(k+1)$

13 **a** $t = 1$ **b** $t = 4, -2$

14 0

15 $f(5) = 125, f(1) = 1, f(-1) = -1$

16 7 **17** −28

18 $f(x+h) - f(x) = 2xh + h^2 - 5h$

19 $4x + 2h + 1$

20 $5(x - c)$ **21** $3k^2 + 5$

22 **a** 2 **b** 0 **c** $n^4 + n^2 + 2$

23 **a** 3

b Denominator cannot be 0 so $x - 3 \ne 0, \therefore x \ne 3$.

c 4

Exercise 3.03

1 **a** x-intercept $\frac{2}{3}$, y-intercept −2

b x-intercept −10, y-intercept 4

c x-intercept 12, y-intercept 4

d x-intercepts 0, −3, y-intercept 0

e x-intercepts ±2, y-intercept −4

f x-intercepts −2, −3, y-intercept 6

g x-intercepts 3, 5, y-intercept 15

h x-intercept $-\sqrt[3]{5}$, y-intercept 5

i x-intercept −3, no y-intercept

j x-intercepts ±3, y-intercept 9

2 **a** $x = 2$

b x-intercept 2, y-intercept −6

3 **a** domain: all real x, range: $[1, \infty)$

b domain: all real x, range: all real y

c domain: $[0, \infty)$, range: $[0, \infty)$

d domain: $[-5, \infty)$, range: $[0, \infty)$

e domain: $[3, \infty)$, range: $(-\infty, 0]$

4 **a** **i** $(0, \infty)$ **ii** $(-\infty, 0)$ **iii** even

b **i** $(-\infty, 2)$ **ii** $(2, \infty)$ **iii** neither

c **i** $(-2, 2)$ **ii** $(-\infty, -2) \cup (2, \infty)$

iii neither

d **i** all real $x, x \ne 0$.

ii none **iii** odd

e **i** none **ii** all real x **iii** neither

5 $f(-x) = (-x)^2 - 2 = x^2 - 2 = f(x)$ so even.

6 **a** $f(x^2) = x^6 + 1$

b $[f(x)]^2 = x^6 + 2x^3 + 1$

c $f(-x) = -x^3 + 1$

d neither odd nor even

e $x = -1$

f x-intercept −1, y-intercept 1

7 $g(-x) = (-x)^8 + 3(-x)^4 - 2(-x)^2 = x^8 + 3x^4 - 2x^2$
$= g(x)$ so even.

8 $f(-x) = -x = -f(x)$ so odd.

9 $f(-x) = (-x)^2 - 1 = x^2 - 1 = f(x)$ so even.

10 $f(-x) = 4(-x) - (-x)^3 = -4x + x^3 = -(4x - x^3)$
$= -f(x)$ so odd.

11 **a** $f(-x) = (-x)^4 + (-x)^2 = x^4 + x^2 = f(x)$ so even

b 0

12 **a** odd **b** neither

c even **d** neither

e neither

13 **a** even values, i.e. $n = 2, 4, 6, \ldots$

b odd values, i.e. $n = 1, 3, 5, \ldots$

14 **a** no value of n

b yes, when n is odd (1, 3, 5, ...)

15 **a** 1 **b** 49 **c** $x = 2$

d x-intercept 2, y-intercept 4

e domain: all real x, range: $y \ge 0$

f $(-x-2)^2$ **g** neither

16

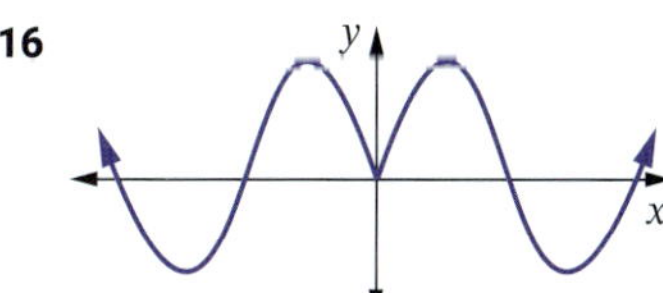

17

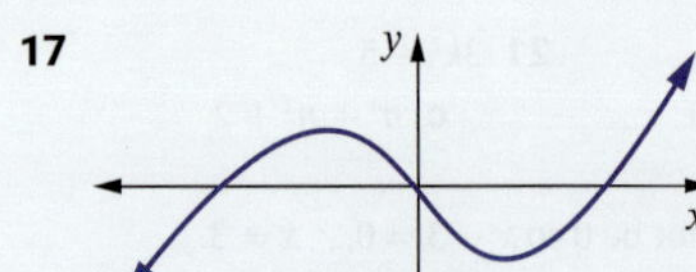

18 a

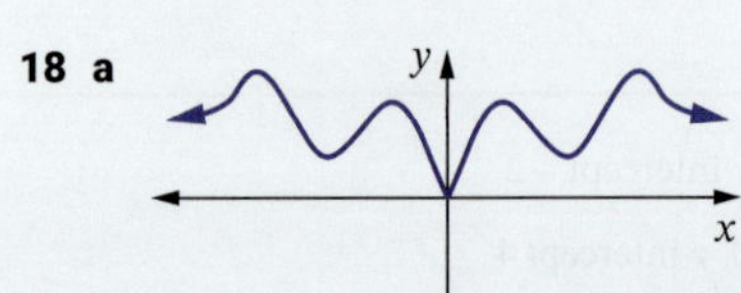

b

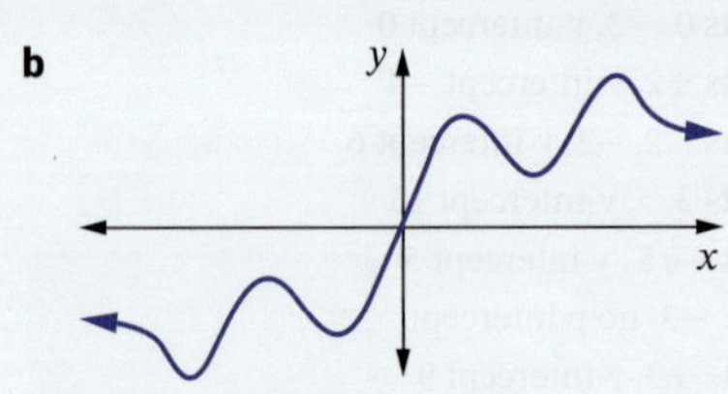

19

	Domain	Range
a	all real x	$y > 0$
b	all real x, $x \neq 2$	all real y, $y \neq 0$
c	$x > -3$	$y > 0$
d	$x < -3, x > 3$	$y > 0$

20 a domain: $[-2, 2]$, range: $[0, 2]$

b domain: $[-2, 2]$, range: $[-2, 0]$

c $y = \pm\sqrt{4 - x^2}$

d domain: $[-2, 2]$, range: $[-2, 2]$

e no

21 domain: all real x, range: $y \geq 0$

22 a domain: all real x, $x \neq -3$, range: $y = 1$

b domain: all real x, $x \neq -3$, range: $y \geq 0$

23 a

	Domain	Range
i	all real x	$y \geq 0$
ii	all real x	$y \geq -2$
iii	all real x	all real y
iv	all real x, $x \neq -\frac{1}{2}$	

24

	i $f(x)$ even	ii $f(x)$ odd
a	even	even
b	even	even
c	even	even
d	even	even
e	odd	even

Exercise 3.04

1 a $N = 12x$ **b** $A = 2n$ **c** $c = 2.5x$
d $y = 4x$ **e** $w = 400x$

2 $c = 850x$

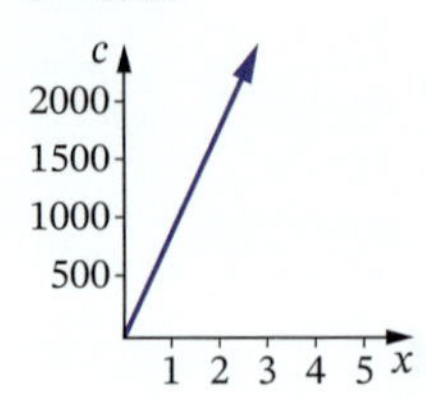

3 a x-intercept 2, y-intercept -2

b x-intercept -3, y-intercept 9

c x-intercept 2, y-intercept 4

d x-intercept $-1\frac{1}{2}$, y-intercept 3

e x-intercept $\frac{4}{5}$, y-intercept -4

f x-intercept $-\frac{1}{2}$, y-intercept 5

g x-intercept 2, y-intercept 2

h x-intercept -2, y-intercept 4

i x-intercept -3, y-intercept 3

j x-intercept $\frac{2}{3}$, y-intercept $-\frac{1}{3}$

4 a

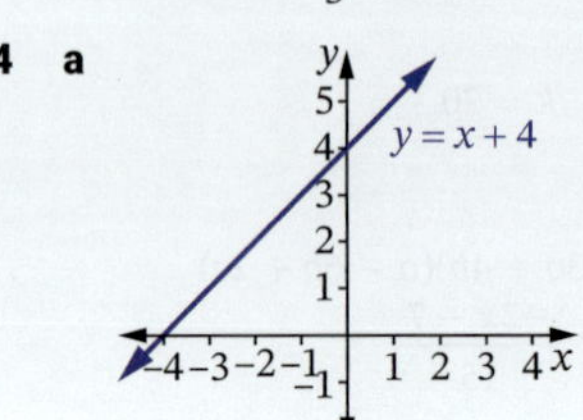

b

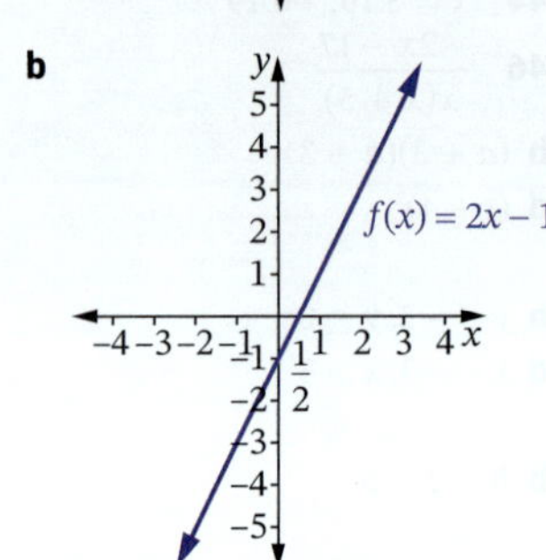

c

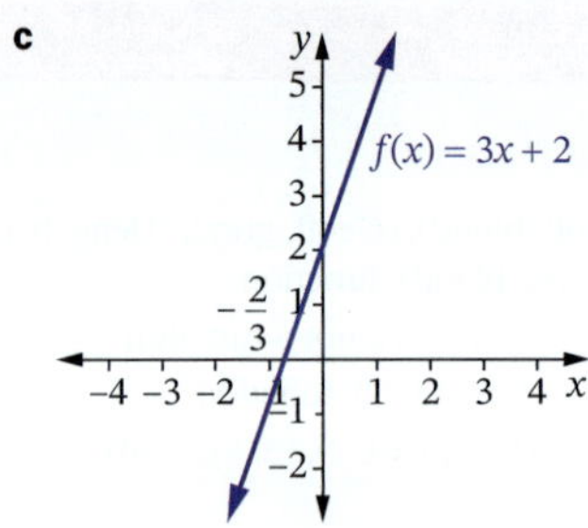

d

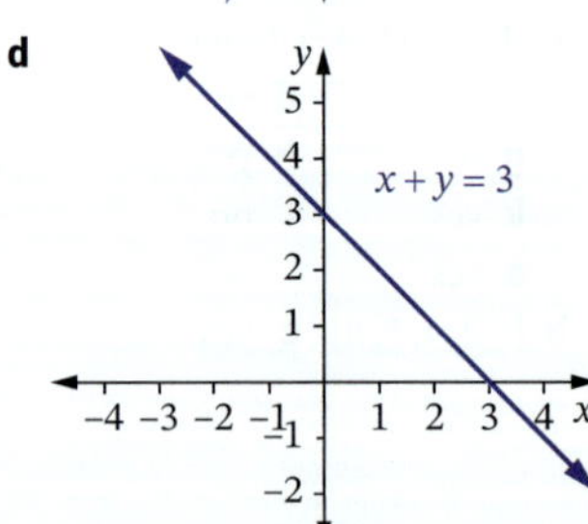

e

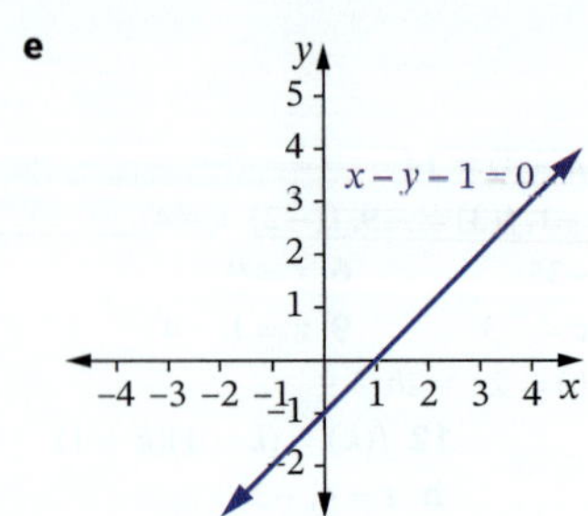

5 **a** domain: all real x, range: all real y
b domain: all real x, range: $[2, 2]$
c domain: $[-4, -4]$, range: all real y
d domain: $[2, 2]$, range: all real y
e domain: all real x, range: $[3, 3]$

6 **a**

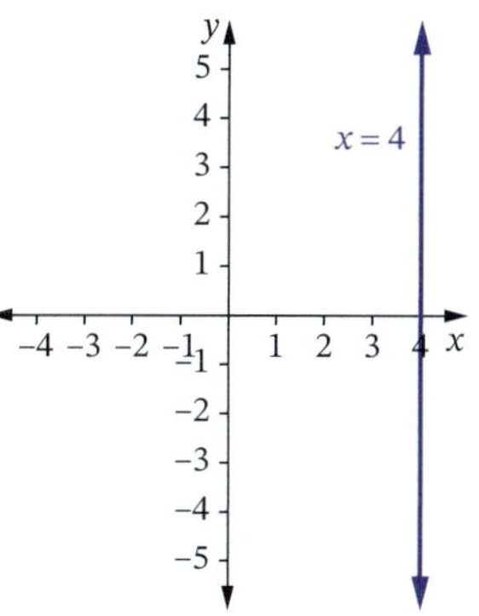

domain: $[4, 4]$, range: all real y

b

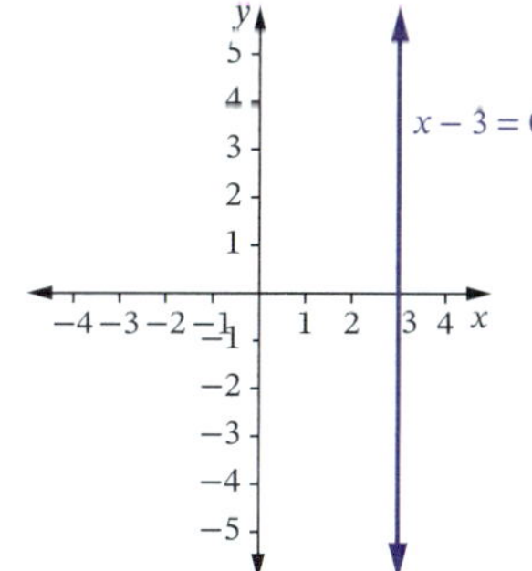

domain: $[3, 3]$, range: all real y

c

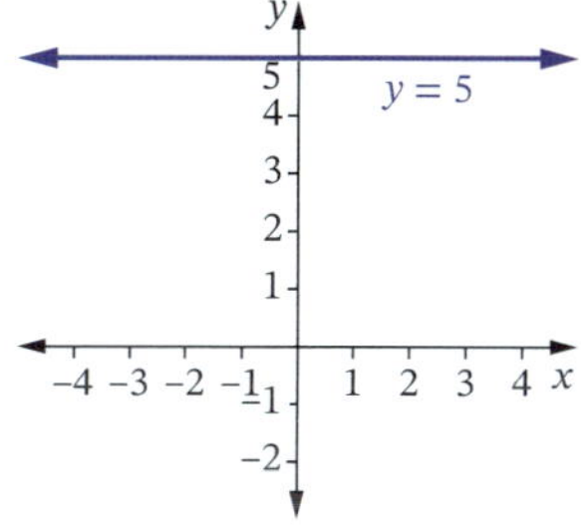

domain: all real x, range: $[5, 5]$

d

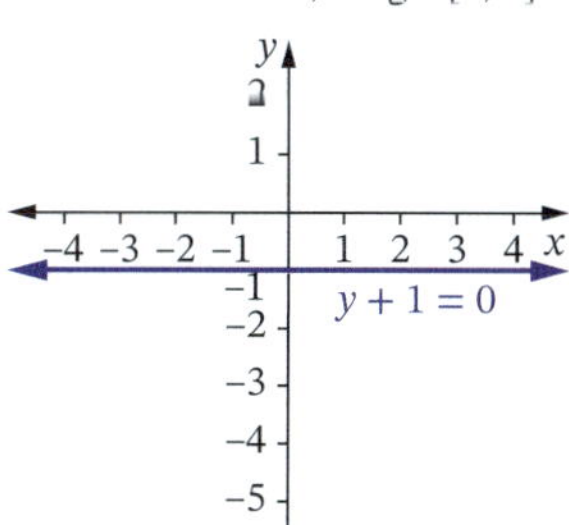

domain: all real x, range: $[-1, -1]$

7 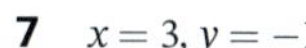
$x = 3, y = -1$

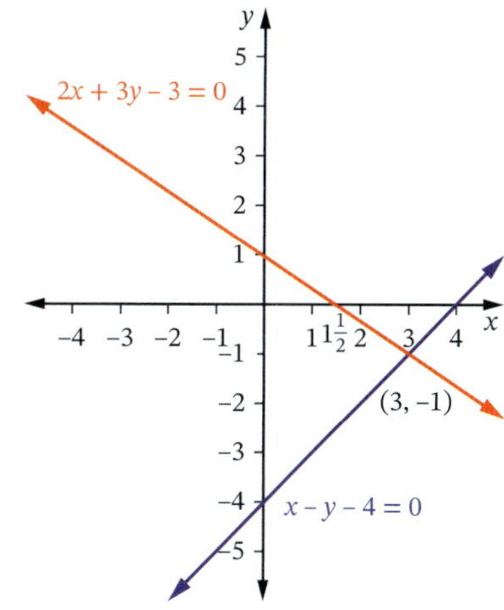

8 **a** $x = -3, y = -5$
b $x < -3$

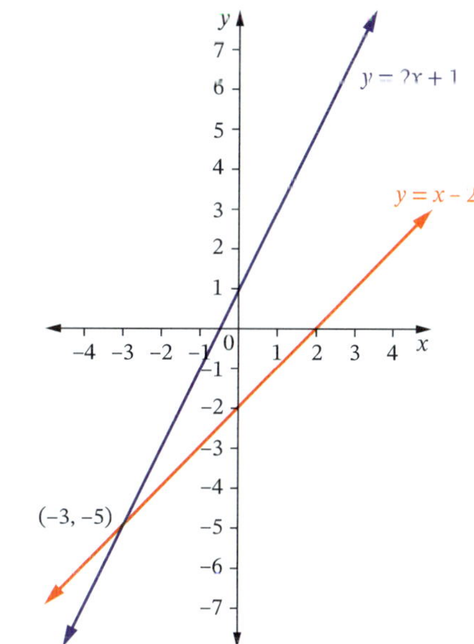

9 **a** $x = -1, y = 5$ **b** $a = 0, b = 4$
c $p = -4, q = 1$ **d** $x = 1, y = -1$
e $x = -1, y = -4$

Exercise 3.05

1 **a** 2 **b** $1\frac{1}{3}$ **c** $-1\frac{1}{3}$
d $-2\frac{2}{5}$ **e** $\frac{2}{3}$ **f** $-\frac{1}{8}$
g $-4\frac{1}{2}$ **h** $-\frac{2}{3}$ **i** $2\frac{1}{4}$

2 **a** 0.5 **b** 7.1 **c** 2.5
d −5.7 **e** −1.2 **f** −0.3

3 **a i** 3 **ii** 5 **b i** 2 **ii** 1
c i 6 **ii** −7 **d i** −1 **ii** 0
e i −4 **ii** 3 **f i** 1 **ii** −2
g i −2 **ii** 6 **h i** −1 **ii** 1
i i 9 **ii** 0

4 **a** $\frac{1}{3}$ **b** $-1\frac{1}{3}$ **c** −2
d $\frac{3}{5}$ **e** $\frac{1}{3}$

5 **a** 63°26′ **b** 59°32′ **c** 80°32′
d 101°19′ **e** 139°38′ **f** 129°48′

6 a i -2 ii 3 b i -5 ii -6

c i 6 ii -1 d i 1 ii 4

e i -2 ii $\frac{1}{2}$ f i 3 ii $1\frac{1}{2}$

7 a -2 b 0 c -1 d -3

e 2 f $-\frac{1}{4}$ g $\frac{1}{5}$ h $\frac{2}{7}$

i $-\frac{3}{5}$ j $-\frac{1}{14}$ k 15 l $-1\frac{1}{2}$

8 $y_1 = 21$ 9 $x = 1.8$ 10 $x = 9$

11 a 350

b The number of phones produced by the factory per day.

12 $m_{AB} = m_{CD} = 1.5$, $m_{AD} = m_{BC} = 0$. Opposite sides are parallel so $ABCD$ is a parallelogram.

Exercise 3.06

1 a $y = 4x - 1$ b $y = -3x + 4$ c $y = 5x$

d $y = 4x + 20$ e $3x + y - 3 = 0$

f $4x - 3y - 12 = 0$

2 a $4x - 3y + 7 = 0$ b $3x - 4y + 4 = 0$

c $4x - 5y + 13 = 0$ d $3x + 4y - 25 = 0$

e $x - 2y + 2 = 0$

3 a -4 b $y = -4x + 8$

4 a $y = 3$ b $x = -1$

5 a 0 b $y = -2x$

6 C

7

	x-intercept	y-intercept
a	3	2
b	4	-5
c	a	b

8 $\frac{x}{2} + \frac{y}{7} = 1$ or $7x + 2y - 14 = 0$

9 $x + y - 2 = 0$

10 See worked solutions.

Exercise 3.07

1 a -3 b $\frac{1}{3}$ c $\frac{3}{4}$ d $1\frac{1}{2}$

e 1 f $-\frac{5}{6}$ g $\frac{1}{3}$ h $\frac{1}{5}$

2 a $x - y + 1 = 0$ b $x - 3y + 16 = 0$

c $x + y - 5 = 0$ d $x + 2y + 5 = 0$

e $x - 2y + 4 = 0$ f $5x - y - 8 = 0$

g $2x + y + 2 = 0$ h $2x - 3y + 16 = 0$

3 $m_1 = m_2 = 3$, so parallel.

4 $m_1 m_2 = -\frac{1}{5} \times 5 = -1$, so perpendicular.

5 $m_1 = m_2 = 1\frac{1}{5}$, so parallel.

6 $m_1 m_2 = -\frac{7}{3} \times \frac{3}{7} = -1$, so perpendicular.

7 $k = -\frac{2}{3}$

8 $m_1 = m_2 = 4$, so parallel.

9 $AB \parallel CD$ ($m_1 = m_2 = 3$) and $BC \parallel AD$ $\left(m_1 = m_2 = -\frac{5}{8}\right)$. So $ABCD$ is a parallelogram.

10 gradient $AC = m_1 = \frac{1}{2}$, gradient $BD = m_2 = -2$.

$m_1 m_2 = \frac{1}{2} \times -2 = -1$ so diagonals are perpendicular.

11 $7x + 6y - 24 = 0$

12 trapezium

Exercise 3.08

1 a $y = x^2 + 2x$ x-intercepts: 0, -2, y-intercept: 0

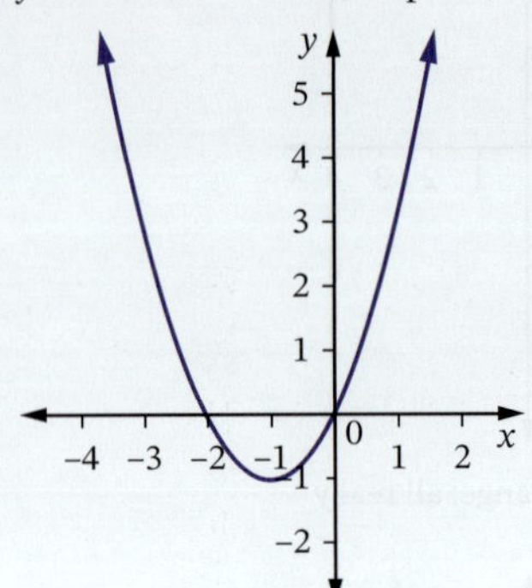

b $y = -x^2 + 3x$ x-intercepts: 0, 3, y-intercept: 0

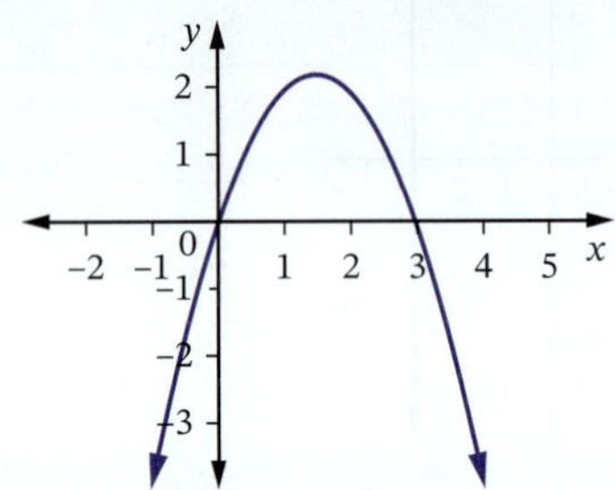

c $f(x) = x^2 - 1$ x-intercepts: ±1, y-intercept: -1

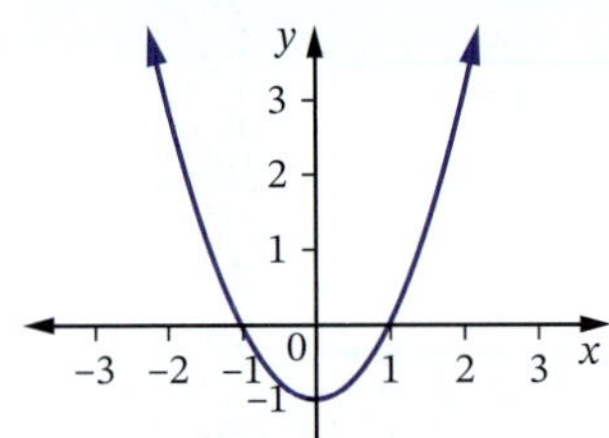

d $y = x^2 - x - 2$ x-intercepts: -1, 2, y-intercept: -2

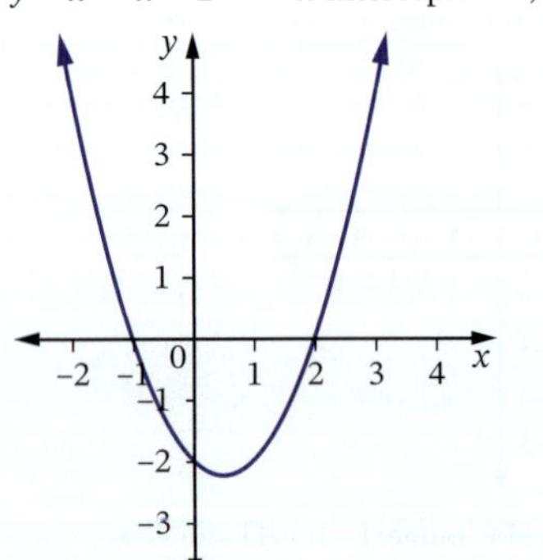

e $y = x^2 - 9x + 8$ x-intercepts: 1, 8, y-intercept: 8

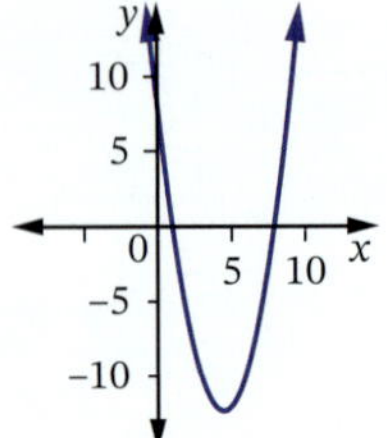

2 $y = x^2 + 5$ is a shift of 5 units up from $y = x^2$. $y = x^2 - 2$ is a shift of 2 units down from $y = x^2$.

3 $y = 4x^2$ is stretched vertically from $y = x^2$. $y = \frac{x^2}{4}$ is shrunk vertically (or stretched horizontally) from $y = x^2$.

4 $y = (x + 3)^2$ is a shift of 3 units to the left from $y = x^2$. $y = (x - 4)^2$ is a shift of 4 units to the right from $y = x^2$.

5 **a**

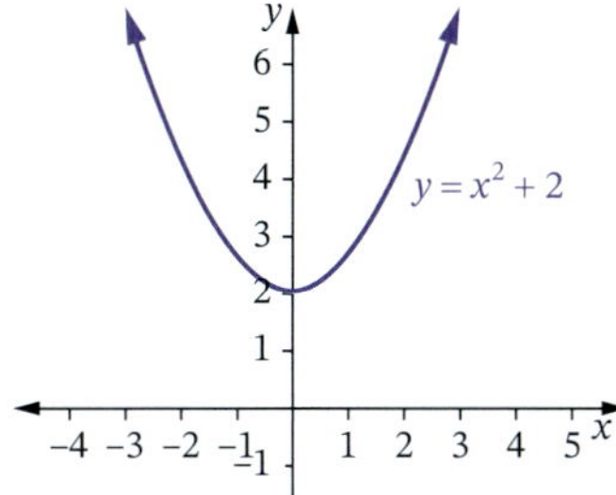

minimum: 2

b

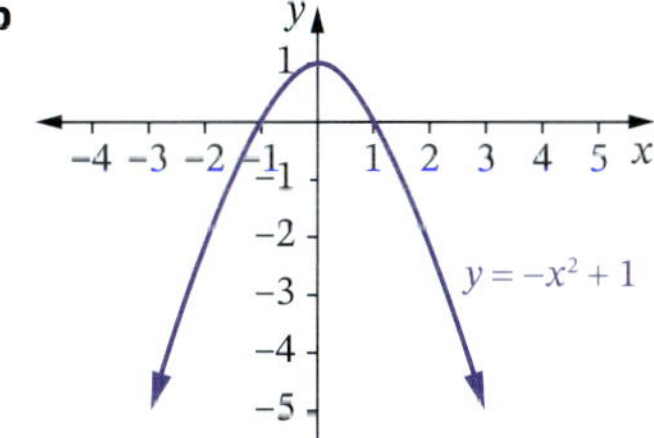

maximum: 1

c

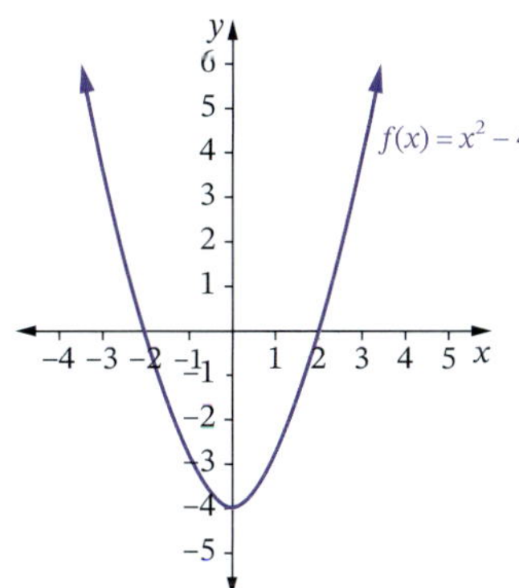

minimum: −4

d

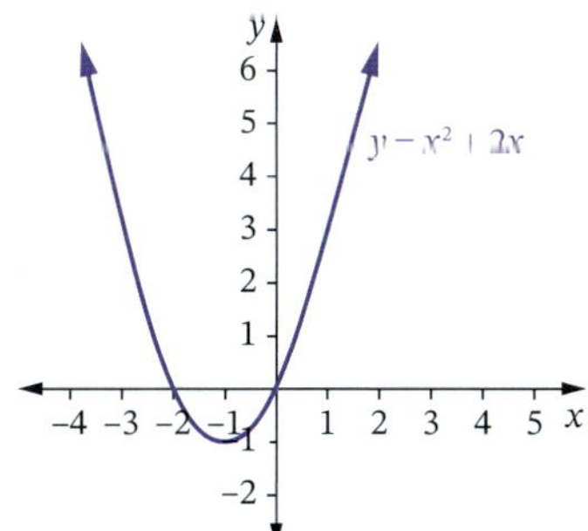

minimum: −1

e

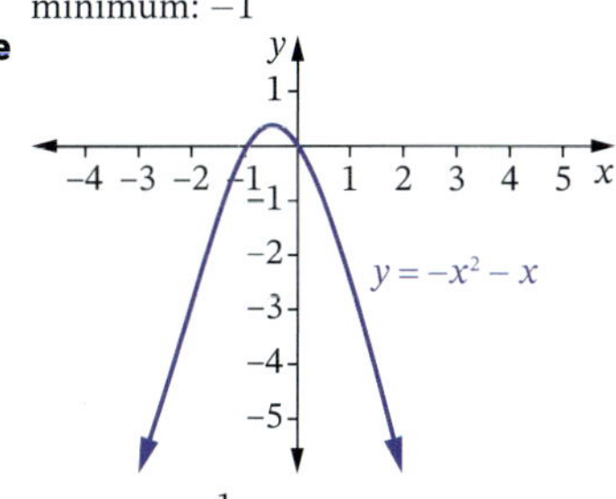

maximum: $\frac{1}{4}$

f

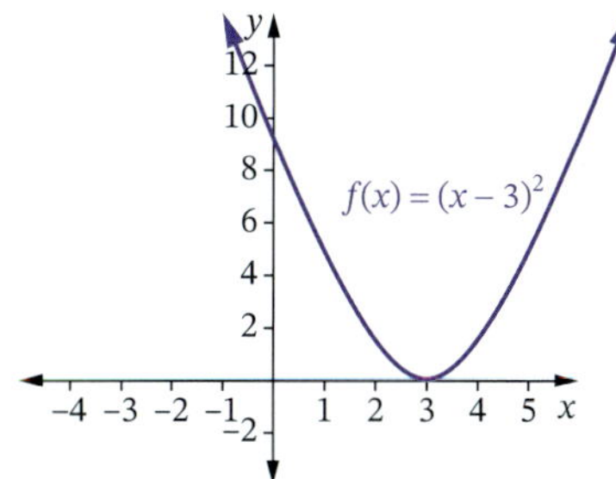

minimum: 0

g

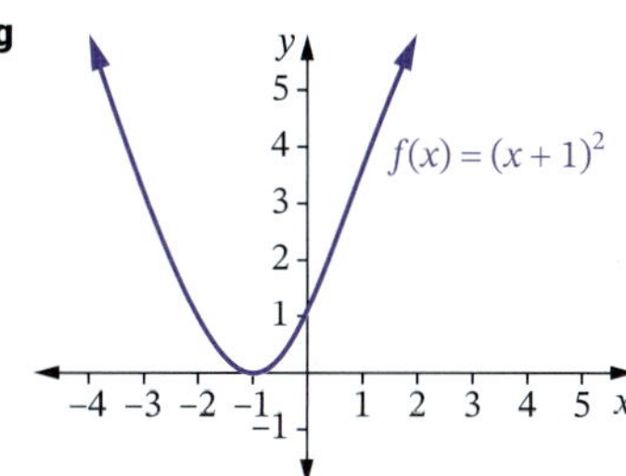

minimum: 0

h

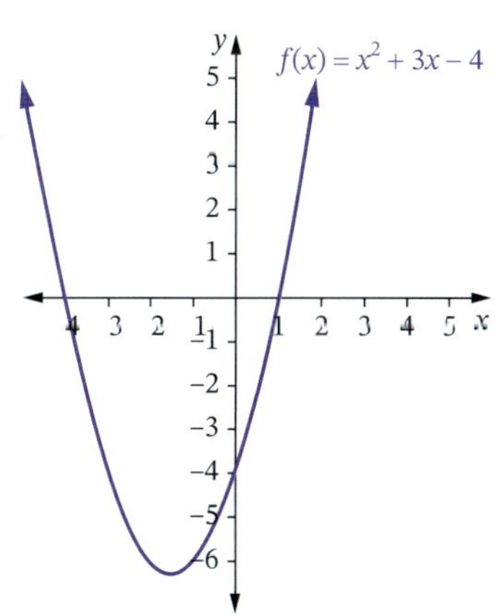

minimum: −6.25

i

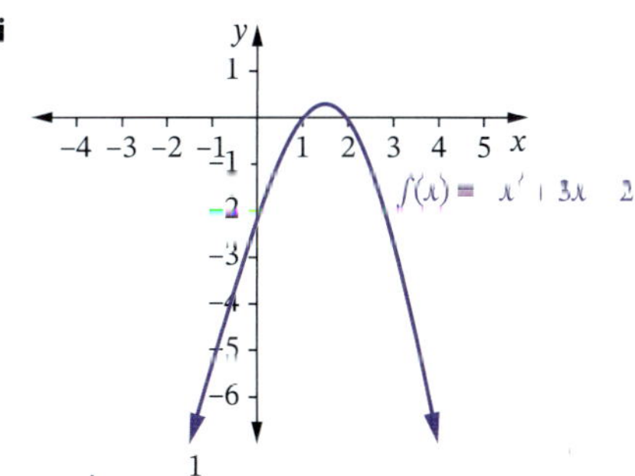

maximum: $\frac{1}{4}$

6 **a** **i** x-intercepts 3, 4, y-intercept 12

ii domain: all real x, range: $y \geq -\frac{1}{4}$

b **i** x-intercepts: 0, −4, y-intercept: 0

ii domain: all real x, range: $y \geq -4$

c **i** x-intercepts: −2, 4, y-intercept: −8

ii domain: all real x, range: $y \geq -9$

d **i** x-intercept: 3, y-intercept: 9

ii domain: all real x, range: $y \geq 0$

e **i** x-intercepts: ±2, y-intercept: 4

ii domain: all real x, range: $y \leq 4$

7 **a** domain: all real x, range: $y \geq -5$

b domain: all real x, range: $y \geq -9$

c domain: all real x, range: $y \geq -2\frac{1}{4}$

d domain: all real x, range: $y \leq 0$

e domain: all real x, range: $y \geq 0$

8 a

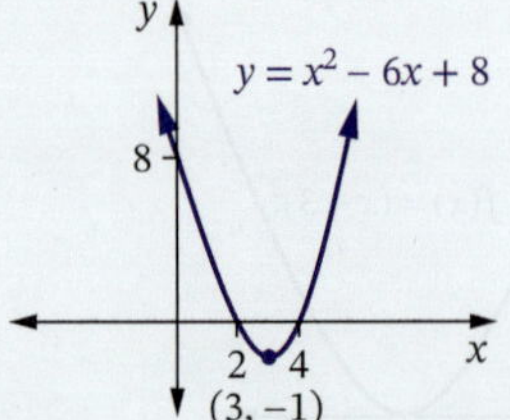

b

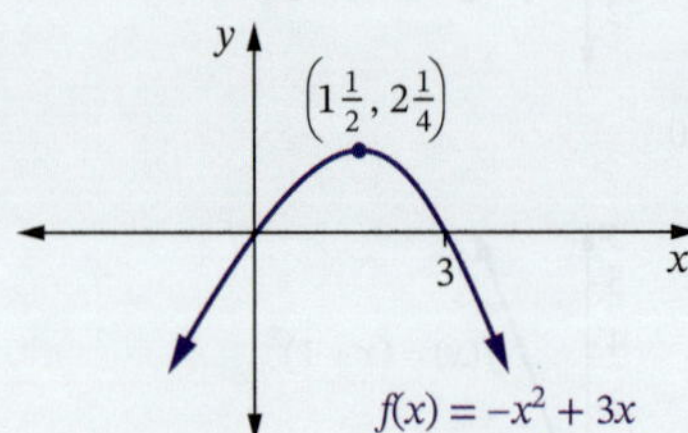

c

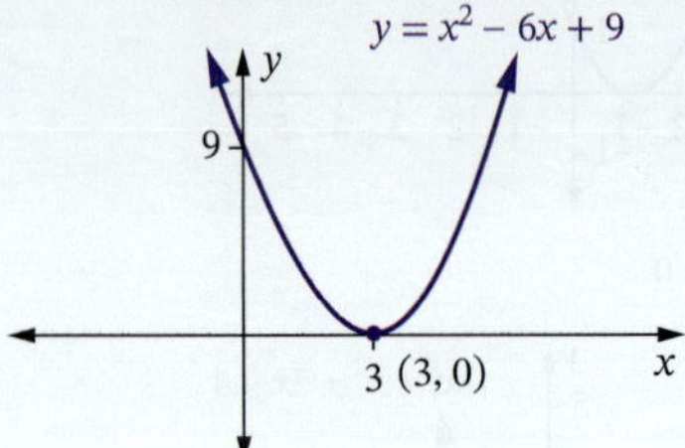

d

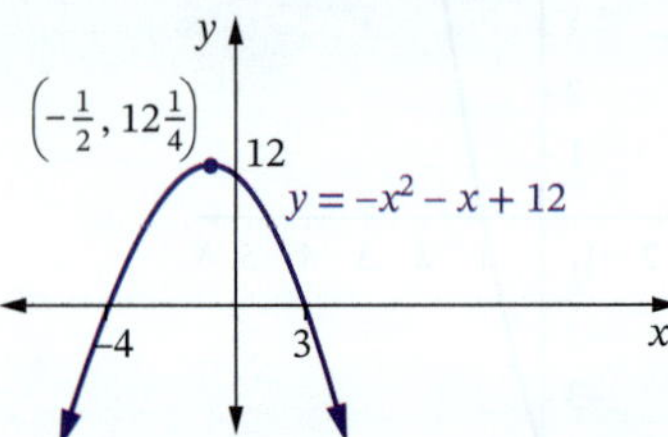

e

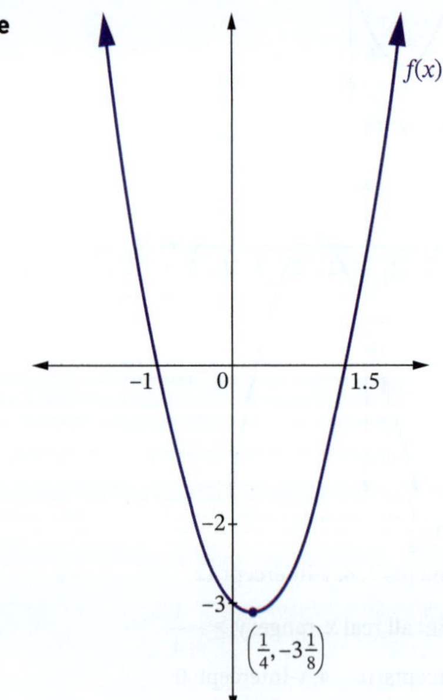

9 **a** $x > 0$ **b** $x < 0$

c $x > 0$ **d** $x < 2$

e $x > -5$

10 $f(-x) = -(-x^2)$

$= -x^2$

$= f(x)$

$\therefore$ even

11 **a** $d = 6\text{ m}$ **b** $w = 2.8\text{ m}$

12 **a** even **b** even **c** even

d neither **e** neither **f** even

g neither **h** neither **i** neither

13 **a** $x = -1, y = -2$ and $x = 2, y = 1$

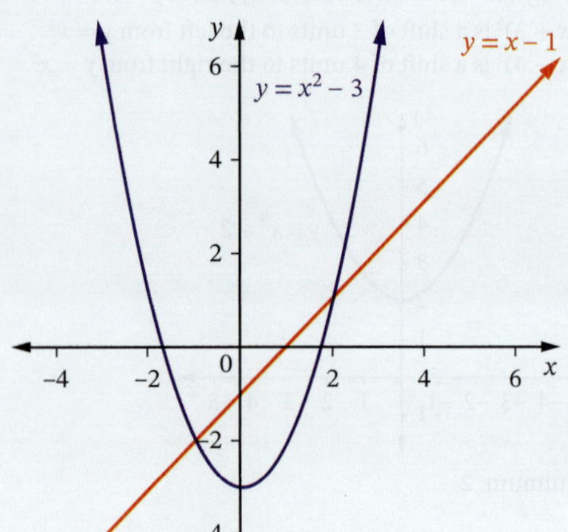

b $x = 0, y = 1$

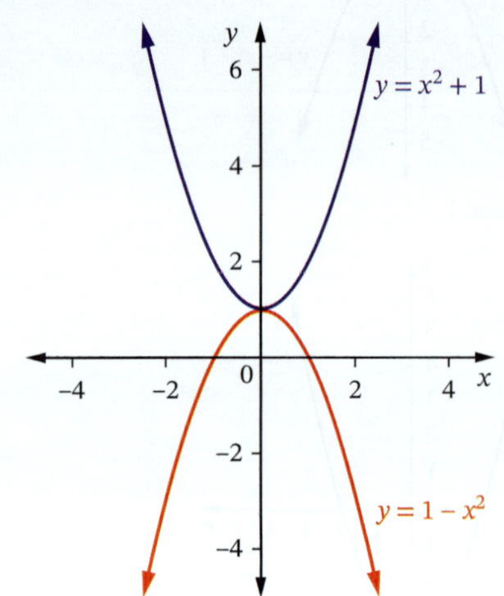

c $x = 1, y = 5$, and $x = 4, y = 11$

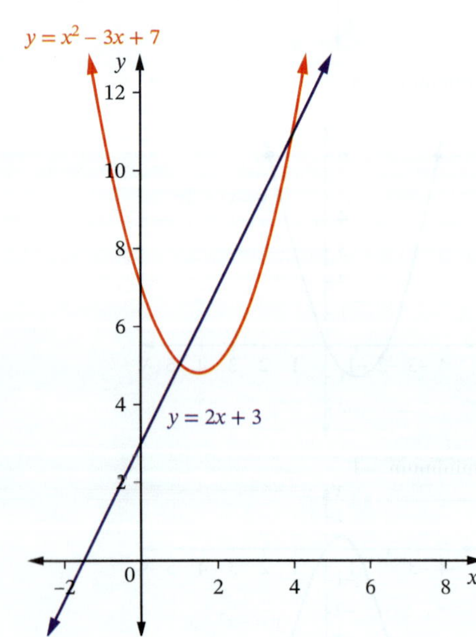

14 a

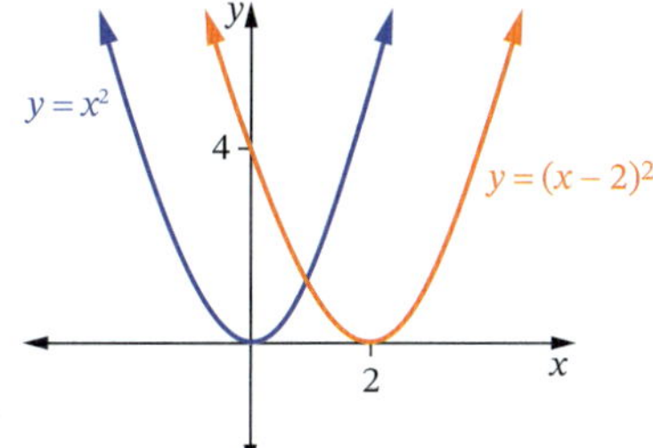

b 1 **c** (1, 1)

15 $x = -1, 2$

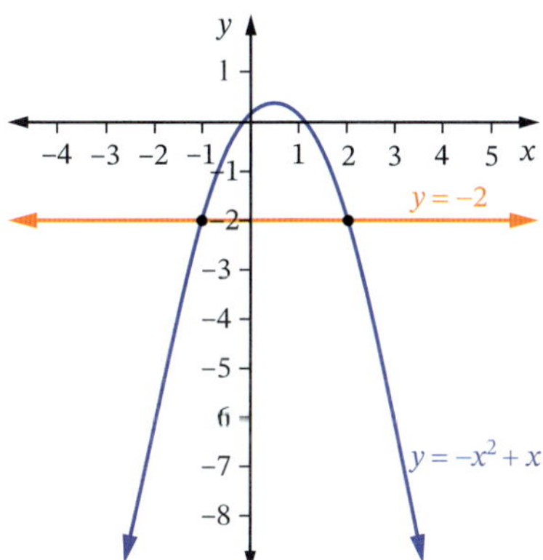

16 $x = 1, 4$

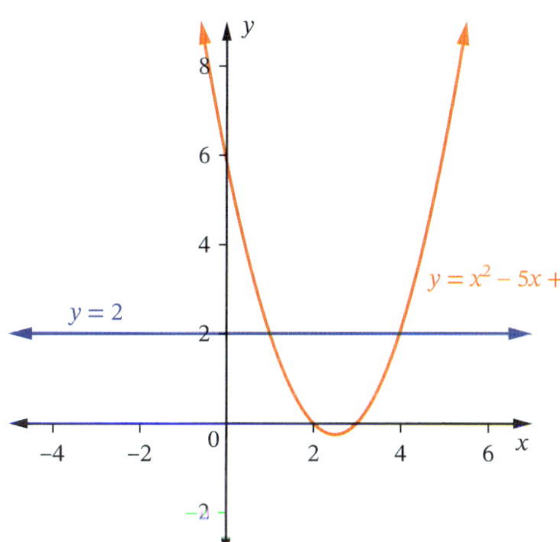

17 $x = -1.5, 2$

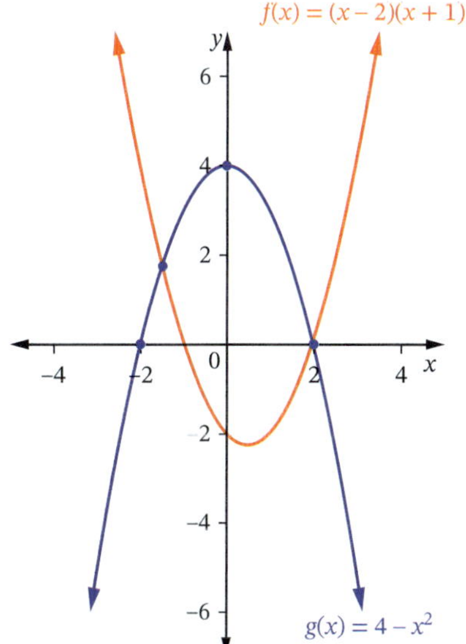

18 nk units

Exercise 3.09

1 $x < -3, x > 3$ **2** $-1 \le x \le 0$
3 $x \le 0, x \ge 2$ **4** $x < -2, x > 2$
5 $0 \le y \le 6$ **6** $0 < t < 2$
7 $x < -4, x > 2$ **8** $p \le -3, p \ge -1$
9 $m < 2, m > 4$ **10** $x \le -3, x \ge 2$
11 $1\frac{1}{2} < h < 2$ **12** $-4 \le x \le 5$
13 $-2\frac{1}{2} \le k \le 7$ **14** $q < 3, q > 6$
15 all real x **16** $n \le -4, n \ge 3$
17 $-3 < x < 5$ **18** $-6 \le t \le 2$
19 $y < -\frac{1}{3}, y > 5$ **20** $x \le -2, x \ge 4$

Exercise 3.10

1 axis of symmetry: $x = -1$, minimum value: -1
2 axis of symmetry: $x = 0$, minimum value: -4
3 axis of symmetry: $x = \frac{3}{8}$, vertex: $\left(\frac{3}{8}, \frac{7}{16}\right)$
4 axis of symmetry: $x = 1$, maximum value: -6
5 axis of symmetry: $x = -1$, vertex: $(-1, 7)$
6 axis of symmetry: $x = -1.5$, minimum value: -0.25
7 a $x = -3, (-3, -12)$ **b** $x = -4, (-4, 17)$
c $x = -3, (-3, -23)$ **d** $x = 1\frac{1}{4}, \left(1\frac{1}{4}, 3\frac{1}{8}\right)$
e $x = -1\frac{1}{4}, \left(-1\frac{1}{4}, -13\frac{1}{4}\right)$
8 a i $x = -1$ **ii** -3 **iii** $(-1, -3)$
b i $x = 1$ **ii** 1 **iii** $(1, 1)$
9 a $(3,0)$ **b** $(-2,0)$ **c** $(-1,3)$
d $(3,0)$ **e** $(-1,8)$ **f** $(5,-1)$
g $(-4,0)$ **h** $(1,-5)$ **i** $(-6,2)$
10 a D **b** G **c** A **d** E
e B **f** H **g** F **h** C
11 a $y = (x+1)^2$, $(-1, 0)$, minimum
b $y = (x-4)^2 - 23$, $(4, -23)$, minimum
c $y = (x+2)^2 - 7$, $(-2, -7)$, minimum
d $y = -(x+1)^2 + 6$, $(-1, 6)$, maximum
e $y = -2(x-2)^2 + 11$, $(2, 11)$, maximum
f $y = -3\left(x - \frac{1}{2}\right)^2 + 7\frac{3}{4}$, $\left(\frac{1}{2}, 7\frac{3}{4}\right)$, maximum
12 a $y = (x-2)^2 - 3$
b

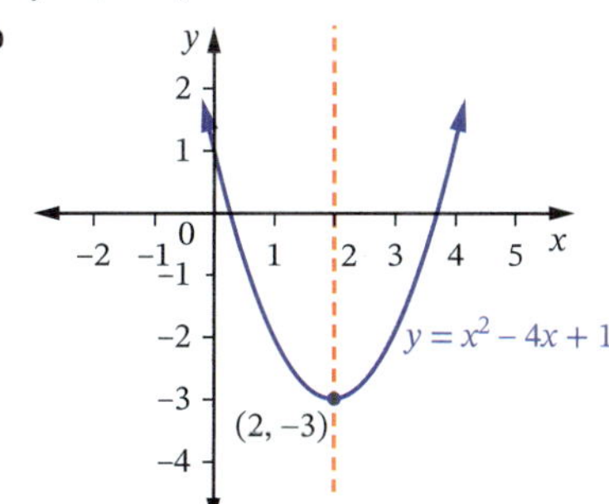

c domain: all real x, range: $y \ge -3$
13 a i -2 **ii** minimum: 0
iii

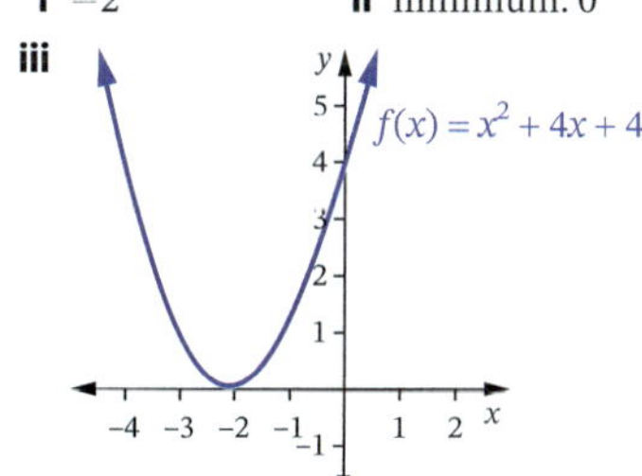

iv $x = -2$

b **i** $-1, 3$ **ii** minimum: -4

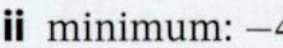

iii

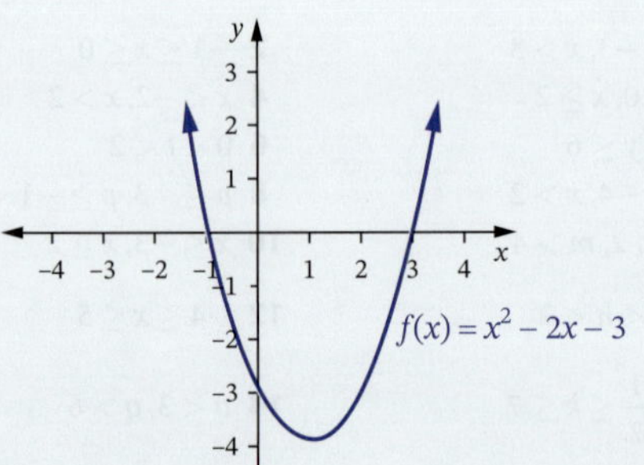

iv $x = 3, -1$

c **i** $3 - 2\sqrt{2}, 3 + 2\sqrt{2}$ **ii** minimum: -8

iii

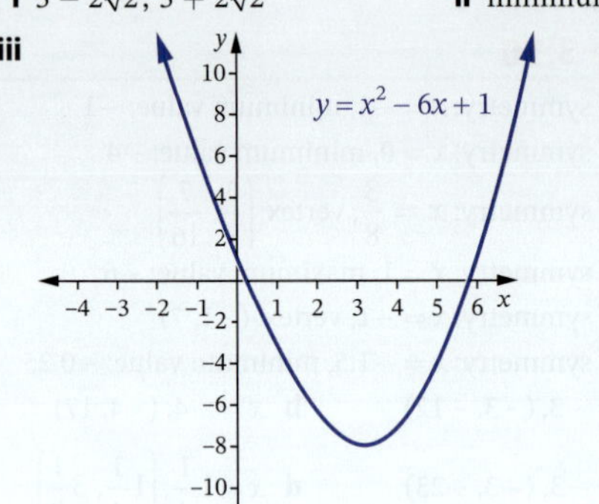

iv $x \approx 5.8, 0.2$

d **i** $-1 - \sqrt{7}, -1 + \sqrt{7}$ **ii** maximum 7

iii

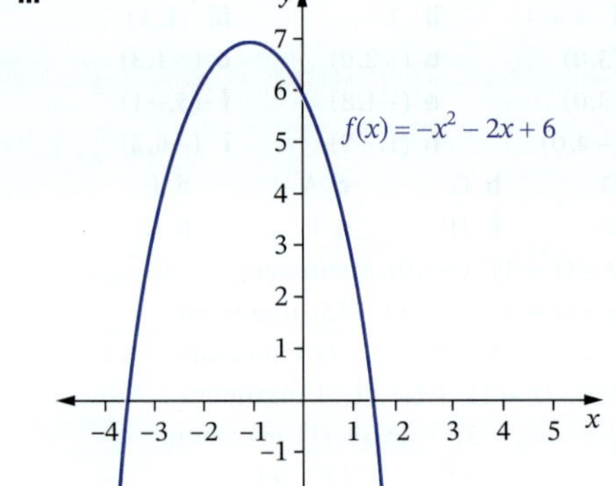

iv $x \approx -3.65, 1.65$

e **i** $\dfrac{-1 - \sqrt{13}}{2}, \dfrac{-1 + \sqrt{13}}{2}$ **ii** maximum: $3\frac{1}{4}$

iii

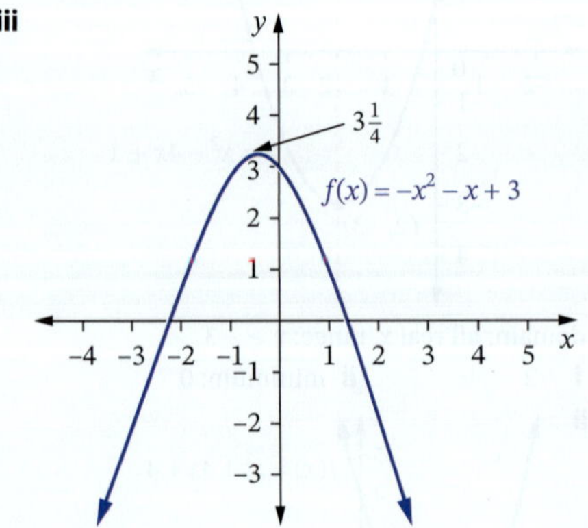

iv $x \approx -2.3, 1.3$

14 **a** 4 **b** none

c

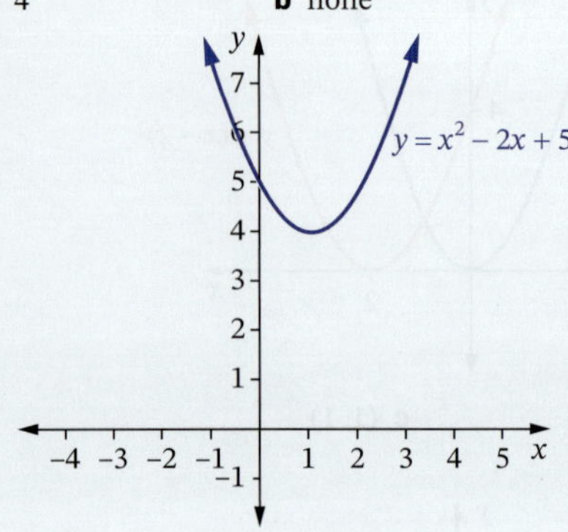

15 **a** $-3\frac{7}{8}$ **b** none

c

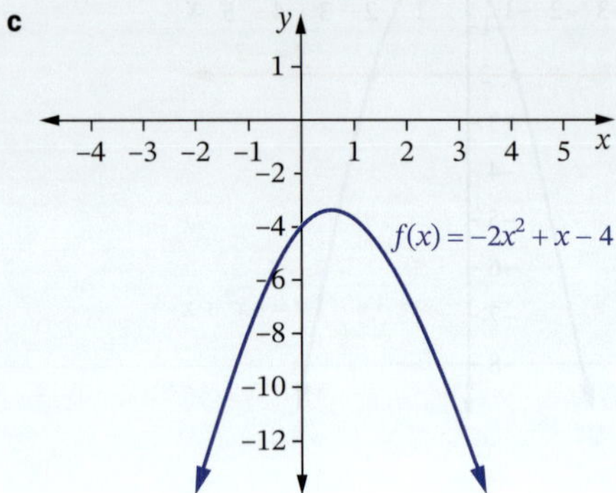

16 D

17 **a** $y = -ax^2 + 3x + 5$ if $a > 0$

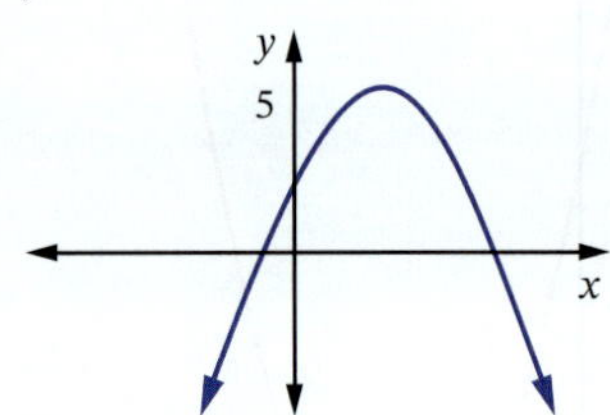

b $f(x) = 3x^2 - 6x + c$ given $c < 0$

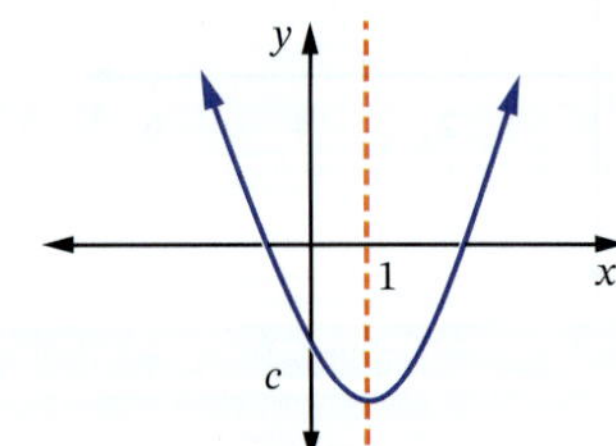

c $y = -x^2 + bx + 2$ if $b > 0$

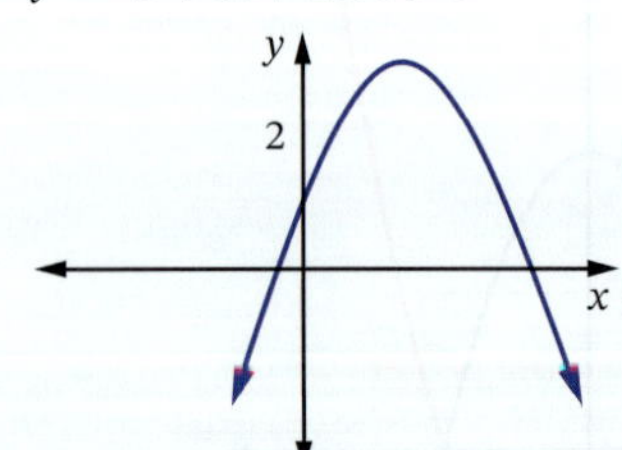

Exercise 3.11

1

	a	b	c
a	1	2	−6
b	2	−11	15
c	1	1	−2
d	1	7	18
e	3	−11	−16
f	4	17	11

2 $m = 2, p = -5, q = 2$

3 $x(x - 2) - 2(x + 1) + 3 + 4$

4 See worked solutions.

5 $A = 1, B = 5, C = -6$

6 $a = 2, b = 1, c = -1$

7 $K = 1, L = 6, M = 7.5$

8 $12(x + 5) + (2x - 3)^2 - 65 - 2$

9 $a = 0, b = -4, c = -21$

10 **a** $y = x^2 - x - 5$ **b** $y = x^2 - 3x$

c $y = 2x^2 - 3x + 7$ **d** $y = x^2 + 4x - 9$

e $y = -x^2 - 2x + 1$

11 **a** $h = -5t^2 + 17.5t + 10$

b $h = 25\,\text{m}$ **c** at $t = 0, 3.5\,\text{s}$

12 **a** $225y = -32x^2 + 1800$

b 6.72 cm **c** 4.6 cm

13 **a** $y = x^2 - 4x$

b **i** $y = 5$ **ii** $y = 32$

c $x = 2$ **d** $x = 2 \pm \sqrt{6}$

14 **a** $f(x) = x^2 + 2x + 7$ **b** $y = 22$

c $a = 1 > 0, \Delta = -24 < 0$

15 $y = 2x^2 - 4x$ **16** $y = -x^2 + 6x + 4$

17 **a** $y = x^2 - 3x - 4$ **b** $y = x^2 + 3x$

c $y = -x^2 + 4x + 5$

18 **a** $y = 3x^2 + 9x - 12$ **b** $y = 2x^2 - 8x - 7$

c $y = -2x^2 - 4x + 3$

Exercise 3.12

1 **a** 20 **b** −47 **c** 12 **d** 49 **e** 9

f −16 **g** 0 **h** 64 **i** 17

2 **a** 17, real distinct irrational roots

b −39, no real roots

c 1, real distinct rational roots

d 0, real equal rational roots

e 33, real distinct irrational roots

f −16, no real roots

g 49, real distinct rational roots

h −116, no real roots

i 1, real distinct rational roots

3 $p = 1$ **4** $k = \pm 2$ **5** $b \leq -\frac{7}{8}$

6 $p > 2$ **7** $k > -2\frac{1}{12}$

8 $a = 3 > 0$

$\Delta = -83 < 0$

9 Solving simultaneously:

$x^2 - 2x - 3 = 0$

$\Delta = 16 > 0$

So there are 2 points of intersection.

10 $\Delta = 68 > 0$

So there are 2 points of intersection.

11 $\Delta = -15 < 0$

So there are no points of intersection.

12 $\Delta = 0$

So there is 1 point of intersection

$\therefore$ the line is a tangent to the parabola.

13 $\Delta = -24 < 0$, graph lies above x-axis.

14 $k \leq -5, k \geq 3$

15 $0 < k < 4$

16 $m < -3, m > 3$

17 **a** 0, 1 or 2 **b** 2 **c** $x = \frac{4 \pm \sqrt{26}}{2}$

Exercise 3.13

1 **a** x-intercept 1, y-intercept −1

b x-intercept 2, y-intercept 8

c x-intercept −5, y-intercept 125

d x-intercept 4, y-intercept 64

e x-intercept −6, y-intercept 1026

f x-intercepts −5, 1, 2, y-intercept 10

2 **a**

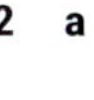

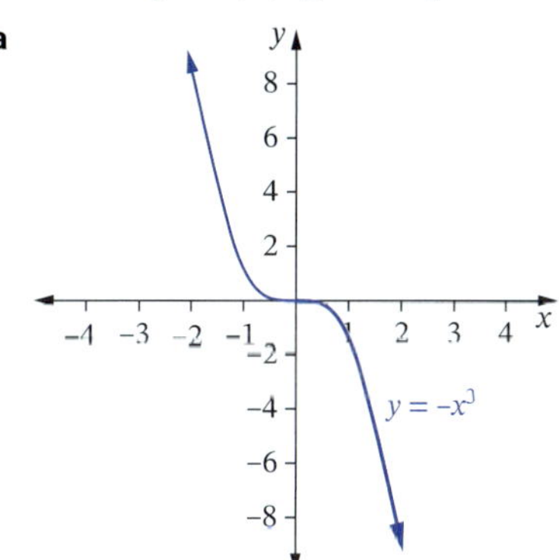

b

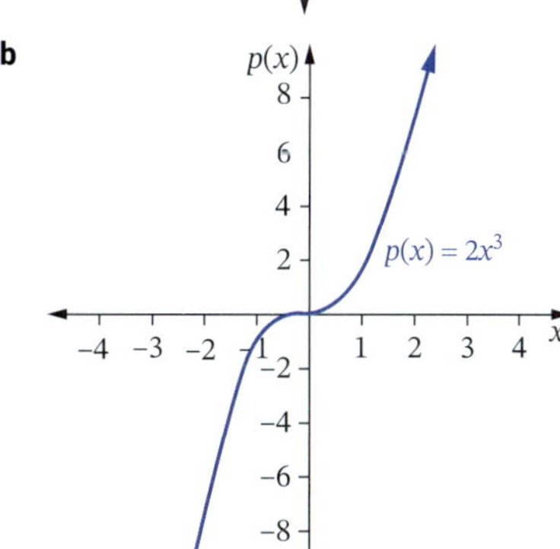

c

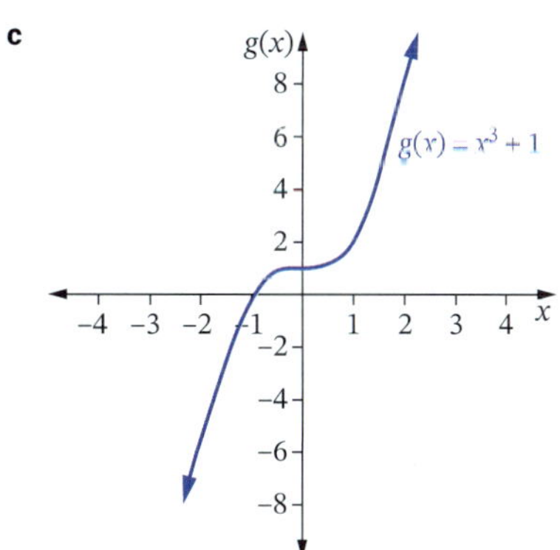

d

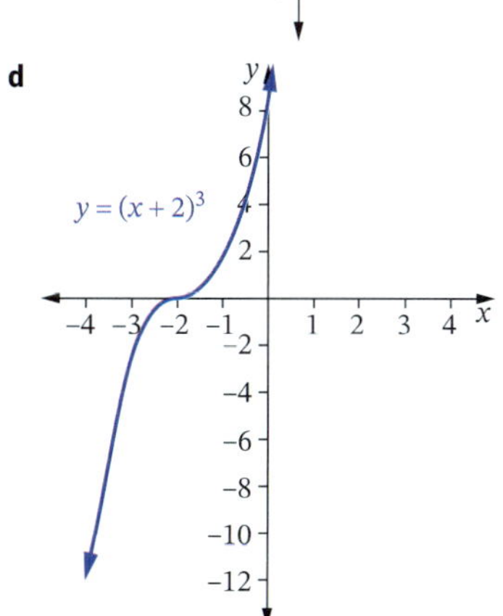

e

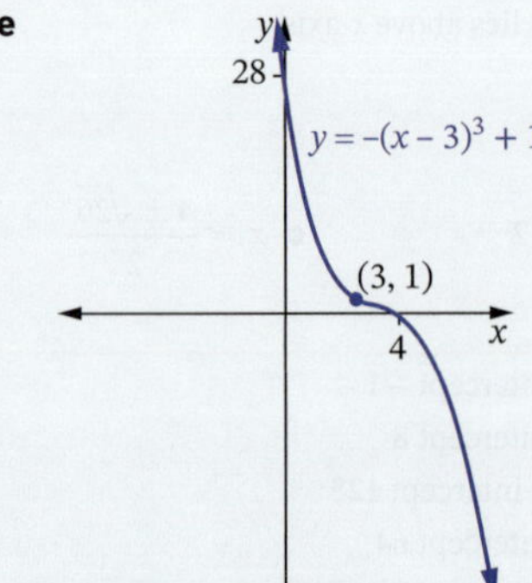

f

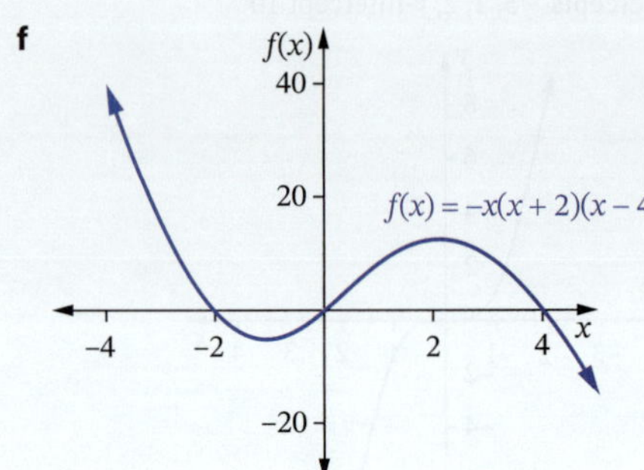

g

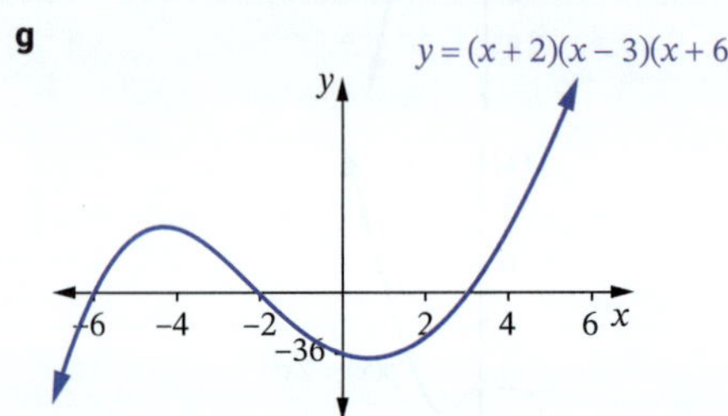

h

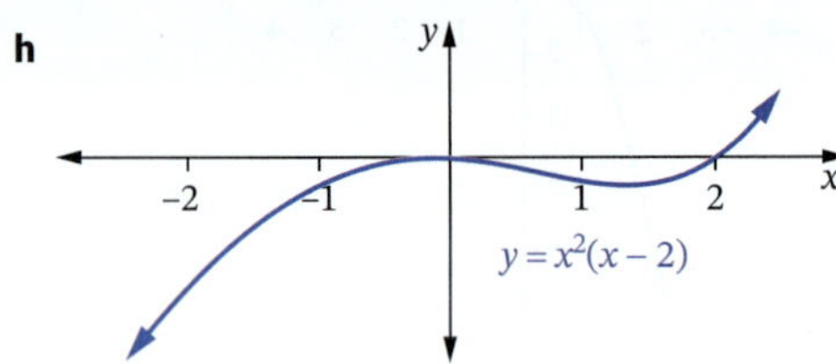

i

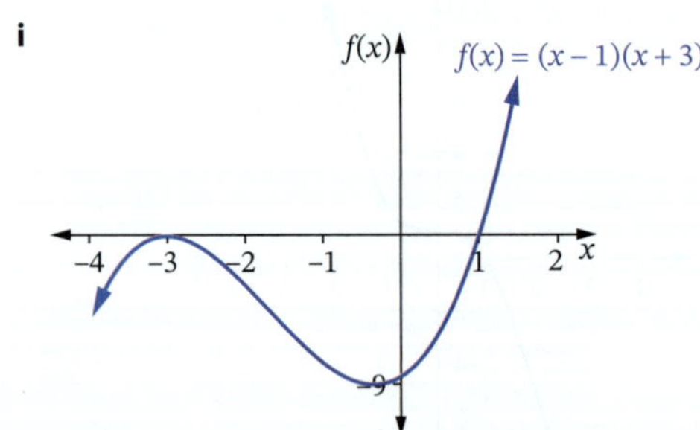

3 **a** (0, 1) **b** (0, 27) **c** (−2, 0)
d (1, −16) **e** (−1, 1)

4 **a** 1.4 **b** −0.3 **c** 0.7
d −1.9 **e** 0.9

5 $f(-x) = -(-x)^3 = x^3 = -f(x)$, so an odd function.

6 a, d

7 **a** Increasing curve with x-intercept 4, point of inflection (0, −64) and y-intercept −64.

b Decreasing curve with x-intercept 3, point of inflection (3, 0) and y-intercept 27.

c Increasing to a maximum turning point between $x = -4$ and $x = -2$, then decreasing to a minimum turning point between $x = -2$ and $x = 0$, then increasing again, x-intercepts −4, −2, 0, y-intercept 0.

d Decreasing to a minimum turning point between $x = -3$ and $x = -1$, then increasing to a maximum turning point between $x = -1$ and $x = 4$, then decreasing again, x-intercepts −3, −1, 4, y-intercept 24.

e Increasing to a maximum turning point at $x = -5$, then decreasing to a minimum turning point between $x = -5$ and $x = 0$, then increasing again, x-intercepts −5, 0, y-intercept 0.

8 **a** $x = 1.7$ **b** $x = -1.3$ **c** $x = 1.7$
d $x = -1.1$ **e** $x = -0.8$ **f** $x = -2, 0, 1$

9 B **10** C

11 $y = x^3 + 3$ is a shift of 3 units up from $y = x^3$, $y = x^3 - 6$ is a shift of 6 units down from $y = x^3$.

12 $y = 3x^3$ is stretched up from $y = x^3$, and $y = \frac{x^3}{2}$ is shrunk down from $y = x^3$.

13 $y = (x+1)^3$ is a shift of 1 unit to the left from $y = x^3$, $y = (x-2)^3$ is a shift of 2 units to the right from $y = x^3$.

Exercise 3.14

1 **a** $P = 15x + 20$
b **i** 380 **ii** 845 **iii** 3725
c **i** 145 **ii** 512 **iii** 843

2 **a** $c = 250n + 7000$
b **i** \$32 000 **ii** \$74 500 **iii** \$307 000
c **i** 180 **ii** 285 **iii** 1440
d 10

3 **a** $A = 450 - 8h$
b **i** 426 L **ii** 258 L
c 56.25 h or 2 days 8 hours 15 minutes

4 **a** $C = 20 - 1.69x$ **b** 11 songs

5 **a** $A = 20\,000 - 320x$
b **i** \$18 400 **ii** \$16 160 **iii** \$800
c 63 months or 5 years 3 months

6 **a** $P = 5x - 100$ **b** \$1400
c 231 **d** (20, 100)

7 71 calculators

8 **a** income: $y = 5x$, costs: $y = x + 264$
b 66 cupcakes **c** \$736 **d** \$64 loss

9 **a** $k = 0.004$ **b** $d = 25.6$ m
c yes ($d = 10$ m) **d** no ($d = 48.4$ m)

10 **a** $A = -x^2 + 3x$
b

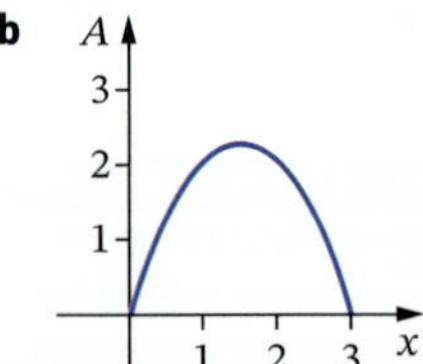

c $x = 1.5$ **d** $A = 2.25$ units^2

11 **a** $I = 2x$ **b** $C = 0.5x + 1500$
c (1000, 2000) **d** \$675

12 **a** $r = 15n$, $c = 3n + 120$
b

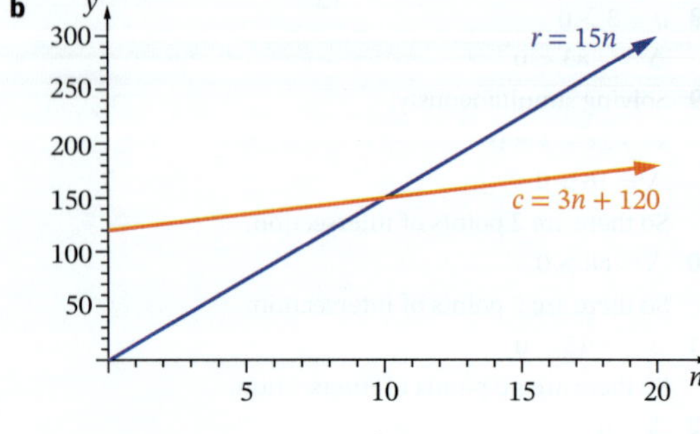

c more than 10 toys **d** (10, 150) **e** \$312

13 a

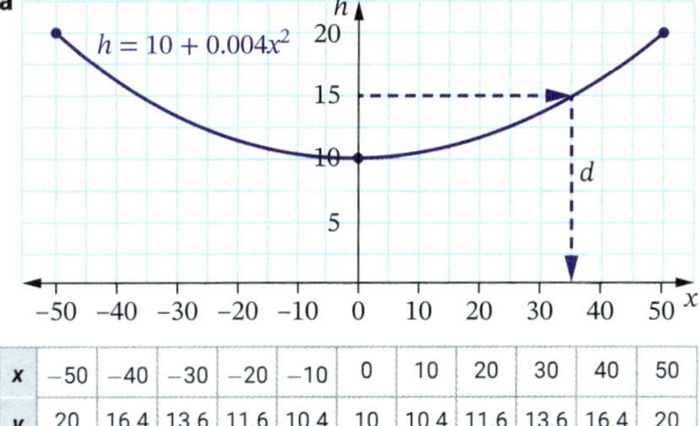

x	−50	−40	−30	−20	−10	0	10	20	30	40	50
y	20	16.4	13.6	11.6	10.4	10	10.4	11.6	13.6	16.4	20

b 20 m **c** 10 m **d** 35 m

14 a 7.2, the height in metres from which the ball is thrown
b 8 m, after 2 s
c 5.2 s
d i 6.75 m **ii** 8 m
e i 1 m **ii** 2.2 m
f 8.3 s

15 a 0, the height of the ball when hit
b 40 m **c** 200 m
d i 39 m and 161 m **ii** 11 m and 189 m

Sample HSC problem

a See worked solutions.
b 200 m **c** 800 m
d 9.875 m **e** 565.7 m

Test yourself 3

1 C **2** A **3** B **4** B **5** D

6 a $f(-2) = 6$ **b** $f(a) = a^2 - 3a - 4$
c $x = 4, -1$

7 a

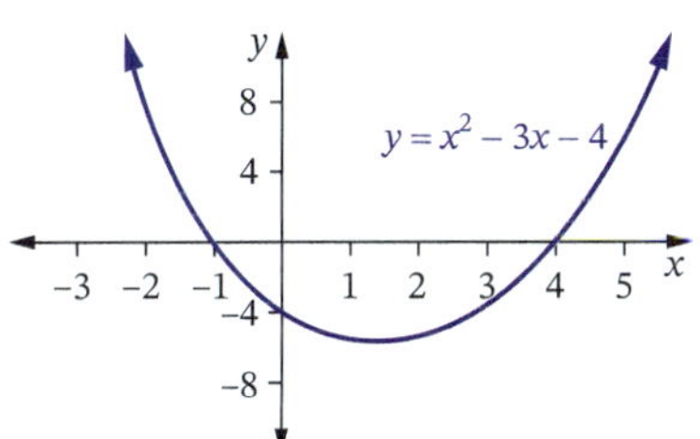

domain: all real x, range: $y \geq -6\frac{1}{4}$

b

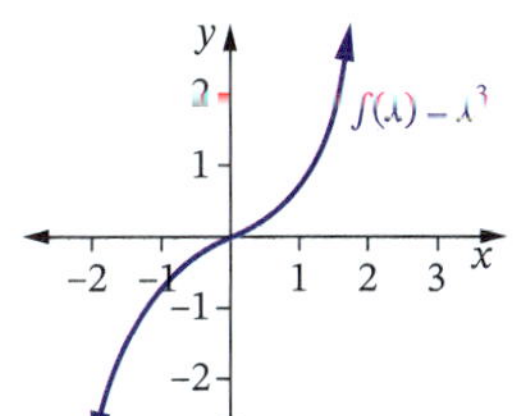

domain: all real x, range: all real y

c

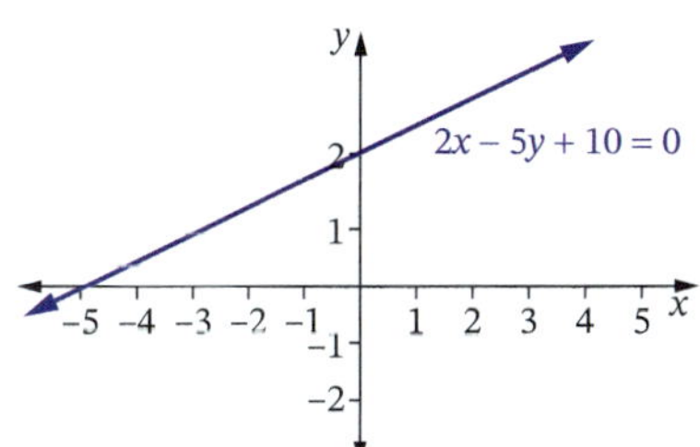

domain: all real x, range: all real y

d

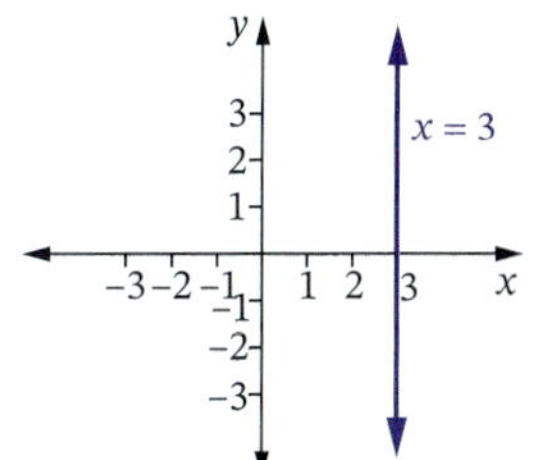

domain: $x = 3$, range: all real y

e

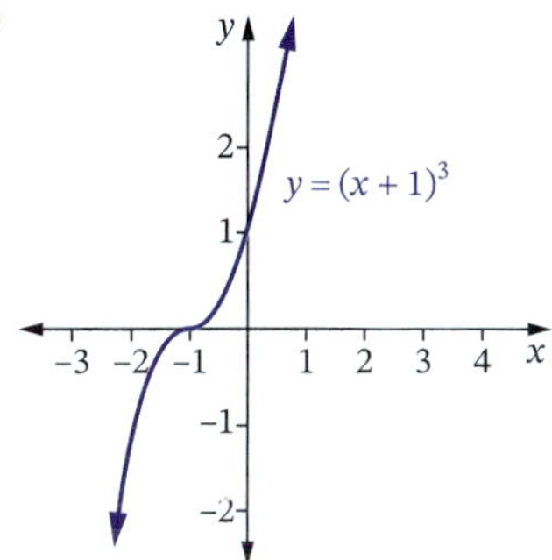

domain: all real x, range: all real y

f

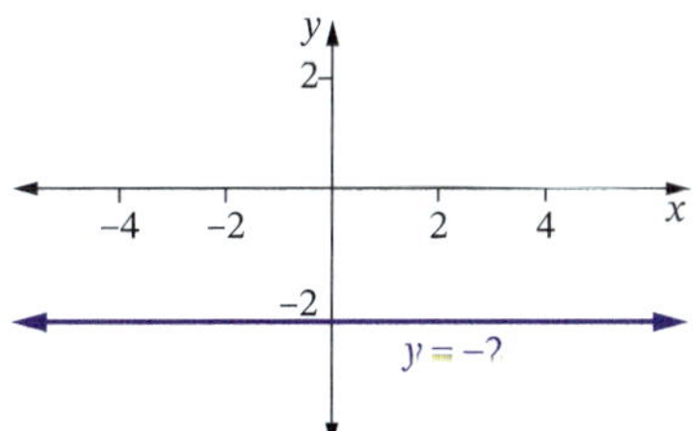

domain: all real x, range: $y = -2$

g

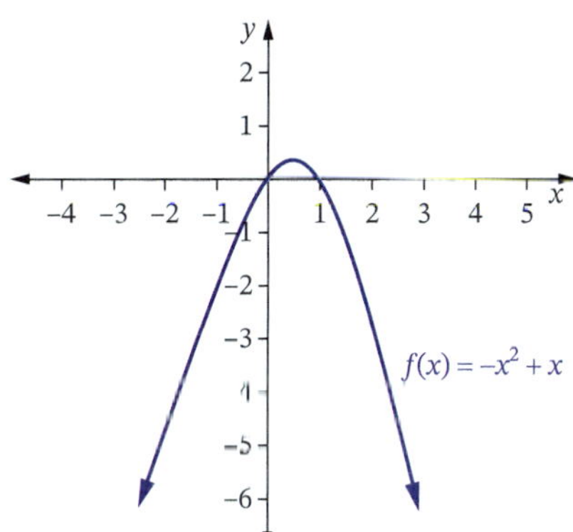

domain: all real x, range: $y \leq \frac{1}{4}$

h

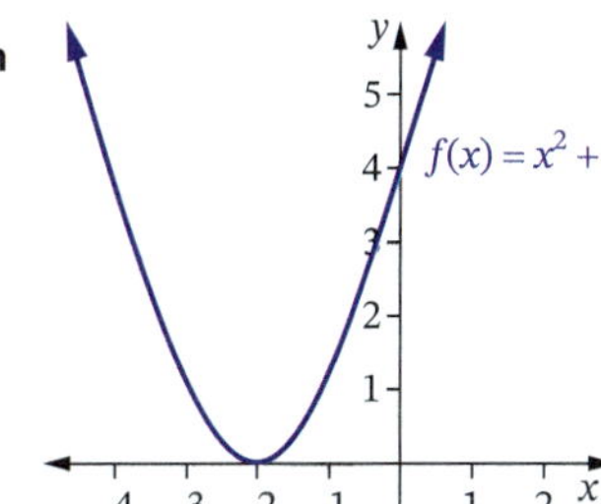

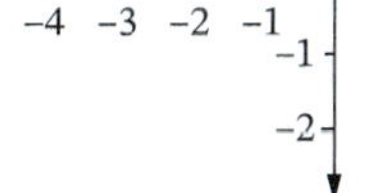

domain: all real x, range: $y \geq 0$

8 a 2 **b** $x = 3\frac{2}{3}$ **c** $x = 1\frac{1}{3}$

9 a $0 \leq x \leq 3$ **b** $n < -3, n > 3$
c $-2 \leq y \leq 2$

ANSWERS

10 a $-1\frac{1}{5}$ **b** 2 **c** $\frac{3}{5}$ **d** 1

11 a $x = 2$ **b** -3

12

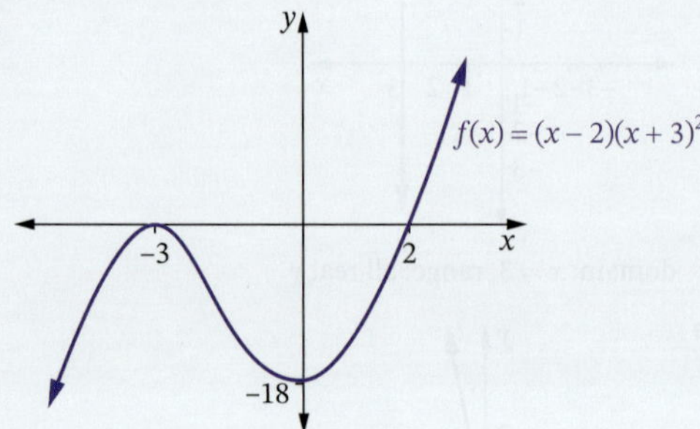

13 a $x = 1, \pm 2$

b x-intercepts 1, ± 2; y-intercept 4

c

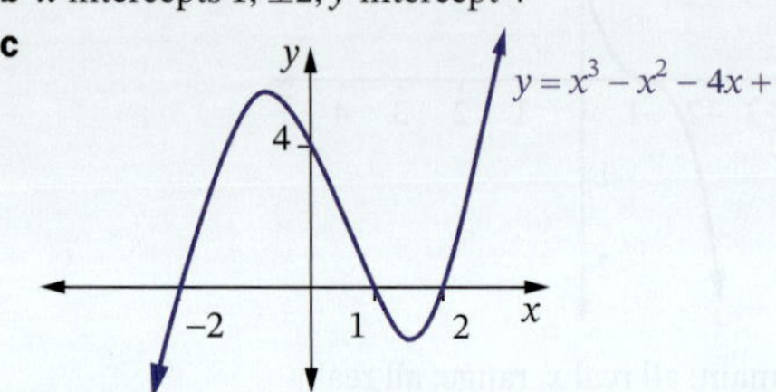

d i 3 **ii** 1

14 a x-intercept -10, y-intercept 4

b x-intercepts $-2, 7$, y-intercept -14

c x-intercept -2, y-intercept 8

d x-intercept 5, y-intercept -2

15 $(-1, 1)$

16 a $x = -\frac{1}{4}$ **b** $6\frac{1}{8}$

17 domain: all real x, range: $y \leq 6\frac{1}{8}$

18 a D **b** B **c** C **d** B **e** A

19 a $7x - y - 11 = 0$ **b** $5x + y - 6 = 0$

c $3x + 2y = 0$ **d** $3x + 5y - 14 = 0$

e $x - 3y - 3 = 0$

20 $a = 2, b = -18, c = 40$

21 a even **b** neither **c** odd

d neither **e** odd

22 $f(-x) = (-x)^3 - (-x) = -x^3 + x = -(x^3 - x) = -f(x)$

23 $m_1 = -\frac{1}{4}, m_2 = 4$ so $m_1 m_2 = -1$

24 $a = -1 < 0$

$\Delta = -7 < 0$

$\therefore -4 + 3x - x^2 < 0$ for all x.

25 a

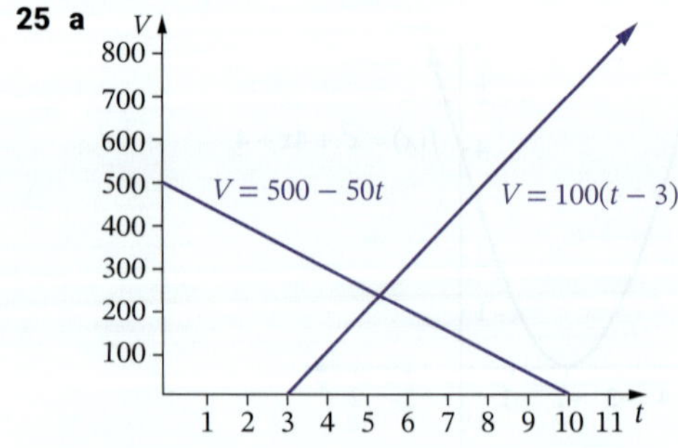

b 5 h **c** 8 h

d i 5.3 h **ii** 233.3 L **iii** 8 h

e 10.5 h

26 $m_1 = m_2 = 5$ so lines are parallel.

27 $x = 4, 5$

28

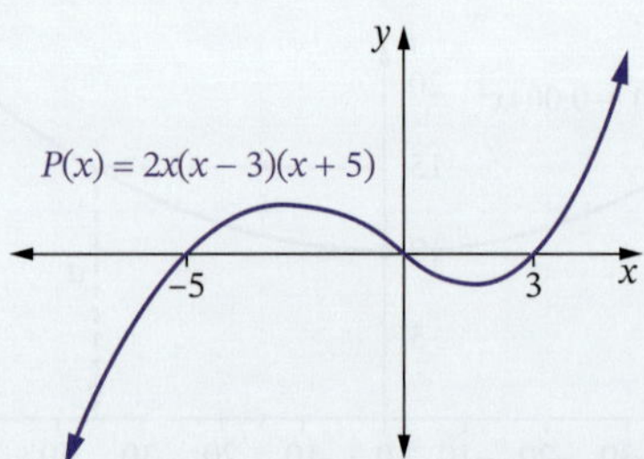

29 $x = 0, 2$ **30** $x = 1\frac{4}{5}$ **31** 15

32 a 4 **b** 5 **c** 9 **d** 3 **e** 2

33 a $y = x^2 - 5x + 4$ **b** $y = -2x^2 + 2x + 12$

34 $k = 1, l = -1, m = 0$

35 a function **b** not a function **c** not a function

d function **e** not a function

36 $f(x) = -3x^3 + 5$

37 $(3, 36)$

38 a $2x + y - 1 = 0$ **b** $\frac{1}{2}$

39 $m < -\frac{9}{16}$

40 a $(1, -1)$ **b** $(1, 0), (2, 2)$

c $(-3, 9), (3, 9)$

41 a $x \leq -1, x \geq 2$ **b** $-3 < x < 1$

42 a

y; 600, 500, 400, 300, 200, 100; $y = 20n$; $y = 5n + 300$; (20, 400); 5 10 15 20 25 30 n

b (20, 400) **c** (20, 400) **d** 20 toys

43 a 25 days **b** $m = 5000 - 200t$

c $m = 500(t - 9)$ **d** 14 days

m; 7000, 6000, 5000, 4000, 3000, 2000, 1000; $m = 500(t - 9)$; $m = 5000 - 200t$; 5 10 15 20 25 30 t

e 14th day **f** 22nd day **g** 29 days

44 a 50 m **b** 10 m

Challenge exercise 3

1 $b = -\frac{2}{3}, 3$

2

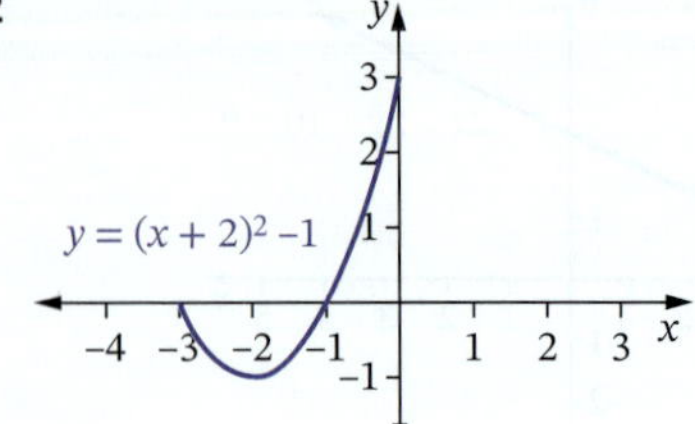

3 $k = -2$ **4** $2x + 3y + 13 = 0$

5 $a = 2, b = 3$

6 **a** $x = 1\frac{2}{3}, \frac{1}{3}$ **b** $x = 0, 2$ **c** $x = 1, 3$ **d** $x = \pm 1$ **e** $x = 1, 2.6, 0.38$

7 $y = 3$

8 $f(3) = 9, f(-4) = 16, f(0) = 1$

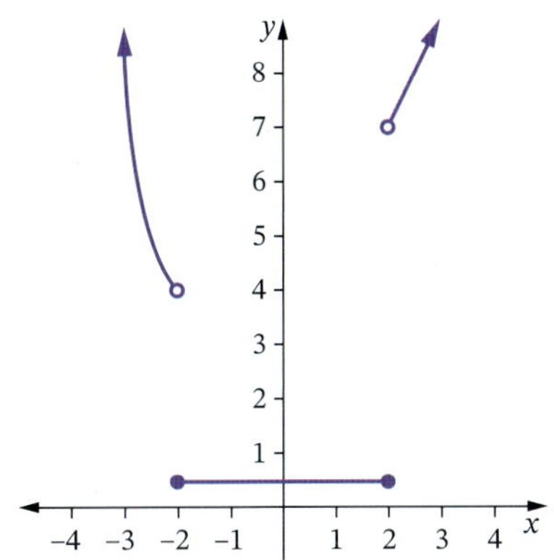

9 $h(2) + h(-1) - h(0) = -3 + 0 - (-1) = -2$

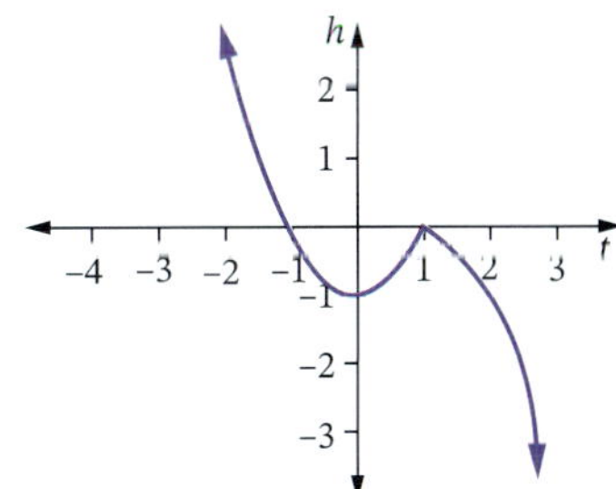

10 $x = 0, 3, -2$

11 $\Delta = (k-4)^2 \geq 0$ and a perfect square $\therefore$ real, rational roots.

12 $p > 0.75$

13 $2x + 5y + 14 = 0$ **14** $(0, 0), (1, 1)$

15 $y = x^3 + 2x^2 - x + 1$

16 $b^2 - 4ac = 0$, so equal roots.

17 $x = \pm 2$

18 $k \leq \frac{1-\sqrt{21}}{2}, k \geq \frac{1+\sqrt{21}}{2}$

19 **a** **i** Substitute points $(0, 113)$, $(251.5, 0)$ and $(-251.5, 0)$ into $h = -0.00179x^2 + 113$. Alternatively, substitute the points into the equation $h = ax^2 + bx + c$ and solve simultaneously to find a, b and c.

ii Substitute points $(0, 134)$, $(251.5, 0)$ and $(-251.5, 0)$ into $h = -0.00212x^2 + 134$. Alternatively, substitute the points into the equation $h = ax^2 + bx + c$ and solve simultaneously to find a, b and c.

b 112.8 m **c** 60.94 m **d** 186.3 m

20 **a** $c = 0.01v^2 + 95$ **b** $C = 97v + \frac{92150}{v}$ **c** \$1995.43 **d** \$4811.49

Chapter 4

Exercise 4.01

1 **a** 47° **b** 82° **c** 19° **d** 77° **e** 52°

2 **a** 47°13′ **b** 81°46′ **c** 19°26′ **d** 76°37′ **e** 52°30′

3 **a** 59°32′ **b** 72°14′ **c** 85°53′ **d** 46°54′ **e** 73°13′

4 **a** 0.635 **b** 0.697 **c** 0.339 **d** 0.928 **e** 1.393

5 **a** 17°20′ **b** 34°20′ **c** 34°12′ **d** 46°34′ **e** 79°10′

6 **a** $x = 6.3$ **b** $y = 5.6$ **c** $b = 3.9$ cm **d** $x = 5.6$ m **e** $m = 2.9$ **f** $x = 13.5$ **g** $y = 10.0$ **h** $p = 3.3$ **i** $x = 5.1$ cm **j** $t = 28.3$ **k** $x = 3.3$ cm **l** $x = 2.9$ cm

7 1.6 m **8** 13.9 m

9 **a** $x = 39°48′$ **b** $\alpha = 35°06′$ **c** $\theta = 37°59′$ **d** $\alpha = 50°37′$ **e** $\alpha = 38°54′$ **f** $\beta = 50°42′$ **g** $x = 44°50′$ **h** $\theta = 30°51′$ **i** $\alpha = 29°43′$ **j** $\theta = 45°37′$ **k** $\alpha = 57°43′$ **l** $\theta = 43°22′$

10 37°57′ **11** 22°14′ **12** 36°52′

13 **a** 18.4 cm **b** 13.8 cm

14 20.3 m

15 $\alpha = 31°58′, \beta = 45°44′$

16 **a** 7.4 cm **b** 6.6 cm **c** 9.0 cm

17 **a** 13 m **b** 65°13′

18 **a** 12.6 cm **b** 22.2 cm

19 **a** 12.9 m **b** 56°37′

20 38 cm

Exercise 4.02

1 10.4 m **2** 21 m **3** 126.9 m

4 72°48′ **5** 12 m **6** 171 m

7 **a**

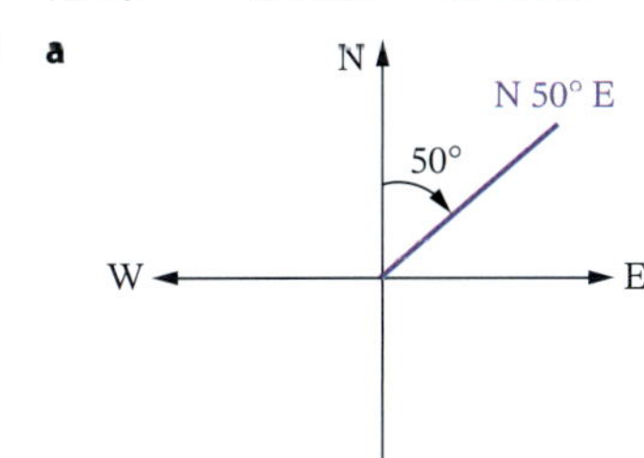

b

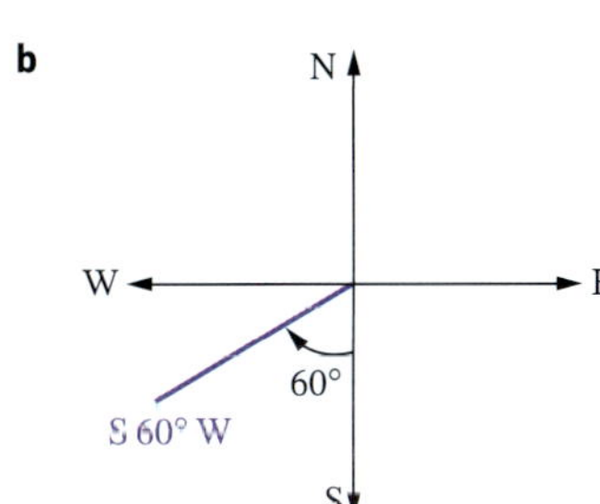

c

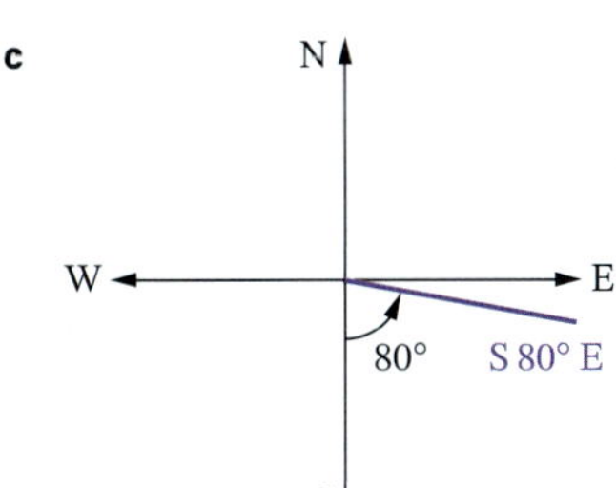

d

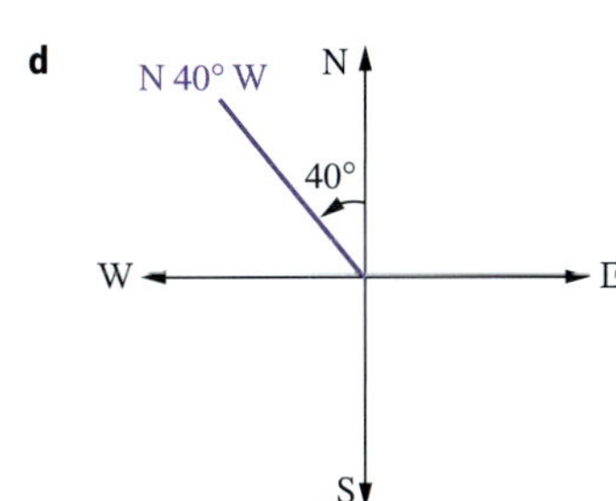

e

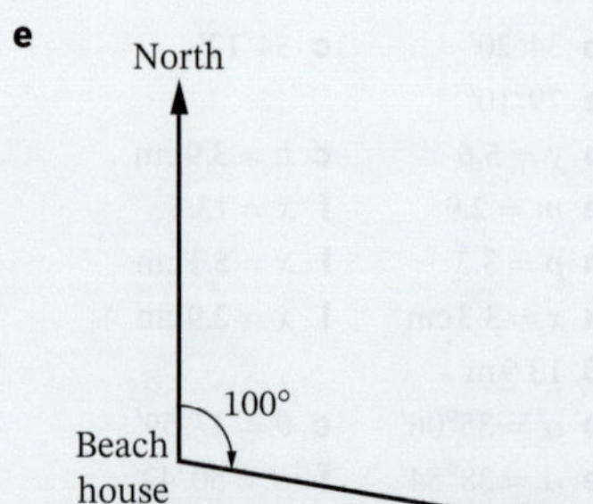

f

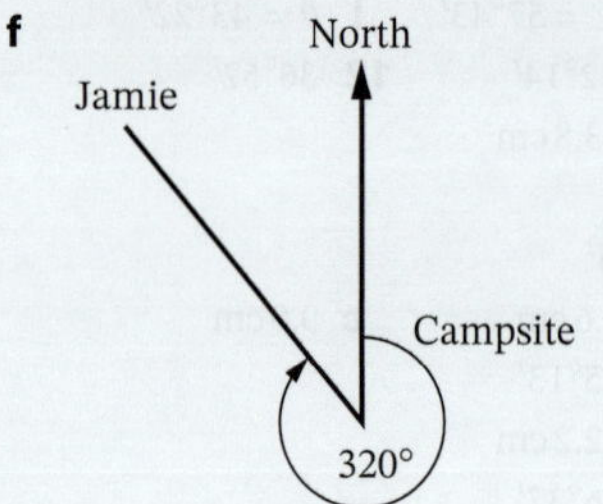

g

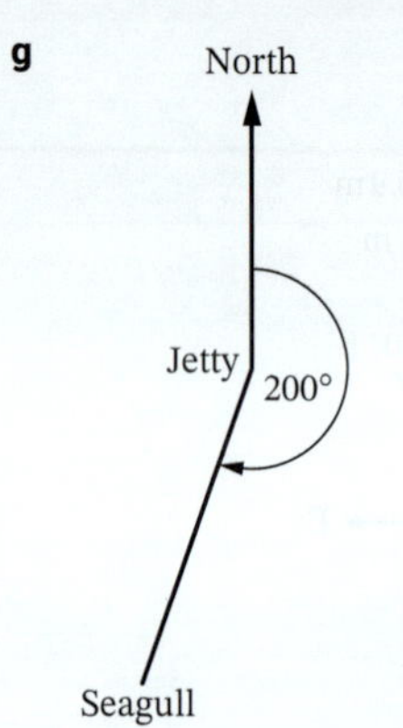

h

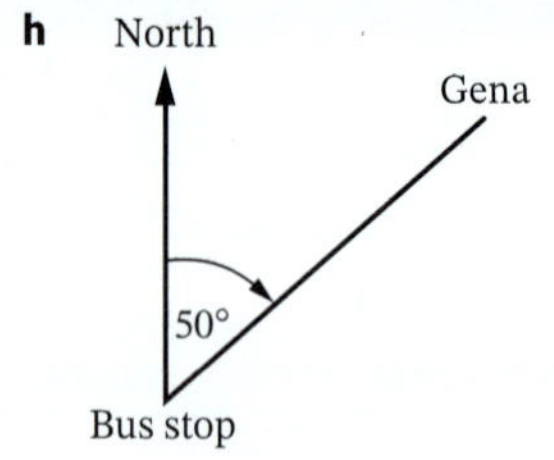

i

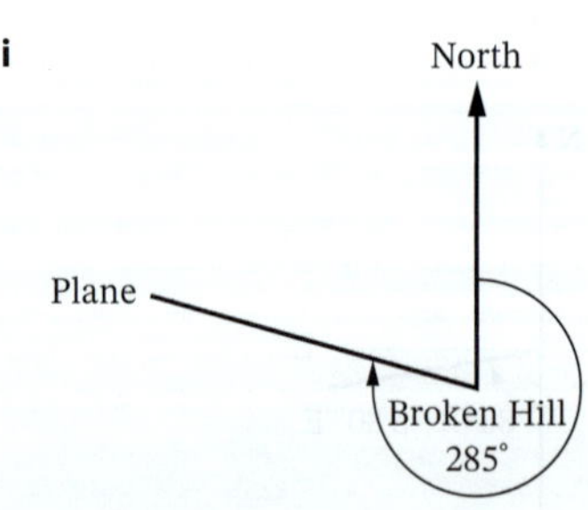

j

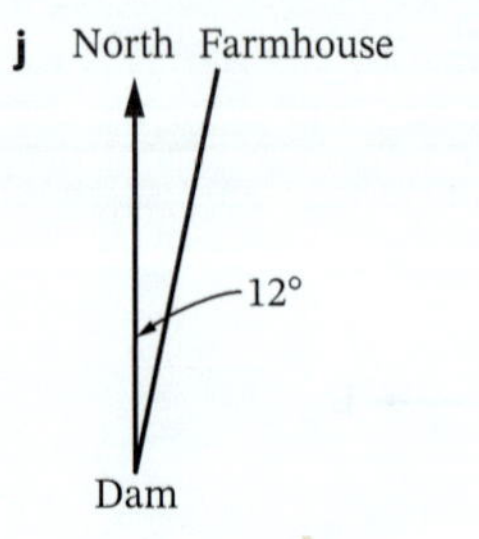

k

North

House 160°

Mohammed

8 **a** **i** S35°E **ii** 145° **b** **i** N80°E **ii** 080°
c **i** N23°W **ii** 337° **d** **i** S **ii** 180°

9 080° **10** 210° **11** 160°

12 **a** 1056.5 km **b** 2265.8 km **c** 245°

13 83.1 m **14** 1.8 km **15** 242°

16 035° **17** 9.2 m

18 9.8 km **19** 51°41′ **20** 2.6 m

21 9°2′ **22** 1931.9 km **23** 35 m

24 149° **25** 198 m **26** 4.8 km

27 9.2 m **28** 217°

29 **a** 1.2 km **b** 7.2 km

30 **a** 13.1 m **b** 50°26′

Exercise 4.03

1 **a** $\sqrt{2}$ **b** 45°
c $\sin 45° = \frac{1}{\sqrt{2}}$, $\cos 45° = \frac{1}{\sqrt{2}}$, $\tan 45° = 1$

2 **a** $\sqrt{3}$
b $\sin 30° = \frac{1}{2}$, $\cos 30° = \frac{\sqrt{3}}{2}$, $\tan 30° = \frac{1}{\sqrt{3}}$
c $\sin 60° = \frac{\sqrt{3}}{2}$, $\cos 60° = \frac{1}{2}$, $\tan 60° = \sqrt{3}$

3 **a** 0 **b** 0 **c** 0
d 0 **e** 0 **f** 1
g 0 **h** 0 **i** undefined
j -1 **k** undefined **l** -1
m 0 **n** 0 **o** -1
p 1

4 **a** $\sin 160° = 0.342$, $\sin 200° = -0.342$, $\sin 340° = -0.342$
b $\cos 305° = 0.574$, $\cos 235° = -0.574$, $\cos 125° = -0.574$
c $\tan 258° = x$, $\tan 102° = -x$, $\tan(-78°) = -x$

5 **a** 25° **b** 40° **c** 60°
d 70° **e** 5° **f** 50°
g 70° **h** 65° **i** 60°
j 80° **k** 60° **l** 40°

6 **a** 1st, 4th **b** 1st, 3rd **c** 1st, 2nd
d 2nd, 4th **e** 3rd, 4th **f** 2nd, 3rd
g 3rd **h** 2nd **i** 4th
j 4th

7 **a** 3rd **b** $-\frac{1}{2}$

8 **a** 4th **b** $-\frac{1}{\sqrt{2}}$

9 **a** 2nd **b** $-\sqrt{3}$

10 **a** 2nd **b** $\frac{1}{\sqrt{2}}$

11 **a** 1st **b** $\frac{\sqrt{3}}{2}$

12 **a** 1 **b** $\frac{1}{\sqrt{2}}$ **c** $-\sqrt{3}$ **d** $\frac{1}{2}$
e $-\frac{1}{2}$ **f** $-\frac{1}{2}$ **g** $\frac{\sqrt{3}}{2}$ **h** $-\frac{1}{\sqrt{3}}$
i $-\frac{\sqrt{3}}{2}$ **j** $-\frac{1}{\sqrt{2}}$

13 **a** $-\frac{\sqrt{3}}{2}$ **b** $\sqrt{3}$ **c** $\frac{\sqrt{3}}{2}$ **d** $\frac{1}{2}$
e $-\frac{1}{2}$ **f** $\sqrt{3}$ **g** $\frac{1}{\sqrt{2}}$ **h** $\frac{1}{\sqrt{2}}$
i -1 **j** $\frac{1}{2}$

14 **a** $-\frac{1}{\sqrt{2}}$ **b** $-\frac{\sqrt{3}}{2}$ **c** $\sqrt{3}$ **d** $-\frac{\sqrt{3}}{2}$
e $-\frac{\sqrt{3}}{2}$ **f** $-\sqrt{3}$ **g** $\frac{1}{2}$ **h** $-\frac{1}{\sqrt{3}}$
i $\frac{1}{\sqrt{2}}$ **j** $-\frac{1}{\sqrt{2}}$

15 $\sin\theta = -\frac{3}{5}, \cos\theta = -\frac{4}{5}$

16 $\cos\theta = -\frac{\sqrt{33}}{7}, \tan\theta = -\frac{4}{\sqrt{33}}$

17 $\cos x = \frac{8}{\sqrt{89}}$

18 $\sin x = -\frac{\sqrt{21}}{5}, \tan x = -\frac{\sqrt{21}}{2}$

19 $\sin x = \frac{5}{\sqrt{74}}, \cos x = -\frac{7}{\sqrt{74}}$

20 $\tan\theta = -\frac{4}{\sqrt{65}}, \cos\theta = \frac{\sqrt{65}}{9}$

21 $\tan x = \frac{\sqrt{55}}{3}, \sin x = -\frac{\sqrt{55}}{8}$

22 **a** $\sin\theta$ **b** $\cos x$ **c** $\tan\beta$
d $-\sin\alpha$ **e** $-\tan\theta$ **f** $-\sin\theta$
g $\cos\alpha$ **h** $-\tan x$ **i** $-\cos\alpha$

Exercise 4.04

1 **a** $x = 8.9$ **b** $y = 9.4$ cm **c** $a = 10.0$
d $b = 10.7$ m **e** $d = 8.0$
2 **a** $\theta = 51°50'$ or $128°10'$
b $\alpha = 61°23'$ or $118°37'$
c $x = 43°03'$ or $136°57'$
d $\theta = 29°4'$ or $150°56'$
e $\alpha = 87°4'$
3 5.7 cm **4** 126°56′
5 **a** 10.3 m **b** 9.4 m
6 **a** 13.5 mm **b** 25 mm
7 **a** 2.6 km **b** 1.6 km
8 075° **9** 284°
10 **a** 14.1 cm **b** 15.6 cm
11 **a** 54.7 mm **b** 35.1 mm
12 **a** 74° or 106° **b** 52° **c** 55° or 125°
13 See worked solutions.

Exercise 4.05

1 **a** $m = 5.8$ **b** $b = 10.4$ m **c** $h = 7.4$ cm
d $n = 16.4$ **e** $y = 9.3$
2 **a** $\theta = 54°19'$ **b** $\theta = 60°27'$ **c** $x = 57°42'$
d $\beta = 131°31'$ **e** $\theta = 73°49'$
3 32.94 mm **4** 11.2 cm and 12.9 cm
5 **a** 11.9 cm **b** 44°11′ **c** 82°12′
6 $\angle XYZ = \angle XZY = 66°10', \angle YXZ = 47°40'$
7 **a** 18.1 mm **b** 78°47′
8 **a** 6.2 cm **b** 12.7 cm
9 12.9 cm
10 **a** 11 cm **b** 30°
11 92.1 km
12 **a** 72° **b** 2.6 cm

Exercise 4.06

1 **a** 7.5 cm^2 **b** 32.3 units^2 **c** 9.9 mm^2
d 30.2 units^2 **e** 6.3 cm^2
2 7.5 cm^2 **3** 15.5 cm^2 **4** 34.8 cm^2
5 **a** 7.8 m **b** 180.8 m^2
6 **a** 5.6 cm **b** 18.5 cm^2 **c** 18.7 cm^2
7 $\frac{25\sqrt{3}}{4}$ cm^2
8 **a** 9.4 mm **b** 33.3 mm^2
9 **a** 60° **b** 3.5 mm **c** 5.3 mm^2
10 122° 52′ **11** 65°, 115°

Exercise 4.07

1 **a** 040° **b** 305°
2 16.4 m **3** 28°
4 **a** 1.21 km **b** 1 minute
5 32 m **6** 107 m
7 $h = 8.5$ **8** 7.7 km
9 5.4 km from A and 5.7 km from B
10 1841 km **11** 89°52′
12 9.9 km **13** 163.5 km
14 3269 km
15 **a** 11.3 cm **b** 44°45′ or 135°15′
16 141°
17 **a** 11.6 cm **b** 73°14′
18 **a** 35°5′ **b** 1.45 m **c** 0.55 m
19 **a** 3.8 km **b** 98°
20 1086.4 cm^2

Exercise 4.08

1 **a** 36° **b** 120° **c** 225° **d** 210°
e 540° **f** 140° **g** 240° **h** 420°
i 20° **j** 50°
2 **a** $\frac{3\pi}{4}$ **b** $\frac{\pi}{6}$ **c** $\frac{5\pi}{6}$ **d** $\frac{4\pi}{3}$
e $\frac{5\pi}{3}$ **f** $\frac{7\pi}{20}$ **g** $\frac{\pi}{12}$ **h** $\frac{5\pi}{2}$
i $\frac{5\pi}{4}$ **j** $\frac{2\pi}{3}$
3 **a** 0.98 **b** 1.19 **c** 2.22
d 5.04 **e** 5.45
4 **a** 0.32 **b** 0.61 **c** 1.78
d 1.54 **e** 0.88
5 **a** 62°27′ **b** 44°0′ **c** 66°28′
d 56°43′ **e** 18°20′ **f** 183°21′
g 154°42′ **h** 246°57′ **i** 320°51′
j 6°18′

ANSWERS

6 **a** 0.34 **b** 0.07 **c** 0.06
d 0.83 **e** −1.14 **f** 0.33
g −1.50 **h** 0.06 **i** −0.73
j 0.16

7 **a** $\frac{1}{\sqrt{2}}$ **b** $\frac{1}{2}$ **c** $\frac{1}{\sqrt{3}}$ **d** $\frac{\sqrt{3}}{2}$
e 1 **f** $\frac{1}{2}$ **g** $\frac{1}{\sqrt{2}}$ **h** $\frac{\sqrt{3}}{2}$
i $\sqrt{3}$ **j** 0

8 **b** 2nd **c** $-\frac{1}{\sqrt{2}}$

9 **b** 2nd **c** $\frac{1}{2}$

10 **b** 4th **c** −1

11 **b** 3rd **c** $-\frac{1}{2}$

12 **b** 4th **c** $-\frac{\sqrt{3}}{2}$

13 **a** **ii** 1st **iii** $\frac{\sqrt{3}}{2}$
b **i** $\frac{1}{\sqrt{2}}$ **ii** $-\sqrt{3}$ **iii** $-\frac{1}{\sqrt{2}}$
iv $\frac{1}{\sqrt{3}}$ **v** $-\frac{\sqrt{3}}{2}$ **vi** $-\frac{1}{2}$

14 **a**

	$\frac{\pi}{3}$	$\frac{2\pi}{3}$	$\frac{4\pi}{3}$	$\frac{5\pi}{3}$	$\frac{7\pi}{3}$	$\frac{8\pi}{3}$	$\frac{10\pi}{3}$	$\frac{11\pi}{3}$
sin	$\frac{\sqrt{3}}{2}$	$\frac{\sqrt{3}}{2}$	$-\frac{\sqrt{3}}{2}$	$-\frac{\sqrt{3}}{2}$	$\frac{\sqrt{3}}{2}$	$\frac{\sqrt{3}}{2}$	$-\frac{\sqrt{3}}{2}$	$-\frac{\sqrt{3}}{2}$
cos	$\frac{1}{2}$	$-\frac{1}{2}$	$-\frac{1}{2}$	$\frac{1}{2}$	$\frac{1}{2}$	$-\frac{1}{2}$	$-\frac{1}{2}$	$\frac{1}{2}$
tan	$\sqrt{3}$	$-\sqrt{3}$	$\sqrt{3}$	$-\sqrt{3}$	$\sqrt{3}$	$-\sqrt{3}$	$\sqrt{3}$	$-\sqrt{3}$

b

	$\frac{\pi}{4}$	$\frac{3\pi}{4}$	$\frac{5\pi}{4}$	$\frac{7\pi}{4}$	$\frac{9\pi}{4}$	$\frac{11\pi}{4}$	$\frac{13\pi}{4}$	$\frac{15\pi}{4}$
sin	$\frac{1}{\sqrt{2}}$	$\frac{1}{\sqrt{2}}$	$-\frac{1}{\sqrt{2}}$	$-\frac{1}{\sqrt{2}}$	$\frac{1}{\sqrt{2}}$	$\frac{1}{\sqrt{2}}$	$-\frac{1}{\sqrt{2}}$	$-\frac{1}{\sqrt{2}}$
cos	$\frac{1}{\sqrt{2}}$	$-\frac{1}{\sqrt{2}}$	$-\frac{1}{\sqrt{2}}$	$\frac{1}{\sqrt{2}}$	$\frac{1}{\sqrt{2}}$	$-\frac{1}{\sqrt{2}}$	$-\frac{1}{\sqrt{2}}$	$\frac{1}{\sqrt{2}}$
tan	1	−1	1	−1	1	−1	1	−1

c

	$\frac{\pi}{6}$	$\frac{5\pi}{6}$	$\frac{7\pi}{6}$	$\frac{11\pi}{6}$	$\frac{13\pi}{6}$	$\frac{17\pi}{6}$	$\frac{19\pi}{6}$	$\frac{23\pi}{6}$
sin	$\frac{1}{2}$	$\frac{1}{2}$	$-\frac{1}{2}$	$-\frac{1}{2}$	$\frac{1}{2}$	$\frac{1}{2}$	$-\frac{1}{2}$	$-\frac{1}{2}$
cos	$\frac{\sqrt{3}}{2}$	$-\frac{\sqrt{3}}{2}$	$-\frac{\sqrt{3}}{2}$	$\frac{\sqrt{3}}{2}$	$\frac{\sqrt{3}}{2}$	$-\frac{\sqrt{3}}{2}$	$-\frac{\sqrt{3}}{2}$	$\frac{\sqrt{3}}{2}$
tan	$\frac{1}{\sqrt{3}}$	$-\frac{1}{\sqrt{3}}$	$\frac{1}{\sqrt{3}}$	$-\frac{1}{\sqrt{3}}$	$\frac{1}{\sqrt{3}}$	$-\frac{1}{\sqrt{3}}$	$\frac{1}{\sqrt{3}}$	$-\frac{1}{\sqrt{3}}$

15

	0	$\frac{\pi}{2}$	π	$\frac{3\pi}{2}$	2π	$\frac{5\pi}{2}$	3π	$\frac{7\pi}{2}$	4π
sin	0	1	0	−1	0	1	0	−1	0
cos	1	0	−1	0	1	0	−1	0	1
tan	0	Not defined	0	Not defined	0	Not defined	0	Not defined	0

Exercise 4.09

1 **a** 4π cm **b** π m **c** $\frac{25\pi}{3}$ cm
d $\frac{\pi}{2}$ cm **e** $\frac{7\pi}{4}$ mm

2 **a** 0.65 m **b** 3.92 cm **c** 6.91 mm
d 2.39 cm **e** 3.03 m

3 1.8 m **4** 7.5 m **5** $\frac{2\pi}{21}$

6 25 mm **7** 1.83 **8** $13\frac{7}{9}$ mm

9 **a** 48.3 mm **b** 25.3 mm

10 $\frac{125\sqrt{35}\,\pi}{648}$ cm^3

11 **a** 1.95 cm **b** 15.64 cm **c** $\frac{40}{360} \times \pi \times 2.8^2$

12 **a** 23.56 cm
b $\frac{3}{4}$, because an arc subtending a central angle of $\frac{3\pi}{2}$ (270°) is $\frac{3}{4}$ of the circumference of the circle.
$\frac{3}{4} \times 2\pi = \frac{3\pi}{2}$

13 **a** $\frac{25\pi}{7}$ cm **b** 9.0 cm **c** 18.28 m^2

14 20.3 m, 32.8 m^2

Exercise 4.10

1 **a** $\frac{3\pi}{2}$ m^2 **b** $\frac{125\pi}{3}$ cm^2
c $\frac{3\pi}{4}$ cm^2 **d** $\frac{245\pi}{8}$ mm^2

2 **a** 0.48 m^2 **b** 6.29 cm^2 **c** 24.88 mm^2
d 7.05 cm^2 **e** 3.18 m^2

3 16.6 m^2 **4** 4.4

5 **a** 88.36 m^2
b $\frac{1}{2}$, because a sector with central angle π is a semicircle.

6 **a** $\frac{35\pi}{6}$ cm **b** $\frac{245\pi}{12}$ cm^2

7 $\frac{6845}{\pi}$ mm^2, 80%, because $\frac{8\pi}{5}$ as a percentage of a revolution 2π is $\frac{\frac{8\pi}{5}}{2\pi} \times 100\% = \frac{4}{5} \times 100\% = 80\%$

8 75 cm^2, $\frac{3}{8}$ **9** 11.97 cm^2 **10** $\frac{\pi}{15}$, 3 cm

11 **a** $\frac{3\pi}{7}$ cm **b** $\frac{9\pi}{14}$ cm^2

12 **a** $\frac{35\pi}{6}$ cm **b** $\frac{175\pi}{12}$ cm² **c** 9.66 cm

13 **a** 10.5 mm **b** 42.9 mm², 12.5%

14 **a** 2π cm² **b** 3.1 cm **c** $4\sqrt{2}$ cm² **d** $2\pi - 4\sqrt{2}$ cm²

15 **a** 77°22′ **b** 70.3 cm² **c** 26.96 cm² **d** 425.43 cm²

16 14.98 cm **17** $\frac{225\pi}{2}$ cm³

18 **a** **i** $A = \pi r^2$ **ii** $A = \frac{1}{2}r^2 \sin\theta$ **iii** $A = \frac{1}{2}r^2\theta$ **iv** $A = \frac{1}{2}r^2(\theta - \sin\theta)$

b $\frac{2\pi - 3\sqrt{3}}{3}$ cm²

Exercise 4.11

1 **a** 8π cm² **b** $\frac{6\pi - 9\sqrt{3}}{4}$ cm² **c** $\frac{125\pi - 75}{3}$ cm² **d** $\frac{3\pi - 9}{4}$ cm²

e $\frac{49\pi - 98\sqrt{2}}{8}$ mm²

2 **a** 0.01 m² **b** 1.45 cm² **c** 3.65 mm² **d** 0.19 cm² **e** 0.99 m²

3 7.2 cm²

4 **a** $\frac{3\pi}{7}$ cm **b** $\frac{9\pi}{14}$ cm² **c** 0.07 cm²

5 134.4 cm

6 **a** 2.6 cm **b** $\frac{5\pi}{6}$ cm **c** 0.29 cm²

7 **a** 10.5 mm **b** 342.04 mm²

8 **a** $\frac{25\pi}{4}$ cm² **b** 0.5 cm²

9 **a** 77°22′ **b** 70.3 cm² **c** 26.96 cm² **d** 425.43 cm²

e area of circle = 452.39 cm², 26.96 + 425.43 = 452.39 cm²

10 37.07 cm² **11** 9.4 cm²

12 **a** 3.84 cm **b** 0.88 cm² **c** 25.84 cm

13 **a** 5 cm² **b** 0.3% **c** 15.6 cm

14 **a** 10π cm **b** 24π cm²

15 **a** $\frac{72 + 20\pi}{9}$ cm **b** 3 : 7

Sample HSC problem

a $\angle PIQ = 110° - 43° = 67°$

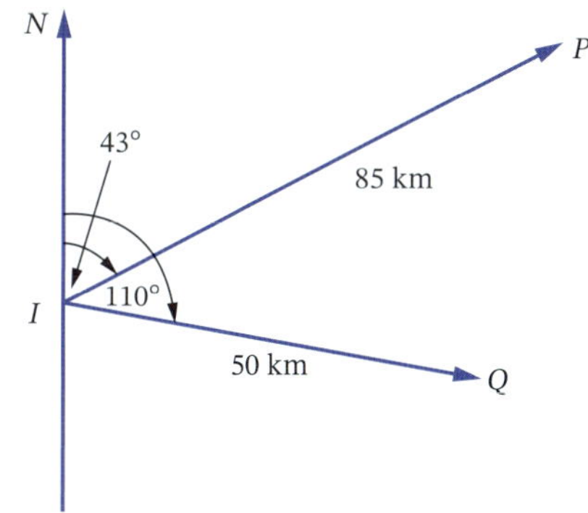

b 80 km **c** 188°

Test yourself 4

1 C **2** A, C **3** C, D **4** D

5 **a** $-\cos\theta$ **b** $-\tan\theta$ **c** $\sin\theta$

6 **a** $\theta = 46°3'$ **b** $\theta = 73°23'$ **c** $\theta = 35°32'$

7 122 km **8** $5\sqrt{3}$

9 **a** 6.3 cm **b** 8.7 m

10 **a** 42°58′ **b** 74°29′ **c** 226°19′ **d** 240°39′ **e** 324°18′

11 **a** $\theta = 65°5'$ **b** $\theta = 84°16'$ **c** $\theta = 39°47', 140°13'$

12 65.3 cm²

13 **a** 209° **b** 029°

14 **a** $AD = \frac{20\sin 39°}{\sin 99°}$ **b** 8.5 m

15 2951 km

16 **a** $\frac{\pi}{3}$ **b** $\frac{\pi}{4}$ **c** $\frac{5\pi}{6}$ **d** π **e** $\frac{\pi}{9}$

17 **a** $\frac{5\pi}{6}$ cm **b** $\frac{25\pi}{12}$ cm² **c** $\frac{25\pi - 75}{12}$ cm²

18 **a** $\sqrt{3}$ **b** $\frac{\sqrt{3}}{2}$ **c** $\frac{1}{\sqrt{2}}$ **d** $\frac{1}{\sqrt{3}}$ **e** $\frac{1}{\sqrt{2}}$ **f** $\frac{1}{2}$ **g** 1 **h** $\frac{1}{2}$ **i** $\frac{\sqrt{3}}{2}$

19 **a** $\frac{8\pi}{7}$ cm² **b** 0.12 cm²

20 54°19′ or 125°41′

21 **a** p **b** $-p$ **c** $-p$ **d** p

22 **a** $\frac{1}{\sqrt{2}}$ **b** $-\frac{\sqrt{3}}{2}$ **c** 0 **d** $-\frac{1}{\sqrt{2}}$ **e** $-\frac{\sqrt{3}}{2}$ **f** $\sqrt{3}$ **g** 1 **h** $\frac{1}{\sqrt{2}}$ **i** $-\frac{1}{\sqrt{3}}$

23 $16\sqrt{3}$ cm²

24 **a** 30° **b** 45° **c** 42° **d** 60°

Challenge exercise 4

1 92°58′ **2** $x = 12.7$ cm

3 **a** $AC = \frac{25.3\sin 39°53'}{\sin 58°53'}$ **b** $h = 19.5$ cm

4 31 m

5 42 cm² **6** 247.7 mm²

7 $\frac{9\pi - 18}{4}$ cm²

8 **a** $-\frac{\sqrt{3}}{2}$ **b** -1 **c** 0

9 $\frac{8\pi - 12\sqrt{3}}{3}$ cm²

10 perimeter 24.73 cm, area 21.89 cm²

ANSWERS

Chapter 5

Exercise 5.01

1 a i domain: all real $x, x \neq 0$, range: all real $y, y \neq 0$

ii no y-intercept

iii

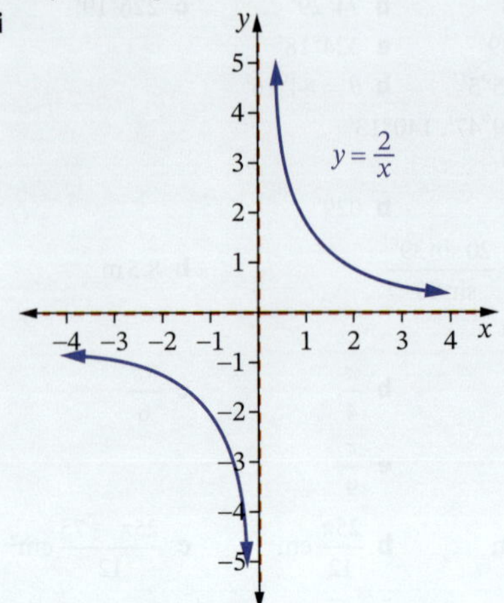

b i domain: all real $x, x \neq 0$, range: all real $y, y \neq 0$

ii no y-intercept

iii

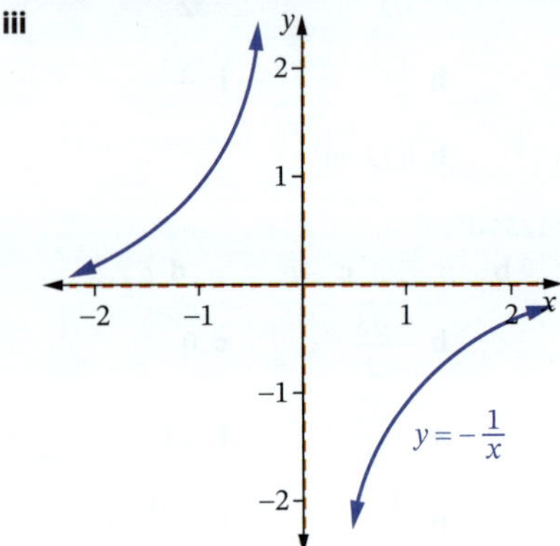

c i domain: all real $x, x \neq -1$, range: all real $y, y \neq 0$

ii 1

iii

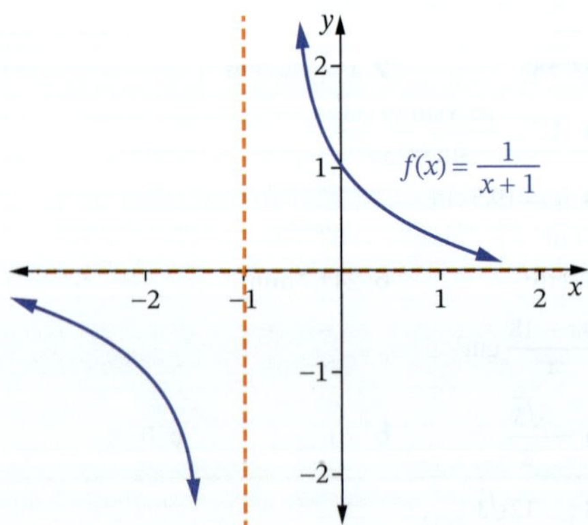

d i domain: all real $x, x \neq 2$, range: all real $y, y \neq 0$

ii $-1\frac{1}{2}$

iii

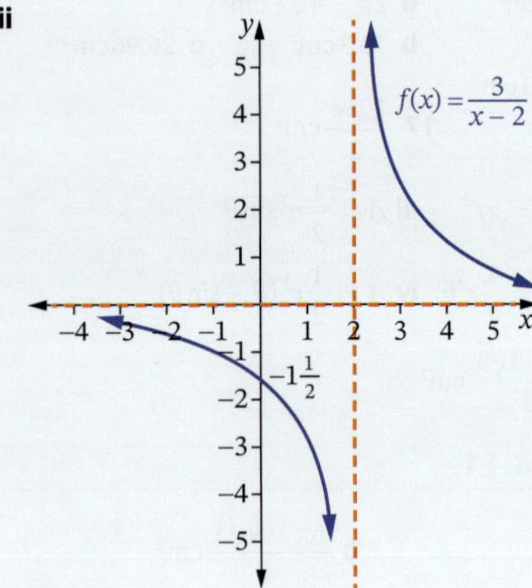

e i domain: all real $x, x \neq -2$, range: all real $y, y \neq 0$

ii $\frac{1}{6}$

iii

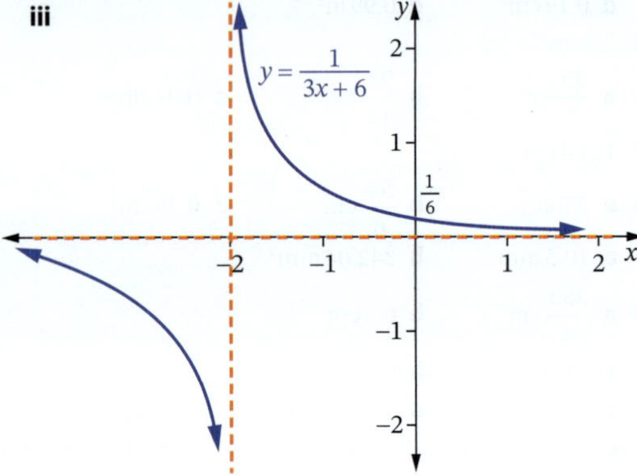

f i domain: all real $x, x \neq 3$, range: all real $y, y \neq 0$

ii $\frac{2}{3}$

iii

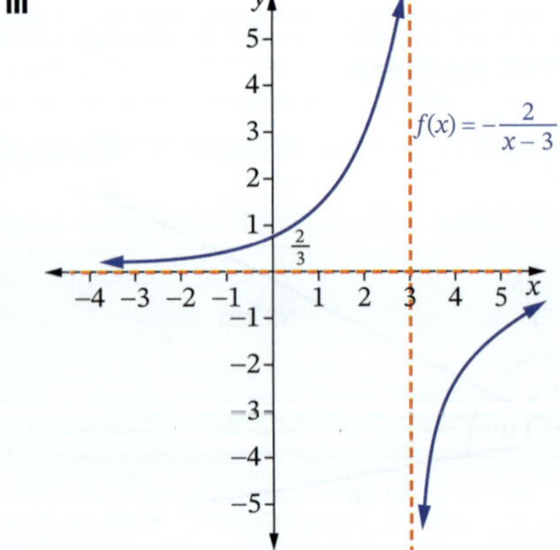

g **i** domain: all real $x, x \neq 1$,
range: all real $y, y \neq 0$

ii -4

iii

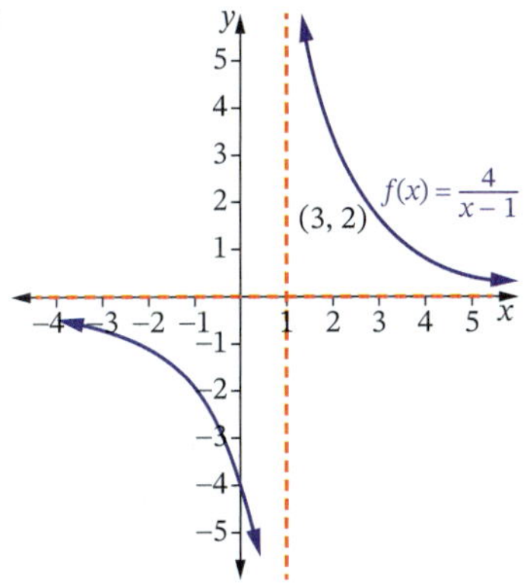

h **i** domain: all real $x, x \neq -1$,
range: all real $y, y \neq 0$

ii -2

iii

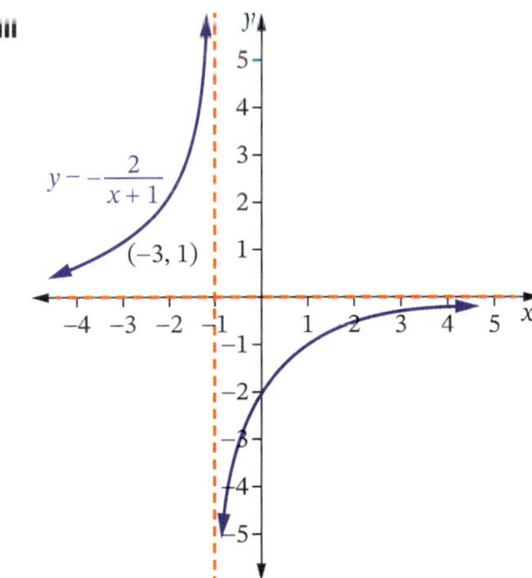

i **i** domain: all real $x, x \neq \frac{1}{2}$,
range: all real $y, y \neq 0$

ii $-\frac{2}{3}$

iii

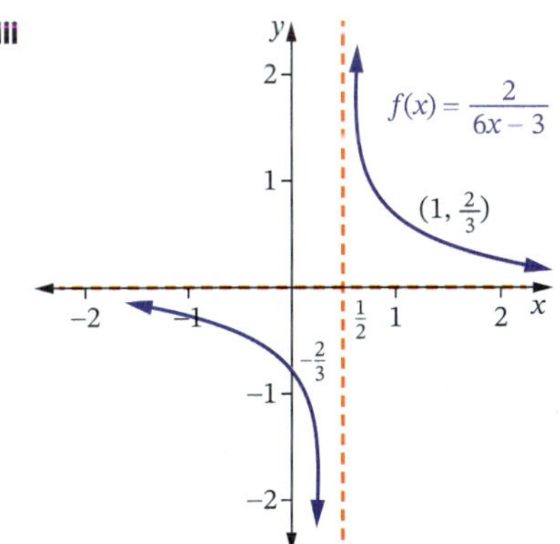

2 $f(-x) = \frac{2}{-x} = -\frac{2}{x} = -f(x)$

$\therefore$ odd function

3 **a** **i** yes **ii** neither **iii** no

b $x = -1, y = 0$

c domain: all real $x, x \neq -1$,
range: all real $y, y \neq 0$

4 **a** continuous **b** discontinuous
c continuous **d** discontinuous

5 not continuous at $x = 0$

6 **a** continuous **b** discontinuous at $x = -1$
c continuous **d** continuous
e discontinuous at $x = \pm 2$
f discontinuous at $x = -1$
g continuous **h** discontinuous at $x = -2$
i continuous

7 **a** $c = x^2 + 150x + 1000$ **b** 40

8 **a** See worked solutions.
b \$766.15
c There are two possible solutions: 28 km/h and 72 km/h

Exercise 5.02

1 **a** $N = 12x$ **b** $A = 2n$ **c** $C = 1.5x$
d $y = 4x$ **e** $w = 400x$

2 $c = 850x$

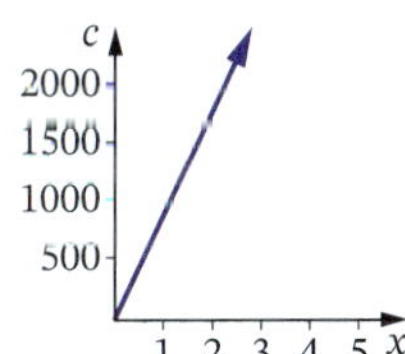

3 **a** $N = 36x$ **b** 1008 **c** 17

4 **a** $P = \frac{d}{3}$ **b** 280 **c** \$171

5 **a** $d = 0.004x^2$
b 40 m
c Yes, its braking distance is 25.6 m, less than 40 m.
d No, its braking distance is 48.4 m, greater than 40 m.
e 91 km/h

6 **a** $A = 5x^2$ **b** 88.2 cm^2 **c** 7.1 cm

7 **a** $C = \frac{176d^3}{49}$ $\left(\text{or } 3.5918...d^3\right)$
b 3592 mL **c** 4.1 cm

8 **a** $P = 18x$ **b** $A = 0.2x^2$ **c** 90 cm

9 **a** $V = 2.7988x^3$ **b** 605 cm^3 **c** 4.47 cm

10 **a** $d = 9t^3$ **b** $P = 5l^6$

Exercise 5.03

1 **a** $D = \frac{160}{x}$ **b** 200 mm **c** 1.4 mm

d

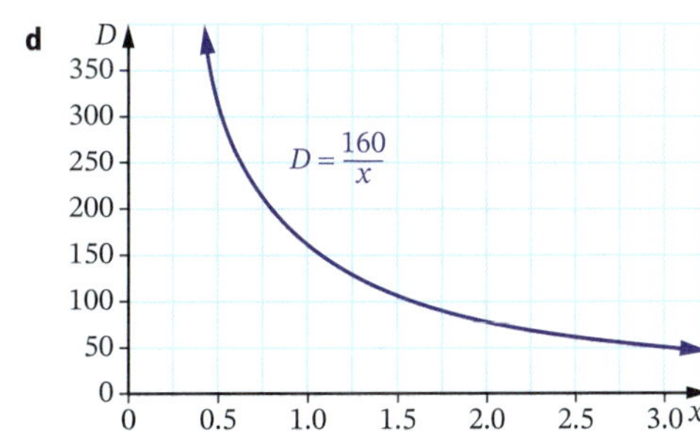

ANSWERS

2 **a** $c = \frac{256}{n}$ **b** \$2.56 **c** 512 boxes

d

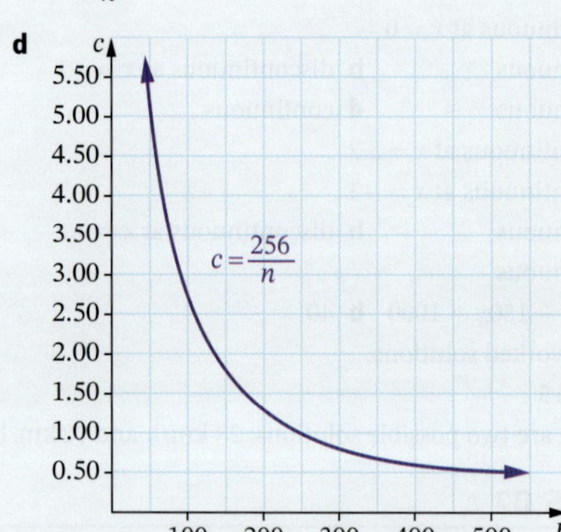

3 **a** $V = \frac{10\,500}{p}$ **b** $262.5\,\text{cm}^3$ **c** $21\,\text{kg/cm}^2$

4 **a** $I = \frac{2.88}{d^2}$ **b** 1.28 lumens **c** 0.44 m

5 **a** $y = \frac{12}{\sqrt{x}}$ **b** $3\sqrt{2}$ **c** 9

d

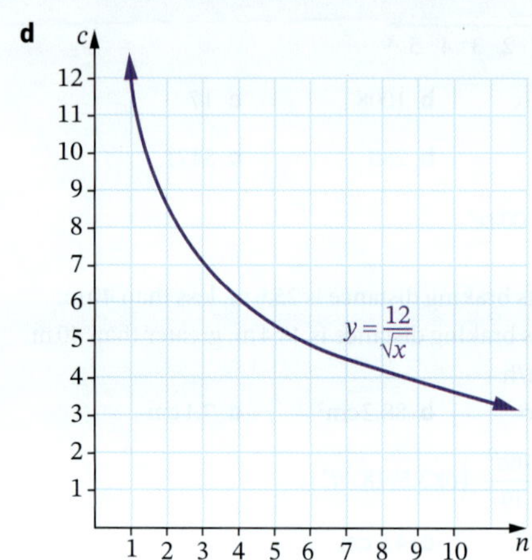

6 **a** $75\,\text{m/s}^2$ **b** 43.3 m

7 **a** $y = \frac{14}{\sqrt{x}}$ **b** $M = \frac{48}{x^4}$

8 6 **9** $y = \frac{38\,880}{x^5}$

Exercise 5.04

1 **a**

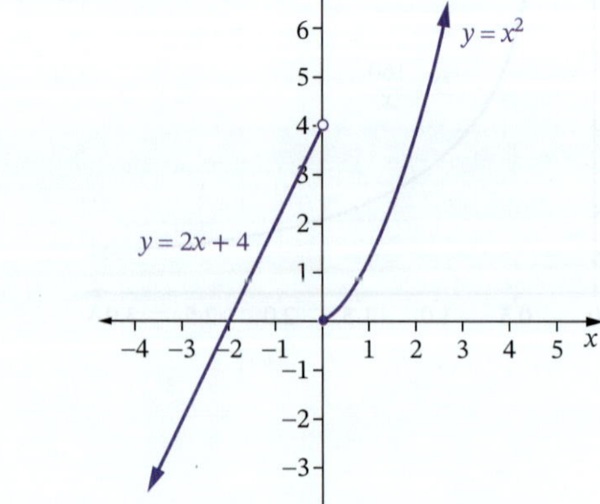

discontinuous at $x = 0$, domain: all real x, range: all real y

b

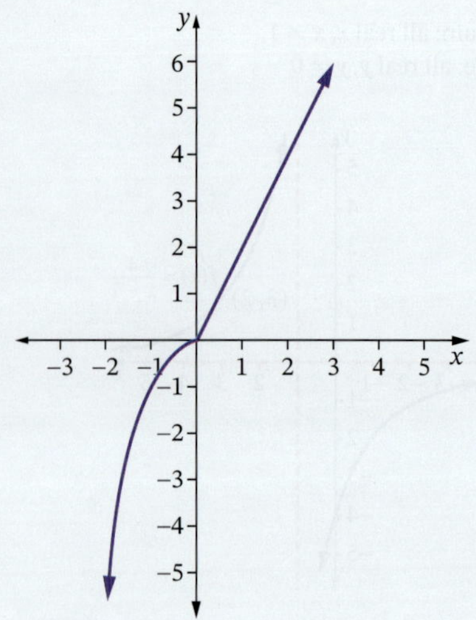

continuous, domain: all real x, range: all real y

c

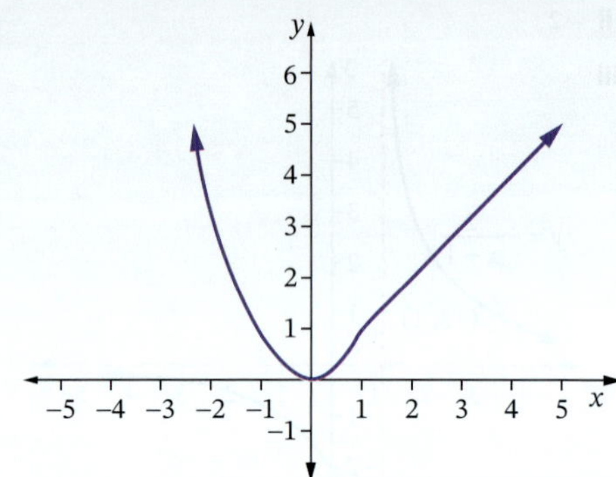

continuous, domain: all real x, range: $y \geq 0$

d

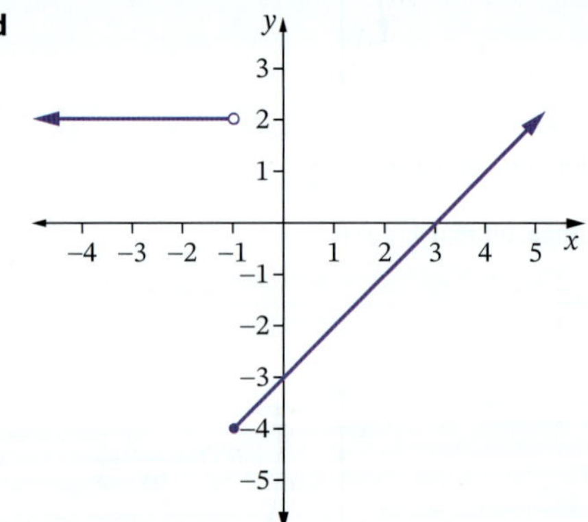

discontinuous at $x = -1$, domain: all real x, range: $y \geq -4$

e

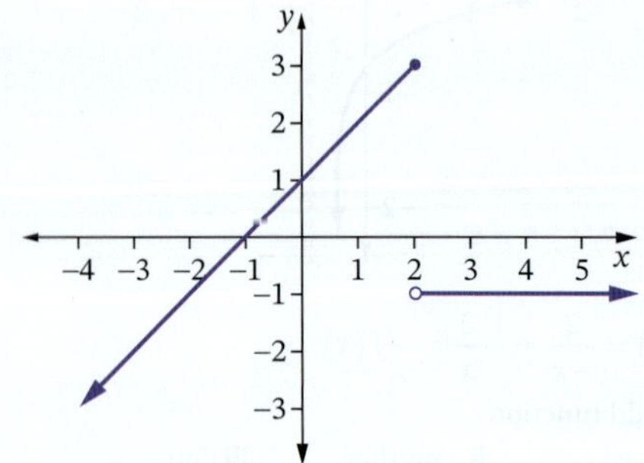

discontinuous at $x = 2$, domain: all real x, range: $y \leq 3$

2 a

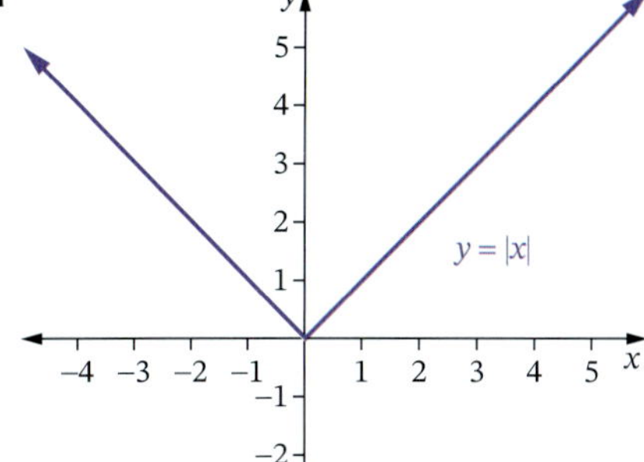

b

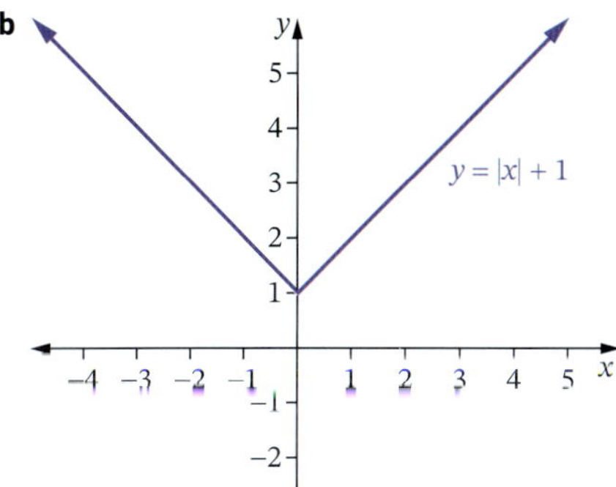

c

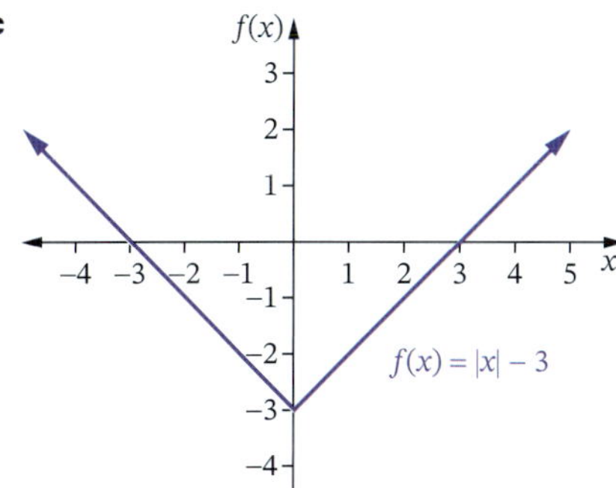

d

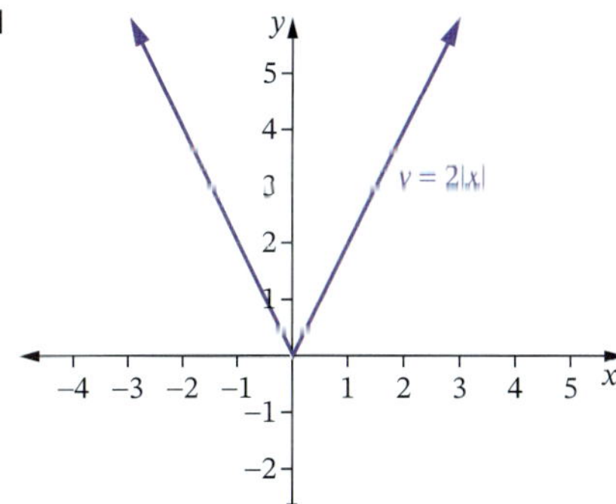

e

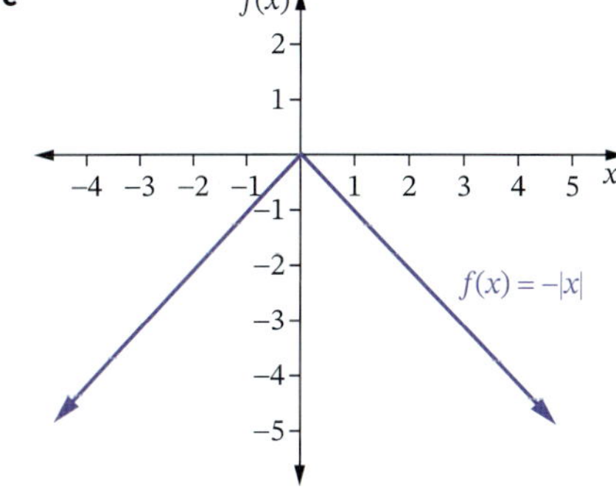

f

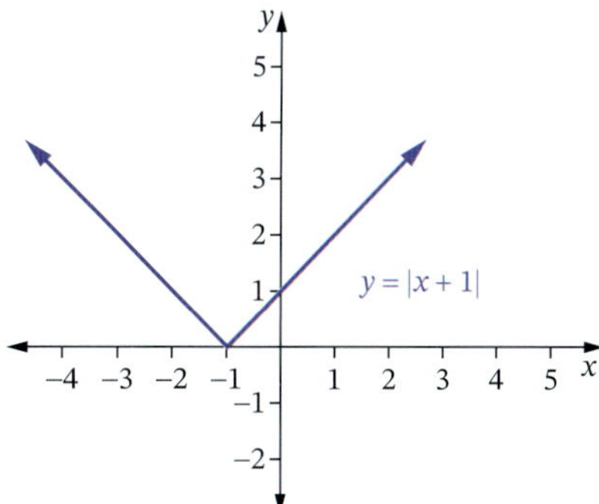

g

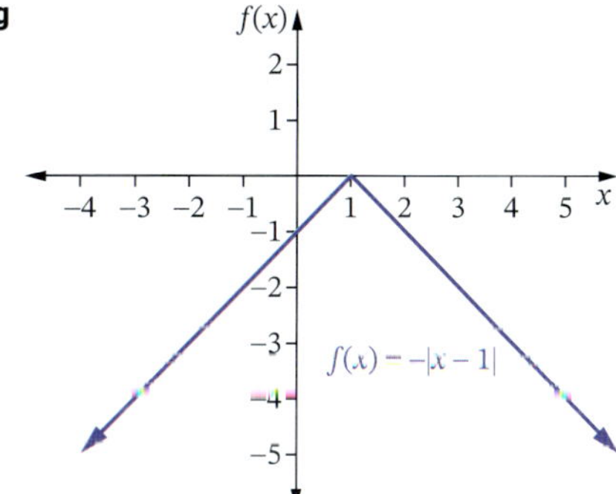

h

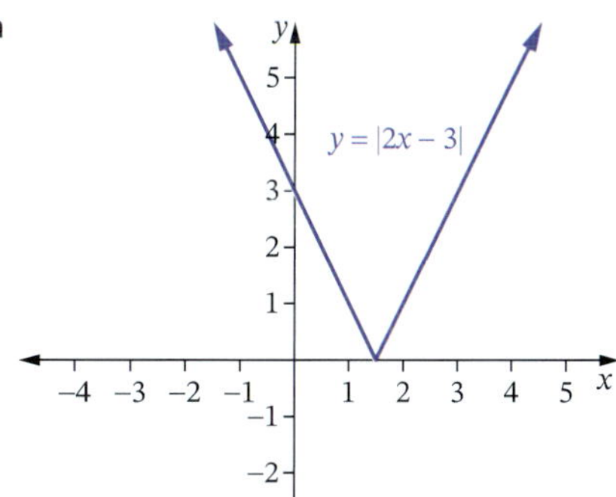

i

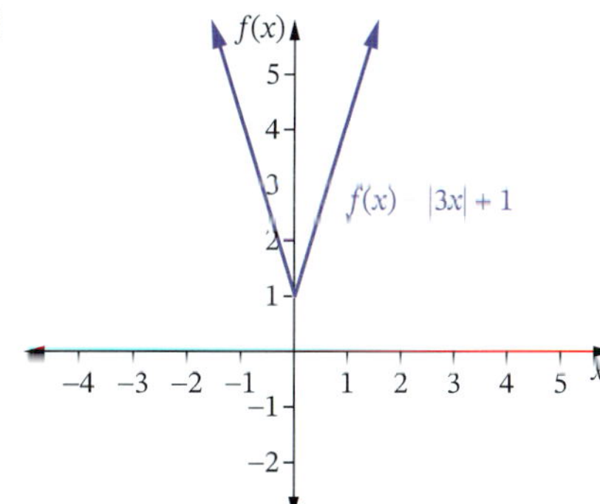

3 $f(-x) = 2|-x|$

$= 2|x|$

$= f(x)$, so even

4

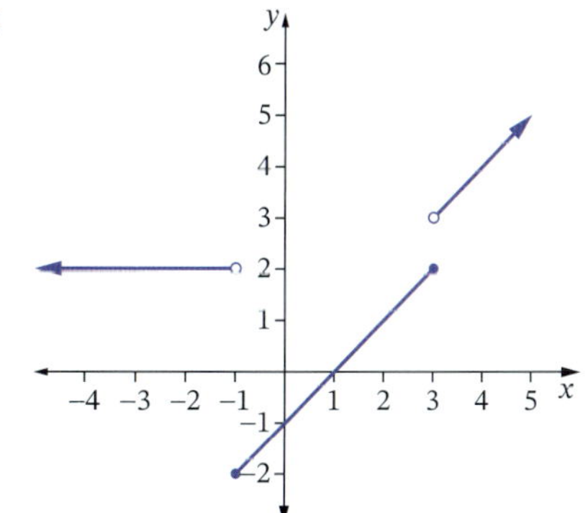

5 $y = |x| + 2$ is 2 units up from $y = |x|$, $y = |x + 2|$ is 2 units to the left of $y = |x|$.

6 **a** $y = |x + 1| - x$

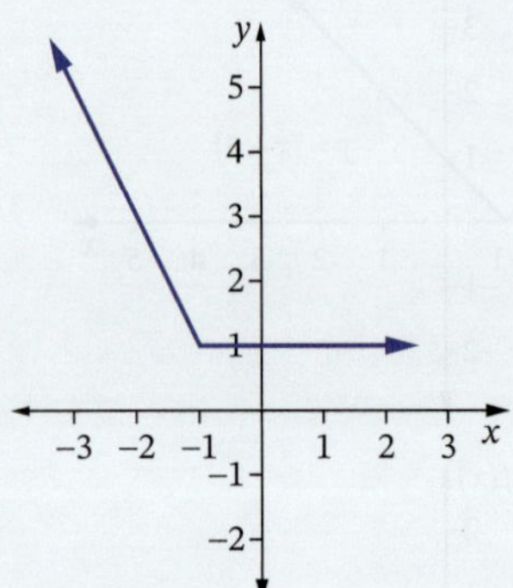

b $y = x^2 + |x|$

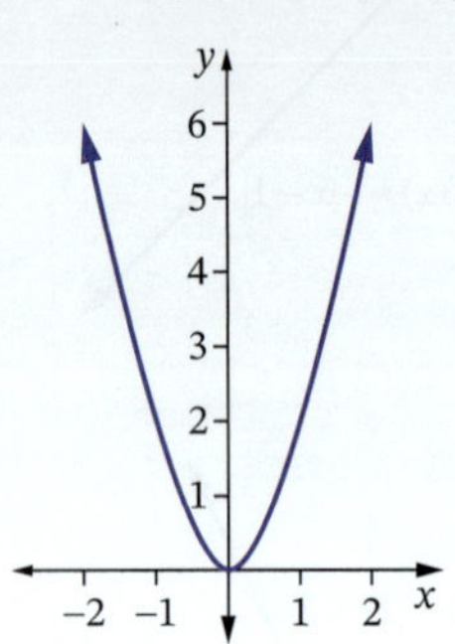

c $y = \frac{|x|}{x}$

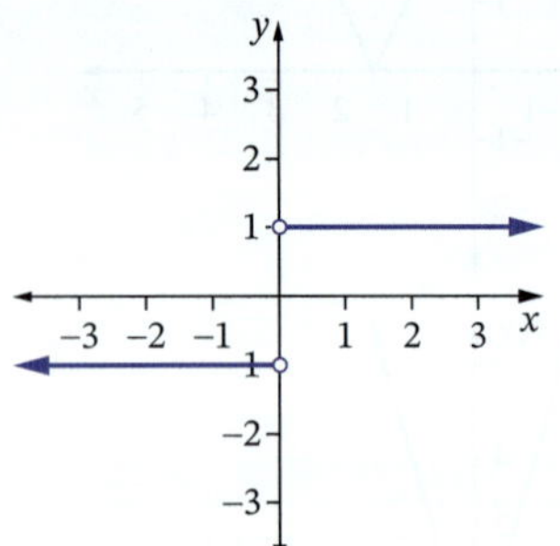

d $f(x) = \frac{|x|}{x^2}$

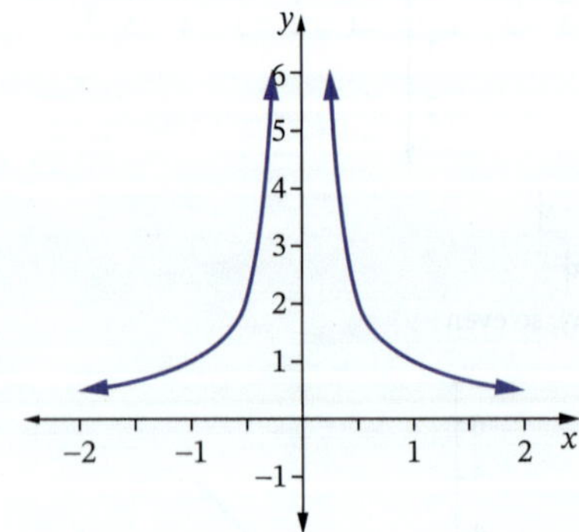

e $f(x) = |3 - ||x| - 2||$

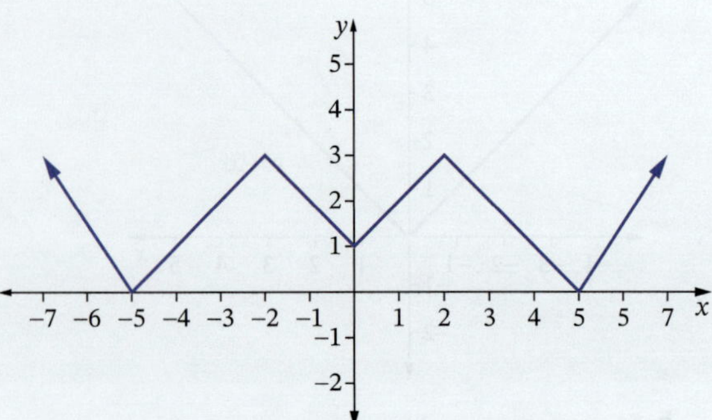

7 **a** $x = \pm 3$ **b** $x = -1, -3$ **c** $x = 3$
d $x = 1, 2$ **e** $x = 4, -7$ **f** $b = 2, -1.2$
g $x = \frac{1}{3}, -1$ **h** $x = 2, -3$ **i** $t = \frac{1}{2}$

8 $x = -2$

9 **a** $|3x - 6| \geq 0$ for all values of x, so if $x > |3x - 6|$, then $x > 0$

b $1\frac{1}{2} < x < 3$

10 $x < -2, -2 < x \leq 1$

Exercise 5.05

1 **a** **i**

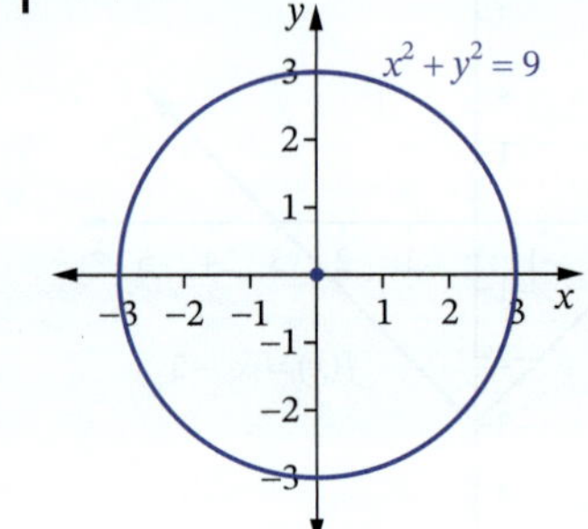

ii domain: $[-3, 3]$, range: $[-3, 3]$

b **i**

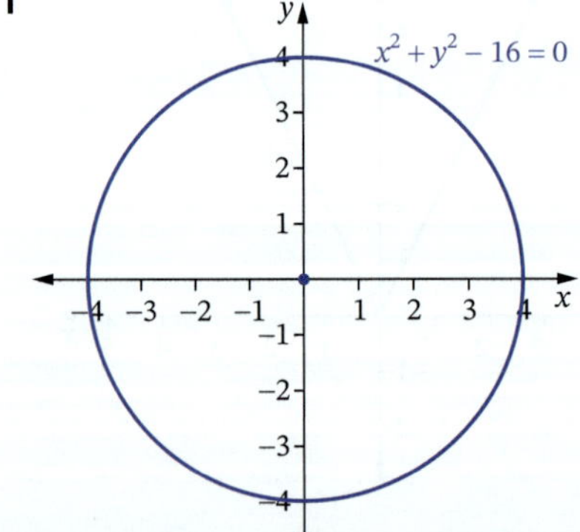

ii domain: $[-4, 4]$, range: $[-4, 4]$

c **i**

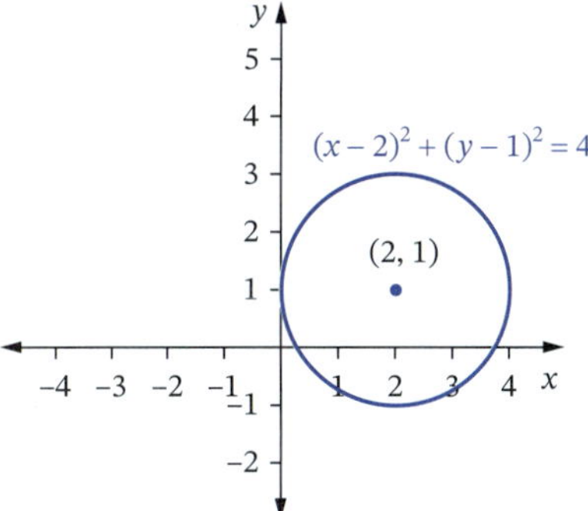

ii domain: [0, 4], range: [−1, 3]

d **i**

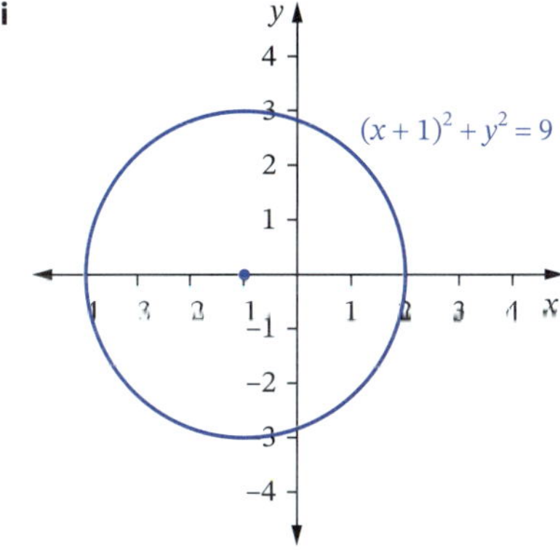

ii domain: [−4, 2], range: [−3, 3]

e **i**

$(x+2)^2+(y-1)^2=1$

(−2, 1)

ii domain: [−3, −1], range: [0, 2]

2 **a** **i** below x-axis

ii

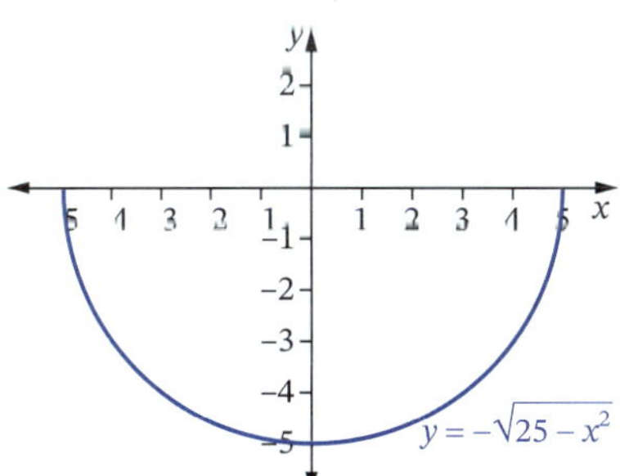

iii domain: [−5, 5], range: [−5, 0]

b **i** above x-axis

ii

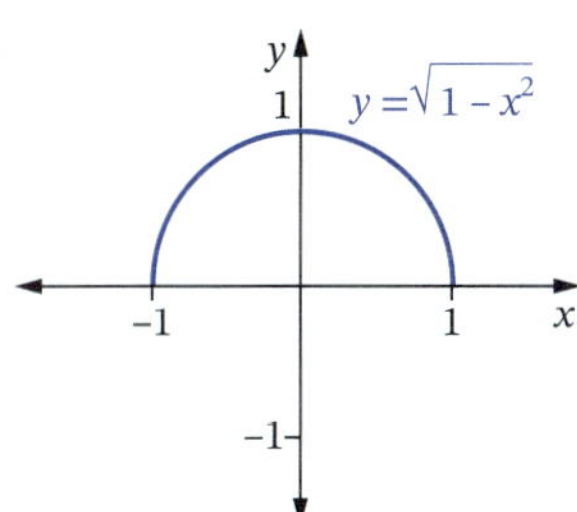

iii domain: [−1, 1], range: [0, 1]

c **i** right of the y-axis

ii

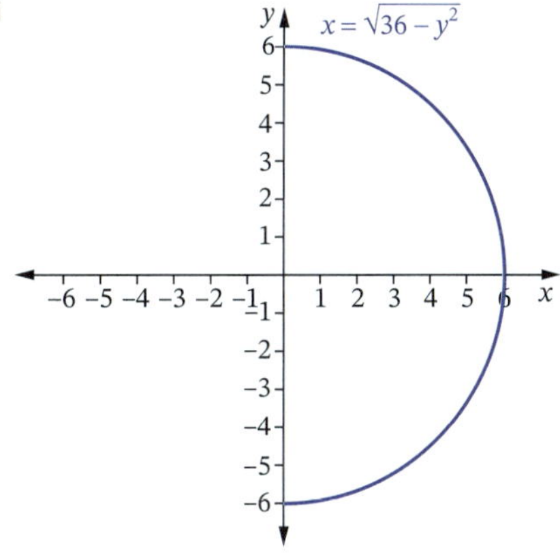

iii domain: [0, 6], range: [−6, 6]

d **i** below x-axis

ii

$y=-\sqrt{64-x^2}$

iii domain: [−8, 8], range: [−8, 0]

e **i** below x-axis

ii

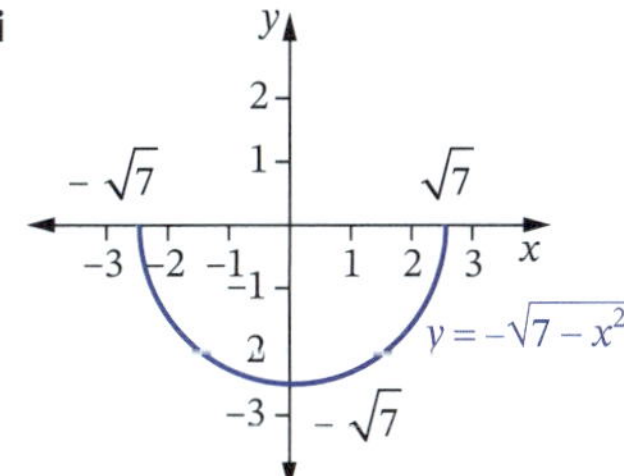

iii domain: $[-\sqrt{7}, \sqrt{7}\,]$, range: $[-\sqrt{7}, 0]$

f **i** left of the y-axis

ii

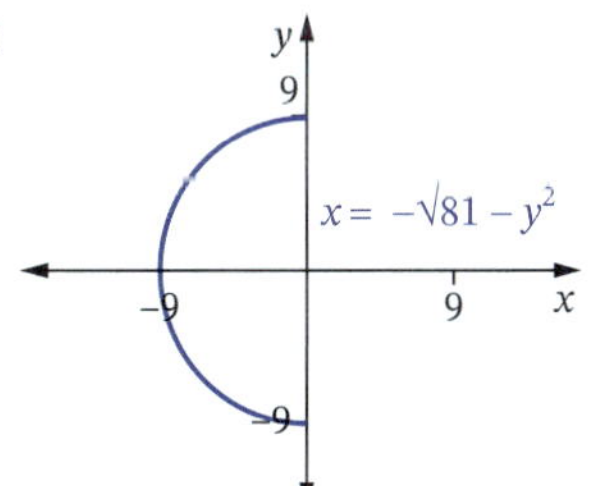

iii domain: [−9, 0], range: [−9, 9]

3 **a** 10, (0, 0) **b** $\sqrt{5}$, (0, 0)
c 4, (4, 5) **d** 7, (5, −6)
e 9, (0, 3)

4 **a** $x^2+y^2=16$
b $x^2-6x+y^2-4y-12=0$
c $x^2+2x+y^2-10y+17=0$
d $x^2-4x+y^2-6y-23=0$
e $x^2+8x+y^2-4y-5=0$
f $x^2+y^2+4y+3=0$
g $x^2-8x+y^2-4y-29=0$
h $x^2+6x+y^2+8y-56=0$
i $x^2+4x+y^2-1=0$
j $x^2+8x+y^2+14y+62=0$

5 **a** 3, (2, 1) **b** 5, (−4, 2)
c 1, (0, 1) **d** 6, (5, −3)
e 1, (−1, 1) **f** 6, (6, 0)
g 5, (−3, 4) **h** 8, (−10, 2)
i 5, (7, −1) **j** $\sqrt{10}$, (−1, −2)

6 **a** centre (−2, 1), radius 2

b

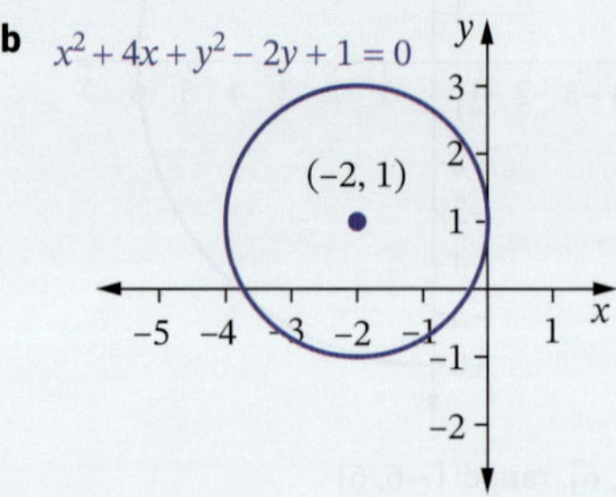

c (0, 1) and (−2, 3)

7 **a** 0, 1 or 2 **b** one point of intersection $\left(-\frac{4}{5}, \frac{3}{5}\right)$

8 $f(-x) = \sqrt{r^2 - (-x)^2} = \sqrt{r^2 - x^2} = f(x)$, so $f(x)$ is even.

9 **a** 66 mm² **b** 22 mm²

Exercise 5.06

1 **a** $y = (x^2 + 1)^2 = x^4 + 2x^2 + 1$
b $y = (5x - 3)^3 = 125x^3 - 225x^2 + 135x - 27$
c $y = (x^2 - 3x + 2)^7$ **d** $y = \sqrt{2x - 1}$
e $y = \sqrt[3]{x^4 + 7x^2 - 4}$ **f** $y = 6x + 3$
g $y = 2x^3 - 7$ **h** $y = 6x^2 - 5$
i $y = 18x^2$ **j** $y = 4x^4 + 24x^2 + 37$

2 **a** domain: all real x, range: $y \geq 0$
b domain: all real x, range: all real y
c domain: $x \geq 2$, range: $y \geq 0$
d domain: $x \geq -3$, range: $y \leq 0$
e domain: [−2, 2], range: [0, 2]
f domain: [−1, 1], range: [−1, 0]

3 **a** $y = \sqrt{x^3}$, 8 **b** $y = (\sqrt{x})^3$, 125

4 **a** $y = \frac{3x^2 + 1}{x^2}, \frac{13}{4}$ **b** $y = \frac{1}{x^2 + 3}, \frac{1}{12}$

5 **a** $f(g(x)) = (x - 2)^3$, neither odd nor even
b $g(f(x)) = x^3 - 2$, neither odd nor even
c −1
d −10

6 **a** $y = f(g(x))$

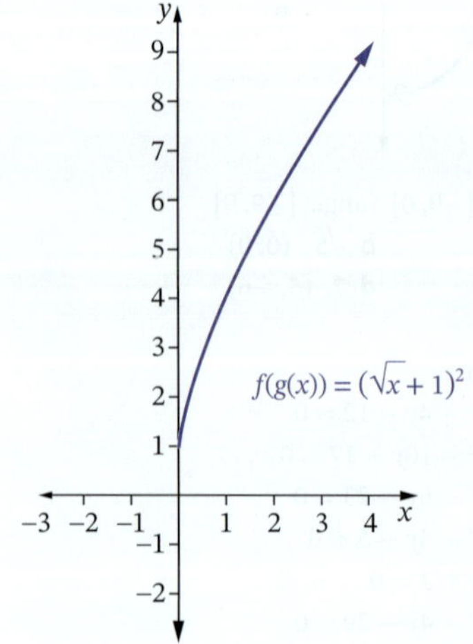

b $y = g(f(x))$

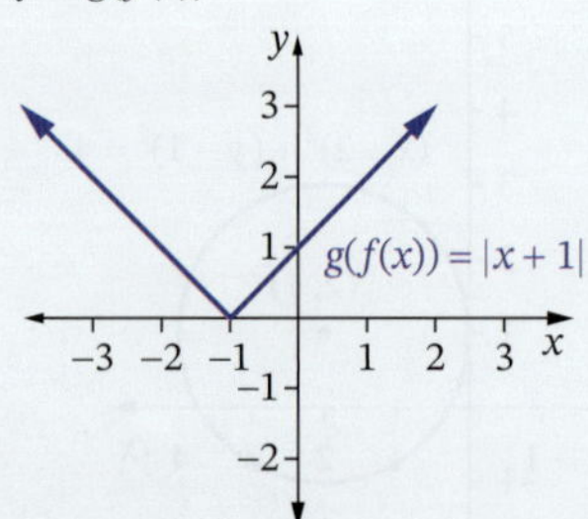

7–8 See worked solutions.

9 D

10 **a** $y = -\frac{1}{|x|}$, domain: all real x, $x \neq 0$, range: $y < 0$

b $y = \left|-\frac{1}{x}\right|$, domain: all real x, $x \neq 0$, range: $y > 0$

Sample HSC problem

a

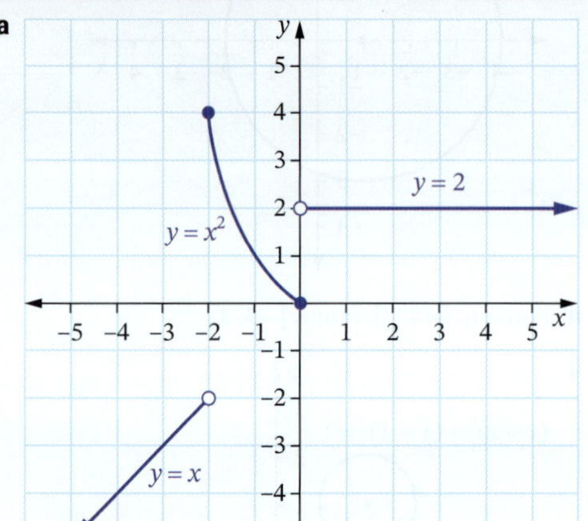

b $x = -2, 0$

c domain: all real x, range: $y < -2$, $0 \leq y \leq 4$

Test yourself 5

1 B **2** A **3** C

4 **a** $A = \frac{150}{n}$

b **i** 15 cm² **ii** 18.75 cm²
c **i** 9 **ii** 6

5 **a**

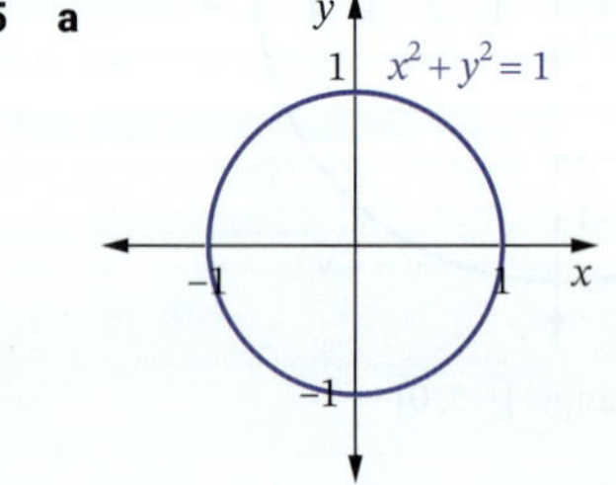

b

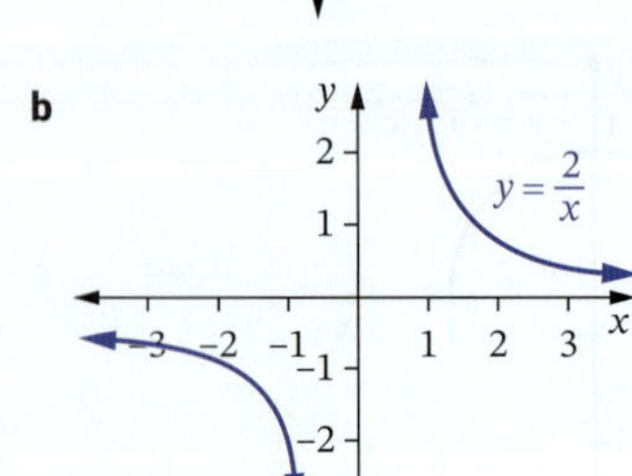

c

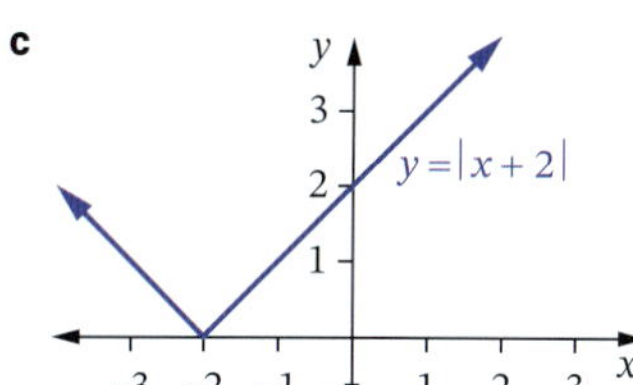

d

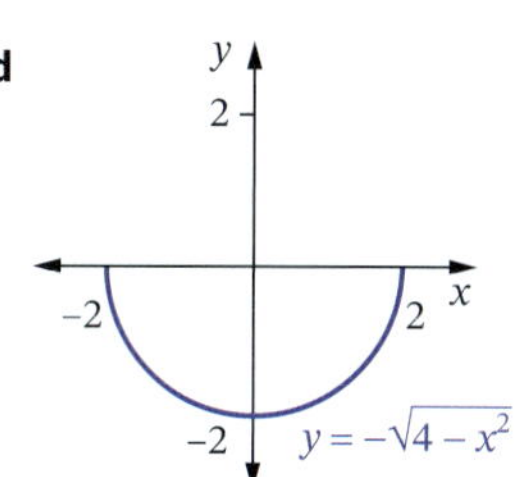

6 radius 4, centre (3, 1)

7 **a** $y = (3x - 1)^3 = 27x^3 - 27x^2 + 9x - 1$

b $y = 3x^3 - 1$

8 **a** no **b** $y = \pm\sqrt{1 - x^2}$

c

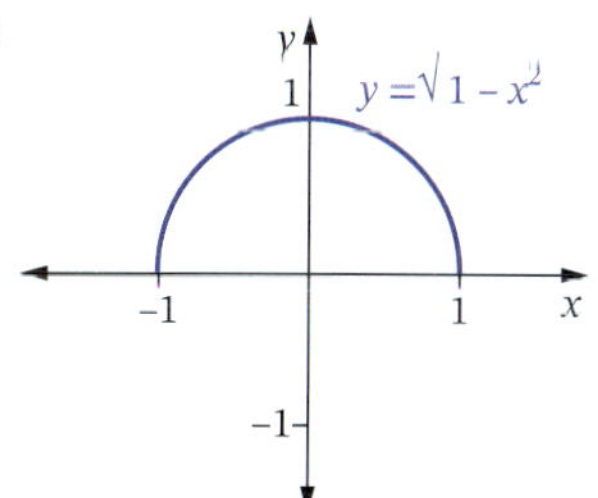

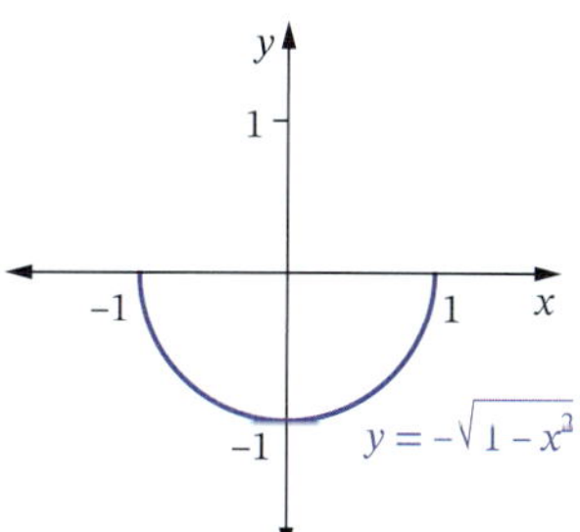

9 **a** domain: $[-4, 4]$, range: $[-4, 4]$

b domain: all real x, $x \neq -2$, range: all real y, $y \neq 0$

c domain: all real x, range: $y \geq 3$

d domain: $[-3, 3]$, range: $[0, 3]$

10 **a** $A = 7x^2$ **b** 700 cm^2 **c** 12.5 cm

11 1.34

12 **a** domain: all real x, $x \neq 3$, range: all real y, $y \neq 0$

b

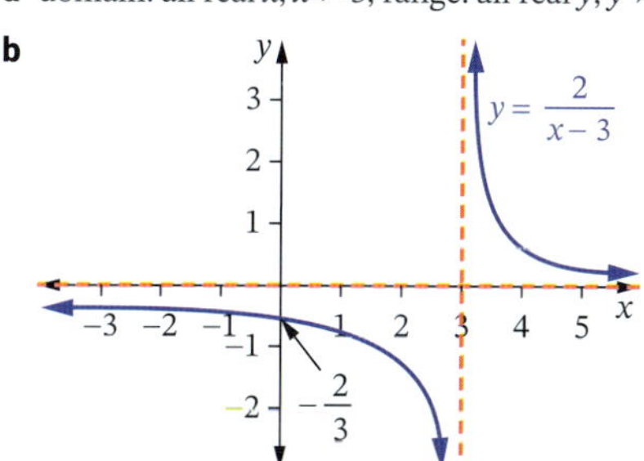

13 **a** radius 10, centre (0, 0)

b radius 11, centre (3, 2)

c radius 3, centre $(-3, -1)$

14 **a** x-intercepts $0, \pm2$, y-intercept 0

b no x-intercepts, y-intercept -2

c x-intercepts ±3, y-intercepts ±3

d x-intercepts ±5, y-intercept 5

e no x-intercepts, y-intercept 5

15 **a**

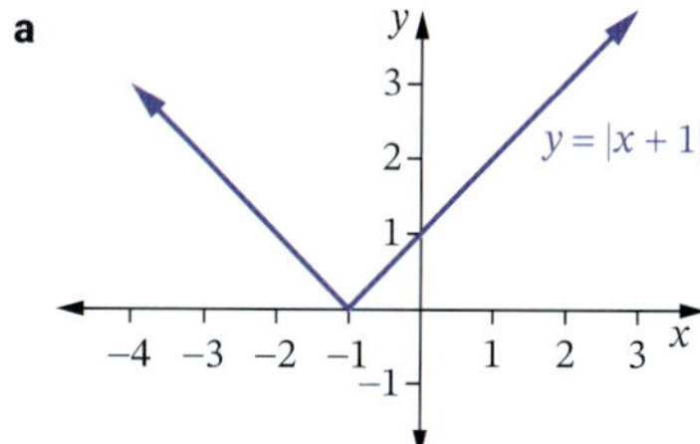

b $x = 2, -4$

Challenge exercise 5

1 $x = 3$

2 **a**

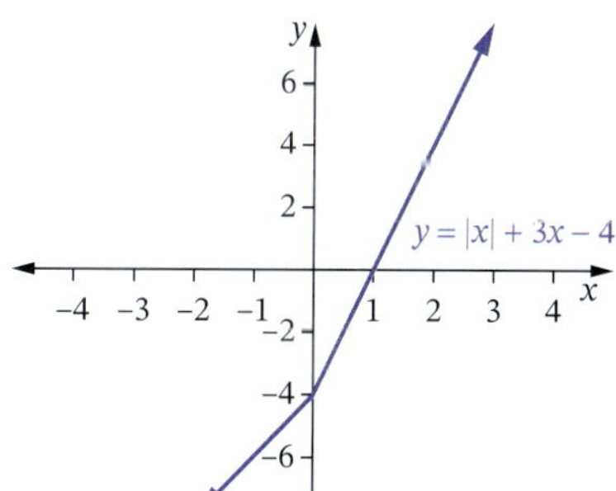

b

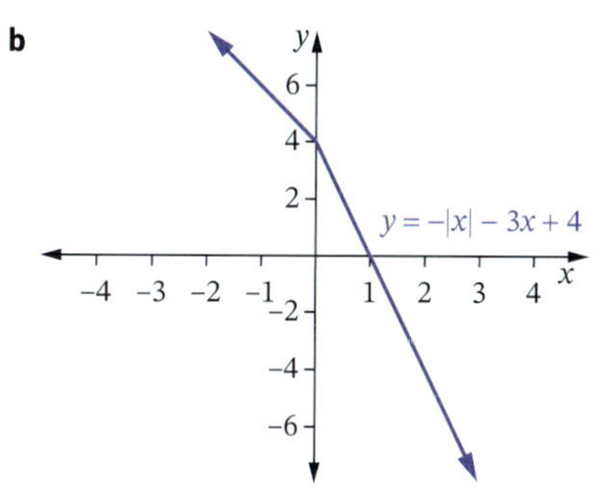

3 centre $\left(-1\frac{1}{2}, 1\right)$, radius $4\frac{1}{2}$

4 $3x + y + 2 = 0$

5

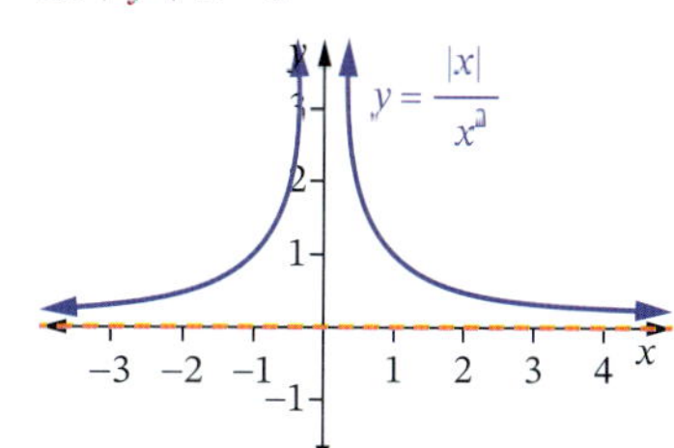

6 **a** See worked solutions.

b domain: all real x, $x \neq -3$, range: all real y, $y \neq 2$

c

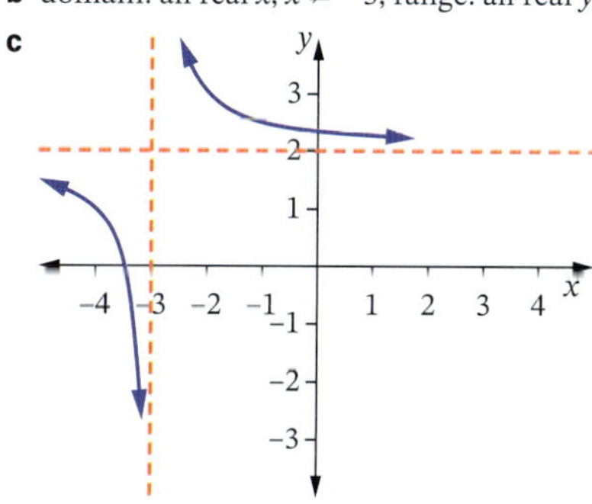

7 Both circles have same centre $(1, -2)$, so concentric.

ANSWERS

8

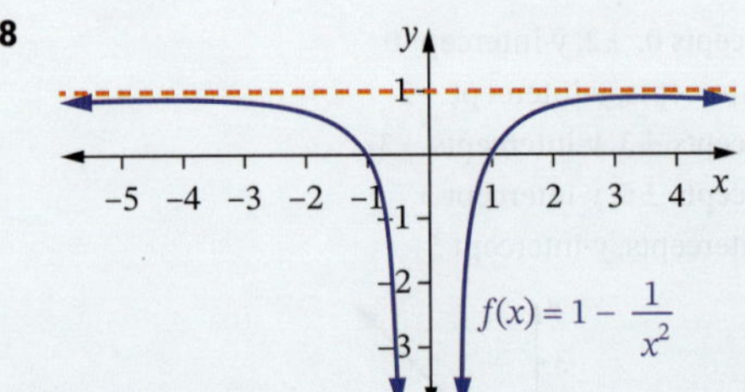

Practice set 2

1 B, D	**2** A	**3** C	**4** D
5 B	**6** A	**7** C	**8** C
9 D	**10** B		

11 a

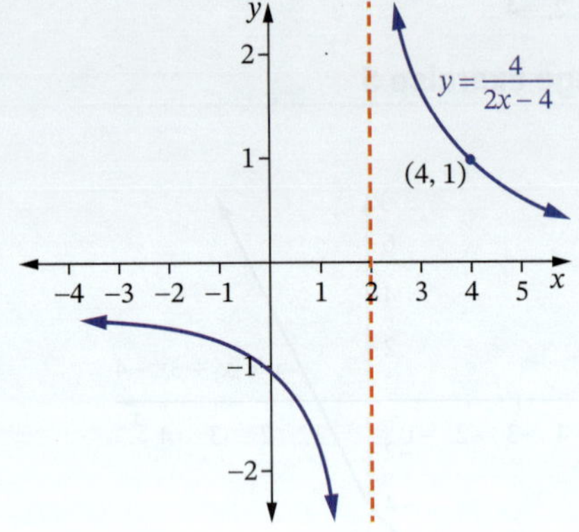

domain: all real x, $x \neq 2$, range: all real y, $y \neq 0$

b

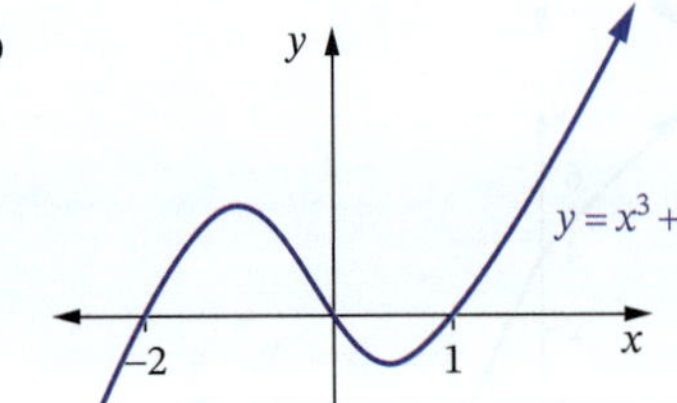

domain: all real x, range: all real y

c

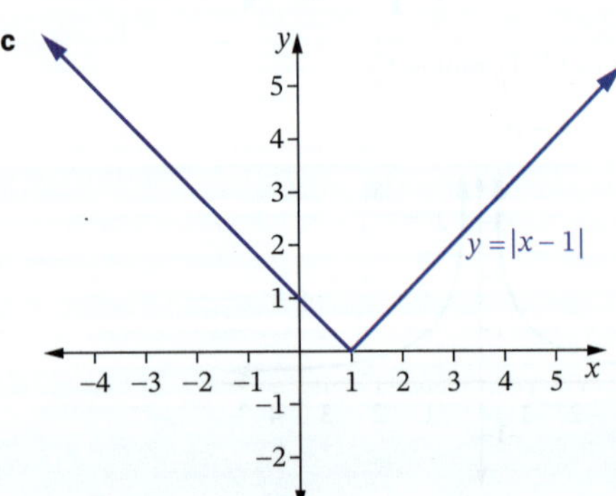

domain: all real x, range: $y \geq 0$

d

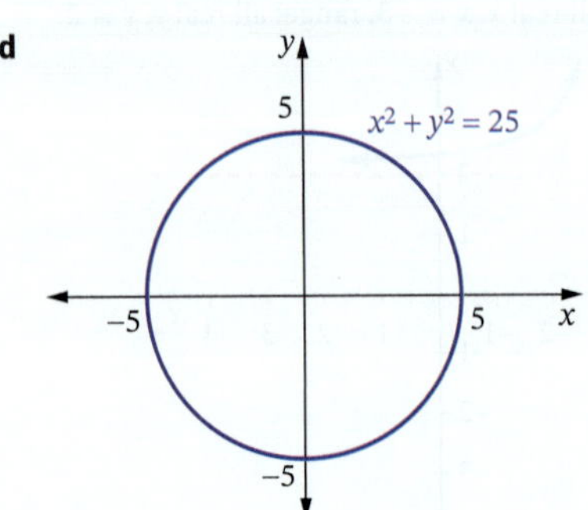

domain: $[-5, 5]$, range: $[-5, 5]$

e

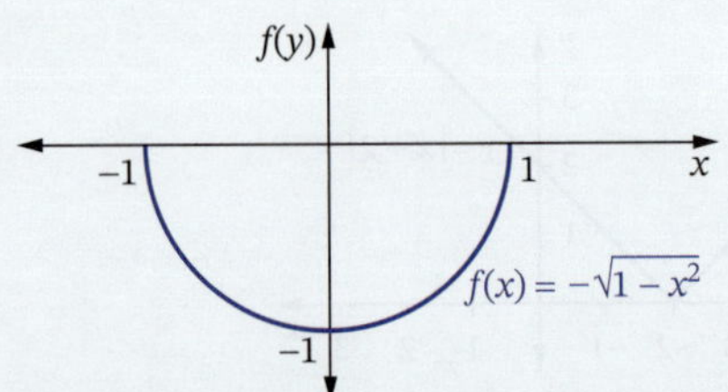

domain: $[-1, 1]$ range: $[-1, 0]$

f

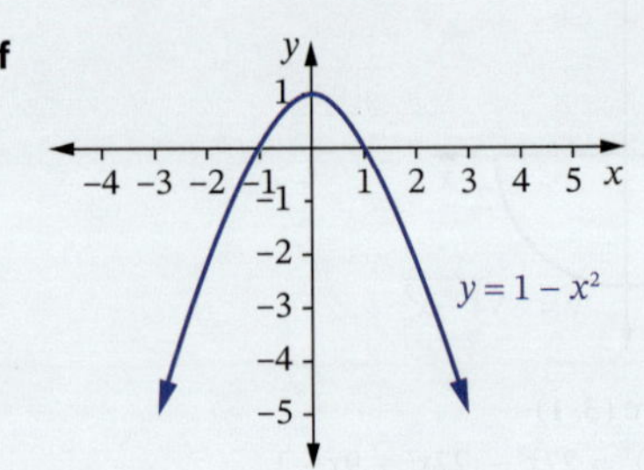

domain: all real x, range: $y \leq 1$

12 a centre $(-1, 3)$, radius 4
b domain: $[-5, 3]$, range: $[-1, 7]$

13 a $52°26'$ **b** 16.6 m^2

14 a $y = -2x + 3$ **b** $x - 5y - 5 = 0$
c $4x - 5y - 8 = 0$ **d** $5x - 4y - 41 = 0$
e $2x + 3y - 3 = 0$ **f** $x - 8y + 15 = 0$
g $x + y - 4 = 0$

15 a 11.25 m^2 **b** 5.6 m or 1.4 m

c

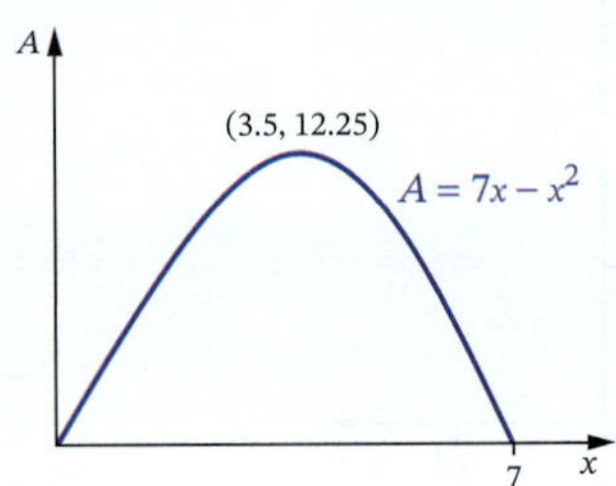

d 12.25 m^2

16 a $\frac{\pi}{3}$ **b** $\frac{5\pi}{6}$ **c** $\frac{\pi}{2}$
d $\frac{\pi}{18}$ **e** $\frac{7\pi}{4}$

17 a

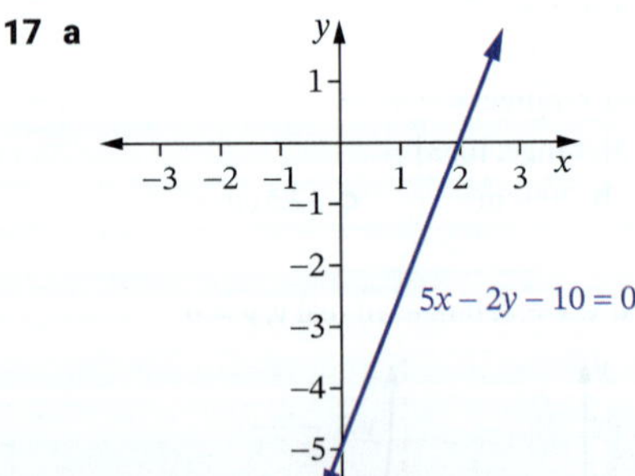

b

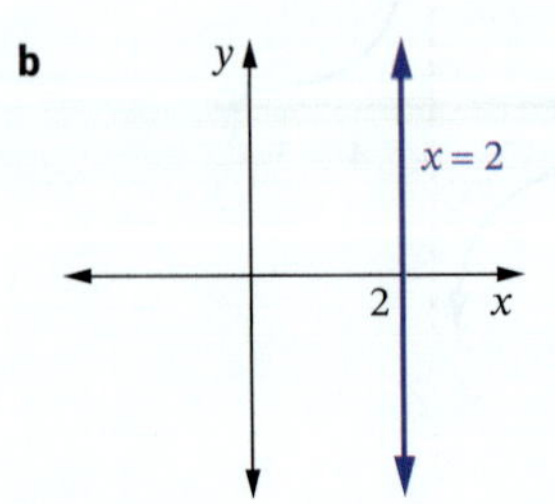

c

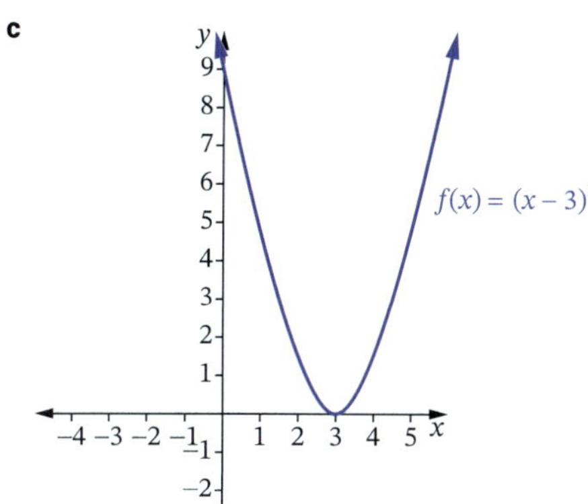

d

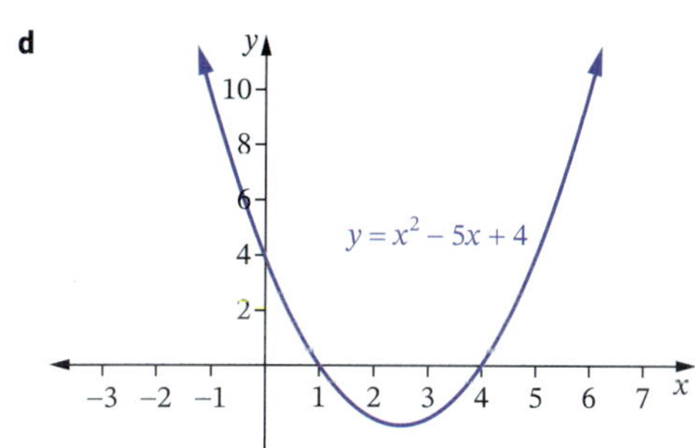

e

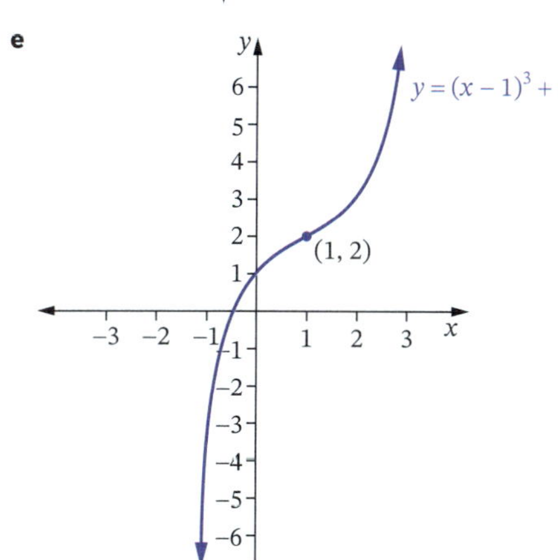

18 **a** 97°24′ **b** 20°38′ **c** 145°32′

19 **a** $m_1m_2 = \frac{3}{4} \times \frac{8}{6} = 1$. So, perpendicular.

b $\left(-1, 1\frac{1}{2}\right)$

20 **a** 0 **b** -2 **c** $|m+1| - 2$

21 **a** $g(2) = 1, g(-3) = -6$

b

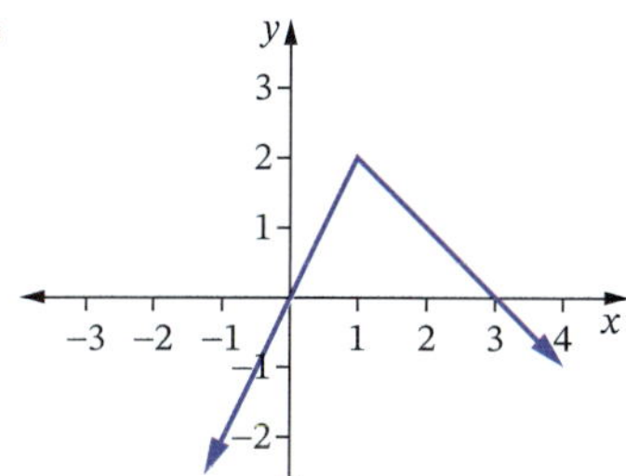

22 $x = 3$ **23** $f(-1) = 7$

24 $\Delta = 0$ so only one point of intersection. See worked solutions.

25 **a** 45° **b** 270° **c** 36°

d 157°30′ **e** 1080°

26 **a** $-\frac{1}{\sqrt{3}}$ **b** $\frac{1}{\sqrt{2}}$ **c** $-\frac{\sqrt{3}}{2}$

d $\frac{1}{\sqrt{2}}$ **e** $-\frac{\sqrt{3}}{2}$ **f** $-\frac{1}{\sqrt{3}}$

27 **a** $a < 0, \Delta < 0$ **b** $a > 0, \Delta < 0$

28 **a** $x = -1, 3$ **b** $2 < x \leq 5$

c $x = 1, -1\frac{2}{3}$ **d** $x \leq -2, x \geq 2$

29 **a** $d = 4.8t$

b **i** 9.6 km **ii** 24 km

c **i** 6.25 h **ii** 3.96 h

d 4.8 km/h

30 **a** $\frac{8\pi}{9}$ cm **b** $\frac{16\pi}{9}$ cm^2

31 $y = x^2 - 2x - 3$

32 $\Delta = 361$ (> 0 and a perfect square)

33 76°52′ **34** 45°49′

35 $x^2 + 4x + y^2 + 6y - 12 = 0$

36 52°37′

37 **a** $2x - y + 4 = 0$ **b** $P(-2, 0), Q(0, 4)$

38 $f(-x) = x^6 - x^2 - 3$ **39** 2°8′

40 **a** $y = (2x+5)^3$ **b** $y = 2x^3 + 5$

c 27 **d** 59

41 **a** $-\tan\theta$ **b** $-\sin\theta$ **c** $\cos\theta$

42 **a** **i** N 50° E **ii** 050°

b **i** S 20° W **ii** 200°

c **i** S 40° E **ii** 140°

d **i** N 50° W **ii** 310°

43 **a** $\frac{5\pi}{6}$ cm **b** $\frac{25\pi}{12}$ cm^2 **c** $\frac{25(\pi-3)}{12}$ cm^2

44 117°56′ **45** $y = 16.5$

46 **a** (4, −1) **b** (3, 9), (−2.5, 6.25)

47 **a** **i** $x = 3$ **ii** minimum (3, −8)

b **i** $x = -1$ **ii** maximum (−1, −1)

48 17.5 m **49** **a** 7 m **b** 27.8 m^2

50 127 m **51** 21π cm^2 **52** 4.9 km

53 **a** domain: all real x, $x \neq -4$, range: all real y, $y \neq 0$

b domain: all real x, range: $y \geq 2$

c domain: $[-2, 2]$, range: $[-2, 0]$

d domain: all real x, range: $y = 4$

e domain: all real x, range: $y \geq -3$

54 **a** 16.85 km **b** 228°

55 $f(-x) = -x^3 + 5x = -(x^3 - 5x) = -f(x)$

56 **a**

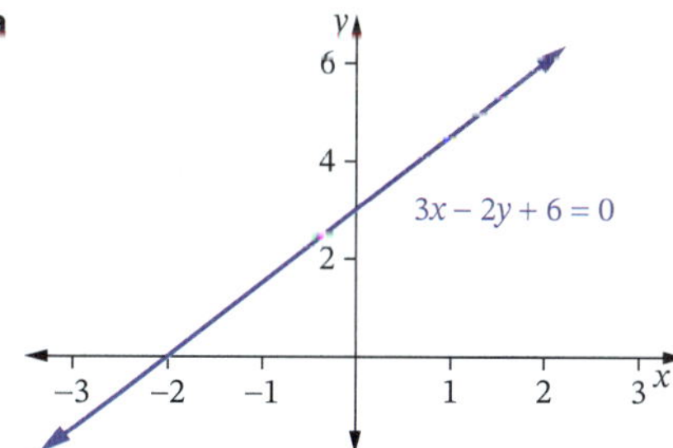

b

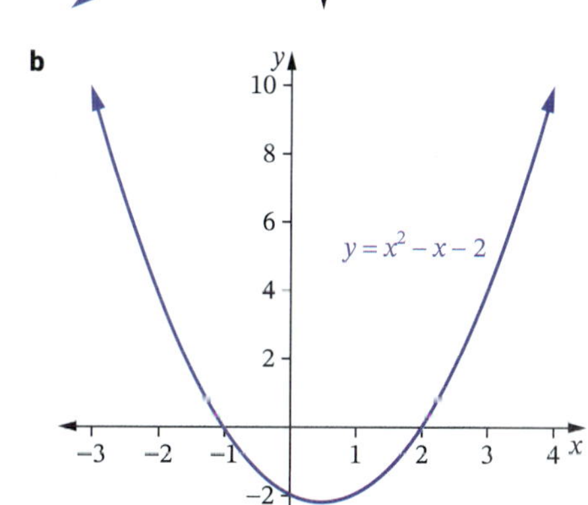

c

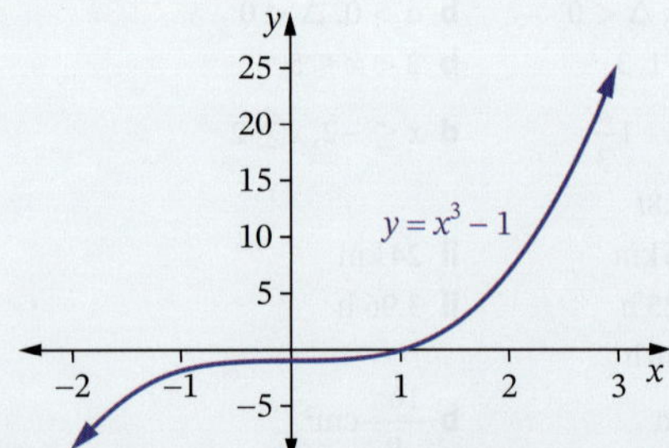

d

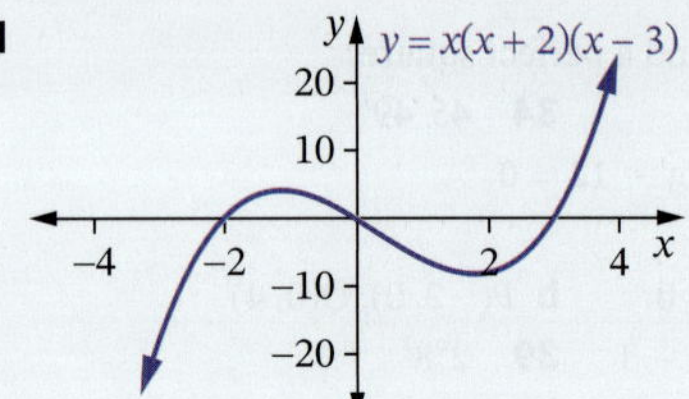

57 1.6 cm²

58 **a** 360° **b** 30° **c** 405°

59 489.3 km

60 **a** $b \geq 2$ **b** $x = 0, 3$
c $n = 2, -7$ **d** $-3 < n < 3$

Chapter 6

Exercise 6.01

1 **a** 4.060401 **b** 3.994004 **c** 4

2 **a** **i** 13.61 **ii** 13.0601 **iii** 12.9401
b 13

3 **a**

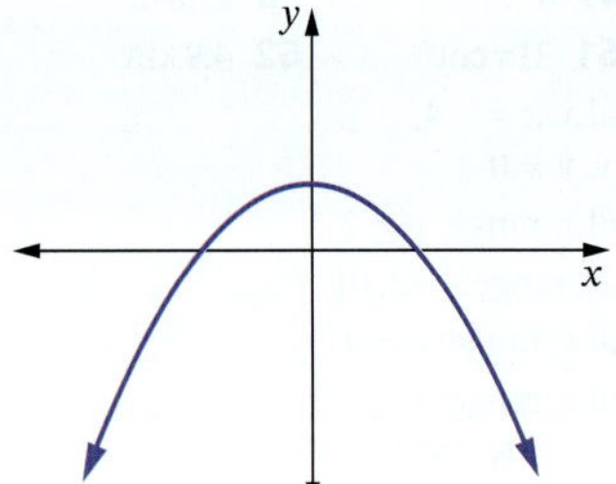

b

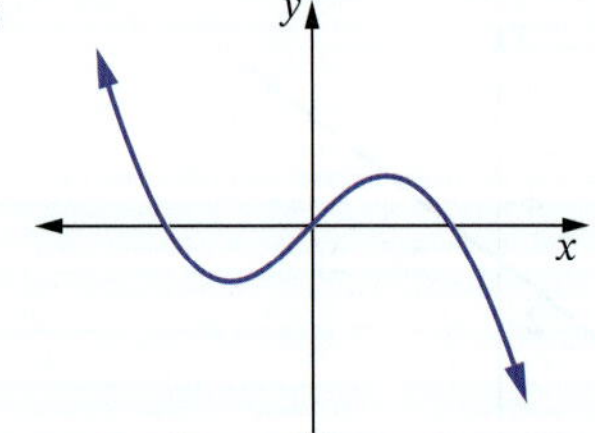

c

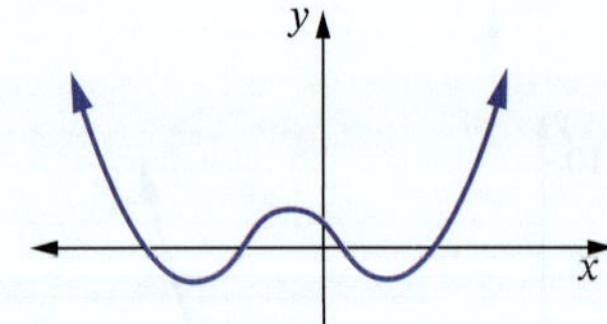

d

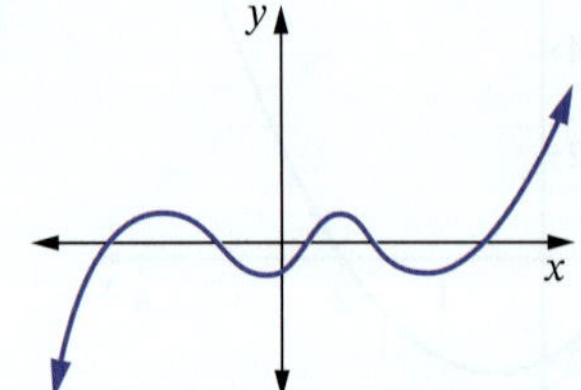

e

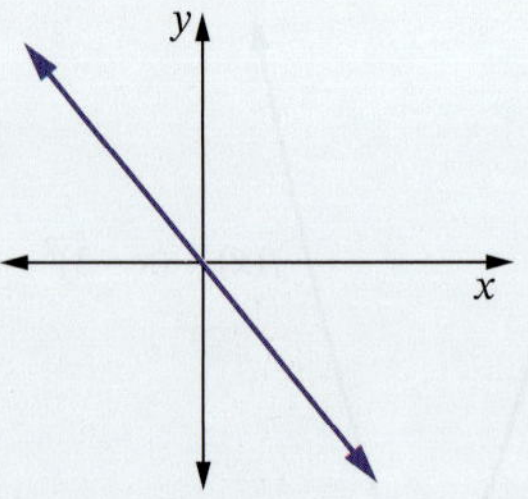

f

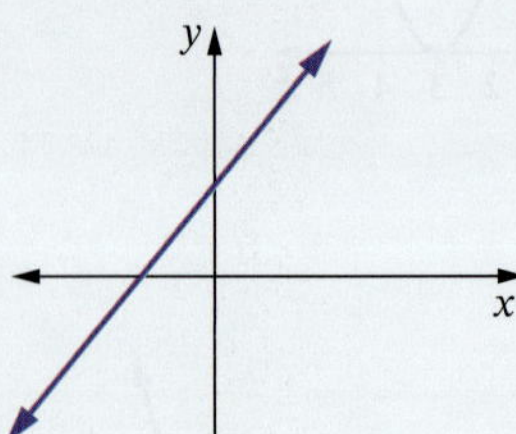

g

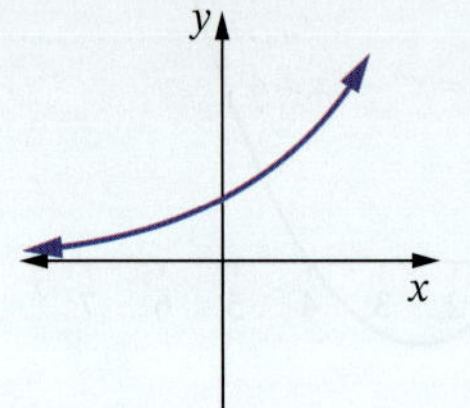

h

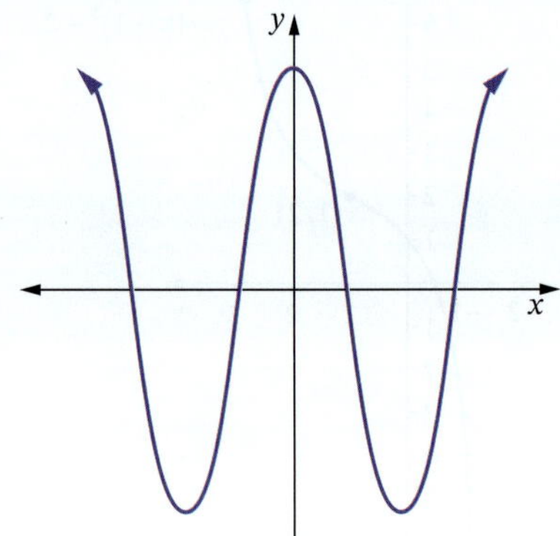

i

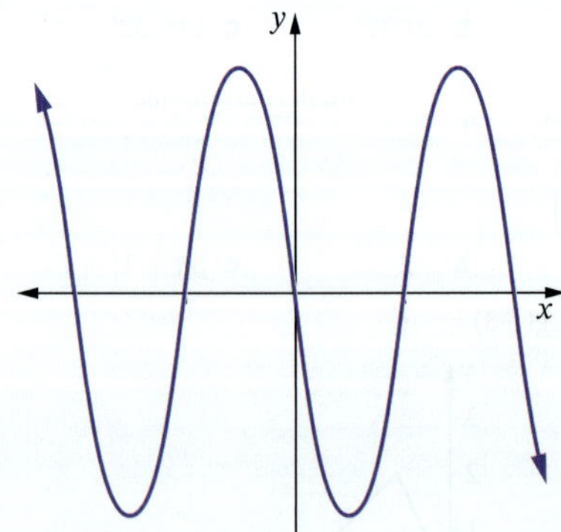

j

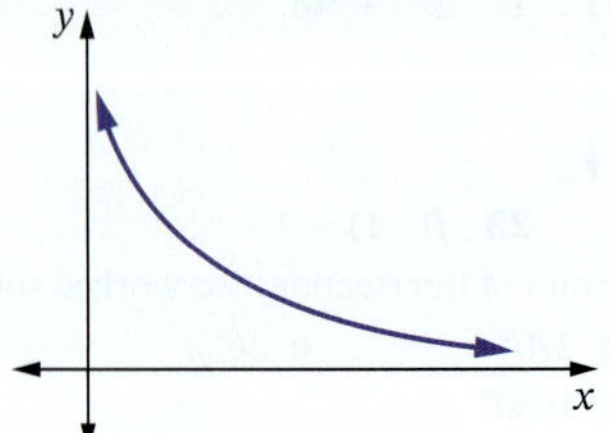

4 **a** 2026 – 2028, 2030 onwards
b 2028 – 2030

5 a i $x < -1, x > 1$ **ii** $-1 < x < 1$ **iii** $x = -1, 1$

b

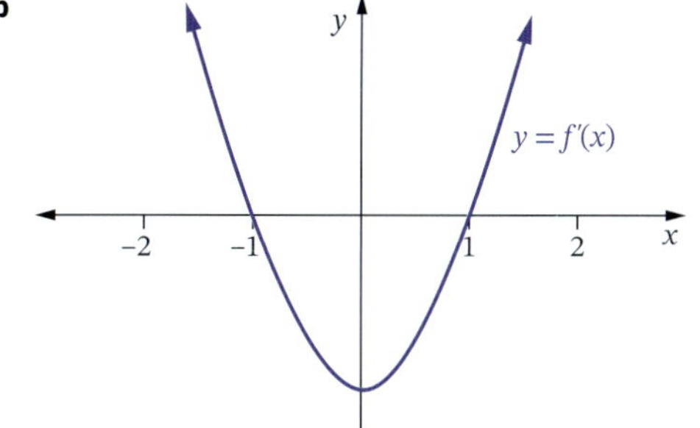

6 a i $x > 0$ **ii** $x < 0$ **iii** $x = 0$

b

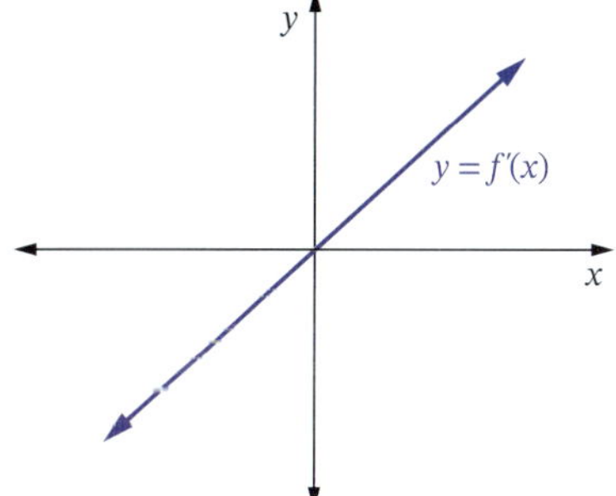

7 a no x values (never increasing)

b $x < 0, x > 0$ **c** $x = 0$

8 The graph is not unique. We only know the gradient of $y = f(x)$ for any value of x, not the actual y value. We could translate this graph vertically and the gradient function/graph would still be the same.

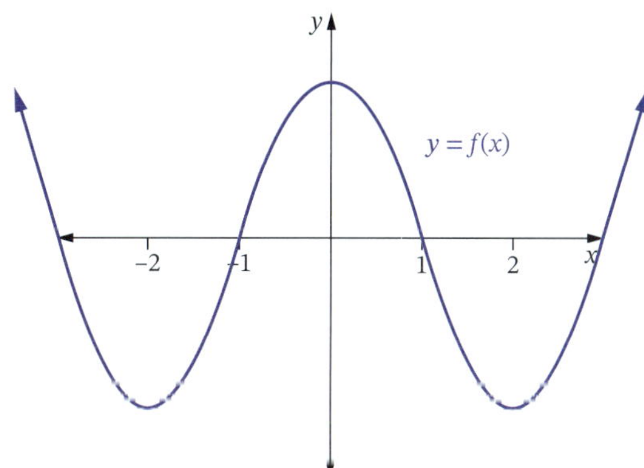

9 a a linear function

b a quadratic function (parabola)

c a constant function, horizontal line of the form $y = k$

10 The derivative $f'(x)$ starts off being positive and constant, then decreases and becomes close to 0 as x becomes large.

11 a no x values (never decreasing)

b $x > 0$

Exercise 6.02

1 6

2 a $x^2 + x + 2xh + h^2 + h + 5$ **b** $2xh + h^2 + h$

c $2x + h + 1$ **d** 5

3 a $4x^2 + 8xh + 4h^2 - 3$ **b** $8xh + 4h^2$

c -8

4 a $x^2 + 2xh + h^2 - 1$ **b** $2xh + h^2$

c 6

5 a $-10x - 3$ **b** 17

6 a $2(x^2 + 2xh + h^2) - 7x - 7h + 3$

b $2x^2 + 4xh + 2h^2 - 7x - 7h + 3 - (2x^2 - 7x + 3)$

c $\dfrac{h(4x + 2h - 7)}{h}$ **d** $4x - 7$

7 a $0, t_2, t_4, t_6$ **b** t_1, t_3, t_5 **c** t_5

d Particle starts at O then moves away (in a positive direction) till t_1, then it turns around and moves back to O by t_2. It then moves the other (negative) direction from O till t_3, then it turns around and moves back to O by t_4.

e The particle moves with positive velocity until it slows down and stops at t_1. It then travels with negative velocity until it slows down and stops at t_3. It travels with positive velocity again to t_4.

8 a 2 cm/s **b** -1 cm/s **c** 0 cm/s

9 a 2 **b** 5 **c** -12 **d** 15 **e** -9

10 a $2x$ **b** $2x + 5$ **c** $8x - 4$ **d** $10x - 1$

11 a 5 cm/s

b 4.5 cm/s

c Make smaller and smaller approximations e.g. average speed between 2.1 and 2, between 2.05 and 2, etc.

d 4 cm/s

12 a $f'(x) = 3x^2 - 2x - 2$

b i 6 **ii** 3

c $x = \dfrac{1 \pm \sqrt{7}}{3}$

d The function is stationary at that point.

13 $f'(x) = -\dfrac{1}{x^2}$

Exercise 6.03

1 a 1 **b** 5 **c** $2x + 3$ **d** $10x - 1$

e $3x^2 + 4x - 7$ **f** $6x^2 - 14x + 7$

g $12x^3 - 4x + 5$ **h** $6x^5 - 25x^4 - 8x^3$

i $10x^4 - 12x^2 + 2x - 2$

2 a $4x + 1$ **b** $8x - 12$ **c** $2x$

d $16x^3 - 24x$ **e** $6x^2 + 6x - 3$

3 a $\dfrac{x}{3} - 1$ **b** $2x^3 - x^2$ **c** $\dfrac{8x^7}{3} - 6x^5$

d $4x$ **e** $\dfrac{1}{4}$ **f** $2x^2 - 2x + 2$

4 $16x - 7$ **5** -56

6 $60x^9 - 40x^7 + 35x^4 - 3$ **7** $10t - 20$ **8** $20x^3$

9 $30t$ **10** $40 - 4t$ **11** $4\pi r^2$ **12** 3

13 a 5 **b** -5 **c** 4

14 a 12 **b** ± 2 **15** 18

16 a $-3x^{-4}$ **b** $1.4x^{0.4}$ **c** $1.2x^{-0.8}$

d $\dfrac{1}{2}x^{-\frac{1}{2}}$ **e** $x^{-\frac{1}{2}} + 3x^{2}$ **f** $x^{-\frac{2}{3}}$

g $6x^{-\frac{1}{4}}$ **h** $x^{-\frac{3}{4}}$

17 a $-\dfrac{1}{x^2}$ **b** $\dfrac{5}{2\sqrt{x}}$ **c** $\dfrac{1}{6\sqrt[6]{x^5}}$

d $-\dfrac{10}{x^6}$ **e** $\dfrac{15}{x^4}$ **f** $-\dfrac{1}{2\sqrt{x^3}}$

g $-\dfrac{3}{x^7}$ **h** $\dfrac{3\sqrt{x}}{2}$ **i** $-\dfrac{2}{3x^2}$

j $-\dfrac{1}{2x^3} - \dfrac{12}{x^5}$

18 $\dfrac{1}{27}$ **19** -3 **20** $\dfrac{1}{32}$ **21** -3

22 $2x + 3\sqrt{x} + 1$ **23** $\dfrac{1}{8}$

24 a $-\dfrac{1}{2\sqrt{x^3}}$ **b** $-\dfrac{1}{16}$

25 $a = 4$ **26** $\left(5, \dfrac{2}{5}\right), \left(-5, -\dfrac{2}{5}\right)$

ANSWERS

Exercise 6.04

1 **a** decreasing at the origin, increasing at $x = -10$
b $(-4, 1221), (5, -1695)$

2 Vertex at $x = 2$, where $\frac{dy}{dx} = 0$, minimum point because $a = 1 > 0$, concave up parabola

3 $(0, 1)$ and $(2, -3)$

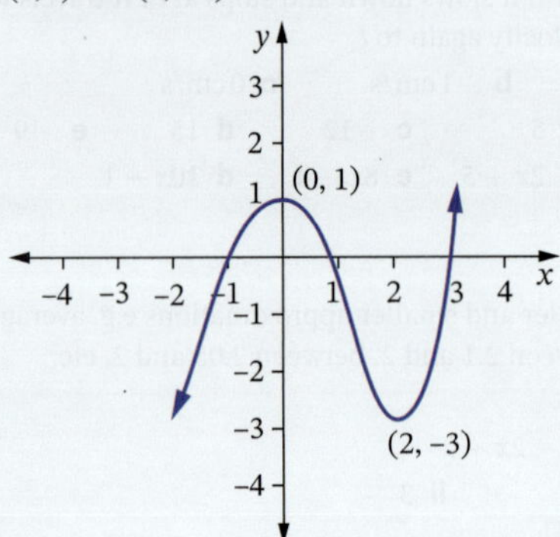

4 **a** 30 **b** 20

5 **a** -8 m/s **b** 13 m **c** 2 s **d** -5 m

6 **a** $x = 2, (2, -4)$
b 0, the minimum point is a stationary point

7 13 L/s

8 **a** $-4 < x < 3$
b $f'(-4) = 0$ and function is increasing between $x = -4$ and $x = 3$, so $x = -4$ must be a minimum point.

9 **a** **i** -7 g/min **ii** -13 g/min
b -19 g/min

10 The object starts moving towards the origin from the negative side, passes the origin and starts to slow down until it turns and moves towards the origin again, increasing in velocity and passing the origin. Throughout its journey, the object's velocity is changing at a constant (steady) rate.

11 **a** The temperature is increasing sharply but then starts to flatten out, still increasing but at a lower rate
b The temperature is increasing slowly from 0, but then its rate of increase starts to grow so the temperature is then rising quickly
c The temperature is decreasing quickly at first, then slowly, flattening out
d The temperature is rising at a constant rate
e The temperature is decreasing slowly at first, and then drops sharply

12 **a** The bird starts flying at a high speed (negative velocity) towards the origin and slows down after it passes the origin, turns and starts flying back to the origin at a constant speed. It goes past where it started, slows down and turns and heads back towards the origin again, again at a higher speed. It passes the origin and repeats the pattern of turning and flying past the origin at a constant speed.
b

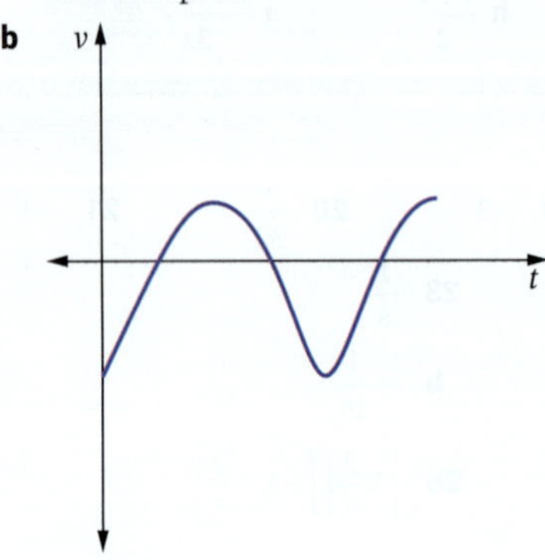

13 **a** $f(x) = x^3 + x^2 - 3x + c$
b $y = x^5 + x^4 - 3x^2 + c$
c $f(x) = \frac{x^4}{4} - \frac{x^3}{3} + \frac{x^2}{2} - 7x + c$

14 $\left(\frac{3}{4}, \frac{\sqrt{3}}{2}\right)$ **15** 165 cm²/s

16 **a** 18 cm/s
b When $t = 0, x = 0$; at 3 s.
c $\sqrt{3}$ s

17 $\frac{dy}{dx} = 2ax + b = 0$

18 **a** 3 m **b** 1 s **c** 4 m
d -2 m/s **e** -4 m/s

19 **a** 15 708 cm³/cm **b** 62 832 cm³/cm

20 **a** 0 s, 7 s, 9 s **b** $1\frac{2}{3}$ s, 9 s
c 65.19 mm, 0 mm

21 $f'(x) = 3ax^2 + 1 > 0$ for all x, where $a > 0$, so $f(x)$ is always increasing.

22 $f'(-3) = f'(3) = -\frac{1}{3}$

23 Proof: see worked solutions.

Exercise 6.05

1 **a** 72 **b** -13 **c** 11 **d** 136
e -4 **f** 149

2 **a** $-\frac{1}{26}$ **b** $\frac{1}{25}$ **c** $\frac{1}{20}$ **d** $-\frac{1}{43}$
e $-\frac{1}{8}$ **f** $-\frac{1}{5}$

3 $(-5, -7)$ **4** $(0, 1)$

5 $(1, 2)$ **6** $\left(-1\frac{3}{4}, -4\frac{15}{16}\right)$

7 **a** $27x - y - 47 = 0$ **b** $7x - y - 1 = 0$
c $4x + y + 17 = 0$ **d** $36x - y - 47 = 0$
e $44t - v - 82 = 0$

8 **a** $(1, -1)$ **b** $6x - y - 7 = 0$

9 $3x + 16y - 8 = 0$

10 **a** $x + 24y - 555 = 0$ **b** $x - 8y + 58 = 0$
c $x - 17y - 516 = 0$ **d** $x - 45y + 3108 = 0$
e $x + 2y - 9 = 0$

11 $x - y + 9 = 0$ **12** $x + 16y - 16 = 0$

13 $(9, 3)$ **14** $x + y + 3 = 0$

15 **a** $108°\ 26'$ at $x = 0$, $71°\ 34'$ at $x = 3$
b $(0, 0)$

16 $\left(4\frac{12}{23}, 9\frac{2}{23}\right)$

17 **a** $12x - y - 16 = 0$ **b** 5.3 units²

18 **a** 6 units **b** 9 units²

19 $5x - y - 7 = 0$, $7x + y + 19 = 0$

Exercise 6.06

1 **a** $4(x + 3)^3$ **b** $6(2x - 1)^2$
c $70x(5x^2 - 4)^6$ **d** $48(8x + 3)^5$
e $-5(1 - x)^4$ **f** $135(5x + 9)^8$
g $4(x - 4) = 4x - 16$
h $6(6x^5 - 4x)(x^6 - 2x^2 + 3)^5$
$= 12x(3x^4 - 2)(x^6 - 2x^2 + 3)^5$
i $\frac{3}{2}(3x - 1)^{-\frac{1}{2}}$ **j** $2(4 - x)^{-3}$
k $-6x(x^2 - 9)^{-4}$ **l** $\frac{5}{3}(5x + 4)^{-\frac{2}{3}}$
m $\frac{3}{2\sqrt{3x + 4}}$ **n** $-\frac{5}{(5x - 2)^2}$

o $-\dfrac{8x}{(x^2+1)^5}$ **p** $-\dfrac{2}{\sqrt[3]{7-3x}}$

q $-\dfrac{5}{2\sqrt{(4+x)^3}}$ **r** $-\dfrac{3}{4\sqrt{(3x-1)^3}}$

s $-\dfrac{4x^3-9x^2+3}{(x^4-3x^3+3x)^2}$ **t** $\dfrac{16\sqrt[3]{4x+1}}{3}$

u $\dfrac{5}{4\sqrt[4]{(7-x)^9}}$

2 9 **3** 40 **4** (4, 1)

5 $x = 2, -1\frac{1}{2}$

6 $16x - y - 15 = 0$ **7** $x + 64y - 1025 = 0$

8 **a** $2x + y + 1 = 0$ **b** $x - 2y + 3 = 0$

9 **a** $0 \le t \le 1.58$

b $v = 12t(2t^2 - 5)^2$; $12t \ge 0$ since $t \ge 0$, $(2t^2 - 5)^2 \ge 0$, so $v \ge 0$

c 216 mm/s

10 **a** $y = \sqrt{25 - x^2}$ **b** $y' = -\dfrac{x}{\sqrt{25-x^2}}$

c Proof: see worked solutions.

11 **a** $5x - y - 9 = 0$

b 1.02 units

c 0.1 units^2

12 $n = -3$ **13** $p = -4, -2$

14–16 Proof: see worked solutions.

Exercise 6.07

1 **a** $8x^3 + 9x^2$ **b** $12x - 1$

c $30x + 21$ **d** $72x^5 - 16x^3$

e $30x^4 - 4x$ **f** $x(5x + 2)(x + 1)^2$

g $8(9x - 1)(3x - 2)^4$ **h** $3x^3(16 - 7x)(4 - x)^2$

i $(10x + 13)(2x + 5)^3$

2 26 **3** 1264 **4** $\dfrac{8\sqrt{7}}{7}$ **5** 176

6 $10x - y - 9 = 0$ **7** $69t - h - 129 = 0$

8 $\dfrac{-6 \pm \sqrt{30}}{3}$

9 **a** $34x - y + 29 = 0$ **b** $\left(\dfrac{11}{12}, 60\dfrac{1}{6}\right)$

10 **a** 0 L **b** [illegible] L/m

c **i** 3675 L/m **ii** 181 875 L/m

11 $x + 304y + 19\,457 = 0$

12 $5x + 106y - 16\,485 = 0$

13 **a** $f'(-4) = 20$: curve is increasing when $x = -4$

b $f'(3) = 55$: curve is increasing when $x = 3$

c $f'(-2) = 0$: curve is stationary when $x = -2$

14 **a** $y' = p(2x - a - b)$

b At $x = a$, $y' = p(a - b)$. At $x = b$, $y' = p(b - a) = -p(a - b)$

c Proof: see worked solutions

Exercise 6.08

1 **a** $\dfrac{-2}{(2x-1)^2}$ **b** $\dfrac{15}{(x+5)^2}$

c $\dfrac{x^4-12x^2}{(x^2-4)^2} = \dfrac{x^2(x^2-12)}{(x^2-4)^2}$

d $\dfrac{16}{(5x+1)^2}$ **e** $\dfrac{-x^2+14x}{x^4} = \dfrac{14-x}{x^3}$

f $\dfrac{11}{(x+3)^2}$ **g** $\dfrac{-2x^2-1}{(2x^2-1)^2}$

h $\dfrac{-3x^2-6x-7}{(3x^2-7)^2}$ **i** $\dfrac{4x^2-12x}{(2x-3)^2} = \dfrac{4x(x-3)}{(2x-3)^2}$

j $\dfrac{-18x}{(x^2-5)^2}$ **k** $\dfrac{2x^3+12x^2}{(x+4)^2} = \dfrac{2x^2(x+6)}{(x+4)^2}$

l $\dfrac{2x^3+9x^2+7}{(x+3)^2}$ **m** $\dfrac{2(x+5)^{\frac{1}{2}} - x(x+5)^{-\frac{1}{2}}}{x+5}$

n $\dfrac{(7x+2)^4 - 28(x-1)(7x+2)^3}{(7x+2)^8} = \dfrac{30-21x}{(7x+2)^5}$

o $\dfrac{3\sqrt{x+1} - \dfrac{3x+1}{2\sqrt{x+1}}}{x+1} = \dfrac{3x+5}{2\sqrt{(x+1)^3}}$

p $\dfrac{\dfrac{2x-3}{2\sqrt{x-1}} - 2\sqrt{x-1}}{(2x-3)^2} = \dfrac{-2x+1}{2\sqrt{x-1}(2x-3)^2}$

2 $\dfrac{1}{8}$ **3** $-1\dfrac{5}{9}$ **4** $x = 0, 1$

5 $x = -9, 3$ **6** $x - 18y + 8 = 0$

7 $17x - 25y - 19 = 0$

8 $x = -2 \pm \sqrt{3}$

9 **a** $120(8x - 1)^4$ **b** $-\dfrac{11}{(2x-1)^2}$

c $(8x + 9)(2x + 9)^2$ **d** $\dfrac{11}{(3t+5)^2}$

e $\dfrac{4h+1}{2h\sqrt{h}}$

10 **a** $x = 1$ **b** $x + 4y - 34 = 0$

Sample HSC problem

a 1 cm/s

b **i** 3 cm/s **ii** 4 cm/s

c $1\frac{1}{3}$ s

Test yourself 6

1 C **2** B **3** D **4** D

5 **a**

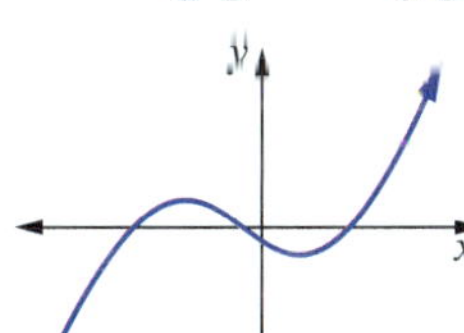

b

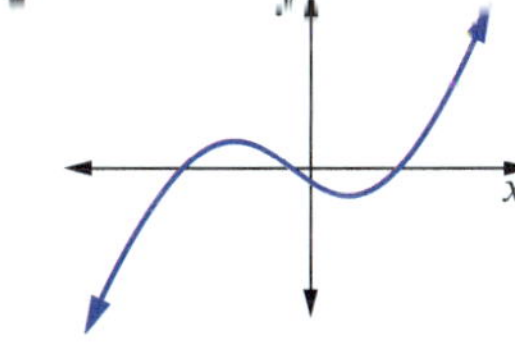

6 $\dfrac{dy}{dx} = 10x - 3$

7 **a** $42x^5 - 9x^2 + 2x - 8$

b $-\dfrac{8}{(x+1)^5}$ **c** $\dfrac{dy}{dx} = \dfrac{5\sqrt{x^3}}{2}$

d $18(x + 2)(x^2 + 4x - 2)^8$

e $\dfrac{7}{(2x+1)^2}$

f $3x^2(3x + 1)^5(9x + 1)$

ANSWERS

8 $4t - 3$ **9** 10 **10** 42

11 a i 10.45 **ii** 9.94 **iii** 9.89

b 10

12 a $-\frac{4}{x^2}$ **b** $\frac{1}{5\sqrt[5]{x^4}}$ **c** $32(4x+9)^3$

d $3(x-1)^2(4x+1)$

e $\frac{4x^3+15x^2+6}{(2x+5)^2}$

13

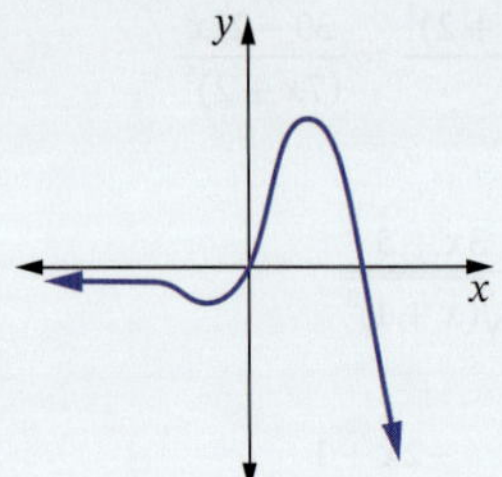

14 $9x - y - 7 = 0$ **15** $(2, 3)$

16 $8\pi r$ **17** -0.9

18 $(-2, 71), (5, -272)$ **19** $4x - y - 6 = 0$

20 a $u + at$ **b** $\frac{1}{5}$

21 $12x + y - 4 = 0$

22 a i 0 m **ii** 4 m **iii** 3 m **iv** 1.75 m

b i 1 m s^{-1} **ii** -1 m s^{-1}

c i 4 m s^{-1} **ii** 0 m s^{-1} **iii** -2 m s^{-1}

23 a $x^2 + 2xh + h^2 - 3x - 3h + 5$

b $2xh + h^2 - 3h$ **c** $2x - 3$

24 a 1 **b** 20 **25** 9

26 a $x < 2, x > 4$ **b** $2 < x < 4$

c $x = 2, 4$

27 $(3, -80), (-2, 45)$

28 a i 5 cm **ii** 6 cm s^{-1}

b 0 cm s^{-1} **c** $t = 1\text{ s}$ **d** 8 cm

29 a t_1, t_3, t_5 **b** t_2, t_4 **c** t_5

d v, t_1, t_2, t_3, t_4, t_5, t

Challenge exercise 6

1 $2x + y = 0, 3x - y - 3 = 0, 6x - y + 12 = 0$

2 a $(2, 2), (-2, -14)$

b $x + 12y - 26 = 0, x + 12y + 170 = 0$

3 $\left(-2\frac{1}{4}, 6\frac{1}{16}\right)$ **4** $n = 8$

5 a $\frac{\pi}{2}, \frac{3\pi}{2}$

b

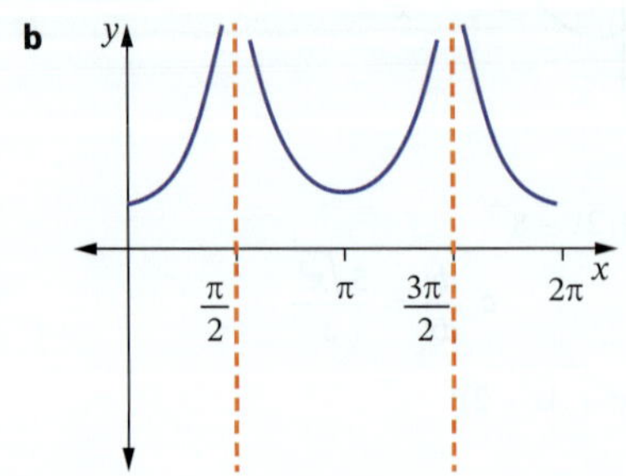

6 $\frac{5\sqrt{22}}{22}$ **7** $2x + y - 25 = 0$

8 a $16x + 32y + 1 = 0, 4x - 2y - 1 = 0$

b $m_1m_2 = -\frac{1}{2} \times 2 = -1$, so perpendicular.

9 $x = 0, 2, 6$ **10** $a = -1, b = 2, c = 4$

11 a $1\text{ m}, 0\text{ m s}^{-1}$

b $(t^3 + 1)^6 = 0$ has solution $t = -1$, but time $t \geq 0$.

Chapter 7

Exercise 7.01

1 a

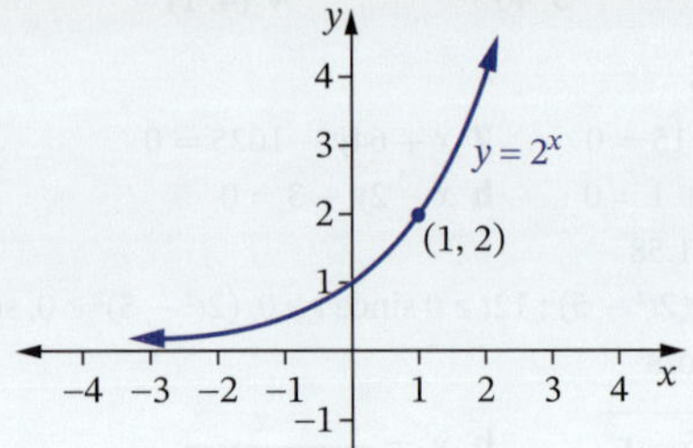

b

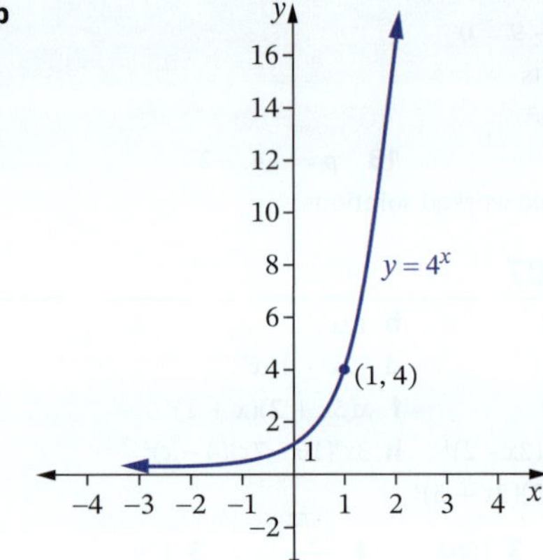

c

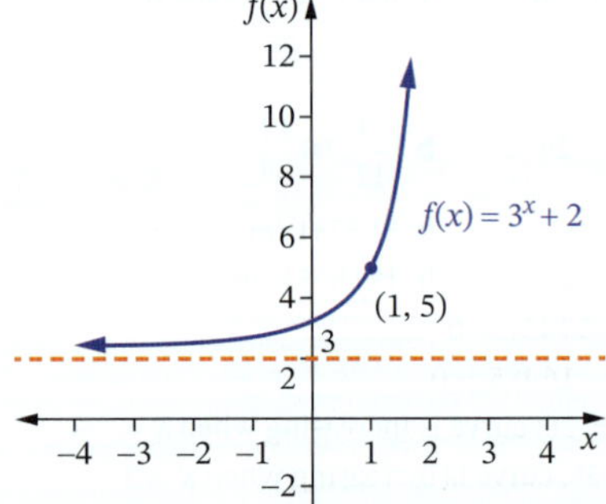

d

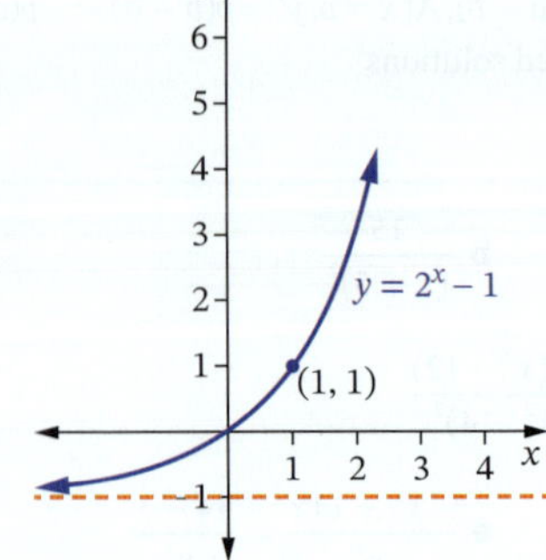

e

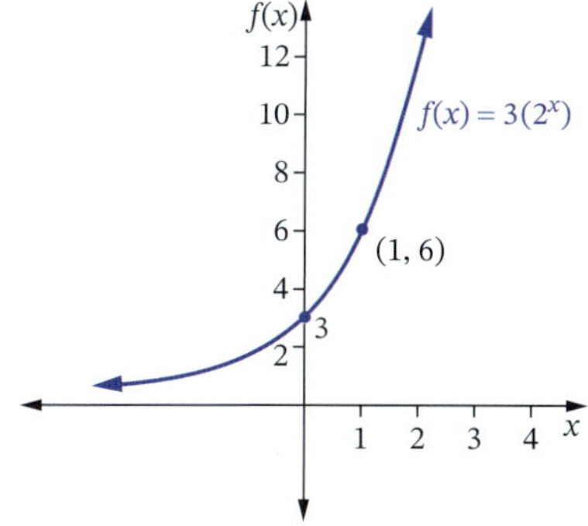

f

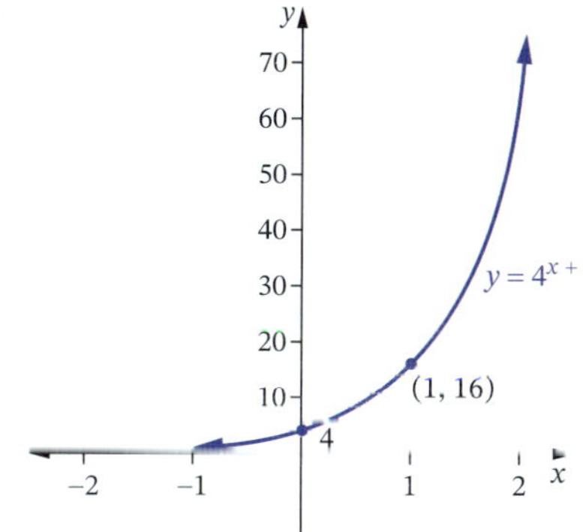

g

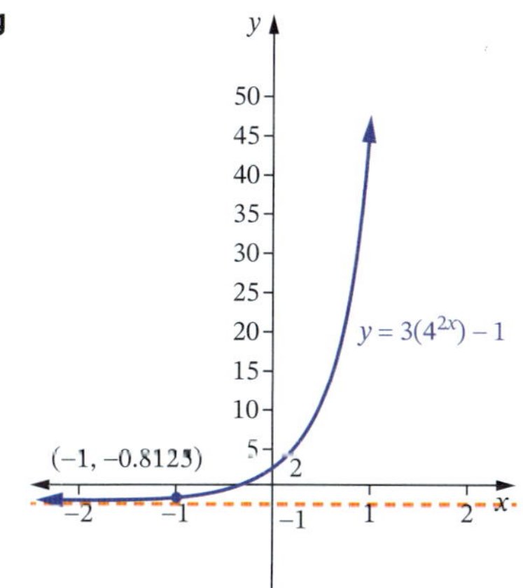

h

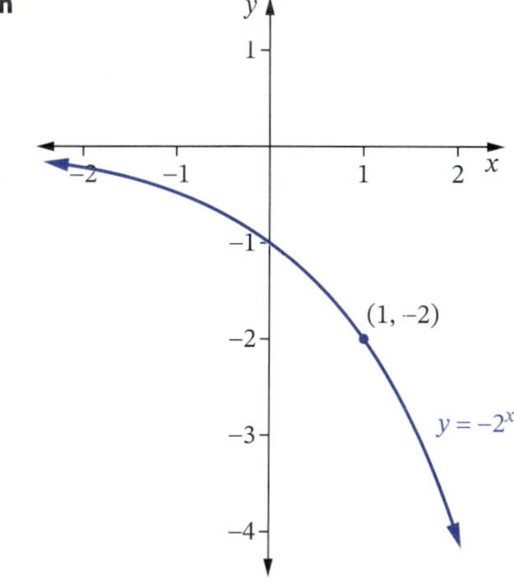

i

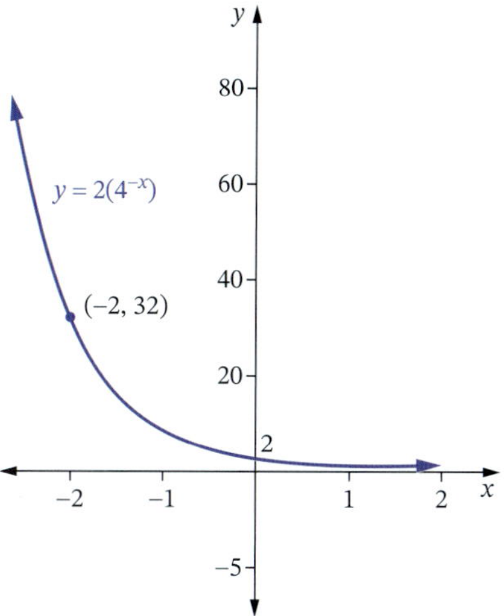

j

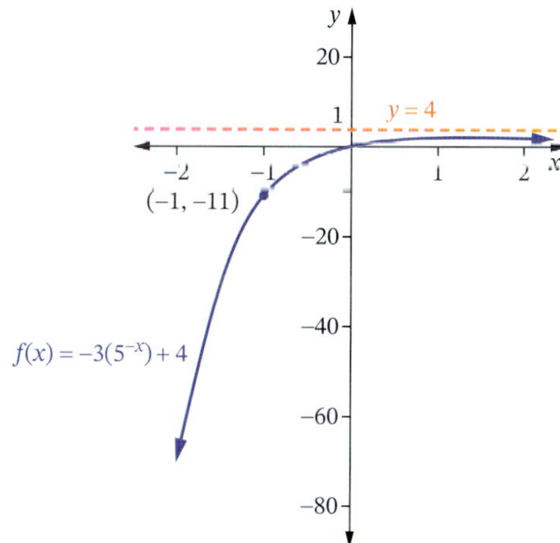

2 **a** domain: all real x, range: $(0, \infty)$
b domain: all real x, range: $(5, \infty)$
c domain: all real x, range: $(0, \infty)$
d domain: all real x, range: $(-\infty, 1)$

3 **a** $y = 2^{3x-4}$ **b** $y = 3(2^x) - 4$

4 **a**

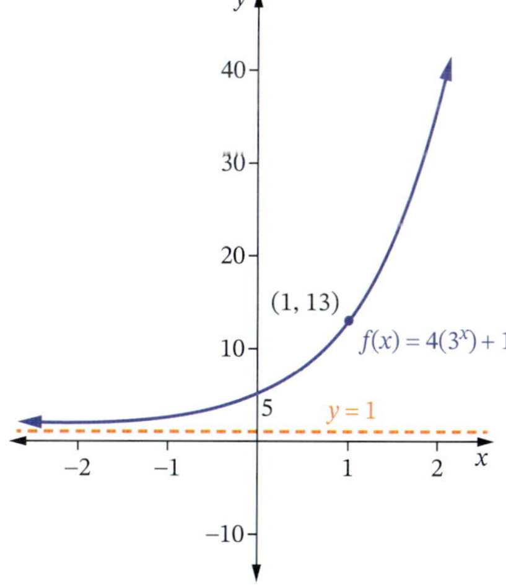

b i

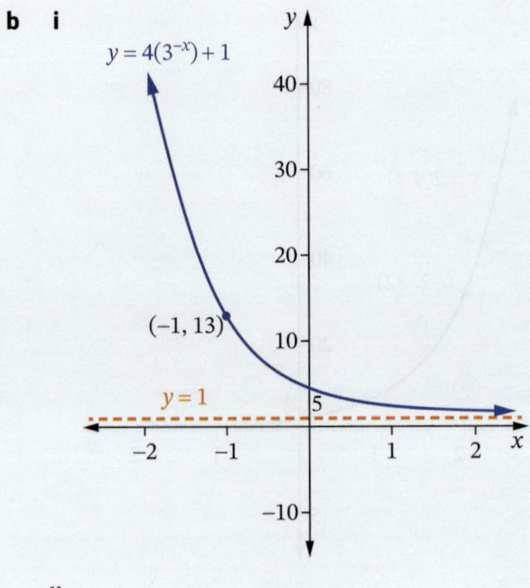

ii

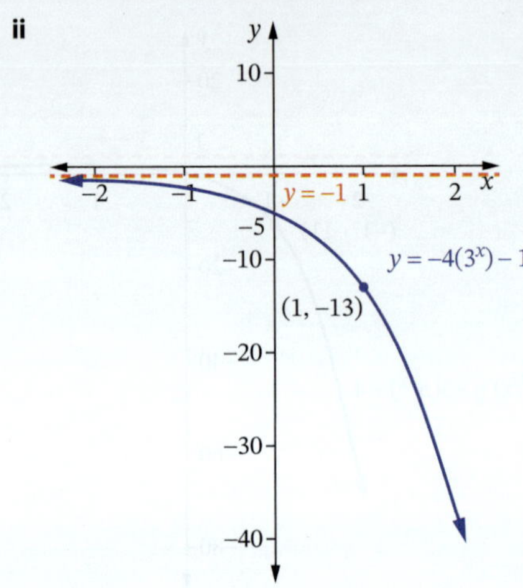

iii

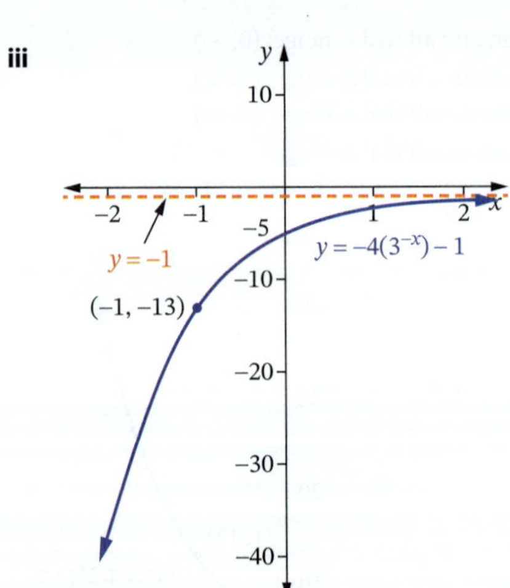

5 a

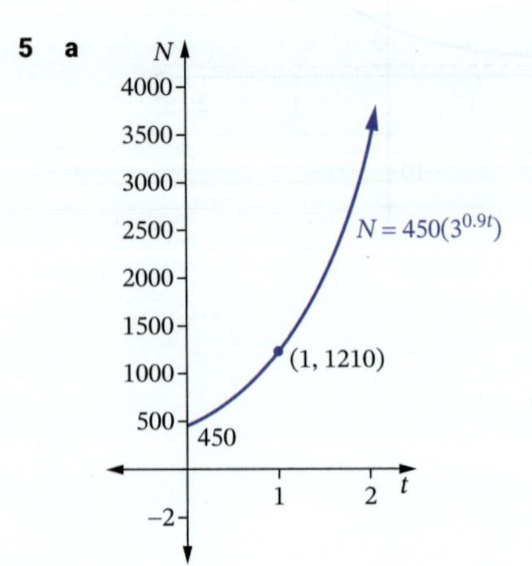

b 450 sales

c i 8739 **ii** 63 133 **iii** 8 857 350

6

	Domain	Range	y-intercept	Asymptote
a	All real x	$y > -3$	−2	$y = -3$
b	All real x	$y < 0$	−1	$y = 0$
c	All real x	$y > 1$	6	$y = 1$
d	All real x	$y > -3$	−2	$y = -3$
e	All real x	$y < 3$	2	$y = 3$

Exercise 7.02

1

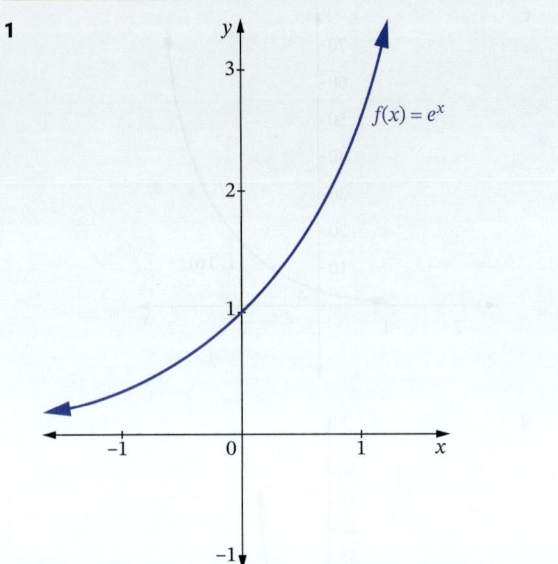

2 **a** 4.48 **b** 0.14 **c** 2.70 **d** 0.05 **e** −0.14

3 a

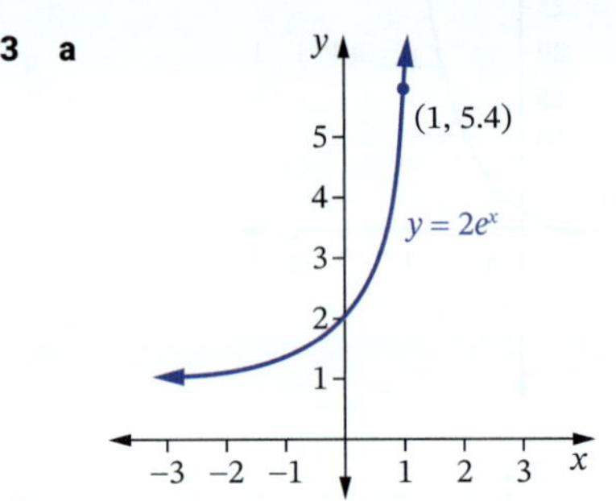

b

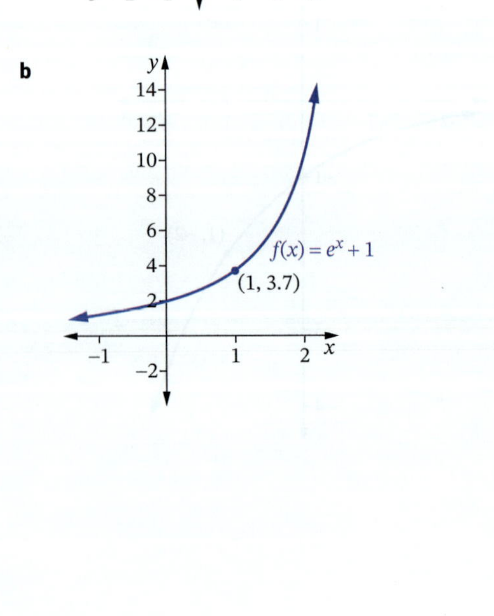

c

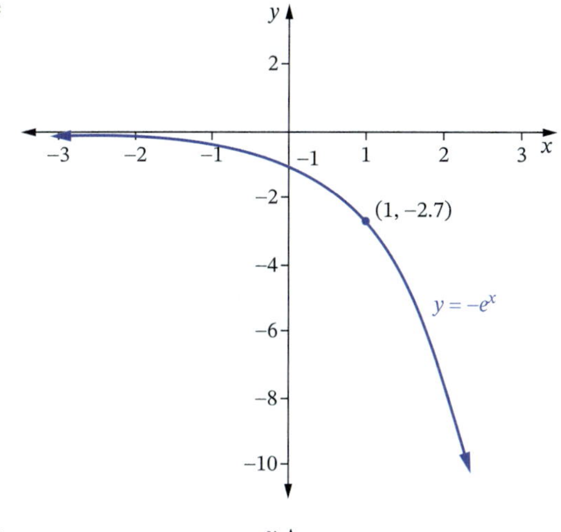

d

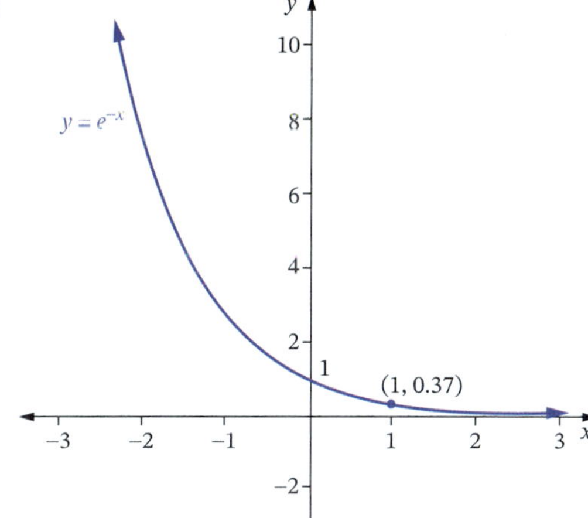

e

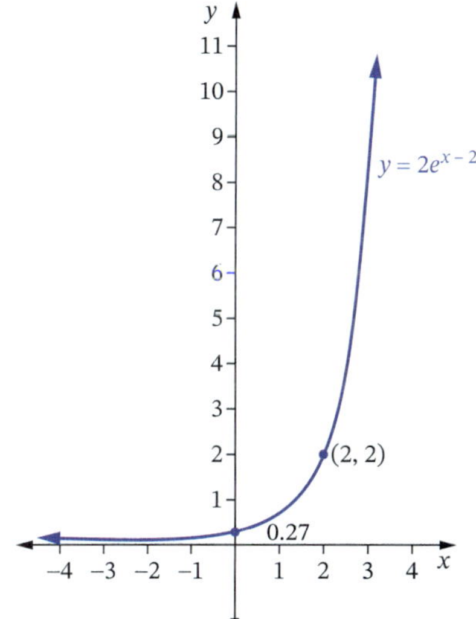

4 domain: all real x, range: $(-2, \infty)$

5 **a** $y = e^{x^3+3}$ **b** $y = e^{3x} + 3$

6 **a**

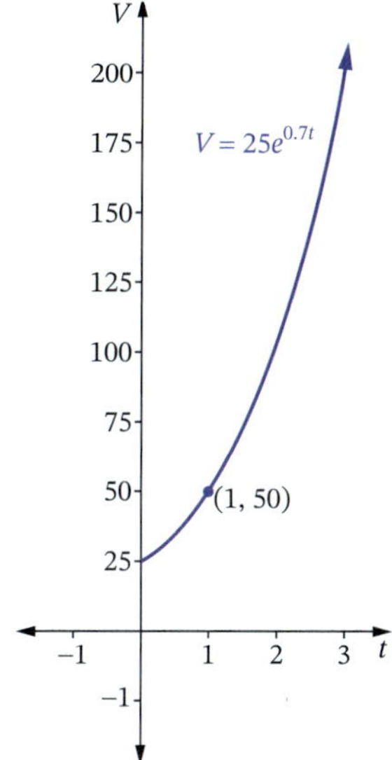

b **i** $204.2\,\text{mm}^3$ **ii** $6760.7\,\text{mm}^3$

c No, because it predicts the volume never stops increasing.

7 **a** 130.4 g **b** 74.5 g **c** 4.5 g

8 **a**

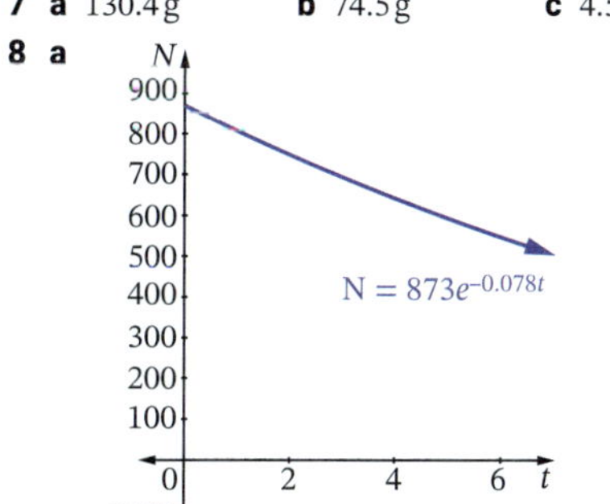

b **i** 873 **ii** 591 **iii** 400

9 **a** 148°C

b **i** 133.9°C **ii** 115.6°C

iii 91.6°C **iv** 23.1°C

c 23°C (room temperature)

10

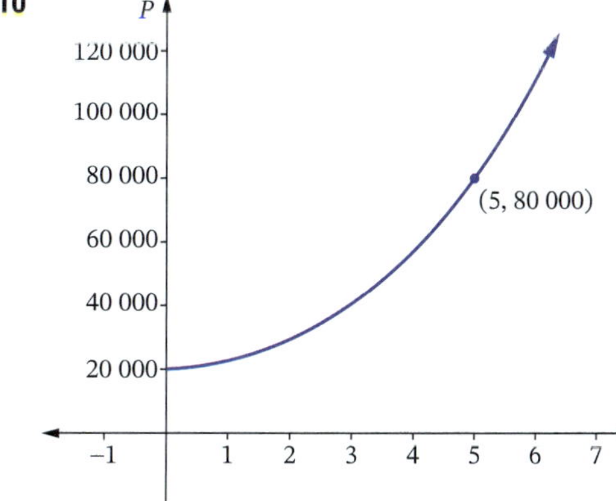

11

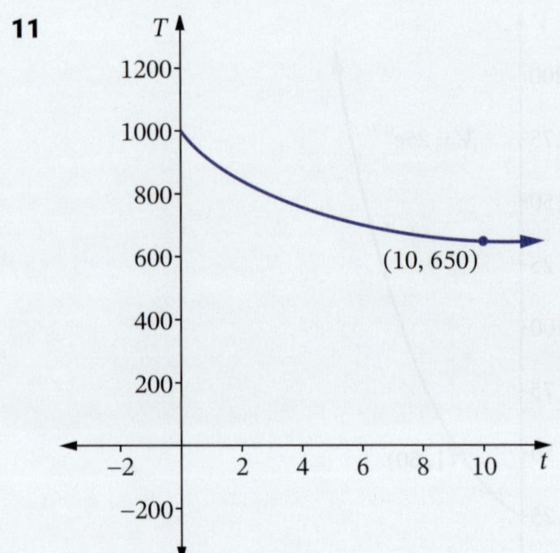

Exercise 7.03

1 **a** $9e^x$ **b** $-e^x$ **c** $e^x + 2x$
d $6x^2 - 6x + 5 - e^x$ **e** $3e^x(e^x + 1)^2$
f $7e^x(e^x + 5)^6$ **g** $4e^x(2e^x - 3)$ **h** $e^x(x + 1)$
i $\frac{e^x(x-1)}{x^2}$ **j** $xe^x(x + 2)$ **k** $e^x(2x + 3)$
l $\frac{e^x(7x-10)}{(7x-3)^2}$ **m** $\frac{5e^x - 5xe^x}{e^{2x}} = \frac{5(1-x)}{e^x}$
n $2e^{2x}$ **o** $-e^{-x}$ **p** $6e^{3x}$
q $-7e^{7x}$ **r** $-6e^{2x} + 2x$
s $2e^{2x} + 2e^{-2x} = 2(e^{2x} + e^{-2x})$ **t** $-5e^{-x} - 3$
u $e^{4x}(4x + 1)$ **v** $\frac{6xe^{3x} + 4e^{3x} + 3}{(x+1)^2}$
w $135e^{3x}(9e^{3x} + 2)^4$
2 $6 - e$ **3** e **4** $-\frac{1}{2e^{10}}$
5 19.81 **6** $y = -ex$
7 $e^3x - y - 3e^3 + e^{-3} = 0$
8 **a** 12 572
b **i** 280 insects/week **ii** 75 742 insects/week
9 **a** 2 m **b** $8e^{40}$ m s^{-1}
10 **a** 1 cm **b** -2.04 cm s^{-1}
c -3.46 cm **d** -0.096 cm s^{-1}
11 **a** **i** 33.1 mm^3 **ii** 163.8 mm^3
b **i** 26.5 mm^3 s^{-1} **ii** 131.0 mm^3 s^{-1}
12 **a** **i** 39 094 **ii** 44 299
b **i** 977 people/year **ii** 1107 people/year
13 **a** **i** 2.95 m **ii** 2.90 m **iii** 2.85 m
b **i** -0.050 m/month **ii** -0.049 m/month
iii -0.048 m/month

Exercise 7.04

1 **a** $\log_3 y = x$ **b** $\log_5 z = x$ **c** $\log_x y = 2$
d $\log_2 a = b$ **e** $\log_b d = 3$ **f** $\log_8 y = x$
g $\log_a y = x$ **h** $\log_e Q = x$
2 **a** $3^x = 5$ **b** $a^x = 7$ **c** $3^b = a$
d $x^y = y$ **e** $a^y = b$ **f** $2^y = 6$
g $3^y = x$ **h** $10^y = 9$ **i** $e^y = 4$
3 **a** $x = 1\,000\,000$ **b** $x = 243$ **c** $x = 7$
d $x = 2$ **e** $x = -1$ **f** $x = 3$
g $x = 44.7$ **h** $x = 10\,000$ **i** $x = 8$
4 **a** 4 **b** 2 **c** 3 **d** 1 **e** 2
f 1 **g** 0 **h** 7 **i** 1
5 **a** 9 **b** 3 **c** -1 **d** 12 **e** 8
f 4 **g** 14 **h** 1 **i** 2
6 **a** -1 **b** $\frac{1}{2}$ **c** $\frac{1}{2}$ **d** -2 **e** $\frac{1}{4}$
f $-\frac{1}{3}$ **g** $-\frac{1}{2}$ **h** $\frac{1}{3}$ **i** $1\frac{1}{2}$ **j** $-1\frac{1}{2}$

7 **a** 3 **b** 4 **c** 29
8 **a** 3.08 **b** 2.94 **c** 3.22 **d** 4.94 **e** 10.40
f 7.04 **g** 0.59 **h** 3.51 **i** 0.43
9 $y = 5$ **10** 44.7 **11** 2.44
12 0, 0 **13** 1, 1
14 **a** 1
b **i** 3 **ii** 2 **iii** 5 **iv** $\frac{1}{2}$
v -1 **vi** 2 **vii** 3 **viii** 5
ix 7 **x** 1 **xi** e
15 **a** 68.4 **b** **i** 58.8 **ii** 46.6
c 8 weeks
16 **a** 2.5, acidic **b** 7, neutral
c 9, alkaline **d** 2, acidic
e 11.9, alkaline **f** 5, acidic
17 **a** $y = \log(2x - 7)$ **b** $y = 2\log x - 7$

Exercise 7.05

1 **a** $\log_a 4y$ **b** $\log_a 20$ **c** $\log_a 4$
d $\log_a\left(\frac{b}{5}\right)$ **e** $\log_x y^3z$ **f** $\log_k 9y^3$
g $\log_a\left(\frac{x^5}{y^2}\right)$ **h** $\log_a\left(\frac{xy}{z}\right)$ **i** $\log_{10} ab^4c^3$
j $\log_3\left(\frac{p^3q}{r^2}\right)$ **k** $-\log_4 n$ **l** $-\log_x 6$
2 **a** 2 **b** 6
3 **a** 1.19 **b** -0.47 **c** 1.55 **d** 1.66 **e** 1.08
f 1.36 **g** 2.02 **h** 1.83 **i** 2.36
4 **a** 2 **b** 6 **c** 2 **d** 3 **e** 1
f 3 **g** 7 **h** $\frac{1}{2}$ **i** -2 **j** 4
5 **a** 1.58 **b** 1.80 **c** 2.41
d 3.58 **e** 2.85 **f** 2.66
g 1.40 **h** 4.55 **i** 4.59
j 7.29
6 **a** $x + y$ **b** $x - y$ **c** $3x$
d $2y$ **e** $2x$ **f** $x + 2y$
g $x + 1$ **h** $1 - y$ **i** $2x + 1$
7 **a** 1.3 **b** 12.8 **c** 16.2
d 9.1 **e** 6.7 **f** 23.8
g -3.7 **h** 3 **i** 22.2
8 **a** $p + q$ **b** $3q$ **c** $q - p$
d $2p$ **e** $p + 5q$ **f** $2p - q$
g $p + 1$ **h** $1 - 2q$ **i** $3 + q$
j $p - 1 - q$
9 **a** $x = 4$ **b** $y = 28$ **c** $x = 48$
d $x = 3$ **e** $k = 6$ **f** $x = 2$
g $x = 3$ **h** $x = 3$
10 **a** $I = 10^{\frac{L}{10}} I_0$ **b** $31\,622.8I_0$
11 **a** Proof: see worked solutions
b **i** 69.9 **ii** 2.2
12 **a** $x = b^{\frac{3}{a}}$ **b** $x = \pm a^{\frac{b}{2}}$

Exercise 7.06

1 **a**

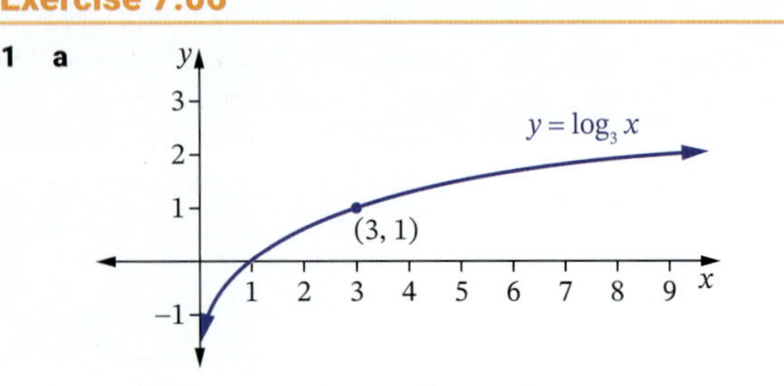

Domain: $x > 0$ and range: $(-\infty, \infty)$

b

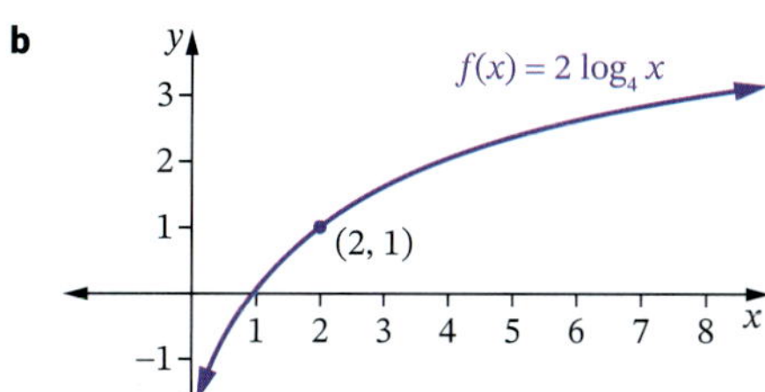

Domain: $x > 0$ and range: $(-\infty, \infty)$

c

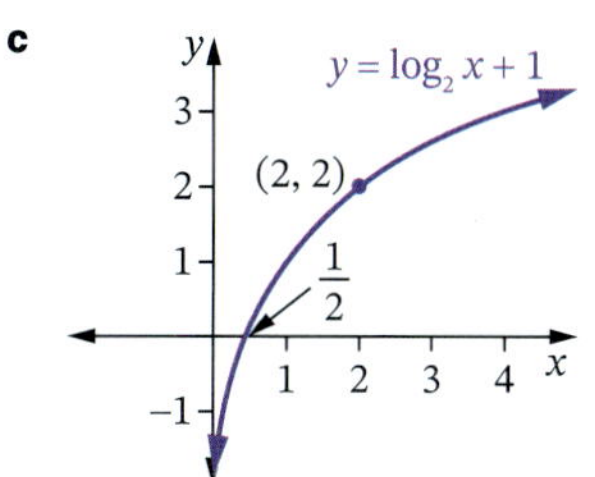

Domain: $x > 0$ and range: $(-\infty, \infty)$

d

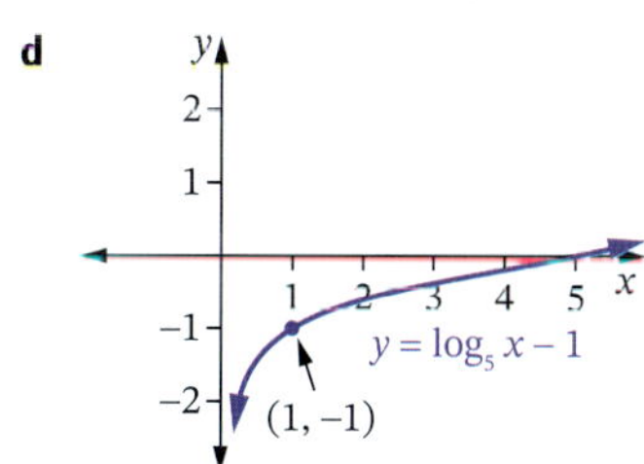

Domain: $x > 0$ and range: $(-\infty, \infty)$

e

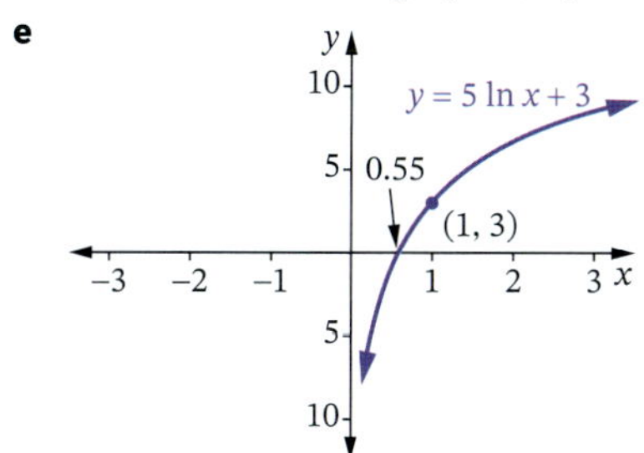

Domain: $x > 0$ and range: $(-\infty, \infty)$

f

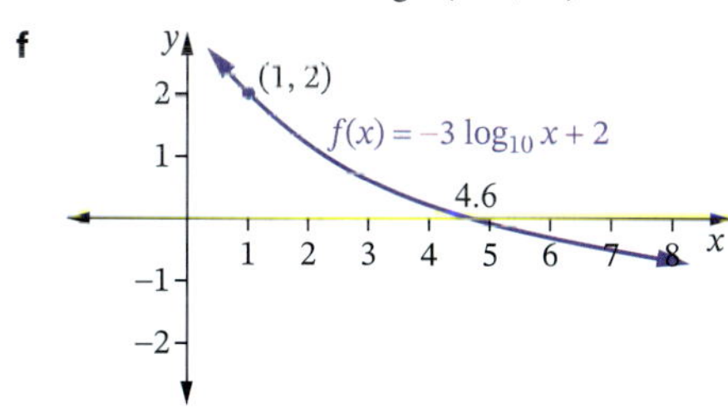

Domain: $x > 0$ and range: $(-\infty, \infty)$

2

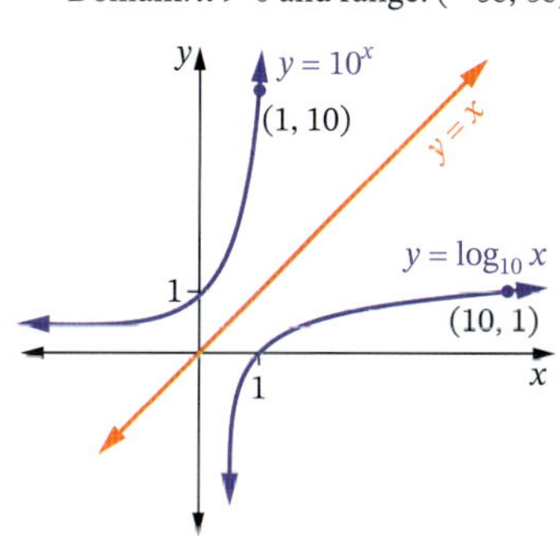

Curves are reflections of each other in the line $y = x$.

3 **a**

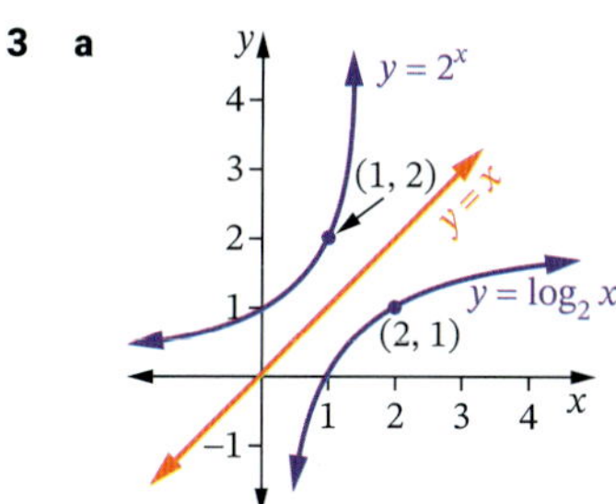

b $y = 2^x$

4

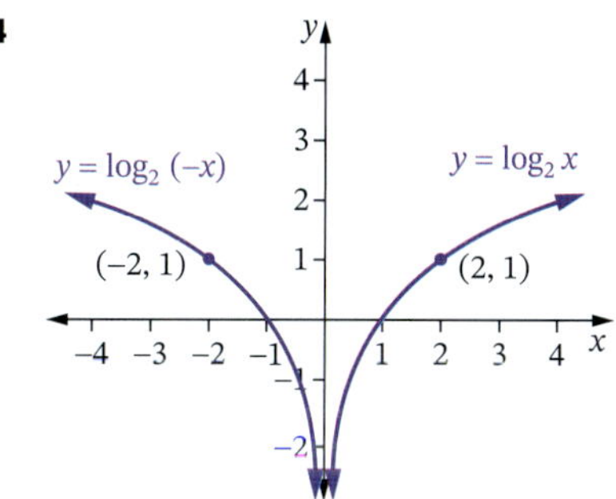

$y = \log_2(-x)$ is a reflection of $y = \log_2 x$ in the y-axis.

5 **a** $y = 7^x$ **b** $y = 9^x$
c $y = e^x$ **d** $y = \log_2 x$
e $y = \log_6 x$ **f** $y = \log_e x$ or $y = \ln x$

6 **a** $x > -1$ **b** $x > \frac{1}{3}$ **c** $x < 2$

Exercise 7.07

1 **a** $x = 1.6$ **b** $x = 1.5$ **c** $x = 1.4$ **d** $x = 3.9$
e $x = 2.2$ **f** $x = 2.3$ **g** $x = 6.2$ **h** $x = 2.8$
i $x = 2.9$ **j** $x = 2.4$

2 **a** $x = 2.58$ **b** $y = 1.68$ **c** $x = 2.73$
d $m = 1.78$ **e** $k = 2.82$ **f** $t = 1.26$
g $x = 1.15$ **h** $p = 5.83$ **i** $x = 3.17$
j $n = 2.58$

3 **a** $x = 0.9$ **b** $n = 0.9$ **c** $x = 6.6$
d $n = 1.2$ **e** $x = -0.2$ **f** $n = 2.2$
g $x = 2.2$ **h** $k = 0.9$ **i** $x = 3.6$

4 **a** $x = 5.30$ **b** $t = 0.536$ **c** $t = 3.62$
d $x = 3.81$ **e** $n = 3.40$ **f** $t = 0.536$
g $t = 24.6$ **h** $k = 67.2$ **i** $t = 54.9$

5 **a** **i** \$850 **ii** \$1010.38 **b** 6.6 years

6 **a** **i** 35 000 **ii** 44 494 **iii** 116 204
b **i** after 34.4 years **ii** after 72.6 years

7 **a** **i** 8900 **ii** 7001 **iii** 309
b **i** 12 years **ii** 79 years

8 **a** 100 g **b** 99.85 g
c 98.5 g **d** 23 105 years

9 **a** 30.1°C **b** 490.4 hours

10 **a** 28 cm **b** $5e^{20}$ cm s^{-1} **c** 2040.1 cm
d 2.52 s **e** 5.30 s

11 **a** 2
b $x = 1, 2, 3, 4, 5, 6, 7, 8$

12 **a** $x = 0, 1$ **b** $x = 1, 2$ **c** $x = 2, 3$

Sample HSC problem

a 3.7 m
b −0.045 metres/month

Test yourself 7

1 D **2** B **3** A, C

4 **a** 3 **b** 1 **c** 3 **d** 2
e 1 **f** 3 **g** $\frac{1}{2}$ **h** −1
i −2 **j** 3

ANSWERS

5 **a** 6.39 **b** 1.98 **c** 3.26 **d** 1.40
e 0.792 **f** 3.91 **g** 5.72 **h** 72.4

6 **a** 6 **b** 2

7 **a** $3^x = a$ **b** $e^y = b$ **c** $10^z = c$

8 **a** 0.92 **b** 1.08 **c** 0.2
d 1.36 **e** 0.64

9 **a** $x = 1.9$ **b** $x = 1.9$ **c** $x = 3$ **d** $x = 36$

10 $t = 18.2$ **11** 0.9

12 **a** 2 **b** 2 **c** 3 **d** 1 **e** 3

13 **a** $\log_a x^5y^3$ **b** $\log_x\left(\frac{k^2p}{3}\right)$

14 **a** 0.65 **b** 1.3

15

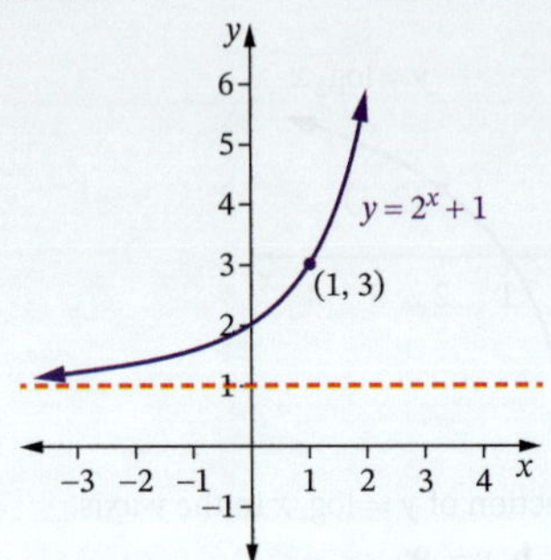

domain: all real x, range: $(1, \infty)$

16 **a** $x = 3.17$ **b** $x = 1.77$ **c** $x = 0.11$
d $y = 0.86$ **e** $n = 0.92$ **f** $x = 2$
g $y = 27$ **h** $n = 49$ **i** $x = 4096$
j $m = 2$

17 **a** $\log_2 y = x$ **b** $\log_5 b = a$ **c** $\log_{10} y = x$
d $\ln z = x$ **e** $\log_3 y = x + 1$

18 **a** $x = 2.7$ **b** $x = 3.1$ **c** $x = 2.1$

19 **a**

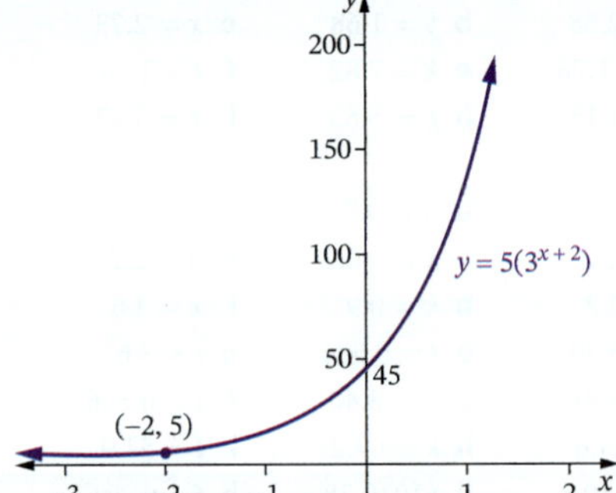

b

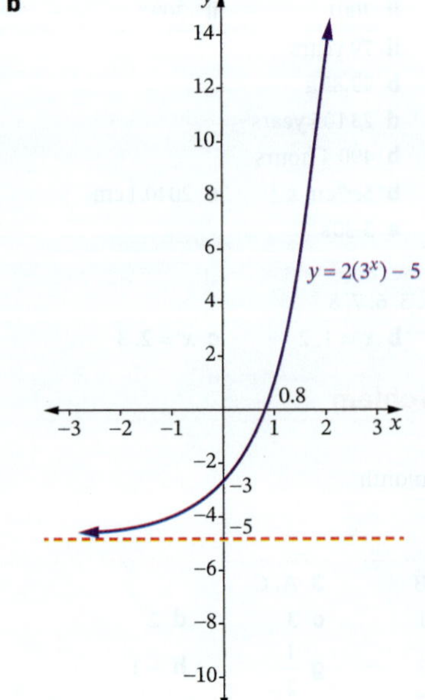

c

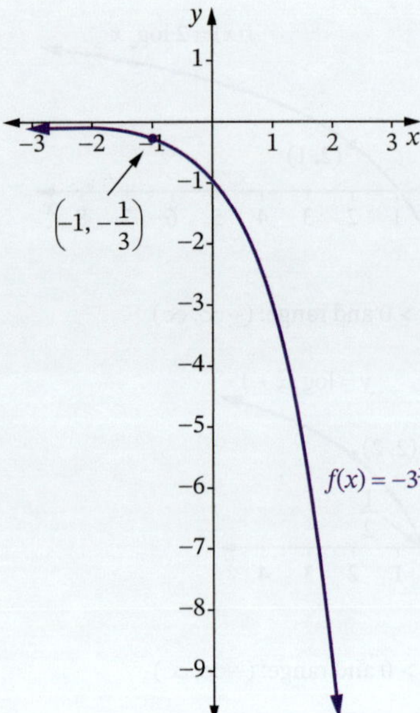

d

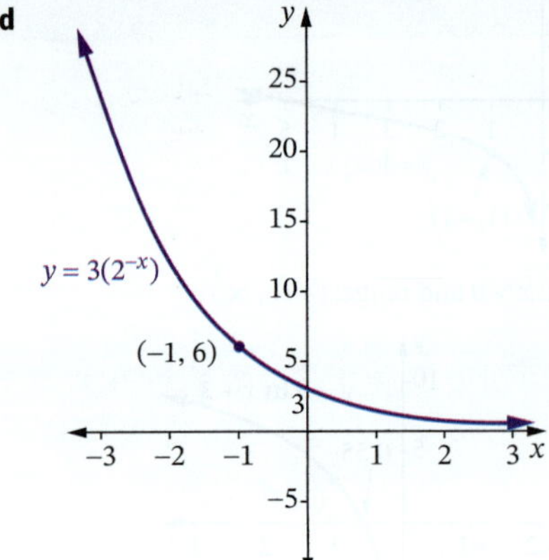

20 **a**

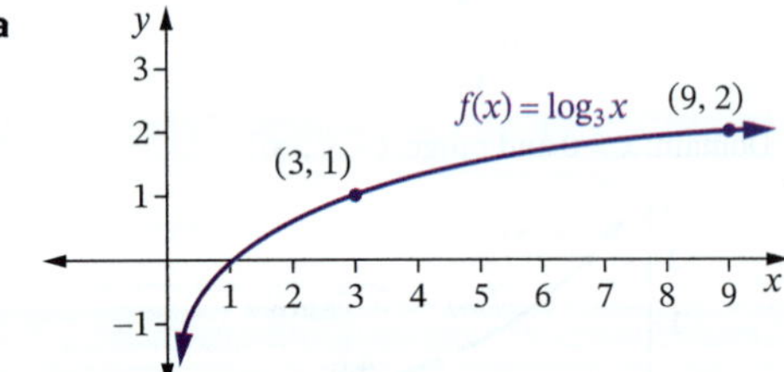

b

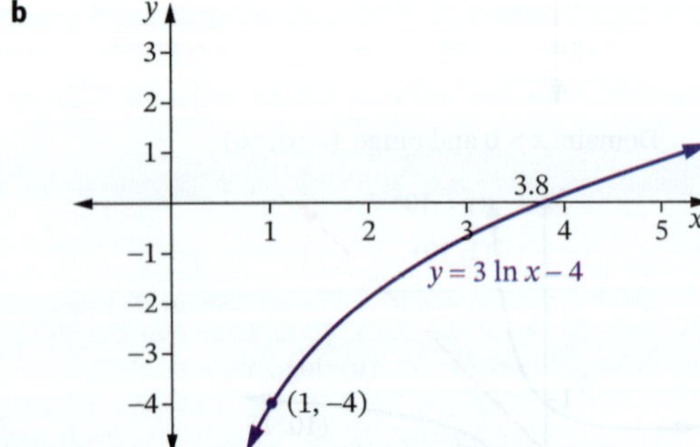

21 **a** $a + b$ **b** $b - a$ **c** $3a$
d $a + 2b$ **e** $3b$

22 **a** $x = 1\frac{1}{3}$ **b** $x = 37$ **c** $x = 1$

23 **a** $-\log_a x$ **b** $-\log_e y$

24 **a** $y = \log_e(6x^2 - 1)$ **b** $y = e^{6x^2 - 1}$
c $y = 6e^{2x} - 1$ **d** $y = x$
e $y = x$ **f** $y = e^x$

25 **a** **i** \$5280 **ii** \$5586.71 **iii** \$5692.86

b **i** 7 years **ii** 34 years

26 **a** $-2e^{-2x}$ **b** $20e^{4x}$

c $-16e^{8x} + 15x^2$ **d** $2xe^{2x}(x+1)$

e $108e^{3x}(4e^{3x}-1)^8$ **f** $\dfrac{1-2x}{e^{2x}}$

27 **a** 1081 **b** 1075

c **i** 88 years **ii** 104 years

28 **a** $e^x + 1$ **b** $-4e^x$ **c** $-3e^{-x}$

d $9e^x(3+e^x)^8$ **e** $3x^4e^x(x+5)$ **f** $\dfrac{e^x(7x-9)}{(7x-2)^2}$

Challenge exercise 7

1 **a** 2.7 **b** 2.8 **c** 2.5

2 (1, 0)

3

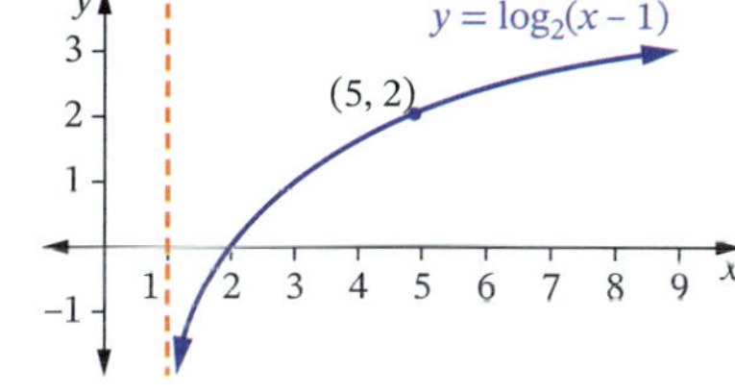

domain: $(1, \infty)$, range: all real y

4 $x = 0.63$

5 $y = 2a^x$ where $0 < a < 1$

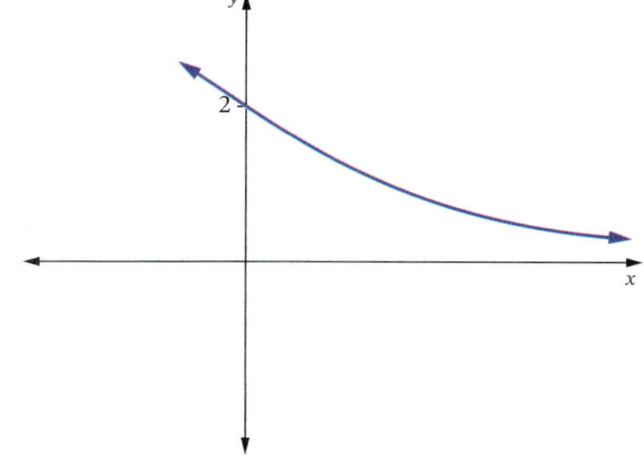

6 **a** Proof includes $x + 2 = 2^{y-8}$ (see worked solutions)

b **i** $y = 10.81$ **ii** $x = -1.99$

7 **a** $3e^2x - y - 3e^4 - 5 = 0$

b $x + 3e^2y - 9e^4 + 15e^2 - 2 = 0$

Practice set 3

1 B 2 D 3 C, D 4 D

5 A 6 C

7 **a** $9x^8 - 8x + 7$ **b** $6x^2 - 2$ **c** $-12x^{-5}$

d $-\dfrac{25}{2x^6}$ **e** $\dfrac{3\sqrt{x}}{2}$ **f** $14(2x+3)^6$

g $-\dfrac{8x}{(x^2-7)^5}$ **h** $\dfrac{5}{3\sqrt[3]{(5x+1)^2}}$ **i** $\dfrac{2(5x^2+15x+1)}{(2x+3)^2}$

8 **a** 100 L **b** 40 L

c −16 (leaking at the rate of 16 L/min)

d 12.2 min

9 **a** $9x - y + 16 = 0$ **b** $x + 9y + 20 = 0$

c $Q = (-20, 0)$

10 **a** $y' = e^x - 1$ **b** $y' = 3e^x$

c $y' = 4e^x(e^x - 2)^3$ **d** $y' = e^x(4x+1)^2(4x+13)$

e $y' = \dfrac{e^x(5x-7)}{(5x-2)^2}$ **f** $y' = 35e^{7x}$

11 $a = 1, b = -3, c = -1$

12 −2 13 $x = \dfrac{1}{2}$

14 **a** $3h^2 + 6xh - 4h$

b Proof involves $\lim\limits_{h \to 0} \dfrac{3h^2 + 6xh - 4h}{h}$; see worked solutions

15 **a** $3x - y - 4 = 0$ **b** $x - y - 2 = 0$

c $x + 3y + 10 = 0$ **d** $R = (-10, 0)$

16 **a**

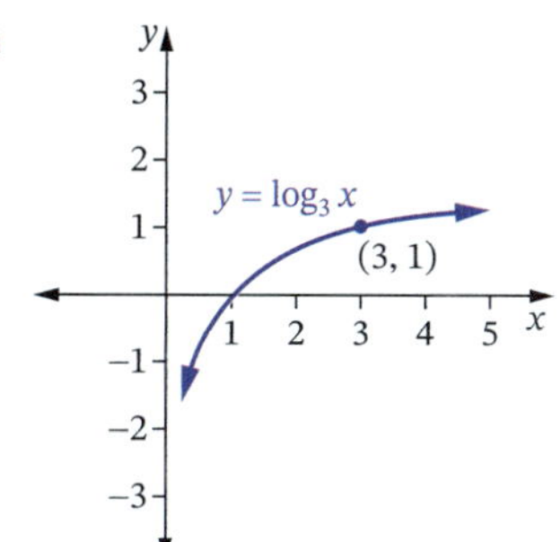

b

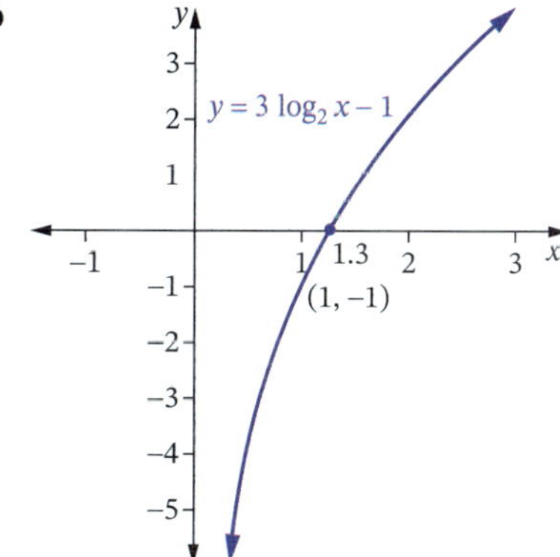

17 **a** $x = e^y$ **b** $x \approx 3.42$

18 $x \approx 0.28$

19

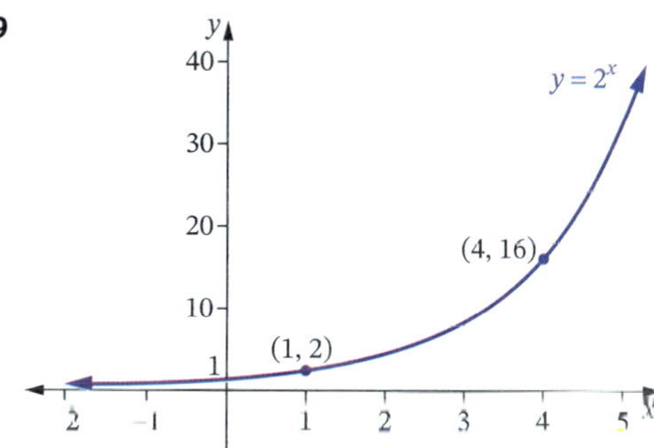

20 $f(-2) = -45, f'(-2) = 48$

21 **a** **i** 24.1202 **ii** 23.8802

b 24

22 **a**

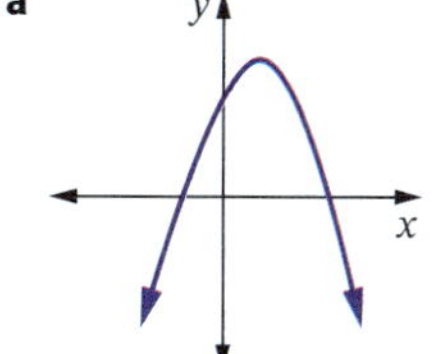

b

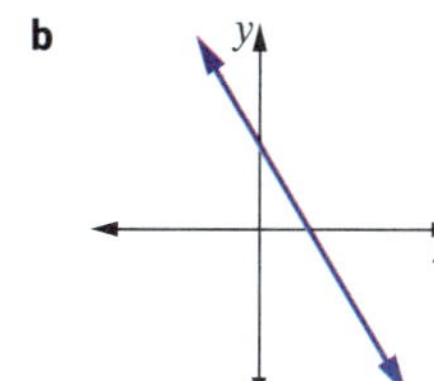

23 **a**

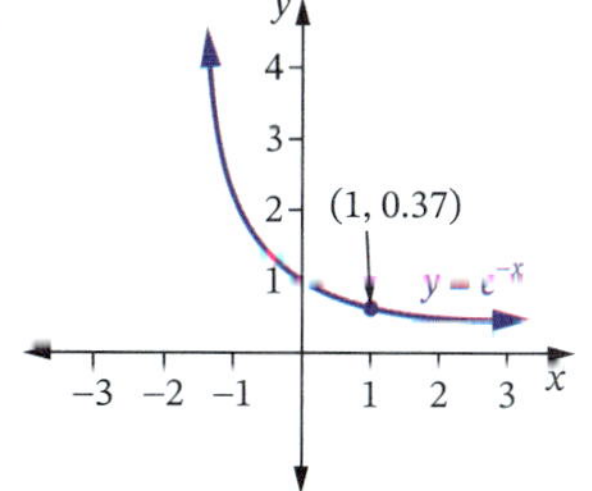

b

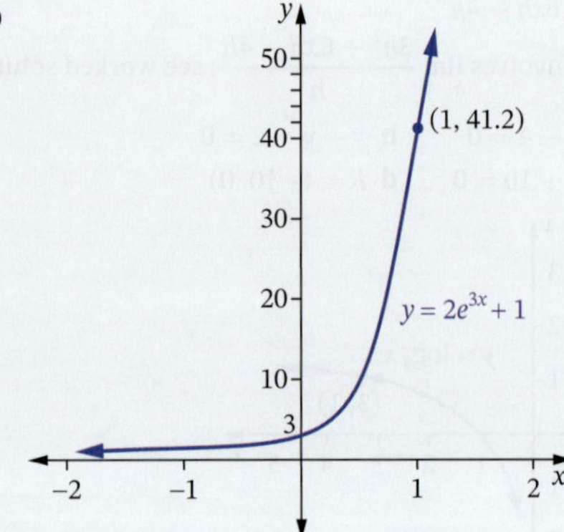

24 **a** 12 cm, 17 cm/s **b** 10 cm, −11 cm/s

25 **a** 4 **b** 1 **c** $\frac{1}{2}$

d 2.04 **e** 3.76 **f** 2.18

26 $y = 6x - 5$

27 **a** $x = 6$ **b** $x = 3$ **c** $x = 3$

d no real solutions **e** $x \approx 0.77$

28 **a** −9, function is decreasing when $x = 2$

b $2 \pm \sqrt{3}$

29 $x + 2y + 1 = 0$

30 **a** $8x^3 - 15x^2 + 6x - 1$ **b** $-\frac{5}{2x^6}$

c $\frac{1}{2\sqrt{x}}$ **d** $14(2x - 3)^6$

e $6x^3(2x - 5)^6(11x - 10)$ **f** $-\frac{31}{(3x - 2)^2}$

31 $y = 5e^2x - 5e^2$

Chapter 8

Exercise 8.01

1 **a** vertical translation 3 units up

b vertical translation 7 units down

c vertical translation 1 unit down

d vertical translation 5 units up

2 **a** vertical translation 1 unit up

b vertical translation 4 units down

c vertical translation 8 units up

3 vertical translation 9 units up

4 **a** $y = x^2 - 3$ **b** $f(x) = 2^x + 8$

c $y = |x| + 1$ **d** $y = x^3 - 4$

e $f(x) = \log_{10} x + 3$ **f** $y = \frac{2}{x} - 7$

5 **a** vertical translation 1 unit down

b vertical translation 6 units up

6 **a** **i** $y = 2x^3 - 2$ **ii** $y = 2x^3 + 6$

b **i** $y = |x| - 3$ **ii** $y = |x| - 6$

c **i** $y = e^x + 1$ **ii** $y = e^x + 5$

d **i** $f(x) = \log_e x + 10$ **ii** $f(x) = \log_e x - 8$

7 **a** (1, −1) **b** (1, −9) **c** $(1, -3 + m)$

8 **a** $P(-1, 1)$ **b** $P(-1, 5)$

9 **a**

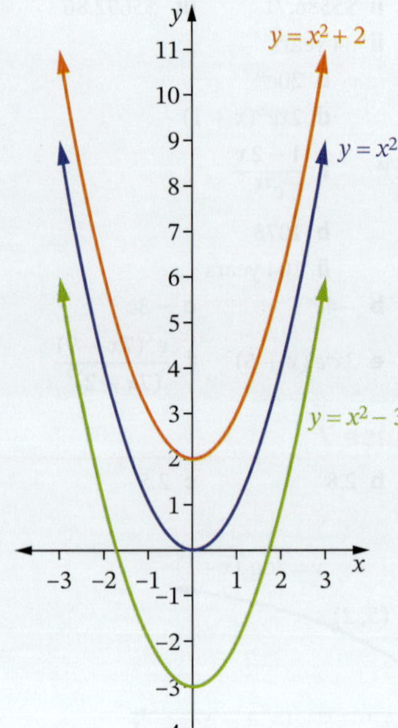

b

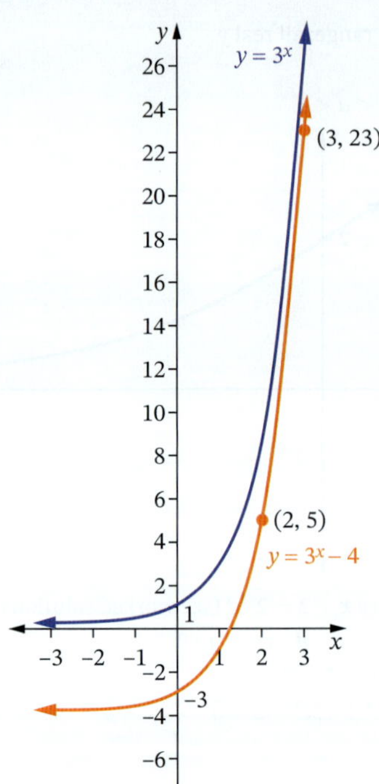

c

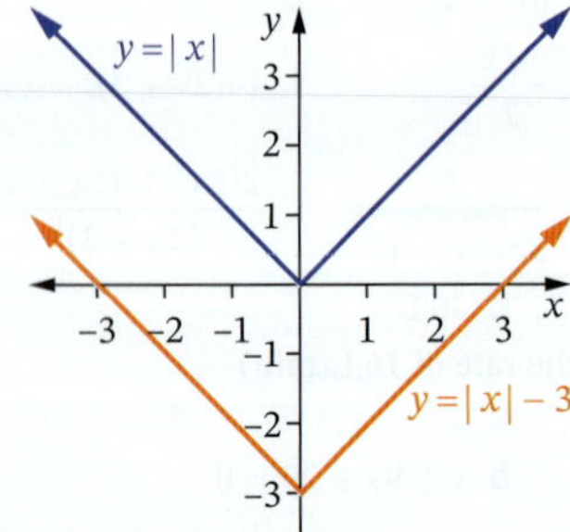

10 **a** $g(x) = f(x) + 5$

b $g(x) = f(x) - 4$

c $g(x) = f(x) - 1$

11 a $y = f(x) - 2$

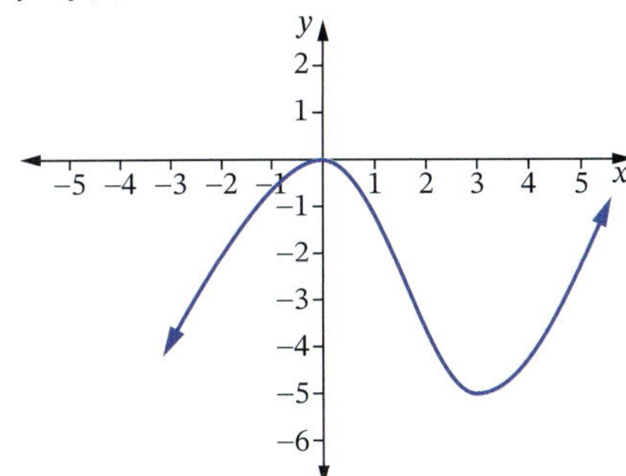

b $y = f(x) + 1$

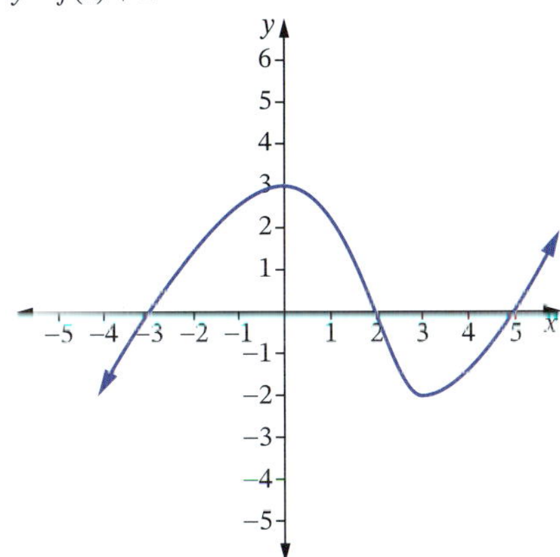

c $y = f(x) - 3$

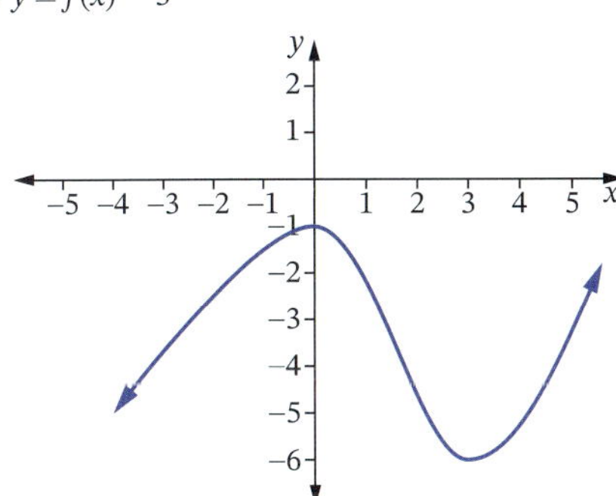

12 a

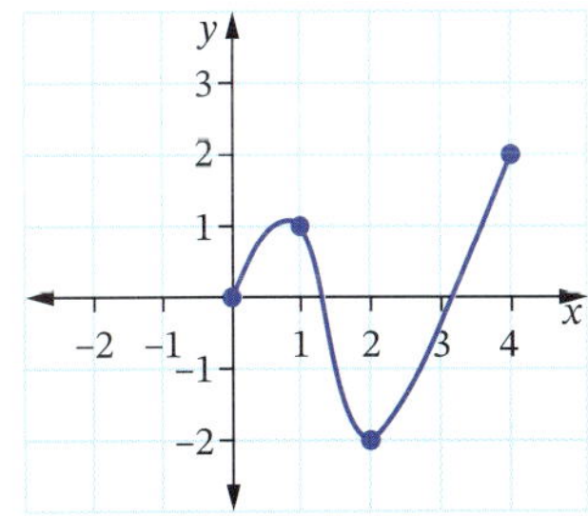

b

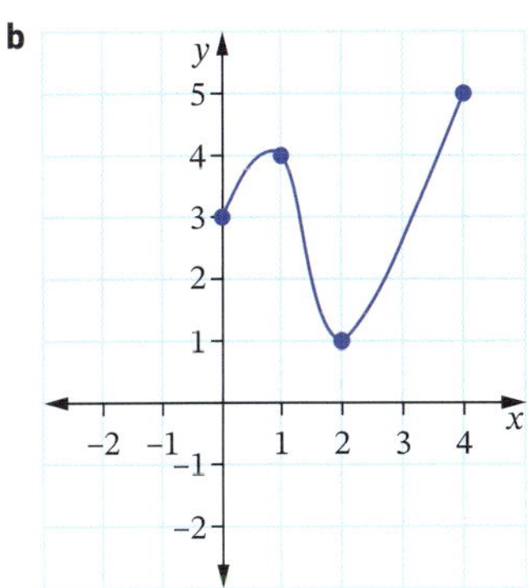

13 $x^2 + (y - 4)^2 = 1$

14 a $(x + 1)^2 + (y + 11)^2 = 16$

b $(x + 1)^2 + (y - 7)^2 = 16$

15 $y = -x^2 + 3x - 2$

16 a $g(x) = f(x) + 3$ **b** $Q(-1,0)$

c $y = -7x - 7$

17 a $h(x) = f(x) - 4$ **b** $B(x, y - 4)$

c $-\frac{1}{5}$

18 a See worked solutions.

b

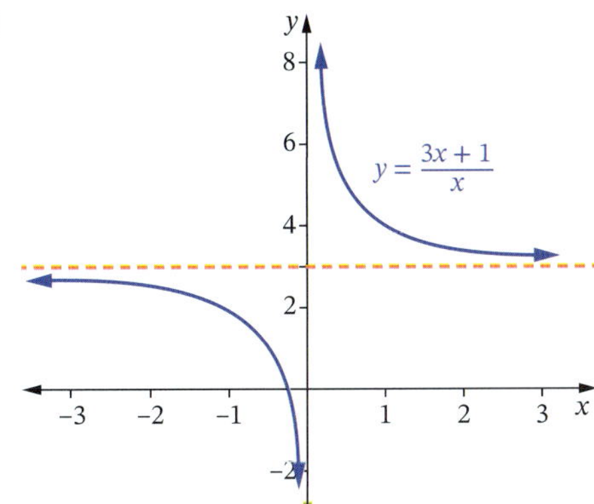

19

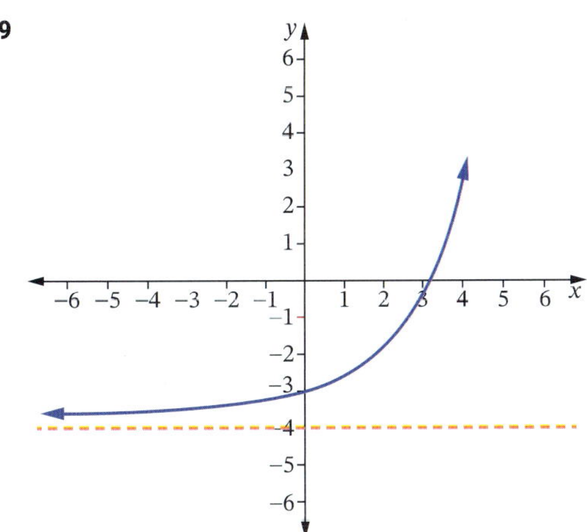

20 radius 4, centre $(-4, -1)$

21 $(x - 4)^2 + (y - 12)^2 = 81$ or $x^2 - 8x + y^2 - 24y + 79 = 0$; radius 9, centre (4, 12)

Exercise 8.02

1 a horizontal translation 4 units to the right

b horizontal translation 2 units to the left

2 a horizontal translation 5 units to the right

b horizontal translation 3 units to the left

3 a $y = (x + 3)^2$ **b** $f(x) = 2^{x-8}$

c $y = |x + 1|$ **d** $y = (x - 4)^3$

e $f(x) = \log_{10}(x + 3)$

4 horizontal translation 3 units to the right

5 a horizontal translation 2 units to the left

b horizontal translation 5 units to the right

6 a i $y = -(x + 4)^2$ **ii** $y = -(x - 8)^2$

b i $y = |x - 3|$ **ii** $y = |x + 4|$

c i $y = e^{x+6}$ **ii** $y = e^{x-5}$

d i $f(x) = \log_2(x - 5)$ **ii** $f(x) = \log_2 x$

7 a $(5, -3)$ **b** $(-8, -3)$ **c** $(1 - t, -3)$

8 a $P(3, 2)$ **b** $P(-9, 2)$

9 a

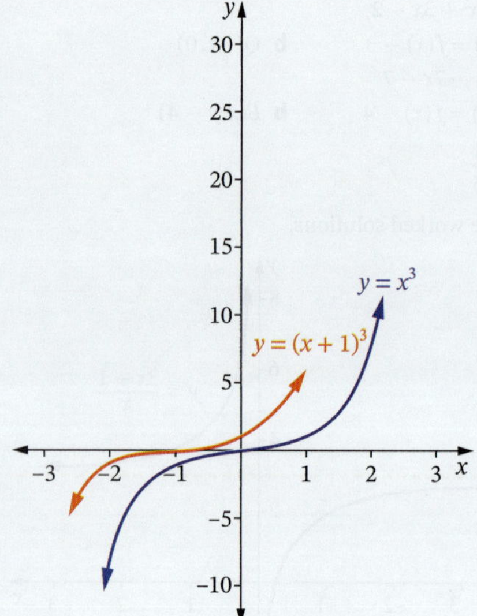

b

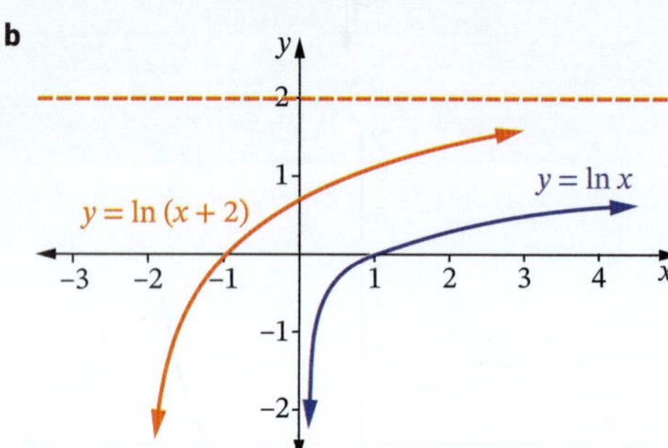

10 a

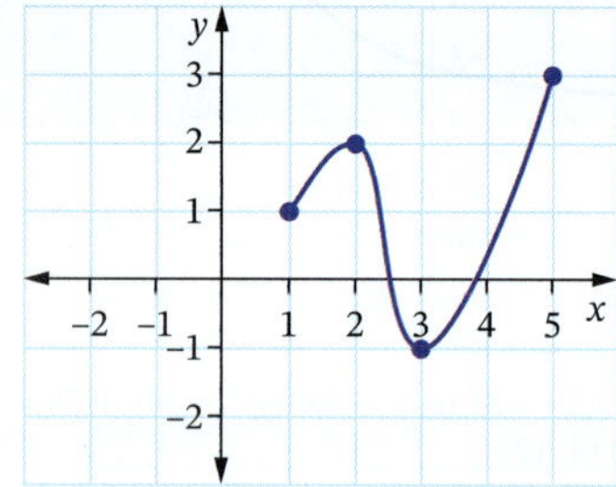

b

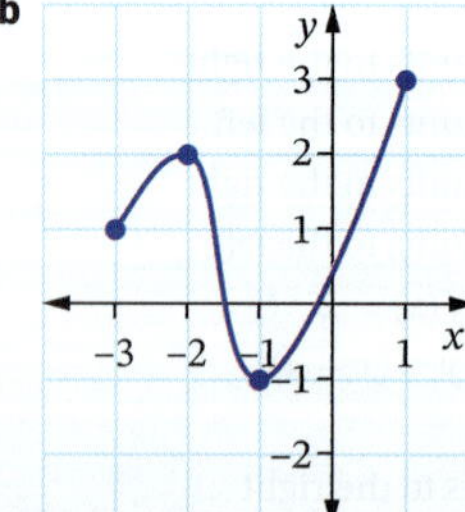

11 a $f(x) = x^5 - 5$ **b** $f(x) = (x - 3)^5$
c $f(x) = x^5 + 2$ **d** $f(x) = (x + 7)^5$

12 $P(7, -2)$

13 a $y = f(x + 7)$ **b** $y = f(x - 6)$

14 $(x + 7)^2 + y^2 = 9$

15 horizontal translation 8 units to the left, vertical translation 3 units down

16 $y = -2x^3 - 20x^2 - 42x$

17 $(x - 2)^2 + (y - 1)^2 = 1$, or $x^2 - 4x + y^2 - 2y + 1 = 0$

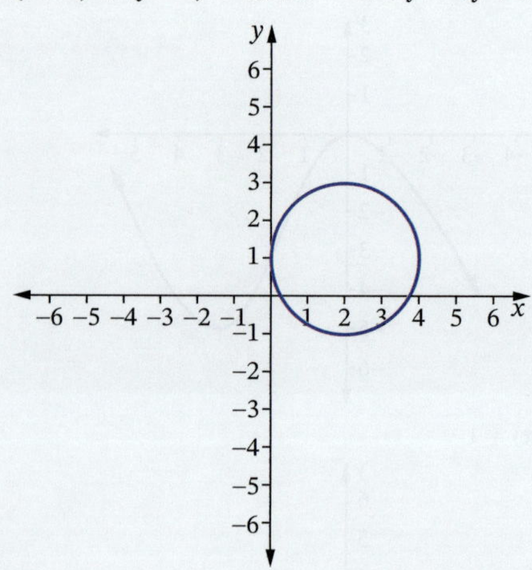

18 $(x + 9)^2 + (y - 1)^2 = 9$, or $x^2 + 18x + y^2 - 2y + 73 = 0$

19 a $g(x) = f(x + 7)$ **b** $(x,y) = (-4, 2)$
c $y = 3x + 14$

20 a $(x + 3)^2 + (y - 2)^2 = 9$ or $x^2 + 6x + y^2 - 4y + 4 = 0$
b $(x - 3)^2 + (y + 2)^2 = 9$ or $x^2 - 6x + y^2 + 4y + 4 = 0$

21 a $V(x) = U(x - 6)$ **b** $(x + 6, y)$ **c** $\frac{1}{2}$

22 radius: 4, centre $= (4, 3)$

Exercise 8.03

1 a $y = -x^2 + 2$ **b** $y = -(x + 1)^3$
c $y = -5x + 3$ **d** $y = -|2x + 5|$
e $y = -\frac{1}{x - 1}$ **f** $y = -3^x$

2 a reflection in x-axis
b reflection in y-axis
c reflection in both x-axis and y-axis

3 a $y = x^2 - 2$ **b** $y = (-x + 1)^3$
c $y = -5x - 3$ **d** $y = |-2x + 5|$
e $y = -\frac{1}{x + 1}$ **f** $y = 3^{-x}$

4

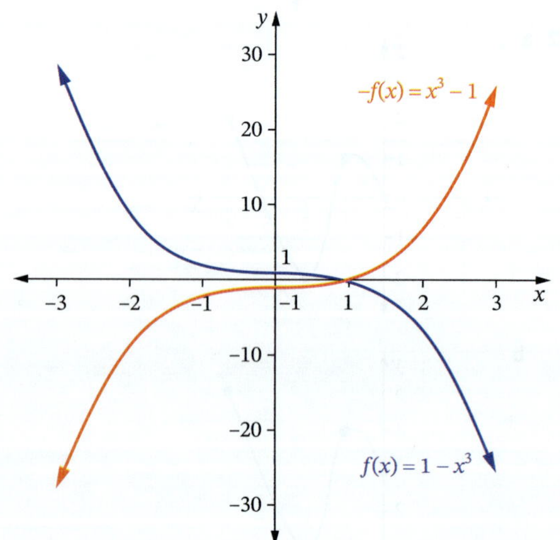

5 **a**

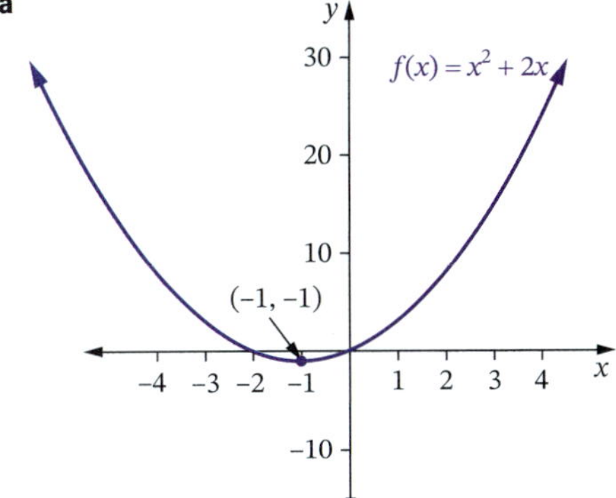

b

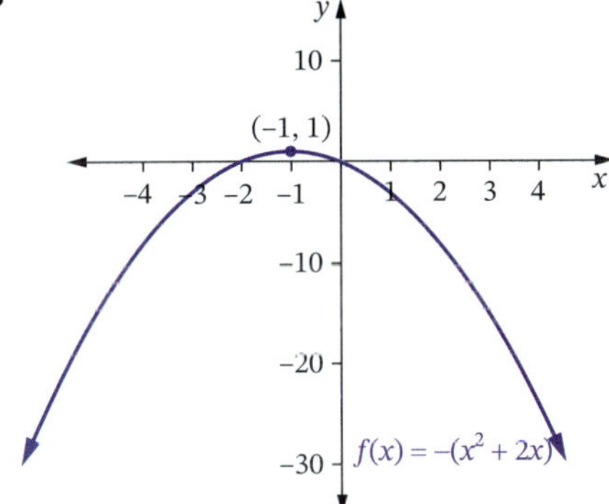

c

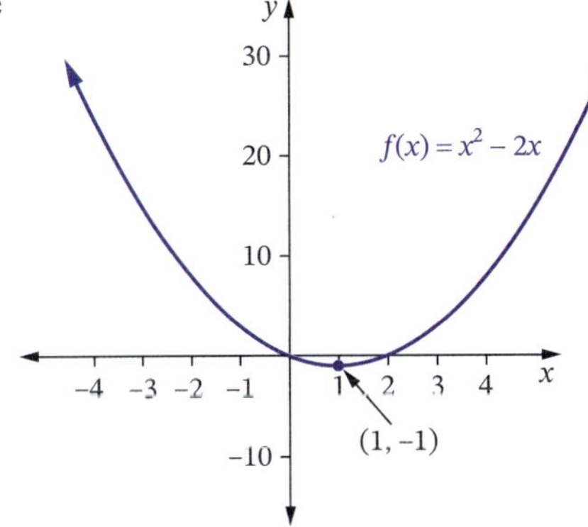

d

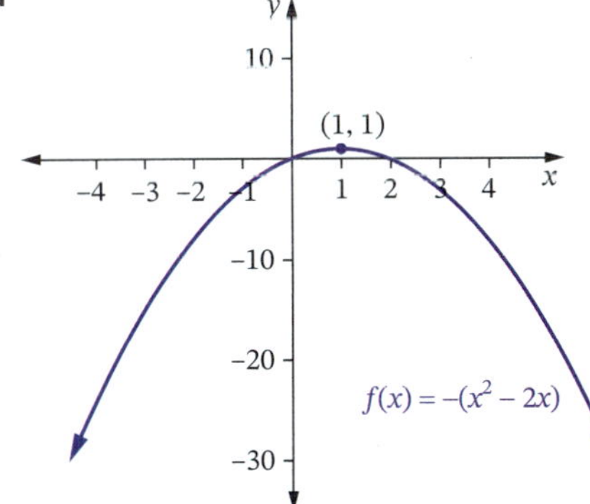

6 **a** $f(-x) = 2(-x)^2 = 2x^2 = f(x)$ so even.

b **i** $y = 2x^2$ **ii** $y = -2x^2$

c

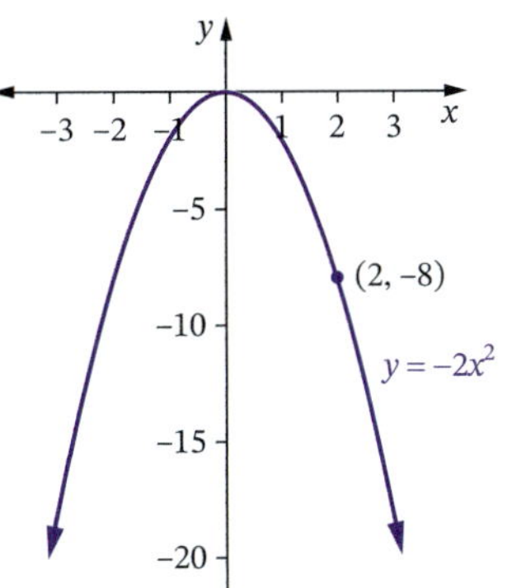

7 **a** $f(-x) = -(-x)^3 = x^3 = -f(x)$ so odd.

b **i** $y = x^3$ **ii** $y = -x^3$

c

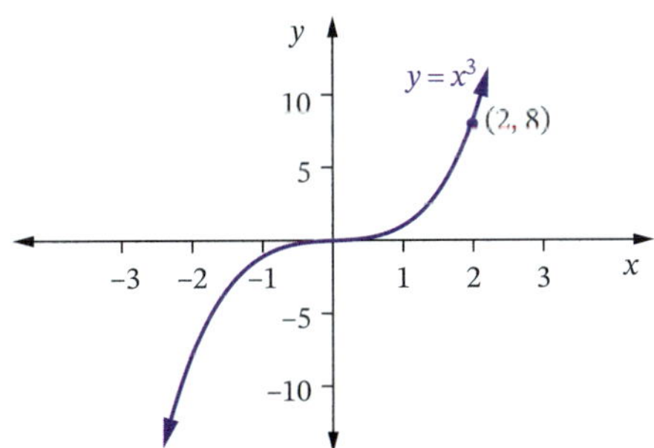

8 **a** x-intercepts 0, 3, 4; y-intercept 0.

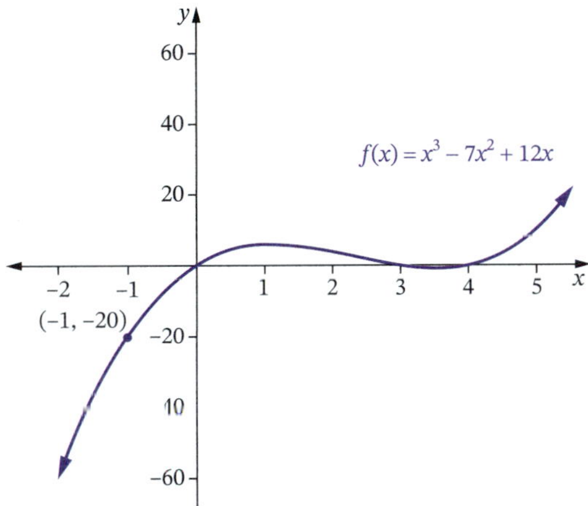

b **i**

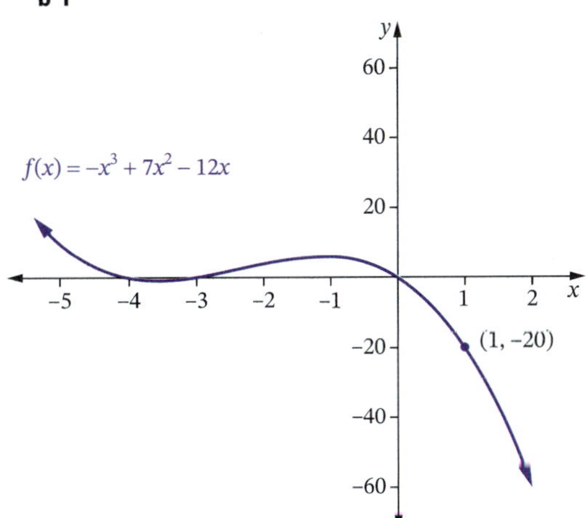

ii

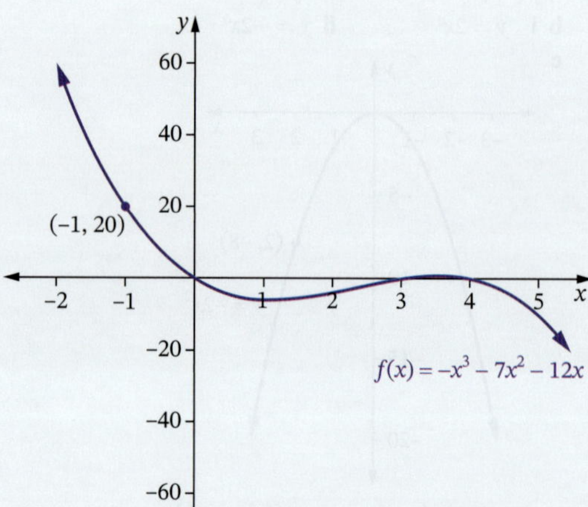

iii

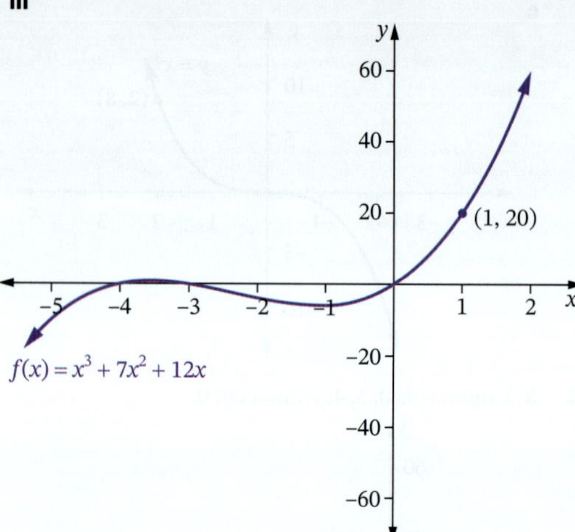

9 **a** $y = -2x - 7$ **b** $y = x^2 + 2x + 5$

c $f(x) = -x^3 + x^2 + 4x + 1$ **d** $f(x) = \dfrac{1}{-x+3}$

e $y = 2^{-x}$ **f** $f(x) = \log_2(-x)$

10

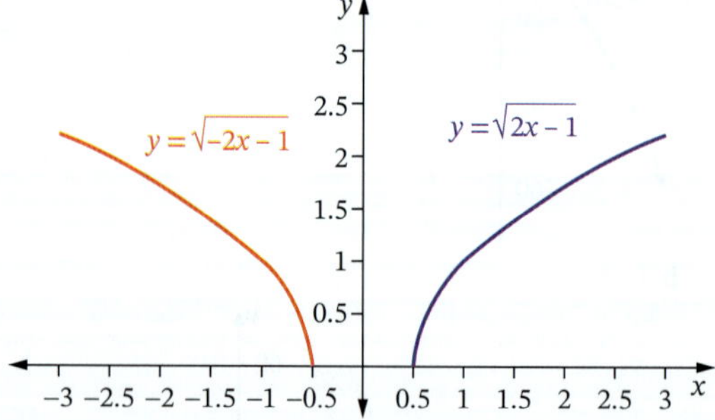

11 **a** $y = x - 1$

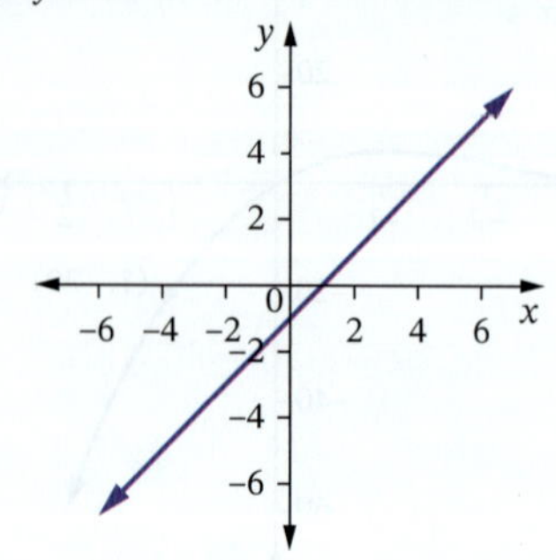

b $y = -x^2 - 3$

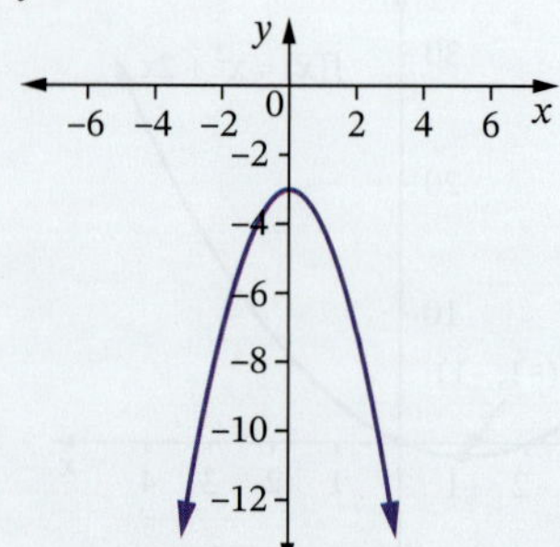

c $f(x) = -x^3 - 4x + 3$

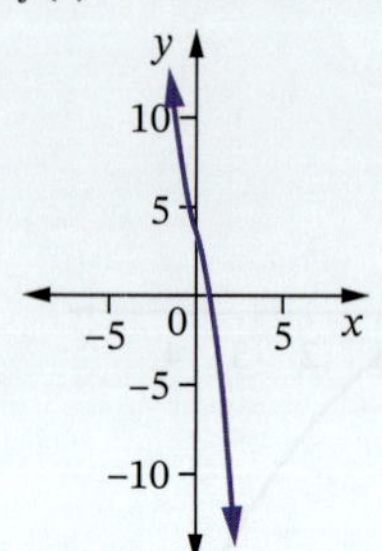

d $f(x) = \dfrac{1}{2x}$

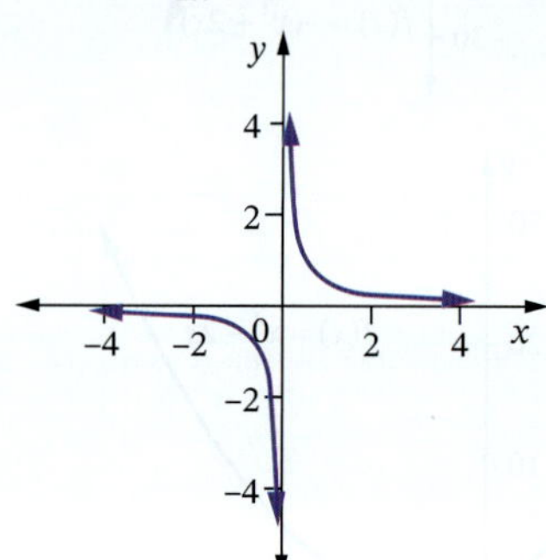

e $y = -5^{3x}$

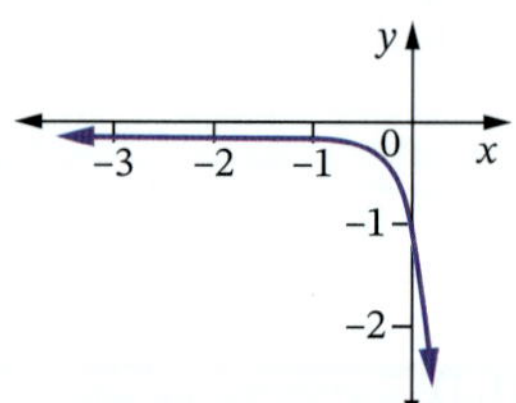

f $f(x) = -\ln x$

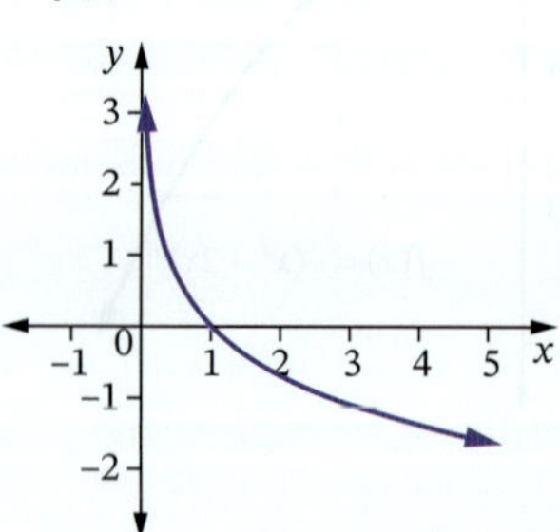

12

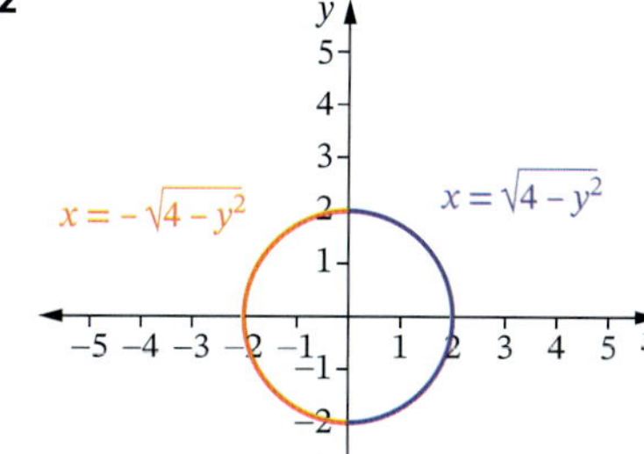

13 The function $y = \log_2 x$ is reflected in both the x- and y-axes.

a $y = -\log_2(-x)$

b

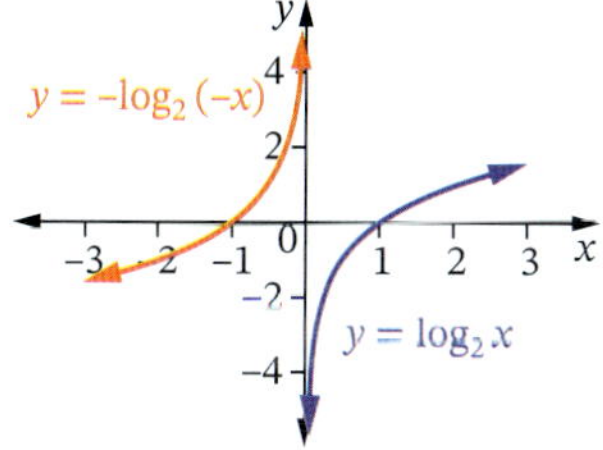

14 a

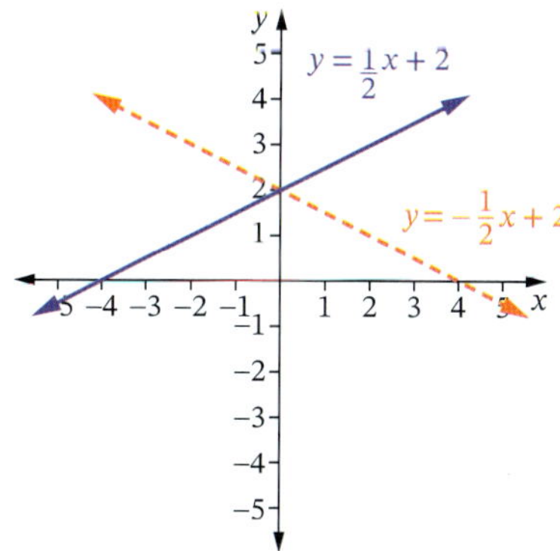

b

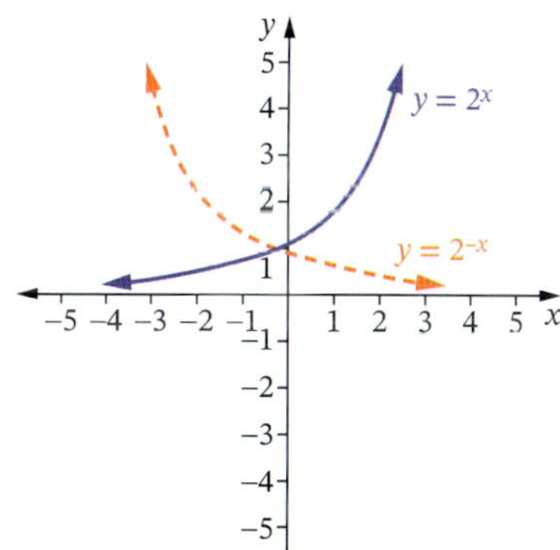

c

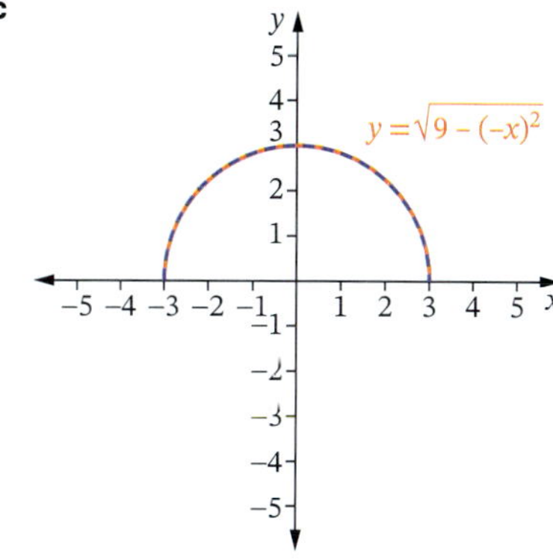

15 a $y = -e^x$

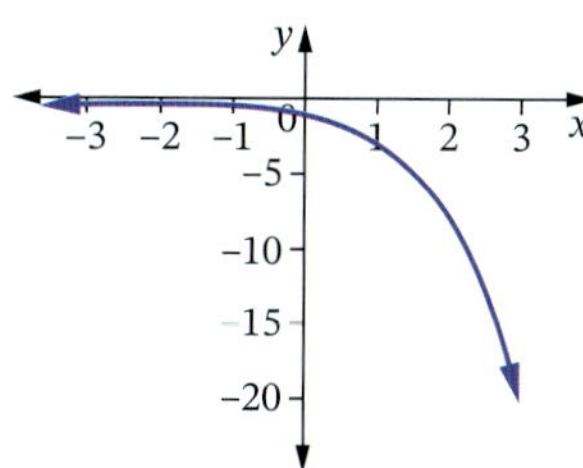

b $y = e^{-x}$

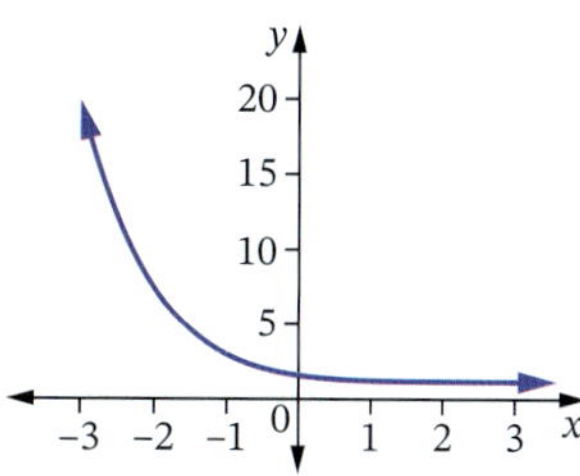

c $y = -e^{-x}$

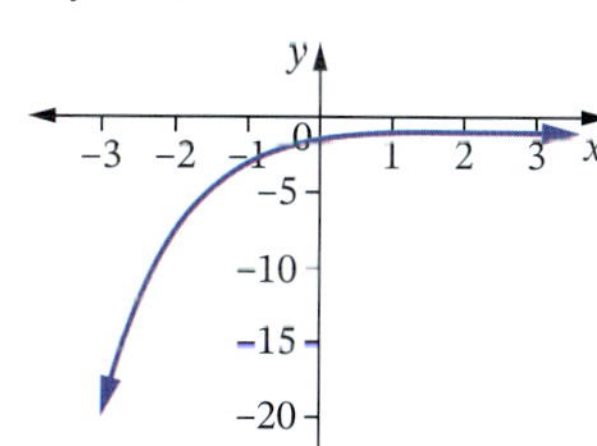

16 a $y = f(-x) + 5$, yes

b $y = -f(x) + 5$, no ($y = -f(x) - 5$)

17 a

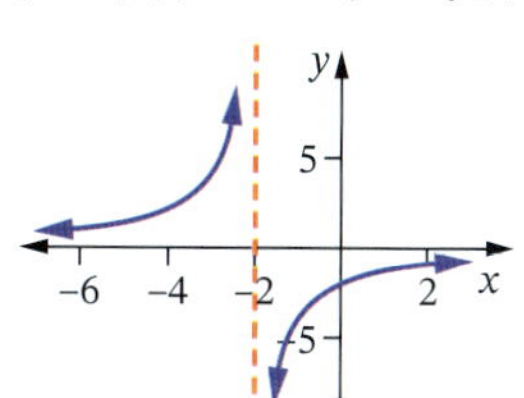

b

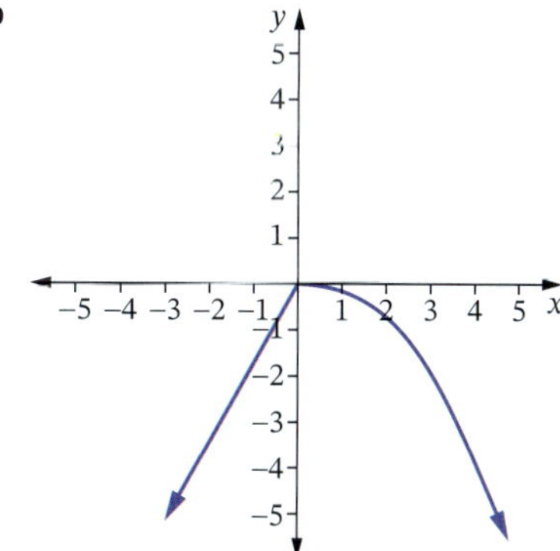

c

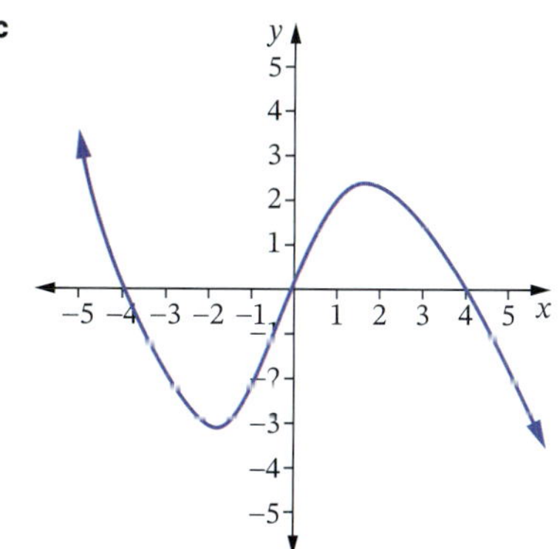

ANSWERS

18 a, b, c $y = -(x + 4)^3 + 2(x + 4)$ or $y = -x^3 - 12x^2 - 46x - 56$

d $y = -(x - 4)^3 + 2(x - 4)$ or $y = -x^3 + 12x^2 - 46x + 56$

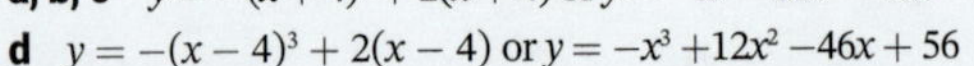

19

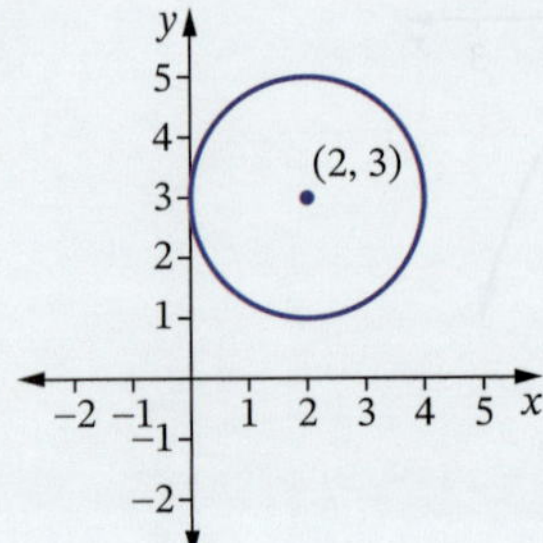

20 $y = -2|-x - 1| - 3$ (or $y = -2|x + 1| - 3$)

Exercise 8.04

1 a i vertical dilation factor 6 (stretched)

ii vertical dilation factor $\frac{1}{2}$ (shrunk)

iii vertical dilation factor -1 (reflection in x-axis)

b i vertical dilation factor 2 (stretched)

ii vertical dilation factor $\frac{1}{6}$ (shrunk)

iii vertical dilation factor -1 (reflection in x-axis)

c i vertical dilation factor 4 (stretched)

ii vertical dilation factor $\frac{1}{7}$ (shrunk)

iii vertical dilation factor $\frac{4}{3}$ (stretched)

d i vertical dilation factor 9 (stretched)

ii vertical dilation factor $\frac{1}{3}$ (shrunk)

iii vertical dilation factor $\frac{3}{8}$ (shrunk)

e i vertical dilation factor 5 (stretched)

ii vertical dilation factor $\frac{1}{8}$ (shrunk)

iii vertical dilation factor -1 (reflection in x-axis)

f i vertical dilation factor: 9 (stretched)

ii vertical dilation factor: -1 (reflection in x-axis)

iii vertical dilation factor: $\frac{2}{5}$ (shrunk)

2 a $y = 6x^2$; domain: all real x, range: $y \geq 0$

b $y = \frac{\ln x}{4}$; domain: $(0, \infty)$; range: all real y

c $f(x) = -|x|$; domain: all real x, range: $y \leq 0$

d $f(x) = 4e^x$; domain: all real x, range: $y > 0$

e $y = \frac{7}{x}$; domain: all real x, $x \neq 0$, range: all real y, $y \neq 0$

3 a $y = 5(3^x)$ **b** $f(x) = \frac{x^2}{3}$

c $y = \frac{1}{2x}$ **d** $y = \frac{2|x|}{3}$

4 a (3, 24) **b** (3, −6) **c** (3, 72) **d** (3, 5)

5 a (4, 4) **b** (4, 6) **c** (4, 36)

d (4, 16) **e** (4, −12)

6 a

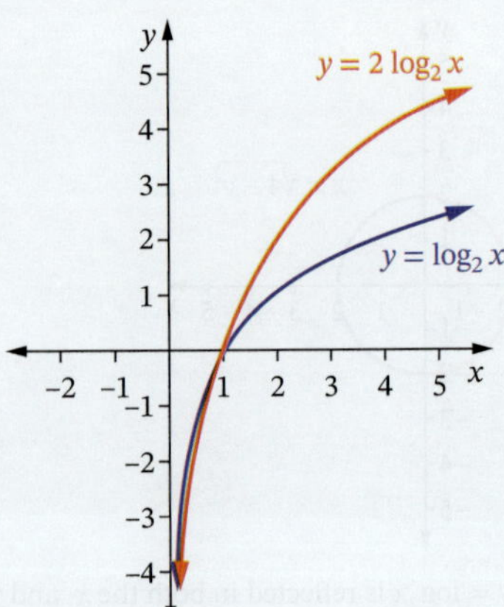

b

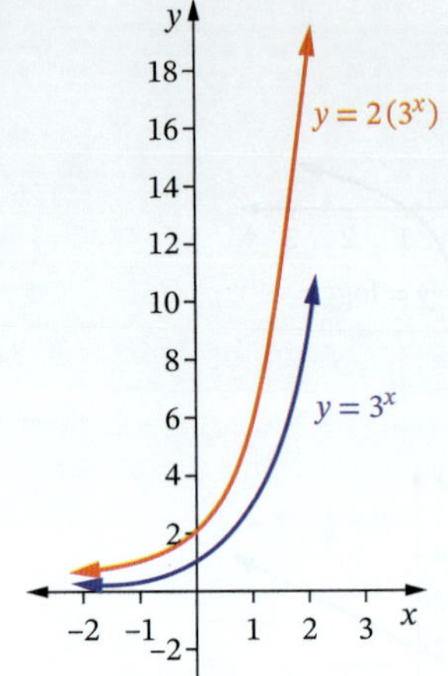

c

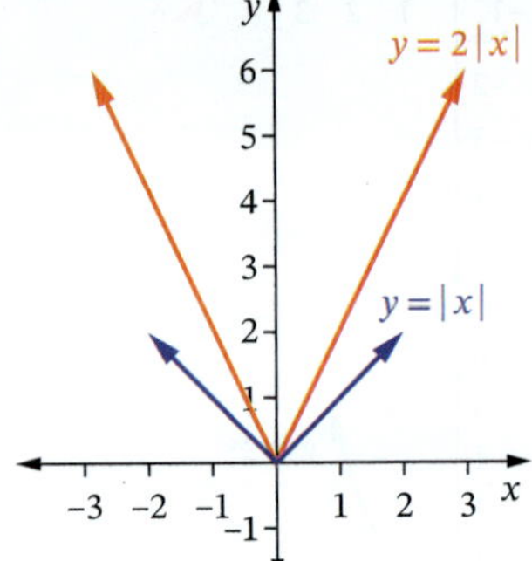

d

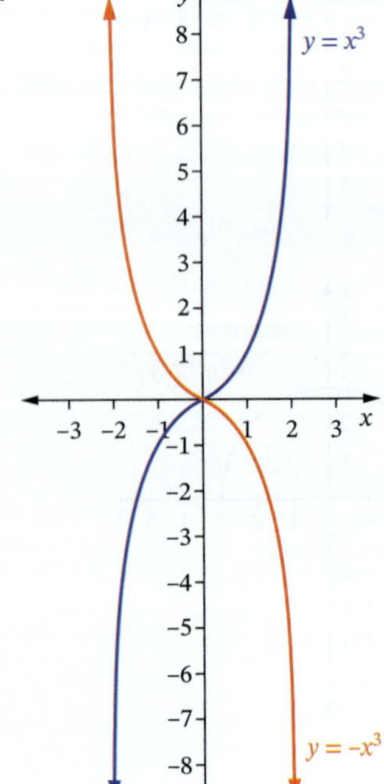

7 **a**

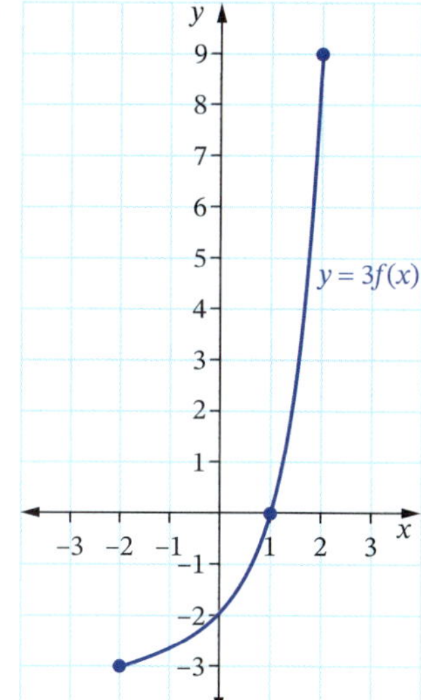

b

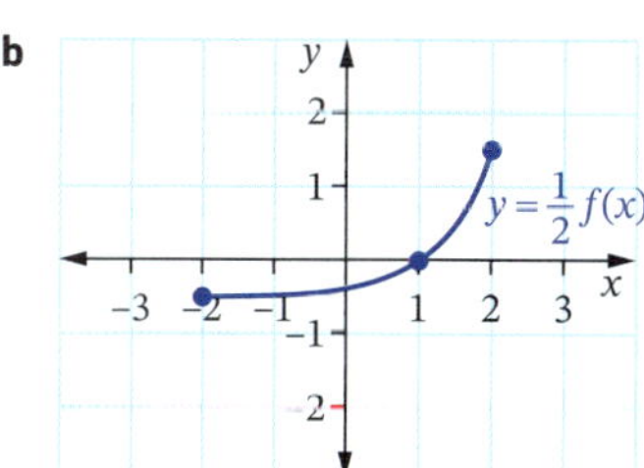

c

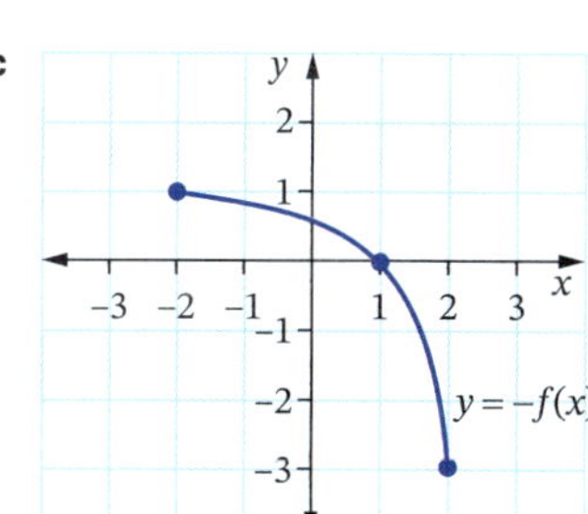

8

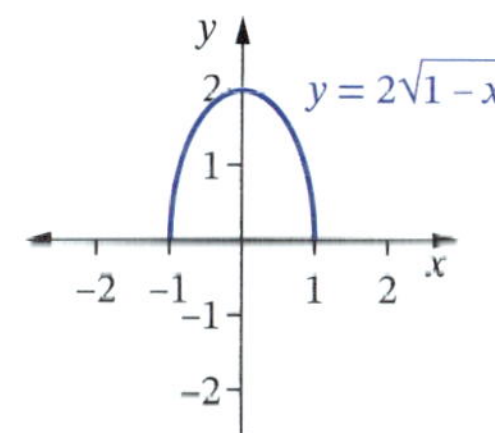

domain $[-1, 1]$ range $[0, 2]$

9 **a** $(-2, 52)$ **b** -6

c -24; vertical dilation stretches the graph, so its gradient is also multiplied by factor 4.

10 $4x^2 + y^2 = 4$ **11** $f(x) = -|x+1|$

12 $4(x-5)^2 + y^2 = 64$ **13** $y = 3x^2 - 6$

14 $(x+2)^2 + 9(y+3)^2 = 1$

Exercise 8.05

1 **a** horizontal dilation factor: $\frac{1}{8}$ (shrunk)

b horizontal dilation factor: 5 (stretched)

c horizontal dilation factor: $\frac{7}{3}$ (stretched)

d horizontal dilation factor: -1 (reflection in y-axis). Note that this is the same graph as $f(x) = x^4$.

2 Answers may vary, depending on choice of dilation.

a **i** horizontal dilation factor: $\frac{1}{2}$ (shrunk), or vertical stretch with dilation factor 4

ii horizontal dilation factor: $\frac{1}{5}$ (shrunk), or vertical stretch with dilation factor 25

iii horizontal dilation factor: 3 (stretched) , or vertical shrink with dilation factor $\frac{1}{9}$

b **i** vertical dilation factor: 4 (stretched), or horizontal shrink with dilation factor $\frac{1}{\sqrt[3]{4}}$

ii horizontal dilation factor: 2 (stretched), or vertical shrink with dilation factor $\frac{1}{8}$

iii horizontal dilation factor: -1 (reflection in y-axis), or reflection in the x-axis

c **i** horizontal dilation factor: $\frac{1}{7}$ (shrunk), or vertical stretch with factor 7^4

ii vertical dilation factor: $\frac{1}{8}$ (shrunk), or horizontal stretch with dilation factor 8^4

iii horizontal dilation factor: $\frac{4}{3}$ (stretched), or vertical shrink with dilation factor $\left(\frac{3}{4}\right)^4$

d **i** horizontal dilation factor: $\frac{1}{5}$ (shrunk)

ii horizontal dilation factor: 2 (stretched)

iii horizontal dilation factor: $\frac{5}{3}$ (stretched)

e **i** horizontal dilation factor: $\frac{1}{3}$ (shrunk)

ii vertical dilation factor: -1 (reflection in x-axis)

iii horizontal dilation factor: 2 (stretched)

f **i** vertical dilation factor: 8 (stretched)

ii horizontal dilation factor: -1 (reflection in y-axis)

iii horizontal dilation factor: 7 (stretched)

3 **a** $f(x) = |5x|$; domain: all real x, range: $[0, \infty)$

b $y = \left(\frac{x}{3}\right)^2$; all real x, range: $[0, \infty)$

c $y = (-x)^3$; domain: all real x, range: all real y

d $y = \frac{e^x}{9}$; domain: all real x, range: $(0, \infty)$

e $y = -\log_4 x$; domain: $x > 0$, range: all real y

4 **a** $(-1, 7)$ **b** $(2, 7)$ **c** $(-6, 7)$

5 **a** $(-72, 1)$ **b** $(-48, 1)$ **c** $(-6, 1)$

6 **a**

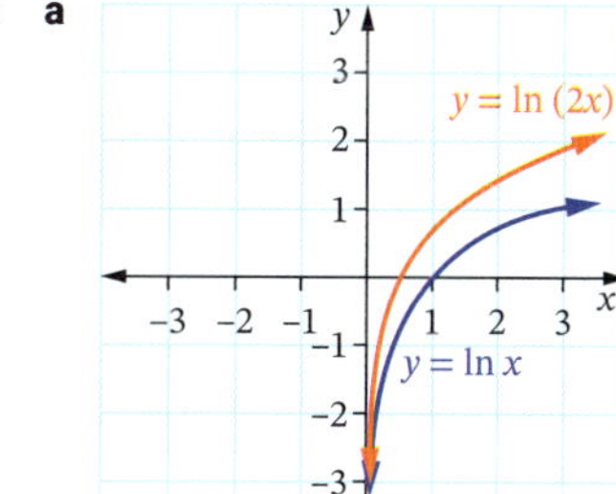

b

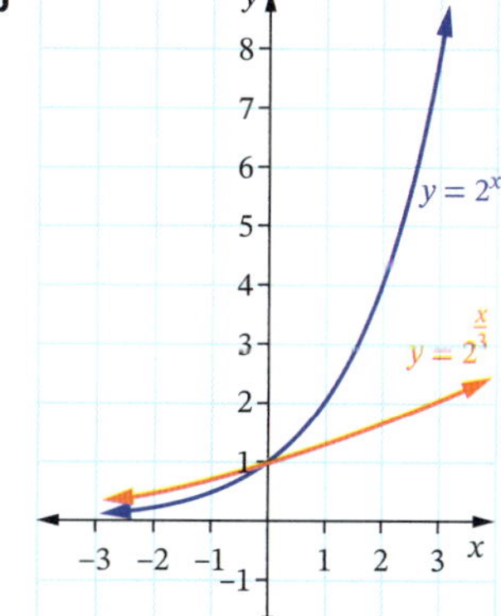

9780170498197

c

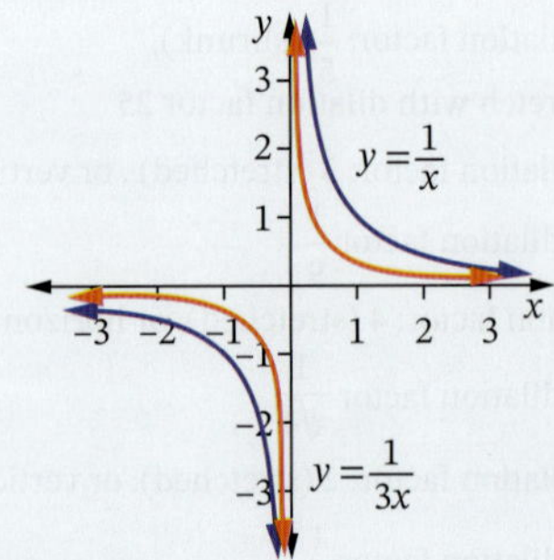

d

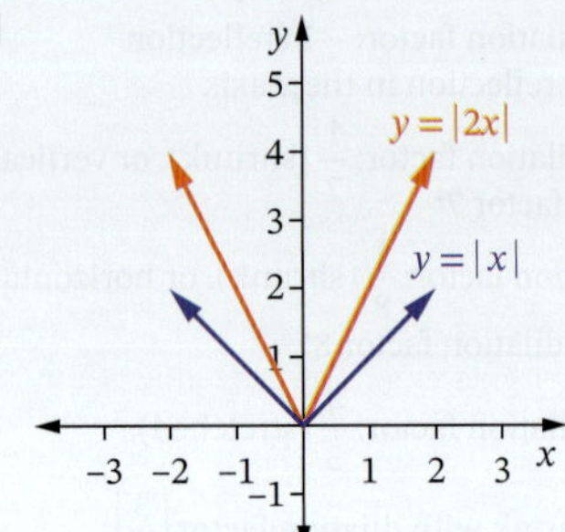

e

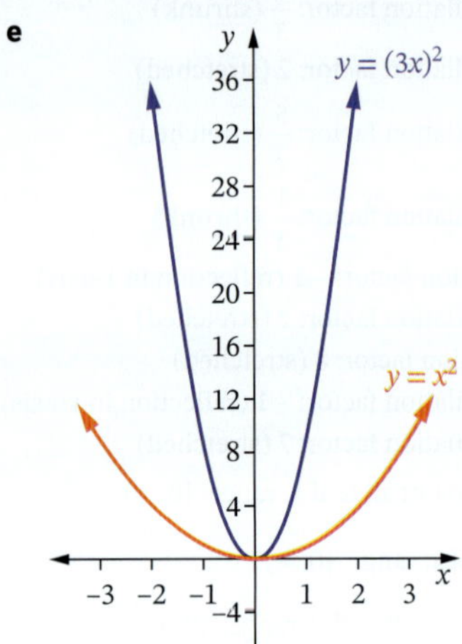

f

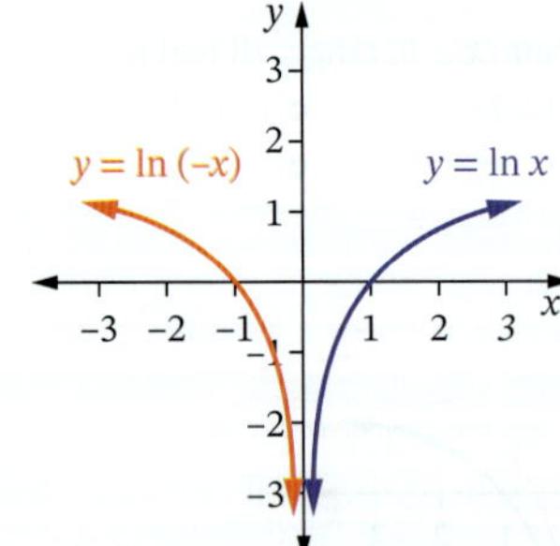

7

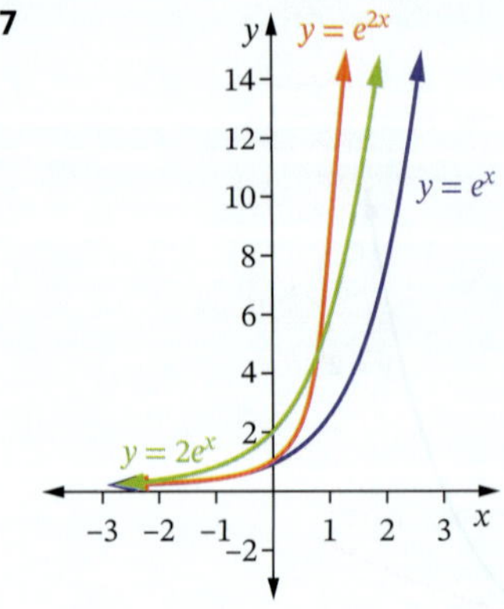

8 A reflection in y-axis transforms $y = f(x)$ into $y = f(-x)$.

a Since $y = x^2$ is an even function, $f(x) = f(-x)$, so a reflection in the y-axis doesn't change the function.

b Since $y = |x|$ is an even function, $f(x) = f(-x)$ so a reflection in the y-axis doesn't change the function.

9 a

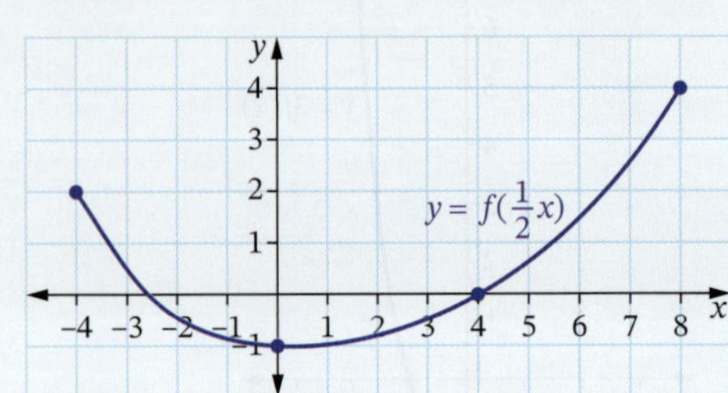

b

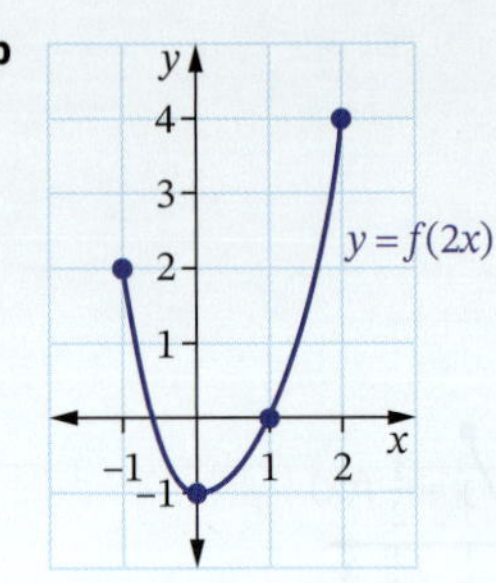

10 a $y = 16x^2 - 12x + 1$ **b** $\left(\frac{1}{2}, -1\right)$

11 a $\left(\frac{1}{2}x\right)^2 + y^2 = 4$ or $x^2 + 4y^2 = 16$

b

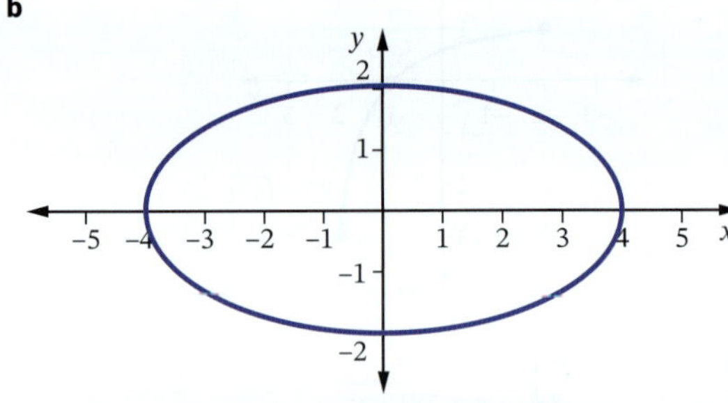

domain: $-4 \le x \le 4$, range: $-2 \le y \le 2$

12 a $4x^2 + 36y^2 = 324$; domain $-9 \le x \le 9$; range $-6 \le y \le 6$

b $x^2 + y^2 = 36$ (an expansion of the original circle with radius twice the length)

c $x^2 + y^2 = \frac{9}{4}$ (a contraction of the original circle with radius half the length)

13 $x^2 + y^2 = 1$

14 a $(1, 2)$ **b** $\frac{1}{2}$

15 a $y = 5|x - 3|$ **b** $y = \left|\frac{x}{2}\right| - 4$

c $y = -\left|\frac{x}{3}\right|$ **d** $y = 6|2x|$

Exercise 8.06

1 a $(5, -11)$ **b** $(-1, -2)$

c $(9, 3)$ **d** $(-2, -17)$

2 a $g(x) = -4x^5$ **b** $g(x) = \left(-\frac{1}{3}x\right)^5$

3 **a** horizontal translation 1 unit to the right, vertical translation 7 units up
b vertical dilation of factor 4, vertical translation 1 unit down
c vertical dilation of factor 5, reflection in x-axis, vertical translation 3 units down
d vertical dilation of factor 2, horizontal translation 7 units to the left
e Rewrite as $y = 6[2(x - 2)]^3 + 5$. Horizontal dilation of factor $\frac{1}{2}$, vertical dilation of factor 6, horizontal translation 2 units to the right, vertical translation 5 units up
f Rewrite as $y = 2[3(x + 3)]^3 - 10$. Horizontal dilation of factor $\frac{1}{3}$, vertical dilation of factor 2, horizontal translation 3 units to the left, vertical translation 10 units down

4 **a** $y = (x + 4)^3 - 3$ **b** $g(x) = |x - 1| + 9$
c $g(x) = 3x - 6$ **d** $y = -e^x + 2$
e $y = (2x)^3 - 5$ **f** $g(x) = \frac{2}{\left(\frac{x}{3}\right)} \equiv \frac{6}{x}$
g $g(x) = 3\sqrt{-2x}$ **h** $y = \ln\left(\frac{x}{3}\right) + 2$
i $g(x) = 3\log_2 4x$ **j** $y = \left(\frac{x}{2}\right)^2 - 3$

5 **a** vertical dilation of factor 2, horizontal translation 3 units to the left, vertical translation 1 unit down
b horizontal dilation of factor $\frac{1}{3}$, reflection in x-axis, vertical translation 9 units up
c horizontal dilation of factor $\frac{1}{5}$, vertical dilation of factor 2, vertical translation 3 units down
d vertical dilation of factor 4, horizontal translation 7 units to the right,vertical translation 1 unit up
e reflection in y-axis, horizontal dilation of factor $\frac{1}{2}$, horizontal translation 1 unit to the left, vertical translation 1 unit down
f horizontal dilation of factor $\frac{1}{2}$ (or vertical dilation of factor $\frac{1}{2}$), reflection in x-axis, vertical translation 8 units up

6 **a** $(9, -31)$ **b** $(4, 5)$
c $(5, -25)$ **d** $(-8, -67)$
e $(6, 21)$

7 **a** $(x + 3, y - 6)$ **b** $(-x, y + 6)$
c $(x - 5, 2y)$ **d** $(3x, y + 5)$
e $\left(\frac{1}{5}(x + 6), -8(y - 1)\right)$

8 **a** $y = f(x + 1) - 2$ **b** $y = f(x - 5) + 3$
c $y = -f(x - 4)$ **d** $y = f(-x) + 2$
e $y = -f\left(\frac{x}{4}\right)$ **f** $y = 2f(x) - 2$

9 **a** domain all real x, range $y \geq 5$
b domain all real x, range $y \geq -2$
c domain all real x, $x \neq 2$, range all real y, $y \neq 1$
d domain all real x, range $y > 2$
e domain $x > 2$ range all real y

10 **a** $y = (x + 1)^2 - 8$
b horizontal translation 1 unit to the left, vertical translation 8 units down

11 horizontal translation 5 units to the right, vertical translation 28 units down

12 -32

13 $40x - y + 46 = 0$

14 $a = 3, b = 5$

Exercise 8.07

1 **a** vertical dilation of factor 3, horizontal translation 1 unit to the left, vertical translation 2 units down
b $y = 3(x + 1)^2 - 2$

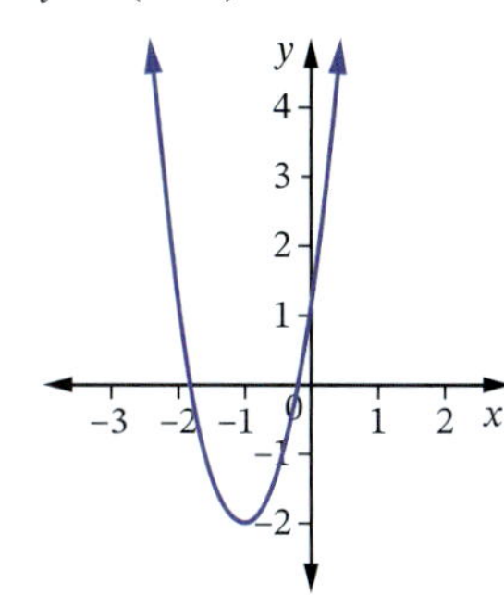

c i $x = 1, -3$ **ii** $-2 \leq x \leq 0$

2

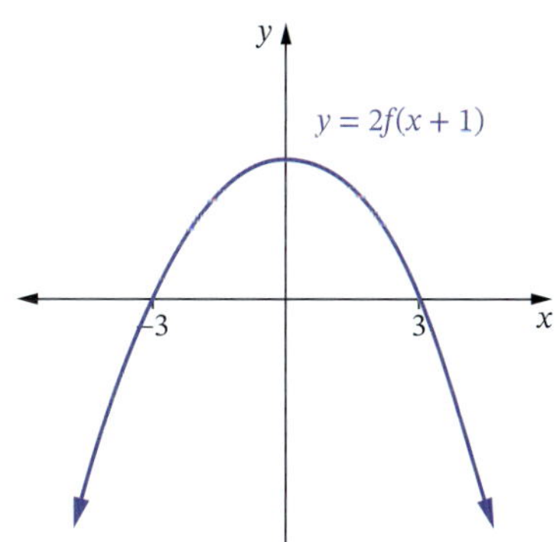

3 **a**

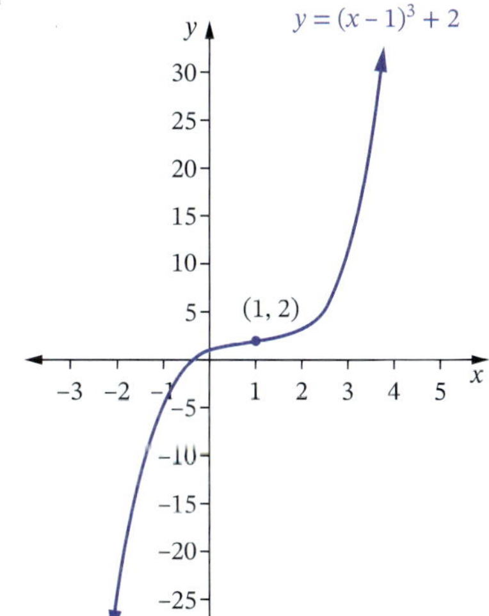

b

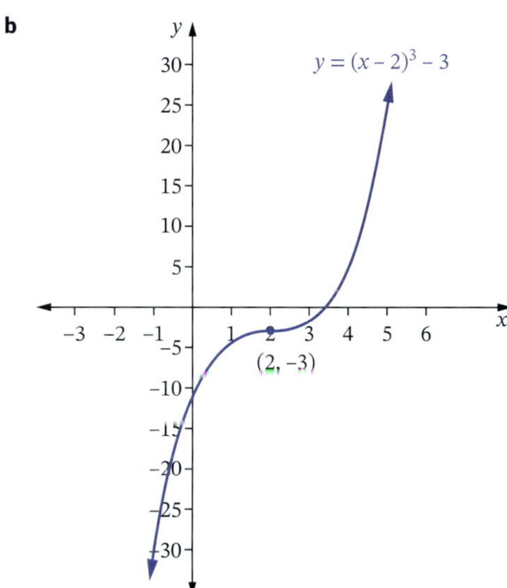

c

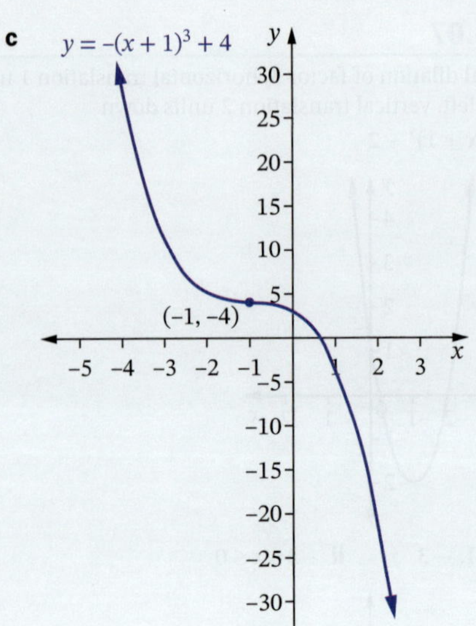

d

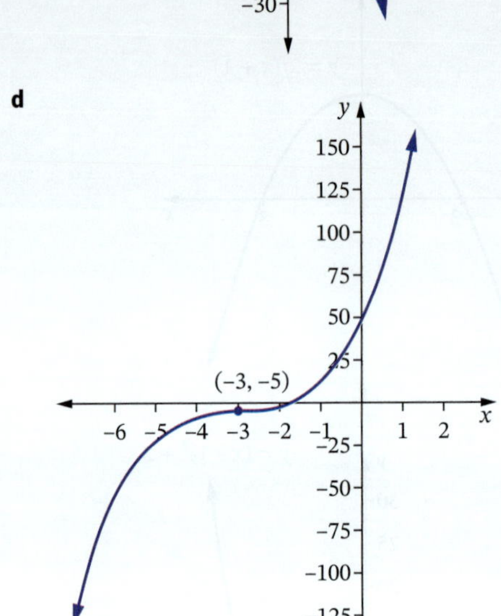

e

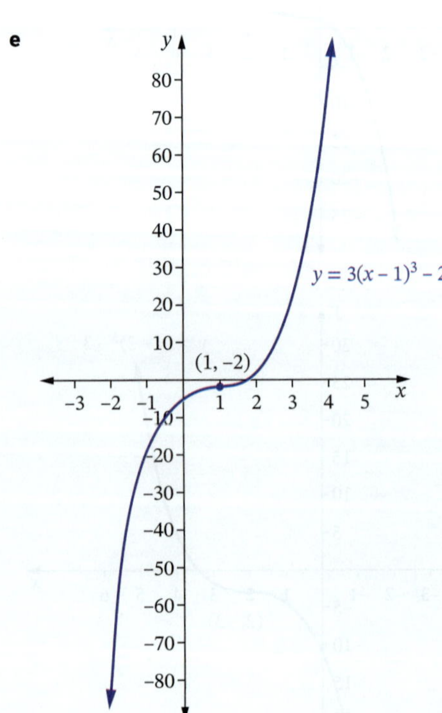

4 a

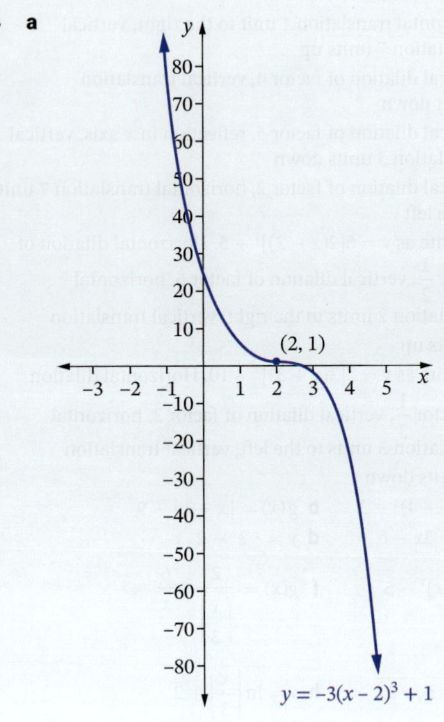

b

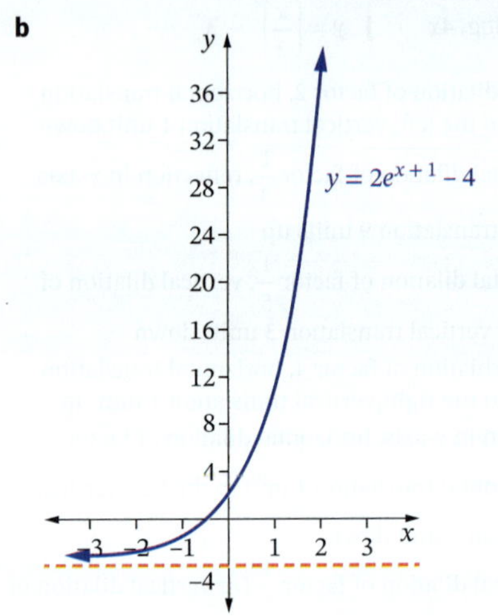

c

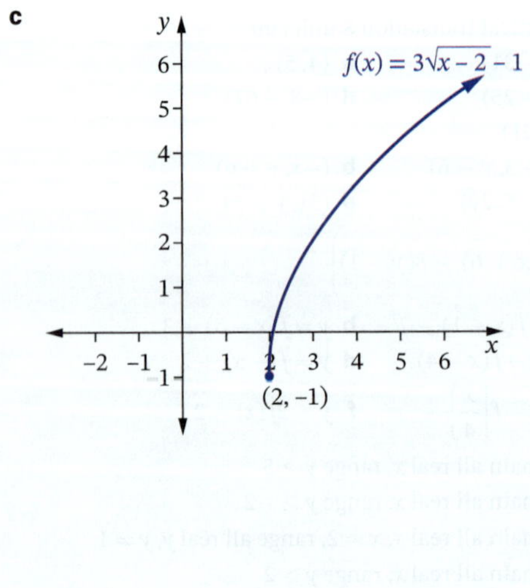

d

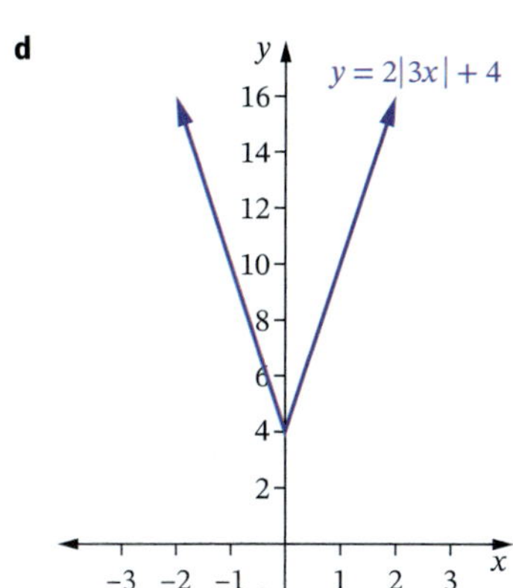

5 a

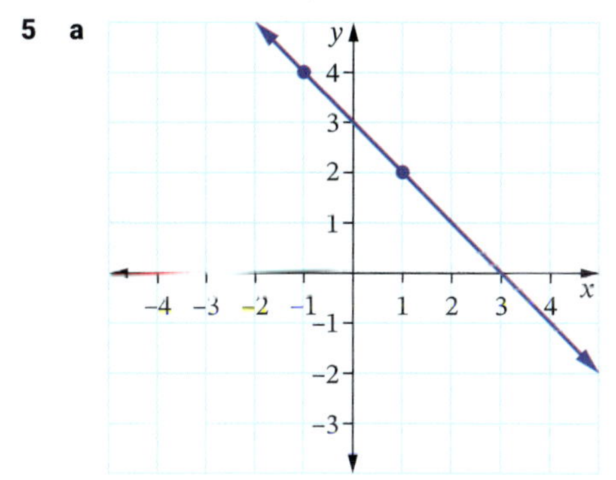

b

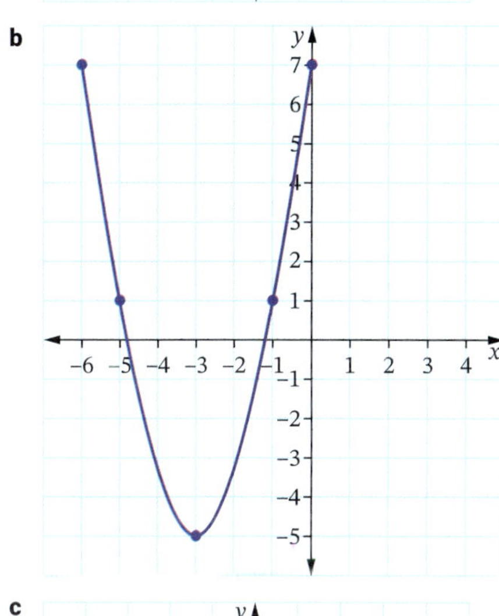

c

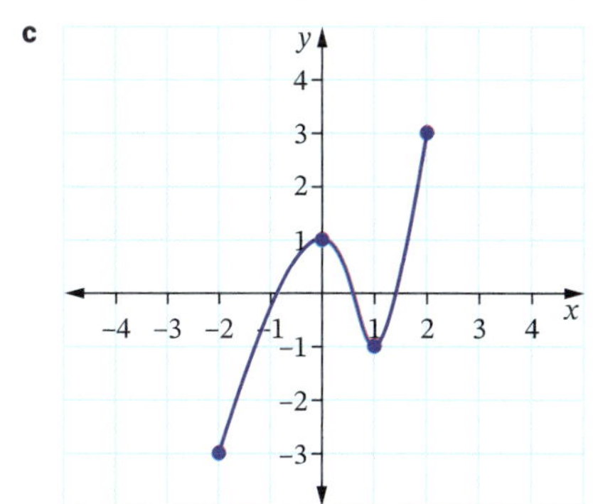

6

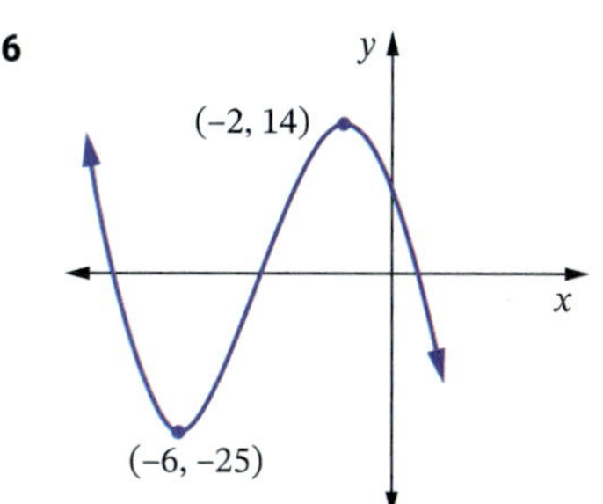

7 a

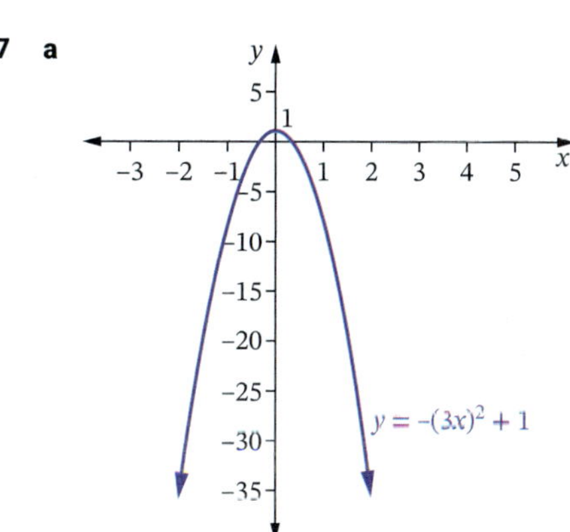

b

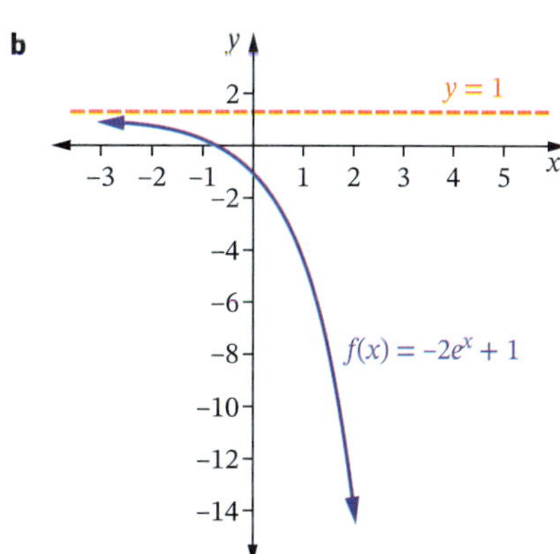

c

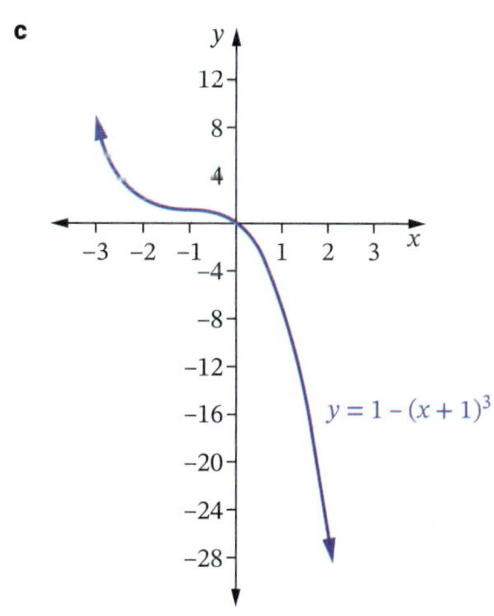

ANSWERS

d

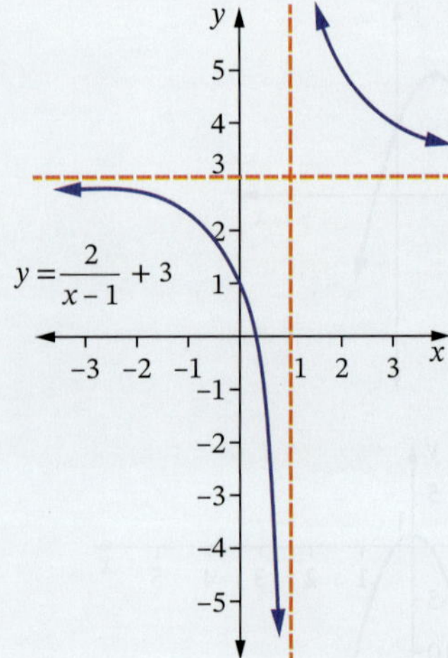

8 a vertical dilation of factor 2, reflection in the x-axis, translation 3 units right, 1 unit up; $(x + 3, -2y + 1)$

b

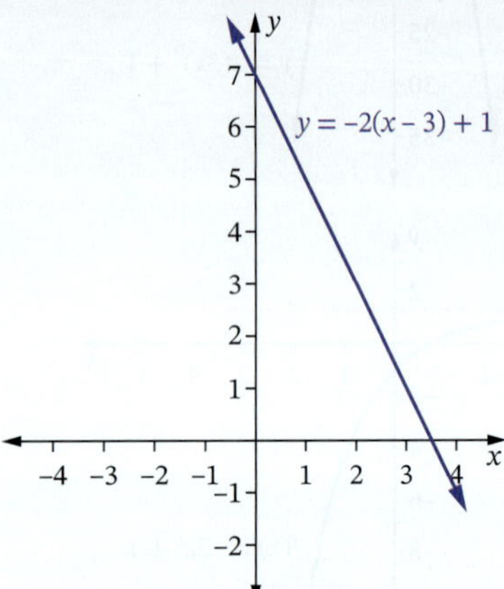

9 a vertical dilation of factor 2, reflection in the x-axis, translation 3 units up; $(x, -2y + 3)$

b

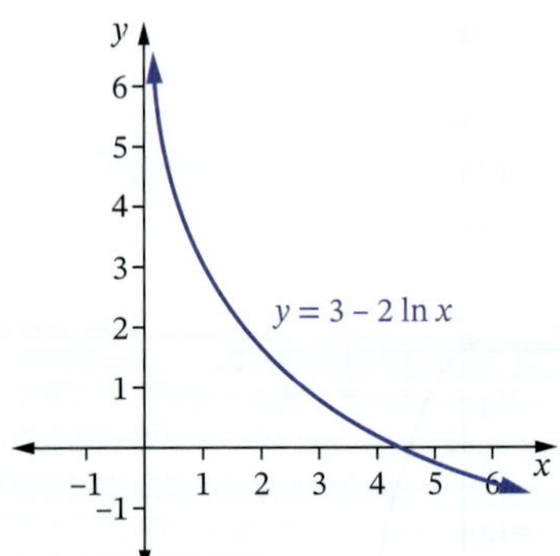

10 a

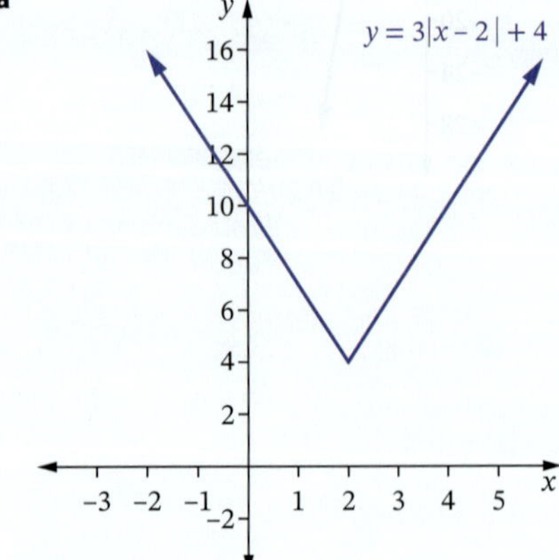

b no solutions **c** $x < 0, x > 4$

11 a See worked solutions.

b vertical dilation of factor 3, reflection in the x-axis, horizontal translation 5 units left, vertical translation 1 unit up.

c and d

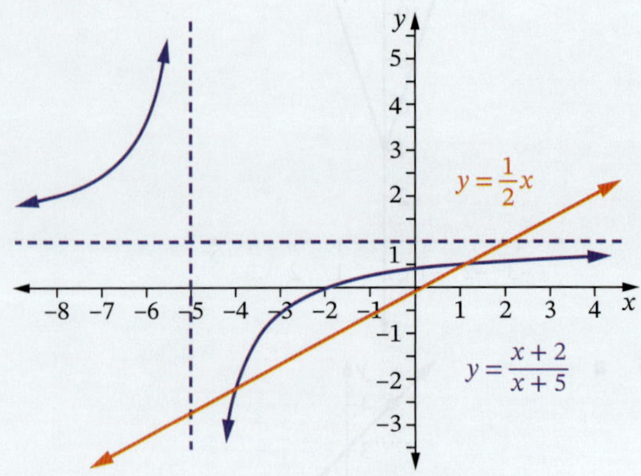

domain: all real x, $x \neq -5$, range: all real y, $y \neq 1$.

e $x < -5, -4 \leq x \leq 1$

12 a $(-5, \frac{1}{3})$

b

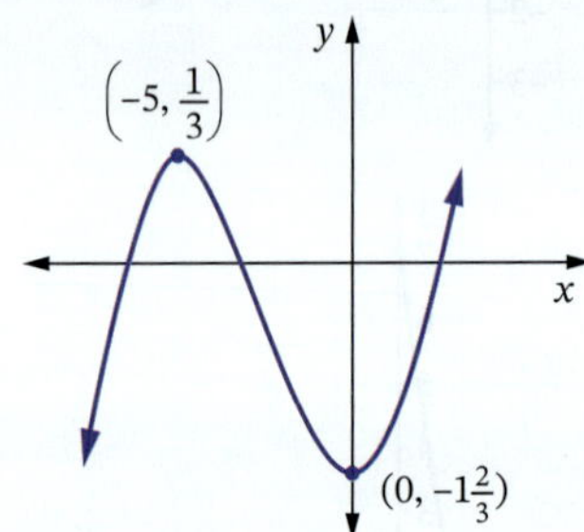

13 a $(-26, 7\frac{2}{3})$

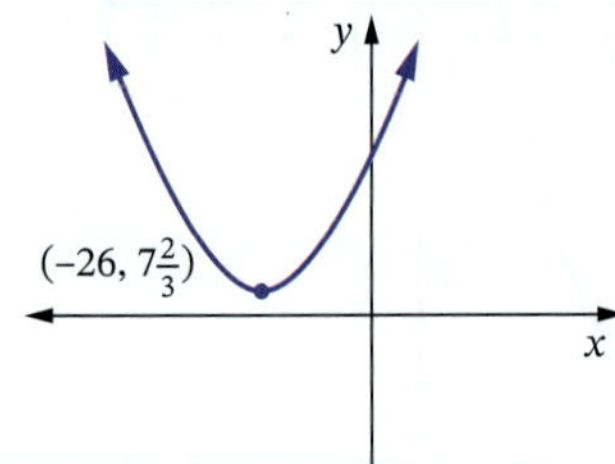

b $(-9, -3.6)$

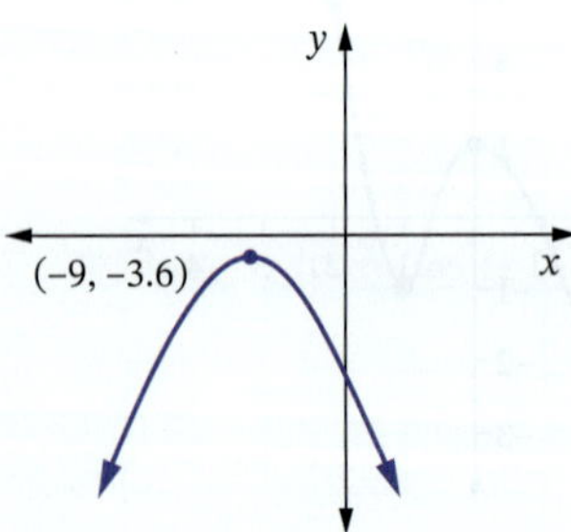

c $(-54, 10\frac{1}{2})$

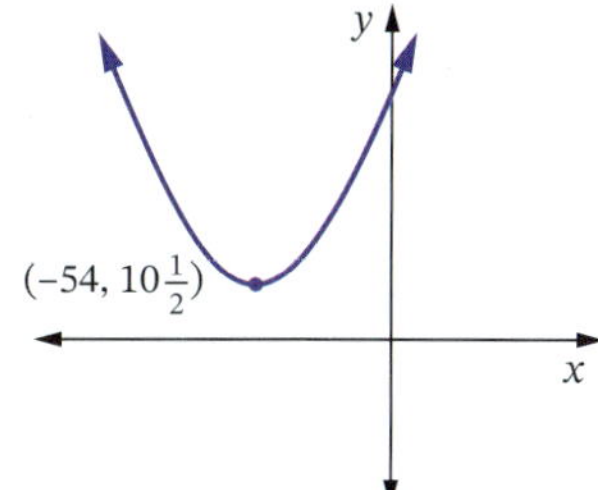

14 a $y = 8|x - 3| - 8$

b $-2\frac{2}{3} < a < 8$

Sample HSC problem

a Proof: see worked solutions.

b vertical dilation of factor 5, horizontal translation 3 units right, vertical translation 1 unit up.

c and **d**

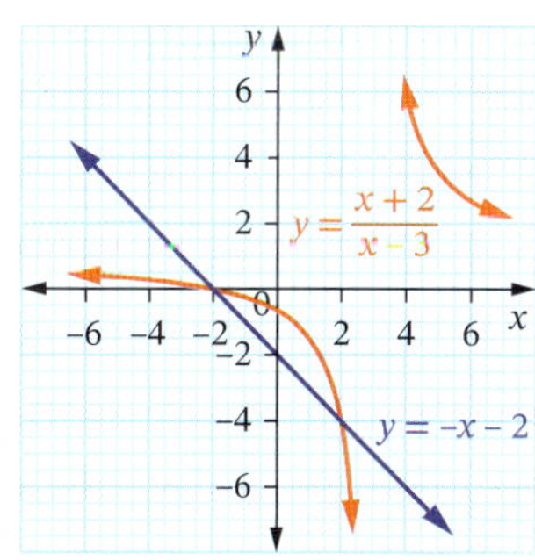

domain: all real x, $x \neq 3$, range: all real y, $y \neq 1$

d $x = -2, 2$

e $x < -2, 2 < x < 3, x > 3$

Test yourself 8

1 B	**2** D	**3** C
4 C	**5** A	**6** B

7 a b is a vertical translation, b units up if $b > 0$ or down if $b < 0$.

a is a horizontal translation, a units to the right if $a > 0$ and to the left if $a < 0$.

l is a vertical dilation, with factor l stretched if $l > 1$ and shrunk if $0 < l < 1$.

k is a horizontal dilation, with factor k stretched if $k > 1$ and shrunk if $0 < k < 1$.

b i reflection in the y-axis

ii reflection in the x-axis

8 $f(x) = -3x^2 + 1$

$f(-x) = -3(-x)^2 + 1$

$= -3x^2 + 1$

$= f(x)$ so even

9 a

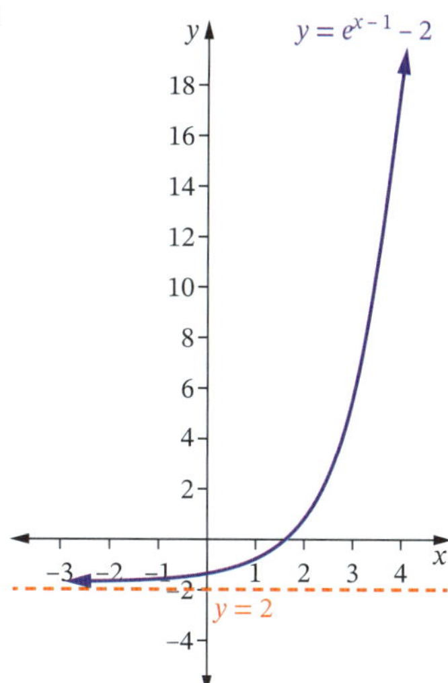

b $x \approx 3.3$ **c** $x = 4.09$

10 a i $y = x^3 + 3$ **ii** $y = (x + 7)^3$

b i $y = 3|x|$ **ii** $y = \left|\frac{x}{2}\right|$

c $f(x) = 5\ln(-x)$ **d** $f(x) = -\frac{1}{x - 4}$

e $f(x) = 9(3^{3(x-2)}) - 6$

11 $f(x) = (x + 4)^4$

12 $(3, -11)$

13 a See worked solutions.

b

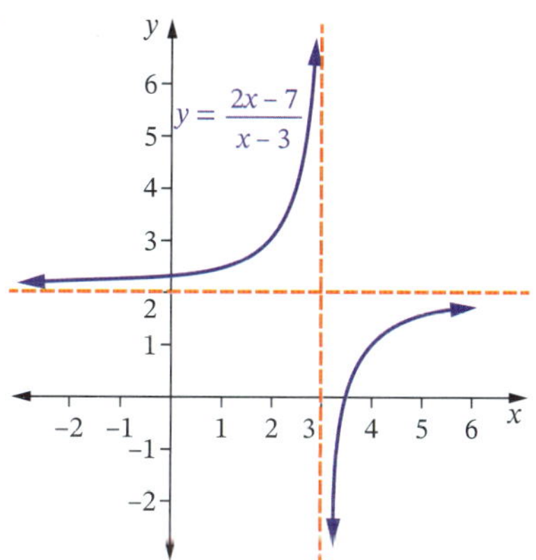

domain: all real x, $x \neq 3$

range: all real y, $y \neq 2$

c i $x < 3, x \geq 3.5$ **ii** $x > 3$

14 a $(x + 3, -7y - 4)$ **b** $x = -6, y = -1$

15 a

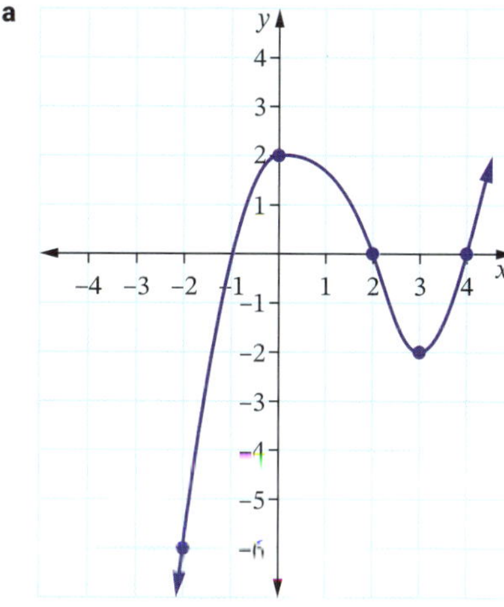

b

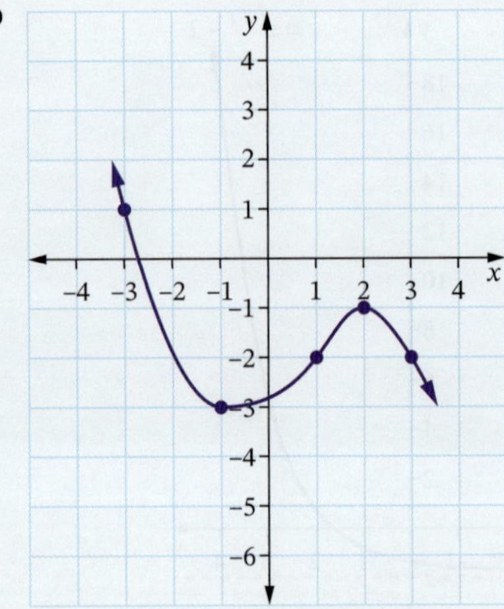

16 a $-3 \le x \le 1$ b $x < -3, x > 1$

17 a

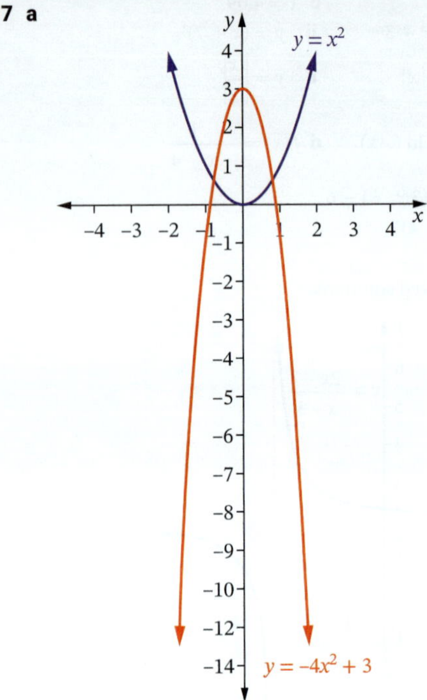

b

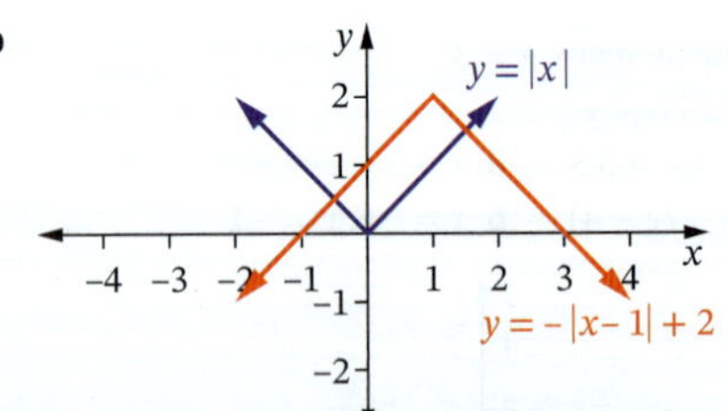

c

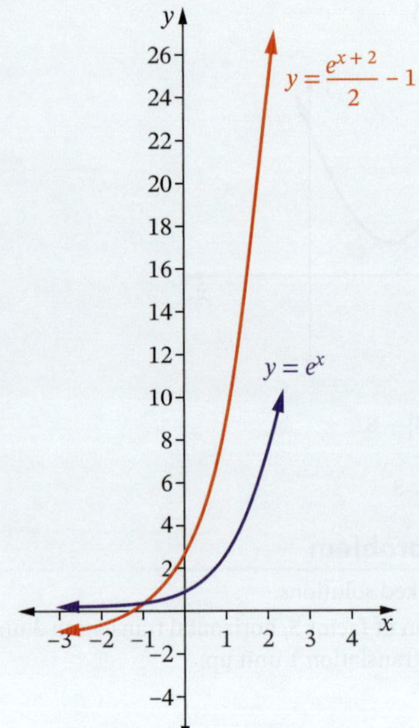

d

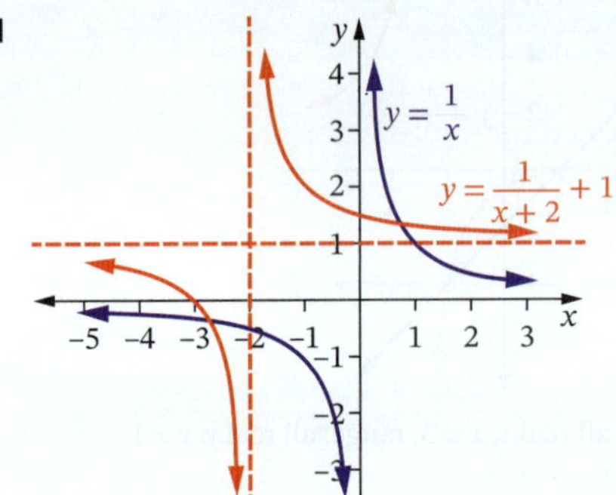

e

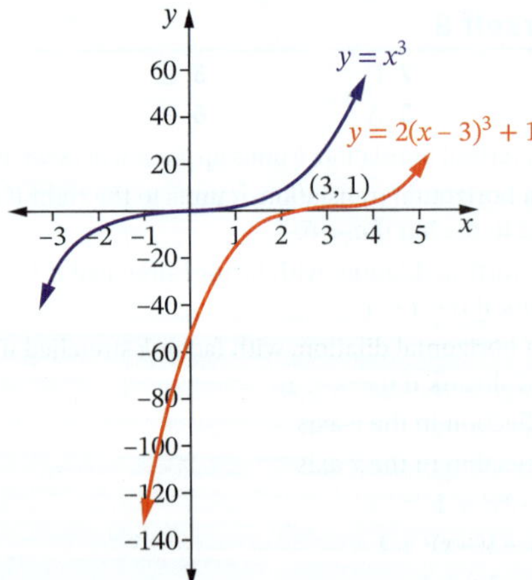

f

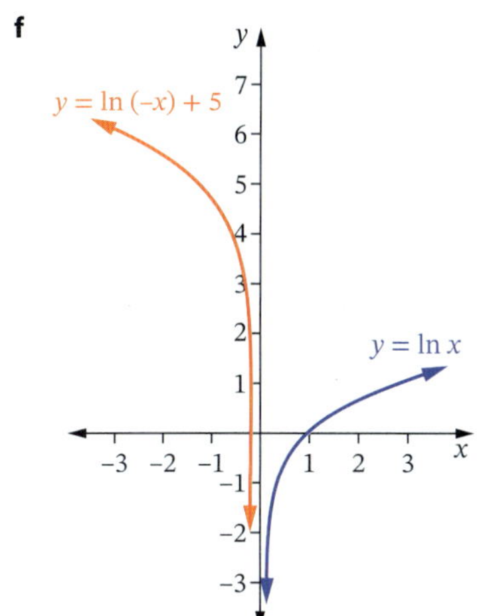

g

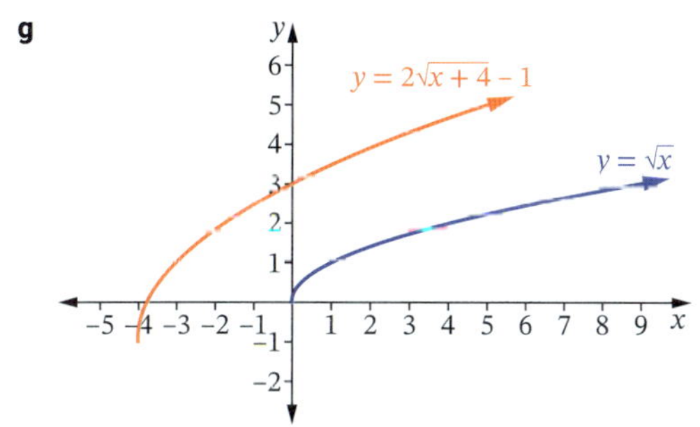

18 **a** shrunk **b** stretched
c shrunk **d** stretched

19 **a** domain: all real x, range: $[-10, \infty)$
b domain: all real x, range: $(-\infty, 2]$
c domain: all real x, $x \neq 3$; range: all real y, $y \neq -5$

Challenge exercise 8

1 $(32, -1)$

2 **a** **i** $P(1, -8)$ **ii** $Q(2, 0)$ **iii** $R(3, -2)$
b $x - 2y - 17 = 0$
c horizontal dilation with factor 2 and horizontal translation 17 units to the right, or vertical dilation with factor $\frac{1}{2}$ and vertical translation $\frac{17}{2}$ units down

3 $x + 3y - 19 = 0$

4 **a** A horizontal dilation with factor $k = 2$:

$$y = \frac{1}{kx} = \frac{1}{\frac{1}{2}x} = \frac{2}{x}$$

A vertical dilation with factor $l = 2$:

$$y = \left(\frac{1}{x}\right) = 2\left(\frac{1}{x}\right) = \frac{2}{x}$$

So these transformations have the same effect on $y = \frac{1}{x}$.

b No. Horizontal dilation gives $y = \frac{4}{x^2}$, vertical dilation gives $y = \frac{2}{x^2}$.

5 **a** $x = -\frac{b}{2a}$ **b** horizontal translation
c **i** $x = -1$ **ii** $x = 3$
iii $x = -b$ **iv** $x = -\frac{b}{a}$

6 $x^2 - 6x + y^2 + 8 = 0$

7 reflection in y-axis, horizontal dilation of factor $\frac{1}{3}$, horizontal translation 2 units to the left, vertical dilation of factor 3, vertical translation 5 units down

8 $y = -x^3 + 3x$

Chapter 9

Exercise 9.01

1 $\sec x = \frac{9}{5}$, $\cot x = \frac{5}{\sqrt{56}}$, $\operatorname{cosec} x = \frac{9}{\sqrt{56}}$

2 $\operatorname{cosec}\theta = \frac{13}{5}$, $\sec\theta = \frac{13}{12}$, $\cot\theta = \frac{12}{5}$

3 $\operatorname{cosec}\theta = \frac{7}{\sqrt{33}}$, $\sec\theta = \frac{7}{4}$, $\cot\theta = \frac{4}{\sqrt{33}}$

4 $\tan\theta = -\frac{\sqrt{11}}{5}$, $\operatorname{cosec}\theta = \frac{6}{\sqrt{11}}$, $\cot\theta = -\frac{5}{\sqrt{11}}$

5 $\sin\theta = -\frac{5}{\sqrt{34}}$, $\operatorname{cosec}\theta = -\frac{\sqrt{34}}{5}$, $\tan\theta = \frac{5}{3}$, $\sec\theta = -\frac{\sqrt{34}}{3}$

6 **a** $\sqrt{2}$ **b** 2 **c** $\sqrt{3}$ **d** $\frac{3}{4}$
e $\frac{1}{2}$ **f** 1 **g** $\sqrt{2}$ **h** 2
i $\frac{4}{3}$ **j** 4 **k** $\frac{\sqrt{3}}{2}$ **l** $\frac{1}{2}$
m $\frac{4}{3}$ **n** 3 **o** $1 - \frac{\sqrt{3}}{2}$

7

	$\sin x$	$\cos x$	$\tan x$	$\operatorname{cosec} x$	$\sec x$	$\cot x$
a 0°	0	1	0	–	1	–
b 90°	1	0	–	1	–	0

8 $\sin 67° = \cos(90° - 67°) = \cos 23°$

9 $\sec 82° = \operatorname{cosec}(90° - 82°) = \operatorname{cosec} 8°$

10 $\tan 48° = \cot(90° - 48°) = \cot 42°$

11 **a** $2\cos 61°$ or $2\sin 29°$
b 0 **c** 0 **d** 1 **e** 2

12 $x = 80$ **13** $y = 22$ **14** $p = 31$

15 $b = 25$ **16** $t = 20$ **17** $k = 15$

18 **a** $\sin\theta$ **b** $\sec\theta$ **c** $\operatorname{cosec} x$
d $\cos^2 x$ **e** $\sin\alpha$ **f** $\operatorname{cosec}^2 x$
g $\sec^2 x$ **h** $\tan^2\theta$ **i** $5\operatorname{cosec}^2\theta$
j $\sin^2 x$ **k** 1 **l** $\sin\theta\cos\theta$

19 **a** $\frac{\cos^2 x}{\sin x}$ **b** $\cos x$ **c** $-\cos^2 3x$
d $\tan 5\alpha$ **e** $\cot^2 2x$ **f** $\cot(3\theta - \alpha)$
g 1 **h** $3\cos\theta$
i $2\sin x\cos x$ (or $1 - \cos 2x$) **j** 1
k $\tan\alpha + \sec\alpha$ **l** -1

20 See worked solutions.

21, 22 Proofs: see worked solutions.

ANSWERS

Exercise 9.02

1 a

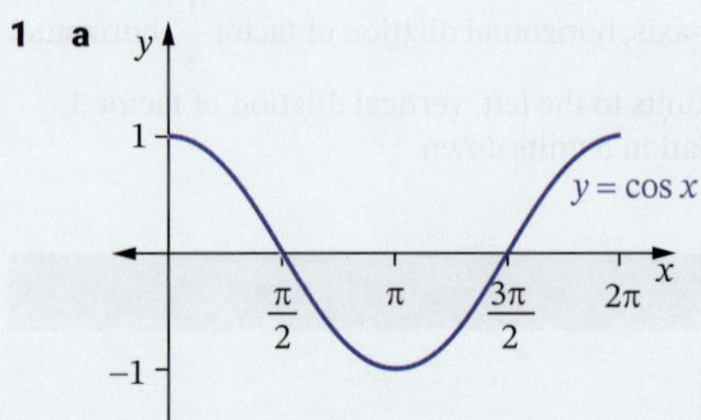

b

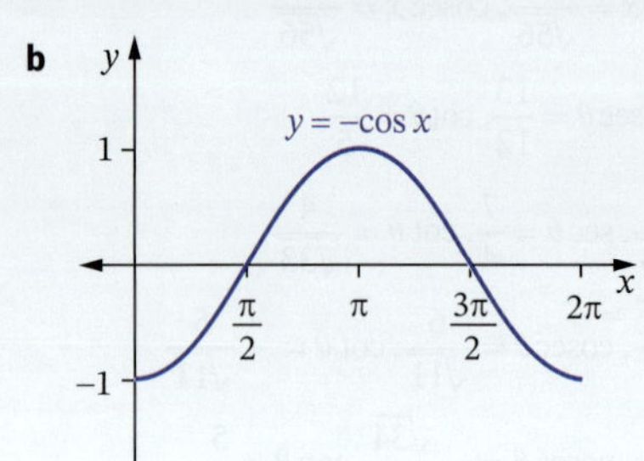

2 a

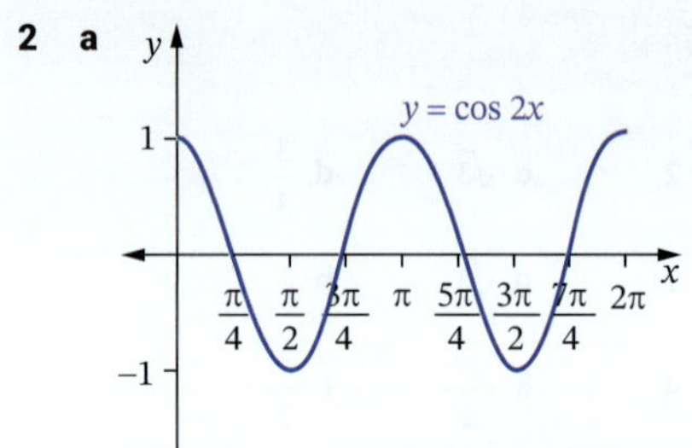

b

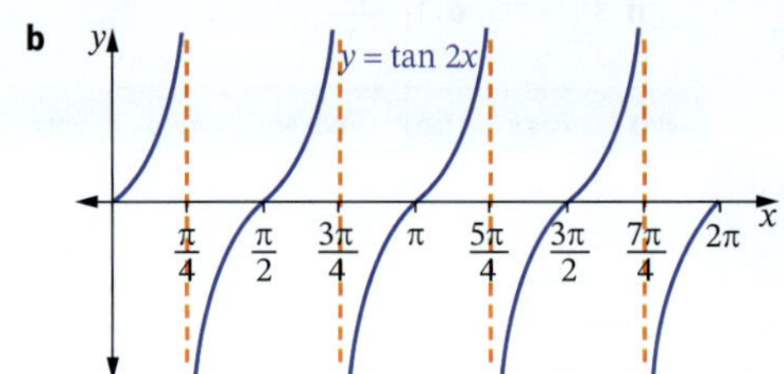

c

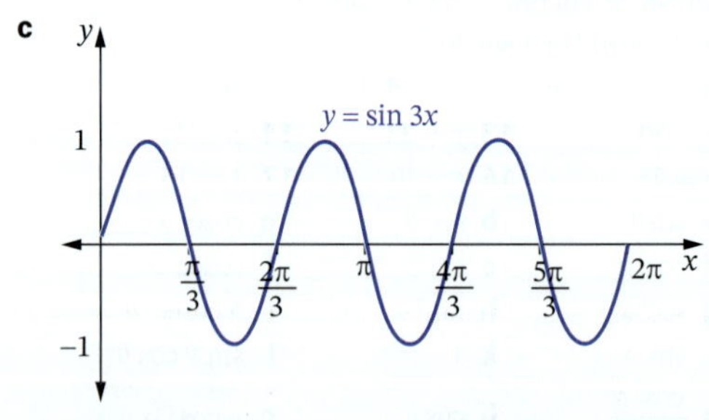

d

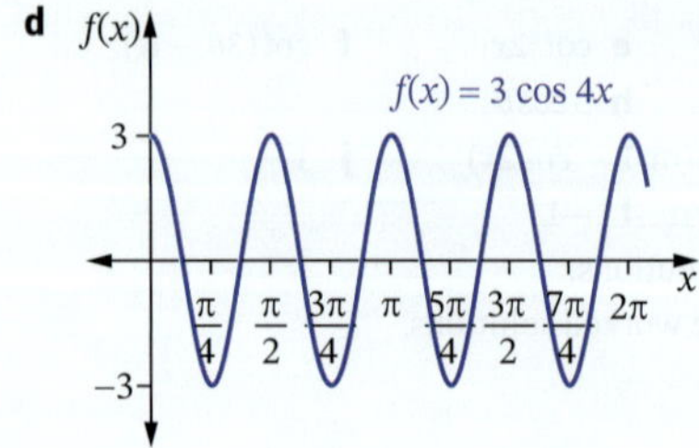

e

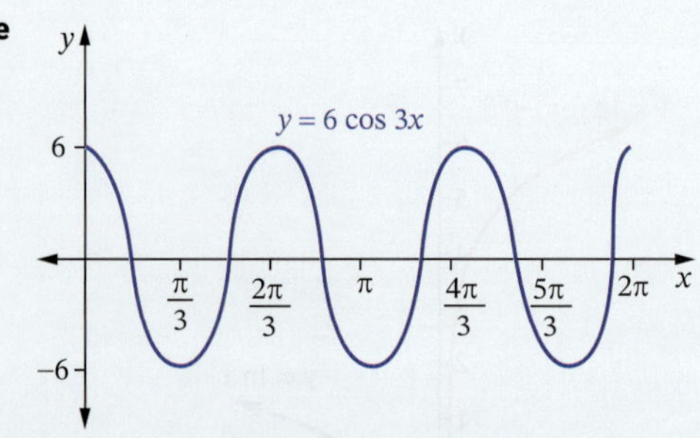

f

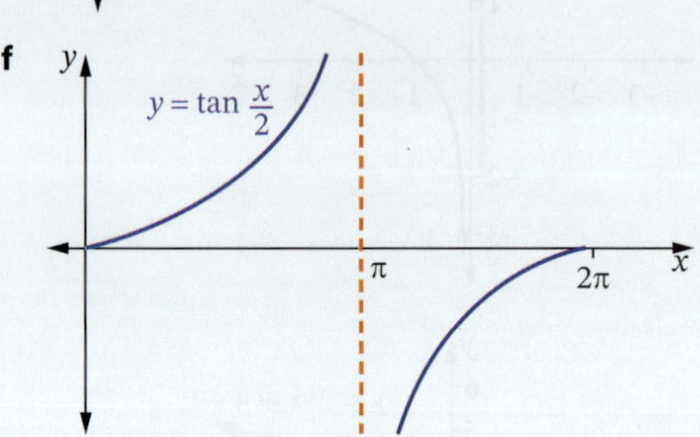

g

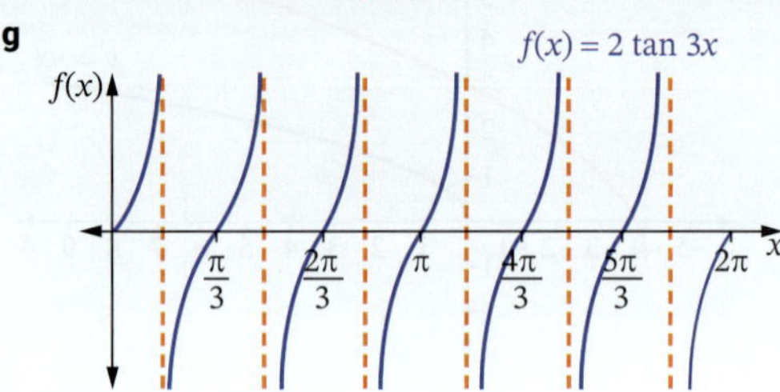

h

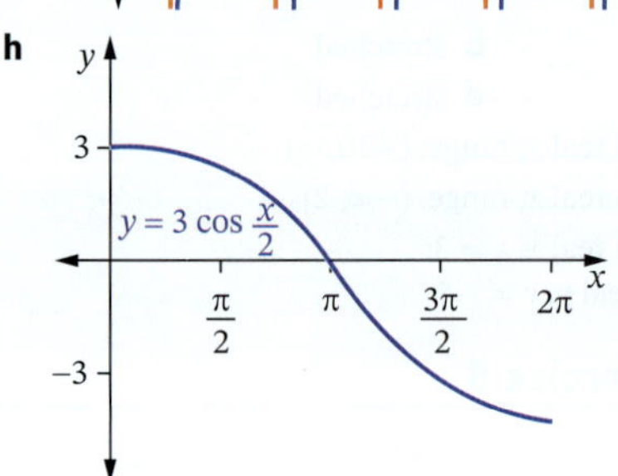

i

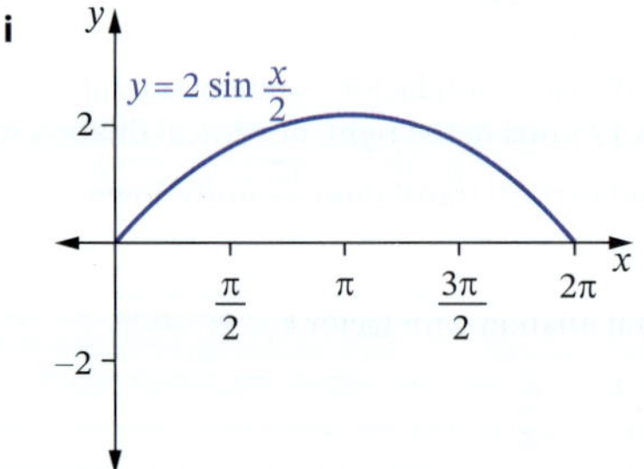

3 a

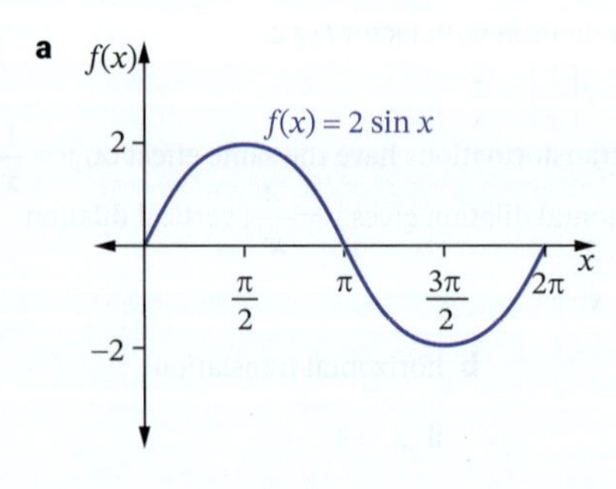

b

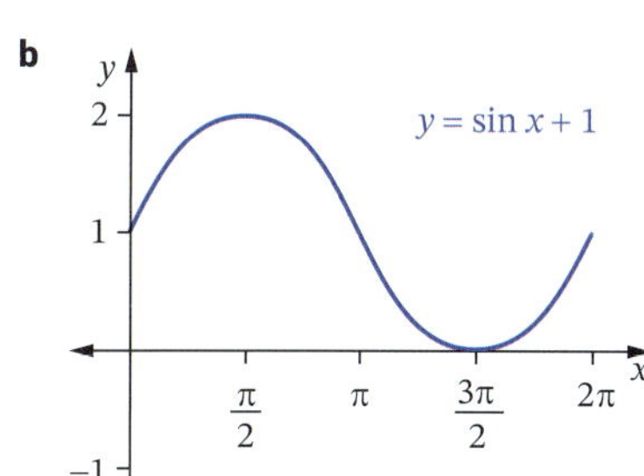

c

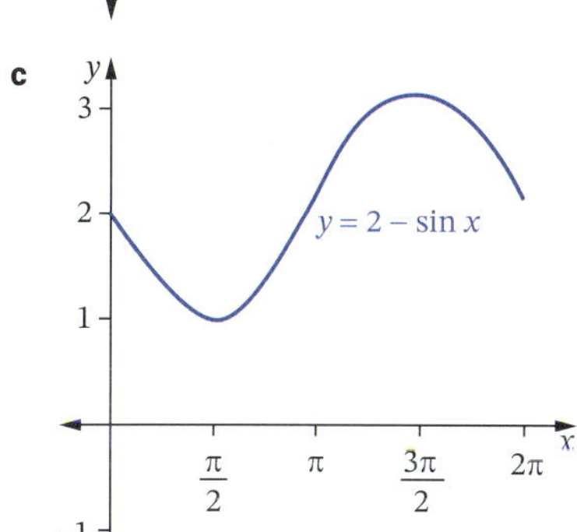

d

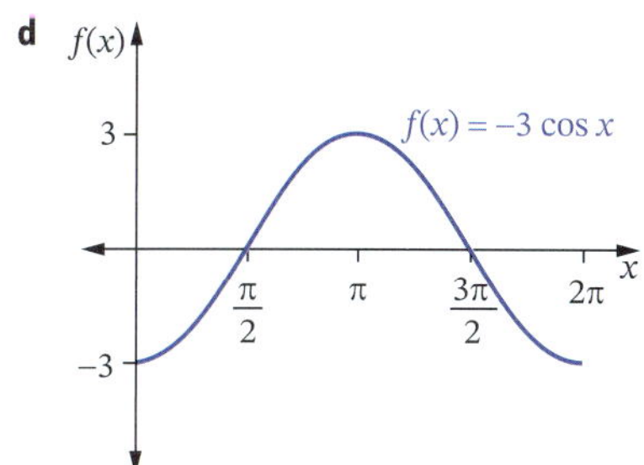

e

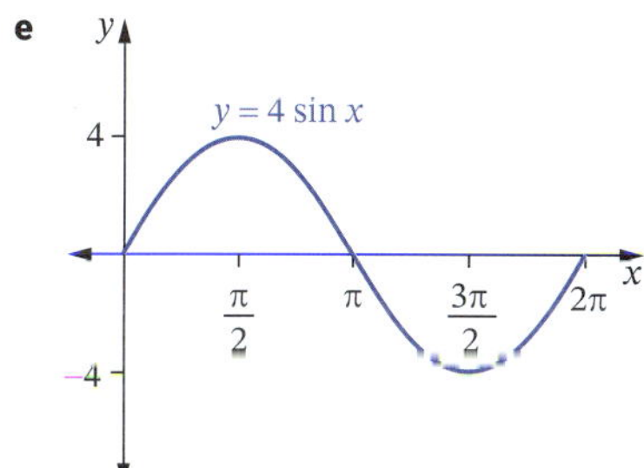

f

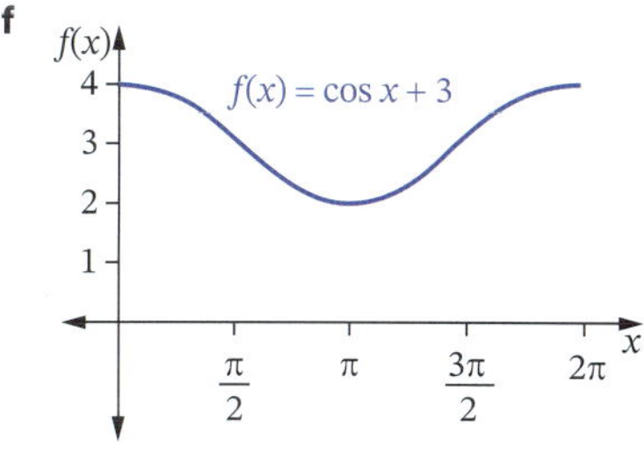

g

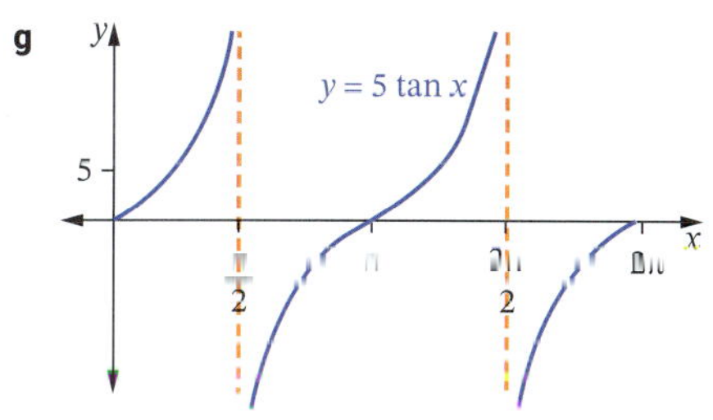

h

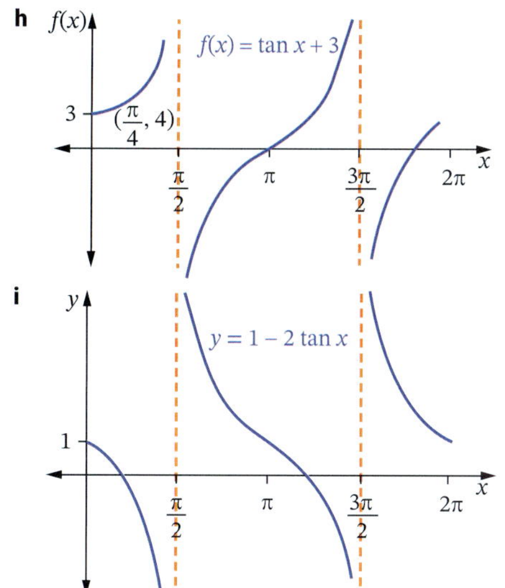

i

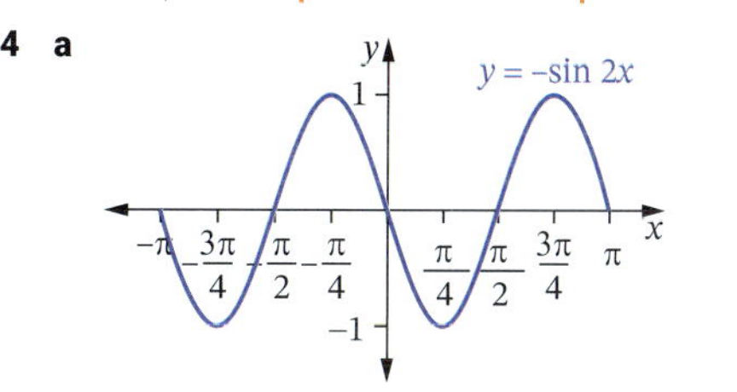

4 a

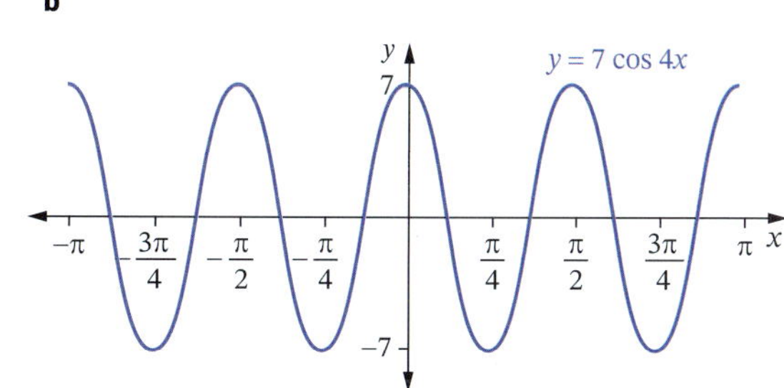

b

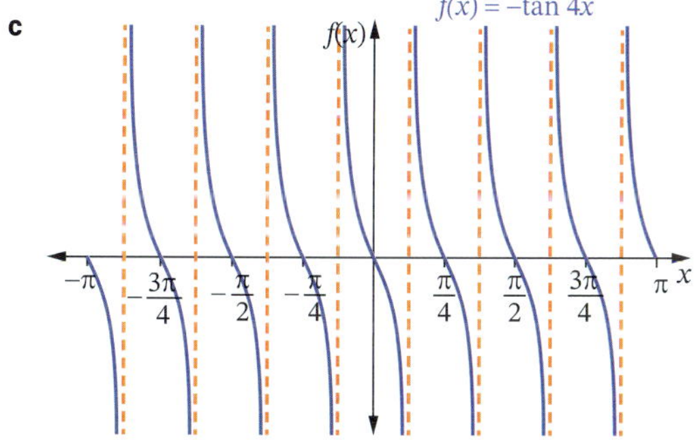

c

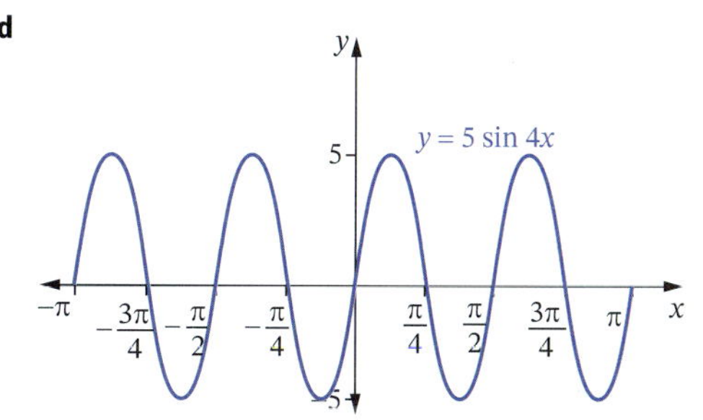

d

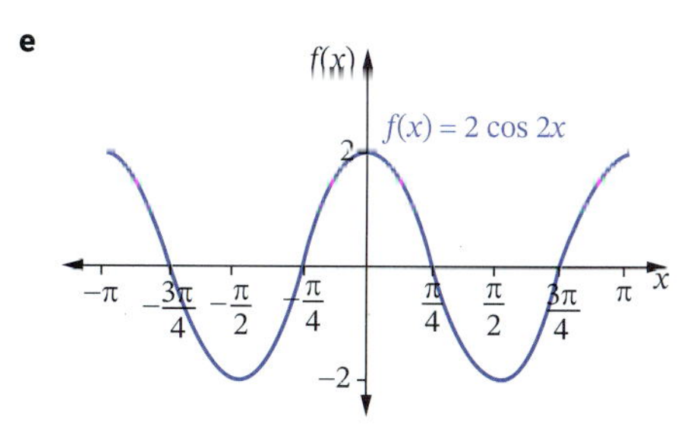

e

f

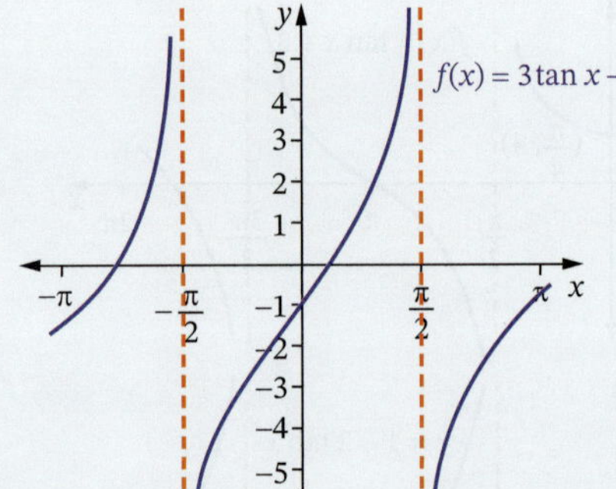

5

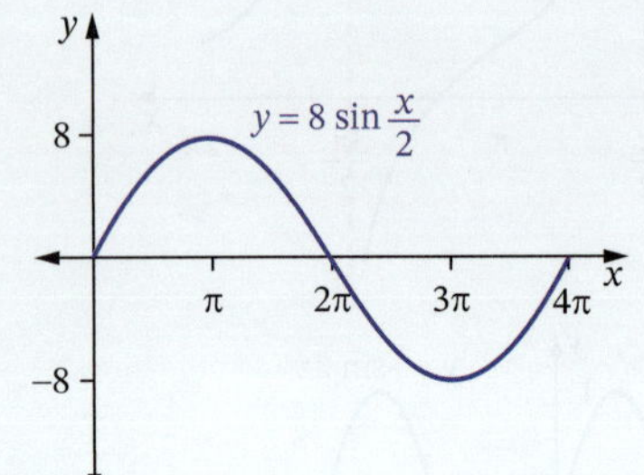

6 a

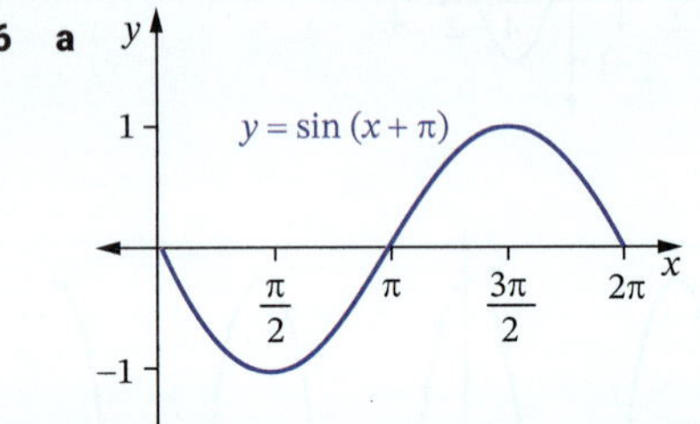

b

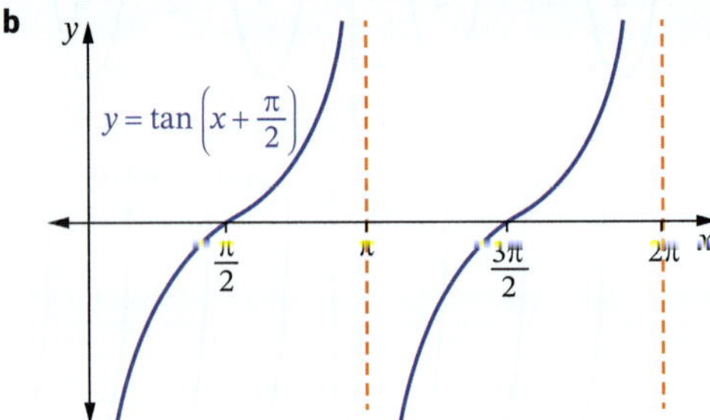

c

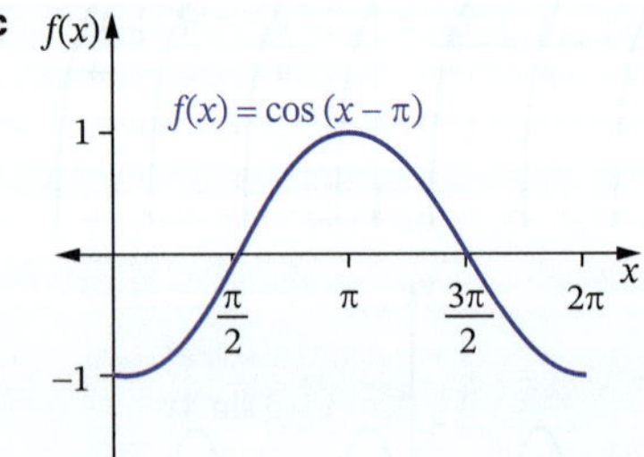

d

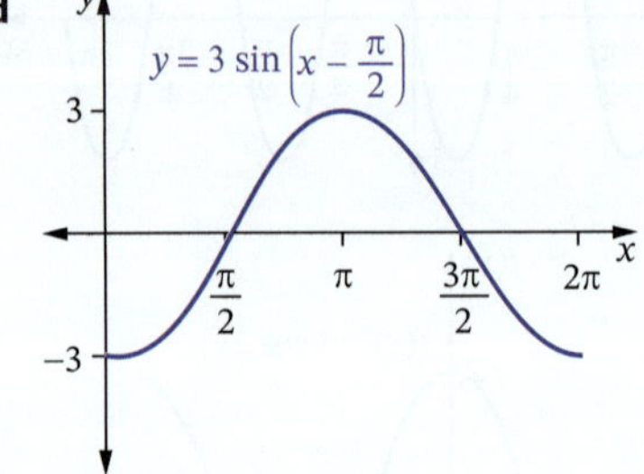

e

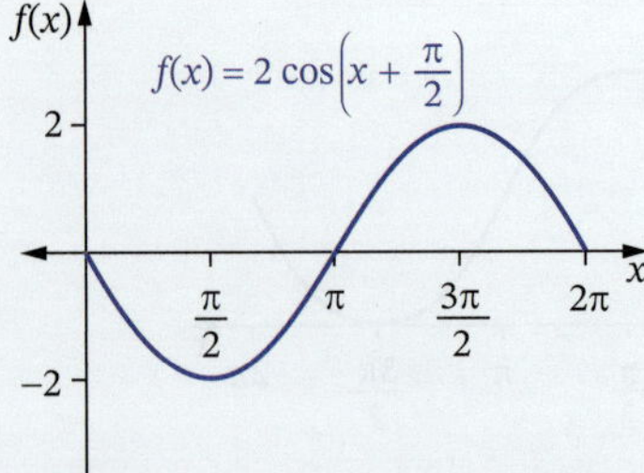

f

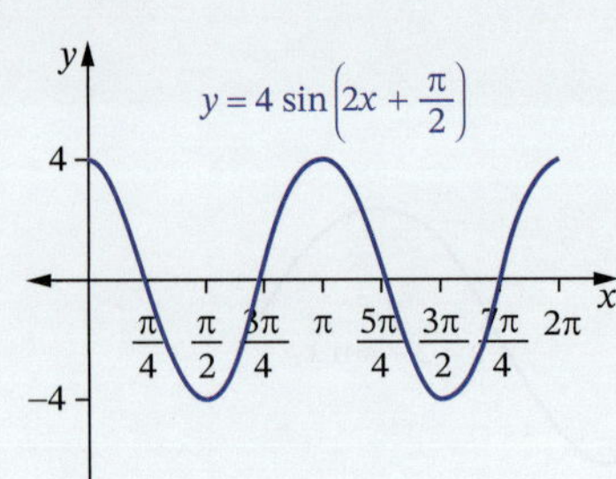

g

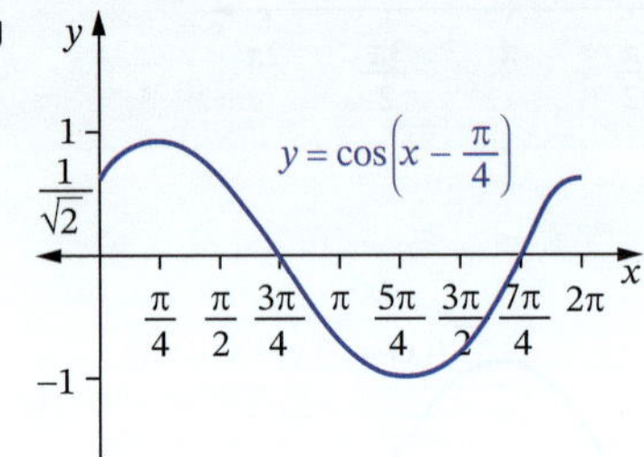

h

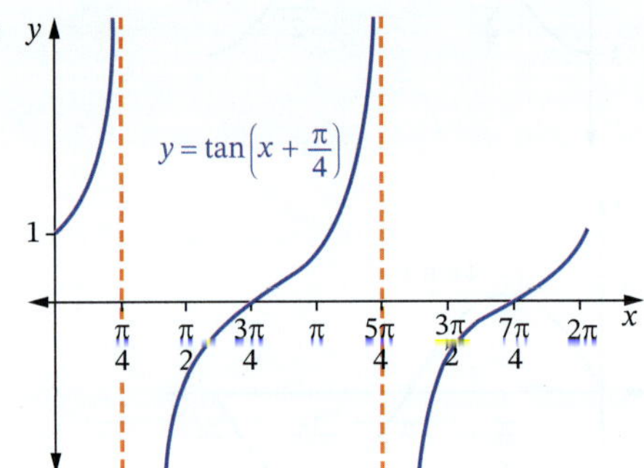

7 a horizontal dilation factor 2, vertical dilation factor 3, horizontal translation 8 units to the right, vertical translation 1 unit down

b $y = 3\sin\left(\frac{x}{2} - 4\right) - 1$

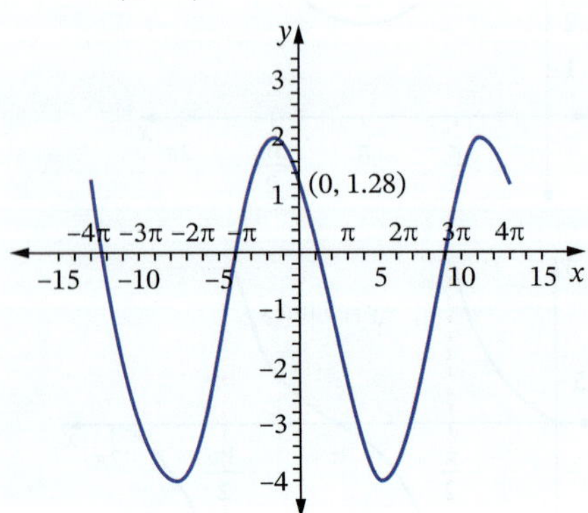

c range: $[-4, 2]$

8 **a**

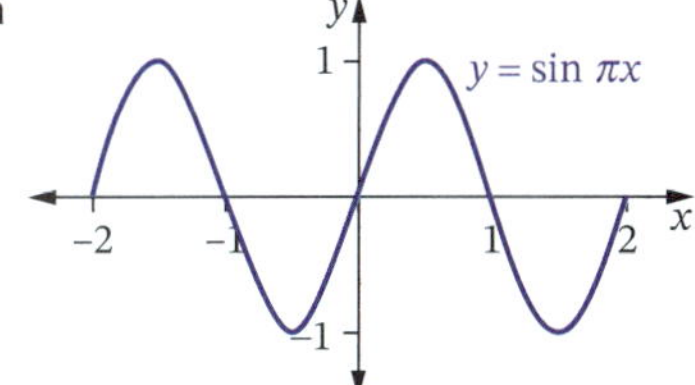

b

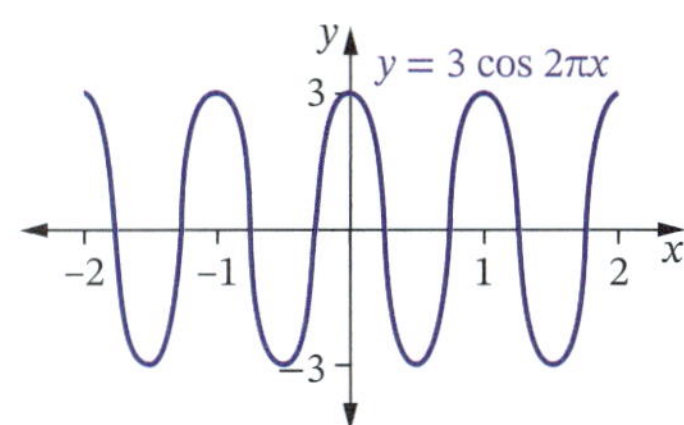

9 **a** **i** 5 **ii** π **iii** 0 **iv** 0

b **i** 1 **ii** 2π **iii** 0 **iv** $-\pi$

c **i** none **ii** $\frac{\pi}{4}$ **iii** -2 **iv** 0

d **i** 3 **ii** 2π **iii** 1 **iv** $\frac{\pi}{4}$

e **i** 8 **ii** 2 **iii** -3 **iv** $-\frac{2}{\pi}$

f **i** none **ii** $\frac{\pi}{5}$ **iii** 2 **iv** $\frac{\pi}{10}$

10 **a** domain: all real x; range: $[-5, 3]$

b domain: all real x, range: $[4, 10]$

c domain: all real x, range: $[-4, 2]$

d domain: all real x, range: $[1, 5]$

11 $a = 2, b = 2, c = 1$

Exercise 9.03

1 **a** $\theta = 20°29', 159°31'$ **b** $\theta = 120°, 240°$

c $\theta = 135°, 315°$ **d** $\theta = 60°, 120°$

e $\theta = 150°, 330°$ **f** $\theta = 30°, 330°$

g $\theta = 30°, 120°, 210°, 300°$

h $\theta = 30°, 150°, 210°, 330°$

i $\theta = 70°, 110°, 190°, 230°, 310°, 350°$

j $\theta = 15°, 45°, 75°, 105°, 135°, 165°, 195°, 225°, 255°, 285°, 315°, 345°$

k $x = 90°, 270°$

l $x = 60°, 90°, 270°, 300°$

2 **a** $\theta = \pm 79°13'$ **b** $\theta = 30°, 150°$

c $\theta = -135°, 45°$ **d** $\theta = -120°, -60°$

e $\theta = -30°, 150°$ **f** $\theta = \pm 30°, \pm 150°$

g $\theta = -45°, 135°$

h $\theta = -157.5°, -67.5°, 22.5°, 112.5°$

3 **a** $x = 0°, 360°$ **b** $x = 270°$

c $x = 0°, 180°, 360°$ **d** $x = 90°$

e $x = 0°, 180°, 360°$

f $x = 0°, 180°, 270°, 360°$

g $x = 0°, 90°, 270°, 360°$

h $x = 0°, 45°, 180°, 225°, 360°$

i $x = 60°, 120°, 240°, 300°$

4 **a** $x = 0, \pi, 2\pi$ **b** $x = 0, \frac{\pi}{2}, \pi, \frac{3\pi}{2}, 2\pi$

c $x = \frac{3\pi}{2}$ **d** $x = 0, 2\pi$ **e** $x = \pi$

5 **a** $x = \frac{\pi}{3}, \frac{5\pi}{3}$ **b** $x = \frac{5\pi}{4}, \frac{7\pi}{4}$

c $x = \frac{\pi}{4}, \frac{5\pi}{4}$ **d** $x = \frac{\pi}{3}, \frac{4\pi}{3}$

e $x = \frac{5\pi}{6}, \frac{7\pi}{6}$

6 **a** $x = \frac{\pi}{3}, \frac{2\pi}{3}$ **b** $x = \pm\frac{\pi}{2}$

c $x = \pm\frac{\pi}{6}, \pm\frac{5\pi}{6}$

7 $x = \pm\frac{2\pi}{3}, \pm\frac{4\pi}{3}$ **8** $x = -90°, 30°, 150°$

9 **a** $x \approx 0.98, 4.12$ **b** $x = \frac{\pi}{3}, \frac{5\pi}{3}$

c $x = \frac{3\pi}{4}, \frac{7\pi}{4}$ **d** $x = \frac{7\pi}{6}, \frac{11\pi}{6}$

e $x = \frac{\pi}{6}, \frac{5\pi}{6}, \frac{3\pi}{2}$ **f** $x \approx 2.2, 5.4$

10 **a** $\theta = \pm\frac{\pi}{3}, \pm\frac{2\pi}{3}$ **b** $\theta = -\frac{5\pi}{6}, -\frac{\pi}{6}, \frac{\pi}{2}$

11 **a** $x = \frac{7\pi}{12}, \frac{19\pi}{12}$ **b** $x = 0, \frac{2\pi}{3}, 2\pi$

c $x = \frac{11\pi}{12}, \frac{17\pi}{12}$

12 **a** $x = 0, \frac{3\pi}{4}, \pi, \frac{7\pi}{4}, 2\pi$ **b** $x = 0, \frac{\pi}{2}, \pi, 2\pi$

c $x = 0, \frac{2\pi}{3}, \frac{4\pi}{3}, 2\pi$ **d** $x = \frac{\pi}{6}, \frac{5\pi}{6}, \frac{7\pi}{6}, \frac{11\pi}{6}$

e $x = 0, \pi, 2\pi$ **f** $x = 0, 2\pi$

g $x = \frac{3\pi}{2}$

h $x = \frac{\pi}{4}, \frac{5\pi}{4}, 1.9, 5$

i $x = \frac{\pi}{18}, \frac{5\pi}{18}, \frac{7\pi}{18}, \frac{11\pi}{18}, \frac{13\pi}{18}, \frac{17\pi}{18}, \frac{19\pi}{18}, \frac{23\pi}{18}, \frac{25\pi}{18}, \frac{29\pi}{18}, \frac{31\pi}{18}, \frac{35\pi}{18}$

13 **a** $\cot^3 x - \cot^2 x - 3\cot x + 3$

b $x = 30°, 45°, 150°, 210°, 225°, 330°$

14 **a** $x = 90°, 210°, 330°$

b $x = 60°, 180°, 300°$

c $x = 60°, 135°, 240°, 315°$

d $x = 30°, 60°, 300°, 330°$

15 $x = 0, \frac{\pi}{4}, \frac{3\pi}{4}, \frac{5\pi}{4}, \frac{7\pi}{4}, 2\pi$

Sample HSC problem

a

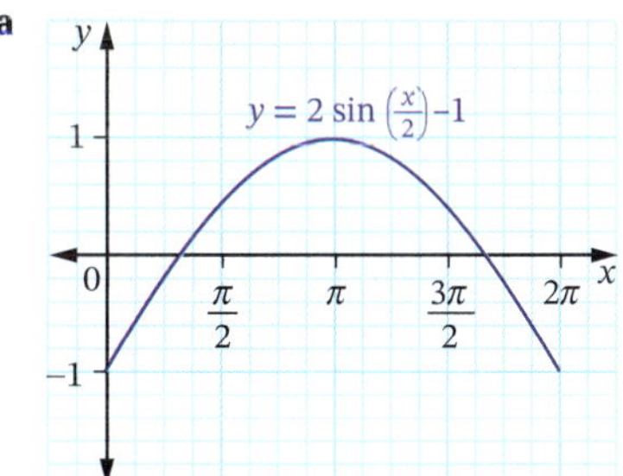

b $x \approx 1.0, 5.2$

c $x = \frac{\pi}{3}, \frac{5\pi}{3}$

ANSWERS

Test yourself 9

1 D **2** A **3** B **4** C **5** B

6 **a** $x = 60°, 120°$ **b** $x = 45°, 225°$
c $x = 120°, 240°$
d $x = 60°, 120°, 240°, 300°$
e $x = 15°, 105°, 195°, 285°$

7 **a** $x = \frac{3\pi}{4}, \frac{7\pi}{4}$ **b** $x = \frac{\pi}{6}, \frac{5\pi}{6}$

c $x = \frac{\pi}{3}, \frac{2\pi}{3}, \frac{4\pi}{3}, \frac{5\pi}{3}$ **d** $x = 0, 2\pi$

e $x = \frac{3\pi}{2}$

8 **a** 

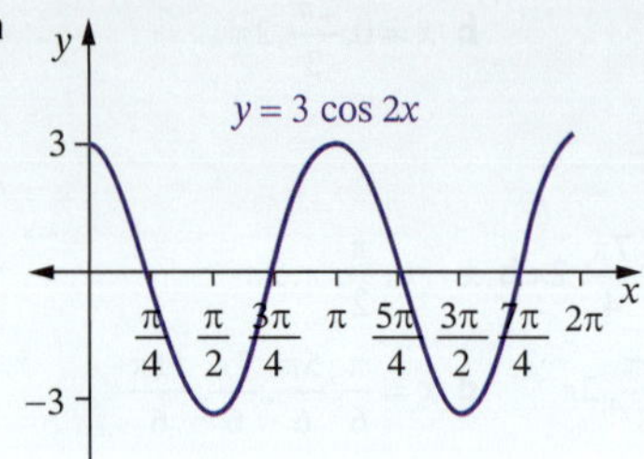

b 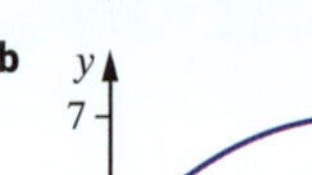
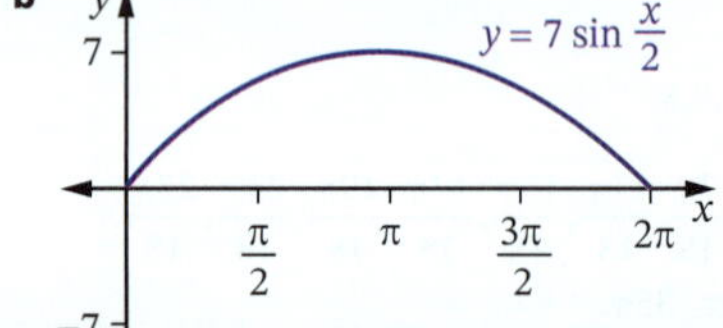

c 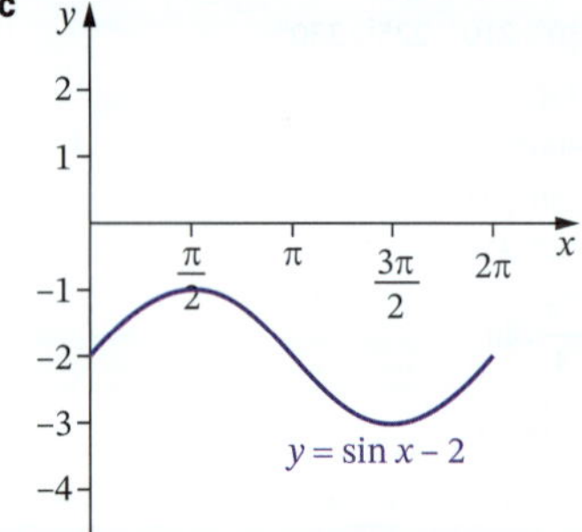

d 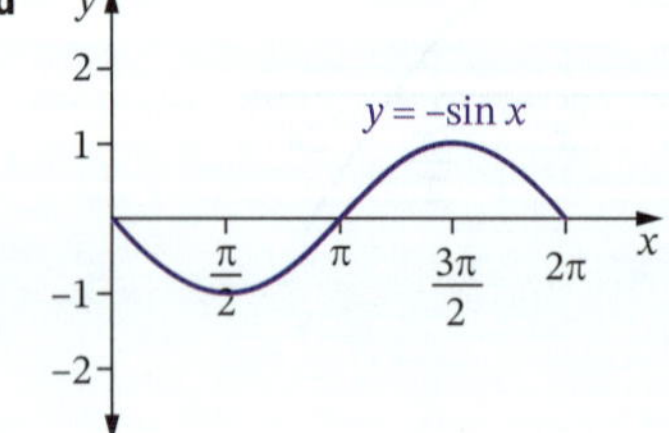

e 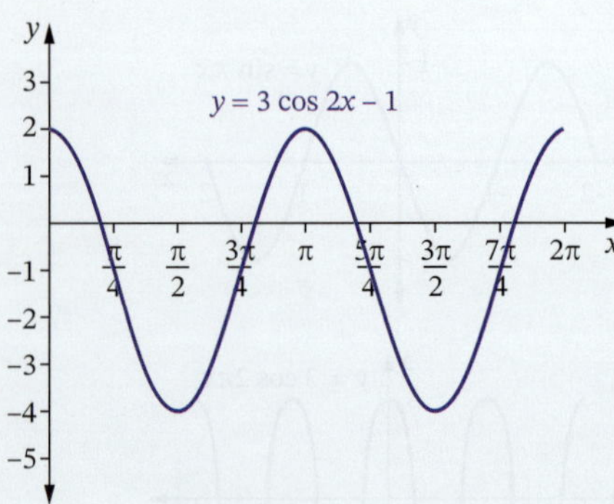

9 **a** $\sin x$ **b** $2\cos A$ **c** $\sin x$
d 1

10 **a** $x = \frac{\pi}{6}, \frac{5\pi}{6}, \frac{7\pi}{6}, \frac{11\pi}{6}$

b $x = \frac{\pi}{12}, \frac{5\pi}{12}, \frac{13\pi}{12}, \frac{17\pi}{12}$

c $x = \frac{3\pi}{2}$

d $x = \frac{\pi}{6}, \frac{\pi}{2}, \frac{7\pi}{6}, \frac{3\pi}{2}$

11 **a** $x = \pm\frac{\pi}{6}, \pm\frac{5\pi}{6}$ **b** $x = \frac{\pi}{12}, -\frac{11\pi}{12}$

c $x = \pm\frac{\pi}{4}$

12 **a** $y = \cos 2x$ **b** $y = 5\cos x$

c $y = -\cos x$ **d** $y = \cos\left(x - \frac{\pi}{6}\right)$
e $y = \cos x + 4$

13 **a** $x = 0°, 45°, 180°, 225°, 360°$
b $x = 60°, 180°, 300°$
c $x = 30°, 150°$
d $x = 0°, 60°, 300°, 360°$

14 See worked solutions.

15 $b = 40$

16 **a** $x = 0°, 180°, 360°$ **b** $x = 155°, 245°$
c $x = 10°\ 54', 59°\ 6', 190°\ 54', 239°\ 6'$

17 **a** $x = -112.5°, -22.5°, 67.5°, 157.5°$
b $x = 90°, -30°, -150°$

18 **a** $\sqrt{2}$ **b** $\sqrt{3}$

c $\frac{2}{\sqrt{3}}$ **d** $\sqrt{3}$

19 **a** domain: all real x, range: $[-1, 11]$
b domain: all real x, range: $[-7, 1]$

20 **a** amplitude 2, period $\frac{2\pi}{3}$, centre -1

b amplitude 1, period 4π, centre 0, phase shift 2π units to the left

c no amplitude, reflection in x-axis, period $\frac{\pi}{5}$, centre 0, phase shift $\frac{\pi}{20}$ to the right

21 **a** $x = 0, \frac{\pi}{2}, \pi, \frac{3\pi}{2}, 2\pi$

b $x = \frac{\pi}{4}, \frac{5\pi}{4}, 2.03, 5.18$

c $x = 0.73, 2.41, 3.79, 5.64$

Challenge exercise 9

1 **a** $-\dfrac{\sqrt{3}}{2}$ **b** -1

2 $\theta = 110°, 230°$

3 **a** period = 2, amplitude = 3

b

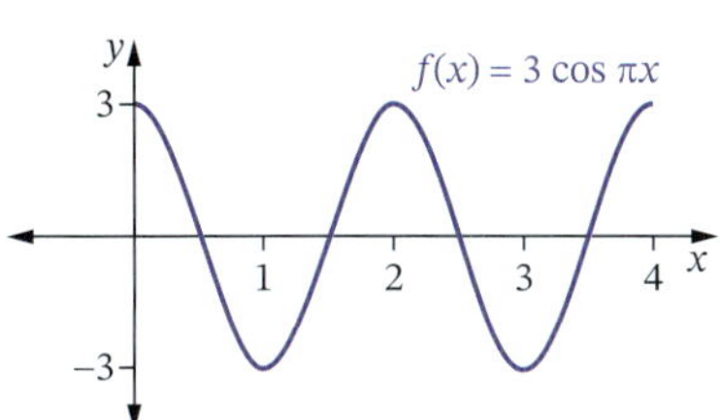

4 **a**

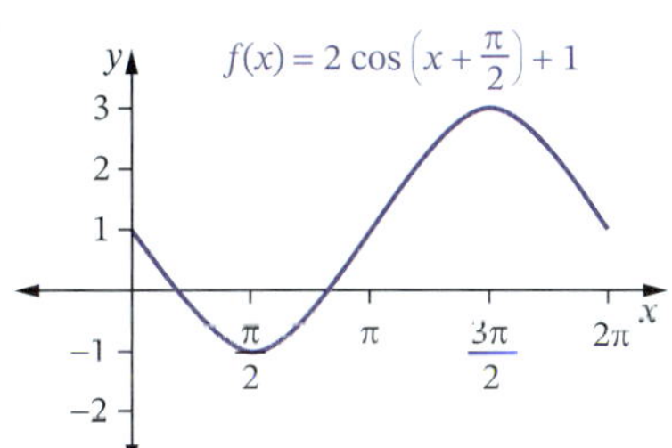

b

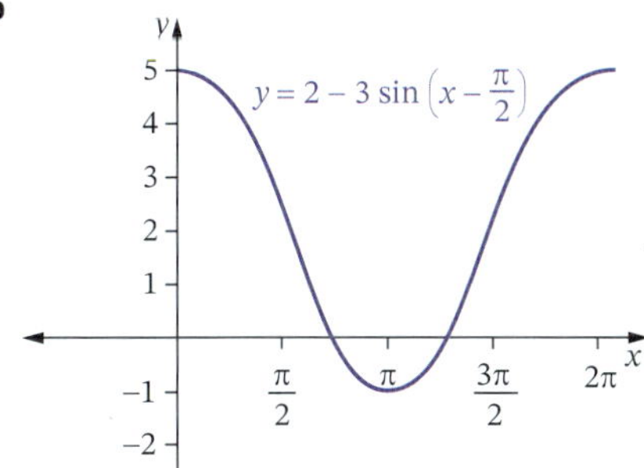

c

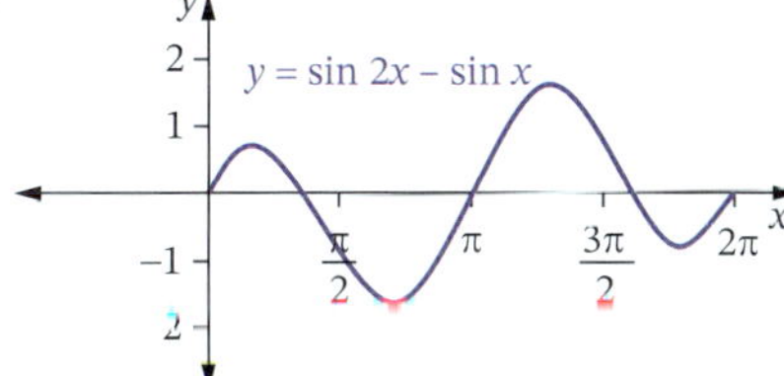

d

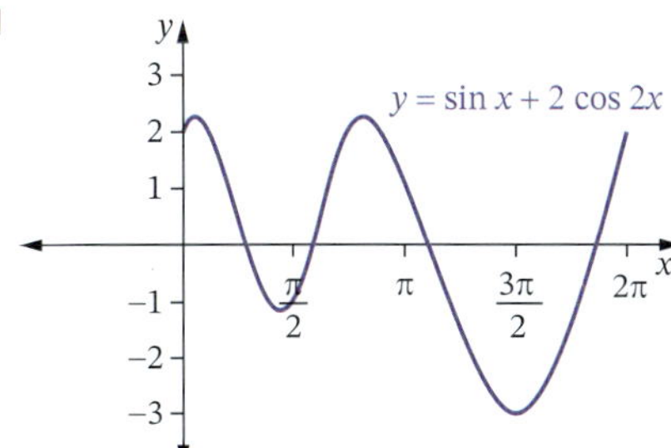

e

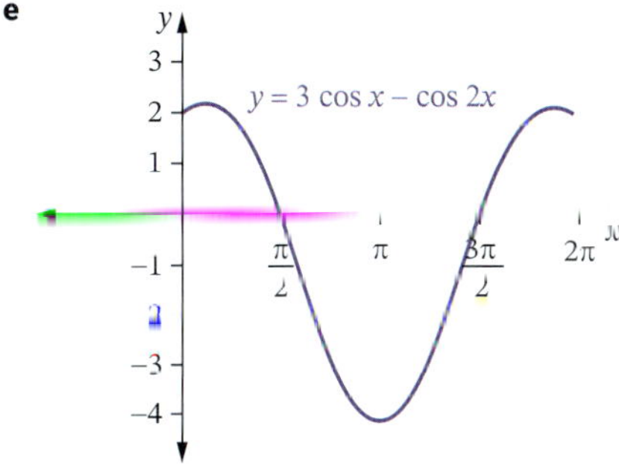

f

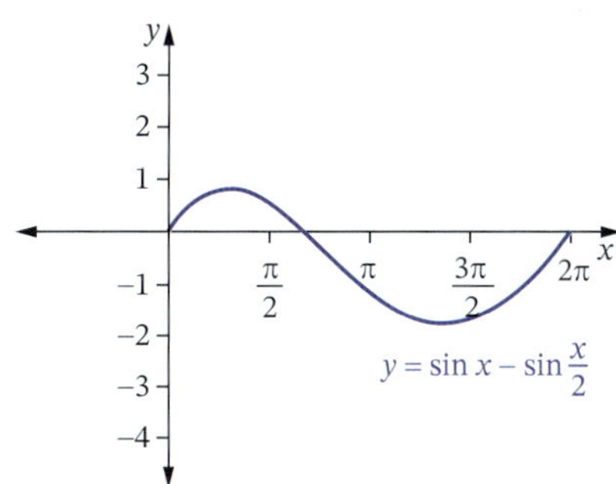

5 **a** $y = 4\cos\left(\frac{1}{3}\left[x - \frac{\pi}{6}\right]\right) - 20$

b amplitude 4, period 6π, centre -20, phase shift $\frac{\pi}{6}$ to the right

6 **a** amplitude 2, period π, phase shift $\frac{\pi}{4}$ units to the right

b $x = \frac{\pi}{6}, \frac{\pi}{3}, \frac{7\pi}{6}, \frac{4\pi}{3}$

7

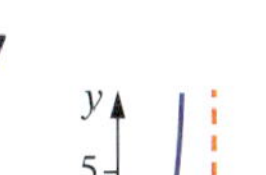

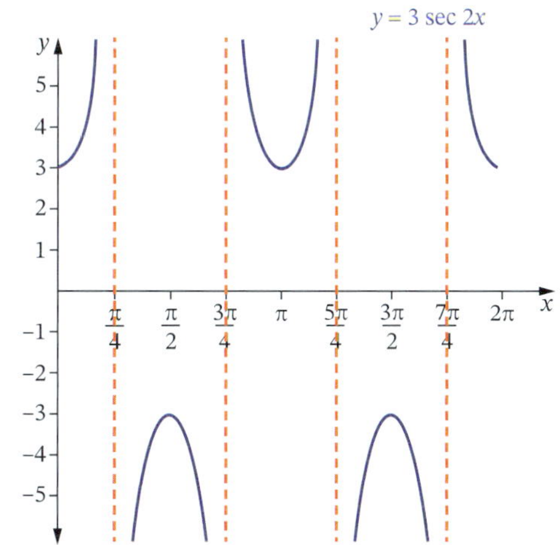

8 $x = \frac{\pi}{8}, \frac{3\pi}{8}, \frac{9\pi}{8}, \frac{11\pi}{8}$

9 See worked solutions.

10 $y = 3\sin 2x - 2$

Chapter 10

Exercise 10.01

1 **a** $n(P) = 5, n(Q) = 9$

b P is a subset of Q

2 **a** **i** {2, 4} **ii** {1, 2, 3, 4, 5, 6}

b **i** {red, white} **ii** {red, yellow, white}

c **i** { } **ii** {4, 5, 6, 7, 8, 9, 10, 11, 12, 15}

d **i** {brown, blue}

ii {blue, green, brown, hazel, grey}

e **i** { } **ii** {1, 2, 3, 4, 5, 6, 7, 8, 9, 10}

3 $A \cap B = \{\ \}$ or $\varnothing$ so A and B are disjoint sets.

4 **a** $\overline{A}$ is the set of all even numbers

b $\overline{B}$ = {all makes of cars except Toyota}

c $\overline{C}$ = {all numbers that are not multiples of 3}

d $\overline{D}$ is the set of tossed coins that come up tails

e $\overline{E}$ = {green, blue, yellow}

f $\overline{F}$ is the set of all losing or drawing teams in a soccer round

5 **a** $A \cup B = \{1, 2, 4, 5, 8, 10, 12, 16, 20\}$
b $A \cap B = \{4, 20\}$
c $\overline{A} = \{3, 6, 7, 8, 9, 11, 12, 13, 14, 15, 16, 17, 18, 19\}$
d $\overline{B} = \{1, 2, 3, 5, 6, 7, 9, 10, 11, 13, 14, 15, 17, 18, 19\}$
e $A \cap \overline{B} = \{1, 2, 5, 10\}$

6 **a**
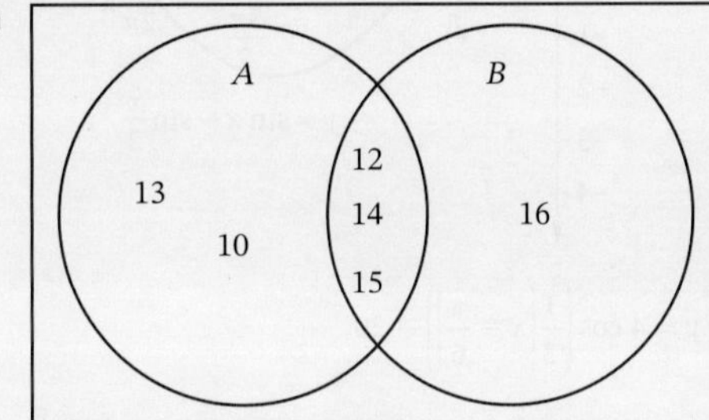

b
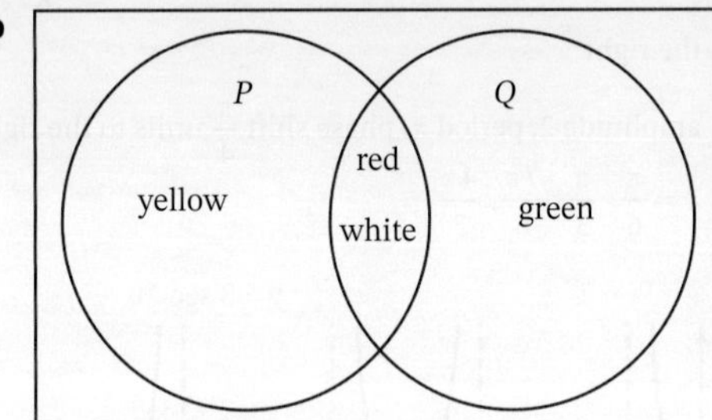

c
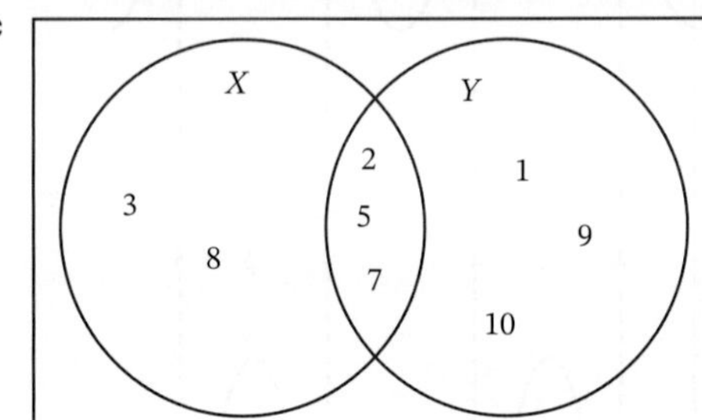

d
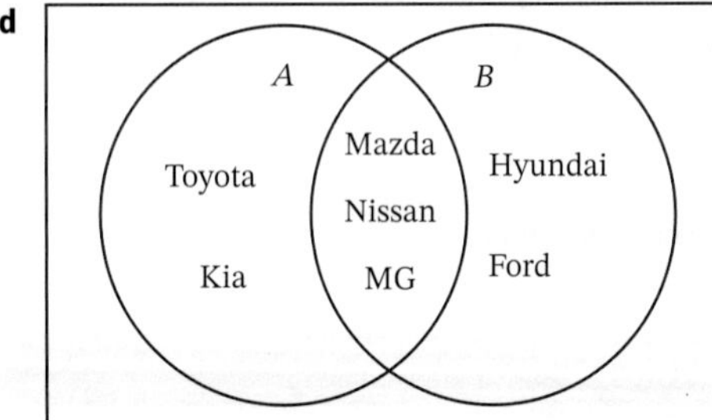

e
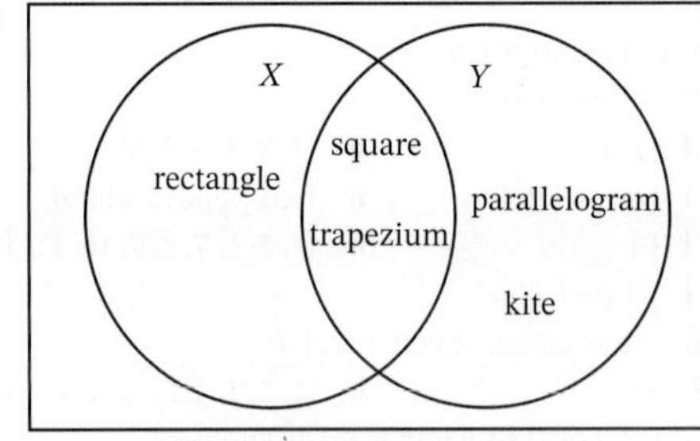

7 **a**
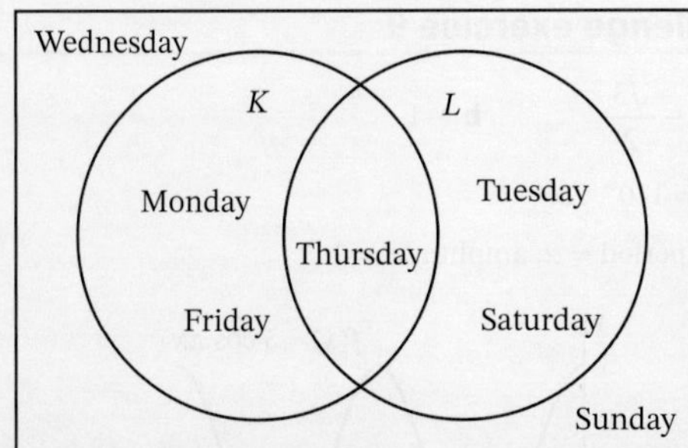

b
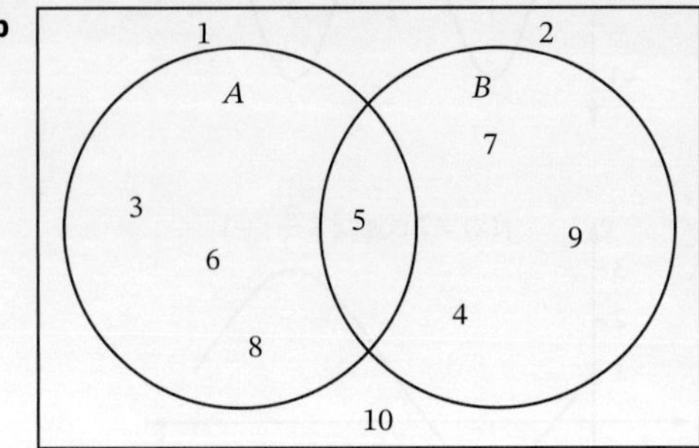

8 **a** 5 **b** 4 **c** 1 **d** 6 **e** 3

9 **a**
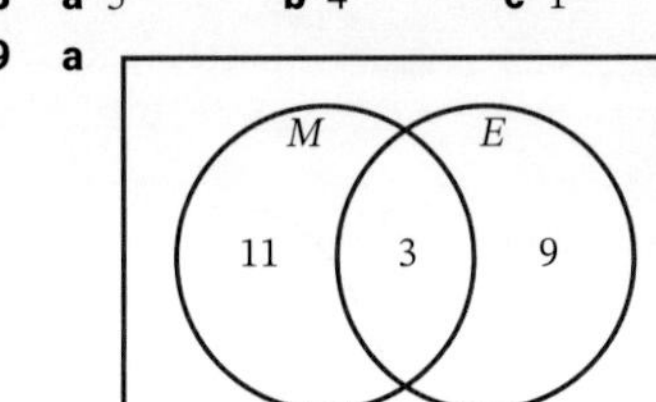

b
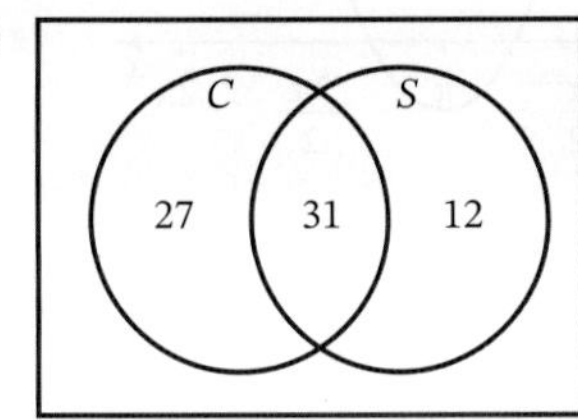

c
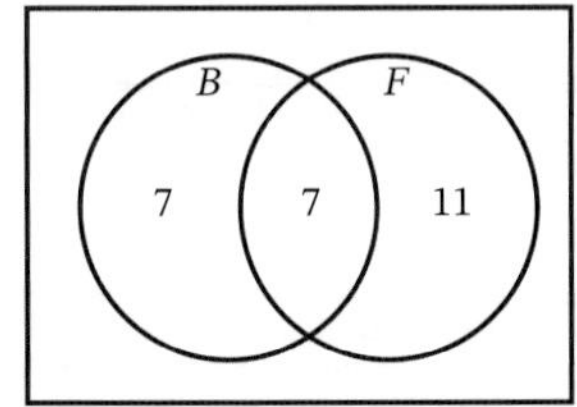

d
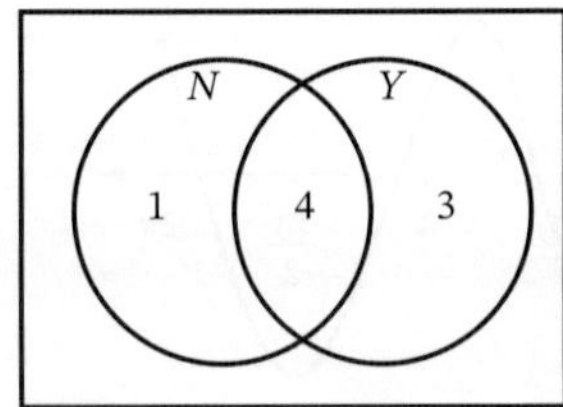

e
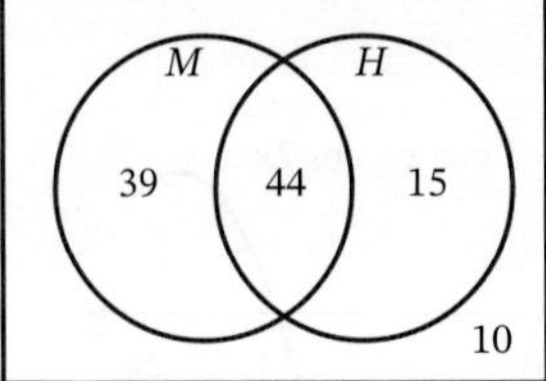

10

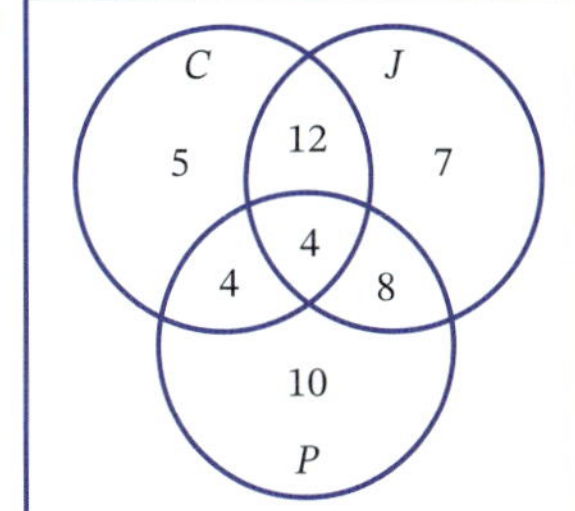

11 **a** 65 **b** 40

Exercise 10.02

1 **a** {heads, tails} **b** {1, 2, 3, 4, 5}
c {1, 2, 3, 4, 5, 6} **d** {red, green, yellow, blue}
e {1, 2, 3, 4, 5, 6, 7, 8}

2 **a** $\frac{1}{52}$ **b** $\frac{51}{52}$

3 **a** $\frac{1}{30}$ **b** $\frac{29}{30}$

4 $\frac{3}{37}$ **5** $\frac{1}{20000}$

6 **a** $\frac{4}{7}$ **b** $\frac{3}{7}$

7 **a** $\frac{11}{20}$ **b** $\frac{3}{4}$

8 **a** $\frac{1}{5}$ **b** $\frac{1}{36}$ **c** $\frac{1}{4}$ **d** $\frac{1}{8}$

9 **a** $\frac{1}{40}$ **b** $\frac{39}{40}$

10 **a** $\frac{2}{9}$ **b** 147

11 99.8%

12 **a** $\frac{1}{6}$ **b** $\frac{1}{2}$ **c** $\frac{1}{3}$

13 **a** $\frac{1}{62}$ **b** $\frac{3}{31}$ **c** $\frac{1}{2}$ **d** $\frac{99}{124}$

14 **a** 98.5% **b** **i** 39 **ii** 2561

15 $\frac{7}{8}$ **16** **a** $\frac{8}{15}$ **b** $\frac{7}{15}$ **c** $\frac{1}{15}$

17 **a** $\frac{1}{2}$ **b** 1

18 **a** $\frac{5}{11}$ **b** 192

19 $\frac{25}{43}$ **20** 34

21 **a** $\frac{7}{31}$ **b** $\frac{7}{31}$ **c** $\frac{12}{31}$

22 $\frac{7}{17}$

23 **a** False, each horse winning is not equally likely. All horses have different abilities and chances.
b False. All competitors have different abilities and chances.
c False. Previous outcomes have no influence.
d False. Previous outcomes have no influence.
e False. All competitors have different abilities and chances.

24 $\frac{1}{3}$

25 **a** $\frac{1}{6}$ **b** $\frac{1}{3}$ **c** $\frac{5}{6}$

26 $\frac{46}{49}$ **27** **a** $\frac{2}{15}$ **b** $\frac{13}{15}$

28 $A \cap B = 0$

Exercise 10.03

1 **a** $\frac{3}{10}$ **b** $\frac{3}{5}$ **c** $\frac{11}{20}$ **d** $\frac{7}{10}$

2 **a** $\frac{1}{5}$ **b** $\frac{1}{2}$ **c** $\frac{3}{5}$ **d** $\frac{3}{5}$
e $\frac{19}{50}$

3 **a** $\frac{5}{26}$ **b** $\frac{9}{26}$ **c** $\frac{12}{13}$

4 **a** $\frac{29}{100}$ **b** $\frac{13}{20}$ **c** $\frac{9}{25}$

5 **a** $\frac{3}{5}$ **b** $\frac{4}{9}$ **c** $\frac{2}{3}$

6 **a** $\frac{3}{14}$ **b** $\frac{13}{28}$ **c** $\frac{9}{28}$

7 **a** $\frac{21}{80}$ **b** $\frac{17}{80}$ **c** $\frac{21}{40}$

8 **a** $\frac{1}{10}$ **b** $\frac{11}{20}$ **c** $\frac{7}{20}$

9 **a** $\frac{7}{25}$ **b** $\frac{2}{15}$ **c** $\frac{44}{75}$

10 **a** $\frac{3}{10}$ **b** $\frac{2}{5}$ **c** $\frac{3}{10}$

11 **a** $\frac{8}{25}$ **b** $\frac{2}{25}$ **c** $\frac{3}{25}$

12 **a** $\frac{19}{112}$ **b** $\frac{3}{28}$

Exercise 10.04

1 $\frac{1}{4}$ **2** $\frac{1}{8}$ **3** $\frac{1}{4}$ **4** $\frac{25}{121}$

5 **a** 0.0441 **b** 0.6241

6 80.4% **7** 32.9%

8 **a** $\frac{29791}{35937}$ **b** $\frac{8}{35937}$ **c** $\frac{6146}{35937}$

9 **a** $\frac{1}{2400}$ **b** $\frac{1}{5760000}$ **c** $\frac{5755201}{5760000}$

10 **a** $\frac{1}{7776}$ **b** $\frac{3125}{7776}$ **c** $\frac{4651}{7776}$

11 **a** $\frac{9}{25000000}$ **b** $\frac{24970009}{25000000}$ **c** $\frac{29991}{25000000}$

12 **a** $\frac{1}{4}$ **b** $\frac{9}{100}$ **c** $\frac{9}{100}$

13 **a** 61.41% **b** 0.34% **c** 99.66%

14 **a** $\frac{1}{2^n}$ **b** $1 - \frac{1}{2^n} = \frac{2^n - 1}{2^n}$

15 **a** $\frac{9}{49}$ **b** $\frac{15}{91}$

16 $\frac{3}{2075}$ **17** $\frac{19}{99}$

18 **a** $\frac{2}{275}$ **b** $\frac{49}{198}$ **c** $\frac{316}{495}$

19 **a** $\frac{253}{861}$ **b** $\frac{57}{287}$ **20** $\frac{14}{95}$

Exercise 10.05

1 **a** $\frac{1}{8}$ **b** $\frac{3}{8}$ **c** $\frac{7}{8}$

2 **a** $\frac{1}{900}$ **b** $\frac{1}{900}$ **c** $\frac{1}{430}$

3 **a** $\frac{25}{169}$ **b** $\frac{80}{169}$

4 **a** 27.5% **b** 23.9% **c** 72.5%

5 **a** $\frac{189}{1000}$ **b** $\frac{441}{1000}$ **c** $\frac{657}{1000}$

6 **a** 0.325 **b** 0.0034 **c** 0.997

7 **a** $\frac{4}{27}$ **b** $\frac{1}{6}$

8 **a** $\frac{1}{25}$ **b** $\frac{1}{825}$ **c** $\frac{64}{825}$ **d** $\frac{152}{165}$ **e** $\frac{13}{165}$

9 **a** $\frac{16}{75}$ **b** $\frac{38}{75}$

10 **a** $\frac{11}{20}$ **b** $\frac{3}{20}$

11 **a** $\frac{84681}{1000000}$ **b** $\frac{912673}{1000000}$ **c** $\frac{27}{1000000}$

12 **a** 17.4% **b** 11.1% **c** 20.9%

13 **a** $\frac{248}{495}$ **b** $\frac{2}{297}$

14 **a** $\frac{22}{425}$ **b** $\frac{368}{425}$ **c** $\frac{7}{425}$

15 **a** $\frac{17}{65}$ **b** $\frac{133}{715}$ **c** $\frac{496}{2145}$

16 **a** $\frac{1}{216}$ **b** $\frac{5}{72}$ **c** $\frac{91}{216}$

17 **a** $\frac{1}{10}$ **b** $\frac{3}{10}$ **c** $\frac{2}{5}$

18 **a** $\frac{5}{18}$ **b** $\frac{5}{9}$ **c** $\frac{13}{18}$

Exercise 10.06

1 **a** $\frac{9}{16}$ **b** $\frac{7}{16}$ **2** $\frac{13}{27}$

3 **a** $\frac{1}{6}$ **b** $\frac{1}{3}$ **4** 75%

5 86.7% **6** 56.25%

7 **a** $\frac{3}{8}$ **b** $\frac{2}{5}$

8 **a** $\frac{9}{19}$ **b** $\frac{4}{19}$ **c** $\frac{5}{19}$

9 **a** $\frac{4}{9}$ **b** $\frac{3}{7}$ **10** **a** $\frac{5}{14}$ **b** $\frac{8}{13}$

11 **a** **i** $\frac{23}{31}$ **ii** $\frac{65}{68}$ **b** **i** $\frac{23}{88}$ **ii** $\frac{8}{11}$

12 **a** $\frac{23}{102}$ **b** $\frac{32}{51}$ **c** $\frac{64}{143}$ **d** $\frac{38}{61}$

13 **a** $\frac{7}{11}$ **b** $\frac{13}{27}$

14 **a** $\frac{17}{61}$ **b** $\frac{105}{278}$ **c** $\frac{85}{138}$ **d** $\frac{34}{155}$ **e** $\frac{102}{295}$

15 **a** 45% **b** 55% **c** 76% **16** 29.9%

17 **a** 79% **b** 17.4% **c** 82.6% **d** 81.8%

18 0.2077

19 $P(L \cap M) = 0.0204$

$P(L)P(M) = 0.17 \times 0.12 = 0.0204$

Since $P(L \cap M) = P(L)P(M)$, L and M are independent.

20 $P(X \cup Y) = P(X) + P(Y) - P(X \cap Y)$

$0.594 = 0.3 + 0.42 - P(X \cap Y)$

$P(X \cap Y) = 0.3 + 0.42 - 0.594$

$= 0.126$

$P(X \cap Y) = P(X)\,P(Y|\,X)$

$0.126 = 0.3 \times P(Y|\,X)$

$0.42 = P(Y|\,X)$

Since $P(Y|\,X) = P(Y) = 0.42$, X and Y are independent.

21 **a** **i** $\frac{3}{10}$ **ii** $\frac{8}{15}$

b $P(W \cap W) = \frac{9}{16} \times \frac{9}{16} = \frac{81}{256}$

$P(W)P(W) = \frac{9}{16} \times \frac{9}{16} = \frac{81}{256}$, so independent since

$P(W \cap W) = P(W)P(W)$

22 **a** No, $P(M \cap C) = \frac{27}{88}$

b $P(M) = \frac{81}{220}$

Exercise 10.07

1 **a** discrete **b** continuous **c** discrete **d** discrete **e** continuous **f** discrete **g** continuous **h** continuous **i** discrete **j** discrete **k** discrete **l** discrete

2 **a** $X = \{0, 1\}$ **b** $X = \{0, 1, 2, \ldots, 10\}$ **c** $X = \{0, 1, 2, 3, \ldots, 20\}$ **d** $X = \{0, 1, 2, \ldots, 31\}$ **e** $X = \{2, 3, 4, \ldots, 12\}$

3 **a** **i**

Score	Frequency
2	1
3	2
4	2
5	5
6	6
7	3
8	5
9	1

ii

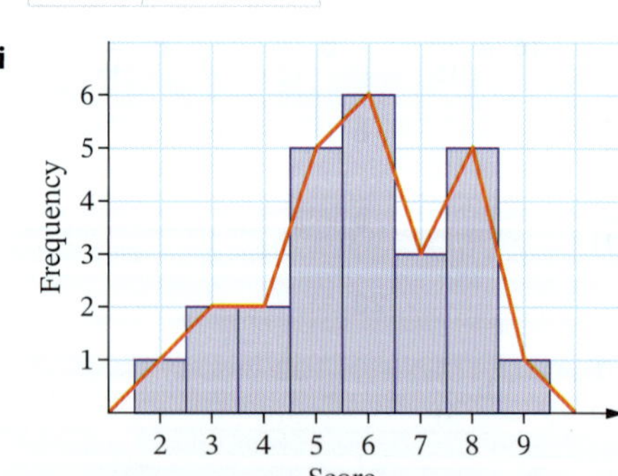

iii highest 9, lowest 2

iv 6

b **i**

Pizzas	Frequency
12	3
13	1
14	1
15	6
16	2
17	4
18	4
19	1

ii

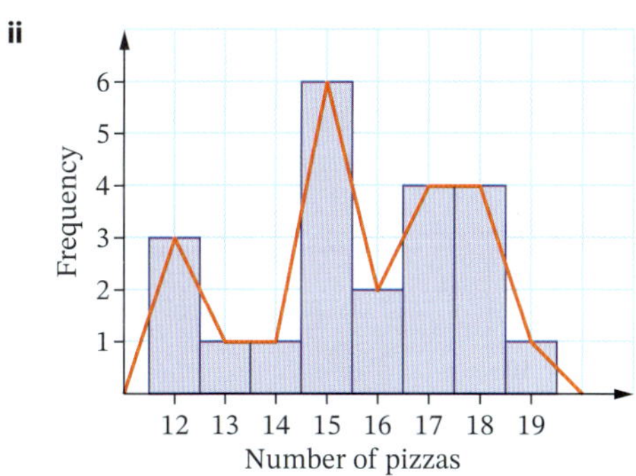

iii highest 19, lowest 12 **iv** 15

c **i**

Gym attendance	Frequency
108	1
109	2
110	5
111	0
112	5
113	4
114	4

ii

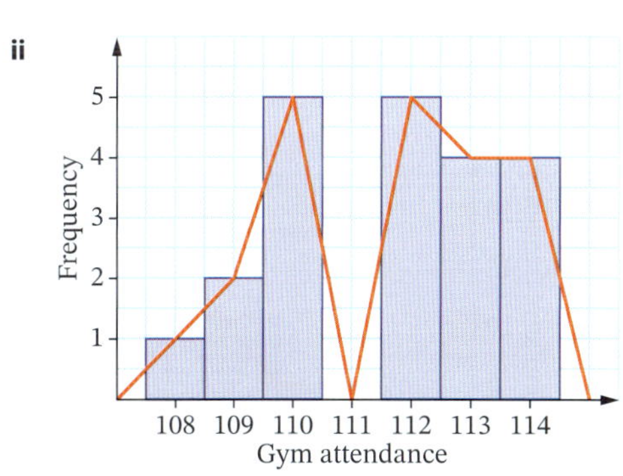

iii highest 114, lowest 108

iv 110 and 112

d **i**

Height (cm)	Class centre	Frequency
155–159	157	5
160–164	162	3
165–169	167	3
170–174	172	6
175–179	177	5
180–184	182	5

ii

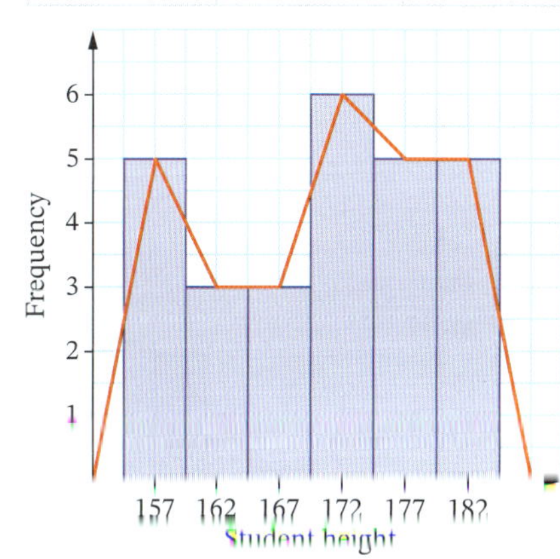

iii highest 182, lowest 157 **iv** 172

4 **a** **i**

Number of cars	Frequency	Cumulative frequency
10	4	4
11	8	12
12	11	23
13	9	32
14	5	37

ii

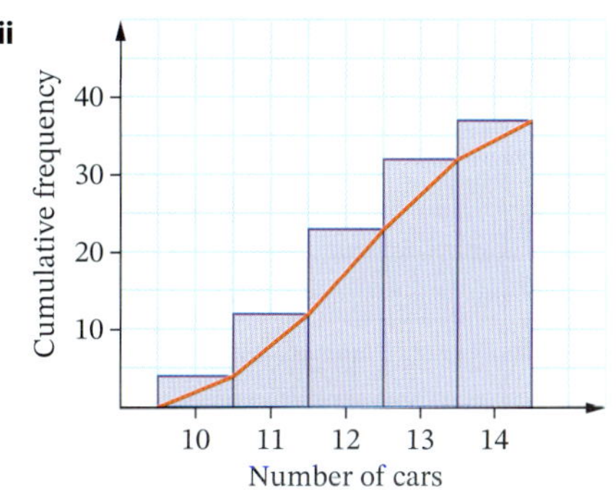

b **i**

Score	Frequency	Cumulative frequency
1	7	7
2	1	8
3	3	11
4	0	11
5	2	13
6	5	18

ii

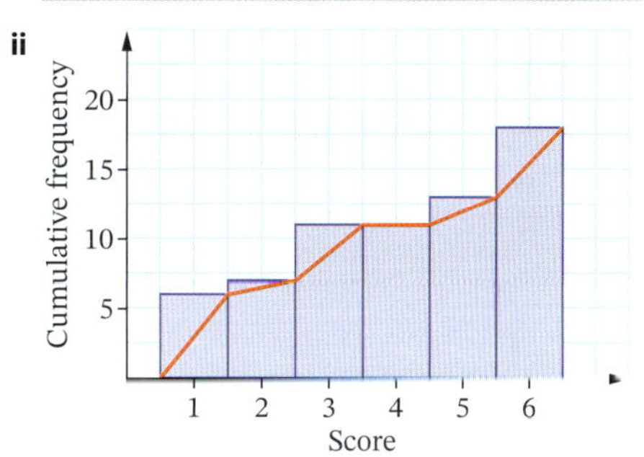

5 **a**

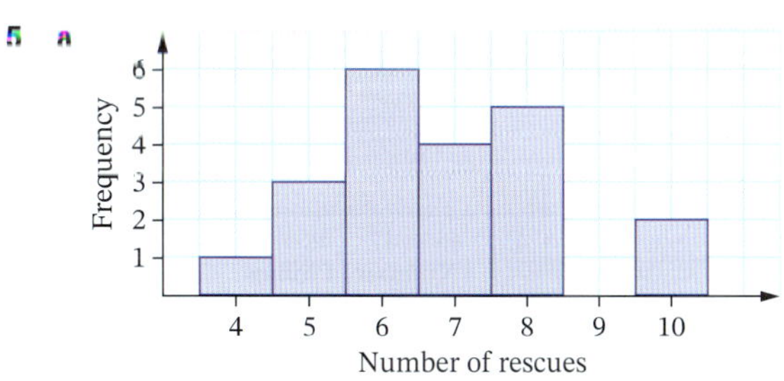

b 11 times **c** 6 rescues

6 **a** **i** 5 **ii** 5 **b** **i** 4 **ii** 5

c **i** 18 **ii** 17 **d** **i** 4 **ii** 5.5

7 **a** **i** 6 **ii** 6 **b** **i** 52 **ii** 51

c **i** 18 **ii** 20 **d** **i** 104 **ii** 104

8 **a** 3 **b** 1.5

c 3.5 **d** 3

9 **a** 3 **b** 1 **c** 2 **d** 2

10 a i

Number of athletes	Frequency	Cumulative frequency
1	5	5
2	6	11
3	4	15
4	8	23
5	5	28
6	2	30

ii

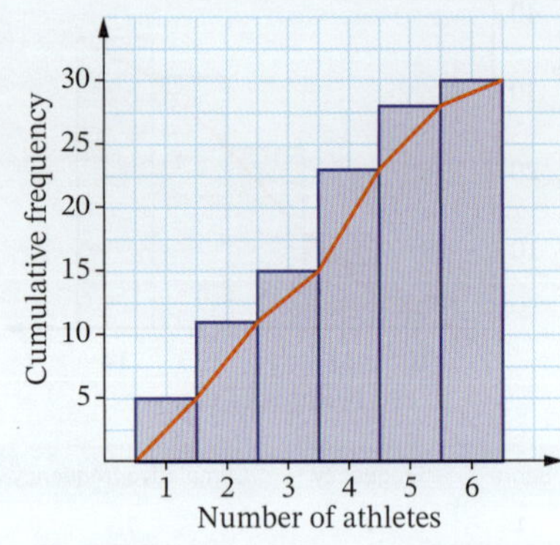

iii 3.5

b i

Number of lollies	Frequency	Cumulative frequency
45	3	3
46	5	8
47	1	9
48	7	16
49	3	19
50	1	20

ii

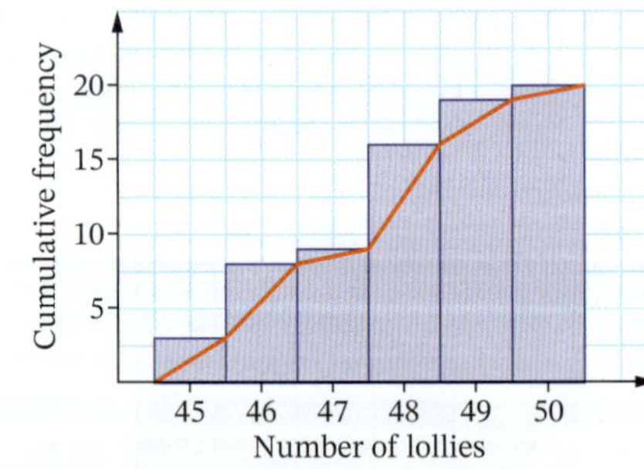

iii 48

c i

Number of hours	Class centre	Frequency	Cumulative frequency
1–5	3	7	7
6–10	8	5	12
11–15	13	3	15
16–20	18	6	21
21–25	23	7	28
26–30	28	2	30

ii

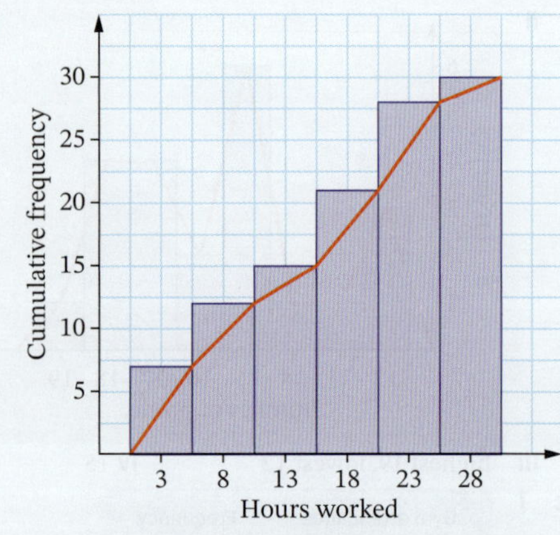

iii 15.5

d i

Time (min)	Class centre	Frequency	Cumulative frequency
2.5–2.8	2.65	3	3
2.9–3.2	3.05	2	5
3.3–3.6	3.45	0	5
3.7–4.0	3.85	6	11
4.1–4.4	4.25	1	12
4.5–4.8	4.65	4	16
4.9–5.2	5.05	4	20

ii

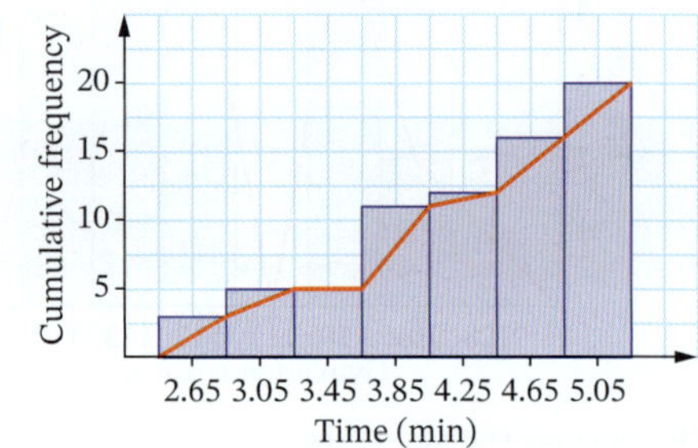

iii 4.0

11 a i

Score	Frequency	Cumulative frequency
2	1	1
3	3	4
4	3	7
5	6	13
6	5	18
7	1	19
8	3	22
9	1	23

ii 5

iii

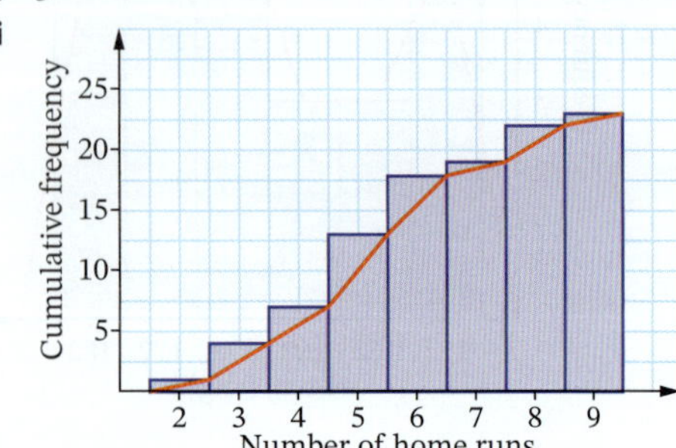

iv 5.5

b i

Number of movies	Frequency	Cumulative frequency
2	1	1
3	0	1
4	2	3
5	3	6
6	4	10
7	1	11
8	3	14
9	0	14
10	2	16

ii 6

iii

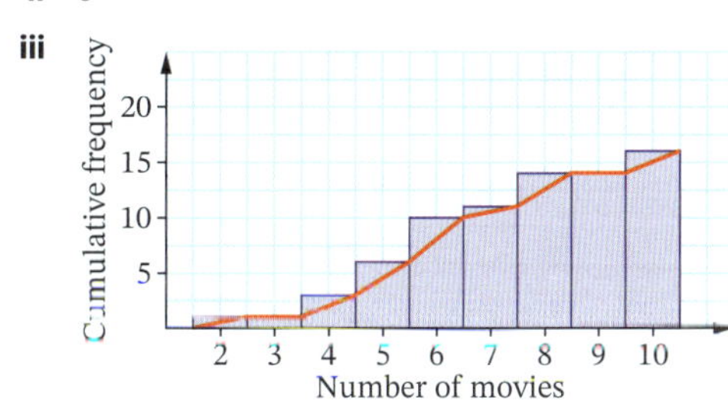

iv 6

12 a 6

b

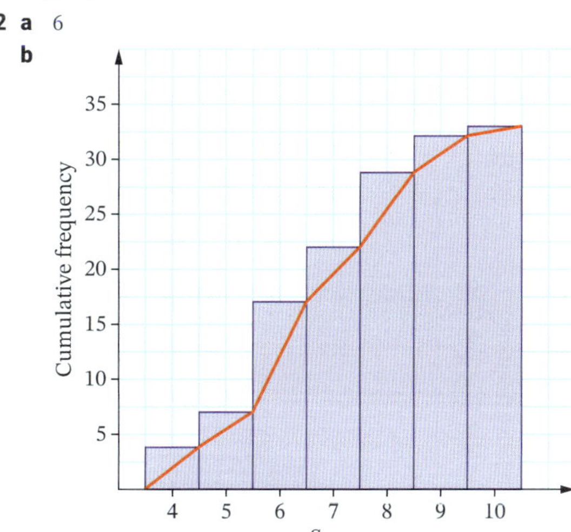

median = 7

13 a

Amount of junk mail items	Frequency
1	6
2	2
3	2
4	2
5	1
6	3
7	4

b

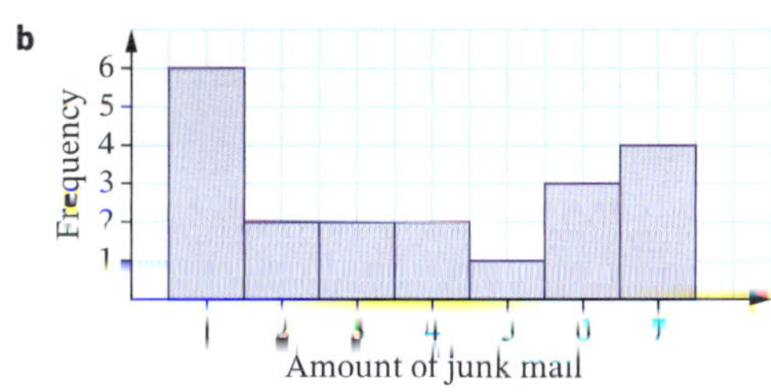

c mode = 1, median = 3.5

Exercise 10.08

1 a

Score	Frequency	Relative frequency
4	6	$\frac{6}{23}$
5	4	$\frac{4}{23}$
6	1	$\frac{1}{23}$
7	7	$\frac{7}{23}$
8	2	$\frac{2}{23}$
9	3	$\frac{3}{23}$

b i $\frac{2}{23}$ **ii** $\frac{11}{23}$ **iii** $\frac{17}{23}$

c i 7 **ii** 6

2 a

Number of days	Frequency	Relative frequency
1	3	15%
2	6	30%
3	1	5%
4	7	35%
5	2	10%
6	1	5%

b 4

c i 15% **ii** 10% **iii** 40%

iv 50% **v** 45%

3 a

Class	Frequency	Relative frequency
0–19	9	0.18
20–39	12	0.24
40–59	18	0.36
60–79	7	0.14
80–99	4	0.08

b i 0.24 **ii** 0.22 **iii** 0.42

4 a

Velocity (m/s)	Frequency	Relative frequency
2–4	2	$\frac{1}{10}$
5–7	7	$\frac{7}{20}$
8–10	4	$\frac{1}{5}$
11–13	1	$\frac{1}{20}$
14–16	6	$\frac{3}{10}$

b i $\frac{7}{20}$ **ii** $\frac{1}{20}$ **iii** $\frac{1}{5}$

iv $\frac{7}{20}$ **v** $\frac{13}{20}$

c i $\frac{9}{20}$ **ii** $\frac{3}{10}$ **iii** $\frac{11}{20}$

5 a 10%

b

Sales/min	Frequency	Relative frequency
0	4	13.3%
1	12	40%
2	6	20%
3	3	10%
4	0	0%
5	5	16.7%

c 1

d i 20% **ii** 16.7% **iii** 26.7%

6 a

Score	Frequency	Relative frequency	Cumulative relative frequency
3	1	$\frac{1}{20}$	$\frac{1}{20}$
4	4	$\frac{1}{5}$	$\frac{1}{4}$
5	3	$\frac{3}{20}$	$\frac{2}{5}$
6	3	$\frac{3}{20}$	$\frac{11}{20}$
7	3	$\frac{3}{20}$	$\frac{7}{10}$
8	2	$\frac{1}{10}$	$\frac{4}{5}$
9	4	$\frac{1}{5}$	1

b

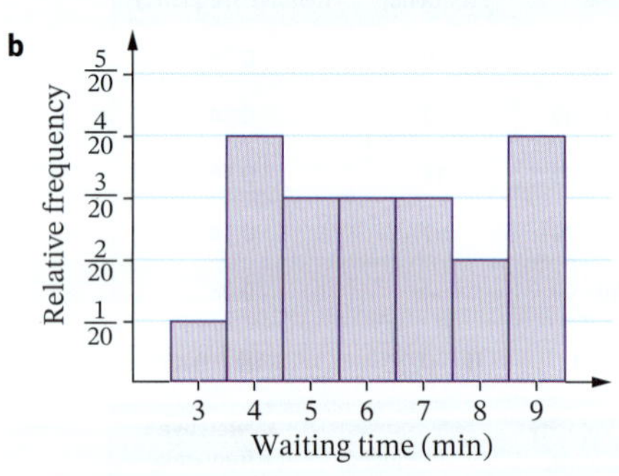

c i $\frac{3}{20}$ **ii** $\frac{3}{10}$ **iii** $\frac{1}{4}$ **iv** $\frac{7}{10}$

7 a

Score	Frequency
4	1
5	5
6	3
7	6
8	2
9	3

b

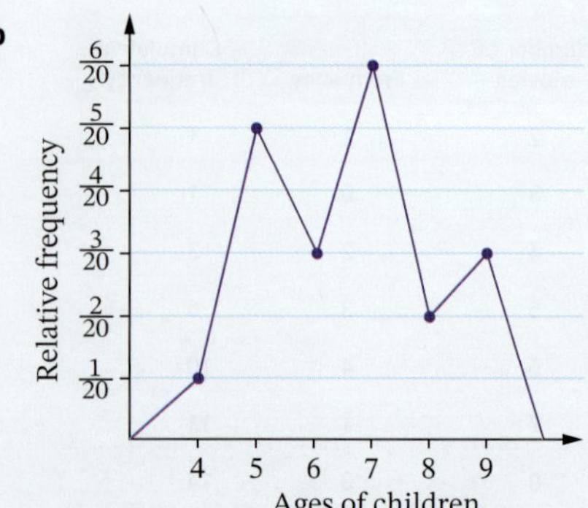

c i 0.1 **ii** 0.7 **iii** 0.25 **iv** 0.95 **v** 0.85

8 a

Relative frequency	Cumulative relative frequency
$\frac{21}{106}$	$\frac{21}{106}$
$\frac{9}{53}$	$\frac{39}{106}$
$\frac{15}{106}$	$\frac{27}{53}$
$\frac{12}{53}$	$\frac{39}{53}$
$\frac{8}{53}$	$\frac{47}{53}$
$\frac{6}{53}$	1

b 106

c

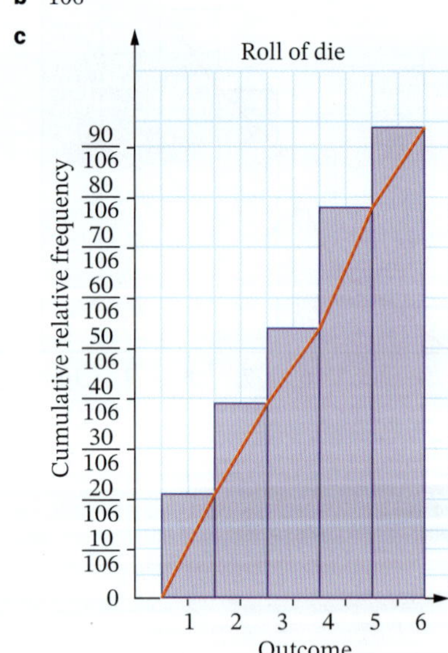

d i $\frac{6}{53}$ **ii** $\frac{39}{106}$ **iii** $\frac{14}{53}$

9 a $\frac{3}{38}$ **b** $\frac{1}{4}$ **c** $\frac{7}{38}$ **d** $\frac{4}{19}$

10 While theoretically $P(H) = P(T) = \frac{1}{2}$, in an experiment anything can happen. Often the probabilities might come close to these in an experiment, but not always.

Sample HSC problem

a $\frac{5}{7}$ **b** $\frac{1}{2}$

Test yourself 10

1 C **2** D **3** B **4** A, D

5 A, C **6** B

7 a 80.4% **b** 1.4% **c** 99.97%

8 a

Outcome	Frequency	Relative frequency	Cumulative relative frequency
1	17	$\frac{17}{100}$	$\frac{17}{100}$
2	21	$\frac{21}{100}$	$\frac{19}{50}$
3	14	$\frac{7}{50}$	$\frac{13}{25}$
4	20	$\frac{1}{5}$	$\frac{18}{25}$
5	18	$\frac{9}{50}$	$\frac{9}{10}$
6	10	$\frac{1}{10}$	1

b i $\frac{7}{50}$ **ii** $\frac{7}{25}$ **iii** $\frac{1}{10}$

iv $\frac{19}{50}$ **v** $\frac{13}{25}$

9 a

	1	2	3	4	5	6
1	0	1	2	3	4	5
2	1	0	1	2	3	4
3	2	1	0	1	2	3
4	3	2	1	0	1	2
5	4	3	2	1	0	1
6	5	4	3	2	1	0

b i $\frac{1}{6}$ **ii** $\frac{1}{6}$ **iii** $\frac{1}{2}$

10 a i $\frac{1}{40}$ **ii** $\frac{39}{40}$ **b** $\frac{39}{796}$

11 a $\frac{4}{15}$ **b** $\frac{1}{10}$ **c** $\frac{3}{17}$ **d** $\frac{3}{11}$

12 False: the events are independent.

13 a $\frac{1}{2}$ **b** $\frac{29}{100}$ **c** $\frac{1}{5}$

d $\frac{11}{25}$ **e** $\frac{16}{25}$

14 a $\frac{2}{5}$ **b** $\frac{7}{15}$ **c** $\frac{2}{15}$

15 a i {3, 5, 6, 7, 8, 9, 10, 11, 12} **ii** {5, 8}

b

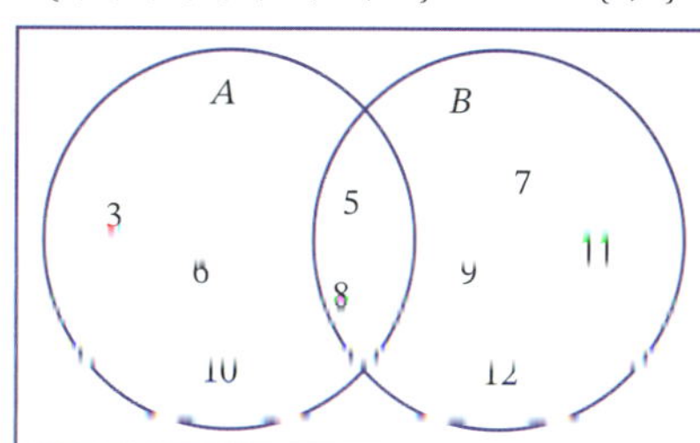

16 a {HH, HT, TH, TT} **b** {red, white, blue}

17 a 0.009% **b** 12.9%

18 a i $\frac{1}{13}$ **ii** $\frac{3}{13}$ **iii** $\frac{5}{26}$

b i $\frac{1}{3}$ **ii** $\frac{1}{4}$

19 a $\frac{5}{12}$ **b** $\frac{1}{3}$

20 a $\frac{9}{40}$

b i $\frac{3}{10}$ **ii** $\frac{27}{160}$ **iii** $\frac{4}{25}$

21 25%

22 a $\frac{1}{15}$ **b** $\frac{4}{5}$

23 a $\frac{1}{50}$ **b** $\frac{147}{7450}$ **c** $\frac{1}{3725}$ **d** $\frac{3577}{3725}$

24 a $\frac{80}{361}$ **b** $\frac{40}{171}$

25 a $\frac{2}{9}$ **b** $\frac{1}{3}$ **c** $\frac{3}{7}$

26 a $\frac{21}{50}$ **b** $\frac{3}{23}$ **c** $\frac{23}{50}$

27 a {0, 1, 2, 3, 4, 5}

b {0, 1, 2, 3, 4, 5, 6, 7, 8, 9, 10}

c {1, 2, 3, 4, 5, ..., 30}

d {0, 1, 2}

e {0, 1, 2, 3, 4, 5, 6, 7, 8, 9}

28 $P(A) \times P(B) = 0.162 = P(A \cap B)$

29 $\frac{14}{29}$

30 a

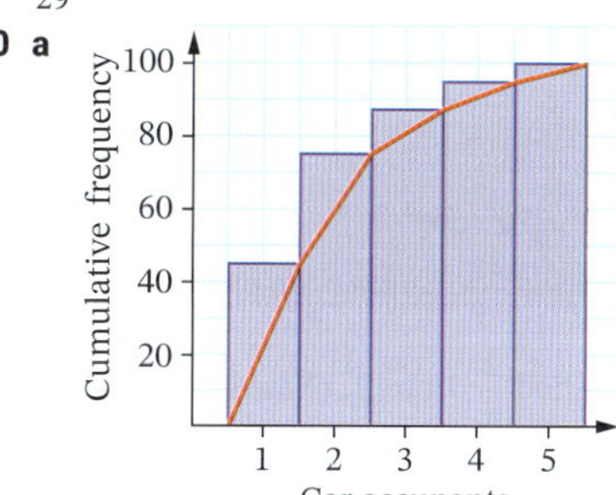

b 2 **c** 1

Challenge exercise 10

1 a $\frac{38}{119}$ **b** $\frac{10}{119}$

2 a 0.04 **b** 0.75 **c** 0.25 **3** $\frac{1}{54145}$

4 a $\frac{5}{8}$ **b** $\frac{2}{5}$

5 No, all combinations are equally likely to win.

6 a $\frac{3}{10}$ **b** $\frac{12}{145}$

7 a $\frac{1}{144}$ **b** $\frac{5}{144}$ **c** $\frac{1}{165\,888}$

8 14.6%

Practice set 4

1 C **2** C **3** C **4** D

5 A **6** D **7** D

8 a

b reflection in x-axis, horizontal translation 2 units to the left

9 a $x = 67°30', 157°30', 247°30', 337°30'$

b $x = 20°, 100°, 140°, 220°, 260°, 340°$

c $x = 150°, 210°$

d $x = 60°, 240°$

e $x = 0°, 90°, 180°, 270°, 360°$

10 a amplitude 4, period $\frac{2\pi}{5}$, centre 0, no phase shift

b amplitude 2, period $\frac{2}{\pi}$ centre 1, phase shift $\frac{\pi}{6}$ units to the right

c period 4π, phase shift 8 units to the left, centre 0

11 a $\frac{3}{25}$ **b** $\frac{8}{25}$ **c** $\frac{8}{25}$

12 a

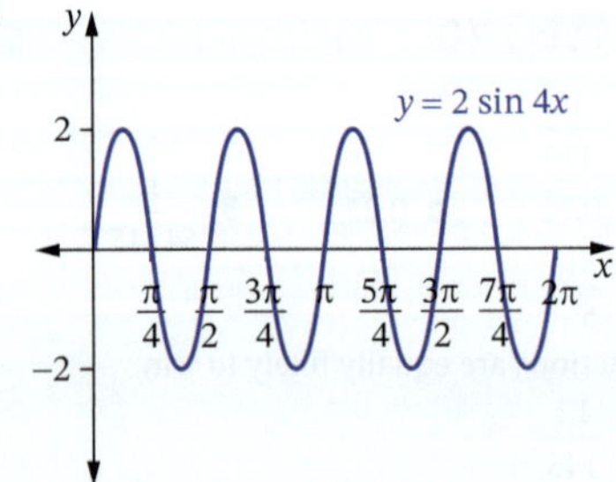

b

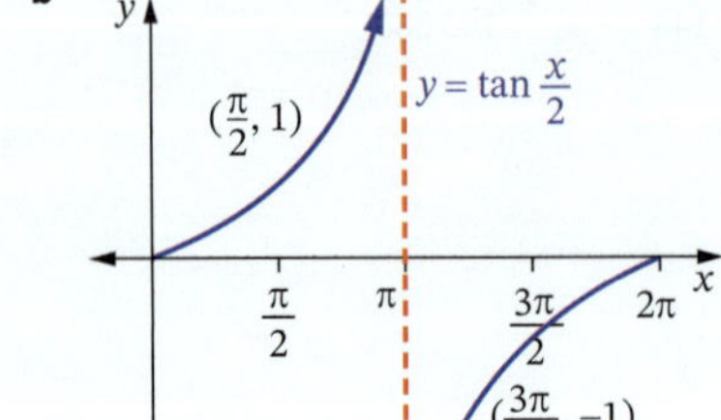

c

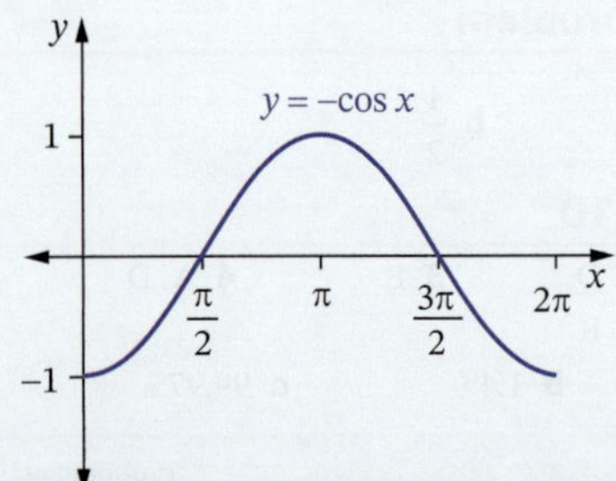

13 a

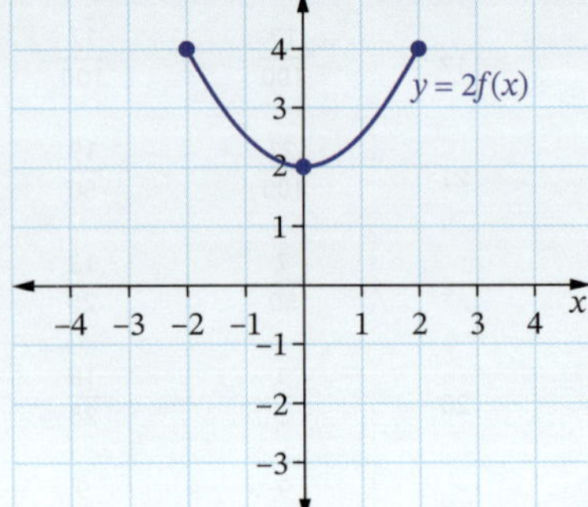

b

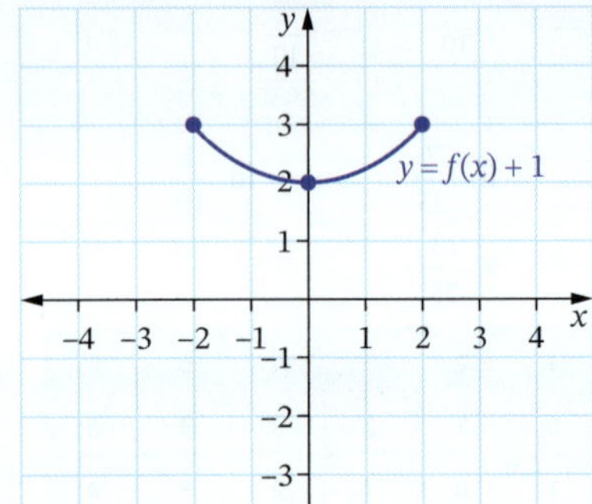

c

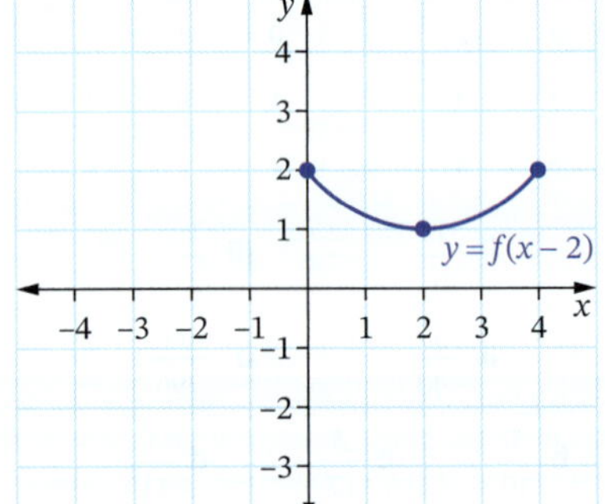

d

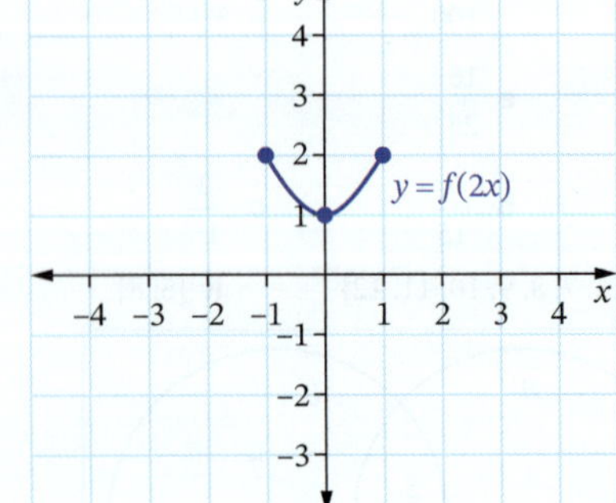

14 **a** $X = \{0, 1, 2, 3, ...\}$
b $X = \{0, 1, 2, ..., 12\}$
c $X = \{0, 1, 2, ..., 31\}$

15 **a** $\frac{1}{36}$ **b** $\frac{1}{6}$ **c** $\frac{11}{36}$
d $\frac{5}{36}$ **e** $\frac{5}{12}$

16 **a** $\frac{7}{50}$ **b** $\frac{11}{20}$

17 $\sin x = -\frac{4}{5}, \cos x = \frac{3}{5}$

18 **a** $x = \frac{2\pi}{3}, \frac{4\pi}{3}$ **b** $x = \frac{\pi}{4}, \frac{3\pi}{4}, \frac{5\pi}{4}, \frac{7\pi}{4}$
c $x = \frac{\pi}{2}, \frac{3\pi}{2}$ **d** $x = \frac{\pi}{12}, \frac{5\pi}{12}, \frac{13\pi}{12}, \frac{17\pi}{12}$

19 **a** $x = 30°, 41° 49', 138° 11', 150°$
b $x = 45°, 105°, 225°, 285°$

20 $y = 4\sqrt{3(x+7)} - 1$

21 vertical dilation, factor 4 (stretch), horizontal translation 1 unit to the right, vertical translation 3 units down

22 **a** $\frac{3}{91}$ **b** $\frac{30}{91}$

23 $\frac{1}{110}$

24 **a** $\frac{5}{36}$ **b** $\frac{5}{12}$ **c** $\frac{1}{6}$ **d** $\frac{7}{36}$ **e** $\frac{1}{2}$

25 **a** $\frac{3}{10}$ **b** $\frac{3}{8}$

26 **a** $\frac{7}{20}$ **b** $\frac{1}{4}$

27 **a** 3.5 **b** 2 **c** 10

28 **a** $x = 135°, 315°$
b $x = 30°, 150°$
c $x = 45°, 135°, 225°, 315°$
d $x = 30°, 120°, 210°, 300°$

29 **a** $\{1, 2, 3, 4, 5, 6\}$
b $\{3, 5\}$
c

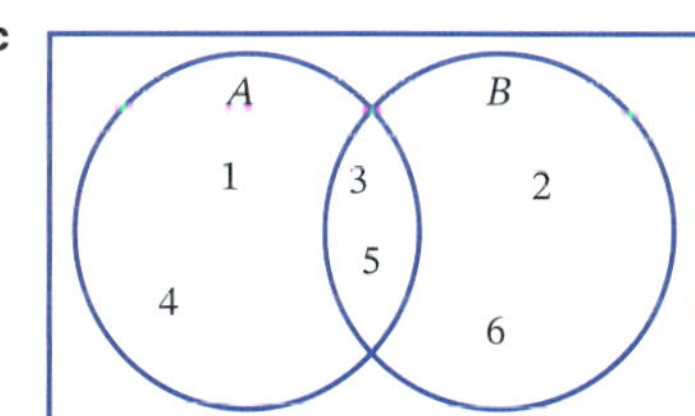

30 $x = 40°$

31 **a**

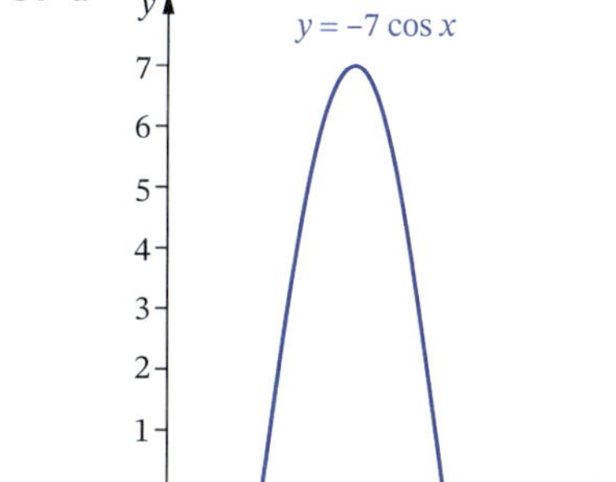

b

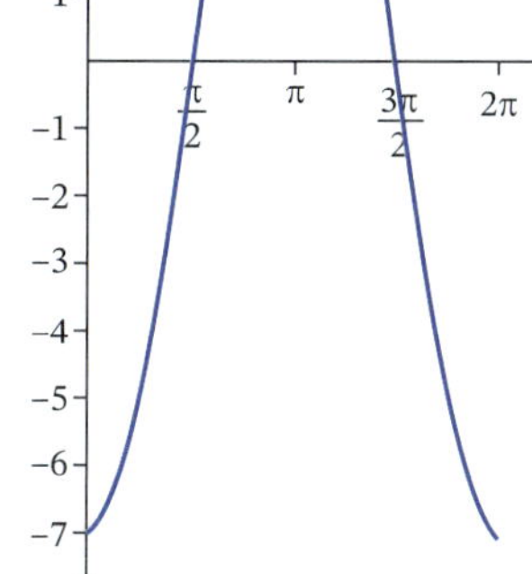

c

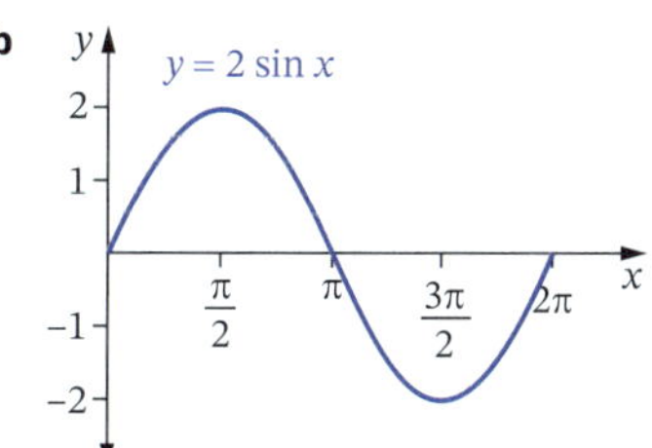

d

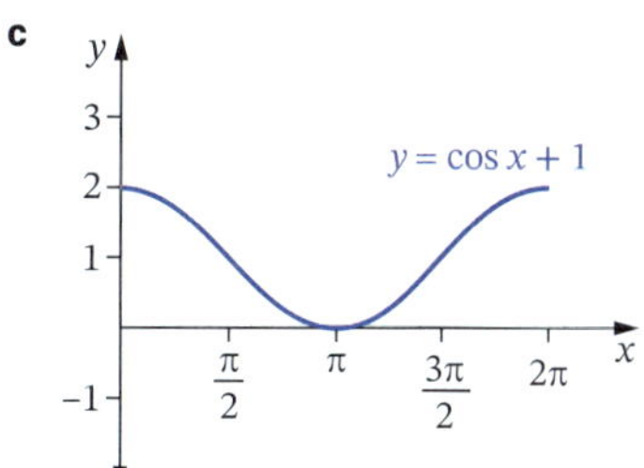

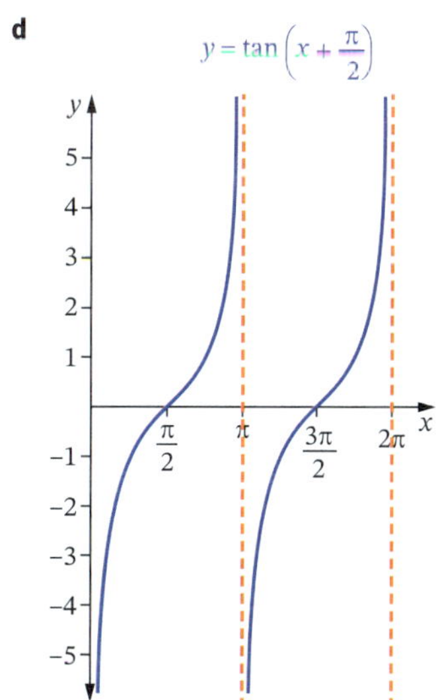

e

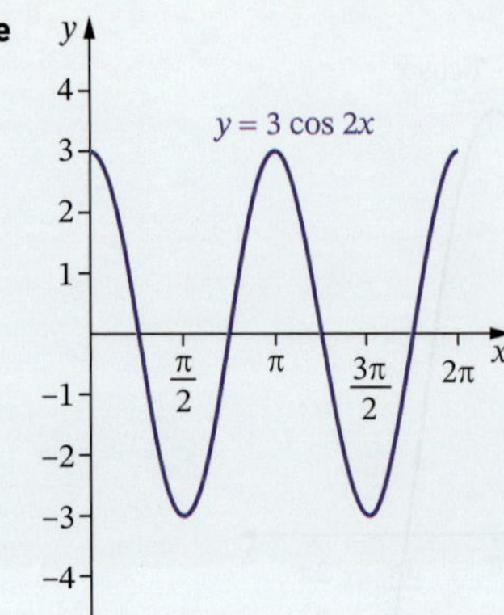

f

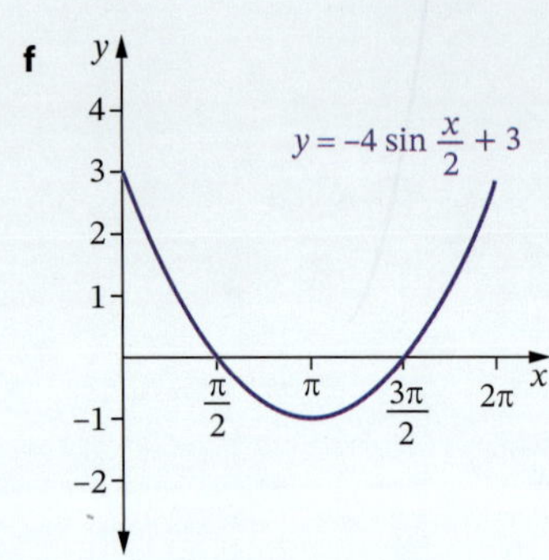

32 Proofs: see worked solutions.

33 **a** $y = 2\sin x$ **b** $y = \sin\left(\frac{x}{2}\right)$

c $y = \sin x - 3$ **d** $y = -5\sin x$

e $y = \sin\left(x + \frac{\pi}{2}\right)$

f $y = 5\sin\left[\frac{1}{3}(x - \pi)\right] + 1$

34 **a** $\frac{1}{12}$ **b** $\frac{1}{4}$

35 Show that $P(Y \mid X) = P(Y) = 0.15$.

36 mutually exclusive

37 75%

38 $P(L)P(M) = 0.054 = P(L \cap M)$

39 **a** $x = 0.9, 5.38$ **b** $x = \frac{\pi}{4}, \frac{3\pi}{4}, \frac{5\pi}{4}, \frac{7\pi}{4}$

c $x = \frac{\pi}{6}, \frac{5\pi}{6}$ **d** $x = \frac{\pi}{2}, \frac{3\pi}{2}$

e $x = \frac{\pi}{3}, \frac{2\pi}{3}, \frac{4\pi}{3}, \frac{5\pi}{3}$ **f** $x = \frac{\pi}{3}, \frac{2\pi}{3}, \frac{4\pi}{3}, \frac{5\pi}{3}$

40 **a** reflection in x-axis, horizontal dilation of factor $\frac{1}{2}$, vertical dilation of factor 2,

horizontal translation 3 units right, vertical translation 5 units up

b $(2, -1)$

c $-\frac{1}{16}$

41 **a** $5\sec^2 x$ **b** $\cot x$

42 vertical dilation of factor 5, horizontal dilation of factor $\frac{1}{3}$, horizontal translation 2 units left, vertical translation 7 units down

GLOSSARY AND INDEX

A

absolute value, $|x|$ The size of a number, x, without sign or direction. Also, the distance of x from 0 on the number line in either direction. (p. 44)

ambiguous case The situation when using the sine rule to find an angle where there may be 2 possible angle sizes: one acute and one obtuse. (p. 163)

amplitude The height from the centre of a sine or cosine function to the maximum or minimum values (peaks and troughs of its graph respectively). For $y = k\sin ax$, the amplitude is k. (p. 388)

angle of depression The angle between the horizontal and the line of sight when looking down to an object below. (p. 147)

angle of elevation The angle between the horizontal and the line of sight when looking up to an object above. (p. 147)

angle of inclination The angle a straight line makes with the positive x-axis, measured anticlockwise. (p. 89)

arc Part of a circle's circumference. (p. 177)

asymptote A line that a curve approaches but doesn't touch. (p. 202)

average rate of change The rate of change between 2 points on a function; the gradient of the line (secant) passing through those points. (p. 240) See also **instantaneous rate of change**.

axis of symmetry A line that divides a shape into halves that are mirror-images of each other. (p. 104)

B

bearing A direction from one point to another, measured in degrees. Bearings may be written as **true bearings** (clockwise from north) or **compass bearings** (from north or south). (p. 148)

binomial A mathematical expression consisting of 2 terms; for example, $x + 3$ and $3x - 1$. (p. 14)

binomial product The product of binomial expressions; for example, $(x + 3)(2x - 1)$. (p. 14)

break-even point The point at which a business' income equals its costs, making neither a profit nor a loss. (p. 122)

C

cardinality of a set The number of elements in the set. The cardinality of set A is written $|A|$ or $n(A)$ (p. 405)

centre The mean value of a trigonometric function that is equidistant from the maximum and minimum values. For $y = k\sin ax + c$, the centre is c. (p. 388)

chain rule A method for differentiating composite functions. (p. 272)

chord An interval inside a circle that joins 2 points on the circle's circumference. (p. 185)

coefficient A number multiplied by a variable in an algebraic term. For example, in ax^3, a is the coefficient of x^3. (p. 18)

compass bearing A bearing expressed as an acute angle measured from north or south, for example, N 20°W or S 67°E. (p. 148) See also **true bearing**.

complement of a set Another set that contains all the elements that are *not* in the original set. (p. 405)

completing the square To add a constant term to a binomial of the form $x^2 + px$ so that the resulting trinomial $x^2 + px + \left(\frac{p}{2}\right)^2$ is a perfect square. (p. 49)

composite function A function of a function, where the output of one function $g(x)$ becomes the input of a second function $f(x)$, written as $f(g(x))$. For example, if $f(x) = x^2$ and $g(x) = 3x + 1$, then $f(g(x)) = (3x + 1)^2$. (p. 222)

conditional probability The probability that an event A occurs when it is known that another event B has occurred. (p. 427)

conjugate surd A surd that matches a binomial surd with the '+' and '−' symbols exchanged. For example, the conjugate surd to $2-\sqrt{7}$ is $2+\sqrt{7}$. (p. 28)

constant term The term in a function that is independent of x. For example, in $y = 2x^4 - x^3 + 3x + 1$, the constant term is 1. (p. 84)

continuous function A function whose graph is smooth and does not have gaps or breaks. (p. 202) See also **discontinuous function**.

cosecant (cosec) $\frac{1}{\sin}$ (p. 379) See also **secant**, **cotangent**.

cosine rule In any triangle, $c^2 = a^2 + b^2 - 2ab\cos C$. See also **sine rule**. (p. 167)

cotangent (cot) $\frac{1}{\tan}$ (p. 379) See also **secant**, **cosecant**.

cubic function A function with x^3 as its highest power. (p. 117)

cumulative frequency A running total of frequencies in a frequency table. (p. 436)

cumulative frequency histogram A histogram of cumulative frequencies. (p. 434)

cumulative frequency polygon (or **ogive**) A frequency polygon of cumulative frequencies. (p. 436)

D

dependent variable The variable in a function whose value depends upon the value of the other variable. In $y = f(x)$, y is dependent upon x, so y is the dependent variable. (p. 69) See also **independent variable**.

derivative function (or **derivative**) The gradient function $y = f'(x)$ of a function $y = f(x)$ obtained through differentiation. (p. 243)

differentiation The process of finding the derivative function. (p. 243)

differentiation from first principles The process of finding the gradient of a tangent to a curve by finding the gradient of the secant between 2 points and finding the limit as one of the points move towards the other point and the secant becomes a tangent. (p. 251)

dilation The process of stretching or shrinking the graph of a function horizontally or vertically from the y- and x-axes respectively. (p. 347) See also **reflection**, **translation**.

dilation factor (or **factor**) The value by which the graph of a function is dilated. (p. 347)

direct variation (or **direct proportion**) A relationship between 2 variables such that as one variable increases, the other variable increases, or as one variable decreases, the other variable decreases. One variable is a constant multiple of the other, with equation $y = kx$ or $y = kx^n$. (p. 205) See also **inverse variation**.

discontinuous function A function whose graph has a gap or break in it; for example, $f(x) = \frac{1}{x}$, whose graph is a hyperbola that is discontinuous at $x = 0$ and $y = 0$. (p. 202). See also **continuous function**.

discriminant The expression $\Delta = b^2 - 4ac$ in the quadratic formula $x = \frac{-b \pm \sqrt{b^2 - 4ac}}{2a}$ for solving quadratic equations of the form $ax^2 + bx + c = 0$. The discriminant determines the number and type of roots of the quadratic equation. (p. 113)

disjoint sets Sets that have no elements in common. (p. 405) See also **mutually exclusive events**.

displacement The distance and direction of an object in relation to the origin, signed distance. (p. 253) See also **velocity**.

domain The set of x values in a function or relation. (p. 69) See also **range**.

E

element The items in a set. (p. 405)

elimination method A method for solving simultaneous equations by adding or subtracting them so that one of the variables is eliminated. (p. 55) See also **substitution method**.

empty set ($\varnothing$ or $\{ \}$) A set with no elements. (p. 405)

equally likely outcomes Outcomes that have the same chance of occurring. (p. 410)

equation A mathematical statement that has a variable or unknown number and an equal sign. An equation can be solved to find the value of the unknown number, for example, $3x + 1 = 7$. (p. 39) See also **inequality**.

Euler's number This number, e, approximately 2.71828, an important constant that is the base of natural exponential and logarithmic functions. (p. 293)

even function A function $f(x)$ such that $f(-x) = f(x)$. Its graph is symmetrical about the y-axis. (p. 79). See also **odd function**.

event A set of one or more outcomes in a chance experiment. (p. 409)

exact ratio A trigonometric ratio of 30°, 45°, 60° and related angles in other quadrants, whose value can be expressed as an exact surd or fraction. (p. 156)

exponential equation An equation where the unknown is the power or index; for example, $2^x = 8$. (p. 312)

exponential function A function of the form $y = a^x$, where $a > 0$ and $a \neq 1$. (p. 289)

expression A mathematical statement involving numbers, variables and operators; for example, $2x - 3$. (p. 6)

F

factor A whole number that divides exactly into another number. For example, 4 is a factor of 28. (p. 16)

factorise To take out the highest common factor in an expression and place the rest in brackets. For example, $2y - 8 = 2(y - 4)$. (p. 16)

frequency histogram A column graph of frequencies where there are no gaps between the columns. (p. 434)

frequency polygon A line graph of frequencies that starts and ends on the horizontal axis. (p. 434)

function A relation where each x value in the domain is matched to exactly one y value in the range. (p. 69)

function notation A way of writing a function as $y = f(x)$, where x is the independent variable and y is the dependent variable. (p. 74)

G

gradient The steepness of a graph at a point on the graph, measured by the ratio $\frac{\text{rise}}{\text{run}}$, or the rate of change in y values compared to the change in x values. (p. 84)

gradient of a secant The gradient of the line between 2 points on a function that measures the average rate of change between the 2 points. (p. 249)

gradient of a tangent The gradient of a line that is a tangent to the curve at a point on a function, that measures the instantaneous rate of change of the function at that point. (p. 243)

H

histogram See **frequency histogram**.

hyperbola The graph of the function $y = \frac{k}{x}$, which is made up of 2 separate curves. (p. 201)

I

identity An equation that shows the equivalence of 2 expressions for all values of the variables. For example, $(a + b)(a - b) = a^2 - b^2$, $\cos^2 x + \sin^2 x = 1$. (p. 381)

included angle The angle between 2 known sides in a triangle. (p. 170)

independent events Events where the occurrence of one event does not affect the probability of another event. (p. 419)

independent variable The variable in a function whose value does not depend on the other variable. In $y = f(x)$, y is dependent upon x so x is the independent variable. (p. 69) See also **dependent variable**.

index (plural: **indices**) The power or exponent of a number. For example, 2^3 has a base of 2 and an index of 3. (p. 5)

inequality A mathematical statement involving an inequality sign with an unknown variable; for example, $x - 7 \leq 12$. (p. 42) See also **equation**.

instantaneous rate of change The rate of change at a particular point on a function; the gradient of the tangent at this point. (p. 242) See also **average rate of change**.

intercepts The values where a graph cuts the x- and y-axes. (p. 77)

intersection of sets ($\cap$) The set of only the elements common to all of the sets. (p. 405) See also **union of sets**.

interval notation A notation that represents an interval by writing its endpoints in square brackets [ ] when they are included and in parentheses ( ) when they are not included. (p. 77)

inverse variation (or **inverse proportion**) A relationship between 2 variables such that as one variable increases, the other variable decreases, or as one variable decreases, the other variable increases. One variable is a multiple of the reciprocal of the other, with equation $y = \frac{k}{x}$ or $y = \frac{k}{x^n}$. (p. 208) See also **direct variation**.

L

limit The value that a function approaches as the independent variable, x, approaches some value. (p. 250)

linear function A function of the form $y = mx + c$ whose graph is a straight line. (p. 84)

logarithm The logarithm of a positive number y is the power to which a given number a, called the base, must be raised to produce the number y. So, $\log_a y = x$ means $y = a^x$. (p. 299)

logarithmic function A function of the form $y = \log_a x$ or $\ln x$. (p. 308)

M

median A measure of centre; the middle value when all the values in a dataset are arranged in order. If there are an even number of values, then the median is the average of the 2 middle values. (p. 436)

mode A measure of centre; the most frequent value(s) in a dataset, with the highest frequency. (p. 436)

mutually exclusive events Events within the same sample space that cannot both occur at the same time; for example, rolling an even number on a die and rolling a 5 on the same die. (p. 411) See also **disjoint sets**.

N

non-mutually exclusive events Events within the same sample space that can occur at the same time; for example, rolling a prime number on a die and rolling an odd number on the same die. (p. 415)

normal A line that is perpendicular to the **tangent** at a given point on a curve. (p. 270)

O

odd function A function $f(x)$ such that $f(-x) = -f(x)$. Its graph has point symmetry about the origin. (p. 80) See also **even function**.

ogive See **cumulative frequency polygon**.

P

parabola The graph of a quadratic function. (p. 97)

perfect square A square number or an algebraic expression that represents one; for example, 64, $(x + 9)^2$, $(a - b)^2$. (p. 20)

period The length of one cycle of a trigonometric function on the x-axis, before the function repeats itself. For $y = k\sin ax$, the period is $\frac{2\pi}{a}$. (p. 388)

periodic function A function that repeats itself regularly, such as a trigonometric function. (p. 388)

phase (or **phase shift**) A horizontal shift (translation). For $y = k\sin[a(x+b)]$, the phase is b, that is, the graph of $y = k\sin ax$ shifted b units to the left. (p. 390)

piecewise function A function that has different functions defined on different intervals. (p. 75)

point of inflection A point on a curve where the concavity changes, such as the turning point on the graph of a cubic function. (p. 118)

probability tree A **tree diagram** that shows the probability on each branch. (p. 423)

product rule A method for differentiating the product of 2 functions. (p. 275) See also **quotient rule**.

power The index or exponent of a number. For example, 2^3 has a base of 2 and a power of 3. (p. 5)

Q

quadrant A quarter of the number plane bounded by the x- and y-axes. (p. 153)

quadratic equation An equation in which the highest power of x is 2. (p. 48)

quadratic formula The formula for solving a quadratic equation of the form $ax^2 + bx + c = 0$. The formula is

$$x = \frac{-b \pm \sqrt{b^2 - 4ac}}{2a}. \text{ (p. 51)}$$

quadratic function A function of the form $y = ax^2 + bx + c$ and whose graph is a parabola. (p. 97)

quadratic inequality An inequality in which the highest power of x is 2. (p. 53)

quotient rule A method for differentiating the quotient of 2 functions. (p. 277) See also **product rule**.

R

radian A unit of angle measurement equal to the size of the angle subtended at the centre of a unit circle by an arc of length 1 unit. (p. 177)

random variable A variable whose value depends on the outcome of a random event, such as highest daily temperature, number of road accidents in an hour. (p. 433)

range The set of y values in a function or relation. (p. 69) See also **domain**.

reciprocal trigonometric ratios The cosecant, secant and cotangent ratios, which are the reciprocals of sine, cosine and tangent respectively. (p. 379)

reflection The process of flipping the graph of a function horizontally and/or vertically across a line without changing its size or shape. (p. 339) See also **dilation, translation**.

relation A mapping between the elements of one set of numbers and another set of numbers. (p. 69) See also **function**.

relative frequency The frequency of an event as a fraction of the total frequency. (p. 442)

root A number that when multiplied by itself a given number of times equals another number. For example, $\sqrt{25} = 5$ because $5^2 = 25$. (p. 5)

root A solution of an equation. (p. 113)

S

sample space The set of all possible outcomes in a chance experiment. (p. 409)

scale factor See **dilation factor**.

secant A line that crosses a curve at 2 points, whose gradient is the average rate of change between the 2 points. (p. 240) See also **tangent**.

secant (sec) $\frac{1}{\cos}$ (p. 379) See also **cosecant**.

sector Part of a circle bounded by 2 radii, a 'pizza slice'. (p. 186)

segment Part of a circle bounded by a chord and an arc. (p. 188)

set A collection of distinct objects called elements or members. For example, set $A = \{1, 2, 3, 4, 5, 6\}$. (p. 405)

simultaneous equations Two or more equations that can be solved together to produce a solution. (p. 55)

sine rule In any triangle, $\frac{a}{\sin A} = \frac{b}{\sin B} = \frac{c}{\sin C}$. (p. 162) See also **cosine rule**.

stationary point A point on the graph of $y = f(x)$ where the tangent is horizontal and its gradient $f'(x)$ is equal to 0. It could be a maximum point, minimum point or a horizontal point of inflection. (p. 245)

subset A part (or all) of a set, containing some (or all) of its elements. (p. 405)

substitution method A method for solving simultaneous equations by making x or y the subject of one of them and then substituting it into the other equation. (p. 55) See also **elimination method**.

surd An irrational root that cannot be simplified; for example, $\sqrt{3}$. (p. 24)

T

tangent A straight line that just touches a curve at one point. The curve has the same gradient or direction as the tangent at that point. (p. 242) See also **secant**.

term A part of an expression containing variables and/or numbers separated by an operation such as $+$, $-$, $\times$ or $\div$. For example, in $2x - 3$ the terms are $2x$ and -3. (p. 14)

transformation A general name for the process of changing the graph of a function by translating, reflecting or dilating it. (p. 325)

translation The process of shifting or sliding the graph of a function horizontally and/or vertically without changing its size or shape. (p. 325) See also **dilation**, **reflection**.

tree diagram A diagram that uses branches to show all possible outcomes for a multi-stage chance experiment. (p. 418) See also **probability tree**.

trinomial An expression with 3 terms; for example, $3x^2 - 2x + 1$. (p. 18)

true bearing (or **three-figure bearing**) A bearing between 000° and 360° that is measured from north in a clockwise direction. (p. 149) See also **compass bearing**.

turning point A maximum or minimum point on a curve, where the curve turns around. (p. 97)

U

union of sets ($\cup$) The combined set containing all elements in the sets. (p. 405) See also **intersection of sets**.

unit circle A circle of radius 1 unit. (p. 153)

V

velocity The rate of change of displacement of an object with respect to time; signed speed. (p. 253)

Venn diagram A diagram that shows the relationship between 2 or more sets using circles (usually overlapping) drawn inside a rectangle. (p. 406)

vertex A turning point. (p. 97) See also **turning point**.

vertical line test A test that checks if a relation is a function: any vertical line drawn on the graph of a function should cut the graph at most once. If the line cuts the graph more than once, then it is not a function. (p. 70)

Z

zero/zeroes An x value of a function or polynomial for which the y value is zero, that is, $f(x) = 0$. (p. 77)